D1727677

# Mathematical and Statistical Methods in Food Science and Technology

The *IFT Press* series reflects the mission of the Institute of Food Technologists — to advance the science of food contributing to healthier people everywhere. Developed in partnership with Wiley Blackwell, *IFT Press* books serve as leading-edge handbooks for industrial application and reference and as essential texts for academic programs. Crafted through rigorous peer review and meticulous research, *IFT Press* publications represent the latest, most significant resources available to food scientists and related agriculture professionals worldwide. Founded in 1939, the Institute of Food Technologists is a nonprofit scientific society with 18,000 individual members working in food science, food technology, and related professions in industry, academia, and government. IFT serves as a conduit for multidisciplinary science thought leadership, championing the use of sound science across the food value chain through knowledge sharing, education, and advocacy.

# Mathematical and Statistical Methods in Food Science and Technology

Edited by

**Daniel Granato**

*Food Science and Technology Graduate Programme, State University of Ponta Grossa, Ponta Grossa, Brazil*

**Gastón Ares**

*Department of Food Science and Technology, Universidad de la República, Montevideo, Uruguay*

WILEY Blackwell

This edition first published 2014 ©2014 by John Wiley & Sons, Ltd

*Registered office:* John Wiley & Sons, Ltd, The Atrium, Southern Gate, Chichester, West Sussex, PO19 8SQ, UK

*Editorial offices:* 9600 Garsington Road, Oxford, OX4 2DQ, UK
The Atrium, Southern Gate, Chichester, West Sussex, PO19 8SQ, UK
111 River Street, Hoboken, NJ 07030-5774, USA

For details of our global editorial offices, for customer services and for information about how to apply for permission to reuse the copyright material in this book please see our website at www.wiley.com/wiley-blackwell.



*Library of Congress Cataloging-in-Publication Data*

Mathematical and statistical methods in food science and technology / edited
by Daniel Granato and Gaston Ares.
pages cm
Includes bibliographical references and index.
ISBN 978-1-118-43368-3 (cloth)
1. Food–Analysis–Stistical methods. 2. Food
contamination–Research–Statistical methods. 3. Food supply–Mathematics.
I. Granato, Daniel, editor of compilation. II. Ares, Gaston, editor of
compilation.
TX541.M377 2013
664′.07–dc23

2013026555

A catalogue record for this book is available from the British Library.

Wiley also publishes its books in a variety of electronic formats. Some content that appears in print may not be available in electronic books.

Cover design and illustration by www.hisandhersdesign.co.uk

Set in 10/12pt, Times by Thomson Digital, Noida, India.
Printed and bound in Malaysia by Vivar Printing Sdn Bhd

1 2014

## ***Titles in the IFT Press series***

- *Accelerating New Food Product Design and Development* (Jacqueline H. Beckley, Elizabeth J. Topp, M. Michele Foley, J.C. Huang, and Witoon Prinyawiwatkul)
- *Advances in Dairy Ingredients* (Geoffrey W. Smithers and Mary Ann Augustin)
- *Bioactive Proteins and Peptides as Functional Foods and Nutraceuticals* (Yoshinori Mine, Eunice Li-Chan, and Bo Jiang)
- *Biofilms in the Food Environment* (Hans P. Blaschek, Hua H. Wang, and Meredith E. Agle)
- *Calorimetry in Food Processing: Analysis and Design of Food Systems* (Gönül Kaletunç)
- *Coffee: Emerging Health Effects and Disease Prevention* (YiFang Chu)
- *Food Carbohydrate Chemistry* (Ronald E. Wrolstad)
- *Food Ingredients for the Global Market* (Yao-Wen Huang and Claire L. Kruger)
- *Food Irradiation Research and Technology*, Second Edition (Christoper H. Sommers and Xuetong Fan)
- *Foodborne Pathogens in the Food Processing Environment: Sources, Detection and Control* (Sadhana Ravishankar, Vijay K. Juneja, and Divya Jaroni)
- *High Pressure Processing of Foods* (Christopher J. Doona and Florence E. Feeherry)
- *Hydrocolloids in Food Processing* (Thomas R. Laaman)
- *Improving Import Food Safety* (Wayne C. Ellefson, Lorna Zach, and Darryl Sullivan)
- *Innovative Food Processing Technologies: Advances in Multiphysics Simulation* (Kai Knoerzer, Pablo Juliano, Peter Roupas, and Cornelis Versteeg)
- *Microbial Safety of Fresh Produce* (Xuetong Fan, Brendan A. Niemira, Christopher J. Doona, Florence E. Feeherry, and Robert B. Gravani)
- *Microbiology and Technology of Fermented Foods* (Robert W. Hutkins)
- *Multiphysics Simulation of Emerging Food Processing Technologies* (Kai Knoerzer, Pablo Juliano, Peter Roupas and Cornelis Versteeg)
- *Multivariate and Probabilistic Analyses of Sensory Science Problems* (Jean-François Meullenet, Rui Xiong, and Christopher J. Findlay
- *Nanoscience and Nanotechnology in Food Systems* (Hongda Chen)
- *Natural Food Flavors and Colorants* (Mathew Attokaran)
- *Nondestructive Testing of Food Quality* (Joseph Irudayaraj and Christoph Reh)
- *Nondigestible Carbohydrates and Digestive Health* (Teresa M. Paeschke and William R. Aimutis)
- *Nonthermal Processing Technologies for Food* (Howard Q. Zhang, Gustavo V. Barbosa-Cánovas, V.M. Balasubramaniam, C. Patrick Dunne, Daniel F. Farkas, and James T.C. Yuan)
- *Nutraceuticals, Glycemic Health and Type 2 Diabetes* (Vijai K. Pasupuleti and James W. Anderson)
- *Organic Meat Production and Processing* (Steven C. Ricke, Ellen J. Van Loo, Michael G. Johnson, and Corliss A. O ' Bryan)
- *Packaging for Nonthermal Processing of Food* (Jung H. Han)
- *Practical Ethics for Food Professionals: Ethics in Research, Education and the Workplace* (J. Peter Clark and Christopher Ritson)
- *Preharvest and Postharvest Food Safety: Contemporary Issues and Future Directions* (Ross C. Beier, Suresh D. Pillai, and Timothy D. Phillips, Editors; Richard L. Ziprin, Associate Editor)
- *Processing and Nutrition of Fats and Oils* (Ernesto M. Hernandez and Afaf Kamal-Eldin)
- *Processing Organic Foods for the Global Market* (Gwendolyn V. Wyard, Anne Plotto, Jessica Walden, and Kathryn Schuett)
- *Regulation of Functional Foods and Nutraceuticals: A Global Perspective* (Clare M. Hasler)
- *Resistant Starch: Sources, Applications and Health Benefits* (Yong-Cheng Shi and Clodualdo Maningat)

- *Sensory and Consumer Research in Food Product Design and Development* (Howard R. Moskowitz, Jacqueline H. Beckley, and Anna V.A. Resurreccion)
- *Sustainability in the Food Industry* (Cheryl J. Baldwin)
- *Thermal Processing of Foods: Control and Automation* (K.P. Sandeep)
- *Trait - Modified Oils in Foods* (Frank T. Orthoefer and Gary R. List)
- *Water Activity in Foods: Fundamentals and Applications* (Gustavo V. Barbosa-Cánovas, Anthony J. Fontana Jr., Shelly J. Schmidt, and Theodore P. Labuza)
- *Whey Processing, Functionality and Health Benefits* (Charles I. Onwulata and Peter J. Huth)

WILEY Blackwell

# Contents

# About the editors

**Daniel Granato** is a Food Engineer. He completed his Master's degree in Food Technology (Federal University of Paraná, Brazil) in 2009 and undertook his Doctoral degree in Food Science (University of São Paulo, Brazil) from 2009–2011. In his Master's dissertation, he used Response Surface Methodology (RSM) to model and optimize physico-chemical and sensory properties of a dairy-free dessert. In his PhD thesis, he used multivariate statistical techniques to assess the influence of the grape variety, origin, price, sensory properties, *in vitro* antioxidant capacity, instrumental colour and chemical composition of red wines from South America on *in vivo* antioxidant activity (using Wistar rats). He has worked at the Adolfo Lutz Institute (São Paulo, Brazil) as a Researcher in the Department of Analysis and Data Processing since 2012. He has authored more than 40 articles in international refereed journals and various presentations at Congresses. He has worked and collaborated with many research groups in the following fields: food safety, food development/optimization using RSM, chemistry, sensory analysis, chemometrics and sensometrics applied in food science and technology. He is an active reviewer of more than 30 scientific peer-reviewed journals.

**Gastón Ares** is a Food Engineer. He received his PhD in chemistry, focusing on sensory and consumer science, from the Universidad de la República (Uruguay) in 2009. He has worked as professor and researcher in the Food Science and Technology Department of the Chemistry Faculty of the Universidad de la República (Uruguay) since 2005. His research has been focused on the application of novel methodologies for the development and evaluation of food products and processes. He has extensive experience in the application of multivariate statistical and mathematical techniques. He has authored more than 80 articles in international refereed journals and numerous presentations in scientific meetings. He was awarded the 2007 Rose Marie Pangborn Sensory Science Scholarship, granted to PhD students in sensory science worldwide. In 2011 he won the Food Quality and Preference Award for a young researcher for his contributions to sensory and consumer science, and the Scopus Uruguay Award in Engineering and Technology. He is member of the Editorial Boards of both the *Journal of Sensory Studies* and the journal *Food Quality and Preference*, as well as associate editor of *Food Research International*.

# List of contributors

**Azila Abdul-Aziz,** Institute of Bioproduct Development, Universiti Teknologi Malaysia, Skudai Johor, Malaysia

**José Manuel Amigo,** Department of Food Science, Faculty of Sciences, University of Copenhagen, Frederiksberg, Denmark

**Gastón Ares,** Department of Food Science and Technology, Facultad de Química, Universidad de la República, Montevideo, Uruguay

**Ramlan Aziz,** Institute of Bioproduct Development, Universiti Teknologi Malaysia, Skudai Johor, Malaysia

**Andreas Baierl,** Department of Statistics and Operations Research, University of Vienna, Vienna, Austria

**Débora Cristiane Bassani,** Department of Biomedicine, Centro Educacional das Faculdades Metropolitanas Unidas, São Paulo, SP, Brazil

**Francis Butler,** School of Biosystems Engineering, Agriculture and Food Science Centre, University College Dublin, Belfield, Dublin 4, Ireland

**Vasco Augusto Pilão Cadavez,** CIMO Mountain Research Centre, School of Agriculture (ESA) of the Polytechnic Institute of Bragança (IPB), Bragança, Portugal

**Shin-Kyo Chung,** School of Food Science & Biotechnology, Kyungpook National University, Daegu, Republic of Korea

**Maria G. Corradini,** Department of Food Science, Rutgers, The State University of New Jersey, New Brunswick, NJ, USA

**Daniel Cozzolino,** School of Agriculture, Food and Wine, Faculty of Sciences, The University of Adelaide, Waite Campus, Glen Osmond, SA, Australia

**Verônica Maria de Araújo Calado,** Escola de Química, Universidade Federal do Rio de Janeiro, Rio de Janeiro, Brazil

**Eva Derndorfer,** University of Applied Sciences Burgenland, University of Salzburg, UMIT Hall in Tirol, and independent sensory consultant, Vienna, Austria

**Rolf Ergon,** Telemark University College, Porsgrunn, Norway

**Tamas Ferenci,** Physiological Control Group, Institute of Information Systems, John von Neumann Faculty of Informatics, Obuda University, Budapest, Hungary

**Ursula Andrea Gonzales-Barron,** CIMO Mountain Research Centre, School of Agriculture (ESA) of the Polytechnic Institute of Bragança (IPB), Bragança, Portugal

**Daniel Granato,** Food Science and Technology Graduate Programme, State University of Ponta Grossa, Ponta Grossa, Brazil

**Aurea Grané,** Statistics Department, Universidad Carlos III de Madrid, Getafe, Spain

**Jos A. Hageman,** Plant Sciences Group, Wageningen UR, Wageningen, The Netherlands

**Noor Hafiza Harun,** Institute of Bioproduct Development, Universiti Teknologi Malaysia, Skudai Johor, Malaysia

**Anthony D. Hitchins,** Rockville, MD, USA [Center for Food Safety and Applied Nutrition, United States Food and Drug Administration (retired)]

**François Husson,** Laboratoire de mathématiques appliqués, Agrocampus Ouest, Rennes Cedex, France

**Agnieszka Jach,** Statistics Department, Universidad Carlos III de Madrid, Getafe, Spain

**Jeffrey E. Jarrett,** University of Rhode Island, Kingston, RI, USA

**Basil Jarvis,** Department of Food and Nutrition Sciences, School of Chemistry, Food and Pharmacy, The University of Reading, Whiteknights, Reading, Berkshire, UK

**Jasenka Gajdos Kljusuric,** Faculty of Food Technology and Biotechnology, University of Zagreb, Zagreb, Croatia

**Levente Kovacs,** Physiological Control Group, Institute of Information Systems, John von Neumann Faculty of Informatics, Obuda University, Budapest, Hungary

**Zelimir Kurtanjek,** Faculty of Food Technology and Biotechnology, University of Zagreb, Zagreb, Croatia

**João A. Lopes,** REQUIMTE, Laboratório de Análises Químicas e Físico-Químicas, Departamento de Ciências Químicas, Faculdade de Farmácia, Universidade do Porto, Porto, Portugal

**Mohd Faizal Muhammad,** Institute of Bioproduct Development, Universiti Teknologi Malaysia, Skudai Johor, Malaysia

**Sulaiman Ngadiran,** Institute of Bioproduct Development, Universiti Teknologi Malaysia, Skudai Johor, Malaysia

**Mark D. Normand,** Department of Food Science, Chenoweth Laboratories, University of Massachusetts, Amherst, MA, USA

**Domingos Sávio Nunes,** Department of Chemistry, Universidade Estadual de Ponta Grossa, Ponta Grossa, PR, Brazil

**Massimo Pacella,** Dipartimento di Ingegneria dell'Innovazione, Università del Salento, Lecce, Italy

**Jérôme Pagès,** Laboratoire de mathématiques appliqués, Agrocampus Ouest, Rennes Cedex, France

**Micha Peleg,** Department of Food Science, Chenoweth Laboratories, University of Massachusetts, Amherst, MA, USA

**Fernando Pérez-Rodríguez,** Department of Food Science and Technology, University of Cordoba – International Campus of Excellence in the AgriFood Sector ceiA3, Campus Rabanales, Edificio Darwin – Córdoba, Spain

**Roshanida Abdul Rahman,** Faculty of Chemical Engineering, Universiti Teknologi Malaysia, Skudai Johor, Malaysia

**Marco S. Reis,** CIEPQPF, Department of Chemical Engineering, University of Coimbra, Coimbra, Portugal

**Åsmund Rinnan,** Department of Food Science, Faculty of Science, University of Copenhagen, Frederiksberg, Denmark

**Ernie Surianiy Rosly,** Institute of Bioproduct Development, Universiti Teknologi Malaysia, Skudai Johor, Malaysia

**Quirico Semeraro,** Dipartimento di Meccanica, Politecnico di Milano, Milan, Italy

**Thomas Skov,** Department of Food Science, Faculty of Science, University of Copenhagen, Frederiksberg, Denmark

**Clara C. Sousa,** REQUIMTE, Laboratório de Microbiologia, Departamento de Ciências Biológicas, Faculdade de Farmácia, Universidade do Porto, Porto, Portugal

**Panagiotis H. Tsarouhas,** Department of Standardization & Transportation of Products – Logistics, Alexander Technological Educational Institute of Thessaloniki, Katerini, Greece

**Grishja van der Veer,** RIKILT Institute of Food Safety, Wageningen UR, Wageningen, The Netherlands

**Saskia M. van Ruth,** Product Design and Quality Group, Wageningen UR and RIKILT Institute of Food Safety, Wageningen UR, Wageningen, The Netherlands

**Paula Varela,** Propiedades físicas y sensoriales de alimentos y ciencia del consumidor, Instituto de Agroquímica y Tecnología de Alimentos (CSIC), Paterna (Valencia), Spain

**Wan Mastura Wan Zamri,** Institute of Bioproduct Development, Universiti Teknologi Malaysia, Skudai Johor, Malaysia

**Harisun Yaakob,** Institute of Bioproduct Development, Universiti Teknologi Malaysia, Skudai Johor, Malaysia

**Hu-Zhe Zheng,** Department of Food & Nutrition Engineering, Jiangsu Food & Pharmaceutical Science College, Jiangsu Huai'an, China

# Acknowledgements

Firstly, I need to thank God that I am here doing what I love: research. All people I love are part of this book and I dedicate this effort to them: Maria Aparecida Granato, Marcos Granato, Gustavo Granato, Felipe Borges, and all my dearest friends. Thank you for understanding and supporting me, no matter what.

I also would like to thank Professor Dr Anderson Sant'Ana for suggesting some of the topics covered in this book and for his support. I want to thank all authors who accepted the invitation to contribute to this book.

Daniel Granato

Thanks to all the authors for joining the project and providing high-quality contributions.

On a personal level I would like to thank my family and friends for their continuous love and support. Thanks for giving me the strength to achieve all my dreams.

Gastón Ares

# Section 1

# 1 The use and importance of design of experiments (DOE) in process modelling in food science and technology

**Daniel Granato[1] and Verônica Maria de Araújo Calado[2]**

[1] *Food Science and Technology Graduate Programme, State University of Ponta Grossa, Ponta Grossa, Brazil*

[2] *Escola de Química, Universidade Federal do Rio de Janeiro, Rio de Janeiro, Brazil*

## ABSTRACT

In the last ten years, the use and applications of mathematical modelling have increased in chemistry and food science and technology. However, it is still common to find researchers using the 'one at a time' approach to test and select variables to develop and optimize products and processes. In this regard, the objectives of this review are to provide some statistical information related to mathematical modelling of processes using design of experiments followed by multiple regression analysis, the so-called response surface methodology (RSM), and to discuss some recent published researches based on RSM optimization of products and processes, with special attention to microbiology, sensory analysis, food development and nutrition.

## INTRODUCTION

The development of food products and/or processes is a complex, expensive and risky multistage process, and special requirements should be considered in this process, such as consumer demands, price, operational conditions and legislation background. To develop or to optimize processes, many companies use statistical approaches, such as response surface methodology (RSM), in their research department in order to achieve the best combination of factors that will render the best characteristic of a product and or process response. In food and chemical companies, RSM has important applications in the design, analysis and optimization of existing products and unit operations, its use decreasing thus the volume of experiments, reagents, time, financial input, energy, among others (Montgomery, 2009).

Mathematical modelling for food development or unit operations to produce a food is increasing and some statistical techniques are being adopted, such as RSM, to solve problems where several independent variables (or factors) influence the response variable value (Nwabueze, 2010). In food systems, the product response of interest to the researcher might include, for example product development, functional and sensory properties, nutritional qualities, antinutritional or toxic levels, shelf life, microbiological quality, packaging performance, processing and media conditions.

It is widely accepted that RSM is a useful tool to analyse results from many different experimental responses (chemical, sensory, physicochemical etc.). Within this context, the objectives of this review are to provide some useful information regarding mathematical modelling by using design of experiments (DOE) followed by response surface methodology, and to discuss some recent published

*Mathematical and Statistical Methods in Food Science and Technology*, First Edition.
Edited by Daniel Granato and Gastón Ares.

researches based on RSM optimization of products and processes, with special attention to microbiology, sensory analysis, development of foods products, and nutrition.

## OVERVIEW OF EXPERIMENTAL DESIGNS

### Types of design

In accordance with Montgomery (2009) and Myers *et al.* (2009), there are several experimental designs that can be applied in food/chemical companies to test ingredients and/or to prepare or reformulate a new food product or even to optimize the conditions to lead to an optimal process. Some of these designs are: full factorial design, fractional factorial design, saturated design; central composite design and mixture design. The use of one of these types depends on the purpose and it is important to note that, in order to achieve a final objective, sometimes it is necessary to use a sequence of two or more designs.

A **full factorial design** is applied when the purpose is to determine which factors (independent variables) are important in the study and tthe range of values (levels) of these factors. This is the only design that can evaluate interaction among all factors. Michel *et al.* (2011) used a two-level full factorial design to assess the effects of factors (extraction time, irradiation power, number of cycles) and their first order interactions on the extraction of antioxidants from sea buckthorn berries by using the pressurized solvent-free microwave assisted extraction technique (PSFME). The best extraction conditions were found and this method was compared to other common extraction techniques, such as pressing, maceration and pressurized liquid extraction; the authors concluded that PSFME leads to the most active and richest extract in phenolic content from buckthorn.

For **two-level factorial designs** ($2^k$), the mathematical model used to describe the relationship between factors and the response variable is linear:

$$Y = \beta_o + \beta_1 x_1 + \beta_2 x_2 + \ldots + \beta_k x_k + \varepsilon \tag{1.1}$$

**Thus, it is not possible to think about optimize this process.** It is common for people use this type of design and find 'optimum values' for the factors selected. Indeed, they are obtaining the best values for the factors, considering the experimental region analysed. However, in several studies, there is an interest in determining which factor level takes the response variable to a maximum or a minimum. Therefore, a more complex model should be proposed to take into consideration the plane curvature formed by the factors and the response variable. In this case, it is possible to work with a three-level factorial design or with a central composite design; in both cases, the parabola is a mathematical model that accomplishes this objective.

In a recent study, Ellendersen *et al.* (2012) used a $2^2$ design to study the influence of temperature and fermentation time on the viability of *Lactobacillus casei* and *L. acidophilus* in apple juice. The best conditions to produce a probiotic apple juice were found to be 10 hours fermentation at 37°C.

Kliemann *et al.* (2009) evaluated the effect of four independent variables (acid, temperature, pH and extraction time) on pectin extraction from passion fruit peel using a $2^4$ factorial design, followed by a central composite design with five levels for the three statistically significant factors (temperature, pH and extraction time); the results were analysed by response surface methodology. The optimal conditions for maximum pectin yield were citric acid at 80°C and pH 1, with an extraction time of 10 minutes, when they considered a model extrapolation. The authors concluded that RSM was a suitable technique to optimize a process that makes good use of a commonly discarded product.

If it is necessary to optimize a process, the design to be used is $3^k$ or central composite design, because they allow quadratic models, as shown by Equation 1.2 for only two factors.

$$Y = \beta_o + \beta_1 x_1 + \beta_2 x_2 + \beta_{12} x_1 x_2 + \beta_{11} x_1^2 + \beta_{22} x_2^2 + \varepsilon \tag{1.2}$$

A **three-level design** ($3^k$) is not the most efficient way to model a quadratic relationship; the central composite design is preferred and requires fewer assays to achieve a better modelling. Gonzalez-Barreiro *et al.* (2000) tested response surface experimental design to optimize the solid phase microextraction (SPME) of the widely used herbicide alachlor. A three-level factorial design ($3^2$) was used to study the effect of extraction time and desorption time on the extraction efficiency and also to optimize the experimental conditions. The extraction time only appeared statistically significant, because the lower level for desorption time (15 min) is long enough to produce the complete desorption of the alachlor extracted by the fibre. No significant interactions were detected.

When there are many factors to be studied and there is not much time or raw materials, it is recommended to use the **fractional factorial design**, aiming at reducing the number of assays. Even though the accuracy of the design is lower, less time and money are spent. Zanariah *et al.* (2012) used a two-level half factorial design for five factors ($HNO_3$ and $H_2O_2$ volumes, sample weight, microwave power and digestion time), which involved 16 experiments, to quantify arsenic in shrimp paste samples treated by a microwave digestion method. They concluded that only two factors (sample weight and microwave power) and their interaction effects were statistically significant. The authors proposed a regression model to predict arsenic concentration, considering the main effects of sample weight and microwave power, the interaction effect between them and the interaction effect between microwave power and digestion time. Because the main effect of this last factor was not statistically important, the authors did not consider it in the mathematical model. **This is a common error made by some researchers; it is necessary to consider the main effects of factors that are not statistically significant if their interaction effects with other factors are.**

A very widely used $3^{k-p}$ fractional factorial design is Box–Behnken, because it considers more experimental points (allowing then more degrees of freedom, which implies a more precise analysis) than the normal fractional factorial, but less than the full factorial design. This type of design is a collection of statistical techniques for designing experiments, building models, evaluating the effects of factors and searching optimum conditions of studied factors for desirable responses (Haaland, 1989). For example Granato *et al.* (2010a) used a $3^2$ design to develop a soy-based guava dessert where guava juice and soy protein were the independent variables, and the responses were the sensory properties and physicochemical characteristics of such products. The authors obtained significant RSM models and concluded that RSM was an adequate approach for modelling the physicochemical parameters and the degree of liking of creaminess of desserts.

A $3^K$ factorial Box–Behnken design was used by Jo *et al.* (2008) to determine the effect of three independent variables (glucose content, pH and temperature) on the hydrogen production rate, and to optimize the process to achieve improved hydrogen production. Thus, by using RSM with the Box–Behnken design, the authors concluded that the maximum hydrogen production rate by *C. tyrobutyricum* JM1 (5089 ml $H_2$ (g dry cell h)$^{-1}$) was obtained under the optimum condition of glucose concentration = 102.08 mM, temperature = 35°C and pH = 6.5.

The extreme case of fractional factorial design is the **saturated design**, where there are not enough degrees of freedom to calculate the interaction effects among some factors, as the number of factors (more than 11, for example) is quite high and the cost and time involved would make the use of factorial designs prohibitive. When there are many factors to be tested, the **Plackett–Burman** design may be an excellent option, once it has been widely used to develop process conditions and to allow the understanding of the effects of various physicochemical, biochemical and sensory variables using a minimum number of experiments. The Plackett–Burman design is widely used in food researches because it allows the screening of main factors from a large number of variables that can be retained in the further optimization process (Siala *et al.*, 2012). For example Siala *et al.* (2012) used a Plackett–Burman design to analyse the effect of various conditions related to the composition of the medium, inoculum size and temperature of fermentation, totalling 11 independent variables. The authors verified that monopotassium phoshphate ($KH_2PO_4$), pH, and temperature were the three most significant factors; then they used a Box–Behnken design of RSM to optimize protease production by *Aspergillus niger* I1.

As mentioned previously, when it is necessary to optimize (to find maximum/minimum values) a response variable, it is necessary to use a $3^k$ or central composite design (CCD), the latter being the better. But first, it is necessary to be sure that the appropriate region of the factors, where the curvature is statistically significant, has been selected. In this case, a **quadratic or second order model**, Equation 1.3 for two factors, should be applied (Nwabueze, 2010).

$$Y = \beta_o + \sum_{j=1}^{k} \beta_j x_j + \sum\sum_{i<j} \beta_{ij} x_i x_j + \sum_{j=1}^{k} \beta_{jj} x_j^2 + \varepsilon \tag{1.3}$$

Second order models are mathematically more complex and used in biochemical reactions and sensory analysis, among others. They would likely be useful as an approximation to the true response surface in a relatively small region. The second order model is very flexible. It can take on a wide variety of functional forms, so it will often work well as an approximation to the true response surface (Keshani *et al.*, 2010).

A central composite design was employed to optimize the extraction conditions of sapodilla juice using hot water extraction (Sin *et al.*, 2006). The independent variables were juice extraction time (30–120 min) and temperature (30–90°C). The combined effects of these variables on juice yield, odour, taste and astringency were investigated. Results showed that the generated regression models adequately explained the data variation and significantly represented the actual relationship between the independent variables and the responses. Higher temperature increased the juice yield, taste and odour but also showed an increase in astringency, which affected the acceptability of the juice. The contour plots showed the relationships among the independent variables and the responses. All regression models were statistically significant ($p < 0.01$) and there was no lack of fit. A superposition of all contour plots allowed the optimum condition to be determined as 60°C for 120 minutes for hot water extraction of sapodilla juice.

The **mixture design** should be used when proposing a new formulation or a new food product. This design allows the determination of the ideal composition of each component in a mixture, with the purpose of achieving a product with the best features (taste, odour, texture, etc.). Several functional and fruit-based products have been developed using a mixture design, including desserts, smoothies, juices and pulp concentrate, among others. Pelissari *et al.* (2012) developed films composed of cassava starch, chitosan and glycerol by blown extrusion using a design for constrained surfaces and mixtures. The effects of the mixture components on the mechanical properties, water vapour permeability (WVP) and opacity of the films were studied. The authors concluded that the design for constrained surfaces and mixtures was a useful tool for this type of study and complexity of film formation conditions.

## Some Considerations

According to Calado and Montgomery (2003), regardless of the design type that will be employed, some considerations should be taken into account prior to collect experimental data:

- Definition of the variables, which can be qualitative (additive type, presence of magnetic agitation, presence of light, etc.) and quantitative (ingredient concentrations, temperature, pressure, etc.).
- Definition of the relevant levels of each independent variable. This can be done by performing an initial experiment.
- Analysis of the results and of the need for relevant changes in the initial design.

A relevant issue to be addressed is the block variable. There are some variables that act as covariates because they indirectly have some influence on response variables, but they should not have. However, imagine that there are many experiments to run and it is not possible to finish all at the same day; the environmental conditions may change day by day. Thus, the day is one of these covariates. Some others are: manufacturer, operator, batch, parts, and so on. There are many examples showing that if significant block variables are not taken into account statistically, the analysis may give the wrong answer, because important factors can be wrongly considered insignificant. If the block variable is identified as not statistically important, it may be considered a replicate, increasing then the degree of freedom. For example imagine that it is necessary to measure the influence of the tip (the only real factor) of an instrument to measure the hardness of a material. There are four different parts of the same material that are supposed to have the same properties. Because it is known that they can be different, the variable 'part' is used as a block. After running the analysis of the experiments, it was concluded that the block was statistically significant as well as the tip. But, if the different parts had been considered replicates, it would have been concluded that the tip was not important for measuring the material hardness, which would be a wrong conclusion.

Regardless of the type of design a researcher uses, it is demanding and essential to test the statistical quality of the results prior to their evaluation. If they are not statistically good, the analysis of the design will lead to misleading conclusions. Herein, the coefficient of variation (CV = standard deviation/mean) for each dependent variable should be calculated and if the results are below 10%, they might be considered excellent, while values up to 20% are considered acceptable. For sensory that uses consumers as panelists, which is subjective by nature, the coefficient of variation may reach values as high as 40% and it can be still considered acceptable. For other applications, such as agriculture, biotechnological processes, microbiology and clinical protocols, the coefficients of variation are high because of a wide dispersion in data. In these cases, it is recommended to establish suitable and acceptable limits. For data that are homogeneous, a CV higher than 30% is considered very bad and the experiments should be repeated.

Once the mathematical model has been selected, it is important to determine its significance by means of a variance analysis (ANOVA). To do that, the standard deviations of the main and the interactions effects of the selected factors should be calculated. If the standard deviations present a lower value than the mean values, it is possible to assume that the mathematical model is significant. If this does not happen, the experimental data should be evaluated in order to not presume that the effect is not significant.

In the evaluation of experimental designs, a mathematical model is provided to relate the response variable with the factor effects. In this regard, the goodness-of-fit of this model needs an assessment and the following criteria should be analysed:

- standard deviation of the estimated parameters and model;
- statistical significance of the estimated parameters;
- regression coefficient;
- value of the objective function;
- significance of the regression (ANOVA);
- analysis of the residuals.

It is considered a good fit to the experimental data when:

- the standard deviation of the parameter presents a lower value than the correspondent effect, indicating that the standard deviation of the proposed mathematical model is low;
- the parameters of a model need to be significant, otherwise they will not contribute to the model;
- it is a myth to consider that if the model presents a regression coefficient ($R^2$) above 90%, then it is considered excellent. This is only one criterion to evaluate the model goodness-of-fit. If a regression

coefficient is low (<70%), the mathematical model is not good and, on the other hand, if its value is high (>90%), it means that other statistical criteria may be used. It is noteworthy emphasizing that depending on the type of analysis, a regression coefficient may be considered good above 70%, such as what happens in sensory evaluation data;

- the value of the objective function should be low;
- the proposed mathematical model must be statistically significant;
- the analysis of the residuals consists in verifying if these residuals (experimental value for a response variable minus predicted value by the mathematical model) have a normal distribution and if the variance is constant. This is a necessary condition for the application of some *post hoc* tests, such as $t$ and $F$. To test the validity of a normal distribution, quantitative tests need to be employed, such as Kolmogorov–Smirnov, Liliefors and Shapiro–Wilks. To test the variance constancy, Levene's test is usually used.

## RESPONSE SURFACE METHODOLOGY: A TOOL FOR ANALYSING AND OPTIMIZING PRODUCTS AND PROCESSES

Response surface methodology consists of a group of mathematical and statistical techniques used in the development of an adequate functional relationship between a response variable ($y$) and a number of associated control variables denoted by $x_1, x_2, \ldots, x_k$. In general, such a relationship is unknown but can be approximated by a low-degree polynomial model of the form

$$y = {}'\boldsymbol{f}(\boldsymbol{x})\boldsymbol{\beta} + \varepsilon \tag{1.4}$$

where $\boldsymbol{x} = (x_1, x_2, \ldots, xk)'$, $\boldsymbol{f}(\boldsymbol{x})$ is a vector function of $p$ elements that consists of powers and cross-products of powers of $x_1, x_2, \ldots, x_k$ up to a certain degree denoted by $d$ ($\geq 1$), $\beta$ is a vector of $p$ unknown constant coefficients referred to as parameters and $\varepsilon$ is a random experimental error assumed to have a zero mean. This is conditional on the belief that a model, which must be significant statistically, provides a suitable representation of the response and the lack of fit is not significant ($p < 0.05$) (Khuri and Mukhopadhyay, 2010).

Simple mathematical models are used to fit experimental data. Usually, linear and quadratic models are sufficient to model sensory, biochemical, physical and physicochemical data (Dutcosky *et al.*, 2006; Capitani *et al.*, 2009; Farris and Piergiovanni, 2009).

The first step in using surface response methodology is to determine a mathematical relationship between the response variable and the independent variables. This relationship is quantitative, covers the entire experimental range tested and includes interactions (if present). Thus, the model can be used to calculate any and all combinations of factors and their effects within the test range (Iwe *et al.*, 2004). The response surfaces are represented mathematically by equations called models, which are similar to the well-known regression equations. First or second order regression models could be used for the analysis of responses $y$ as a function of independent variables. A brief summary of all steps that should be taken to build a response surface and then a mathematical model is presented in Figure 1.1 and Figure 1.2.

It is clear that the first model to be considered should be a straight line, as it is the simplest one. Linear behaviours usually occur in physicochemical analysis of ingredients mixture, such as pH, water activity, instrumental colour and titratable acidity. Equation 1.1 represents a **first order model**, as presented before.

First order models may not be able to adequately predict the response if there is a complex relationship between a dependent (response) variable and the independent (process) variables. If there is a curvature in the plane formed by a response variable and two other factors, then a polynomial with higher degree, such as a **quadratic or second order model** (Equation 1.3), should be applied (Nwabueze, 2010).

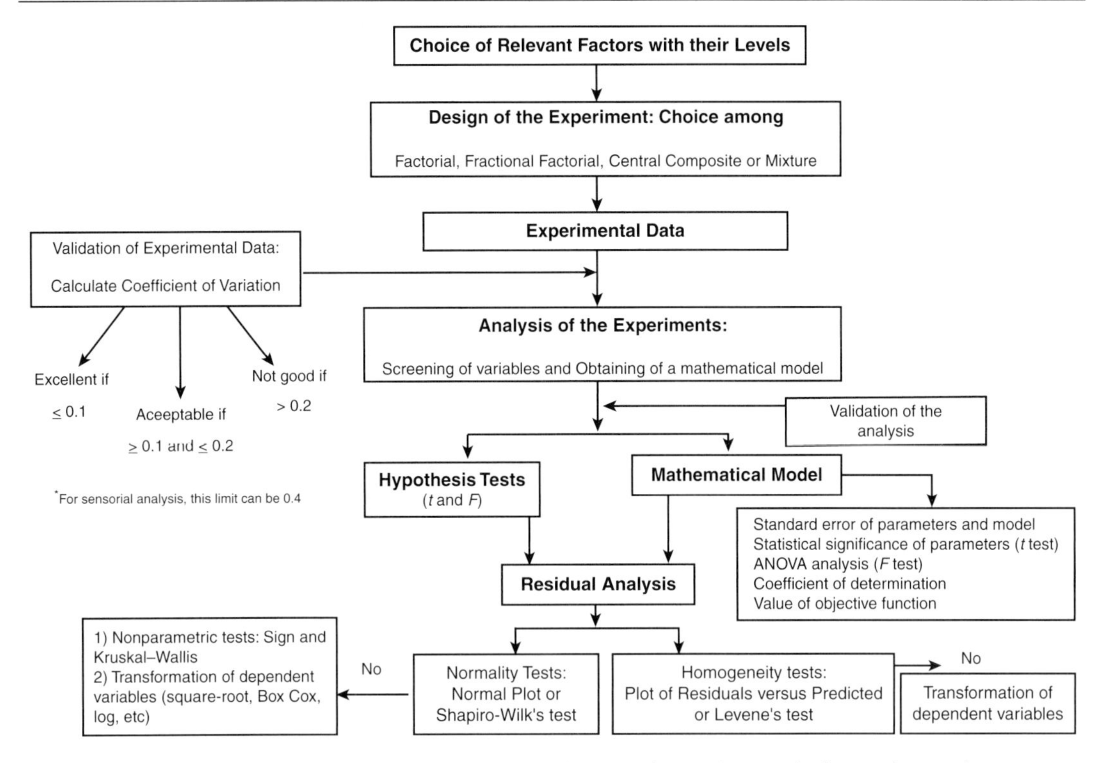


**Figure 1.1** Summary of the recommended statistical procedures used to analyse results from a design of experiments.

In accordance with Khuri and Mukhopadhyay (2010), the objectives of a mathematical model generated by RSM are:

- to determine a statistical significance of all factors whose levels are represented by $x_1, x_2, \ldots, x_k$;
- to establish a relationship between y and $x_1, x_2, \ldots, x_k$ that can be used to predict response values for a given set of control variables;

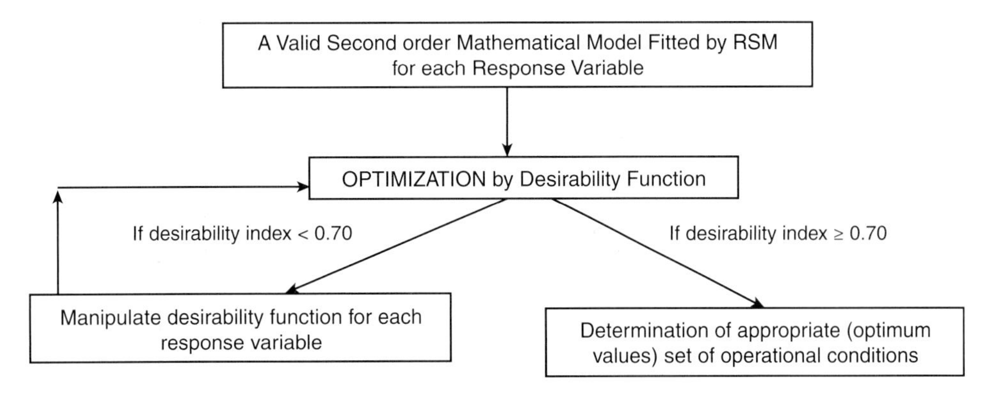


**Figure 1.2** Steps to obtain optimized food/process conditions.

- to determine the optimum set of $x_1, x_2, \ldots, x_k$ that results into a maximum (or minimum) response over a certain region of interest by means of a simultaneous optimization of the selected response variables. This gives information on the direction and magnitude of the influence of the factors and their combined effects on the product characteristics (Nwabueze, 2010).

By using an appropriate estimation method applied to the chosen model, the regression coefficients will be obtained and the estimated response can be easily calculated. Because the relation among the response and independent variables are usually not known a priori, different models should be tested in order to better fit the experimental data (Bas and Boyaci, 2007). For verification of the model adequacy, several techniques are used. Some of these techniques are residual analysis, scaling residuals, prediction of error sum of squares residuals and tests of lack of fit (Granato *et al.*, 2010a). The lack of fit is a measure of a model failure in representing data in the experimental domain (Montgomery, 2009). If there is a significant lack of fit, as indicated by a low probability value ($p < 0.05$), the response predictor is discarded. The overall predictive capability of the model is commonly explained by the regression coefficient ($R^2$), but this coefficient alone does not measure the model accuracy. $R^2$ is defined as the ratio of the explained variation to the total variation and is a measure of the degree of fit (Myers and Montgomery, 2002).

Many researchers use different critical regression coefficient values to determine whether the mathematical models can be considered good and predictive or just cannot be used for prediction purposes. Henika (1982) stated that for sensory data the regression coefficient must be above 85% to be considered satisfactory; however, Granato *et al.* (2010a, 2010b) established that a value $\geq 70\%$ was considered good for sensory, colorimetric and physicochemical results, while Joglekar and May (1987) suggested that for a good model fit, $R^2$ should be at least 80%. For the models that present a regression coefficient below 70%, it must be considered that there is a failure of the models to represent the data in the experimental domain (Myers and Montgomery, 2002). However, in many food science and technology applications, such as enzymology, kinetic studies and sensory evaluation, it is not surprising if no mathematical model can be adequately fitted to the experimental data. For example the affective tests, used to determine consumer acceptance of a food, is extremely subjective depending on the sensory method that is applied (Shihani *et al.*, 2006; Nikzadeh and Sedaghat, 2008); hence, there is a great variance among the scores given by the assessors and no mathematical model can be successfully used to model the scores.

## PROCESS OPTIMIZATION

In accordance with Bas and Boyaci (2007), the development of a food product and the evaluation and optimization of a process are affected by numerous factors (chemical, operational, physical, sensory and physicochemical). Once it is not possible to identify the effects of all factors, it is necessary to select those ones that have major effects. Screening experiments are useful to identify the independent variables (factors) and factorial designs may be used to achieve this objective. After identifying the important factors, the improvement direction is determined and the levels of the factors are identified. Determining these levels is important because the success of an optimization process is directly related to them. Mistakenly chosen levels result in an unsuccessful optimization.

To determine the best conditions (factor levels that result in the desirable values to a response variable) to develop a product, some researchers optimize only one factor at a time, keeping constant the remaining ones. This procedure is called 'one factor at a time' optimization. That is, the ideal level of one factor, which provided the best (maximum/minimum) value to a response variable, is defined. Another factor is then studied, keeping constant the others. This procedure continues until all factors have been analysed. Besides laborious, this procedure is erroneous, as it does not take into account the interactions among factors (Box and Draper, 1987). This means that the supposed 'ideal' level of a factor is determined based on certain levels of the others. If other levels were chosen for these other factors, the result could completely change; that is, the 'ideal' levels of these factors would be different, resulting in a different and maybe wrong value for the response variables (Khuri and Mukhopadhyay, 2010). Therefore, all factors must be simultaneously varied, with a minimum number of assays, according to the design methodology. The major disadvantage of this technique – one factor each time – is that it does not include interactive effects among the variables and, eventually, it does not depict the complete effects of the factors on the process (Bas and Boyaci, 2007). To overcome this problem, optimization studies using RSM can be performed to obtain optimum conditions.

## Simultaneous optimization of response variables

The main objective of optimization is to determine the levels of independent variables that lead to the best characteristics of a particular product, such as physicochemical, colorimetric, sensory and nutritional properties, without extending excessively the experiment time with a large number of assays. These procedures can be performed using a RSM approach. One of the main objectives of RSM is to determine optimum settings of the control variables that result in a maximum (or a minimum) response over a certain region of interest. This requires having a 'good' fitting model that provides an adequate representation of the mean response, because such a model is used to determine the value of the optimum (Khuri and Mukhopadhyay, 2010). For food processes, and especially for food development, optimization is a way to obtain ideal conditions to achieve a desired quality (physicochemical, chemical and sensory, for example) (Myers *et al.*, 2009). Optimization of a product is an effective strategy of accomplishing its successful development. If a food cannot be re-engineered or modified to fulfil consumer specifications, it will not succeed in the market. Hence, optimization is required and well established in many food companies.

Simultaneous optimization techniques are used when there is more than one response variable and it is necessary to find the 'optimum points' of the factors that fulfil all requirements for all response variables at the same time. This is an optimization problem with restriction and nonlinear programming techniques are usually used. Most researchers use a graphical approach of superimposing the different response surfaces and finding the experimental region that gives the desired values for all response variables simultaneously. This methodology, although visually attractive, is inefficient and cannot be automated.

For first order models, the method of steepest ascent (or descent) is a viable technique for sequentially moving toward the optimum response (Myers *et al.*, 2009).

For second order models, which are the most used for food development purposes, simultaneous optimization using the desirability function technique is the recommended tool (Granato, 2010b; Cruz *et al.*, 2010). It is based on the idea that the 'quality' of a product or process has multiple quality characteristics (Reis *et al.*, 2008). The desirability approach, proposed initially by Derringer and Suich (1980), seems very promising for optimizing simultaneous response variables, besides being easily performed (Reis *et al.*, 2008).

The general approach consists in first converting each response variable into a desirability function $d_i$, that varies from 0 to 1 (Calado and Montgomery, 2003). That is, if it is necessary to find the factor levels that take to a maximum response variable value, it is necessary to set $d_i = 1$ for high values and $d_i = 0$ for low values of this response variable. In the case a minimum response variable value is required, it is necessary to set $d_i = 0$ for high values and $d_i = 1$ for low values of this response variable. The idea is that this desirability function acts as a penalty function that leads the algorithm to regions where the desired response variable values can be found. The factor levels that take to a maximum or a minimum of the response variable are called 'optimum points'.

Equation 1.4 expresses the global desirability function, $D$, defined as the geometric mean of the individual desirability functions. The algorithm should search for response variable values where D tends to 1.

$$D = (d_1 \cdot d_2 \cdot \ldots \cdot d_m)^{1/m} \quad (1.5)$$

where $m$ is the number of response variables.

This approach is not recommended for simultaneous optimization of more than four response variables because of constraints to achieve all expected results (Reis *et al.*, 2008). Thus, for a large number of response variables, it is necessary to select those ones that characterize the product in a more specific way and possess a relation with the main quality features (Yi *et al.*, 2009).

Before starting the searching process for optimum values of the independent variables, the appropriate model should be determined in order to describe each response variable as a function of the factors and then to find an appropriate set of operational conditions that optimize all response variables (Calado and Montgomery, 2003).

## RSM application to foods/process development/optimization

The use of optimization procedures to improve the process conditions and to obtain a product with certain characteristics is essential to companies when a high yield, low production costs and low use of energy are desired. A search for significant effects on quality parameters is demanding and, in this context, factorial designs (to eliminate some independent variables that are not statistically significant) followed by RSM (for the remaining factors) are tools that help the researchers to estimate the influence of variables on the process, by eliminating the variables that do not seem to contribute to the final product's quality, and also to optimize the process conditions (pressure, temperature, agitation velocity, energy input, moisture etc.) in order to obtain an improved product.

Herein, some examples of RSM applications are presented for food science and technology, microbiology, food development, sensory evaluation and nutrition.

### *Applications of RSM to food science and technology*

Working with açaí powder obtained from spray drying, Tonon *et al.* (2008) proposed a central composite design to study the influence of inlet air temperature, feed flow rate and maltodextrin concentration on the spray drying process yield, moisture content, hygroscopicity and anthocyanin retention. By using 17 experiments, the authors concluded that inlet temperature showed a significant effect on all the response variables.

Tiwary *et al.* (2012) used a central composite design to study the effect of pectinase concentration, cellulase concentration, hemicellulase concentration, temperature and incubation time on the stevioside extraction from *Stevia rebaudiana* leaves. The authors tested 26 experimental conditions and optimized them by using graphical and numerical approaches. A second order quadratic equation was then fitted to the data by multiple regression procedure and authors obtained a $R^2 = 0.9776$ and a $p$-value $= 0.007$ for the regression, showing the significance of the DOE followed by RSM to model the experimental data.

The effects of enzyme concentration (0.16–0.84 mg/100 g guava pulp), incubation temperature (36.6–53.4°C), and incubation time (0.95–11 h) on guava juice yield were evaluated by Kaur *et al.* (2009). A central composite design was applied to and analysed by response surface methodology, showing that enzyme concentration was the most significant factor affecting the juice yield. A quadratic model was fitted to the results and presented no significant lack of fit ($p > 0.05$) and a satisfactory regression coefficient ($R^2 = 85.0\%$). Optimum juice yield was obtained with 0.70 mg/100 g guava pulp, 7.27 hours of incubation time at 43.3°C, reiterating the importance and suitability of DOE and RSM on process optimization.

Jing *et al.* (2011) used a multipleresponse surface methodology in order to optimize the anthocyanin content, glucosinolate content and clarity when working with chitosan to remove glucosinolates from radish anthocyanin extracts. The factors were the purification conditions: pH, chitosan concentration and treatment duration. The authors used a Box–Behnken model design and optimized using a desirability function approach. A second order polynomial model was applied to adjust the three response variables. The optimal purification condition was: pH = 3.92, chitosan concentration = 1.59 g/100 ml, and treatment duration = 2.74 hours. Equation 1.4 predicts a desirability value of 0.87. The authors carried out a triplicate experiment to validate the selected purification conditions and concluded that the experimental data were in good agreement with the predicted values. Relative errors between predicted and actual values were 0.9%, 0.5% and 3.0%, respectively, for anthocyanin content, clarity and glucosinolate content, indicating that the selected processing parameters could produce radish anthocyanin extracts with high quality.

### *Microbiology*

Recent investigations have been performed in the microbiology field to address some issues related to optimization of experimental protocols, development of new culture media and processes and also to test new ingredients to produce culture media. In the past, the optimization of media compounds by the traditional 'one-variable-at-a-time' strategy involving changing one independent variable was the most frequently used operation in biotechnology, but it is known that this approach is extremely time consuming, expensive and incapable of detecting the true optimum conditions, especially because of the interactions among the factors (Calado and Montgomery, 2003; Siala *et al.*, 2012), as already discussed. Nowadays, design of experiments and RSM have been more used to optimize culture conditions and medium composition of fermentation processes, conditions of enzyme reaction and processing parameters in the production of food, drugs and enzymes by fungi, bacteria and yeasts.

Ramirez *et al.* (2001) studied the influence of some culture conditions on the final concentration of astaxanthin (a pigment of the carotenoid family) by using a $2^5$ factorial design with four central points. The five factors were: temperature, carbon concentration, nitrogen concentration, pH and inoculum. The authors used two different culture media – Yucca and YM. The statistically significant factors depended on the medium. The higher astaxanthin concentration was obtained for Yucca medium and the significant main effects were pH and carbon; the interaction effects were carbon × temperature and inoculum × temperature. Thus, the authors decided to apply central composite design to optimize the astaxanthin concentration by considering carbon and temperature (the main effect was not important but the interaction effects involving temperature were very important) as the only factors; the others were established at 5% of inoculum, 6.0 of pH and 0.5 g/l of nitrogen, because at these levels they obtained the highest astaxanthin concentration. The predicted optimum factor levels were carbon concentration equals to 11.25 g/l and temperature equals to 19.7°C; the maximum astaxanthin concentration was then 7823 μg/l. The adjusted $R^2$ equalled 0.985. The typical production conditions before optimization procedure were: temperature = 22°C, pH = 5.0, carbon concentration = 6.0 g/l, nitrogen concentration = 1.0 g/l and inoculum = 5%. Under these conditions, astaxanthin concentration was only 4200 μg/l, a value 54% less than the optimum one, showing the importance of optimizing a process.

In order to optimize the pectin hydrolysis by pectolytic enzymes produced by *Aspergillus niger*, Busto *et al.* (2007) used a central composite design, totalling 46 assays, and used enzyme concentration, substrate concentration, pH, temperature and reaction time as independent variables at five different levels (2 axial points), whereas the response variables were the reactor conversion, reducing sugar concentration, endopectolytic productivity and enzymatic depolymerization productivity. The authors used full second order polynomial models to explain the experimental data and results showed that the model did not present lack of fit and the $R^2$ ranged from 0.96 to 0.99. The optimization procedure was performed using the graphical approach and the optimum conditions were found to be 0.03% and 0.7% of enzyme and substrate concentrations, respectively, at 46°C, 1 hour of incubation time and pH 4.8.

Singhal and Bule (2009) used a Plackett–Burman design to assess the effect of medium component on the production of ubiquinone-10 (CoQ10), a vitamin-like lipophilic component with recognized antioxidant and anticancer effects. The authors varied the concentrations of glycerol, yeast extract, calcium carbonate and magnesium sulfate (independent variables) on the production of CoQ10, in a total of 30 different combinations. Data were subjected to RSM and the experimental results were fitted into a second order regression equation, which presented a $R^2 = 0.979$ and the regression was very significant ($p < 0.0001$). The optimal concentrations for the independent variables obtained from the model were 40, 17.72, 1.57 and 0.23 g/l for glycerol, yeast extract, calcium carbonate ($CaCO_3$) and magnesium sulfate ($MgSO_4$), respectively.

The kinetic activity of cellulolytic enzymes produced by *Aspergillus niger* during the solid state fermentation of potato peels was investigated by Santos *et al.* (2012). For this purpose, author used a $2^{3-1}$ fractional factorial design added with four central points to evaluate the influence of temperature, water content and time on the enzymatic activity of some enzymes. Pareto charts and 3D response surface plots were built to explain the influence of the factors on the responses, and the polynomial equations seemed to be suitable to describe the results once there was not significant lack of fit, the regression was deemed statistically significant, and the $R^2$ values were above 0.87. The desirability function was used to optimize the experimental conditions to maximize the kinetic activity of xylanase and the best combination of factors was: 81.92 hours of fermentation at 28.85°C, water content of 50.72%.

### *Food development*

Although it is still common to find researchers using the 'one at a time' approach to develop food products, this method has been put aside once it fails to optimize properties of a food product or even the best combination and levels of ingredients to enhance the desired properties, such as sensory appeal, nutritional profile and cost, among others. Nowadays, food companies have attempted to use RSM to develop food products to enhance product characteristics and to optimize industrial process to obtain a desired property. By using RSM, it is possible to check the significance of each ingredient and also the interaction of ingredients on each response, which is clearly an advantage towards the 'one factor at a time' approach. There are numerous publications regarding the development of new foods, beverages and ingredients, but here only a few are analysed.

A low/no added pork sausage formulation was developed by Murphy *et al.* (2004), where the effects of added surimi (0–40%), fat (5–30%) and water (10–35%) on the physical, textural and sensory properties were analysed by RSM. In order to accomplish the objective, the authors employed a central composite rotatable design containing five levels of each factor, totalling 15 formulations. Data were fitted with second order polynomial equations and results showed that the mathematical models were highly significant ($p < 0.05$) for protein and moisture contents, hardness, water-holding capacity and shear force. The authors did not provide any other quality parameter of the RSM models, inhibiting the full evaluation of the proposed models. Peaks in RSM three-dimensional plots and contour plots were used to extrapolate the optimum level of the three variables (surimi, fat and water). Extrapolation is not a suitable technique to optimize a food product. It would be more appropriate to use the simultaneous optimization to render a potential optimized formulation.

Wadikar *et al.* (2010) used a central composite design to develop ginger-based ready-to-eat appetizers. The formulation varied in relation to the content of raisins, red sugar, and ginger powder; samples were analysed in terms of sensory acceptability and total sugars. The data were subjected to multiple regression analysis and 3D surface plots were built to explain the experimental results. The quadratic polynomial equations were significant ($p < 0.05$) and the $R^2$ and adjusted $R^2$ were 0.9232/0.7849 and 0.9898/0.9716 for the sensory score and total content of sugars, respectively, showing that such models describe the actual data well. The food product was optimized by the numerical optimization procedure in order to maximize its sensory acceptability and authors observed that the optimized formulation had a shelf-life of eight months in metalized polyester pouches and contained 6.8 g/100 g of proteins, 5 g/100 g of crude fibre and 37 mg/100 g of vitamin C. This study showed that it is possible to develop new food products with enhanced functionality by using a response surface approach.

Dutcosky *et al.* (2006) developed tasty cereal bars with prebiotic functional properties using three sources of fibres: inulin, oligofructose and gum acacia. The authors used a simplex-centroid design, considering these three components. The response variables were degree of liking and the attributes selected (brightness, dryness of cereals flakes, banana volatile odour, cinnamon volatile odour, banana flavour, sweetness, crunchiness, hardness, chewiness). Applying the optimization technique of Derringer–Suich, two optimal formulations were detected: 50% inulin, 50% oligofructose and 0% gum acacia and/or 8.46% inulin, 66.16% oligofructose, and 25.38% gum acacia.

### *Sensory evaluation*

Deshpande *et al.* (2008) developed and optimized the overall acceptability of a chocolate-flavoured, peanut–soy beverage by using a three-component constrained mixture design, using peanut, soy (flour or protein isolate) and chocolate syrup as independent variables. The authors tested 28 formulations and data were subjected to multiple regression analysis; the graphical optimization technique was used to maximize the consumer acceptability of the final product. The optimal combination of factors was found to be 34.1–45.5 g/100 g peanut, 31.2–42.9 g/100 g soy flour and 22.4–24.1 g/100 g chocolate syrup.

Pepper-based appetizers, developed in the form of convenient beverage mixes, were developed by Wadikar *et al.* (2008). They used a central composite rotatable design without any blocking. The authors tested the effect of black-gram flour, milk powder, salt and pepper powder on the overall acceptability of test samples. The experimental data were used to fit a second order polynomial equation and results showed that the regression was significant ($p < 0.05$) and the $R^2$ and $R^2_{adj}$ were 72.76 and 59.96%, respectively, indicating the model was not so suitable to express the actual results, once it presented a low adjusted regression coefficient.

With the objective of optimizing the roasting of robusta coffee (*Coffea canephora* conillon), Mendes *et al.* (2001) employed a two factor central composite design (3 central points, 2 levels of axial points, totalling 11 samples) to optimize the settings for roasting time and the initial internal temperature of the roaster drum on response variables of sensory attributes (aroma, flavour and colour). The models for beverage aroma, flavour and colour presented no lack of fit ($p \geq 0.05$) and $R^2$ of 80%, 70% and 96%, respectively. The $R^2$ for the predictive model of beverage colour is quite high, although those referring to the predictive models for the acceptance of aroma and flavour are also satisfactory, considering that the response variables are hedonic sensory measurements, which often show a high variation.

### *Nutrition*

Numerous studies have demonstrated that spices have considerable antioxidant properties, mostly because of the amount and variety of polyphenolic compounds present in those plant extracts. In this regard, Hossain *et al.* (2011) used a central composite design to investigate the effects of methanol concentration and extraction temperature on the phenolic compounds and antioxidant activity measured by the FRAP (ferric reducing antioxidant power) assay. Data were fitted into a second order polynomial equation and the authors obtained high $R^2$ values – ranging from 0.952 to 0.99 for both variables. In addition, the lack of fit results were not significant ($p > 0.05$) and the regression models were highly significant ($p < 0.0001$), proving the suitability of RSM to analyse and model the extraction of antioxidant polyphenols from spices.

Trevisan and Areas (2012) worked with a production of corn–flaxseed snacks aiming at obtaining the maximum expansion ratio (ER), as the sensory quality and the acceptance of snack foods depend mainly on this variable and texture parameters. They analysed the effects of three independent extrusion parameters (variables), moisture content (x1), temperature (x2) and flaxseed flour content (x3) on the ER. By using a centre composite design, the authors concluded that the factor levels that resulted in a maximum expansion ratio (3.93) were: humidity = 19%, temperature = 123°C, and flaxseed content = 25%.

Martínez *et al.* (2004) used a second order fractional factorial design including three levels for each factor (carrot, rice, pea/potato, chicken/veal liver) to develop infant foods (beikosts) with a goal of achieving low amounts of antinutritive substances and high trace element content. The results were subjected to response surface methodology and the authors verified that carrot was the main source of tannins in beikosts and was the key factor in controlling antinutritive substances, whereas rice and potato were key ingredients for controlling phytic acid content in the formula. None of the vegetable ingredients exerted major effects on trace element content in the final product, with the exception of a significant effect of rice on manganese content and pea on copper content. From this study, it is possible to state that the development of foods with special nutritional requirements is feasible by using a statistical approach.

## STATISTICAL PACKAGES

In order to design and analyse experimental data, there are some free (the well-known are *R* and Action for Microsoft Excel) and commercial statistical packages, such as SAS (*Statistical Analysis Software*), SPSS (*Statistical Package for Social Science*), Statistica, Statgraphics, Minitab, Design-Expert and Prisma, among others. Among these, Minitab and Statistica are the most used packages for design of experiments. They both have a friendly interface, although Statistica seems more complete and has a magnific graphics output. Action software, developed by Brazilian scientists, is also free to download and presents the DOE features. This software also has suitable graphics output and is the first statistical system that utilizes the *R* platform together with Microsoft Excel.

## FINAL REMARKS AND PERSPECTIVES

The use of DOE and RSM for food development and process optimization in Food Science and Technology has increased in the last 10 years. In this paper, some types of experimental designs and examples of how DOE and RSM may be applied in microbiology, sensory tests, process optimization and nutrition were reported to provide experimental information to future experimenters. Within this context, the authors believe that readers can take into consideration this information in order to build and to analyse experimental designs to help them to obtain the right answers for their problems.

## REFERENCES

Bas, I., and Boyaci, I. S. (2007) Modelling and optimization I: Usability of response surface methodology. *Journal of Food Engineering*, **78**, 836–845.

Box, G. E. P., and Draper, N. (1987) *Empirical model-building and response surface*, 1st edn. John Wiley & Sons, Inc., New York.

Busto, M. D. Rodríguez-Nogales, J. M., Ortega, N. and Perez-Mateos, M. (2007) Experimental design and response surface modelling applied for the optimisation of pectin hydrolysis by enzymes from *A. niger* CECT 2088. *Food Chemistry*, **101** 634–642.

Calado, V., and Montgomery, D. C. (2003) *Planejamento de Experimentos usando o Statistica*, 1st edn. E-papers Serviços Editoriais Ltda, Rio de Janeiro. www.e-papers.com.br (last accessed 2 July 2013).

Capitani, C., Carvalho, A. C., Botelho, P. P. *et al.* (2009) Synergism on antioxidant activity between natural compounds optimized by response surface methodology. *European Journal of Lipid Science and Technology*, **111**(11), 1100–1110.

Cruz, A. G., Faria, J. A. F., Walter, E. H. M. *et al.* (2010) Processing optimization of probiotic yogurt supplemented with glucose oxidase using response surface methodology. *Journal of Dairy Science*, **93**, 5059–5068.

Deshpande, R. P., Chinnan, M. S. and McWatters, K. H. (2008) Optimization of a chocolate-flavored, peanut–soy beverage using response surface methodology (RSM) as applied to consumer acceptability data. *LWT – Food Science and Technology*, **41**, 1485–1492.

Dutcosky, S. D., Grossmann, M. V. E., Silva, R. S. and Welsch, A. K. (2006) Combined sensory optimization of a prebiotic cereal product using multicomponent mixture experiments. *Food Chemistry*, **98**, 630–638.

Ellendersen, L. S. N., Granato, D., Guergoletto, K. B. and Wosiacki, G. (2012) Development and sensory profile of a probiotic beverage from apple fermented with *Lactobacillus casei*. *Engineering in Life Sciences*, **12**, 475–485.

Farris, F. and Piergiovanni, L. (2009) Optimization of manufacture of almond paste cookies using response surface methodology. *Journal of Food Process Engineering*, **32**, 64–87.

Gonzalez-Barreiro, C., Lores, M., Casais, M. C. and Cela, R. (2000) Optimisation of alachlor solid-phase microextraction from water samples using experimental design. *Journal of Chromatography A*, **896**, 373–379.

Granato, D., Bigaski, J., Castro, I. A. and Masson, M. L. (2010a) Sensory evaluation and physicochemical otimisation of soy-based desserts using response surface methodology. *Food Chemistry*, **121**(3), 899–906.

Granato, D., Castro, I. A., Ellendersen, L. S. N. and Masson, M. L. (2010b) Physical stability assessment and sensory optimization of a dairy-free emulsion using response surface methodology. *Journal of Food Science*, **73**, 149–155.

Haaland, P. D. (1989) Statistical problem solving. In: *Experimental Design in Biotechnology* (ed. P. D. Haaland). Marcel Dekker, New York, NY, pp. 1–18.

Henika, R. B. (1982) Use of response-surface methodology in sensory evaluation. *Food Technology*, **36**, 96–101.

Hossain, M. B., Barry-Ryan, C., Martin-Diana, A. B. and Brunton, N. P. (2011) Optimisation of accelerated solvent extraction of antioxidant compounds from rosemary (*Rosmarinus officinalis* L.), marjoram (*Origanum majorana* L.) and oregano (*Origanum vulgare* L.) using response surface methodology. *Food Chemistry*, **126**, 339–346.

Iwe, M. O., van Zuilichem, D. J., Stolp, W. and Ngoddy, P. O. (2004) Effect of extrusion cooking of soy–sweet potato mixtures on available lysine content and browning index of extrudates. *Journal of Food Engineering*, **62**, 143–150.

Jing, P., Ruan, Si-Y., Dong, Y. *et al.* (2011) Optimization of purification conditions of radish (*Raphanus sativus* L.) anthocyanin-rich extracts using chitosan. *LWT – Food Science and Technology*, **44**, 2097–2103.

Jo, J. H., Lee, D. S., Park, D. and Park, J. M. (2008) Statistical optimization of key process variables for enhanced hydrogen production by newly isolated *Clostridium tyrobutyricum* JM1. *International Journal of Hydrogen Energy*, **33**, 5176–5183.

Joglekar, A. M. and May, A. T. (1987) Product excellence through design of experiments. *Cereal Foods World*, **32**, 857–868.

Kaur, S., Sarkar, B. C. and Sharma, H. K. (2009) Optimization of enzymatic hydrolysis pretreatment conditions for enhanced juice recovery from guava fruit using response surface methodology. *Food and Bioprocess Technology*, **2**, 96–100.

Khuri, A. I. and Mukhopadhyay, S. (2010) Response surface methodology. *WIREs Computational Statistics*, **2**, 128–149.

Kliemann, E., De Simas, K. N., Amante, E. R. *et al.* (2009) Optimization of pectin acid extraction from passion fruit peel (*Passiflora edulis flavicarpa*) using response surface methodology. *International Journal of Food Science and Technology*, **44**, 476–483.

Martínez, B., Rincón, F., Ibáñez, M. V. and Abellán, P. (2004) Improving the nutritive value of homogenized infant foods using response surface methodology. *Journal of Food Science*, **69**, 38–43.

Mendes, L. C., Menezes, H. C. and Silva, M. A. A. P. (2001) Optimization of the roasting of robusta coffee (*C. canephora* conillon) using acceptability tests and RSM. *Food Quality and Preference*, **12**, 153–162.

Myers, R. H. and Montgomery, D. C. (2002) *Response Surface Methodology*, 2nd edn. John Wiley & Sons, Inc., New York.

Michel, T., Destandau, E. and Elfakir, C. (2011) Evaluation of a simple and promising method for extraction of antioxidants from sea buckthorn (*Hippophaë rhamnoides* L.) berries: pressurised solvent-free microwave assisted extraction. *Food Chemistry*, **126**, 1380–1386.

Montgomery, D. C. (2009) *Design and analysis of experiments*, 5th edn. John Wiley & Sons, Inc., New York.

Murphy, S. C., Gilroy, D., Kerry, J. F. *et al.* (2004) Evaluation of surimi, fat and water content in a low/no added pork sausage formulation using response surface methodology. *Meat Science*, **66**, 689–701.

Myers, R. H., Montgomery, D. C. and Anderson-Cook, C. M. (2009) *Response Surface Methodology: process and product optimization using designed experiments*, 3rd edn. John Wiley & Sons, Inc., New York.

Nikzadeh, V. and Sedaghat, N. (2008) Physical and sensory changes in pistachio nuts as affected by roasting temperature and storage. *American-Eurasian Journal of Agriculture and Environmental Sciences*, **4**(4), 478–483.

Nwabueze, T. U. (2010) Basic steps in adapting response surface methodology as mathematical modelling for bioprocess optimization in the food systems. *International Journal of Food Science and Technology*, **45**(9), 1768–1776.

Pelissari, F. M., Yamashitaa, F., Garciab, M. A. *et al.* (2012) Constrained mixture design applied to the development of cassava starch–chitosan blown films. *Journal of Food Engineering*, **108**, 262–267.

Ramirez, J., Gutierrez, H. and Gschaedler, A. (2001) Optimization of astaxanthin production by *Phaffia rhodozyma* through factorial design and response surface methodology, *Journal of Biotechnology*, **88**, 259–268.

Reis, F. R., Masson, M. L., Waszczynskyj, N. (2008) Influence of a blanching pretreatment on color, oil uptake and water activity of potato sticks, and its optimization. *Journal of Food Process Engineering*, **31**(6), 833–852.

Santos, T. C., Gomes, D. P. P., Bonono, R. C. F. and Franco, M. (2012) Optimisation of solid state fermentation of potato peel for the production of cellulolytic enzymes. *Food Chemistry*, **133**, 1299–1304.

Shihani, N., Kumbhar, B. K. and Kulshreshtha, M. (2006) Modeling of extrusion process using response surface methodology and artificial neural networks. *Journal of Engineering Science and Technology*, **1**(1), 31–40.

Siala, R., Frikha, F., Mhamdi, S. *et al.* (2012) Optimization of acid protease production by *Aspergillus niger* I1 on shrimp peptone using statistical experimental design. *The Scientific World Journal*, **2012**, 1–11. doi:10.1100/2012/564932.

Sin, H. N., Yusof, S., Hamid, S. A. and Rahman, A. (2006) Optimization of hot water extraction for sapodilla juice using response surface methodology. *Journal of Food Engineering*, **74**, 352–358.

Singhal, R. S. and Bule, M. V. (2009) Use of carrot juice and tomato juice as natural precursors for enhanced production of ubiquinone-10 by *Pseudomonas diminuta* NCIM 2865. *Food Chemistry*, **116**, 302–305.

Tiwary, A. K., Puri, M., Sharma, D. and Barrow, C. J. (2012) Optimisation of novel method for the extraction of steviosides from Stevia rebaudiana leaves. *Food Chemistry*, **132**, 1113–1120.

Tonon, R. V., Brabet, C. and Hubinger, M. D. (2008) Influence of process conditions on the physicochemical properties of açai (*Euterpe oleraceae* Mart.) powder produced by spray drying. *Journal of Food Engineering*, **88**, 411–418.

Trevisan, A. J. B. and Areas, J. A. G. (2012) Development of corn and flaxseed snacks with high-fibre content using response surface methodology (RSM), *International Journal of Food Sciences and Nutrition*, **63**(3), 362–367.

Wadikar, D. D., Majumdar, T. K., Nanjappa, C. *et al.* (2008) Development of shelf stable pepper based appetizers by response surface methodology (RSM). *LWT – Food Science and Technology*, **41**, 1400–1411.

Wadikar, D. D., Nanjappa, C., Premavalli, K. S. and Bawa, A. S. (2010) Development of ginger based ready-to-eat appetizers by response surface methodology. *Appetite*, **55**, 76–83.

Yi, C., Shi, J., Xue, S. J. *et al.* (2009) Effects of supercritical fluid extraction parameters on lycopene yield and antioxidant activity. *Food Chemistry*, **113**(4), 1088–1094.

Zanariah, C. W., Ngah, C. W. and Yahya, M. A. (2012) Optimisation of digestion method for determination of arsenic in shrimp paste sample using atomic absorption spectrometry. *Food Chemistry*, **134**(4), 2406–2410.

# 2 The use of correlation, association and regression to analyse processes and products

**Daniel Cozzolino**

*School of Agriculture, Food and Wine, Faculty of Sciences, The University of Adelaide, Waite Campus, Glen Osmond, SA, Australia*

## ABSTRACT

No single technique can solve all analytical issues in the food processing industries. However, modern instrumental techniques such as visible (Vis), near infrared (NIR), and mid infrared (MIR) spectroscopy, electronic noses (EN), electronic tongues (ET) and other sensors (e.g. temperature, pressure) combined with multivariate data analysis (MVA) have many advantages over chemical, physical and other classic instrumental methods of analysis. This chapter provides a general description of the methods and applications of regression analysis based on MVA used during the process analysis in the food industries. Most of the examples and applications given in this chapter are related with the use of infrared spectroscopy. However, the same basic principles can be applied to a different number of methods and instruments currently used by the food industry.

## INTRODUCTION

No single technique can solve all analytical issues in the food processing industries. However, modern instrumental techniques such as visible (Vis), near infrared (NIR), and mid infrared (MIR) spectroscopy, electronic noses (EN), electronic tongues (ET) and other sensors (e.g. temperature, pressure) have many advantages over chemical, physical and other classic instrumental methods of analysis (Arvantoyannis *et al.*, 1999; Blanco and Villaroya, 2002; McClure, 2003; Cozzolino *et al.*, 2006a, 2011a, 2011b; Cozzolino, 2007, 2009, 2010, 2011, 2012; Roggo *et al.*, 2007; Hashimoto and Kameoka, 2008; Huang *et al.*, 2008; Cozzolino and Murray, 2012). Traditionally, much of the research into analytical and food chemistry has been conducted in a manner that can be described as 'univariate' in nature, since it was focused only into the examination of the effects (responses) of a single variable on the overall matrix (Munck *et al.*, 1998; Jaumot *et al.*, 2004; Cozzolino and Murray, 2012). At the time that many statistical methods were developed around the 1920s (Bendell *et al.*, 1999), samples were considered cheap and measurements expensive. Since that time, the nature of technology has changed; at present, samples are considered expensive while measurements cheap (Gishen *et al.*, 2005; Cozzolino and Murray, 2012).

The analysis of the effects of one variable at a time by the application of classical statistical analysis (e.g. analysis of variance) can provide useful descriptive information. However, specific information about relationships among variables and other important relationships in the entire matrix might be lost (Wold, 1995; Martens and Martens, 2001; Munck, 2007; Munck *et al.*, 2010). Multivariate analysis (MVA) or chemometrics was developed in the late 1960s, and introduced by a number of research groups in chemistry,

*Mathematical and Statistical Methods in Food Science and Technology*, First Edition.
Edited by Daniel Granato and Gastón Ares.

mainly in the fields of analytical, physical and organic chemistry, in order to deal with the introduction of instrumentation, which provides multiple responses (e.g. peaks, wavelengths) for each sample analysed, as well as with the increasingly availability and use of computers (Wold, 1995; Munck *et al.*, 1998; Otto, 1999). With modern chemical measurements, we are often confronted with so much data that the essential information might not be readily evident. Certainly that can be the case with chromatographic or spectral data for which many different observations (peaks or wavelengths) have been collected in a single analysis (Wold, 1995; Munck *et al.*, 1998; Otto, 1999; Munck and Moller, 2011).

Traditionally as analysts, we strive to eliminate matrix interference in our methods by isolating or extracting the analyte we wish to measure, making the measurement apparently simple and certain (Wold, 1995; Munck *et al.*, 1998; Otto, 1999). However, this scientific approach ignores the possible effects of chemical and physical interactions between the large amounts of constituents present in the sample – this will be especially evident for such complex materials as food. Univariate models do not consider the contributions of more than one variable source and can result in models that could be an oversimplification (Wold, 1995; Munck *et al.*, 1998; Otto, 1999). Therefore, we need to look at the sample in its whole and not just at a single component if we wish to unravel all the complicated interactions between the constituents as well as to understand their combined effects on the whole matrix (Wold, 1995; Geladi, 2003). Multivariate methods provide the means to move beyond the one-dimensional (univariate) world (Wold, 1995; Geladi, 2003). In many cases, MVA can reveal constituents that are important through the various interferences and interactions (Wold, 1995; Geladi, 2003). Today, many food quality measurement techniques are multivariate and based on indirect measurements of the chemical and physical properties by the application of modern instrumental methods and techniques (Esbensen, 2002; Geladi, 2003; Hashimoto and Kameoka, 2008; Woodcock *et al.*, 2008; Blanco and Bernardez, 2009; Cozzolino and Murray, 2012).

A typical characteristic of many of the most useful of these instrumental techniques is that, paradoxically, the measurement variable might not have a direct relationship with the property of interest; for instance, the concentration of a particular chemical in the sample – that is, the technique is a correlative method. The explanation for this is often found in chemical and physical interferences. For example spectroscopy techniques provide the possibility to obtain more information from a single measurement, because they can record responses at many wavelengths simultaneously, and it becomes essential to use MVA in order to extract the information. Specific details of the numerous algorithms, formulas and procedures used in multivariate analysis can be found in more specialized literature (Massart *et al.*, 1988; Wold, 1995; Munck *et al.*, 1998; Otto, 1999; Esbensen, 2002; Geladi, 2003; Mark and Workman, 2003).

This chapter provides a general description of the methods and applications of regression analysis based on multivariate data analysis used during the process analysis in the food industries. Most of the examples and applications given are related to the use of infrared spectroscopy. However, the same basic principles can be applied to a different number of methods and instruments currently used by the food industry.

## PROCESS ANALYSIS

Over the past 30 years, on/in/at line analysis, the so called process analytical technologies (PAT), has provided to be one of the most efficient and advanced tools for continuous monitoring, controlling the processes and the quality of raw ingredients and products in several fields, including food processing, petrochemical and pharmaceutical industries (Beebe *et al.*, 1993; Workman *et al.*, 1999, 2001; Liu *et al.*, 2001; Blanco and Villaroya, 2002; Huang *et al.*, 2008; Karande *et al.*, 2010). In this new context the sample becomes an integral component of the system (Figure 2.1). We are moving from the laboratory to the process (Esbensen, 2002; Kueppers and Haider, 2003).

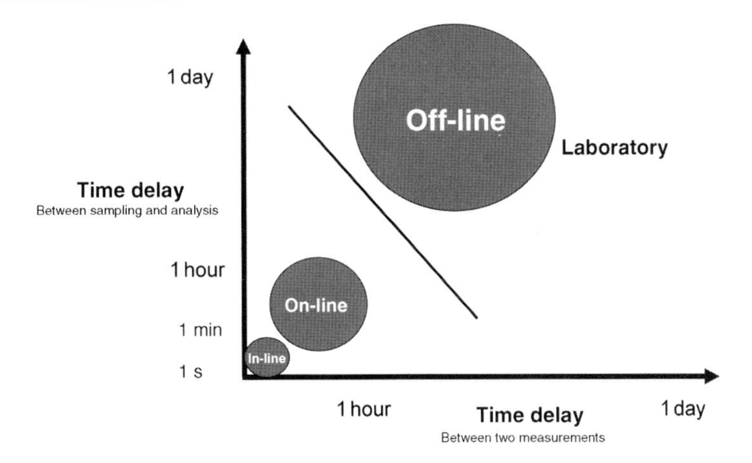


**Figure 2.1** Sampling: from the laboratory to the 'real world'. Adapted from Kuppers & Haider, 2003.

Numerous applications can be found on the use and implementation of instrumental methods such as those based on infrared (IR), temperature, moisture, gas, and pressure sensors for on-line and at-line analysis. The PAT tools are categorized into four areas, namely multivariate data acquisition and analysis tools, modern process analyser or process analytical tools, process and endpoint monitoring and control tools, and continuous improvement and knowledge management tools (Workman *et al.*, 1999, 2001; Skibsted, 2005). On-line applications based on IR spectroscopy started around the end of the 1980s. They were developed primary due to the availability of fibre optics (Workman *et al.*, 1999, 2001; Skibsted, 2005). Most of such applications were related with the control and monitoring of liquids, mainly because the application of fibre optics was easier to use with such type of samples. Simple systems are also used based on single sensors that collect changes in temperature, pressure, pH and in the mixture of gases, among others.

In today's manufacturers environment there is an increasing need for real process analytical chemistry. The driving force behind this transition from traditional laboratory analysis to process analysis is the need for more repaid process control information, as well as economical, safety, and environmental factors (Workman *et al.* 1999, 2001; Skibsted, 2005). Therefore, complex or multiplex systems (combination of several sensors) and techniques are used (Blanco and Villaroya, 2002; McClure, 2003; Roggo *et al.*, 2007; Blanco and Bernardez, 2009; Huang *et al.*, 2008; Cozzolino, 2009). The overall requirements on the use of MVA methods for on line applications are: (i) can be assembled in the production line and take place under realistic environment; (ii) early detection of possible failures; (iii) permanent monitoring of the conditions; and (iv) assessment of the conditions at any desired time (Roggo *et al.*, 2007; Blanco and Bernardez, 2009; Huang *et al.*, 2008).

## MULTIVARIATE METHODS

Multivariate statistical techniques (or chemometrics) are therefore required to extract the information about the attributes measured that is buried in the signal produced by the instrumental method applied ('model calibration'). Essentially, this process involves regression techniques coupled with pre-processing methods (Brereton, 2000, 2008; Næs *et al.*, 2002; Cozzolino *et al.*, 2006a, 2011a, 2011b; Nicolai *et al.*, 2007). In many applications, it is expensive, time consuming or difficult to measure a property of interest directly. Such cases require the analyst to predict something of interest based on related properties that are easier to measure (Næs *et al.*, 2002; McClure, 2003; Nicolai *et al.*, 2007; Cozzolino *et al.*, 2006, 2011; Walsh and

Kawano, 2009). Hence, the goal of MVA regression analysis is to develop a calibration model which correlates the information in the set of known measurements to the desired property.

Calibration is the process by which the mathematical relationship between the values provided by a measuring instrument or system and those known for the measured material object is established. The mathematical expression relating analytical responses or signals to concentrations is known as the calibration equation or calibration model (Mark and Workman, 2003; Blanco and Bernardez, 2009). Most analytical techniques use a straight line for calibration on account of its straightforward equation and its ability to illustrate a direct relationship between measured signals and concentrations (univariate calibration) (Mark and Workman, 2003; Blanco and Bernardez, 2009). However, a linear calibration model can only be useful for quantitation purposes if the analytical signal depends exclusively on the concentration of the specific analyte for which the model has been developed. Such exclusive dependence is the exception rather than the rule in the analysis of complex samples by spectroscopic techniques, such as IR spectroscopy, as well as in other instrumental analytical methods (Mark and Workman, 2003; Nicolai *et al.*, 2007; Blanco and Bernardez, 2009).

Calibration models are usually constructed by least squares regression (LSR) of the absorbance (or apparent absorbance) values for a set of standards against their concentrations. Multiple linear regression (MLR) is subject to two major restrictions. One is the dimension of matrix **X**; thus, the number of variables used cannot exceed that of samples. The other is that no two $X$ variables should be mutually related; otherwise, the matrix (**X** T **X**) cannot be inverted. In real-world applications, where data are typically noisy, variables are highly unlikely to be fully correlated; however, a substantial degree of correlation between variables can lead to an unstable inverted matrix (Haaland and Thomas, 1988; Mark and Workman, 2003; Blanco and Bernardez, 2009). Other common MVA algorithms for performing regression include partial least squares (PLS) and principal component regression (PCR) (Haaland and Thomas, 1988). These regression methods are designed to avoid issues associated with noise and correlations (collinearity) in the data (Geladi and Kowalski, 1986; Wold *et al.*, 2001; Næs *et al.*, 2002; Walsh and Kawano, 2009). Table 2.1 summarizes some of the most common algorithms used to develop calibrations (Blanco and Bernardez, 2009).

Because spectroscopy is a relative technique, the samples used for calibration must be previously analysed with adequate accuracy and precision (Blanco and Bernardez, 2009). This entails using an instrument capable of remaining operational for a long time and a simple, robust enough model capable of retaining its predictive ability for new samples over long periods. Constructing a multivariate calibration model is a complex, time consuming process that requires careful selection of variables in

**Table 2.1** Common algorithms used to develop calibrations. Adapted from Blanco, M., Bernardez, M. (2009).

| Method | Characteristics |
|---|---|
| Least squares (LS) | Easy to calculate and understand, it is used for simple sample sets (e.g. pure compounds). Requires isolated variables (e.g. spectral bands), large prediction errors. |
| Classic least squares (CLS) | Based in Beers' Law, it uses a large number of variables, not suitable for mixture of compounds, susceptible to baseline effects |
| Inverse least squares (ILS) | It is a flexible method used for indirect calibration only and unrestricted as regards the number of variables, it is the basis for multivariate regression |
| Principal component regression (PCR) | It is a flexible, full-spectrum method. variable-compression, it uses inverse regression, optimization requires knowledge of PCA |
| Partial least squares regression (PLSR) | Flexible, full-spectrum, variable-compression method compatible with inverse and indirect calibration, combines ILS and CLS, calibrations are robust, effect of collinearity |
| Artificial neural networks (ANNs) | Flexible, compression method using a restricted number of input variables and compatible with inverse and indirect calibration |

order to ensure accurate prediction of unknown samples. This requires knowledge not only of the target samples but also of MVA techniques, in order to obtain a model retaining its predictive ability over time and amenable to easy updating. Because the model will usually be applied by unskilled operators, it should deliver analytical information in an easily interpreted manner (Blanco and Bernardez, 2009). The process of obtaining a robust model involves the following steps:

(i) choosing the samples for inclusion in the calibration set;
(ii) determining the property to be predicted by using an appropriate method to measure such samples;
(iii) obtaining the analytical instrumental signal (e.g. spectra);
(iv) constructing the model;
(v) validation;
(vi) pre-processing of the data;
(vii) prediction of unknown samples.

Selection of calibration samples is one of the most important steps in constructing a calibration model and involves choosing a series of samples, which ideally should encompass all possible sources of physical and chemical variability in the samples to be subsequently predicted (Blanco and Bernardez, 2009). The model will only operate accurately if both the calibration samples and the prediction samples belong to the same population. Usually, the set or population of available samples is split into two subsets, called the calibration set (used to construct the model) and the validation set (Næs *et al.*, 2002; McClure, 2003; Cozzolino *et al.*, 2006a, 2011a, 2011b; Nicolai *et al.*, 2007; Walsh and Kawano, 2009). The samples included in the calibration set should span the whole variability in both calibration and validation; thus, the selected samples should be uniformly distributed throughout the calibration range in the multidimensional space defined by spectral variability. One simple method for selecting samples based on spectral variability uses a scatter plot obtained from a principal component analysis (PCA) applied to the whole set of available spectra. Inspecting the most significant PCs in the graph allows the distribution of the sample spectra to be clearly envisaged; those to be included in the calibration set are chosen from both the extremes and the middle of the score maps obtained and simultaneously checked to uniformly encompass the range spanned by the quantity to be determined (Figure 2.2). This method is effective when the first two or three PCs contain a high proportion of the total variance (Mark, 1991; Fearn, 1997, 2002; Næs *et al.*, 2002; McClure, 2003; Cozzolino *et al.*, 2006a, 2006b, 2011a, 2011b; Nicolai *et al.*, 2007; Blanco and Bernardez, 2009; Walsh and Kawano, 2009). As a rule of thumb, samples used to build a calibration model should be selected from samples similar to those that will be analysed in the future (Murray, 1993, 1999; Murray and Cowe, 2004). The calibration samples should be subjected exactly to the same handling process to be adopted with future samples. It is also recommended to obtain the widest range in composition not only for the property of interest but for all sources of possible variation to be encounter in the future. Obtain sufficient extreme samples to emphasize the tail of the distribution (Murray, 1993, 1999; Murray and Cowe, 2004).

As previously described, before building a calibration model for a given analyte using spectra (e.g. NIR and MIR), a series of steps needs to be considered. Spectral pre-processing techniques are used to remove any irrelevant information that cannot be handled by the regression techniques. Several pre-processing methods have been developed for this purpose and several references can be found elsewhere (Brereton, 2000; Martens and Martens, 2000; Mark and Workman, 2003; Næs *et al.*, 2002; Blanco and Bernardez, 2000; Karoui *et al.*, 2010). This can include averaging over spectra, which is used to reduce the number of wavelengths or to smooth the spectrum (Nicolai *et al.*, 2007), moving average filters and the Savitzky–Golay algorithm (Næs *et al.*, 2002; Mark and Workman, 2003; Nicolai *et al.*, 2007; Blanco and Bernardez, 2009). Another pre-processing technique commonly used is standardization. Standardization means dividing the spectrum at every wavelength by the standard deviation of the spectrum at this

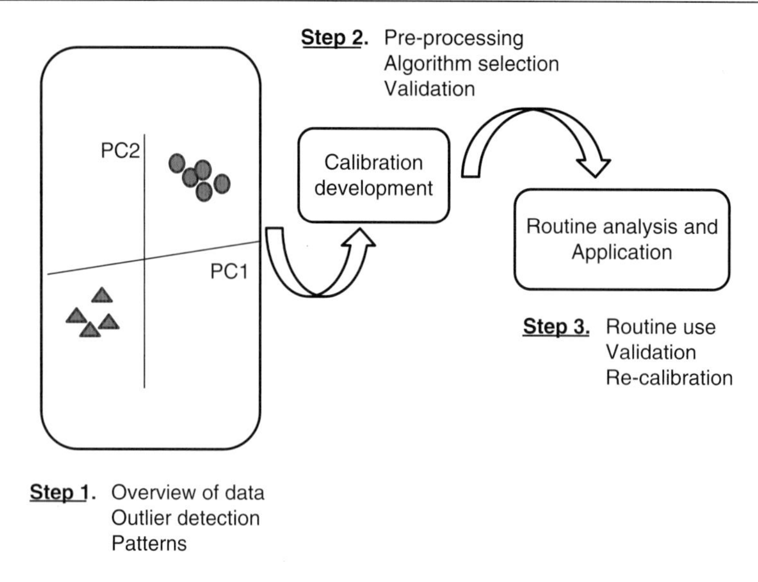


**Figure 2.2** Steps during the application of multivariate data methods in food processing.

wavelength. Typically variances of all wavelengths are standardized to one, which results in an equal influence of the variables in the model. Other standardization procedures are possible as well (Næs *et al.*, 2002; Mark and Workman, 2003; Blanco and Bernardez, 2009). Standardization is commonly used when variables are measured in different units or have different ranges. While most chemometrics software packages offer several normalization methods, multiple scatter correction (MSC) and standard normal variate correction (SNV), are the most popular normalization technique (Dhanoa *et al.*, 1994; Næs *et al.*, 2002; Mark and Workman, 2003; Blanco and Bernardez, 2009). MSC is used to compensate for additive (baseline shift) and multiplicative (tilt) effects in the spectral data, which are induced by physical effects (scattering, particle size and the refractive index) (Mark and Workman, 2003; Nicolai *et al.*, 2007; Blanco and Bernardez, 2009). In SNV, each individual spectrum is normalized to zero mean and unit variance (Dhanoa *et al.*, 1994). Derivation is used to remove baseline shifts and superposed peaks (Duckworth, 2004; Næs *et al.*, 2002).

## OUTLIER DETECTION

Outliers may be induced by typing errors, file transfer, interface errors, sensor malfunctions and fouling, poor sensor calibration, bad sampling or sample presentation, among other factors. A sample can be considered as an outlier according to the X-variables only (spectra), to the Y-variables only (reference), or to both. It might also not be an outlier for either separate sets of variables but become an outlier when the X–Y relationship is considered. During calibration, outliers related to the spectra show up in a graphical scale or diagram called a PCA scores plot as points outside the normal range of variability. Alternatively, the leverage of a spectrum might be calculated as the distance to the centre of all spectra relative to the variability in its particular direction. Additionally, X-residuals plots can be constructed. Y-outliers can be identified as extreme values in the Y-residuals plot. In practice, however, only those outliers that have an effect on the regression model have to be removed (Adams, 1995; Wise and Gallagher, 1996; Mark and Workman, 2003; Nicolai *et al.*, 2007; Brereton, 2008; Cozzolino *et al.*, 2011a, 2011b).

## MODEL ACCURACY AND VALIDATION

Once the calibration model has been developed, its ability to predict unknown samples not present in the calibration set (e.g. not used to construct the model) should be assessed. This involves applying the model to a limited number of samples not included in the calibration set, for which the target property to be predicted by the model is previously known. The results provided by the model are directly compared with the reference values; if the two are essentially identical, the model will afford accurate predictions and be useful for determining the target property in future (e.g. unknown samples). In order to assess the accuracy of the calibration model and to avoid overfitting, validation procedures have to be applied; a calibration model without validation is nonsense. In feasibility studies, cross validation is a practical method to demonstrate that NIR spectroscopy can predict something but the actual accuracy must be estimated with an appropriate test set or validation set (Otto, 1999; Brereton, 2000; Næs *et al.*, 2002). As mentioned previously, the predictive ability of the method needs to be demonstrated using an independent validation set. Independent means that samples need to come from experiments, harvest times, or new batches with spectra all taken at a time different from the calibration spectra (Fearn, 1997; Cozzolino *et al.*, 2011a, 2011b); for example samples obtained from a different orchard, different season or different region or environment.

Many statistics are reported in the literature to interpret a calibration. The prediction error of a calibration model is defined as the root mean square error for cross validation (RMSECV) when cross validation is used or the root mean square error for prediction (RMSEP) when internal or external validation is used (Næs *et al.*, 2002; Walsh and Kawano, 2009). This value gives the average uncertainty that can be expected for predictions of future samples. The results of future predictions with a 95% confidence interval can be expressed as the predicted value $Y = \pm 1.96 \times \text{RMSEP}$ (Næs *et al.*, 2002; Walsh and Kawano, 2009). The number of latent variables (terms) in the calibration model is typically determined as that which minimizes the RMSECV or RMSEP. In some publications the standard error of prediction (SEP) is reported instead of the RMSEP (Adams, 1995; Brereton, 2000; Næs *et al.*, 2002). The quality of the model can be also evaluated if it provides an RMSEP value not exceeding 1.4 times the standard error of the laboratory (SEL).

Some other useful statistics are commonly used to interpret calibrations, namely residual predictive deviation (RPD) (Williams, 2001; Fearn, 2002). The RPD is defined as the ratio of the standard deviation of the response variable to the RMSEP or RMSECV (some authors use the term SDR). An RPD between 1.5 and 2 means that the model can discriminate low from high values of the response variable; a value between 2 and 2.5 indicates that coarse quantitative predictions are possible; and a value between 3 and 5 or above corresponds to good and excellent prediction accuracy, respectively (Williams, 2001; Fearn, 2002). The coefficient of determination ($R^2$) represents the proportion of explained variance of the response variable in the calibration or validation sets. Correlation (r) addresses what the strength of the (usually linear) relationship is between two variables. Thus, correlation cannot sufficiently explain what the 'cause' of a relationship is; it therefore says nothing about the specificity of a calibration. In the same way, as with correlation, predictions cannot sufficiently explain what the 'cause' of a relationship is; it therefore says nothing about the specificity of a calibration. 'Relationships' means that there is some structured association (linear, quadratic etc.) between **X** and **Y**. Note, however, that even though causality implies association, association does not imply causality (causality is not proved by association). Correlation is a measure of the strength of the (usually linear) relationship between two variables. The validity of a prediction is the degree to which a measure predicts the future behaviour or results it is designed to predict. Finally, bias is a measure of the difference between the average or expected value of a distribution (i.e. NIR predictions) and the true value (i.e. assay). As with correlation and prediction, bias cannot sufficiently explain what the 'cause' of a relationship is; therefore it too says nothing about the specificity of a calibration (Norris and Ritchie, 2008).

## OVERFITTING AND UNDERFITTING

When using any multivariate data technique, it is important to select an optimum number of variables or components (Adams, 1995; Næs *et al.*, 2002). If too many are used, too much redundancy in the X-variables is modelled and the solution can become overfitted – the model will be very dependent on the data set and will give poor prediction results. On the other hand, using too few components will cause underfitting and the model will not be large enough to capture the variability in the data. This 'fitting' effect is strongly dependent on the number of samples used to develop the model; in general, more samples give rise to more accurate predictions (Næs *et al.*, 2002).

## ROUTINE ANALYSES AND APPLICATIONS

A validated calibration model is fit for use in routine analyses and can be used unaltered over long periods provided a reference method is used from time to time to analyse anecdotal samples in order to check whether it continues to produce accurate and precise results (Blanco and Bernardez, 2009). Likewise, the instrument should be monitored over time in order to detect any alteration in its response or performance. By using control graphs for the results obtained over time, deviations in the model or instrument can detected and appropriate corrective measures taken. If the instrument is found to operate as scheduled, it can be suspected that deviations in the results are due to some failure in the model, which can be checked by analysing the samples concerned with the reference method, or due to a change in the target samples caused by the presence of a new source of variability, which will reflect in an expanded confidence range for the results and call for recalibration of the model (Blanco and Bernardez, 2009).

The advantages of using MVA techniques combined with instrumental methods enable detection of quality changes of raw materials and final product under steady or process conditions. To determine the efficacy of using any analytical technology in a process, several critical factors must be considered. For example is continuous real time information necessary for process control? The type of process (continuous vs batch) as well as the process characteristics define the type of process control information required (Wise and Gallagher, 1996). Process measurement can be preferable to laboratory analysis if there is an issue in obtaining a representative sample for analysis. For many industrial processes, there are safety related issues to sample collection, which can make laboratory measures undesirable. The cost associated with routine laboratory analysis may justify process measurements. For example the requirements for 24 h/day laboratory testing needs personal are available 24 h/day to collect samples and perform the analysis.

Taking in mind that quality can be defined as fitness for purpose; the first consideration in any application of the quality measurement of a material is to determine the exact purpose of the analysis. The fitness for purpose will be related to the sample selected. When developing an on-line/in-line method, a representative process sample must be identified and collected to further develop calibrations for either quantitative or qualitative applications. The sample must not only cover the range of analyte levels expected in the process but must include all the other variables, such as temperature, changes in particle size, physical changes in the sample and equipment, among others. Any analysis is only as good as the sample taken. Inadequate sampling invalidates the efforts of the analysis if the sample taken is not representative of the bulk from which is taken (Murray and Cowe, 2004). It is important to remember that the overall error obtained will be the sum of the multiplicative errors of sampling and analytical method.

$$\text{Overall ERROR} = \text{Sampling ERROR} + \text{Analytical ERROR}$$

## SUMMARY

The combination of sensors and multivariate data analysis techniques is applicable to many foods and agricultural commodities to predict and to monitor chemical composition with high accuracy. The main attractions of this methodology are to reduce time and the speed of analysis. Therefore, a mathematical relationship between the instrument and the analyte needs to be found, in a process called calibration. Although, as it seems this can be considered as a purely mathematical or statistical exercise, calibration development can be considered as complex process that implies the understanding of a *system* constituted by the sample, instrument, multivariate data analysis method and the final user. In addition, some elements of that system need to be considered and understood before, during and after calibration, such as knowing and understanding the reference laboratory method (standard error of the laboratory method, reference method, limitations), knowing the physics and chemical basis of the spectra, knowing the interactions between the sample and the instrument, as well as the interpretation of the calibration or mathematical relationships. It is, therefore, important that the individual that developed the calibrations has knowledge in order to produce a method that can be reliable. In this context, infrared spectroscopy technology has been successfully applied for composition analysis, product quality assessment and in production control. The infrared spectrum can give a global signature of composition (fingerprint) which, with the application of MVA techniques (e.g. PCA, PLS regression and discriminant analysis), can be used to elucidate particular compositional characteristics not easily detected by targeted chemical analysis. The main advantages of these new analytical approaches over the traditional chemical and chromatographic methods are the ease of use in routine operations and that they require minimal or no sample preparation.

Without doubt one of the biggest challenges when multivariate methods are combined with instrumental techniques to trace, monitor and predict chemical composition will be the interpretation of the complex models obtained. Although much time was devoted to the interpretation of such models through MVA, the knowledge of the fundamentals (e.g. molecular spectroscopy, chemistry, biochemistry, physics) involved is still the main barrier to understanding the basis and functionality of the models developed, in order to applied them efficiently in routine analysis. Nowadays the importance of food quality in human health and well-being cannot be overemphasized. The consumer is less concerned with gross composition issues, such as protein or fat or moisture. However, more sophisticated questions about the wholesomeness of food produced, such as its freedom from hormone and antibiotic residues, animal welfare issues and honesty in production (origin of foods, traceability, authenticity) become more and more relevant in the food industry.

At the beginning of the second decade of the twenty-first century, the combination of instrumental methods (e.g. infrared spectroscopy, electronic noses) with multivariate data techniques has a role in the production plant and for surveillance at critical points in the food chain. The greatest impact of vibrational spectroscopy in the food industry so far has been its use for the measurement of many compositional parameters. The advantages of ability to predict multiple parameters and speed of analysis mean that vibrational spectroscopy has a revolutionized the food industry. The future development of such applications will provide the industry with a very fast and nondestructive method to monitor composition or changes and to detect unwanted problems, providing a rapid means of qualitative rather than quantitative analysis. The potential savings, reduction in time and cost of analysis, the environmentally friendly nature of the technology, positioned rapid instrumental techniques as the most attractive techniques with a bright future in the field of the analysis.

However, one of the main constraints that faces the application of these methodologies is the lack of formal education in both instrumental techniques and multivariate data methods applied to foods. These are still a barrier for the wide spread of this technology as an analytical tool for the analysis of foods during processing.

## REFERENCES

Adams, M.J. (1995) *Chemometrics in Analytical Spectroscopy*. The Royal Society of Chemistry, Cambridge, UK.

Arvantoyannis, I., Katsota, M.N., Psarra, P. *et al.* (1999) Application of quality control methods for assessing wine authenticity: Use of multivariate analysis (chemometrics). *Trends Food Science and Technology* **10**, 321–336.

Beebe, K.R., Blaser, W.W., Bredeweg, R.A. *et al.* (1993) Process Analytical Chemistry. *Analytical Chemistry* **65**, 199R–216R.

Bendell, A., Disney, J. and McCollin, C. (1999) The future role of statistics in quality engineering and management. *The Statistician* **48**, 299–326.

Blanco, M. and Bernardez, M. (2009) Multivariate calibration for quantitative analysis. In: *Infrared Spectroscopy for Food Quality Analysis and Control* (ed. D.W. Sun). Elsevier, Oxford, UK.

Blanco, M. and Villaroya, I. (2002) NIR spectroscopy: a rapid-response analytical tool. *Trends Analytical Chemistry* **21**, 240–250.

Brereton R.G. (2000) Introduction to multivariate calibration in analytical chemistry. *The Analyst* **125**, 2125–2154.

Brereton R.G. (2008). *Applied Chemometrics for Scientists*. John Wiley & Sons Ltd, Chichester, UK.

Cozzolino, D. (2007) Application of near infrared spectroscopy to analyse livestock animal by-products. In: *Near Infrared Spectroscopy in Food Science and Technology* (eds Y. Ozaki, W.F. McClure and A.A. Christy). John Wiley & Sons Ltd, Chichester, UK.

Cozzolino, D. (2009) Near infrared spectroscopy in natural products analysis. *Planta Medica* **75**, 746–757.

Cozzolino, D. (2011) Infrared methods for high throughput screening of metabolites: food and medical applications. *Communications Chemistry High Throughput Screening* **14**, 125–131.

Cozzolino, D. (2012) Recent trends on the use of infrared spectroscopy to trace and authenticate natural and agricultural food products. *Applied Spectroscopy Reviews* **47**, 518–530.

Cozzolino, D. and Murray, I. (2012) A review on the application of infrared technologies to determine and monitor composition and other quality characteristics in raw fish, fish products, and seafood. *Applied Spectroscopy Reviews* **47**, 207–218.

Cozzolino, D., Cynkar, W., Janik, L. *et al.* (2006a) Analysis of grape and wine by near infrared spectroscopy – a review. *Journal of Near Infrared Spectroscopy* **14**, 279–289.

Cozzolino, D., Parker, M., Dambergs, R.G. *et al.* (2006b) Chemometrics and visible–near infrared spectroscopic monitoring of red wine fermentation in a pilot scale. *Biotechnology and Bioengineering* **95**, 1101–1107.

Cozzolino, D., Cynkar, W.U., Dambergs, R.G. *et al.* (2009) Multivariate methods in grape and wine analysis. *International Journal of Wine Research* **1**, 123–130.

Cozzolino, D., Shah, N., Cynkar, W. and Smith, P. (2011a) A practical overview of multivariate data analysis applied to spectroscopy. *Food Research International* **44**, 1888–1896.

Cozzolino, D., Shah, N., Cynkar, W. and Smith, P. (2011b) Technical solutions for analysis of grape juice, must and wine: the role of infrared spectroscopy and chemometrics. *Analytical and Bioanalytical Chemistry* **401**, 1479–1488.

Dhanoa, D.J.M.S., Lister, S.J., Sanderson, R. and Barnes, D.J. (1994) The link between Multiplicative Scatter Correction (MSC) and Standard Normal Variate (SNV transformations of NIR spectra. *Journal of Near Infrared Spectroscopy* **2**, 43–50.

Duckworth, J. (2004) Mathematical data processing. In: *Near Infrared Spectroscopy in Agriculture* (eds C.A. Roberts, J. Workman and J.B. Reeves). American Society of Agronomy, Crop Science Society of America, Soil Science Society of America. Madison, WI, USA, pp. 115–132.

Esbensen, K.H. (2002) Multivariate data analysis in practice. CAMO Process AS, Oslo, Norway.

Fearn, T. (2002) Assessing calibrations: SEP, RPD, RER and R2. *NIR News* **13**, 12–14.

Fearn, T. (1997) Validation. *NIR News* **8**, 7–8.

Geladi, P. and Kowalski, B.R. (1986) Partial least-squares regression: a tutorial. *Analytica Chimica Acta* **185**, 1–15.

Geladi, P. (2003) Chemometrics in spectroscopy. Part I. Classical chemometrics. *Spectrochimica Acta Part B*. **58**, 767–782.

Gishen, M., Dambergs, R.G. and Cozzolino, D. (2005) Grape and wine analysis – enhancing the power of spectroscopy with chemometrics. A review of some applications in the Australian wine industry. *Australian Journal of Grape and Wine Research* **11**, 296–305.

Haaland, D.M. and Thomas, E.V. (1988) Partial least-squares methods for spectral analysis. 1. Relation to other quantitative calibration methods and the extraction of qualitative information. *Analytical Chemistry* **60**, 1193–1198.

Hashimoto, A. and Kameoka, T. (2008) Applications of infrared spectroscopy to biochemical, food, and agricultural processes. *Applied Spectroscopy Reviews* **43**, 416–451.

Huang, H., Yu, H., Xu, H. and Ying, Y. (2008) Near infrared spectroscopy for on/in-line monitoring of quality in foods and beverages: a review. *Journal of Food Engineering* **87**, 303–313.

Jaumot, J., Vives, M. and Gargallo, R. (2004) Application of multivariate resolution methods to the study of biochemical and byophysical processes. *Analytical Biochemistry* **327**, 1–13.

Karande, A.D., Sia Heng, P.W. and Liew, C.V. (2010) In-line quantification of micronized drug and excipients in tablets by near infrared (NIR) spectroscopy: Real time monitoring of tabletting process. *International Journal of Pharmaceutics* **396**, 63–74.

Karoui, R., Downey, G. and Blecker, Ch. (2010) Mid-infrared spectroscopy coupled with chemometrics: a tool for the analysis of intact food systems and the exploration of their molecular structure–quality relationships – A Review. *Chemical Reviews* **110**, 6144–6168.

Kueppers, S. and Haider, M. (2003) Process analytical chemistry – future trends in industry. *Analytical and Bioanalytical Chemistry* **376**, 313–315.

Liu, Y.-C., Wang, F.-S. and Lee, W.-C. (2001) On-line monitoring and controlling system for fermentation processes. *Biochemical Engineering Journal* **7**, 17–25.

Mark, H. (1991) *Principles and Practice of Spectroscopic Calibration*. John Wiley & Sons Ltd, Toronto, Canada.

Mark, H. and Workman, J. (2003) *Statistics in Spectroscopy*, 2nd edn. Elsevier, London, UK.

Martens, H. and Martens, M. (2001) *Multivariate Analysis of Quality. An Introduction*. John Wiley & Sons Ltd, Chichester, UK.

Massart, D.L., Vandegiste, B.G.M., Deming, S.N. *et al.* (1988) *Chemometrics: A Textbook*. Elsevier, Amsterdam, The Netherlands.

McClure, F.W. (2003) 204 years of near infrared technology: 1800–2003. *Journal of Near Infrared Spectroscopy* **11**, 487–498.

Munck, L. (2007) A new holistic exploratory approach to systems biology by near infrared spectroscopy evaluated by chemometrics and data inspection. *Journal of Chemometrics* **21**, 406–426.

Munck, L. and Møller, J.B. (2011) Adapting cereal plants and human society to a changing climate and economy merged by the concept of self-organization. In Barley: Production, Improvement, and Uses (ed. S.E. Ullrich). John Wiley & Sons, Inc., Hoboken, NJ, USA, pp. 563–602.

Munck, L., Norgaard, L., Engelsen, S.B. *et al.* (1998) Chemometrics in food science: a demonstration of the feasibility of a highly exploratory, inductive evaluation strategy of fundamental scientific significance. *Chemometrics and Intelligent Laboratory Systems* **44**, 31–60.

Munck, L., Møller, J.B., Rinnan, Å. *et al.* (2010) A physiochemical theory on the applicability of soft mathematical models – experimentally interpreted. *Journal of Chemometrics* **24**, 481–495.

Murray, I. (1993) Forage analysis by near infrared spectroscopy. In Sward Management Handbook (eds A. Davies, R.D. Baker, S.A. Grant and A.S. Laidlaw). British Grassland Society, UK, pp. 285–312.

Murray, I. (1999) NIR spectroscopy of food: simple things, subtle things and spectra. *NIR News* **10**, 10–12.

Murray, I. and Cowe, I. (2004) Sample preparation. In: *Near Infrared Spectroscopy in Agriculture* (eds C.A. Roberts, J. Workman and J.B. Reeves). American Society of Agronomy, Crop Science Society of America, Soil Science Society of America. Madison, WI, USA, pp. 75–115.

Naes, T., Isaksson, T., Fearn, T. and Davies, T. (2002) *A User-Friendly Guide to Multivariate Calibration and Classification*. NIR Publications, Chichester, UK.

Nicolai, B.M., Beullens, K., Bobelyn, E. *et al.* (2007). Non-destructive measurement of fruit and vegetable quality by means of NIR spectroscopy: a review. *Post Harvest Biology and Technology* **46**, 99–118.

Norris, K.H. and Ritchie, G.E. (2008) Assuring specificity for a multivariate near-infrared (NIR) calibration: The example of the Chambersburg Shoot-out 2002 data set. *Journal of Pharmacology Biomedical Analysis* **48**, 1037–1041.

Otto, M. (1999) *Chemometrics: Statistics and Computer Application in Analytical Chemistry*. John Wiley & Sons Ltd. Chichester, UK.

Roggo, Y., Chalus, P., Maurer, L. *et al.* (2007) A review of near infrared spectroscopy and chemometrics in pharmaceutical technologies. *Journal of Pharmacology Biomedical Analysis* **44**, 683–700.

Skibsted, E. (2005) PAT and beyond. Academic Report from the University of Amsterdam, The Netherlands.

Walsh, K.B. and Kawano, S. (2009) Near infrared spectroscopy. In Optical Monitoring of Fresh and Processed Agricultural Crops (ed. M. Zude). CRC Press, Boca Raton, FL, USA, pp. 192–239.

Williams, P.C. (2001) Implementation of near-infrared technology. In Near Infrared Technology in the Agricultural and Food Industries (eds P.C. Williams, and K.H. Norris). American Association of Cereal Chemists, St Paul, MN, USA, pp. 145–169.

Wise, B.M. and Gallagher, N.B. (1996) The process chemometrics approach to process monitoring and fault detection. *Journal of Processing and Control* **6**, 329–348.

Wold, S. (1995) Chemometrics; what do we mean with it, and what do we want from it? *Chemometrics and Intelligent Laboratory Systems* **30**, 109–115.

Wold, S., Sjöstrom, M. and Eriksson, L. (2001) PLS-regression: a basic tool of chemometrics. *Chemometrics and Intelligent Laboratory Systems* **58**, 109–130.

Woodcock, T., Downey, G. and O'Donnell, C.P. (2008) Better quality food and beverages: the role of near infrared spectroscopy. *Journal of Near Infrared Spectroscopy* **16**, 1–29.

Workman, J.J., Veltkamp, D.J., Doherty, S. *et al.* (1999) Process Analytical Chemistry. *Analytical Chemistry* **71**, 121R–180R.

Workman, J.J., Creasy, K.E., Doherty, S. *et al.* (2001) Process analytical chemistry. *Analytical Chemistry* **73**, 2705–2718.

# 3 Case study: Optimization of enzyme-aided extraction of polyphenols from unripe apples by response surface methodology

**Hu-Zhe Zheng[1] and Shin-Kyo Chung[2]**

[1] *Department of Food & Nutritional Engineering, Jiangsu Food & Pharmaceutical Science College, Jiangsu Huai'an, China*

[2] *School of Food Science & Biotechnology, Kyungpook National University, Daegu, Republic of Korea*

## ABSTRACT

Unripe apples from manual thinning out and falling contain a high amount of antioxidant polyphenols. To enhance the extraction yield of polyphenols from the unripe apples, an optimization study of Viscozyme L aided polyphenol extraction has been carried out with response surface methodology (RSM). The optimum reaction condition was determined: 1.95 FBG (fungal beta-glucanase unit), 47.12°C, and 12.52 h. The experimental values of total phenolic content (TPC) and caffeic acid content (CAC) of Viscozyme L treatment at optimum conditions showed a good agreement with the predicted values. The TPC and CAC of Viscozyme L treatment at optimum conditions was 2 and 13-fold higher than those of the control treatment. The proposed optimum condition could be expected for the enhancement of unripe apple polyphenols extraction.

## INTRODUCTION

Apples are one of the most widely cultivated and consumed fruits in the world due to their palatability and health-benefit constituents, such as organic acid, dietary fibre, vitamins and polyphenols (Andreas *et al.*, 2001). The global apple production increased by approximately 10%, from 58 to 69 million metric tonnes, between 2003 and 2010; Korea (south) produced about 460 000 metric tonnes of apples in 2010 (FAOSTAT, 2012).

Among their constituents, the beneficial effects have been attributed mainly to the biological actions of polyphenols, such as free radical scavenging activity (Adil *et al.*, 2007), metal chelation (Chien *et al.*, 2007), anti-allergic activity (Kojima *et al.*, 2000; Akiyama *et al.*, 2005), anticancer activity (Yanagida *et al.*, 2002) and anti-arteriosclerosis activity (Stefania *et al.*, 2007). The main classes of polyphenols in apple are flavan-3-ols (catechin, epicatechin, proanthocyanidin), hydroxycinnamic acids (chlorogenic acid, *p*-coumaroyl quinic acid), dihydrochalcone (phloridzin, phloretin glucoside) and flavonols (quercetin, rutin) (Andreas *et al.*, 2001; Rosa *et al.*, 2005). Interestingly, these polyphenols contents are inversely correlated with the maturity of apples (Akiyama *et al.*, 2005; Wu *et al.*, 2007, Zheng *et al.*, 2012). Unripe apples which have resulted

*Mathematical and Statistical Methods in Food Science and Technology*, First Edition.
Edited by Daniel Granato and Gastón Ares.

from manual thinning out or falling were typically discarded in the orchard, although they were estimated to account for about 25% of Korea's total apple production (Park *et al.*, 2004). A small portion of them has been used as animal feed and fertilizer, in spite of causing environmental contamination (Sudha *et al.*, 2007).

Agricultural by-products have attracted considerable attention from scientists as potential resources of natural antioxidants that can be processed for pharmaceutical, cosmetic and food products. However, the low extraction efficiency of polyphenols has been the major hindrance to their use (Wang and Weller, 2006; Makris *et al.*, 2007). Carbohydrate-hydrolysing enzymes, for example pectinase, cellulase, hemicellulase, gluconase and Viscozyme L, have recently been used to increase the extraction efficiency of cell-wall bounded polyphenols from plants (Landbo and Meyer, 2001; Barberousse *et al.*, 2009). These enzymes might have seemed to take a role in disintegrating the plant cell wall matrix in order to facilitate the polyphenol extraction (Renard *et al.*, 2001; Garrote *et al.*, 2004; Manuel *et al.*, 2006; Stalikas, 2007; Bourvellec *et al.*, 2009).

Zheng and coworkers have also investigated the polyphenols extraction yields of several carbohydrate-hydrolysing enzymes with unripe apples; Viscozyme L was selected as the most efficient enzyme (Zheng *et al.*, 2009). Viscozyme L is a multi-enzyme complex containing a wide range of carbohydrate-hydrolysing enzymes that has been used with oat bran (Xiao and Yao, 2008) and wheat (Sørensen *et al.*, 2005). Some processing variables, such as the ratio of enzymes, substrates, solvents, temperature and pH values, may affect the enzyme-aided polyphenol extraction efficiency from plant material (Meyer *et al.*, 1998). Response surface methodology (RSM) was used for the optimization of several natural compound extraction techniques, such as supercritical fluid extraction (Adil *et al.*, 2007), microwave extraction (Hayat *et al.*, 2009) and solvent extraction (Pompeu *et al.*, 2009). In order to approach more efficient conditions for introducing Viscozyme L, the total phenolic content (TPC) and extraction yield have been examined previously with respect to the main variables, such as the ratio of enzymes to substrate, reaction pH, temperature and time.

In the present study, Viscozyme L aided hydrolysis parameters, such as the ratio of Viscozyme L to substrate, the reaction temperature and time, were optimized using RSM by employing a five level, three variable central composite rotatable design, in order to obtain the optimum conditions for the extraction of polyphenols from unripe apples.

## EXPERIMENTS

### Materials, chemicals and instruments

Unripe apples (*Malus pumila* cv. Fuji) were collected at 85 days after full bloom from the orchard of Kyungpook National University in Daegu, Korea, in 2009, and stored in a freezer (–70°C) until the experiment. Viscozyme L (from *Aspergillus aculeatus*, 100 fungal beta-glucanase units (FBG)/ml, Novozymes, Bagsvaerd, Denmark) was used in the study. The FBG unit is determined based on the Christian *et al.* (2005) assay with some modification, and defined as the quantity of Viscozyme L liberating one μmol of glucose from $\beta$-glucan per minute. Folin–Ciocalteu phenol reagent, caffeic acid, chlorogenic acid, *p*-coumaric acid, ferulic acid, phloretin,and phloridzin were obtained from Sigma Co. (St Louis, MO). All organic solvents and other chemicals were analytical grade from Duksan Co. (Seoul, Korea), except for the high performance liquid chromatography solvents (HPLC, J.T. Baker, Phillipsburg, NJ). An ultraviolet-visible (UV-vis) spectrophotometer (UV 1601 PC, Shimadzu Co., Kyoto, Japan) and HPLC (LC-10A, Shimadzu Co., Kyoto, Japan) were used for the determination of total phenolic content (TPC) and caffeic acid content (CAC).

## Viscozyme L aided polyphenol extraction

Two hundred grams of unripe apples were blanched at 80°C for 10 minutes for polyphenol oxidase (PPO) inhibition (Buckow *et al.*, 2009), cut into cylindrical shapes without peeling, added to a half volume of distilled buffer (w/v), and then homogenized. The variable of enzyme hydrolysis buffer pH (3.7) was kept at constant value. For the enzyme-aided hydrolysis reaction, 10 g of unripe apple homogenate was put into a 30 ml vial, Viscozyme L solution was added, with the ratio of Viscozyme L to substrate from 0 to 0.03 (0–3 FBG unit), and then incubated at a selected temperature (30–70°C) for different times (4–20 h).

## Experimental design for the RSM procedure

Based on the preliminary experiments, three independent variables – the ratio of Viscozyme L to substrate, reaction temperature and time – were selected (Table 3.1). The RSM used a five level, three variable central composite rotatable design (CCRD) consisting of 16 experimental runs (Cochran and Cox, 1957) to study the response pattern and to determine the optimum combination of Viscozyme L aided hydrolysis reaction variables. The effect of the independent variables $X_1$ (the ratio of enzyme to substrate, *En/S*, v/w), $X_2$ (Reaction temperature, *Rt*, °C) and $X_3$ (Reaction time, *h*) at five variation levels were selected (Table 3.1). Experiments were randomized in order to maximize the effects of unexplained variability in the observed responses due to extraneous factors.

The variables were coded according to Equation 3.1:

$$x_i = (X_i - \bar{X}_i)/\Delta X_i \tag{3.1}$$

where $x_i$ = dimensionless value of an independent variable; $X_i$ = real value of an independent variable; $\bar{X}_i$ = real value of an independent variable at the center point; and $\Delta X_i$ = step change.

The specific codes are:

$$x_1 = (En/S - 0.015)/0.0075 \tag{3.2}$$

$$x_2 = (Rt. - 50)/10 \tag{3.3}$$

$$x_3 = (h - 12)/4 \tag{3.4}$$

Data from the central composite experimental design were subjected to second order polynomial regression analysis using least square regression methodology to obtain the parameters of the mathematical models. Experimental data were fitted to the second order polynomial model shown in Equation 3.5:

$$Y_i = \beta_0 + \sum_{i=1}^{k} \beta_i X_i + \sum_{i=1}^{k} \beta_{ii} X_i^2 + \sum_{\substack{i=!\\ i<j}}^{k-1} \sum_{j=2}^{k} \beta_{ij} X_i X_j \tag{3.5}$$

**Table 3.1** Independent variable values of the process and their corresponding levels.

| Independent variables | Symbol | | Levels | | | | |
|---|---|---|---|---|---|---|---|
| | Uncodified | Codified | −2 | −1 | 0 | 1 | 2 |
| Ratio of Viscozyme L to substrate (v/w) | $X_1$ | $x_1$ | 0 | 0.0075 | 0.015 | 0.0225 | 0.03 |
| Reaction temperature (°C) | $X_2$ | $x_2$ | 30 | 40 | 50 | 60 | 70 |
| Reaction time (h) | $X_3$ | $x_3$ | 4 | 8 | 12 | 16 | 20 |

**Table 3.2** Experimental design of the five-variable central composite, TPC and CAC of Viscozyme L aided hydrolysis reaction from unripe apples (Zheng *et al.*, 2010). Reproduced with permission from The Korean Society for Applied Biological Chemistry.

| Run NO. | Factor values[a] | | | Response values | |
|---|---|---|---|---|---|
| | $X_1$ | $X_2$ | $X_3$ | TPC (mg GAE/100 g) | CAC (mg/kg) |
| 1 | 0.0075 (−1) | 40 (−1) | 8 (−1) | 85.88 ± 0.60[b] | 31.59 ± 0.36 |
| 2 | 0.0225 (1) | 40 (−1) | 8 (−1) | 99.46 ± 0.43 | 36.05 ± 0.37 |
| 3 | 0.0075 (−1) | 40 (−1) | 16 (1) | 89.49 ± 0.56 | 35.84 ± 0.31 |
| 4 | 0.0225 (1) | 40 (−1) | 16 (1) | 100.78 ± 0.44 | 35.26 ± 0.35 |
| 5 | 0.0075 (−1) | 60 (1) | 8 (−1) | 90.16 ± 0.09 | 20.51 ± 0.25 |
| 6 | 0.0225 (1) | 60 (1) | 8 (−1) | 102.42 ± 0.70 | 23.24 ± 0.22 |
| 7 | 0.0075 (−1) | 60 (1) | 16 (1) | 91.83 ± 0.60 | 18.91 ± 0.17 |
| 8 | 0.0225 (1) | 60 (1) | 16 (1) | 98.52 ± 0.44 | 21.63 ± 0.26 |
| 9 | 0.015 (0) | 50 (0) | 12 (0) | 114.90 ± 0.73 | 43.21 ± 0.45 |
| 10 | 0.015 (0) | 50 (0) | 12 (0) | 115.85 ± 1.07 | 42.80 ± 0.44 |
| 11 | 0.015 (0) | 30 (−2) | 12 (0) | 90.37 ± 0.15 | 32.99 ± 0.30 |
| 12 | 0.015 (0) | 70 (2) | 12 (0) | 92.36 ± 0.37 | 13.48 ± 0.19 |
| 13 | 0.015 (0) | 50 (0) | 4 (−2) | 101.02 ± 0.70 | 24.93 ± 0.35 |
| 14 | 0.015 (0) | 50 (0) | 20 (2) | 106.86 ± 0.73 | 31.97 ± 0.37 |
| 15 | 0 (−2) | 50 (0) | 12 (0) | 51.85 ± 0.44 | 3.14 ± 0.22 |
| 16 | 0.03 (2) | 50 (0) | 12 (0) | 98.25 ± 0.73 | 20.87 ± 0.42 |

[a]Numbers in parentheses are coded symbols for levels of independent parameters, $X_1$; ratio of Viscozyme L to substrate, $X_2$; reaction temperature (°C), $X_3$; reaction time (h)
[b]Means ± standard deviation ($n = 3$).

where $X_1, X_2, \ldots, X_k$ are the independent variables affecting the responses $Y_i$; $\beta_0$, $\beta_i$ ($i = 1, 2, \ldots, k$), $\beta_{ii}$($i = 1, 2, \ldots, k$), and $\beta_{ij}$ ($i = 1, 2, \ldots, k$; $j = 1, 2, \ldots, k$) are the regression coefficients for intercept, linear, quadratic, and interaction terms, respectively; $k$ is the number of variables. The responses obtained from the experimental design set (Table 3.2) were subjected to multiple nonlinear regression analysis using the Statistical Analysis System (version 9.1, Institute Inc., Cary, NC, USA), to obtain the coefficients of the second order polynomial model. The quality of the fit of the polynomial model was expressed by $R^2$, adjusted $R^2$, and its statistical significance was checked using an $F$-test.

## Determination of the optimum conditions and evaluation of the model

The optimum Viscozyme L aided hydrolysis condition for both TPC and CAC was estimated by superimposing the four-dimensional response surfaces for both components using the Mathematica program (version 7.0, Wolfram Research, Inc., USA). To verify the significance of a regression equation, the optimal value estimated by setting up the hydrolysis conditions at any point within the estimated ranges and then applying those to the regression equation, was compared with the actual values from a real extraction experiment.

## Determination of TPC and CAC

TPC was determined using Folin–Ciocalteu reagent with some modifications (Singleton *et al.*, 1999) and expressed as gallic acid equivalent (mg GAE/100 g). Twenty μl of unripe apples polyphenol solution was injected into a HPLC after filtration (0.45 μm) with an ODS-HG-5 (Develosil, 150 × 4.6 mm, i.d.) column, in a mobile phase of 2% acetic acid in water (solvent A), 0.5% acetic acid and 45.5% acetonitrile in water (solvent B), with a flow rate of 1.0 ml/min, and monitored at 290 nm. The CAC was determined by the linear regression equation used for standard caffeic acid solutions ranges from 5 to 50 mg/kg.

### Statistical analysis

All determinations were carried out in triplicate and the experimental results obtained were expressed as means ± SD. Statistical analysis was performed by using the Statistical Analysis System (version 9.1, Institute Inc., Cary, NC, USA). Data were analysed by one-way analysis of variance and the mean values were considered significantly different when $p < 0.05$. The optimal extraction conditions were estimated through four-dimensional response surface analyses of the three independent variables and each dependent variable.

## RESULTS AND DISCUSSION

### Modelling of the Viscozyme L aided polyphenol extraction reaction

In general, the efficiency of the enzyme-aided phytochemical compounds extraction was influenced by multiple parameters including but not limited to enzyme type and concentration, the reaction solution pH, the reaction temperature and time, and their effects were either independent or interactive (Meyer *et al.*, 1998; Landbo and Meyer, 2001; Sørensen *et al.*, 2005; Manuel *et al.*, 2006; Zheng *et al.*, 2009). In order to optimize the extraction process with reference to the extraction of TPC and CAC from unripe apples using Viscozyme L aided extraction, a central composite design was developed as represented in Table 3.2. The experimental values of TPC and CAC of unripe apples extracts at various experimental conditions are also presented in Table 3.2. The results of the analysis of variance, goodness of fit, and the adequacy of the models are summarized in Table 3.3. The data showed a good fit with Equation 3.5, which was statistically acceptable at $p < 0.05$ and adequate with satisfactory $R^2$ values. The values of coefficients presented in Table 3.3 were used in final predictive equations. On the basis of these equations

**Table 3.3** Regression coefficients and of *t*-value predicted quadratic polynomial models for the response TPC and CAC of Viscozyme L aided hydrolysis reaction from unripe apples (Zheng *et al.*, 2010). Reproduced with permission from The Korean Society for Applied Biological Chemistry.

| Term | TPC | | CAC | |
|---|---|---|---|---|
| | coefficient | *t*-value | coefficient | *t*-value |
| $\beta_0$ | −151.6025 | $-2.78^c$ | −137.111250 | $-3.68^b$ |
| Linear | | | | |
| $\beta_1$ | 7401.5 | $4.55^b$ | 4628.250000 | $4.16^b$ |
| $\beta_2$ | 6.489625 | $4.35^b$ | 4.569188 | $4.48^b$ |
| $\beta_3$ | 6.127188 | $1.90^d$ | 7.039844 | $3.19^c$ |
| Quadratic | | | | |
| $\beta_{11}$ | −179222 | $-7.62^a$ | −137778 | $-8.57^a$ |
| $\beta_{22}$ | −0.060025 | $-4.54^b$ | −0.049425 | $-5.46^b$ |
| $\beta_{33}$ | −0.178828 | $-2.16^d$ | −0.227422 | $-4.02^b$ |
| Cross-product | | | | |
| $\beta_{12}$ | −9.866667 | −0.40 | 2.616667 | 0.15 |
| $\beta_{13}$ | −32.75 | −0.53 | −21.041667 | −0.49 |
| $\beta_{23}$ | −0.022375 | −0.48 | −0.020844 | −0.65 |
| $R^{2e}$ | $0.9470^b$ | | $0.9555^b$ | |
| Lack of fit | 0.0879 | | 0.0555 | |

[a]Significant at $p < 0.001$,
[b]$p < 0.01$,
[c]$p < 0.05$,
[d]$p < 0.1$,
[e]Coefficient of multiple determination.

**Table 3.4** Polynomial equations calculated by RSM program for extraction conditions of extract from unripe apples.

| Response | Second order polynomials[a] | $R^2$ | $R^2$-adjusted |
|---|---|---|---|
| TPC (mg GAE/100 g) | $Y_{TPC} = -151.57375 + 7401.5X_1 + 6.489625X_2 + 6.122813X_3 - 9.866667X_1X_2 - 32.75X_1X_3 - 0.022375X_2X_3 - 179222X_1^2 - 0.060025X_2^2 - 0.178672X_3^2$ | 0.9470 | 0.9205 |
| CAC (mg/kg) | $Y_{CAC} = -137.111250 + 4628.25X_1 + 4.569188X_2 + 7.039844X_3 + 2.616667X_1X_2 - 21.041667X_1X_3 - 0.020844X_2X_3 - 137778X_1^2 - 0.049425XX_2^2 - 0.227422X_3^2$ | 0.9555 | 0.9333 |

[a]$X_1$; ratio of Viscozyme L to substrate, $X_2$; reaction temperature (°C), $X_3$; reaction time (h).

(Table 3.4), four-dimensional plots were constructed to predict the relationships between independent variables and dependent variables.

## Effect of Viscozyme L aided hydrolysis variables on TPC

The TPC of the unripe apples extracts obtained by Viscozyme L aided extraction and based on the central composite design are shown in Table 3.2. The analysis variance of TPC showed that the linear and quadratic effects of ratio of Viscozyme L to substrate and reaction temperature were significant ($p < 0.01$) (Table 3.4). Xiao and Yao (2008) suggested that, for a good fit of a model, $R^2$ should be at least 0.80. In the results, the multiple coefficients of correlation $R^2$ (0.9470) indicated a close agreement between the experimental and predicted values of the TPC yield. The corresponding variables were more significant when the absolute $t$-value became larger and the $p$-value became smaller (Fu *et al.*, 2006). The factor $t$-test value (7.62) corresponded to the ratio of Viscozyme L to substrate ($\beta_{11}$), while the $t$-test values for reaction time ($\beta_3$) were smaller at 1.90, but the $p$-values were still significant at 0.1. It was noted that the variable with the largest effect was the quadratic term of $\beta_{11}$, followed by linear term of $\beta_1$, quadratic term of reaction temperature ($\beta_{22}$), linear term of $\beta_2$, quadratic term of $\beta_{33}$, and linear term of $\beta_3$, but the cross-product term was not significant (Table 3.3).

Table 3.4 demonstrates the response surface regression analysis of the results. The $R^2$ of the regression equation for the TPC was 0.9470 and the level of significance was at $p < 0.01$. Figure 3.1 illustrates the four-dimensional response surfaces drawn using the response surface regression equation of the TPC yield. The TPC was affected chiefly by the ratio of Viscozyme L to substrate, reaction temperature (Tables 3.3 and 3.5). The TPC increased as the ratio of Viscozyme L to substrate and reaction temperature increased up to a maximum (the ratio of Viscozyme L to substrate 0.018 (1.8 FBG), reaction temperature 50.27°C and reaction time 12.33 h); the estimated stationary point was at the maximum (116.40 mg GAE/100 g) (Table 3.5). The TPC decreased as the ratio of Viscozyme L to substrate, reaction temperature and reaction time were increased above these maximum values. These results suggested that an increase of reaction temperature might favour Viscozyme L activity and enhance the decomposition activity during the enzyme reaction. Nevertheless, an excessive decrease or increase of the reaction temperature partly inhibited Viscozyme L activity, thereby decreasing the TPC. Similar results were reported by Zheng *et al.* (2008), Landbo and Meyer (2001), and Pinelo *et al.* (2006), who found that an optimum given temperature may enhance the extraction efficiency of antioxidants, while too high a temperature had a significantly negative effect. Furthermore, an excessive increase of reaction time reduced TPC, which was possibly due to the decomposition of antioxidants during long Viscozyme L reaction times (Manuel *et al.*, 2008; Zheng *et al.*, 2008, 2009). Tables 3.2 to 3.6 and Figure 3.1 show that the TPC of the unripe apples was significantly ($p < 0.01$) affected by the linear and quadratic terms of the ratio of Viscozyme L to substrate ($p < 0.001$) and reaction temperature ($p < 0.05$), and that reaction time was not significant (Zheng *et al.*, 2010).

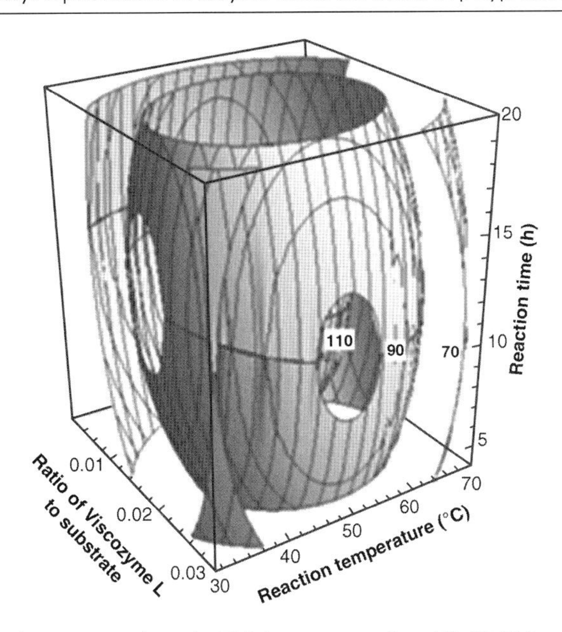


**Figure 3.1** Four-dimensional response surfaces for TPC (at constant values, 70-90-110 mg GAE/100 g) of extracts from unripe apples as functions of the ratio of Viscozyme L to substrate, reaction temperature, and reaction time in the Viscozyme L aided extraction (Zheng *et al.*, 2010).

**Table 3.5** Analysis of variance of the factors obtained from RSM, and the optimum extraction conditions by superimposing response surfaces of TPC and CAC from unripe apples in the Viscozyme L aided extraction.

| Response variables | *F*-value | *p*-value | Optimum extraction conditions[a] | | | Maximum | Morphology |
|---|---|---|---|---|---|---|---|
| | | | $X_1$ | $X_2$ | $X_3$ | | |
| TPC (mg GAE/100 g) | 11.92[b] | 0.0035 | 0.018 | 50.27 | 12.33 | 116.40 | Maximum |
| CAC (mg/kg) | 14.32[b] | 0.0021 | 0.016 | 43.97 | 12.71 | 45.70 | Maximum |

[a]$X_1$; ratio of Viscozyme L to substrate, $X_2$; reaction temperature (°C), $X_3$; reaction time (h).
[b]Significant at $p < 0.01$.

## Effect of Viscozyme L aided hydrolysis variables on CAC

Chlorogenic acid, a major phenolic compound of apples was hydrolysed to caffeic acid by carbohydrate-hydrolysing enzymes (Zheng *et al.*, 2008). Hence, caffeic acid has been suggested as a marker to detect whether cleavage of the quinic esters of chlorogenic acid by carbohydrate-hydrolysing enzymes has occurred (Benoit *et al.*, 2006; Zheng *et al.*, 2009). Although caffeic acid and chlorogenic acids can act as antioxidants *in vitro*, caffeic acid is efficiently absorbed through the small intestine, increasing the total antioxidant status of plasma in mammals (Nardini *et al.*, 2002; Gonthier *et al.*, 2006). Therefore, the hydrolysis of chlorogenic acids to caffeic acid before consumption or during digestion is very meaningful to the bioavailability. CAC from unripe apples under various hydrolysis conditions using Viscozyme L is presented in the Table 3.2. Experimental data were subjected to regression analysis, and the coefficients of the estimate are presented in Table 3.3. The response surface regression equation for CAC according to extraction conditions is demonstrated in Table 3.4, and the $R^2$ of the regression

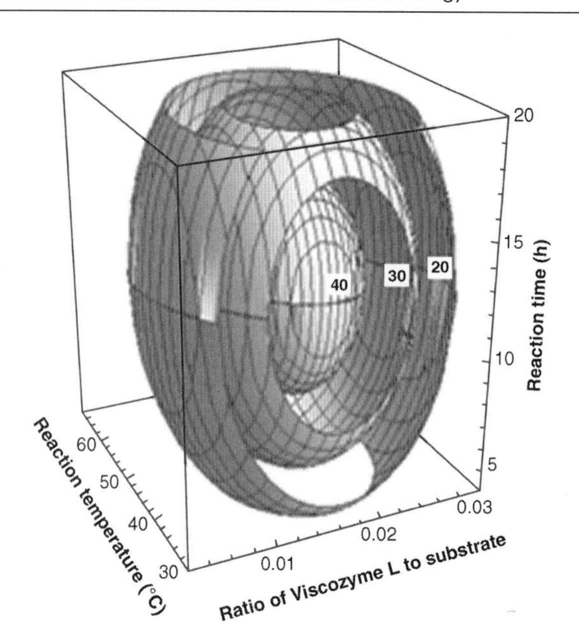


**Figure 3.2** Four-dimensional response surfaces for CAC (at constant values, 20-30-40 mg/kg) of extracts from unripe apples as functions of the ratio of Viscozyme L to substrate, reaction temperature, and reaction time in the Viscozyme L aided extraction (Zheng *et al.*, 2010). Reproduced with permission from The Korean Society for Applied Biological Chemistry.

equation was 0.9555 and the level of significance was at $p < 0.01$. Figure 3.2 demonstrates the four-dimensional response surfaces for CAC obtained from the regression equation. The CAC was considerably affected by the ratio of Viscozyme L to substrate, reaction temperature and reaction time (Tables 3.3 and 3.5). The CAC increased as the ratio of Viscozyme L to substrate, reaction temperature and reaction time increased up to a maximum (the ratio of Viscozyme L to substrate 0.016 (1.6 FBG), reaction temperature 43.97°C and reaction time 12.71 h); the estimated stationary point was at the maximum (45.70 mg/kg) (Table 3.5). These results were supported by other research conducted on olive oil by-products (Bouzid *et al.*, 2005); coffee pulp and apple marc (Benoit *et al.*, 2006) to release simple phenolic compounds such as *p*-coumaric and caffeic acids using fungal enzymes. However, the influence of reaction variables on the production of caffeic acid from plant byproducts has not yet been reported. This research might be the first attempt to optimize enzyme aided caffeic acid production in apple by RSM.

## Estimation and validation of optimum hydrolysis condition

To establish the extraction conditions for the effective substance form unripe apples, the research examined the extraction characteristics for TPC and CAC to estimate the range of the optimum extraction conditions by superimposing the response surface of the representative quality factors TPC and CAC. Therefore, the response surfaces for 110 mg GAE/100 g of TPC and 40 mg/kg of CAC were superimposed to reveal the overlapped portions, which had the highest response values in the Viscozyme L aided extraction (Figure 3.3). The ranges of Viscozyme L aided extraction conditions that were assumed to maximize both components in the Viscozyme L extraction were 0.018–0.02 the ratio of

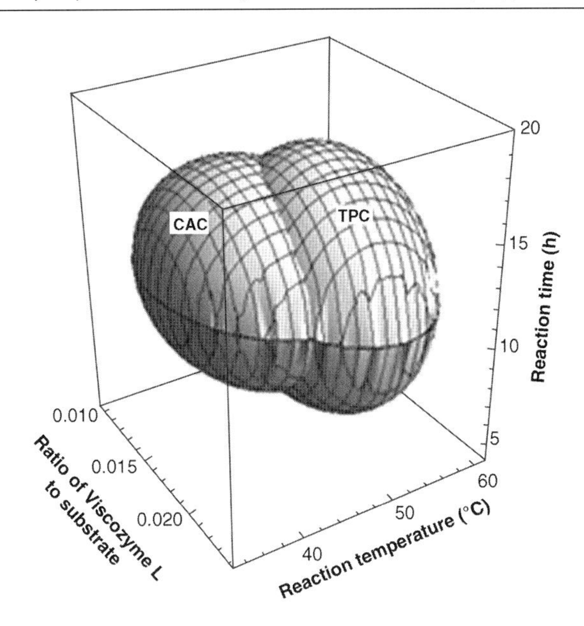


**Figure 3.3** Superimposed response surface for optimization of TPC and CAC (at constant values of 110 mg GAE/100 g TPC and 40 mg/kg CAC) of extracts from unripe apples as functions of the ratio of Viscozyme L to substrate, reaction temperature, and reaction time in the Viscozyme L aided extraction (Zheng *et al.*, 2010).

Viscozyme L to substrate (1.8–2.0 FBG), 40–50°C reaction temperature, 11–13 hours reaction time. To validate the estimated optimum extraction conditions for both components, an optimum point for each condition was selected within the ranges, that is the ratio of Viscozyme L to substrate 0.0195 (1.95 FBG), reaction temperature 47.12°C and reaction time 12.52 hours (Table 3.6). Under these conditions, RSM models predicted the yield of TPC and CAC to be 115.52 mg GAE/100 g and 43.77 mg/kg of unripe apples, respectively (Table 3.6). These values showed good agreement with the experimental values (110.52 mg GAE/100 g of TPC and 43.13 mg/kg of CAC) that were executed and determined at the optimum condition, which reflects the fitness of the optimization. It was noteworthy that the TPC and CAC with the Viscozyme L treatment was 2.2- and 13.4-fold higher than that of the control, respectively (Table 3.6). The composition of unripe apple polyphenol with Viscozyme L treatment at the optimized

**Table 3.6** Comparison between the predicted value and experimental value for the response variables at the given optimum conditions for unripe apples.

| Response variables | Predicted condition[a] | | | Maximum value | | B/A × 100 (%) | Control[b] |
|---|---|---|---|---|---|---|---|
| | $X_1$ | $X_2$ | $X_3$ | Predicted value (A) | Experimental values (B) | | |
| TPC (mg GAE/100 g) | 0.0195 | 47.12 | 12.52 | 115.52 | 110.52 ± 0.30[c] | 95.67 | 49.83 ± 0.11 |
| CAC (mg/kg) | | | | 43.77 | 43.13 ± 0.45 | 98.54 | 3.21 ± 0.05 |

[a]$X_1$; ratio of Viscozyme L to substrate, $X_2$; reaction temperature (°C), $X_3$; reaction time (h).
[b]Without Viscozyme L treatment.
[c]Means ± SD($n = 3$).

condition were investigated by HPLC; 15 kinds of polyphenol compounds, including three kinds of flavan-3-ols, four kinds of hydroxycinnamic acids and two kinds of dihydrochalcones were identified (Zheng *et al.*, 2009). Viscozyme L treatment enhanced the contents of caffeic acid content, *p*-coumaric acid and ferulic acid about 44, 8, and 4 times more than control treatment, respectively.

## CONCLUSION

The optimum polyphenol extraction conditions from unripe apple with Viscozyme L treatment were determined: 1.95 FBG at 47.12°C for 12.52 hours. The TPC and CAC of Viscozyme L treatment at optimum condition were 2.2- and 13.4-fold higher than those of the control, respectively. The proposed optimum conditions could be adapted for the unripe apple polyphenols extraction.

## REFERENCES

Adil, İ.H., Çetin, H.İ., Yener, M.E. and Bayındırlı, A. (2007) Subcritical (carbon dioxide+ethanol) extraction of polyphenol from apple and peach pomaces, and determination of the antioxidant activities of the extracts. *The Journal of supercritical fluids* **43**(1), 55–63.

Akiyama, H., Yuji, S., Takahiro, W. *et al.* (2005) Dietary unripe apple polyphenol inhibits the development of food allergies in murine models. *FEBS Letters* **579**(20), 4485–4491.

Andreas, S., Keller, P. and Carle, R. (2001) Determination of phenolic acids and flavonoids of apple and pear by high-performance liquid chromatography. *Journal of Chromatography A*, **910**, 265–273.

Barberousse, H., Kamoun, A., Chaabouni, M. *et al.* (2009) Optimization of enzymatic extraction of ferulic acid from wheat bran, using response surface methodology, and characterization of the resulting fractions. *Journal of the Science of Food and Agriculture* **89**(10), 1634–1641.

Benoit, I., Navarro, D., Marnet, N. *et al.* (2006) Feruloyl esterases as a tool for the release of phenolic compounds from agro-industrial by-products. *Carbohydrate Research* **341**(11), 1820–1827.

Bourvellec, C.L., Guyot, S. and Renard, C.M.G.C. (2009) Interactions between apple (Malus x domestica Borkh.) polyphenol and cell walls modulate the extractability of polysaccharides. *Carbohydrate Polymers* **75**, 251–261.

Bouzid, O., Navarro, D., Roche, M. *et al.* (2005) Fungal enzymes as a powerful tool to release simple phenolic compounds from olive oil by-product. *Process Biochemistry* **40**(5), 1855–1862.

Buckow, R., Weiss, U. and Dietrich, K. (2009) Inactivation kinetics of apple polyphenol oxidase in different pressure-temperature domains. *Innovative Food Science and Emerging* **10**(4), 441–448.

Chien, P.J., Sheu, F., Huang, W.T. and Su, M.S. (2007) Effect of molecular weight of chitosans on their antioxidative activities in apple juice. *Food Chemistry* **102**(4), 1192–1198.

Christian, R., Serge, H., Marc, O. *et al.* (2005) Development of an enzymatic assay for the determination of cellulose bioavailability in municipal solid waste. *Biodegradation* **16**, 415–422.

Cochran, W.G. and Cox, G.M. (1957) Some methods for the study of response surfaces. In: *Experimental Designs*, 2nd edn. John Wiley & Sons, Inc., New York, pp. 335–375.

FAOSTAT (2012) FAO Statistics Division. http://faostat.fao.org (last accessed 6 June 2013).

Fu, C.L., Tian, H.J., Li, Q.H. *et al.* (2006) Ultrasound-assisted extraction of xyloglucan from apple pomace. *Ultrasonics Sonochemistry* **13**(6), 511–516.

Garrote, G., Cruz, J.M., Moure, A. *et al.* (2004) Antioxidant activity of byproducts from the hydrolytic processing of selected lignocellulosic materials. *Trends in Food Science and Technology* **15**, 191–200.

Gonthier, M.P., Remesy, A.S., Cheynier, V. *et al.* (2006) Microbial metabolism of caffeic acid and its esters chlorogenic and caftaric acids by human faecal microbiota *in vitro*. *Biomedicine and Pharmacotherapy* **60**(9), 536–540.

Hayat, K., Hussain, S., Abbas, S. *et al.* (2009) Optimized microwave-assisted extraction of phenolic acids from citrus mandarin peels and evaluation of antioxidant activity *in vitro*. *Separation and Purification Technology* **70**(1), 63–70.

Kojima, T., Akiyama, S., Taniuchi, S. *et al.* (2000) Anti-allergic effect of apple polyphenol on patients with atopic dermatitis: a pilot study. *Allergology International* **49**(1), 69–73.

Landbo, A.K. and Meyer, A.S. (2001) Enzyme-assisted extraction of antioxidative phenols from black currant juice press residues (*Ribes nigrum*). *Journal of Agricultural and Food Chemistry* **49**(7), 3169–3177.

Makris, D.P., Boskou, G. and Andrikopoulos, N.K. (2007) Polyphenolic content and in vitro antioxidant characteristics of wine industry and other agri-food solid waste extracts. *Journal of Food Composition and Analysis* **20**(2), 125–132.

Manuel, P., Arnous, A. and Meyer, A.S. (2006) Upgrading of grape skins: Significance of plant cell-wall structural components and extraction techniques for phenol release. *Trends in Food Science and Technology* **17**, 579–590.
Manuel, P., Zornoza, B. and Meyer, A.S. (2008) Selective release of phenols from apple skin: Mass transfer kinetics during solvent and enzyme-assisted extraction. *Separation and Purification Technology* **63**, 620–627.
Meyer, A.S., Jepsen, S.M., Sørensen, N.S. (1998) Enzymatic release of antioxidants for human low-density lipoprotein from grape pomace. *Journal of Agricultural and Food Chemistry* **46**(7), 2439–2446.
Nardini, M., Cirillo, E., Natella, F. and Scaccini, C. (2002) Absorption of phenolic acids in humans after coffee consumption. *Journal of Agricultural and Food Chemistry* **50**(20), 5735–5741.
Park, M.W., Park, Y.K. and Kim, E.S. (2004) Properties of phenolic compounds in unripe apples. *Journal of the East Asian Society of Dietary Life* **14**(4), 343–347.
Pinelo, M., Arnous, A. and Meyer, A.S. (2006) Upgrading of grape skins: significance of plant cell-wall structural components and extraction techniques for phenol release. *Trends in Food Science and Technology* **17**(11), 579–590.
Pompeu, D.R., Silva, E.M. and Rogez, H. (2009) Optimisation of the solvent extraction of phenolic antioxidants from fruits of *Euterpe oleracea* using Response Surface Methodology. *Bioresource Technology* **100**(23), 6076–6082.
Renard, C.M.G.C., Baron, A., Guyot, S. and Drilleau, J.F. (2001) Interactions between apple cell walls and native apple polyphenol: quantification and some consequences. *International Journal of Biological Macromolecules* **29**(2), 115–125.
Singleton, V.L., Orthofer, R. and Lamuela-Raventos, R.M. (1999) Analysis of total phenolic and other oxidation substrates and antioxidants by means of Folin–Ciocalteu reagent. *Methods in Enzymology* **299**, 152–178.
Sørensen, H.R., Pedersen, S., Anders, V.N. and Meyer, A.S. (2005) Efficiencies of designed enzyme combinations in releasing arabinose and xylose from wheat arabinoxylan in an industrial ethanol fermentation residue. *Enzyme and Microbial Technology* **36**(5-6) 773–784.
Stalikas, C.D. (2007) Extraction, separation, and detection methods for phenolic acids and flavonoids. *Journal of Separation Science* **30**(18), 3268–3295.
Stefania, D.A., Amelia, C., Marianna, R. *et al.* (2007) Effect of reddening-ripening on the antioxidant activity of polyphenol extracts from Cv. 'Annurca' apple fruits. *Journal of Agricultural and Food Chemistry* **55**(24), 9977–9985.
Sudha, M.L., Baskaran, V. and Leelavathi, K. (2007) Apple pomace as a source of dietary fiber and polyphenol and its effect on the rheological characteristics and cake making. *Food Chemistry* **104**(2), 686–692.
Wang, L. and Weller, C.L. (2006) Recent advances in extraction of nutraceuticals from plants. *Trends in Food Science and Technology* **17**(6), 300–312.
Wu, J.H., Gao, H.Y., Zhao, L. *et al.* (2007) Chemical compositional characterization of some apple cultivars. *Food Chemistry* **103**(1), 88–93.
Xiao, G. and Yao, H. (2008) Optimization of Viscozyme L-assisted extraction of oat bran protein using response surface methodology. *Food Chemistry* **106**(1), 345–351.
Yanagida, A., Kanda, T., Tanabe, M. *et al.* (2002) Inhibitory effects of apple polyphenols and related compounds on cariogenic factors of mutans streptococci. *Journal of Agricultural and Food Chemistry* **48**(11), 5666–5671.
Zheng, H.Z., Lee, H.R., Lee, S.H. *et al.* (2008) Pectinase assisted extraction of polyphenol from apple pomace. *Chinese Journal of Analytical Chemistry* **36**(3), 306–310.
Zheng, H.Z., Hwang, I.W. and Chung, S.K. (2009) Enhancing polyphenol extraction from unripe apples by carbohydrate-hydrolyzing enzymes. *Journal of Zhejiang University SCIENCE B* **10**(12), 912–919.
Zheng, H.Z., Hwang, I.W., Kim, S.K. *et al.* (2010) Optimization of carbohydrate-hydrolyzing enzyme aided polyphenol extraction from unripe apples. *Journal of the Korean Society for Applied Biological Chemistry* **53**(3), 342–350.
Zheng, H.Z., Kim, Y.L. and Chung, S.K. (2012) A profile of physicochemical and antioxidant changes during fruit growth for the utilisation of unripe apples. *Food Chemistry* **131**, 106–110.

# 4 Case study: Statistical analysis of eurycomanone yield using a full factorial design

**Azila Abdul-Aziz[1], Harisun Yaakob[1], Ramlan Aziz[1], Roshanida Abdul Rahman[2], Sulaiman Ngadiran[1], Mohd Faizal Muhammad[1], Noor Hafiza Harun[1], Wan Mastura Wan Zamri[1] and Ernie Surianiy Rosly[1]**

[1] *Institute of Bioproduct Development, Universiti Teknologi Malaysia, Skudai Johor, Malaysia*

[2] *Faculty of Chemical Engineeering, Universiti Teknologi Malaysia, Skudai Johor, Malaysia*

## ABSTRACT

*Eurycoma longifolia*, also known as Tongkat Ali, is popular in Asia due to its aphrodisiac properties. The root extract also exhibits cytotoxic, anti-ulcer, antitumour, antipyretic, antischistosomal and antimalarial activities. This work focuses on the development of experimental design to find a set of optimum process variables for water extraction of *Eurycoma longifolia* roots. A two level full factorial design has been successfully employed to generate a linear regression model for the process variables (solvent to raw material ratio, extraction time and particle size) with eurycomanone yield as the response. The regression model was statistically significant with $R^2$ of 0.98; the residuals of the model are normally distributed. The factor that presented has the strongest influence on yield was solvent-to-raw material ratio, followed by extraction time. The interaction effect between ratio and particle size on eurycomanone yield was also significant in this model. The validity of this analysis is limited to water extraction at 100°C with an agitation rate of 400 rpm.

## INTRODUCTION

*Eurycoma longifolia* or Tongkat Ali is amongst the top ranking medicinal and aromatic plants (MAPs) in Malaysia for its alleged testosterone-enhancing properties. It is a flowering shrub native to Malaysia and other Southeast Asia countries, such as Indonesia, Vietnam, Thailand and Laos. *Eurycoma longifolia* is part of the *Simaroubaceae* family and is also known as Long Jack, Pasak Bumi (Indonesia), Cay ba Binh (Vietnam), Penawar Pahit, Bedara Pahit, Tongkat Baginda, Petala Bumi and Setunjang Bumi (Ismail *et al.*, 1999).

*Eurycoma longifolia* is traditionally used as a health tonic by decoction of the roots or bark to cure a variety of illnesses, such as diarrhoea, glandular swelling, bleeding, dropsy, persistent cough and hypertension, and relief of pains in the bones. Recently, investigations and phytochemical screenings conducted by researchers have shown that *Eurycoma longifolia* extracts, particularly from the roots, exhibit aphrodisiac, cytotoxic, anti-ulcer, antitumour, antipyretic, antischistosomal and antimalarial activities (Perry and Metzger, 1980; Ismail *et al.*, 1999; Jagananth and Ng, 2000).

Various bioactive constituents have been isolated and characterized from *Eurycoma longifolia*, and the major portions are from the groups of alkaloids, quassinoids and saponins (Bhat and Karim, 2010).

*Mathematical and Statistical Methods in Food Science and Technology*, First Edition.
Edited by Daniel Granato and Gastón Ares.

The phytochemicals that are responsible for the pharmacological activities and therapeutic effects of *Eurycoma longifolia* include quassinoids (Chan *et al.*, 1991, 1992; Morita *et al.*, 1993a), tirucallene-type triterpenes (Itokawa *et al.*, 1992), squalene derivatives (Itokawa *et al.*, 1991; Morita *et al.*, 1993b), biphenyl-neo-lignans (Morita *et al.*, 1992), anthraquinones (Lin *et al.*, 2001), canthin-6-one-alkaloids (Kardono *et al.*, 1991), β-carboline alkaloids (Kardono *et al.*, 1991; Kuo *et al.*, 2003) and dimeric dihydrobenzofuran (Kuo *et al.*, 2004).

Quassinoid is the characteristic phytochemical of the *Simaroubaceae* family that imparts its bitter taste. The major components of quassinoids in *Eurycoma longifolia* include eurycomanone, 13α(21)-epoxyeurycomanone, eurycomanol, eurycomanol-2-O-β-D-dglucopyranoside and 13,21-dihydroeurycomanone (Teh *et al.*, 2011). Thus far, eurycomanone is known as the target marker for *Eurycoma longifolia* extract due to its high yield and significant phytochemical bioactivities (Kumaresan and Sarmidi, 2003; Low *et al.*, 2005). This active ingredient is believed to be responsible for the aphrodisiac effects of *Eurycoma longifolia* extract. Eurycomanone has also been shown to demonstrate anticancer activities towards the leukemia cells, KB IV and P388 (Zakaria *et al.*, 2009) and a cytotoxic effect on other cancer cells, such as lung cancer, skin cancer (melanoma), colon cancer, breast cancer and fibrosarcoma (Tada *et al.*, 1991).

The Malaysian government has developed a Malaysian Standard for *Eurycoma longifolia* extract for commercial purposes. The amount of standardized eurycomanone in freeze dried water extract of *Eurycoma longifolia* is specified to be between 0.8 and 1.5% (w/v) (Malaysian Standard, 2010).

Studies on the effect of processing parameters on *Eurycoma longifolia* water extraction have been pursued by a few groups of researchers (Kumaresan and Sarmidi, 2003; Low *et al.*, 2005). However, there is still limited information on using statistical analysis on the yield of eurycomanone or other bioactive compounds from *Eurycoma longifolia* extract. Mohamad *et al.* (2010) have developed a mass transfer model of the *Eurycoma longifolia* solid–liquid extraction process and used the model to predict the yield of phytochemical components. The model was simulated using Matlab programming. Meanwhile, Athimulam *et al.* (2006) have modelled and optimized the processes for the pilot scale production of *Eurycoma longifolia* by proposing three debottlenecking schemes using SuperPro Designer®; they focused on the throughput and economic analysis.

Compared to classical experimental techniques, statistical mathematical methods result in increased research efficiency (Lazic, 2004). Statistical design of experiments is a quick and cost-effective technique where the outcomes can be applied further for optimization of any process, including the extraction system (Antony and Roy, 1999). Analysis of the design of experiments (DOE) is built on the basis of the analysis of variance (ANOVA), that is a collection of statistical models that can be used to test for significant difference between means of groups under the assumption that the sampled populations are normally distributed.

In this sense, this work focused on the development of experimental designs for finding a set of process variables for water extraction of *Eurycoma longifolia* roots. The effects of processing parameters, namely solvent to raw material ratio, duration of extraction and raw material particle size, on eurycomanone yield were investigated using statistical methodology based on design of experiments using Design Expert (Version 6.0.8) software.

## MATERIALS AND METHODS

### Materials

*Eurycoma longifolia* roots were purchased from Felda Agricultural Services Sdn. Bhd., Malaysia. The plants were planted in Jengka, Pahang, Malaysia, by the Federal Land Development Authority

(FELDA) and the age of the plants was approximately five years. The average growth performance of the cultivated plants was recorded as the following: 6 cm stem diameter at 15 cm above ground and plant height of 470 cm. The growth performance of the plants met the requirements for production of *Eurycoma longifolia* extracts. The harvested plants of *Eurycoma longifolia* (roots) were then subjected to the pre-processing activities of cleaning, drying, grinding and sieving before extraction.

## Methods

Experimental design using statistical analysis is a suitable and valuable tool for simultaneously finding out individual and interaction effects of all parameters that may affect the response of the studied system. *Eurycoma longifolia* extraction is influenced by both quantitative (duration of extraction, agitation rate, temperature and solvent to raw material ratio) and qualitative (solvent type and particle size) factors. The design that has been used to study the combinations of factors for *Eurycoma longifolia* extraction is two levels full factorial design. It consists of a $2^k$ experiment in which k is the factor for each experiment at two levels.

### *Solid-liquid extraction*

Extraction is the key component in production of *Eurycoma longifolia* roots. Three factors affecting *Eurycoma longifolia* extraction on eurycomanone yield were investigated in this study. The factors were solvent to raw material ratio, duration of extraction and particle size of raw material. The range of each factor is summarized in Table 4.1.

The type of extractor used in this study was a batch solid–liquid extractor that can accommodate up to two litres volume. The volume of the total solution for each run was fixed at 500 ml, whereas the amount of solvent and *Eurycoma longifolia* roots was employed at the designated solvent to raw material ratios (weight per weight).

### *Determination of eurycomanone yield*

To measure the yield of eurycomanone in the *Eurycoma longifolia* extract, liquid chromatography analysis was performed using a Waters Alliance 2690 HPLC system with 2487 DAD detector (Milford, MA). Separation was achieved using Ascentris™ RP-Amide 5 μm column with the dimensions 250 × 4.6 mm. The column was isocratically eluted with water–acetonitrile–*ortho*-phosphoric acid. The mobile phase was 90% of 0.05% phosphoric acid and 10% acetonitrile at a flow rate of 1 ml/min (Malaysian Standard, 2010) with the injection volume of 20 μl. The low mobile phase flow rate was chosen to allow the peaks to separate more distinctly. Eurycomanone was monitored at 254 nm (Malaysian Standard, 2010) and the yield calculated by referring to an external standard curve constructed using authentic standard.

**Table 4.1** The processing parameters for *Eurycoma longifolia* roots water extraction.

| Factor | Range | References |
|---|---|---|
| Solvent to raw material ratio (g/g) | 10 : 1, 25 : 1, 40 :1 | Sim, 2004 Kumaresan, 2008 |
| Duration of extraction (h) | 1, 2, 3 | Sim, 2004 Kumaresan, 2008 |
| Particle size of raw material (mm) | 0.25–0.5, 1.0–2.0 | Sim, 2004 |

**Table 4.2** Extraction factors and their coded level in full factorial design.

| Factor | Unit | Symbol | Level | |
|---|---|---|---|---|
| | | | **−1 (Low)** | **+1 (High)** |
| Solvent to raw material ratio | w/w | A | 10 | 40 |
| Duration | Hour | B | 1 | 3 |
| Particle size | Mm | C | S1 (0.25–0.5) | S3 (1.0–2.0) |

### *Two level full factorial design*

To improve the extraction system, the three variables (or experimental factors) used are solvent to raw material ratio ($A$), duration ($B$) and particle size ($C$). Meanwhile, the response variable is eurycomanone yield ($Y$) of the *Eurycoma longifolia* extract. A full factorial design of $2^3$ with four centre points was applied in this study, resulting in twelve experimental runs. Two-level designs are denoted as 'high' and 'low' or '+1' and '−1', respectively. Two replicates were made at each setting in order to obtain an average value of the response (or idea of consistency). A $p$ value of lower than 0.05 was considered to be significant. Design Expert (Version 6.0.8) was used to analyse the data using ANOVA with 95% reliability. The range and coded level of extraction variables studied are listed in Table 4.2. In this work, the half-normal probability plot was used to select the model.

Running the full complement of all factors can give the estimate of all main and interaction effects. The main effect is referred to as the average response when the factor changes from one level to another level, such as from high to low level, while the interaction effect is the difference in response between the levels of one factor (Montgomery, 2009). Centre points are used to estimate the pure error and curvature in the model. The full factorial design model can be presented using the following mathematical equation:

$$Y = \beta_0 + \beta_1 A + \beta_2 B + \beta_3 C + \beta_{12} AB + \beta_{13} AC + \beta_{23} BC + \beta_{123} ABC + \varepsilon \qquad (4.1)$$

where $\beta_0$ is the intercept, $\beta_1, \beta_2, \beta_3$ are the coefficients for three main effects, $\beta_{12}, \beta_{13}, \beta_{23}$ are the coefficients of two factors and $\beta_{123}$ is the three-factor interaction coefficient; $ABC$ represent the factors in the model. Meanwhile, $\varepsilon$ is the experimental error. From this linear regression equation, the factors that have the greatest impact on eurycomanone yield can be determined and any significant interaction between the factors can be detected. Then, the one-factor interaction and 3D surface plots were explored for regions of interest in this factorial analysis.

## RESULTS AND DISCUSSION

To get high quality *Eurycoma longifolia* extract, the extraction process has to be optimized. Extraction of a plant compound is achieved when solvent diffuses into the solid matrix of the plant material and then solubilizes the targeted compounds with similar polarity (Handa, 2006; Ncube *et al*., 2008). Water extraction is found to be the best technique for plant extracts that are intended to be consumed orally, because water is safe for consumption and is nontoxic even at reasonably high dosages.

To obtain high quality, efficient and safe *Eurycoma longifolia* extract, effective processing techniques using appropriate technology are crucial. The engineering activities, that is the extraction process, should be optimized in order to extract the maximum amount of active phytochemicals in *Eurycoma longifolia* and to ensure that the extract retains a phytochemical profile that is as close as possible to the *Eurycoma longifolia* plant.

In the extraction process, the most important criterion to be evaluated is the extraction yield (quantity in percentage), namely the total yield or the yield of a certain target compound or compounds. The parameters that affect extraction yield are:

(i) *Solvent characteristics.* The characteristics that should be considered include the type of solvent used, its polarity, pH and concentration. *Eurycoma longifolia* roots are mostly extracted using water by most manufacturers (Sambandan *et al.*, 2006; Malaysian Standard, 2010). Water was used as the solvent for this work because it is known as the generally recognized as safe (GRAS) solvent that is usually applied in natural medicinal plant production (Takeuchi *et al.*, 2008).

(ii) *Duration of extraction.* The duration or length of the extraction period was investigated in order to estimate the time required to extract the maximum extractable quantity of solute at equilibrium. Duration of extraction of 1–3 h was studied in this work.

(iii) *Solvent to raw material ratio.* Ratio study was employed to evaluate a viable extraction ratio of solvent to raw *Eurycoma longifolia* (weight per weight). The ratios studied in this work were 10 : 1, 25 : 1 and 40 : 1.

(iv) *Particle size.* Reduction in particle size is important for better extraction performance due to the increase in surface area, which will enhance the contact between solvent and solute. However, excessively small grain size can result in sample agglutination and hinders efficient solvent permeation (Kumaresan, 2008). The grain sizes investigated in this work were 0.25–0.5 mm (S1), and 1.0–2.0 mm (S3).

(v) *Temperature.* Heat treatment is employed to enhance the mechanism of the diffusion process during extraction. However, increasing temperature at continuous exposure can also generate unwanted reactions, such as degradation of thermo labile compounds and others. In previous studies, the optimum temperature for water extraction of *Eurycoma longifolia* roots was found to be 100°C (Sim, 2004; Kumaresan, 2008). Thus, in this work, the boiling temperature of water at atmospheric pressure was chosen as the extraction temperature.

(vi) *Mechanical actions.* Two criteria that are important in mechanical action are pressure and agitation rate. Since the extraction process was performed under atmospheric pressure, only agitation rate was emphasized in this study. Increasing the agitation speed will increase the mass transfer coefficient as well as produce a high yield of extraction. Mohamad *et al.* (2010) had suggested that the best agitation rate for the extraction of *Eurycoma longifolia* was 400 rpm. According to Kumaresan (2008), 400 rpm was chosen as the optimal agitation rate for his work, as this value was sufficient to mix the particles and maintain them suspended without creating a vortex. Thus the agitation speed employed in this study was at 400 rpm.

The goal of the statistical analysis in this study was to identify the main effects of and the interaction effects among the processing parameters during *Eurycoma longifolia* extraction process. The complete design matrix of the $2^3$ full factorial experiments and the corresponding output response on eurycomanone yield are shown in Table 4.3.

Analysis of variances (ANOVA) is employed to determine the significance of each factor based on effects calculations. ANOVA is a statistical technique that partitions the whole variation into its component parts, each of which is allied with a different source of variation (Kim *et al.*, 2003). Results of the significance of each independent variable of *Eurycoma longifolia* extraction process and their interaction effects are presented in Table 4.4. The $p$ value, which is the lowest level of significance leading to the rejection of the null hypothesis ($H_0$), shows that the generated model (with the $p$ value of 0.0037) is significant.

**Table 4.3** The design matrix of the two level full factorial design and the corresponding output response of solvent to raw material ratio, duration and particle size on eurycomanone yield.

| Trial No. | Independent variables | | | Response |
|---|---|---|---|---|
| | A | B | C | Y |
| | Solvent to raw material ratio (w/w) | Duration (h) | Particle size (mm) | Yield of eurycomanone (%) |
| 1 | 10 | 1 | S1 | 0.0897 |
| 2 | 40 | 1 | S1 | 0.1708 |
| 3 | 10 | 3 | S1 | 0.0793 |
| 4 | 40 | 3 | S1 | 0.1675 |
| 5 | 10 | 1 | S3 | 0.1236 |
| 6 | 40 | 1 | S3 | 0.1421 |
| 7 | 10 | 3 | S3 | 0.0826 |
| 8 | 40 | 3 | S3 | 0.1277 |
| 9 | 25 | 2 | S1 | 0.1390 |
| 10 | 25 | 2 | S3 | 0.1478 |
| 11 | 25 | 2 | S1 | 0.1406 |
| 12 | 25 | 2 | S3 | 0.1497 |

Data are means of two replicates.

**Table 4.4** ANOVA test results of *Eurycoma longifolia* extraction process.

| Source of variation | Sum of squares | Degrees of freedom | Mean Square | F-ratio | p value (Prob > F) |
|---|---|---|---|---|---|
| *Model* | *0.0091* | 6 | *0.0015* | 25.60 | *0.0037 significant* |
| A – ratio | 0.0068 | 1 | 0.0068 | 113.85 | 0.0004 |
| B – duration | 0.006 | 1 | 0.006 | 10.02 | 0.0340 |
| C – particle size | 0.000015 | 1 | 0.000015 | 0.25 | 0.6425 |
| AB | 0.00014 | 1 | 0.00014 | 2.38 | 0.1975 |
| AC | 0.0014 | 1 | 0.0014 | 23.45 | 0.0084 |
| BC | 0.00022 | 1 | 0.00022 | 3.65 | 0.1287 |
| Pure error | 0.000003 | 2 | 0.0000015 | — | — |
| Cor Total | 0.011 | 11 | — | — | — |

The half-normal probability plot (Figure 4.1) shows that the main effects of *A* (ratio) and *B* (duration), and the interaction effect of *AC* fall well out on the tail of the normal curve (points furthest to the right). Thus, these three biggest effects of *A, AC* and *B* (from the greatest to smallest, respectively) on eurycomanone yield are statistically significant with 95% confidence level. However, the other effects, namely *C* (particle size), *AB* and *BC* which fall in line (points nearest zero) are found to be insignificant in this model.

The regression equation that represents the model of eurycomanone yield from the solvent to material ratio, duration and particle size within the investigated ranges is:

$$Y = 0.12 + 0.029A - 0.0086B - 0.0011C + 0.0042AB - 0.013AC - 0.0052BC \quad (4.2)$$

This function describes how the experimental variables and their interactions influence the eurycomanone yield. A positive value of the quantitative factor indicates that the increase in the parameter will increase the yield. On the contrary, negative values of the effects suggest a decrease in response. The

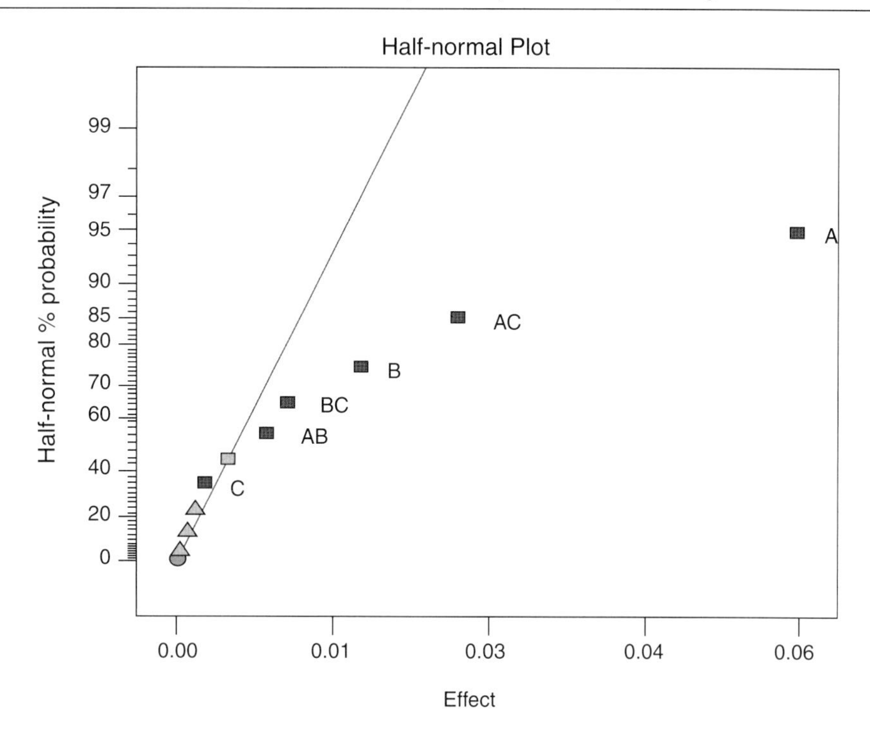


**Figure 4.1** Half Normal plot of the effect of processing on eurycomanone yield.

equation shows that the solvent to raw material ratio ($A$) has a positive impact on eurycomanone yield, while the extraction duration ($B$) and combination of $AC$ produce negative effects.

The parity plot (Figure 4.2) compares the eurycomanone yield of actual (experimental) values with the predicted (model) values. The coefficient of determination ($R^2$) value is approximately 0.98, suggesting an adequate model for the response variable, that is eurycomanone yield. The adjusted $R^2$ ($R^2_{adj}$) value is 0.94. This value indicates the increment when a new variable is added if there is enough reduction in the residual sum of squares ($SS_E$) to balance for the loss of a degree of freedom. The $p$ value of lack of fit for the model was 0.013.

Another hypothesis implicit in this analysis is that the residuals have a normal distribution. This can be checked with a normality plot of the residuals, as shown in Figure 4.3. The normal probability plot is a graphical representation for finding out whether data are distributed normally or not (Pokhrel and Viraraghvan, 2006). The residual values elucidate the difference between predicted values (model) and the observed values (experimental). Figure 4.3 shows that the experimental points follow the theoretical distributions verifying that the residuals are approximately normally distributed.

The probability of $p$ value is used to find out the statistically significant effect in the model. Figures 4.4 and Figure 4.5 show the main effects of and interaction effects between the three factors on the yield of eurycomanone. The generated interaction plots with the dotted line represent the higher level of the factors (with 95% Least Significant Difference (LSD) bars). Note that the factors used in this study have never been evaluated simultaneously using full factorial designs and were chosen for their importance. The results obtained in a statistical factorial design largely depend on the ranges of the factors studied. These plots use the solvent to raw material ratio of 10 : 1, duration of 1 h and particle size of 0.25–0.5 mm levels as the baseline.

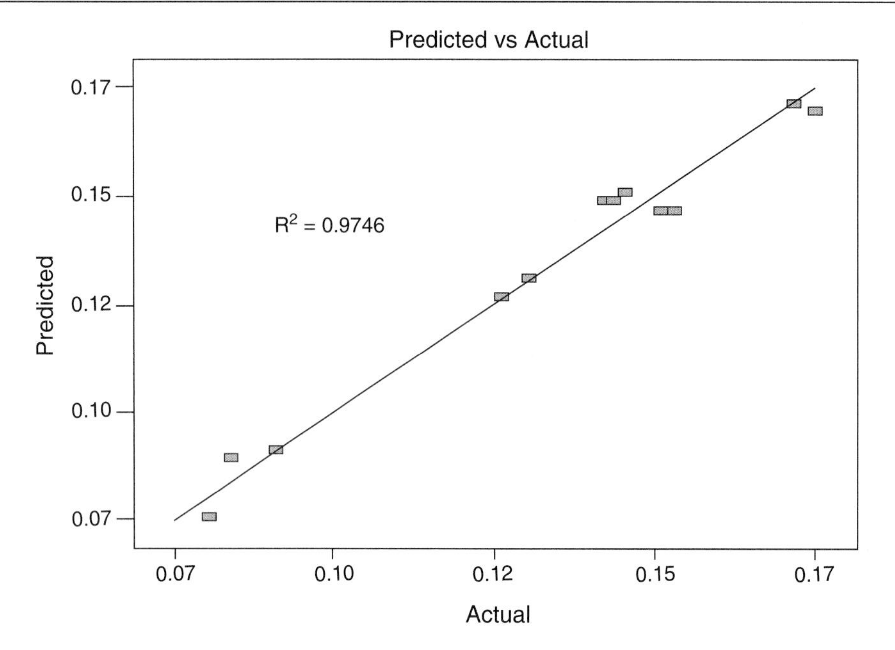


**Figure 4.2** Predicted-actual values plot for eurycomanone yield.

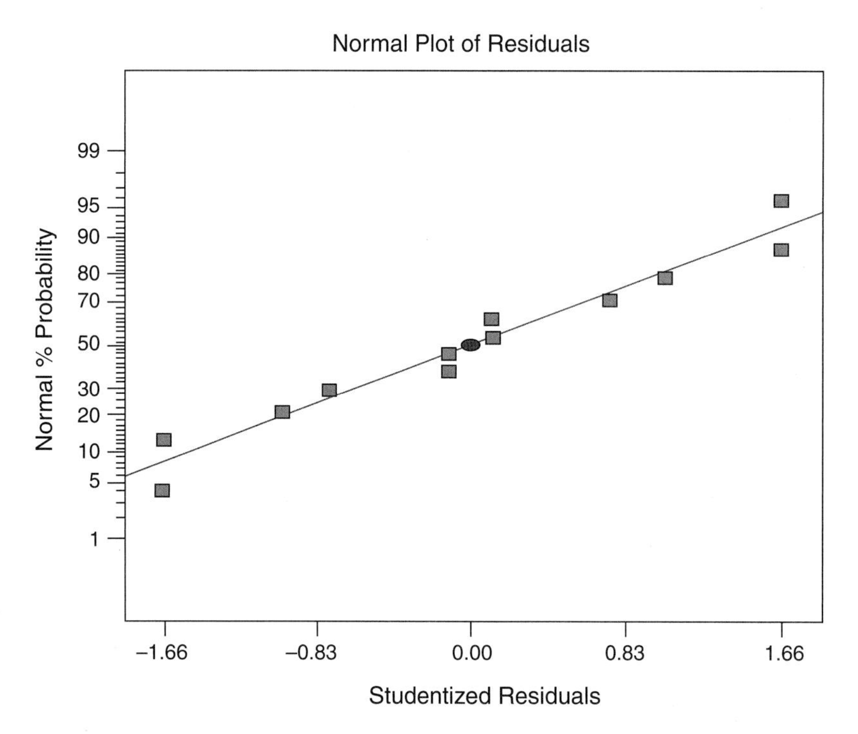


**Figure 4.3** Normal probability for plot residuals for eurycomanone yield.

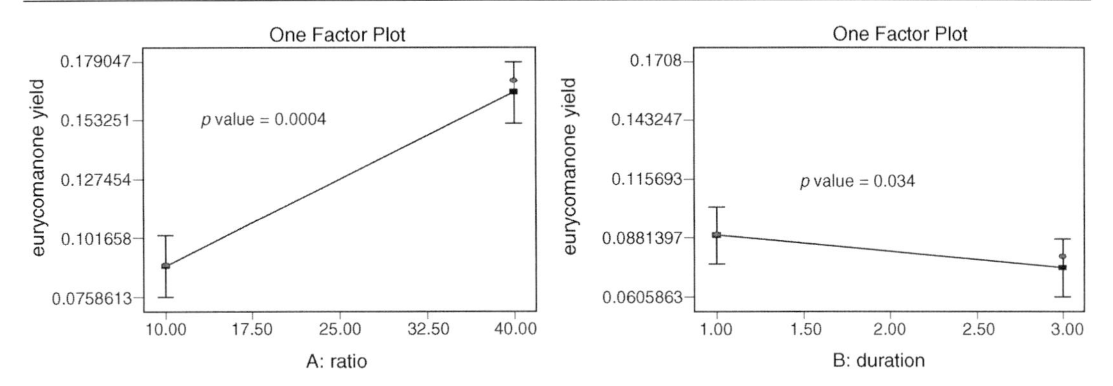


**Figure 4.4** Main effects of the processing parameters on eurycomanone yield.

Figure 4.4 shows that the solvent to raw material ratio ($A$) is predicted to have the greatest effect on yield. The plot illustrates that increasing the ratio will increase eurycomanone yield, that is the positive effect is observed when the factor changes from low to high level. The model suggests that higher eurycomanone yield can be obtained at solvent to raw material ratio (g/g) of 40 : 1 compared to 25 : 1 and 10 : 1 at the conditions of this work.

Another significant effect in this work was the duration of extraction ($B$). An increase in the extraction duration from 1 to 3 h is expected to slightly decrease the percentage of eurycomanone yield from 0.0897 to 0.074% (w/w). This 'duration effect' behaviour had been similarly detected in other related works (Sim, 2004; Kumaresan, 2008). An extraction time that is too long may result in the extraction

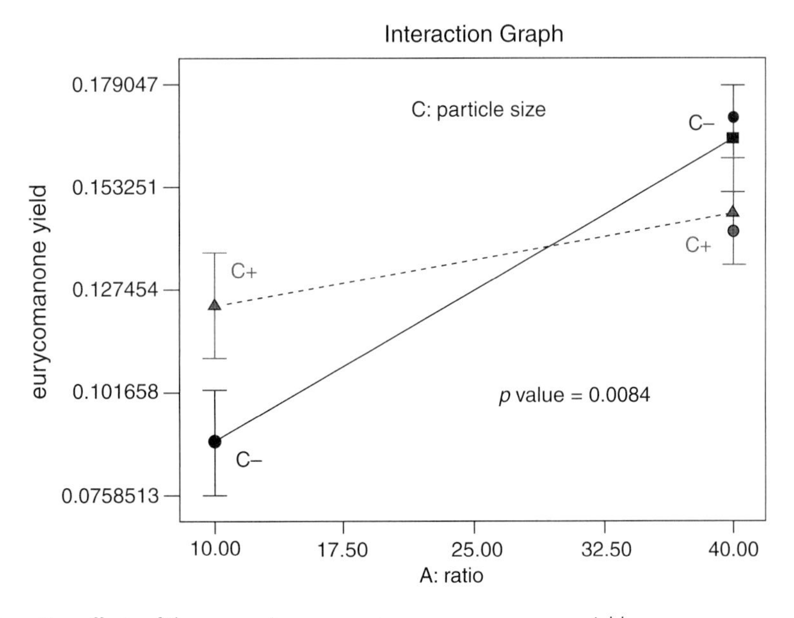


**Figure 4.5** Interaction effects of the processing parameters on eurycomanone yield.

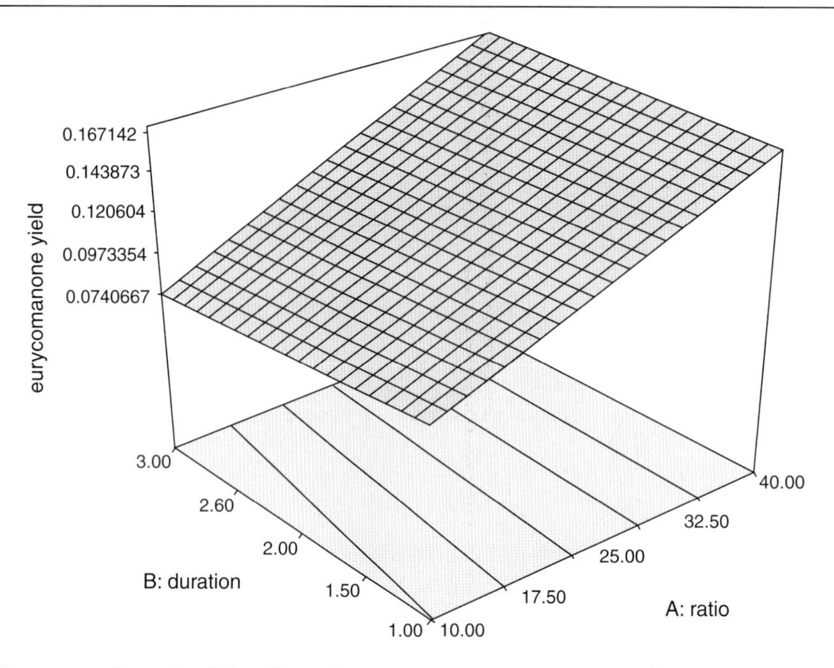


**Figure 4.6** Response surface plot of the effects of parameters on eurycomanone yield.

of unwanted constituents. As a result, the model anticipates that the lower extraction time improves the yield of eurycomanone.

Figure 4.5 shows the interaction effect of *AC* on eurycomanone yield, where the effect of solvent to raw material ratio depends on the level of particle size. An interaction is effective when the change in the response of a factor from low to high levels is dependent on the level of another factor (Mathialagan and Viraraghavan, 2005). The graph predicts that using bigger particle size (*C*) of *Eurycoma longifolia*, leads to a small change in eurycomanone yield and the effect is less affected by the solvent to raw material ratio (*B*). However, when using smaller particle size, the eurycomanone yield changed substantially from 0.089 to 0.166% (w/w). The LSD bars overlap at the high solvent to raw material ratio end of the interaction graph, implying that there is no significant difference in eurycomanone yield at the high level of solvent to raw material ratio.

The surface plot of the response function is useful in understanding both the main and interaction effects of the factors (Arbizu and Luis Pérez, 2003; Cojocaru and Trznadel, 2007). The surface plot (Figure 4.6) of the response, that is eurycomanone yield, shows the same outcomes as observed in the effects plots, that is the yield increases with increasing solvent to raw material ratio and slightly decreases with increasing extraction duration.

## CONCLUSIONS

A full factorial experimental design has been successfully employed to generate a linear regression model for the factors that affect eurycomanone yield in the production of *Eurycoma longifolia* root extract. The solvent to raw material ratio and duration of extraction have been shown to be significant in the production of *Eurycoma longifolia* extract on account of eurycomanone yield. The factor that has the greatest effect on yield is solvent to raw material ratio, which is predicted to have a positive influence on eurycomanone yield, that is increasing the ratio will increase eurycomanone yield. Duration of extraction affects eurycomanone yield negatively, that is increasing the duration of extraction will decrease the yield. The interaction effect of ratio and particle size on eurycomanone yield is also significant in this model. The validity of this analysis is limited to water as the solvent, an extraction temperature of 100 °C and an agitation rate of 400 rpm.

## REFERENCES


Antony, J. and Roy, R.K. (1999) Improving the process quality using statistical design of experiments: a case study. *Quality Assurance* **6**, 87–95.

Arbizu, I.P. and Luis Pérez, C.J. (2003) Surface roughness prediction by factorial design of experiments in turning processes. *Journal of Materials Processing Technology* **143-144**, 390–396.

Athimulam, A, Kumaresan, S., Foo, D.C.Y. *et al.* (2006) Modelling and optimization of Eurycoma longifolia water extract production. *Food and Bioproducts Processing* **84**(C2), 139–149.

Bhat, R. and Karim, A.A. (2010) Tongkat Ali (*Eurycoma Longifolia* Jack): A review on its ethnobotany and pharmacological importance. *Fitoterapia* **81**, 669–679.

Chan, K.L., Iitaka, Y., Noguchi, H. *et al.* (1992) 6α-hydroxyeurycomalactone, a quassinoid from *Eurycoma longifolia. Phytochemistry* **31**(12), 4295–4298.

Chan, K.L., Lee, S.P., Sam, T.W. *et al.* (1991) 13β, 18-dihydroeurycomanol, a quassinoid from *Eurycoma longifolia. Phytochemistry* **30**(9), 3138–3141.

Cojocaru, C. and Trznadel, G.Z. (2007) Response surface modeling and optimization of copper removal from aqua solutions using polymer assisted ultrafiltration. *Journal of Membrane Science* **298**, 56–70.

Kim, H.M., Kim, J.G., Cho, J.D. and Hong, J.W. (2003) Optimization and characterization of UV-curable adhesives for optical communication by response surface methodology, *Polymer Testing* **22**, 899–906.

Handa, S.S., Rakesh, D.D. and Vasisht, K. (2006) *Compendium of Medicinal and Aromatic Plants ASIA*, Volume II. ICS-UNIDO, Vienna, Austria, pp. 75–84.

Ismail, Z., Ismail, N. and Lassa, J. (1999) *Malaysian herbal monograph*, Volume 1. Malaysian Monograph Committee, Institut Penyelidikan Perubatan, Kuala Lumpur, Malaysia, pp. 1–93.

Itokawa, H., Kishi, E., Morita, H. and Takeya, K. (1992) Cytotoxic quassinoids and tirucallane-type triterpenes from the woods of *Eurycoma longifolia. Chemical and Pharmaceutical Bulletin* **40**(4), 1053–1055.

Itokawa, H., Kishi, E., Morita, H. *et al.* (1991) Eurylene, a new squalene-type triterpene from *Eurycoma longifolia. Tetrahedron Letters* **32**(15), 1803–1804.

Jagananth, J.B. and Ng, L.T. (2000) *Herbs – the Green Pharmacy of Malaysia*. Vinpress Sdn. Bhd. and Malaysian Agricultural Research and Development Institute (MARDI), Kuala Lumpur, Malaysia, pp. 45–46.

Kardono, L.B.S., Angerhofer, C.K., Tsauri, S. *et al.* (1991) Studies on Indonesian medicinal plants. IV Cytotoxic and antimalarial constituents of the roots of Eurycoma longifolia. *Journal of Natural Products* **54**(5), 1360–1367.

Kumaresan, S. (2008) A process engineering approach to the standardization of eurycomanone in *Eurycoma longifolia* water extract. PhD Thesis in Chemical Engineering, Universiti Teknologi Malaysia, Skudai Johor, Malaysia.

Kumaresan, S. and Sarmidi, M.R. (2003) A preliminary study into the effect of processing on *Eurycoma longifolia* water extract yield. In: *Proceedings of the International Conference on Chemical and Bioprocess Engineering*, University Malaysia Sabah, Kota Kinabalu, pp. 750–754.

Kuo, P.C., Damu, A. G., Lee, K.H. and Wu, T.S. (2004) Cytotoxic and antimalarial constituents from the roots of *Eurycoma longifolia. Bioorganic and Medicinal Chemistry* **12**(3), 537–544.

Kuo, P.C., Shi, L.S., Damu, A.G. *et al.* (2003) Cytotoxic and antimalarial β-carboline alkaloids from the roots of *Eurycoma longifolia. Journal of Natural Products* **66**, 1324–1327.

Lazic, Z.R. (2004) *Design of Experiments in Chemical Engineering: A Practical Guide*. Wiley-VCH Verlag GmbH, Weinhem, pp. 157–443.

Lin, L. C., Peng, C. Y., Wang, H. S. *et al.* (2001) Reinvestigation of the chemical constituents of *Eurycoma longifolia. Chinese Pharmaceutical Journal* **53**(2), 97–106.

Low, B.S., Ng, B.H., Choy, W.P. *et al.* (2005) Bioavailability and pharmacokinetic studies of eurycomanone from *Eurycoma longifolia. Planta Medica* **71**, 803–807.

Malaysian Standard (2010) MS No. 07U001R0, Phytopharmaceutical Aspect of Freeze Dried Water Extract from Tongkat Ali Roots – Specification. Department of Standards Malaysia, pp. 1–13.

Mathialagan, T. and Viraraghavan, T. (2005) Biosorption of pentachlorophenol by fungal biomass from aqueous solutions: a factorial design analysis. *Environmental Technology* **6**, 571–579.

Mohamad, M., Ali, M.W. and Ahmad, A. (2010) Modelling for extraction of major phytochemical components from *Eurycoma longifolia*. *Journal of Applied Sciences* **10**(21), 2572–2577.

Montgomery, D.C. (2009) *Design and Analysis of Experiments*, 7th edn. John Wiley and Sons (Asia) Pte org.

Morita, H., Kishi, E., Takeya, K. and Itokawa, H. (1992) Biphenylneolignans from wood of *Eurycoma longifolia*. *Phytochemistry* **31**(11), 3993–3995.

Morita H., Kishi, E., Takeya, K. *et al.* (1993a) Highly oxygenated quassinoids from *Eurycoma longifolia*. *Phytochemistry* **33**(3), 691–696.

Morita, H., Kishi, E., Takeya, K. *et al.* (1993b) Squalene derivatives from *Eurycoma longifolia*. *Phytochemistry* **34**(3), 765–771.

Ncube, N.S., Afolayan, A.J. and Okoh, A.I. (2008) Assessment techniques of antimicrobial properties of natural compounds of plant origin: current methods and future trends. *African Journal of Biotechnology* **7**(12), 1797–1806.

Perry, L.M. and Metzger, J. (1980) *Medicinal Plants of East and Southeast Asia: Attributed Properties and Uses*. Massachusetts Institute of Technology Press, MA, USA.

Pokhrel, D. and Viraraghvan, T. (2006) Arsenic removal from aqueous solution by iron oxide coated fungal biomass: a factorial design analysis. *Water, Air, & Soil Pollution* **173**, 195–208.

Sambandan, T.G., Saad, J.M., Rha, C.K. *et al.* (2006) *Bioactive Fraction of Eurycoma longifolia*. US Patent No. 7 132 117 B2.

Sim, C.C. (2004) Mass Transfer Modelling of *Eurycoma longifolia* Batch Extraction Process, Thesis of Bachelor Degree in Chemical Engineering, Universiti Teknologi Malaysia, Skudai Johor, Malaysia.

Tada, H., Yasuda, F., Otani, K. *et al.* (1991) New antiulcer quassinoids from *Eurycoma longifolia*. *European Journal of Medicinal Chemistry* **26**, 345–349.

Takeuchi, T.M., Pereira, C.G., Braga, M.E.M. *et al.* (2008) Low-pressure solvent extraction (solid–liquid extraction, microwave assisted and ultrasound assisted) from condimentary plants. In: *Extracting Bioactive Compounds for Food Products: Theory and Applications* (ed. M. AngelaA. Meireles). Taylor & Francis, pp. 140–150.

Teh, C.H., Murugaiyah, V. and Chan, K.L. (2011) Developing a validated liquid chromatography–mass spectrometric method for the simultaneous analysis of five bioactive quassinoid markers for the standardization of manufactured batches of Eurycoma longifolia Jack extract as antimalarial medicaments. Journal of Chromatography A **1218**, 1861–1877.

Zakaria, Y., Rahmat, A., Pihie, A.H.L. *et al.* (2009) Eurycomanone induce apoptosis in HepG2 cells via up-regulation of p53. *Cancer Cell International*. **9**(16), doi: 10.1186/1475-2867-9-16.

# Section 2

# 5 Applications of principal component analysis (PCA) in food science and technology

**Aurea Grané and Agnieszka Jach**
*Statistics Department, Universidad Carlos III de Madrid, Getafe, Spain*

## ABSTRACT

In this chapter a brief introduction is given to principal component analysis (PCA), a multivariate statistical technique widely used by applied researchers. The main objectives of PCA are presented followed by a review of some theoretical aspects, such as its definition, computation and some mathematical properties. Finally, some advanced topics, such as the study of the component sensitivity and the use of PCA in multivariate regression, are treated. Throughout the chapter, theoretical results are accompanied with illustrative examples. R-scripts are also provided, so that the reader can easily apply the explained methodologies. The complete collection of data sets and R-scripts with comments can be found in the supplementary materials available at http://www.est.uc3m.es/agrane/eng/public.html.

## INTRODUCTION

Principal component analysis (PCA) is a classical multivariate analysis technique. It was pioneered in 1891 by Karl Pearson, as a way of adjusting planes via orthogonal least squares, and developed further, in 1933, by Harold Hotelling in covariance and correlation analyses. Since then its popularity has increased well beyond the borders of the statistical community. Nowadays, PCA is one of the most widely used techniques, especially in those fields that deal with large data sets, where a dimensionality reduction is sought. The mathematics involved in principal component analysis boils down to the spectral decomposition of symmetrical and positively-definite matrices. This posed a real challenge in PCA in the pre-computer era but nowadays the computational aspects are only a minor issue, which allows the analyst to focus on the statistical questions. Comprehensive surveys on PCA can be found elsewhere (Jollife, 1986; Jackson, 1991; Basilevsky, 1994), as well as general introductions to PCA (Jonhson and Wichern, 1998; Rencher, 1998).

## GOAL

The main purpose of principal component analysis is to solve the problem of *dimensionality reduction*:

> Is it possible to capture the 'information' contained in a data set with a smaller set of variables than the observed one?

*Mathematical and Statistical Methods in Food Science and Technology*, First Edition.
Edited by Daniel Granato and Gastón Ares.

From an intuitive point of view, the situation is the following: if one variable is a function of other variables, then it contains redundant information, and is thus superfluous. We would like to establish the same when there is stochastic dependence instead of functional dependence. If the observed variables are highly correlated, it should be possible to replace them by a smaller number of new variables without a great loss of information. If that is the case, the dimensionality reduction will allow us to:

- Simplify posterior analyses, that are based on these few new variables.
- Carry out a graphical visualization/representation in low dimension (1, 2 or 3).
- Examine and interpret some relations between the observed variables.

Additionally, these nonredundant new variables often have interpretations that are useful for applied researchers (size and shape in biometry, socioeconomic dimensions in econometrics, speed and acceleration in physics/chemistry, among others). This is why principal component analysis has many of applications in biology, physics, engineering, psychology, economy, agriculture, sociology and so on, although they may exist under different names. For example in physics or engineering, principal components are called *empirical orthogonal functions* (Jolliffe, 1986) or the *Karhunen–Loeve expansion* (Watanabe, 1967).

Principal components are often the first step before performing the main analysis, like regression or multivariate analysis of variance (MANOVA). For example in data sets with a large number of observed variables, it can be useful to replace those variables by a smaller set of principal components.

## DEFINITION

Let $\mathbf{X}$ be a $nxp$ data matrix, containing the information on $p$ quantitative variables, $X_1, \ldots, X_p$, measured on a set of $n$ individuals. Matrix $\mathbf{X}$ can be seen in three different ways: as a $1xp$ vector of observed variables, as an $nx1$ vector of individuals and, finally, as a $nxp$ data matrix. The following notation will be used:

$$\mathbf{X} = [X_1, \ldots, X_p] = \begin{pmatrix} \mathbf{x}_1' \\ \vdots \\ \mathbf{x}_n' \end{pmatrix} = \begin{pmatrix} x_{11} & \cdots & x_{1p} \\ \vdots & \ddots & \vdots \\ x_{n1} & \cdots & x_{np} \end{pmatrix}$$

Henceforth, we will assume that $\mathbf{X}$ is a centred matrix, otherwise we apply the transformation $\mathbf{HX}$, where $\mathbf{H} = \mathbf{I} - \frac{1}{n}\mathbf{11}'$ is the centring matrix, $\mathbf{I}$ is the $nxn$ identity matrix and $\mathbf{1}$ is an $nx1$ vector of ones, which leads to a centred data matrix. (Remark: The centring matrix $\mathbf{H}$ is symmetrical, $\mathbf{H}' = \mathbf{H}$, and idempotent, $\mathbf{H}^2 = \mathbf{H}$). The principal components of $\mathbf{X}$ are $p$ composite variables that are linear combinations of $X_1, \ldots, X_p$, that is:

$$\begin{cases} Y_1 = X_1 t_{11} + \cdots + X_p t_{p1} \\ Y_2 = X_1 t_{12} + \cdots + X_p t_{p2} \\ \vdots \\ Y_p = X_1 t_{1p} + \cdots + X_p t_{pp} \end{cases}$$

equivalently, in matrix notation, they are:

$$Y_j = \mathbf{X}\mathbf{t_j}, \quad j = 1, \ldots, p,$$

where $\mathbf{X} = [X_1, \ldots, X_p]$ is a row vector containing the observed variables, and each $\boldsymbol{t_j} = (t_{1j}, \ldots, t_{pj})'$ is a column vector of coefficients. These $p$ vectors of coefficients, and hence the $Y_j$ variables, are determined according to the following recursive construction:

(i) The vector $\mathbf{t_1}$ containing the coefficients of

$$Y_1 = X_1 t_{11} + \cdots + X_p t_{p1}$$

is the one that maximizes the variance $var(Y_1)$ among all the linear combinations of the $X_j$s. Additionally, it is required that $\mathbf{t_1'} \mathbf{t_1} = 1$, because otherwise the problem is indeterminate: the variance $var(Y_1)$ can be artificially increased by multiplying vector $\mathbf{t_1}$ by a constant.

(ii) The vector $\mathbf{t_2}$ containing the coefficients of

$$Y_2 = X_1 t_{12} + \cdots + X_p t_{p2}$$

is the one that maximizes the variance $var(Y_2)$ among all the linear combinations of the $X_j$s that are uncorrelated with variable $Y_1$. As before, it is required that $\mathbf{t_2'} \mathbf{t_2} = 1$.

(iii) Given variables $Y_1, \ldots, Y_k$, for $k < p$, the vector $\mathbf{t_{k+1}}$ containing the coefficients of

$$Y_{k+1} = X_1 t_{1,k+1} + \cdots + X_p t_{p,k+1}$$

is the one that maximizes the variance $var(Y_{k+1})$ among all the linear combinations of the $X_j$s that are uncorrelated with variables $Y_1, \ldots, Y_k$ and verifying that $\mathbf{t_{k+1}'} \mathbf{t_{k+1}} = 1$.

## EFFECTIVE COMPUTATION

Let $\mathbf{S}$ be the covariance matrix of $\mathbf{X}$, that is, $\mathbf{S} = var(\mathbf{X}) = \frac{1}{n} \mathbf{X}' \mathbf{H} \mathbf{X}$. The problem of finding the maximum of function $V(\mathbf{t_1}) = \mathbf{t_1'} \mathbf{S} \mathbf{t_1}$, with the constraint that $\mathbf{t_1'} \mathbf{t_1} = 1$, is equivalent to obtaining the eigenvector of $\mathbf{S}$ of maximum eigenvalue, whose norm is equal to 1. This eigenvalue is the maximum of that function, in other words, the variance of $Y_1$. When a variational problem leads to an eigenvalue/eigenvector problem, the successive maximization problems are solved simultaneously by the spectral decomposition of the (symmetrical and positive semi-definite) matrix

$$\mathbf{S} = \mathbf{T} \boldsymbol{\Lambda} \mathbf{T}'$$

where $\boldsymbol{\Lambda} = diag(\lambda_1, \ldots, \lambda_p)$ is a diagonal matrix that contains the ranked eigenvalues of $\mathbf{S}$, $\lambda_1 > \ldots > \lambda_p$ and $\mathbf{T} = [\mathbf{t_1}, \ldots, \mathbf{t_p}]$ is an orthogonal matrix (that is, $\mathbf{T}'\mathbf{T} = \mathbf{T}\mathbf{T}' = \mathbf{I}$) with the corresponding eigenvectors (Mardia et al., 1979). The principal components of $\mathbf{X}$ are obtained as the new variables:

$$Y_j = \mathbf{X} \mathbf{t_j}, \quad j = 1, \ldots, p.$$

Notice that, for each $j$, the new variable $Y_j$ is constructed from the $j$-th eigenvector of $\mathbf{S}$ and the $j$-th eigenvalue gives the variance $var(Y_j) = \lambda_j$.

## SOME PROPERTIES

(i) By construction, the principal components have decreasing variances, that is:

$$\begin{aligned} var(Y_1) &= var(\mathbf{Xt_1}) = \mathbf{t'_1St_1} = \lambda_1\mathbf{t'_1t_1} = \lambda_1, \\ var(Y_2) &= var(\mathbf{Xt_2}) = \mathbf{t'_2St_2} = \lambda_2\mathbf{t'_2t_2} = \lambda_2, \\ &\vdots \\ var(Y_p) &= var(\mathbf{Xt_p}) = \mathbf{t'_pSt_p} = \lambda_p\mathbf{t'_pt_p} = \lambda_p, \end{aligned}$$

with $\lambda_1 > \cdots > \lambda_p$. This means that the first principal component is the one that captures the highest percentage of information of the data set. The second principal component is the second highest, and so on.

(ii) They are pairwise uncorrelated, that is:

$$cov(Y_k, Y_j) = cov(\mathbf{Xt_k}, \mathbf{Xt_j}) = \mathbf{t'_kSt_j} = \lambda_j\mathbf{t'_kt_j} = 0, \quad \text{when } k \neq j$$

because **T** is an orthogonal matrix. This property will be very useful for graphical purposes, because the principal components form a coordinate system of orthogonal axes. Properties 1 and 2 are particularly useful for dimensionality reduction. We come back to this point later.

(iii) The covariances between each principal component and the observed variables $X_1, \ldots, X_p$ are given by the row vector:

$$\text{cov}(Y_j, [X_1, \ldots, X_p]) = \lambda_j\mathbf{t'_j} \quad \text{for } j = 1, \ldots, p$$

To this end, defining $\mathbf{Y} = \mathbf{XT}$, implies that:

$$cov(\mathbf{Y}, \mathbf{X}) = \frac{1}{n}\mathbf{Y'HX} = \frac{1}{n}\mathbf{T'X'HX} = \mathbf{T'S} = \mathbf{T'T\Lambda T'} = \mathbf{\Lambda T'}$$

Therefore, the $j$-th row of $\mathbf{\Lambda T'}$ contains the covariances between $Y_j$ and the observed variables $X_1, \ldots, X_p$. For instance, the covariance between $Y_1$ and $X_1, \ldots, X_p$ is given by $\lambda_1\mathbf{t'_1}$.

(iv) The correlation between $Y_j$ and the observed variable $X_i$ is given by

$$corr(Y_j, X_i) = \frac{cov(Y_j, X_i)}{\sqrt{var(Y_j)var(X_i)}} = \frac{\lambda_j t_{ij}}{\sqrt{\lambda_j s_i^2}} = \frac{t_{ij}}{s_i}\sqrt{\lambda_j},$$

where $t_{ij}$ is the $i$-th element of the $\mathbf{t_j}$ vector and $s_i^2 = var(X_i)$. This property is useful for interpreting the axes, since it gives the weight of the observed variables in the new coordinate system.

## REPRESENTATION OF THE INDIVIDUALS: A GEOMETRICAL INTERPRETATION

One of the main purposes of principal components is the graphical visualization of the set of individuals in a low dimensional pictorial representation. As we have seen, principal components define a coordinate system of orthogonal axes. Now we are interested in obtaining the coordinates of each individual in this coordinate system.

Remember that the $i$-th row of matrix $\mathbf{X}$, that is, $\mathbf{x}'_i = (x_{i1}, \ldots, x_{ip})$ contains the measurements of $p$ variables on the $i$-th individual. The coordinates of this individual in the new coordinate system are given by

$$\mathbf{y}'_i = \mathbf{x}'_i\mathbf{T} = \left(\mathbf{x}'_i\mathbf{t}_1, \ldots, \mathbf{x}'_i\mathbf{t}_p\right)$$

and the whole transformed data matrix is $\mathbf{Y} = \mathbf{XT}$, which represents the 'measurements' of the new variables (the principal components) on the set of $n$ individuals. This transformation can be geometrically interpreted considering the $n$ individuals as $n$ points of $\mathbb{R}^p$.

Let us consider the squared Euclidean distance between the $i$-th and $j$-th individuals in the new coordinate system, that is:

$$d^2(i,j) = (\mathbf{y}_i - \mathbf{y}_j)'(\mathbf{y}_i - \mathbf{y}_j) = \left(\mathbf{x}'_i\mathbf{T} - \mathbf{x}'_j\mathbf{T}\right)(\mathbf{T}'\mathbf{x}_i - \mathbf{T}'\mathbf{x}_j) = \left(\mathbf{x}'_i - \mathbf{x}'_j\right)\mathbf{TT}'(\mathbf{x}_i - \mathbf{x}_j) = (\mathbf{x}_i - \mathbf{x}_j)'(\mathbf{x}_i - \mathbf{x}_j)$$

Therefore, ignoring any orientation, this transformation can be seen as a rotation in $\mathbb{R}^p$, which searches for orthogonal directions of maximum dispersion. These directions are the new axes (the principal components), which are linear combinations of the observed variables. Hence, the first of the new axes (the first principal component) is the direction along which the dispersion of the individuals is maximum (remember that the first principal component was defined so that $var(Y_1)$ was maximum). Successively, each of the subsequent principal components is the direction, orthogonal to the previous ones, along which the individuals show more dispersion.

## Example 5.1

Let us see the geometrical interpretation of the section above through a numerical example. Consider the following data set given by Table 5.1.

Suppose we are interested in computing the principal components from the information contained in variables $X_1$ and $X_2$.

**Table 5.1** Data set for Example 5.1.

| | $X_1$ | $X_2$ |
|---|---|---|
| **1** | 8.84 | 8.47 |
| **2** | 3.36 | 3.03 |
| **3** | −0.05 | −0.08 |
| **4** | 3.46 | 3.37 |
| **5** | 3.43 | 0 |
| **6** | 0.56 | 0.56 |
| **7** | 0 | −0.55 |
| **8** | 0.78 | 0.75 |
| **9** | −3.03 | −4.8 |
| **10** | 5.91 | 5.35 |
| **11** | 8.27 | 6.03 |
| **12** | 1.65 | 0 |
| **13** | −1.32 | −1.33 |
| **14** | 3.24 | 3.24 |
| **15** | 2.93 | 2.93 |
| **16** | 6.13 | 6.13 |
| **17** | 0.57 | −0.32 |
| **18** | −6.06 | −6.67 |

So, the first step is to take matrix $\mathbf{X}$ to be the $18 \times 2$ data matrix defined by $X_1$ and $X_2$ columns of Table 5.1. If $\mathbf{X}$ is not centred, we can center it by considering the transformation $\mathbf{X} - \frac{1}{n}\mathbf{1}'\bar{\mathbf{x}}'$, where $\bar{\mathbf{x}} = (2.1483 \quad 1.4506)'$ is the mean vector. This is not compulsory but it is convenient for representation purposes. Next, we compute the covariance matrix

$$\mathbf{S} = \begin{pmatrix} 13.4275 & 13.1227 \\ 13.1227 & 13.7018 \end{pmatrix}$$

and obtain the principal components from the spectral decomposition of $\mathbf{S} = \mathbf{T}\boldsymbol{\Lambda}\mathbf{T}'$, where:

$$\mathbf{T} = \begin{pmatrix} 0.7034 & -0.7108 \\ 0.7108 & 0.7034 \end{pmatrix}, \quad \boldsymbol{\Lambda} = \begin{pmatrix} 26.6881 & 0 \\ 0 & 0.4412 \end{pmatrix}.$$

As we have seen in the section *Effective Computation*, the columns of $\mathbf{T}$ characterize the principal components, which are given by the equations:

$$Y_1 = 0.7034X_1 + 0.7108X_2,$$

$$Y_2 = -0.7108X_1 + 0.7034X_2.$$

The correlations between the $X$ variables and the $Y$s are given by:

$$corr(Y_1, X_1) = \frac{t_{11}}{s_1}\sqrt{\lambda_1} = \frac{0.7034}{\sqrt{13.4275}}\sqrt{26.6881} = 0.9917,$$

$$corr(Y_1, X_2) = \frac{t_{21}}{s_2}\sqrt{\lambda_1} = \frac{0.7108}{\sqrt{13.7018}}\sqrt{26.6881} = 0.9920,$$

$$corr(Y_2, X_1) = \frac{t_{12}}{s_1}\sqrt{\lambda_2} = \frac{-0.7108}{\sqrt{13.4275}}\sqrt{0.4412} = -0.1288,$$

$$corr(Y_2, X_2) = \frac{t_{22}}{s_2}\sqrt{\lambda_2} = \frac{0.7034}{\sqrt{13.7018}}\sqrt{0.4412} = 0.1262.$$

Let us analyse the effect of transformation $\mathbf{T}$ on the individuals of Table 5.1. Figure 5.1 shows the representation of these individuals. In the panel on the left, individuals are plotted as coordinates in $X_1$ and $X_2$, whereas the panel on the right shows the individuals on the plane defined by the principal components. Notice that now we are looking at the *same* cloud of points, representing the individuals, in a different coordinate system.

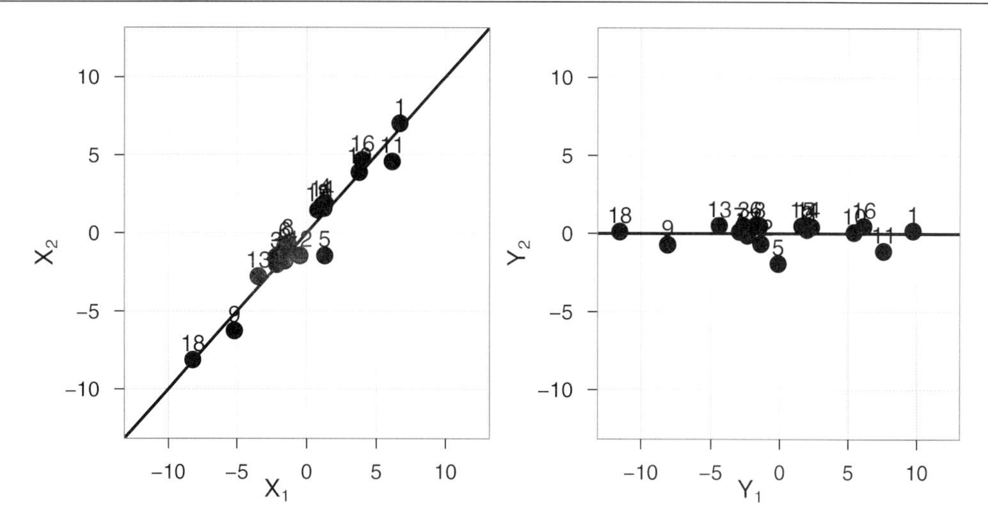


**Figure 5.1** Geometrical interpretation of the principal components.

## DIMENSIONALITY REDUCTION

The global dispersion of the data set is measured in terms of the *total variation* of the data matrix **X** and is interpreted as the amount of information available in the data set. The total variation of **X** is defined as:

$$TVar(\mathbf{X}) = tr(var(\mathbf{X})) = tr(\mathbf{S}) = \sum_{j=1}^{p} \lambda_j$$

Since the transformation matrix **T** is an orthogonal matrix, the total variation of $\mathbf{Y} = \mathbf{XT}$ is the same as $TVar(\mathbf{X})$:

$$TVar(\mathbf{Y}) = tr(var(\mathbf{Y})) = tr\left(\frac{1}{n}\mathbf{T}'\mathbf{X}'\mathbf{X}\mathbf{T}\right) = tr(\mathbf{T}'\mathbf{S}\mathbf{T}) = tr(\mathbf{S}) = \sum_{j=1}^{p} \lambda_j$$

As can be seen, each principal component explains a certain percentage of the total variability. For instance, the percentage of variability explained by the $k$-th principal component is given by:

$$\frac{var(Y_k)}{TVar(\mathbf{Y})} \times 100 = \frac{\lambda_k}{\sum_{j=1}^{p} \lambda_j} \times 100$$

Therefore, the dimensionality reduction is achieved by selecting the first $q < p$ principal components along which the dispersion of the individuals is maximum, or equivalently in ignoring the last $p - q$ axes along which the dispersion is minimum. The percentage of variability explained by the first $q < p$ principal components is:

$$P_q = \frac{\sum_{j=1}^{q} \lambda_j}{\sum_{j=1}^{p} \lambda_j} \times 100, \quad q < p$$

When the percentage $P_q$ is near 100%, it is said that variables $Y_1, \ldots, Y_q$ can replace variables $X_1, \ldots, X_p$ without a great loss of information, in terms of the *total variation*.

## Example 5.2

The 2011 Global Food Policy Report prepared by the International Food Policy Research Institute (IFPRI, 2012) emphasizes the important connection between agriculture, nutrition and health. Although the three sectors seek to improve human well-being, agriculture has rarely been explicitly deployed as a tool to address nutrition and health challenges. However, the rise of agriculture to a higher position in the Global Agenda highlights its potential (as a supplier of food, a source of income and an engine of growth) to reduce malnutrition and ill-health for the world's most vulnerable people. One essential element in increasing agricultural productivity and, hence, in promoting agricultural growth is greater investment in agricultural research and development.

Table 5.2 contains some information about the public spending on agricultural research and development for several Sub-Saharan African countries. Data come from the Agricultural Science & Technology Indicators (ASTI) initiative, which is one of the few sources of information on agricultural science and technology statistics for low- and middle-income countries (www.asti.cgiar.org).

Suppose that we are interested in constructing an index of public research using the information of Table 5.2, that is the four variables $X_1$ = public research spending 2005 PPP dollars (millions), $X_2$ = public spending as a share of agricultural GDP (%), $X_3$ = FTE public researchers and $X_4$ = women as share of total researchers (%). The first principal component is computed by taking the four numerical columns of Table 5.2 as the data matrix **X**, centring it and computing the covariance matrix

$$\mathbf{S} = \begin{pmatrix} 7279.4342 & 4.9228 & 32214.2814 & 257.9399 \\ 4.9228 & 0.9637 & -55.0799 & 4.4238 \\ 32214.2814 & -55.0799 & 202178.7060 & 944.6526 \\ 257.9399 & 4.4238 & 944.6526 & 100.2188 \end{pmatrix}$$

and obtaining the principal components as:

$$\begin{aligned} Y_1 &= 0.1590X_1 - 0.0003X_2 + 0.9873X_3 + 0.0047X_4 \\ Y_2 &= 0.9859X_1 + 0.0066X_2 - 0.1590X_3 + 0.0521X_4 \\ Y_3 &= 0.0524X_1 - 0.0442X_2 - 0.0037X_3 - 0.9976X_4 \\ Y_4 &= 0.0042X_1 - 0.9990X_2 - 0.0011X_3 + 0.0445X_4 \end{aligned} \qquad (5.1)$$

where the percentage of explained variability of the first component is 98.96%, which looks quite good. However, looking more carefully at component $Y_1$, we see that the weight of variable $X_3$ is extremely large (0.9873) compared to those of the other variables. On the other hand, if we order the $X$ variables according to their variances, $X_3$ is the one with greatest variance, followed by $X_1$, $X_4$ and $X_2$, which coincides with the variables that have more weight in each principal component. This is not a by chance! We will come back to this example later.

The R-script shown in Box 5.1 computes the principal components from the covariance matrix of a given data set. (The complete collection of data sets and R-scripts with comments can be found in the supplementary materials available at http://www.est.uc3m.es/agrane/eng/public.html.)

**Box 5.1 R-script for the computation of the principal components from the covariance matrix**

```
PrinCompCov<-function(X)
  {
    #auxilary routine to find covariance matrix of a column-centered X
    my.cov<-function(X){
    X=as.matrix(X);t(X)%*% scale(diag(nrow(X)),scale=FALSE)%*%X/nrow(X)}
    #obtain cov matrix, eigen values and vectors
    S<-my.cov(X)
    outS<-eigen(S)
    T1<-outS$vectors
    D1<-outS$values
    #correct the signs in columns of T1
    if((sum(sign(T1[,1])) < 0))
      T1<-sweep(T1,2,FUN="*", rep(-1,ncol(T1))^(1:ncol(T1)))
    if( (sum(sign(T1[,2])) < 0))
      T1<-sweep(T1,2,FUN="*", rep(-1,ncol(T1))^(1+(1:ncol(T1))))
    #reconstruct (center columns in X)
    Y1<-as.matrix(scale(X,scale=FALSE))%*%T1
    #get the %s
    percl<-round(100*D1/sum(D1),4)
    cpercl<-cumsum(percl)
    rownames(Y1)<-rownames(X)
    list(T=T1,Y=Y1,cperc=cpercl)
  }
```

## COVARIANCE OR CORRELATION MATRIX?

We have learned to compute the principal components of a data matrix $\mathbf{X}$ from its covariance matrix $\mathbf{S}$. However, these principal components have an important drawback: they are not scale invariant. This means that the eigenvalues and eigenvectors of the original variables do not transform directly to the eigenvalues and eigenvectors of scaled variables. In applications, variables are usually scaled. For example when we change the measurement units we are re-scaling the variables.

Scaling involves transformations based on a diagonal matrix. If we consider a diagonal matrix $\mathbf{\Phi}$ and we transform the original data matrix according to $\mathbf{\Phi}$, that is $\mathbf{X_\Phi} = \mathbf{\Phi X}$, then the covariance matrix of $\mathbf{X_\Phi}$ is $var(\mathbf{X_\Phi}) = \mathbf{\Phi S \Phi}$, which obviously is different from $var(\mathbf{X}) = \mathbf{S}$. As a result the eigenvectors and eigenvalues computed from $var(\mathbf{X_\Phi})$ will be different from those computed from $var(\mathbf{X})$.

Another important drawback is the following: if there are few original variables with moderately large variances with respect to the others, their weights on the principal components will be artificially large, although the correlations between these original variables and the components might be small. For example if $X_1$ has a variance much greater than the other variables $X_2, \ldots, X_p$, then the first principal component $Y_1$ will be nearly equivalent to $X_1$.

To summarize, we can say that it is recommended to compute the principal components of a data matrix $\mathbf{X}$ from its covariance matrix $\mathbf{S}$ whenever the observed variables $X_1, \ldots, X_p$ are commensurate,

**Table 5.2** Public sector agricultural research and development (R&D) spending and staffing in some Sub-Saharan African countries. (Latest data available 2008.)[a]

| | Country | Public research spending 2005 PPP dollars (millions) | Public spending as a share of agricultural GDP (%) | FTE public researchers | Women as share of total researchers (%) |
|---|---|---|---|---|---|
| **1** | Benin | 21.6 | 0.57 | 115.4 | 16.9 |
| **2** | Botswana | 19 | 4.32 | 97.4 | 29.8 |
| **3** | Burkina Faso | 19.4 | 0.43 | 239.9 | 11.6 |
| **4** | Burundi | 9.6 | 1.78 | 97.8 | 14.8 |
| **5** | Congo, Rep. | 4.6 | 0.85 | 93.8 | 16.2 |
| **6** | Côte d'Ivoire | 42.6 | 0.54 | 122.6 | 16.8 |
| **7** | Eritrea | 3 | 0.45 | 121.9 | 32 |
| **8** | Ethiopia | 68.6 | 0.27 | 1318.3 | 6.8 |
| **9** | Gabon | 1.6 | 0.2 | 61.4 | 22.4 |
| **10** | Gambia, The | 2.5 | 0.5 | 37.7 | 13.7 |
| **11** | Ghana | 95.4 | 0.9 | 537.1 | 17 |
| **12** | Guinea | 4 | 0.18 | 229.2 | 3.2 |
| **13** | Kenya | 171.5 | 1.3 | 1011.5 | 26.8 |
| **14** | Madagascar | 11.9 | 0.27 | 212.4 | 29.8 |
| **15** | Malawi | 21.4 | 0.68 | 126.5 | 15.7 |
| **16** | Mali | 24.7 | 0.57 | 312.7 | 13.4 |
| **17** | Mauritania | 6.4 | 1.16 | 73.7 | 4.9 |
| **18** | Mauritius | 22.1 | 3.92 | 158.3 | 41.4 |
| **19** | Mozambique | 17.7 | 0.38 | 263.3 | 29 |
| **20** | Namibia | 21.6 | 2.03 | 70.2 | 16.5 |
| **21** | Niger | 6.2 | 0.17 | 93.4 | 8.3 |
| **22** | Nigeria | 403.9 | 0.42 | 2062 | 21.3 |
| **23** | Rwanda | 18.1 | 0.53 | 104.2 | 14.6 |
| **24** | Senegal | 25.4 | 0.87 | 141.1 | 9.9 |
| **25** | Sierra Leone | 5.9 | 0.31 | 66.6 | 5.2 |
| **26** | South Africa | 272.1 | 2.02 | 783.9 | 40.1 |
| **27** | Sudan | 51.5 | 0.27 | 1020.5 | 36.2 |
| **28** | Tanzania | 77.1 | 0.5 | 673.5 | 21.3 |
| **29** | Togo | 8.7 | 0.47 | 62.7 | 9.9 |
| **30** | Uganda | 87.7 | 1.24 | 298.7 | 21.5 |
| **31** | Zambia | 8.1 | 0.29 | 208.5 | 22.9 |

*PPP = purchasing power parity. **FTE = full-time equivalent.

Note: Table includes only countries where ASTI has conducted survey rounds since 2002. Public agricultural research and development (R&D) includes government, higher education, and not-for-profit agencies but excludes the private sector. Purchasing power parities (PPPs) measure the relative purchasing power of currencies across countries by eliminating national differences in pricing levels for a wide range of goods and services. PPPs are relatively stable over time, whereas exchange rates fluctuate considerably. Measuring researchers in full-time equivalents (FTEs) takes into account the proportion of time researchers spend on R&D activities. For example four university professors who spend 25% of their time on research would individually represent 0.25 FTEs and collectively be counted as one FTE.

[a]IFPRI, 2011. Reproduced with permission from the International Food Policy Research Institute.

that is similar in measurement scale and in variance (Rencher, 1998). When this is not the case, it is recommended to compute them from the correlation matrix

$$\mathbf{R} = corr(\mathbf{X}) = \mathbf{S_0}^{-1/2}\mathbf{S}\mathbf{S_0}^{-1/2}$$

where $\mathbf{S_0} = diag(s_1^2, \ldots, s_p^2)$ and $s_i^2 = var(X_i)$, for $i = 1, \ldots, p$. Since we are assuming that matrix $\mathbf{X}$ is a centred matrix, to obtain the principal components from matrix $\mathbf{R}$ is equivalent to standardizing the observed variables to zero mean and unit variance.

Analogously as has been done in the section *Effective Computation*, principal components from $\mathbf{R}$ are obtained though the spectral decomposition of $\mathbf{R} = \tilde{\mathbf{T}}\tilde{\mathbf{\Lambda}}\tilde{\mathbf{T}}'$, where $\tilde{\mathbf{T}}'\tilde{\mathbf{T}} = \tilde{\mathbf{T}}\tilde{\mathbf{T}}' = \mathbf{I}$ and $\tilde{\mathbf{\Lambda}} = diag(\tilde{\lambda}_1, \ldots, \tilde{\lambda}_p)$ with $\tilde{\lambda}_1 > \ldots > \tilde{\lambda}_p$. The columns of $\tilde{\mathbf{T}} = [\tilde{\mathbf{t}}_1, \ldots, \tilde{\mathbf{t}}_\mathbf{p}]$ are the vectors of coefficients of the principal components $\tilde{Y}_j$s, and the eigenvalues are the corresponding variances.

The representation of the individuals now is given by (notice the difference from the section *Representation of the Individuals: A Geometrical Interpretation*):

$$\tilde{\mathbf{Y}} = \mathbf{X}\mathbf{S_0}^{-1/2}\tilde{\mathbf{T}}$$

where $\mathbf{S_0} = diag(s_1^2, \ldots, s_p^2)$ and $s_i^2 = var(X_i)$, for $i = 1, \ldots, p$.

As might be expected, the eigenvalues and eigenvectors obtained from $\mathbf{R}$ will be different from those obtained from $\mathbf{S}$ and so too the principal components. Some authors (Naik and Khattree, 1996) prefer to obtain $\mathbf{S}$-based principal components to $\mathbf{R}$-based for use in further computations like multiple regression.

The R-script shown in Box 5.2 computes the principal components from the correlation matrix of a given data set.

**Box 5.2 R-script for the computation of the principal components from the correlation matrix.**

```
PrinCompCor<-function(X)
 {
  R<-cor(X)
  outR<-eigen(R)
  T2<-outR$vectors
  D2<-outR$values
  #correct the signs in columns of T2
  if((sum(sign(T2[,1])) < 0))
    T2<-sweep(T2,2,FUN="*", rep(-1,ncol(T2))^(1:ncol(T2)))
  if( (sum(sign(T2[,2])) < 0))
    T2<-sweep(T2,2,FUN="*", rep(-1,ncol(T2))^(1+(1:ncol(T2))))
  #reconstruct (standardize columns in X)
  Y2<-as.matrix(scale(X))%*%T2
  #get the %s
  perc2<-round(100*D2/sum(D2),4)
  cperc2<-cumsum(perc2)
  rownames(Y2)<-rownames(X)
  list(T=T2,Y=Y2,cperc=cperc2)
 }
```

## DETERMINING THE NUMBER OF COMPONENTS

When performing a principal component analysis, the researcher must decide how many components to use to represent the individuals (and the remaining ones are discarded). There are several methods to decide on the number of components to include. Some of them are:

(i) **Percentage of explained variability**. This is the simplest and easiest method. The number of components to select is based on the percentage of explained variability accounted for. It is desirable to achieve a relatively high percentage of explained variability, for example 70–90%. If principal components are used as a first-step analysis before performing the main analysis, say regression or MANOVA, it may be useful to keep more components than if one is merely interested in a low-dimension representation.

(ii) **Kaiser's criterion**. Exclude those components whose eigenvalues are smaller than the average eigenvalue $\bar{\lambda} = \frac{1}{p}\sum_{j=1}^{p} \lambda_j$, or else smaller than $\bar{\lambda} = 1$ if they are computed from matrix **R**. This method often works well in practice.

(iii) **Jollife's criterion**. When the number of variables $p \leq 20$ Kaiser's criterion tends to include few components. Jollife's criterion excludes those components whose eigenvalues are smaller than $0.7\bar{\lambda} = \frac{0.7}{p}\sum_{j=1}^{p} \lambda_j$, or smaller than $\bar{\lambda} = 0.7$ if they are computed from matrix **R**.

(iv) **Cattell's scree graph**. The aim of this procedure is to find a visual break between the 'large' eigenvalues and the 'small' ones. The term *scree*, suggested by Cattell (1966), refers to the geological term for the debris at the bottom of a rocky cliff. It is a graphical tool based on a Cartesian representation in $\mathbb{R}^2$. The number of the principal component $(1, 2, \ldots, p)$ is represented in the horizontal axis and the percentage of explained variability (or the eigenvalues) in the vertical one. Finally, the plotted points are joined by a line. The first $q < p$ components with the steepest slopes are considered (Figure 5.2).

(v) **Dimensionality test**. This test assumes random sampling from a multivariate normal distribution and is applicable to the covariance matrix **S**. Note that principal components provide valuable descriptive information in many situations where variables are not normally distributed. To determine the number of components to retain, we test the following null hypothesis:

$$H_{0k} : \lambda_1 > \cdots > \lambda_k > \lambda_{k+1} = \cdots = \lambda_p$$

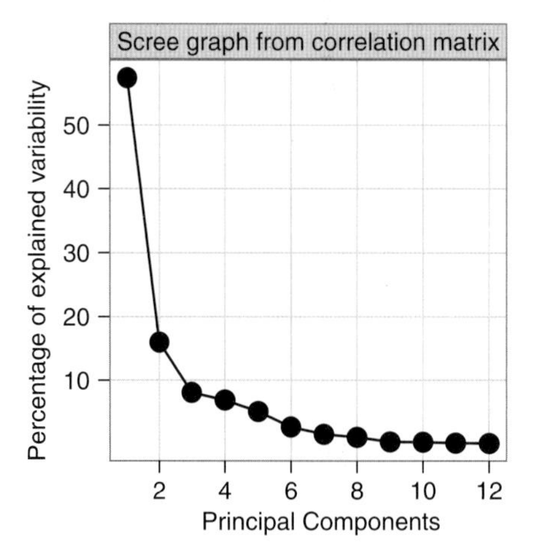


**Figure 5.2** Catell's srcee graph.

that means that there are $k$ significantly different eigenvalues, that is $k$ principal components, or equivalently that we can discard the last $p$–$k$ principal components. To test $H_{0k}$ the following statistic is used:

$$u = \left(n - \frac{2p+11}{6}\right)\left((p-k)\ln\left(\frac{\hat{\lambda}_{k+1} + \ldots + \hat{\lambda}_p}{p-k}\right) - \sum_{i=k+1}^{p} \ln \hat{\lambda}_i\right)$$

where $\hat{\lambda}_1 > \cdots > \hat{\lambda}_k > \hat{\lambda}_{k+1} > \cdots > \hat{\lambda}_p$ are the estimated eigenvalues from the covariance matrix **S**. When the null hypothesis $H_{0k}$ is true, the test statistic $u$ is asymptotically distributed as a $\chi^2$ with $(p-k-1)(p-k+2)/2$ degrees of freedom and $H_{0k}$ is rejected for large values of $u$.

The test is performed sequentially for $H_{0,p-2}, H_{0,p-3}, H_{0,p-4} \ldots$ until $H_{0k}$ is rejected for some value of $k = p-2, \ldots, 0$. This procedure tends to retain one or more 'smaller' principal components that are not important descriptors. A Bonferroni adjustment to preserve the overall significance level is helpful in this regard (Rencher, 1998).

However, if the eigenvalues are computed from the correlation matrix, it is considerably difficult to test the above null hypothesis, mainly because the test statistic is not asymptotically chi-squared distributed (Mardia *et al.*, 1979).

## Example 5.2 (revisited)

Looking carefully at the four observed variables of Table 5.2, we can see, firstly, that the measurement units are not the same (PPPs, %, proportions, . . . ) and, secondly, their variances are not of the same magnitude. Hence, according to the section *Covariance or Correlation Matrix?*, it is not recommended to compute the principal components from the covariance matrix **S**. If we do not consider this advice and proceed by computing the principal components from **S**, the first component will be dominated by the variable with largest variance (see the previous part of this example). In this case, it is more appropriate to compute the components from the correlation matrix:

$$\mathbf{R} = \begin{pmatrix} 1.0000 & 0.0588 & 0.8397 & 0.3020 \\ 0.0588 & 1.0000 & -0.1248 & 0.4501 \\ 0.8397 & -0.1248 & 1.0000 & 0.2099 \\ 0.3020 & 0.4501 & 0.2099 & 1.0000 \end{pmatrix}$$

In doing so we obtain:

$$\begin{aligned} Y_1 &= 0.6582X_1 + 0.1407X_2 + 0.6241X_3 + 0.3967X_4 \\ Y_2 &= 0.1783X_1 - 0.7415X_2 + 0.3320X_3 - 0.5552X_4 \\ Y_3 &= -0.2296X_1 - 0.6378X_2 - 0.0787X_3 + 0.7310X_4 \\ Y_4 &= 0.6944X_1 - 0.1538X_2 - 0.7029X_3 + 0.0082X_4 \end{aligned} \tag{5.2}$$

where the percentage of explained variability of the first two components is 84.24% (49.77% for $Y_1$ and 38.47% for $Y_2$). For ease of reference, the notation $\tilde{Y}'$s of the section *Covariance or Correlation Matrix?* are omitted; here it is clear that principal components of Equation 5.2 are different from those of Equation 5.1. Figure 5.3 shows the principal component representation of the individuals of Table 5.2 computed from matrix **R**. Looking at the component $Y_1$ of Equation 5.2, we see that all $X$ variables contribute positively, meaning that the higher the values of a country in the $X$ variables, the higher its values in $Y_1$. Although $X_1$ (public research spending 2005 in PPP million dollars), $X_3$ (full-time equivalent public researchers) and $X_4$ (percentage of women as share of total researchers) are those with higher weights, this principal component can be interpreted as a global index of public research and development in the agricultural sector. For countries situated on the right of the horizontal axis, such as Nigeria and South Africa, the public sector agricultural R&D spending and staffing is greater than for those situated on the left of the horizontal axis, such as Sierra Leone, Burundi and Gabon, among others.

In Box 5.3 an R-script for plotting the principal component representation and Catell's scree-graph is provided.

**Box 5.3 R-script for the principal component representation and Catell's scree-graph**

```
PlotPrinComp<-function(T,Y,cperc,type="covariance")
 {
   perc<-c(cperc[1],diff(cperc))
   pca2.df<-data.frame(PC1=Y[,1],PC2=Y[,2],label=rownames(Y),
                       type=rep(type,nrow(Y)))
   label<-paste("PCA from ",type," matrix (",cperc[2],"%)",sep="")
   pca2.df$type<-factor(pca2.df$type,levels=rev(levels(pca2.df$type)),
                        labels=label)
   p2<-ggplot(data=pca2.df,aes(x=PC1,y=PC2))
   p2<-p2+geom_point()+facet_wrap(~type,scales="free",ncol=1)+
   scale_y_continuous("2nd Principal Component")+
   scale_x_continuous("1st Principal Component")
   p2 <- p2 + geom_text(data=pca2.df, hjust=+0.5,vjust=-0.5,
                     aes(x=PC1, y=PC2, label=label), size=4) +theme_bw()
   scree.df<-data.frame(index=1:ncol(T),percent=perc,type=rep(type,nrow(T)))
   label=paste("Scree graph from ",type," matrix",sep="")
   scree.df$type<-factor(scree.df$type,levels=rev(levels(scree.df$type)),
                        labels=label)
   s<-ggplot(data=scree.df,aes(x=index,y=percent))
   s<-s+geom_point(size=4)+geom_line()+facet_wrap(~type,scales="free",ncol=1)
   s<-s+scale_x_continuous("Principal Components")
   s<-s+ scale_y_continuous("Percentage of explained variability")+theme_bw()
   list(p=p2,s=s)
 }
 #Usage:
 library(ggplot2); outCor<-PrinCompCor(X)
 T<-outCor$T; Y<-outCor$Y
 cperc<-outCor$cperc
 outPlot<-PlotPrinComp(T,Y,cperc,type="correlation")
 grid.newpage()
 pushViewport(viewport(layout = grid.layout(1, 2)))
 print(outPlot$p, vp = viewport(layout.pos.row = 1, layout.pos.col = 1))
 print(outPlot$s, vp = viewport(layout.pos.row = 1, layout.pos.col = 2))
```

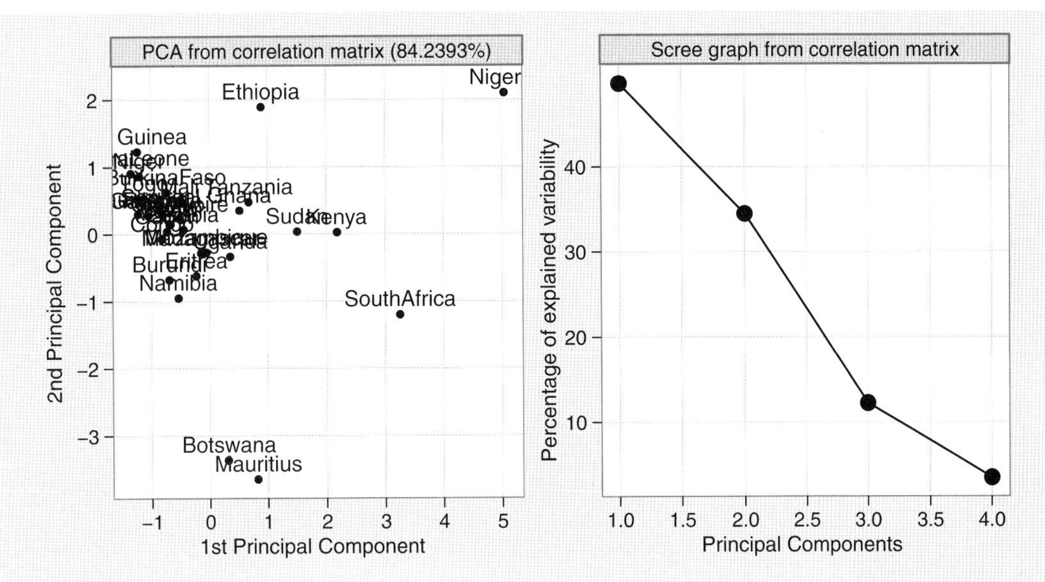


**Figure 5.3** Principal component representations and scree graphs for the data of Example 5.2.

The second principal component basically counterbalances $X_2$ (percentage of public spending as a share of agricultural GDP) and $X_4$ (percentage of women as share of total researchers) to $X_3$ (full-time equivalent public researchers). Hence, countries that spend more resources on research and with a greater presence of women in the research field are situated at the bottom of the vertical axis. This may be also an indicator of the development of the country. Mauritius and Botswana are in this situation. The contrary will happen for those countries situated on the top of the vertical axis, such as Nigeria and Ethiopia. Therefore, a good place to be is the bottom right corner, where there is only South Africa.

## Example 5.3

Table 5.3 contains observations of 12 variables measuring the nutritional contribution (per 100 grams) of some dairy products on 30 samples. We will apply several methods of choosing the number of components.

Once again, these variables are not commensurate. They are different with respect to scale and variance. Therefore, we will compute the principal components from the correlation matrix. Table 5.4 shows only the principal components $Y1$–$Y4$ jointly with their variances (eigenvalues):

As seen in the section *Determining the Number of Components*, the easiest method to discard components is to fix a percentage of explained variability. For example if we want to explain 70% of the variability, we should take two components. Notice that, since the trace of the correlation matrix is $tr(\mathbf{R}) = 12$, the percentage of explained variability from the first two principal components is $(6.8909 + 1.9152)/12 \times 100 = 73.3845\%$. According to Kaiser's criterion, we would also take two components (because there are two eigenvalues $>1$). The scree graph in the right panel of Figure 5.4 also suggests taking two components. However, Jollife's criterion implies four (because the variance of the fifth component is $0.6121 < 0.7$). Notice that we cannot perform the dimensionality test

**Table 5.3** Nutritional information of some dairy products.

| | | | kcal | Protein (g) | Fat (g) | Sodium (mg) | Calcium (mg) | Iron (mg) | Phosphorus (mg) | Potassium (mg) | Vitamin A (U.I.) | Vitamin B1 (mg) | Vitamin B2 (mg) | Vitamin B3 (mg) |
|---|---|---|---|---|---|---|---|---|---|---|---|---|---|---|
| **Milk** | Whole | 1 | 60 | 3 | 3 | 30 | 110 | 0.2 | 80 | 140 | 200 | 0.1 | 0.2 | 0.2 |
| | Semi skimmed | 2 | 45 | 3 | 1.5 | 0 | 110 | 0.1 | 85 | 0 | 200 | 0.1 | 0.2 | 0.2 |
| | Skimmed | 3 | 31 | 2.9 | 1 | 0 | 120 | 0.1 | 100 | 0 | 150 | 0.02 | 0.2 | 0 |
| | Condensed | 4 | 320 | 8.2 | 8.2 | 100 | 250 | 0.1 | 200 | 300 | 360 | 0.1 | 0.3 | 0.2 |
| | Chocolate milk | 5 | 80 | 3.3 | 2.5 | 50 | 100 | 0.2 | 90 | 140 | 80 | 0.05 | 0.15 | 0.1 |
| **Yoghurt** | Creamy | 6 | 110 | 3 | 6 | 0 | 130 | 0 | 90 | 0 | 0 | 0 | 0 | 0 |
| | Full fat | 7 | 85 | 2.8 | 3.3 | 60 | 150 | 0.1 | 100 | 190 | 1000 | 0.05 | 0.2 | 0.1 |
| | Low fat | 8 | 75 | 3 | 1.5 | 0 | 0 | 0 | 0 | 0 | 0 | 0 | 0 | 0 |
| | Fat free | 9 | 40 | 4 | 0.1 | 0 | 130 | 0 | 90 | 0 | 0 | 0 | 0 | 0 |
| **Pourable** | Low fat | 10 | 80 | 2.9 | 1.5 | 0 | 115 | 0 | 0 | 0 | 0 | 0 | 0 | 0 |
| **Whole grain** | Low fat | 11 | 120 | 5.2 | 1.8 | 0 | 180 | 0 | 0 | 0 | 0 | 0 | 0 | 0 |
| **Milk** | Bioactive full fat | 12 | 95 | 4.1 | 2.6 | 0 | 145 | 0 | 0 | 0 | 0 | 0 | 0 | 0 |
| **Milk** | Bioactive fat free | 13 | 35 | 3.5 | 0.1 | 0 | 0 | 0 | 0 | 0 | 0 | 0 | 0 | 0 |
| **Cheese** | Hard | 14 | 150 | 11 | 10 | 70 | 150 | 0 | 0 | 0 | 0 | 0 | 0 | 0 |
| **Hard** | Low fat | 15 | 110 | 12 | 4.5 | 250 | 150 | 0.5 | 150 | 90 | 15 | 0.02 | 0.2 | 0.1 |
| **Hard** | Full fat | 16 | 300 | 18 | 25 | 470 | 300 | 1 | 400 | 80 | 500 | 0.03 | 0.4 | 0.1 |
| | Melted | 17 | 300 | 15 | 25 | 450 | 300 | 0 | 0 | 0 | 0 | 0 | 0 | 0 |
| **Soft** | Full fat | 18 | 300 | 24 | 23 | 450 | 800 | 0 | 600 | 0 | 0 | 0 | 0 | 0 |
| **Soft** | Low fat | 19 | 240 | 28 | 14 | 450 | 800 | 0 | 600 | 0 | 0 | 0 | 0 | 0 |
| | Mozzarella | 20 | 240 | 20 | 16 | 750 | 75 | 0.3 | 200 | 110 | 400 | 0.03 | 0.2 | 0.1 |
| | Camembert | 21 | 300 | 18 | 25 | 900 | 200 | 0.2 | 300 | 100 | 1000 | 0.04 | 0.6 | 0.9 |
| | Cheddar | 22 | 400 | 25 | 31 | 700 | 750 | 1 | 500 | 90 | 1300 | 0.03 | 0.4 | 0.1 |
| | Processed | 23 | 250 | 16 | 17 | 600 | 400 | 0.5 | 330 | 0 | 0 | 0 | 0 | 0 |
| | Edam | 24 | 320 | 25 | 20 | 700 | 700 | 0.6 | 500 | 80 | 500 | 0.06 | 0.3 | 0.1 |
| | Emmental | 25 | 400 | 27 | 28 | 450 | 1000 | 0.3 | 600 | 100 | 600 | 0.05 | 0.3 | 0.2 |
| | Gouda | 26 | 370 | 25 | 29 | 700 | 700 | 0.5 | 440 | 100 | 400 | 0.03 | 0.2 | 0.1 |
| | Gruyere | 27 | 400 | 28 | 28 | 380 | 1000 | 0 | 600 | 100 | 0 | 0.05 | 0.3 | 0.1 |
| | Parmesan | 28 | 390 | 34 | 25 | 700 | 1100 | 0.8 | 800 | 130 | 1000 | 0.02 | 0.6 | 0.2 |
| | Provolone | 29 | 390 | 28 | 28 | 1100 | 900 | 0.5 | 650 | 70 | 300 | 0.5 | 0.3 | 0.2 |
| | Roquefort | 30 | 370 | 21 | 30 | 1800 | 500 | 0.5 | 360 | 90 | 1200 | 0.04 | 0.6 | 0.8 |

**Table 5.4** Principal components of data set described in Table 5.3.

| | $Y_1$ | $Y_2$ | $Y_3$ | $Y_4$ | $Y_5$ ... |
|---|---|---|---|---|---|
| **eigenvalues** | **6.8909** | **1.9152** | **0.9734** | **0.8278** | **0.6121** ... |
| $X_1$ | 0.3503 | −0.1682 | 0.0179 | 0.0236 | ⋮ |
| $X_2$ | 0.3371 | −0.2998 | 0.0661 | 0.0088 | |
| $X_3$ | 0.3473 | −0.1653 | 0.1431 | −0.1130 | |
| $X_4$ | 0.3257 | 0.0526 | 0.1412 | −0.4171 | |
| $X_5$ | 0.3027 | −0.3705 | −0.0808 | 0.1057 | |
| $X_6$ | 0.2704 | 0.0829 | 0.0819 | 0.3532 | |
| $X_7$ | 0.3287 | −0.2752 | −0.0804 | 0.1189 | |
| $X_8$ | 0.1668 | 0.3790 | −0.3869 | 0.5311 | |
| $X_9$ | 0.2765 | 0.3794 | 0.1888 | 0.1538 | |
| $X_{10}$ | 0.1376 | 0.0160 | −0.8552 | −0.3541 | |
| $X_{11}$ | 0.3098 | 0.3622 | 0.0194 | 0.0527 | |
| $X_{12}$ | 0.2123 | 0.4613 | 0.1342 | −0.4782 | |

described in the section *Determining the Number of Components*, because principal components are computed from the correlation matrix.

Nearly all methods coincide in selecting two components. We can interpret them in the following way. The weights of the first component are all positive. Hence, it can be seen as a weighted average of all the variables. Therefore, those dairy products with more calories, more fat and so on, will be situated on the right of the horizontal axis. Looking at Figure 5.4, we can see that milks, yoghurts and soft cheeses are on the left, whereas stronger cheeses are on the right. On the other hand, the second

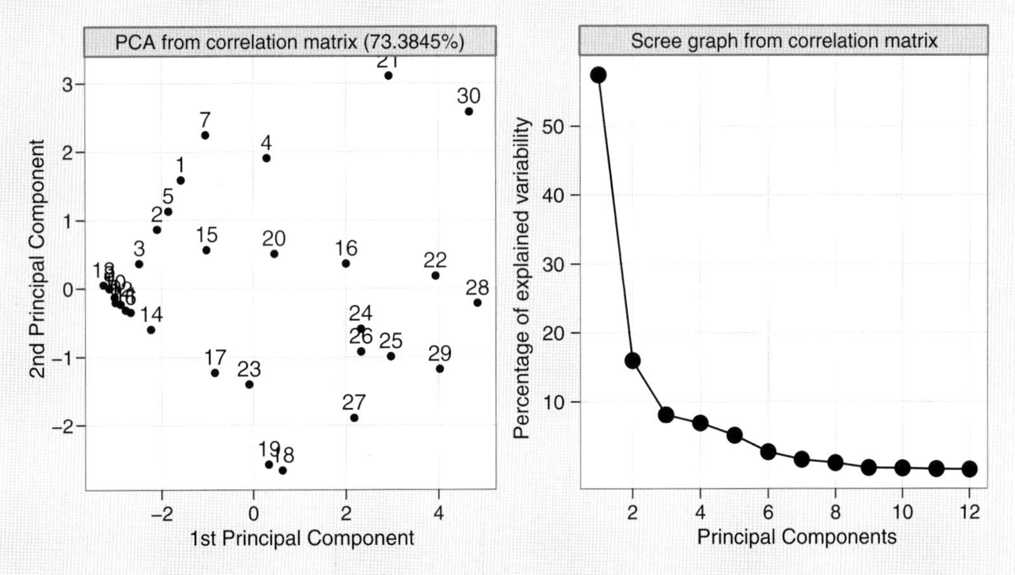


**Figure 5.4** Principal component representation and scree graph for data from Example 5.3.

component seems to confront vitamins (A, B2 and B3) with kcal, proteins, fat, calcium and phosphorus. Thus, those dairy products with more vitamins and less kcal, proteins, fat, calcium and phosphorus will be situated on the top of the vertical axis. Products such as camembert and roquefort cheeses, full fat yoghurt or condensed milk are in those regions.

## Example 5.4

To illustrate the dimensionality test described in the section *Determining the Number of Components*, consider the subgroup of cheeses described in Table 5.5.

In this case, the selected variables are commensurate in scale, hence we can compute the principal components from the covariance matrix **S** and implement the dimensionality test by using the R-script shown in Box 5.4.

**Table 5.5** Nutritional properties of some cheeses.

| | kcal | sodium (mg) | calcium (mg) | phosphorus (mg) | potassium (mg) | Vitamin A (U.I.) |
|---|---|---|---|---|---|---|
| Mozzarella | 240 | 750 | 75 | 200 | 110 | 400 |
| Camembert | 300 | 900 | 200 | 300 | 100 | 1000 |
| Cheddar | 400 | 700 | 750 | 500 | 90 | 1300 |
| Processed | 250 | 600 | 400 | 330 | 0 | 0 |
| Edam | 320 | 700 | 700 | 500 | 80 | 500 |
| Emmental | 400 | 450 | 1000 | 600 | 100 | 600 |
| Gouda | 370 | 700 | 700 | 440 | 100 | 400 |
| Gruyere | 400 | 380 | 1000 | 600 | 100 | 0 |
| Parmesan | 390 | 700 | 1100 | 800 | 130 | 1000 |
| Provolone | 390 | 1100 | 900 | 650 | 70 | 300 |
| Roquefort | 370 | 1800 | 500 | 360 | 90 | 1200 |

### Box 5.4 R-script for the dimensionality test

```
dimtest<-function(X)
 {
   dims<-dim(X)
   n<-dims[1]
   p<-dims[2]
   S<-cov(X)
   D<-eigen(S,only.values=TRUE)$values
   logD<-log(D)
   result<-NULL
   for(k in seq(p-2,0,-1))
   {
     t<-(n-(2*p+11)/6)*((p-k)*log(sum(D[(k+1):p])/(p-k))-sum(logD[(k+1):p]))
     df<-(p-k-1)*(p-k+2)/2
     result<-rbind(result, c(p-k, round( c(t,1-pchisq(t,df)), 4)))
   }
 return(result)
 }
```

Table 5.6 contains the results of the dimensionality test. We can see that for a 5% significance level (1% with Bonferroni correction), we would select three different components.

In Figure 5.5 the scree graph for this data set is plotted. We can observe that the eigenvalues of the last three principal components are very small and do not differ much from each other.

**Table 5.6** Dimensionality test for data set described on Table 5.5.

| Null hypothesis $H_{0k}$ | Discard *k* last components | Test statistic | *p* value |
|---|---|---|---|
| $\lambda_1 > \lambda_2 > \lambda_3 > \lambda_4 > \lambda_5 = \lambda_6$ | 2 | 3.1673 | 0.2052 |
| $\lambda_1 > \lambda_2 > \lambda_3 > \lambda_4 = \lambda_5 = \lambda_6$ | 3 | 8.7461 | 0.1196 |
| $\lambda_1 > \lambda_2 > \lambda_3 = \lambda_4 = \lambda_5 = \lambda_6$ | 4 | 60.3165 | 0.0000 |
| $\lambda_1 > \lambda_2 = \lambda_3 = \lambda_4 = \lambda_5 = \lambda_6$ | 5 | 76.9951 | 0.0000 |
| $\lambda_1 = \lambda_2 = \lambda_3 = \lambda_4 = \lambda_5 = \lambda_6$ | 6 | 89.9939 | 0.0000 |

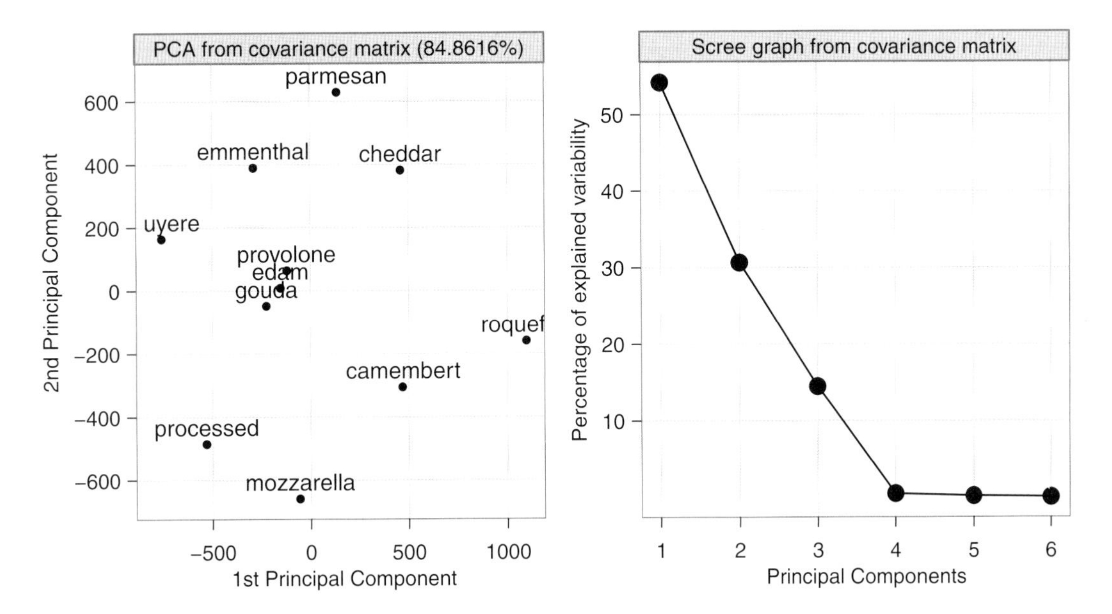


**Figure 5.5** Scree graph for data set described on Table 5.5.

## SOME PATTERNS IN R OR IN S AND THEIR INTERPRETATION

If all correlations in **R** or all covariances in **S** are positive, then the Perron–Frobenius theorem states that all elements of the first eigenvector are positive, that is the first principal component has positive weights. This means that the first principal component is a weighted average of the original variables, sometimes referred to a *size* component, because an increase of the original variables implies an increase of the component. Regarding the other $p$–1 eigenvectors, since they are orthogonal to the first one, they have both positive and negative elements. Hence, the next $p$–1 principal components have both positive and negative weights, and are called *shape* components. This means that an increase in those components is produced by increases in some of the original variables and decreases in other ones. The size and shape characterizations are typical in allometry; nevertheless, they can be generalized to other fields in a wide sense. As an example, we revisit the data set of Table 5.3 (Nutritional information of some dairy products.). The reader can check that the elements of the covariance matrix (or the correlation matrix) are all positive. As expected, the first principal component has positive weights (Table 5.4) and can be seen as a generalized size component. The next components have positive and negative weights and can be interpreted as generalized shape components.

Krzanowski (1984a, 1984b) suggested performing a *sensitivity analysis* in order to investigate the stability of the principal components, stating that confidence can only be expressed in the interpretation of an analysis if the components remain stable under small departures from optimality of the criterion function. In particular, Krzanowski (1984b) studied the effect on the principal component coefficients produced by small changes (perturbations) in the variance of the principal component. To illustrate sensitivity analysis, we decrease the variance of the $j$-th principal component, that is $var(Y_j) = var(\mathbf{Xt_j}) = \lambda_j$, by a small quantity $\varepsilon > 0$, and by $\mathbf{t}_j^*$ we denote the vector that differs as much as possible from $\mathbf{t_j}$, subject to the constrain that $var\left(Y_j^*\right) = var\left(\mathbf{Xt}_j^*\right) = \lambda_j - \varepsilon$. Krzanowski (1984b) showed that

$$\mathbf{t}_j^* = \left\{ \mathbf{t_j} \pm \mathbf{t_{j+1}} \left( \frac{\varepsilon}{\lambda_j - \lambda_{j+1}} \right)^{1/2} \right\} \Big/ \left\{ 1 + \left( \frac{\varepsilon}{\lambda_j - \lambda_{j+1}} \right)^{1/2} \right\} \quad (5.3)$$

and the angle $\theta$ between $\mathbf{t_j}$ and $\mathbf{t}_j^*$ is given by:

$$\cos(\theta) = \left( 1 + \frac{\varepsilon}{\lambda_j - \lambda_{j+1}} \right)^{-1/2} \quad (5.4)$$

Thus, if $\lambda_j$ is close to $\lambda_{j+1}$, the angle $\theta$ will be large, and the $j$-th principal component $Y_j = \mathbf{Xt_j}$ will be more unstable than if $\lambda_{j+1}$ is further away from $\lambda_j$. Similarly, if the variance $var(Y_j)$ is increased by $\varepsilon$, the component $Y_j$ becomes unstable if $\lambda_{j-1} - \lambda_j$ is small. Thus the stability of the component $Y_j$ depends on the distance of $\lambda_j$ from both $\lambda_{j-1}$ and $\lambda_{j+1}$. Roughly speaking, isolated components with large variances should therefore be fairly stable, unlike the remaining ones.

## Example 5.5

Let us study the stability of the principal components of the data set given by Table 5.2. Recall that in Example 5.2, it was argued that, for this particular case, it is preferable to compute the principal components from the correlation matrix. Hence, the stability study refers to the eigenvalues and eigenvectors of that matrix. Equations 5.3 and 5.4 can be easily implemented with the R-script shown in Box 5.5.

**Box 5.5 R-script for the stability study of principal components**

```
PrinCompStability<-function(X,alpha=0.05)
 {
  dims<-dim(X)
  S<-cov(X)
  R<-cor(X)
  outS<-eigen(S)
  outR<-eigen(R)
  T1<-outS$vectors
  T2<-outR$vectors
  D1<-outS$values
  D2<-outR$values
  #auxilary routine
  find.perturbed<-function(T0,D0,alpha=alpha)
```

```
  {
    epsilon<-alpha*D0
    p<-length(D0)
    Tp<-theta<-Dp<-NULL
    for(j in 1:(p-1))
    {
      Tp<-cbind(Tp, (T0[,j]+T0[,j+1]*sqrt(epsilon[j]/(D0[j]-
      D0[j+1])))/sqrt(1+epsilon[j] / (D0[j]-D0[j+1])))
      theta<-c(theta,acos(1/sqrt(1+epsilon[j]/(D0[j]-D0[j+1]))))
      Dp<-c(Dp, D0[j]-epsilon[j])
    }
    theta<-theta*360/(2*pi)
    list(Tp=Tp,Dp=Dp,theta=theta)
  }
  outS<-find.perturbed(T1,D1,alpha=alpha)
  outR<-find.perturbed(T2,D2,alpha=alpha)
  list(DpS=outS$Dp,TpS=outS$Tp,thetaS=outS$theta,
  DpR=outR$Dp,TpR=outR$Tp,thetaR=outR$theta)
}
```

Table 5.7 contains the sensitivity study for this data set for three different values of the perturbation parameter $\varepsilon$. For better understanding, in Figure 5.6 the effect of these perturbations on the principal component coefficients (columns $X_1$—$X_4$ of Table 5.7) is plotted.

The first component is the one showing greatest deviation in terms of angular separation. From Table 5.7 we can see how the first component weights of $X_2$ and $X_4$ progressively decrease as the perturbation parameter value increases. These effects can be also observed in Figure 5.6.

**Table 5.7** Sensitivity analysis of data set described in Table 5.2 (Public sector agricultural research and development (R&D) spending and staffing in some Sub-Saharan African countries).

| Component | Variance | Principal component coefficients | | | | Angle (in degrees) |
|---|---|---|---|---|---|---|
| | | $X_1$ | $X_2$ | $X_3$ | $X_4$ | |
| 1, No perturbation | 1.9908 | 0.6582 | 0.1407 | 0.6241 | 0.3967 | |
| 1, Perturbed 1% | 1.9709 | 0.6794 | 0.0068 | 0.6732 | 0.2919 | 10.2 |
| 1, Perturbed 5% | 1.8913 | 0.6771 | −0.1469 | 0.7030 | 0.1603 | 22.0 |
| 1, Perturbed 10% | 1.7917 | 0.6601 | −0.2451 | 0.7066 | 0.0696 | 29.7 |
| 2, No perturbation | 1.3787 | 0.1783 | −0.7415 | 0.3320 | −0.5552 | |
| 2, Perturbed 1% | 1.3650 | 0.1485 | −0.8146 | 0.3197 | −0.4606 | 7.1 |
| 2, Perturbed 5% | 1.3098 | 0.1102 | −0.8854 | 0.2987 | −0.3387 | 15.6 |
| 2, Perturbed 10% | 1.2409 | 0.0818 | −0.9236 | 0.2800 | −0.2487 | 21.5 |
| 3, No perturbation | 0.4898 | −0.22296 | −0.6378 | −0.0787 | 0.7310 | |
| 3, Perturbed 1% | 0.4849 | −0.1463 | −0.6514 | −0.1609 | 0.7269 | 6.8 |
| 3, Perturbed 5% | 0.4653 | −0.0442 | −0.6559 | −0.2561 | 0.7087 | 14.8 |
| 3, Perturbed 10% | 0.4408 | 0.0285 | −0.6512 | −0.3203 | 0.6874 | 21.5 |

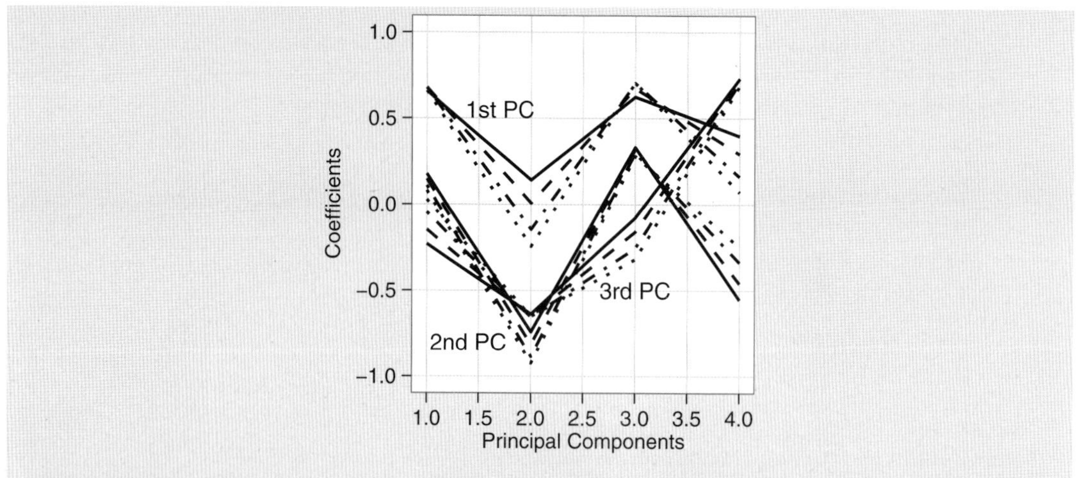


**Figure 5.6** Effect of three perturbations on the principal component coefficients (solid line – original eigenvector, dashed line – perturbed 1%, dash-dotted line – perturbed 5%, dotted line – perturbed 10%).

## RELATIONSHIP WITH THE ORTHOGONAL REGRESSION

Consider the following problem in $\mathbb{R}^2$: Given the observations $(x_i, y_i)$, $i = 1, \ldots, n$, we are interested in finding a line that is a 'good approximation' to these pairs of $n$ points, without giving any preference to any of the two coordinates.

In this case, the function to optimize is the *orthogonal* distance between the pairs of $n$ points and a line of generic equation $Ax + By = C$, where $(A, B)$ denotes the orthogonal vector (Figure 5.7). Notice that

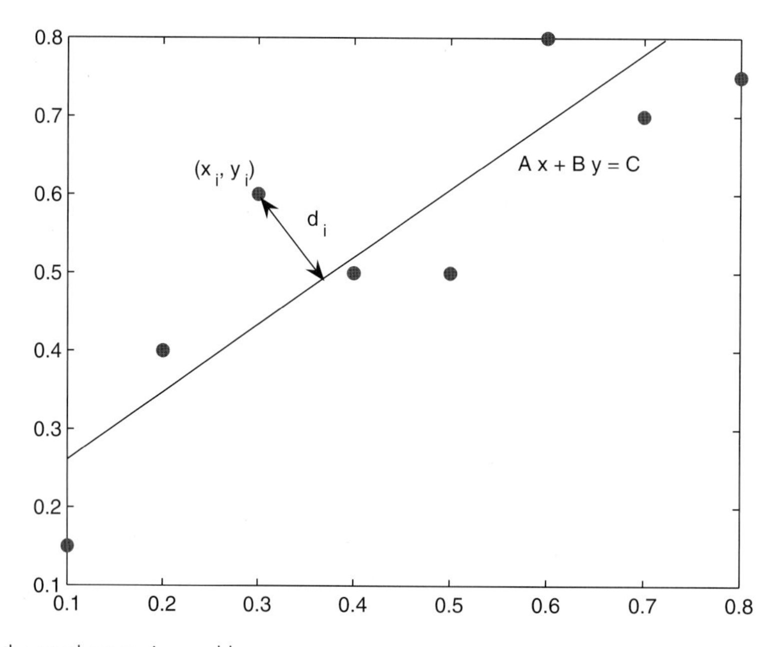


**Figure 5.7** Orthogonal regression problem.

this situation is different from the simple linear regression context. In addition, we will impose that the vector $(A, B)$ has unit norm, so that the problem can be solved. Hence, the equation of the line is given by:

$$\alpha x + \beta y = \gamma, \tag{5.5}$$

where $\alpha = \frac{A}{\sqrt{A^2+B^2}}, \beta = \frac{B}{\sqrt{A^2+B^2}}, \gamma = \frac{C}{\sqrt{A^2+B^2}}$.

The function to optimize is the average of the squared orthogonal distances between the points $(x_i, y_i)$, $i = 1, \ldots, n$, and the line $\alpha x + \beta y = \gamma$, that is:

$$\overline{d^2} = \frac{1}{n}\sum_{i=1}^{n}(\alpha x_i + \beta y_i - \gamma)^2 = \alpha^2\overline{x^2} + \beta^2\overline{y^2} + \gamma^2 + 2\alpha\beta\overline{xy} - 2\alpha\gamma\bar{x} - 2\beta\gamma\bar{y}, \tag{5.6}$$

where $\bar{x} = \frac{1}{n}\sum_{i=1}^{n} x_i, \bar{y} = \frac{1}{n}\sum_{i=1}^{n} y_i, \overline{xy} = \frac{1}{n}\sum_{i=1}^{n} x_i y_i, \overline{x^2} = \frac{1}{n}\sum_{i=1}^{n} x_i^2, \overline{y^2} = \frac{1}{n}\sum_{i=1}^{n} y_i^2$.

Using the sample variances and covariance, Equation 5.6 can be written as:

$$d^2 = \alpha^2 s_x^2 + \beta^2 s_y^2 + 2\alpha\beta s_{xy} + (\alpha\bar{x} + \beta\bar{y} - \gamma)^2 \tag{5.7}$$

where $s_x^2 = \overline{x^2} - \bar{x}^2$, $s_y^2 = \overline{y^2} - \bar{y}^2$ and $s_{xy} = \overline{xy} - \bar{x}\bar{y}$. The problem now consists in finding those values of $\alpha$, $\beta$ and $\gamma$ such that function $\overline{d^2}$ of Equation 5.7 is minimum.

Since the first part of Equation 5.7 does not depend on $\gamma$ and $(\alpha\bar{x} + \beta\bar{y} - \gamma)^2 \geq 0$, the minimum will be attained when $\gamma = \alpha\bar{x} + \beta\bar{y}$. Substituting this value of $\gamma$ in Equation 5.6, we have that the regression line passes through the gravity center, that is:

$$\alpha(x - \bar{x}) + \beta(y - \bar{y}) = 0$$

With the restriction that $\gamma = \alpha\bar{x} + \beta\bar{y}$, and using matrix notation, the problem of optimizing $\overline{d^2}$ of Equation 5.7 is equivalent to finding the extremes of the quadratic form

$$\overline{d^2} = \begin{pmatrix}\alpha & \beta\end{pmatrix}\begin{pmatrix} s_x^2 & s_{xy} \\ s_{xy} & s_y^2 \end{pmatrix}\begin{pmatrix}\alpha \\ \beta\end{pmatrix}.$$

The eigenvector $(\alpha, \beta)$ of the covariance matrix of maximum eigenvalue will be the maximum of function $\overline{d^2}$ and the eigenvector of the covariance matrix of minimum eigenvalue will be the minimum that we are searching for. The latter gives the direction orthogonal to the line $Ax + By = C$, whereas the former gives the direction of the line $Ax + By = C$. Therefore, this line is the direction along which the dispersion is maximum, or equivalently its direction vector (the eigenvector of maximum eigenvalue) determines the coefficients of the first principal component. Analogously, the direction orthogonal to the line $Ax + By = C$ is the direction along which the dispersion is minimum, that is the vector orthogonal to this line (the eigenvector of minimum eigenvalue) determines the coefficients of the second principal component.

## MULTIPLE REGRESSION ON PRINCIPAL COMPONENTS

In this section we consider the problem of multiple regression with one continuous response $y$ and $q$ continuous regressors $x$. In particular, we are interested in the centred model for multiple regression, whose equation is:

$$y_i = \alpha + \beta_1(x_{i1} - \bar{x}_1) + \cdots + \beta_q(x_{iq} - \bar{x}_q) + u_i, \qquad i = 1, 2, \ldots, n, \tag{5.8}$$

where $u_i$ is a random error term. There are some situations where the least square estimator of the $\beta_j$ parameters, that is the $\hat{\beta}_j$s, becomes unstable, in the sense that the variances of the $\hat{\beta}_j$s are very large. This may typically happen when the number of regressors is close to the sample size, that is $q \approx n$, or when the regressors are near-linear dependent.

A possible solution to the problem of *multicollinearity* is to regress the response $y$ on the principal components of the xs. In doing so, if we use the first $m < q$ principal components, the large variances of the $\hat{\beta}_j$s due to multicollinearity will be reduced, but some bias to the new $\hat{\beta}_j$s is introduced. However, this may be still worthwhile if the multicollinearity is severe. Other alternatives may be ridge regression (Hoerl and Kennard, 1970a, 1970b) or the 'lasso' method (Tibshirani, 1996).

In the following we obtain the estimators of $\beta_j$ parameters via principal component regression. In matrix notation, the model of Equation 5.8 can be written as

$$\mathbf{y} = [\mathbf{1} \quad \mathbf{X}]\begin{pmatrix} \alpha \\ \boldsymbol{\beta} \end{pmatrix} + \mathbf{u}, \tag{5.9}$$

where $\mathbf{y}$ is the $n$x1 vector of the response variable, $\mathbf{X}$ is the $n$x$q$ matrix of the centred regressors, $\mathbf{1}$ is a $n$x1 vector of ones, $\boldsymbol{\beta}$ is a $q$x1 vector of parameters and $\mathbf{u}$ is a $n$x1 random vector with null expectation and covariance matrix $var(\mathbf{u}) = \sigma^2\mathbf{I}$, where $\mathbf{I}$ is the $n$x$n$ identity matrix, $\sigma^2 > 0$.

The least square estimator of $\boldsymbol{\beta}$ is given by:

$$\hat{\boldsymbol{\beta}} = (\mathbf{X}'\mathbf{X})^{-1}\mathbf{X}'\mathbf{y}$$

whose covariance matrix is $var\left(\hat{\boldsymbol{\beta}}\right) = \sigma^2(\mathbf{X}'\mathbf{X})^{-1}$. Note that (i) the variances of $\hat{\beta}_j$s are the diagonal elements of this matrix, (ii) the eigenvalues of $(\mathbf{X}'\mathbf{X})^{-1}$ are the inverse of the eigenvalues of $\mathbf{X}'\mathbf{X}$, and (iii) the eigenvalues of $\mathbf{X}'\mathbf{X}$ are proportional to the eigenvalues of covariance matrix of $\mathbf{X}$, that is if $\lambda_j$ is an eigenvalue of $\mathbf{S} = \mathbf{X}'\mathbf{X}/n$, then $n\lambda_j$ is an eigenvalue of $\mathbf{X}'\mathbf{X}$ (remember that $\mathbf{X}$ is a centred matrix).

Hence, the total variance $\sum_{j=1}^{q} var(\hat{\beta}_j)$ can be written as:

$$\sum_{j=1}^{q} var(\hat{\beta}_j) = \sigma^2 tr(\mathbf{X}'\mathbf{X})^{-1} = \frac{\sigma^2}{n}\sum_{j=1}^{q}\frac{1}{\lambda_j} \tag{5.10}$$

where $\lambda_1 > \lambda_2 > \ldots > \lambda_q$ are the eigenvalues of the covariance matrix $\mathbf{S}$. From Equation 5.10 we observe that small eigenvalues inflate the total variance and also induce multicollinearity among the regressors. To show that, let $\lambda_q \approx 0$ be the smallest eigenvalue of the covariance matrix $\mathbf{S}$ and $\mathbf{t_q}$ the corresponding eigenvector, such that $\mathbf{t}'_\mathbf{q}\mathbf{t_q} = 1$. Then, since

$$\mathbf{S}\mathbf{t_q} = \lambda_q\mathbf{t_q}$$

multiplying on the left by $\mathbf{t}'_\mathbf{q}$ gives

$$\mathbf{t}'_\mathbf{q}\mathbf{S}\mathbf{t_q} = \lambda_q\mathbf{t}'_\mathbf{q}\mathbf{t_q} = \lambda_q \approx 0$$

and since $\mathbf{S} = \mathbf{X}'\mathbf{X}/n$, we have that

$$\frac{1}{n}\mathbf{t}_\mathbf{q}'\mathbf{X}'\mathbf{X}\mathbf{t}_\mathbf{q} = \mathbf{t}_\mathbf{q}'\mathbf{S}\mathbf{t}_\mathbf{q} \approx 0 \Rightarrow \mathbf{X}\mathbf{t}_\mathbf{q} \approx 0. \quad (5.11)$$

Finally, since $\mathbf{Xt_q}$ is a linear combination of the columns of $\mathbf{X}$, Equation 5.11 says that there is near-linear dependency between the columns of $\mathbf{X}$.

We have seen that small eigenvalues that inflate the total variance are also responsible of inducing multicollinearity among the regressors. Since these eigenvalues correspond to the last principal components of $\mathbf{S}$, we can reduce the total variance by considering an estimator based on the first principal components.

Consider the principal components computed from the covariance matrix $\mathbf{S}$. That is, we compute the spectral decomposition of $\mathbf{S} = \mathbf{T}\boldsymbol{\Lambda}\mathbf{T}'$, where $\mathbf{T}$ is the orthogonal matrix whose columns are the eigenvectors of $\mathbf{S}$ and $\boldsymbol{\Lambda}$ is a diagonal matrix containing the eigenvalues of $\mathbf{S}$ ordered in decreasing order, and compute the principal components as $\mathbf{Z} = \mathbf{XT}$. Then Equation 5.9 can be written as:

$$\mathbf{y} = \alpha\mathbf{1} + \mathbf{X}\boldsymbol{\beta} + \mathbf{u} = \alpha\mathbf{1} + \mathbf{Z}\mathbf{T}'\boldsymbol{\beta} + \mathbf{u} = \alpha\mathbf{1} + \mathbf{Z}\boldsymbol{\gamma} + \mathbf{u} \quad (5.12)$$

where $\boldsymbol{\gamma} = \mathbf{T}'\boldsymbol{\beta}$ is a $q$x1 vector of parameters (remember that $\mathbf{T}$ is an orthogonal matrix, that is $\mathbf{T}'\mathbf{T} = \mathbf{T}'\mathbf{T} = \mathbf{I}$). Now parameter $\boldsymbol{\gamma}$ can be estimated by:

$$\hat{\boldsymbol{\gamma}} = (\mathbf{Z}'\mathbf{Z})^{-1}\mathbf{Z}'\mathbf{y}$$

whose covariance matrix is given by:

$$var(\hat{\boldsymbol{\gamma}}) = \sigma^2(\mathbf{Z}'\mathbf{Z})^{-1} = \sigma^2(\mathbf{T}'\mathbf{X}'\mathbf{X}\mathbf{T})^{-1} = \sigma^2(n\mathbf{T}'\mathbf{S}\mathbf{T})^{-1} = \frac{\sigma^2}{n}\boldsymbol{\Lambda}^{-1} = \frac{\sigma^2}{n}\begin{pmatrix} 1/\lambda_1 & \dots & 0 \\ \vdots & \ddots & \vdots \\ 0 & \dots & 1/\lambda_q \end{pmatrix}. \quad (5.13)$$

From Equation 5.13 we see that we can reduce the total variance of the $\hat{\gamma}_j$s by considering only those components with large $\lambda_j$s, for example the first $m < q$ principal components. Therefore, the estimator based on the first $m < q$ principal components is given by:

$$\hat{\boldsymbol{\gamma}}_\mathbf{m} = \frac{\sigma^2}{n}\boldsymbol{\Lambda}_\mathbf{m}^{-1}\mathbf{T}_\mathbf{m}'\mathbf{X}'\mathbf{y}$$

where $\boldsymbol{\Lambda}_\mathbf{m}^{\mathbf{-1}} = diag\left(\frac{1}{\lambda_1}, \dots, \frac{1}{\lambda_m}\right)$ and $\mathbf{T_m} = [\mathbf{t}_1, \dots \mathbf{t_m}]$.

If the principal components have a natural intuitive interpretation it may be better to leave the regression equation (and coefficient estimates) in terms of the components. If this is not the case, it may be more convenient to transform back to the original variables. The following formula recovers the estimators for the regression coefficients for the original $x$s as:

$$\tilde{\boldsymbol{\beta}} = \mathbf{T_m}\hat{\boldsymbol{\gamma}}_\mathbf{m} = \frac{\sigma^2}{n}\mathbf{T_m}\boldsymbol{\Lambda}_\mathbf{m}^{-1}\mathbf{T}_\mathbf{m}'\mathbf{X}'\mathbf{y} \quad (5.14)$$

whose covariance matrix is

$$var(\tilde{\boldsymbol{\beta}}) = \mathbf{T_m}var(\hat{\boldsymbol{\gamma}}_\mathbf{m})\mathbf{T}_\mathbf{m}' = \frac{\sigma^2}{n}\mathbf{T_m}\boldsymbol{\Lambda}_\mathbf{m}^{-1}\mathbf{T_m}' \quad (5.15)$$

These equations have their analogues when principal components are computed from the correlation matrix $\mathbf{R}$. In this case, matrix $\mathbf{X}$ is replaced by $\mathbf{X_s} = \mathbf{X}\mathbf{S_0}^{-1/2}$, where $\mathbf{S_0} = diag(s_1^2, \ldots, s_q^2)$ is a diagonal matrix containing the variances of the regressors. Calling $\tilde{\mathbf{T}}_\mathbf{m}$ and $\tilde{\mathbf{\Lambda}}_\mathbf{m}$ the matrices containing the first $m$ eigenvectors and eigenvalues of $\mathbf{R}$, then the analogues of Equations 5.14 and 5.15 are:

$$\tilde{\boldsymbol{\beta}}^* = \frac{\sigma^2}{n}\tilde{\mathbf{T}}_\mathbf{m}\tilde{\mathbf{\Lambda}}_\mathbf{m}^{-1}\tilde{\mathbf{T}}_\mathbf{m}'\mathbf{X}_\mathbf{s}'\mathbf{y}$$

$$var\left(\tilde{\boldsymbol{\beta}}^*\right) = \frac{\sigma^2}{n}\tilde{\mathbf{T}}_\mathbf{m}\tilde{\mathbf{\Lambda}}_\mathbf{m}^{-1}\tilde{\mathbf{T}}_\mathbf{m}'$$

In selecting the number of principal components to consider, Rencher (1998) also proposes studying those components that correlate better with the response $y$. We know that keeping components with smaller eigenvalue will increase the total variance of the estimated coefficients, but Rencher suggested that the increase may be offset by the gain in predictability.

## Example 5.6

Ares *et al.* (2009) developed a sensory quality index for the appearance and odour of strawberries based on consumer perception through multiple regression on the first two principal components. Table 5.8 contains the analysed data set. Let us study one of the authors' procedures.

Suppose that our interest is to explain 'consumer acceptability' on strawberries in terms of several sensory attributes determined by a trained assessors' panel. In doing so, we choose the first seven numerical columns of Table 5.8 to be the regressors and 'consumer acceptability' would be the response.

Let $\mathbf{X}$ be the $6 \times 7$ centred data matrix containing the regressors. We compute the correlation matrix and obtain:

$$\mathbf{R} = \begin{pmatrix} 1.0000 & -0.5819 & 0.7631 & 0.2844 & 0.8258 & -0.5661 & -0.5823 \\ -0.5819 & 1.0000 & -0.3424 & 0.4602 & -0.3468 & 0.6667 & 0.1187 \\ 0.7631 & -0.3424 & 1.0000 & 0.5419 & 0.9117 & -0.2496 & -0.9090 \\ 0.2844 & 0.4602 & 0.5419 & 1.0000 & 0.6627 & 0.0706 & -0.7462 \\ 0.8258 & -0.3468 & 0.9117 & 0.6627 & 1.0000 & -0.4719 & -0.9154 \\ -0.5661 & 0.6667 & -0.2496 & 0.0706 & -0.4719 & 1.0000 & 0.3025 \\ -0.5823 & 0.1187 & -0.9090 & -0.7462 & -0.9154 & 0.3025 & 1.0000 \end{pmatrix}$$

**Table 5.8** Average scores for the evaluated sensory attributes for strawberries with different sensory qualities, as evaluated by a trained assessors' panel and average consumer acceptability.

| Sample | Off-odour | Strawberry odour | Browning on the sepals | Red colour | Dark bruises | Gloss | Surface evenness | Consumer acceptability |
|---|---|---|---|---|---|---|---|---|
| **Unripe** | 1.9 | 0.9 | 2.6 | 2.9 | 0.3 | 2.1 | 8.7 | 4.8 |
| **Ripe** | 1.3 | 8.3 | 1.6 | 7.7 | 0.5 | 6.7 | 9.1 | 7.7 |
| **Overripe** | 1.8 | 7.5 | 6.8 | 9.4 | 4.0 | 6.7 | 1.3 | 5.6 |
| **A** | 1.4 | 7.8 | 2.8 | 8.8 | 2.6 | 1.2 | 4.3 | 6.1 |
| **B** | 3.5 | 2.8 | 5.7 | 9.2 | 6.1 | 1.5 | 1.7 | 5.5 |
| **C** | 8.5 | 1.8 | 8.3 | 9.1 | 7.9 | 0.3 | 0.7 | 1.9 |

A: ripe strawberries stored at 5°C for 6 d; B: ripe strawberries stored at 0°C for 5 d and at 25°C for 1 d; C: ripe strawberries stored at 0°C for 2 d and at 25°C for 4 d.

**Table 5.9** Principal components for the strawberries sensory attributes (Example 5.6).

| | $Y_1$(60.02%) | corr($Y_1$, $X_i$) | $Y_2$(28.03%) | corr($Y_2$, $X_i$) |
|---|---|---|---|---|
| $X_1$ = off-odour | 0.4213 | 0.8636 | −0.2052 | −0.2874 |
| $X_2$ = strawberry odour | −0.2073 | −0.4249 | 0.6163 | 0.8634 |
| $X_3$ = browning on the sepals | 0.4532 | 0.9289 | 0.0938 | 0.1314 |
| $X_4$ = red colour | 0.2813 | 0.5767 | 0.5486 | 0.7686 |
| $X_5$ = dark bruises | 0.4828 | 0.9896 | 0.0675 | 0.0945 |
| $X_6$ = gloss | −0.2596 | −0.5321 | 0.4569 | 0.6400 |
| $X_7$ = surface evenness | −0.4411 | −0.9042 | −0.2344 | −0.3284 |

where we can observe that there are some variables with high correlations. Variable $X_1$ = 'off-odour' has correlation coefficients of 0.7631 with variable $X_3$ = 'browning on the sepals' and 0.8258 with variable $X_5$ = 'dark bruises'. Variable $X_3$ is highly correlated with $X_4$ = 'red colour' (0.9117) and $X_7$ = 'surface evenness' (−0.9090). finally, variable $X_5$ is highly correlated with $X_7$ (−0.9154).

Moreover, the determinant of the correlation matrix is $\det(\mathbf{R}) = 7.6459 \times 10^{-36}$, indicating the presence of multicollinearity, meaning that least squared estimators will be unstable. As we have seen, a possible solution is to obtain the regression coefficients via principal component regression.

We start by computing the principal components from the correlation matrix. In Table 5.9 the first two principal components only are reproduced jointly with the percentage of explained variability and their correlations with the original variables.

From Table 5.9 we can see that the first principal component mainly contrasts positively with off-odour, browning sepals and dark bruises and negatively with surface evenness. All these attributes are sensory defects to decay, fermentation, bruising and shrivelling. Therefore, $Y_1$ can be considered an index of sensory deterioration. Those individuals with large values of $Y_1$ will correspond to deteriorated strawberries. The second principal component is positively correlated with strawberry odour, red colour (both considered indicators of ripeness) and gloss (an indicator of freshness). Component $Y_2$ is related to attributes that are desirable sensory characteristics of strawberries. Thus, it can be considered a desirable attributes index of strawberries. Individuals with large values of $Y_2$ will correspond to ripe strawberries. Figure 5.8 contains the principal component representation for the strawberry sensory attributes.

From Figure 5.8 we can see that individuals are split in different groups according to their sensory quality. Hence, individuals located in the second quadrant (negative values of $Y_1$ and positive values of $Y_2$) correspond to ripe strawberries with little sensory deterioration. On the other hand, individuals located in the fourth quadrant (positive values of $Y_1$ and negative values of $Y_2$) correspond to deteriorated strawberries. The reader is referred to Ares *et al.* (2009) for a more detailed interpretation.

The first two principal components explain 88.01% of the variability of the strawberry sensory attributes. However, since we are interested in establishing a regression model that explains 'consumer acceptability' in terms of the strawberry sensory attributes, it is important to select those principal components that are more correlated with the response variable. Table 5.10 contains those correlations, where we can see that the $Y_1$ and $Y_2$ have the largest correlations (in absolute value).

To determine how the evaluated sensory attributes affect consumer acceptability, we can establish the following regression model on the two first principal components:

$$y_i = \alpha + \beta_1 Y_{i1} + \beta_2 Y_{i2} + u_i$$

where $u_i$ is a random error term with zero expectation and variance $\sigma^2$. The estimated equation is (in R software, we would write `lm(y~Y1+Y2))`:

$$y = 5.2667 - 0.6789\, Y_1 + 0.5922\, Y_2$$

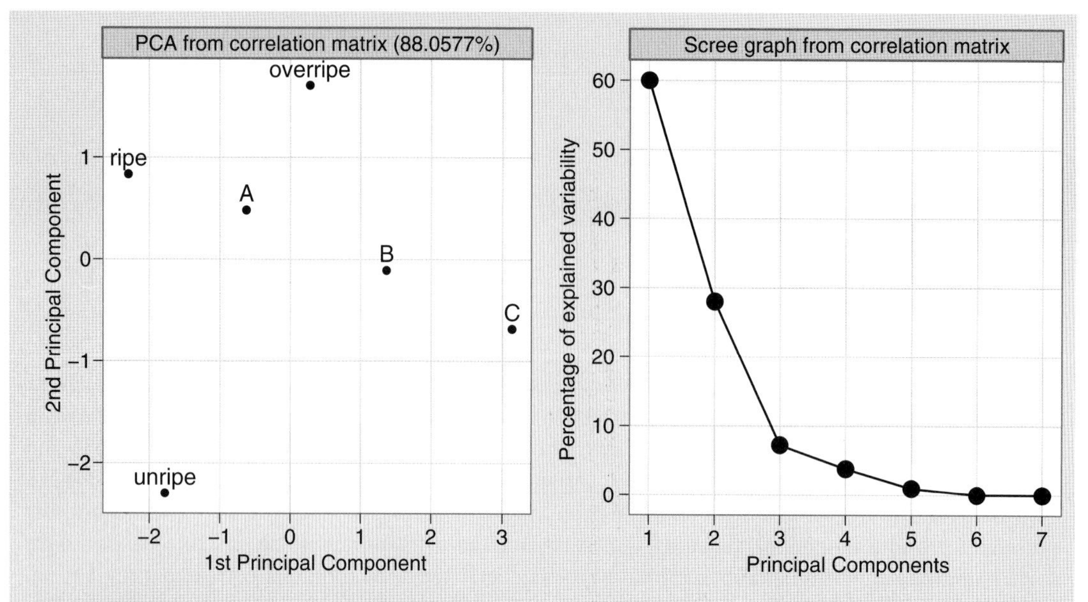


**Figure 5.8** Principal component representation for the strawberries sensory attributes (Example 5.6).

**Table 5.10** Correlations between all principal components and the response variable (Example 5.6).

| | $Y_1$ | $Y_2$ | $Y_3$ | $Y_4$ | $Y_5$ | $Y_6$ | $Y_7$ |
|---|---|---|---|---|---|---|---|
| y = 'consumer acceptability' | −0.7961 | 0.4745 | 0.1661 | −0.0775 | −0.3278 | 0.3253 | −0.4196 |

with estimated $\sigma^2$ equal to $s^2 = 0.8619$ and determination coefficient of 0.859, meaning that the previous model is able to capture or explain 85.9% of the variability of the data. As can be seen, consumer acceptability decreases when the sensory deterioration index increases (and the other index remains constant) and consumer acceptability increases when the desirable attributes index increases (and the other index remains constant).

To finish with this example, let us analyse what would happened if we estimated the regression coefficients via least squares, without correcting for multicollinearity. In Table 5.11 the regression coefficients and variances of the ordinary least squares estimators are compare with those obtained from the principal components. As can be seen, the estimator variances computed via least squares make no sense. This is due to the near-singularity of the correlation matrix of the regressors.

**Table 5.11** Comparison of least squares and principal component regression coefficients and variances (Example 5.6).

| **Standardized variable** | **Least squares** | | **Principal component** | |
|---|---|---|---|---|
| | **Coefficient** | **Variance/$s^2$** | **Coefficient** | **Variance/$s^2$** |
| $X_1$ = off odour | −2.6602 | $-0.4152 \times 10^{14}$ | −1.0277 | 0.0106 |
| $X_2$ = strawberry odour | 0.1535 | $-0.1924 \times 10^{14}$ | 4.4919 | 0.0340 |
| $X_3$ = browning on the sepals | −0.0121 | $0.1267 \times 10^{14}$ | 1.3269 | 0.0089 |
| $X_4$ = red colour | 0.9780 | $0.6383 \times 10^{14}$ | 4.6169 | 0.0287 |
| $X_5$ = dark bruises | 1.7880 | $-0.5854 \times 10^{14}$ | 1.1625 | 0.0096 |
| $X_6$ = gloss | 0.4103 | $0.0058 \times 10^{14}$ | 3.1891 | 0.0204 |
| $X_7$ = surface evenness | 3.6822 | $-1.0685 \times 10^{14}$ | −2.3985 | 0.0124 |

## REFERENCES


Ares, G., Barrios, S., Lareo, C. and Lema, P. (2009) Development of a sensory quality index for strawberries based on correlation between sensory data and consumer perception. *Postharvest Biology and Technology* **52**, 97–102.

Basilevsky, A. (1994) *Statistical Factor Analysis and Related Methods*. John Wiley & Sons, Inc., New York.

Cattell, R. B. (1966) The Scree test for the number of factors. *Multivariate Behavioral Research* **1**, 245–276.

Hoerl, A. E. and Kennard, R. W. (1970a) Ridge regression: Biased estimation for nonorthogonal problems. *Technometrics* **12**, 55–67.

Hoerl, A. E. and Kennard, R. W. (1970b) Ridge regression: Applications to nonorthogonal problems. *Technometrics* **12**, 69–82.

IFPRI (2012) *2011 Global Food Policy Report*. International Food Policy Research Institute, Washington, DC. doi: 10.2499/9780896295476.

Jackson, J. E. (1991) *A User's Guide to Principal Components*. John Wiley & Sons, Inc., New York.

Johnson, R. A. and Wichern, D. W. (1998) *Applied Multivariate Statistical Analysis*, 4th edn. Prentice Hall, NJy.

Jolliffe, I. T. (1986) *Principal Component Analysis*. Springer, New York.

Krzanowski, W. J. (1984a) Principal component analysis in the presence of group structure. *Applied Statistics* **33**, 164–168.

Krzanowski, W. J. (1984b) Sensitivity of principal components. *Journal of the Royal Statistical Society Series B* **46**, 558–863.

Mardia, K. V., Kent, J. T. and Bibby, J. M. (1979) *Multivariate Analysis*. Academic Press, London.

Naik, D. N. and Khattree, R. (1996) Revisiting Olympic track records: Some practical considerations in principal component analysis. *American Statistician* **50**, 140–144.

Rencher, A. C. (1998) *Multivariate Statistical Inference and Applications*. John Wiley & Sons, Inc., New York.

Tibshirani, R. (1996) Regression shrinkage and selection via lasso. *Journal of the Royal Statistical Society Series B* **58**, 267–288.

Watanabe, S. (1967) Karhunen–Loeve expansion and factor analysis. In: *Transactions of the Fourth Prague Conference*. Czechoslovakia Academy of Science, Prague.

# 6 Multiple factor analysis: Presentation of the method using sensory data

**Jérôme Pagès and François Husson**
*Laboratoire de mathématiques appliqués, Agrocampus Ouest, Rennes Cedex, France*

## ABSTRACT

Multiple factor analysis (MFA) is devoted to data tables in which a set of individuals is described by several groups of variables. It balances the influence of these groups in a single analysis and provides results describing all the aspects of the comparison between groups of variables. Described here are the main features of the method from an application in the field of sensory analysis: ten wines were described by two panels, the sensory descriptions of each panel constituting a group of variables. Tables and graphs were obtained using the R package FactoMineR (free).

## INTRODUCTION

Multiple factor analysis (MFA) is an analysis applied to tables in which a set of individuals is described by several groups of both categorical and quantitative variables (Escofier and Pagès 2008).

There are many application examples in sensory analysis. In the two following classic situations, the individuals are food products:

- In the first case, sensory information for each product was provided by several panels. Each panel's descriptors (which varied from one panel to the next) represent a group of variables. This kind of example will be used to present MFA in this chapter (the description of another example, oriented toward the results rather than the methodology, can be found in Pagès *et al.*, 2007).
- In the second case there is only one panel, but in addition there are also physicochemical variables. The set of variables is therefore arranged into two groups: sensory and physicochemical.

Another classic example, which is beyond the framework of sensory analysis, is a set of individuals which is the object of multiple repeated measurements over time. In this case, the measurements conducted at one given time make up a group.

For the user, the crucial point is: what do we really mean by 'taking into account the group structure of a set of variables'? The presentation of MFA in this chapter is organized with the aim of answering this question. The first problem we encounter is thus that of balancing the groups. Once this balance is achieved, it is possible to compare the groups of variables with one another according to a number of aspects, that is to say mainly the images of the individuals with which they are associated, along with their principal components.

*Mathematical and Statistical Methods in Food Science and Technology*, First Edition.
Edited by Daniel Granato and Gastón Ares.

To illustrate the way in which MFA answers these questions, a small example was chosen: ten individuals described by two groups of variables made up of 31 and 15 quantitative variables. This size is handy for explaining the method in a reduced format while still taking into account its advantages, which become more and more valuable the more data we work with.

Due to the large number of situations in which we want to simultaneously analyse multiple groups of variables, and the suitability of MFA to these situations, this method is now well established, particularly in the food industry and sensory analysis sectors. There are many different software packages available to users with which they can implement such an analysis: SPAD, XLSTAT, UNIWIN PLUS, Ade4 (package R), FactoMineR (Package R), to name but a few. The latter of these pieces of software, which is free, as are all R packages, was developed by the Applied Mathematics Department at Agrocampus, Rennes, France ($LMA^2$). This is the software that will be used in this chapter (a presentation of classic data analysis methods using the FactoMineR package can be found in Husson *et al.*, 2010). The data set used is available on the $LMA^2$ website; readers can therefore easily compute themselves all of the results presented here, should they wish to.

# DATA

## Ten Touraine white wines

Within the context of research on Loire Valley wines, the typical characteristics of Chenin vines compared with Sauvignon vines were studied. Ten Touraine wines were chosen for this study. Of these wines, five of them (numbered 1–5) are officially labelled Touraine (Sauvignon grape variety). The remaining five (numbered 6–10) belong to the Vouvray appellation (Chenin grape variety). In the latter group, two wines (7 and 10) are cask-aged (and thus have a woody taste) and one (number 6) has residual sugars (7 g/l). On the graphs, the wine's labels begin with their number, followed by a letter indicating variety (S or C).

## Two panels

These wines were subjected to two descriptive tastings.

A first panel was made up of eight experts in both wine tasting and Loire Valley wines. The experts evaluated the ten wines during a single session according to a questionnaire of forty descriptors.

The second panel was composed of twelve food industry students who followed a short training course of two sessions in wine tasting. The final tasting took place over three sessions, during which each taster evaluated each wine according to a simple questionnaire with only twenty descriptors.

In both cases, the order in which the wines were presented was defined using Williams Latin squares.

## Pre-processing

An analysis of variance was conducted for each panel and each descriptor, introducing within the model the wine effect, the taster effect and the wine × taster interaction effect wherever possible (student panel case). In this way it was possible to choose the descriptors with significant effects. The usual 5% threshold was chosen for the 'student' panel, whereas for the 'expert' panel a threshold of 10% was chosen due to the scarcity of the data. With these thresholds, no 'important' descriptors were discarded. Thirty one descriptors were finally retained for the experts and 15 for the students.

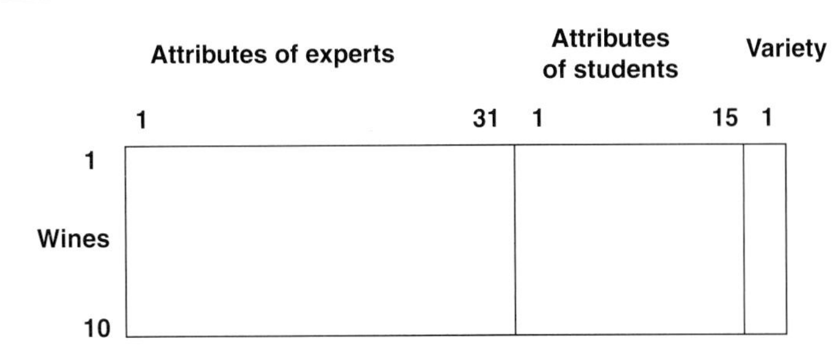


**Figure 6.1** The wine data set.

From these data, we calculated the mean table for the student panel, which has no missing data (and which, at the intersection of row $i$ and column $k$, contains the mean scores obtained by product $i$ for descriptor $k$).

On the other hand, there were some missing data from the expert panel. Also, for this panel, the mean table contained the adjusted means for the analysis of variance model: score = judge + product.

## Analysis of the table

To compare the evaluations provided by the two panels, all of the data are brought together in a single table (Figure 6.1), in which:

- the rows are the wines;
- the columns are the descriptors organized into two groups;
- an additional column indicates the grape variety.

In this MFA, we chose to centre and reduce all of the descriptors in order to assign the same weight to each.

# WEIGHTING TO BALANCE GROUPS OF VARIABLES

The first function of MFA is to balance the groups of variables within a global analysis. In this example, the 'expert' group includes twice as many variables as the 'student' group. Without balancing, that is to say by merely submitting the table juxtaposing the variables of the two groups to a simple Principal Component Analysis (PCA), the expert panel would 'automatically' dominate the construction of the dimensions. This can be seen in Table 6.1, where the 'expert' group contributes 68.91% (64.15% respectively) to the first axis (and second axis, respectively) of this PCA (the contribution of a group of variables is the sum of the contributions of the variables in the group).

**Table 6.1** PCA performed on the 46 variables. Decomposition of the inertia by axis and by group of variables.

| | Global inertia | F1 | F2 | F1 (%) | F2 (%) |
|---|---|---|---|---|---|
| PCA whole | 46 | 24.95 | 10.02 | 54.24 | 21.79 |
| Expert Group | 31 | 17.19 | 6.43 | 68.91 | 64.15 |
| Student Group | 15 | 7.76 | 3.59 | 31.09 | 35.85 |

**Table 6.2** Decomposition of inertia by dimension and by group. Rows 1 and 2: Separate PCA. Rows 3 to 5: MFA.

| | Global inertia | F1 | F2 | F1 (%) | F2 (%) |
|---|---|---|---|---|---|
| PCA Expert | 31 | 17.57 | 6.16 | 56.66 | 19.86 |
| PCA Student | 15 | 8.61 | 2.96 | 57.37 | 19.72 |
| MFA | 3.51 | 1.89 | 0.77 | 53.98 | 22.04 |
| Expert Group | 1.76 | 0.95 | 0.39 | 49.98 | 50.66 |
| Student Group | 1.74 | 0.95 | 0.38 | 50.02 | 49.34 |

To balance the influence of the groups, the MFA assigns a weight to each variable. These weights are constant within the group: the structure of the group therefore does not change. The maximum axial inertia of each group is worth 1. In other words, if with these weights we conduct a PCA on the variables of a single group, the first eigenvalue is 1 by construction. Technically, to obtain this property, the weight assigned to each variable of a group is equal to the inverse of the first eigenvalue of the PCA of that group. In the example, the weight of each variable from the 'expert' group (and 'student' group respectively) is thus worth 1/17.57 (and 1/8.61 respectively; the eigenvalues of the separate PCA figure in the first two rows in Table 6.2).

A PCA with these weights is at the heart of the MFA. Thus, in the MFA applied to this data, both groups contribute equally to the first axis (Table 6.2 rows 4 and 5; 49.98% and 50.02%). This is the result we would expect. Furthermore, in this example, both groups contribute equally to the second axis (50.66% and 49.34%).

## Representing wines and descriptors

MFA provides representations of the individuals (wines) and variables (descriptors) which are read in the same way as PCA representations. In particular, MFA benefits from the transition relationships of principal axes method. More precisely, for a given axis, an individual's coordinate can be calculated from the variables' coordinates and the individual's values. According to this relationship (given in Appendix 6.A), an individual appears on the side of the variables for which it carries a high value and opposite the variables for which it carries a low value. Thus, the first axis opposes (Figure 6.2 and Figure 6.3):

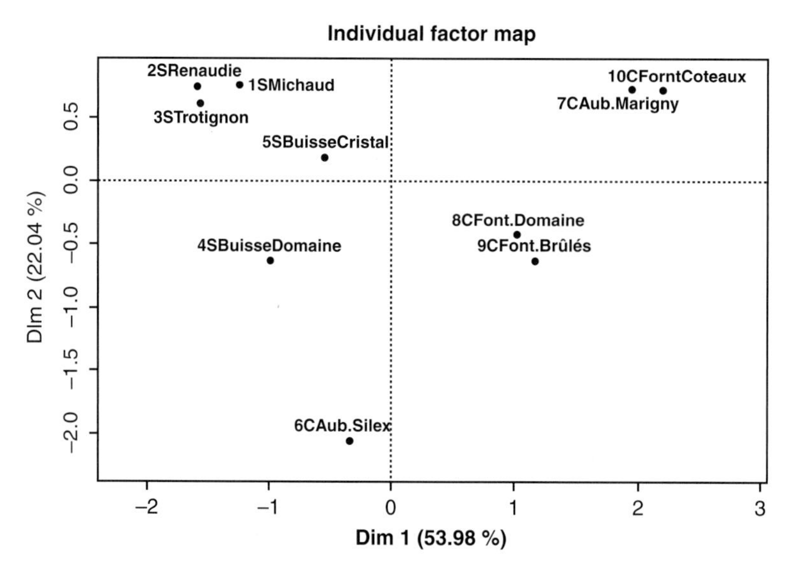


**Figure 6.2** Representation of wines on the first plane of the MFA.

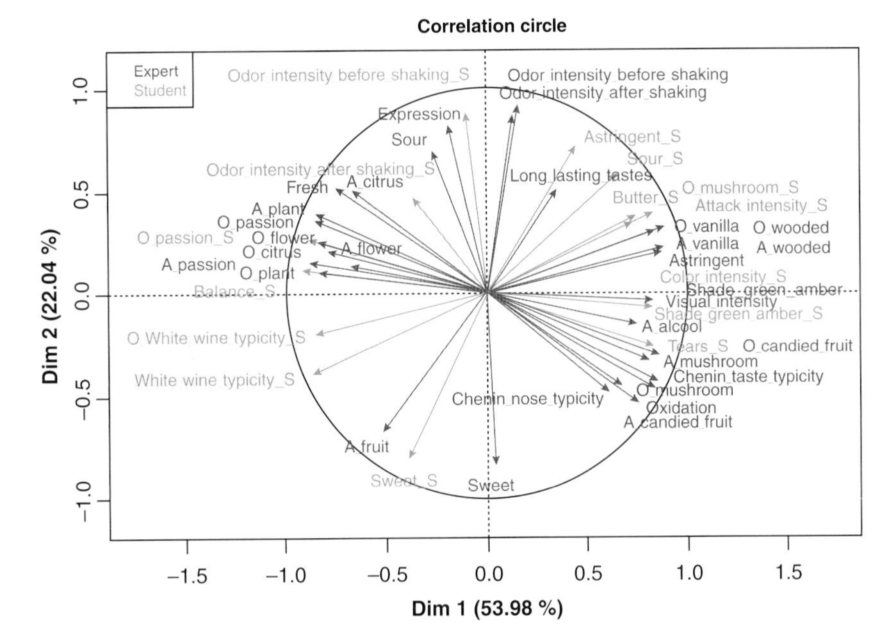


**Figure 6.3** Representation of descriptors on the first plane of the MFA. (See colour version of this figure in colour plate section.)

- amber-coloured wines, well characterized by the experts (high scores for 'vanilla' and 'woody');
- slightly green wines, the smell of which was described as 'passion fruit' by both panels, and in greater depth by the experts (flowery, vegetal, citrusy).

From the point of view of the individuals, this axis opposes the two grape varieties. The olfactory characteristics classically attributed to Sauvignon (particularly 'passion fruit') are well represented here; amongst other things, only wines 7 and 10 were cask-aged, which explains the high scores for the descriptors 'woody' and 'vanilla'.

Furthermore, wines 1, 2 and 3 seem more typical of Sauvignon than the other two Sauvignons (4 and 5); the woody Chenin wines, 7 and 10, are the most different from the Sauvignons.

The second axis mainly opposes wine 6 from all the others (contribution: 54%). This wine is characterized by both panels as sweet (it contains 7 g/L of residual sugars) and the odour is characterized as not intense (it is opposed to wines 1, 2, 3, 7 and 10, all of which have a strong odour, even if the odour differs for each).

## SUPERIMPOSED REPRESENTATION OF THE WINES ANALYSED BY EACH PANEL

Partial individuals are an important concept in MFA. In the example, two images can be assigned to each wine, each of which corresponds to the descriptions given by one single panel. In considering all of the individuals, we construct clouds of individuals which are said to be 'partial'. There are as many partial clouds as there are groups of individuals.

Comparing partial clouds is a big part of studying a multiple table. In this example we will, therefore, focus on comparing the wine descriptions provided by both panels. For this comparison the difficulty is that the descriptors are not the same for both panels. We will also approach this comparison with

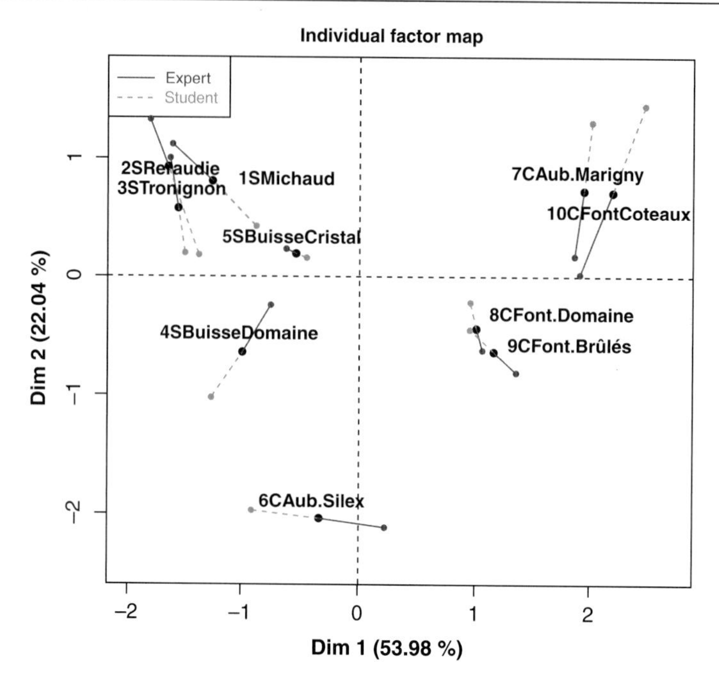


**Figure 6.4** Representation of the mean and partial individuals on the first plane of the MFA. (See colour version of this figure in colour plate section.)

questions such as: 'Are two wines which are close for one panel also close for the other (without imposing the variables which bring them together)?' 'Is wine 6, which is unusual when we analyse both panels simultaneously (Figure 6.2), also unusual when we consider only one panel?'

More precisely, the cloud of individuals for one group is that which is considered in the separate PCA for this group (prior to the projection on MFA principal components, that is to say the whole space). In the example, we therefore wish to compare the cloud of wines described by the experts with that of the wines described by the students. To compare these two clouds, the MFA superimposes them onto the representation of the mean cloud (Figure 6.4). Technically, this representation is calculated using the aforementioned transition relationship but is limited to the variables of a single group (this relationship is explained in detail in Appendix 6.A). The result is a natural rule of interpretation. Thus, for a given axis, a partial individual in group $j$ is found on the side of the variables of group $j$ for which it carries high values and opposite the variables of group $j$ for which it carries low values.

For each wine, Figure 6.4 includes three points: the overall point (identical to Figure 6.2) and two partial points. By construction, the overall point is at the barycentre of the corresponding partial points, which is why they are called 'mean' points. It is helpful to link each partial point to the corresponding mean point using a segment (known as a star graph).

Overall, the two partial representations for the same wine are alike: the two clouds (of points) look similar. However, when we look more closely, there are some deviations which merit further investigation:

- Wines 7 and 10 (woody Chenins) were judged as being more unusual by the students than by the experts. In comparison with the students, wine 6 was considered by the experts to be the closest to the other Chenins: as a result the Chenins are much more homogeneous according to the experts' descriptions.
- Wines 1, 2 and 3 were judged as being more unusual by the experts than by the students.

Finally, the deviation between Sauvignon and Chenin was the most significant for the experts.

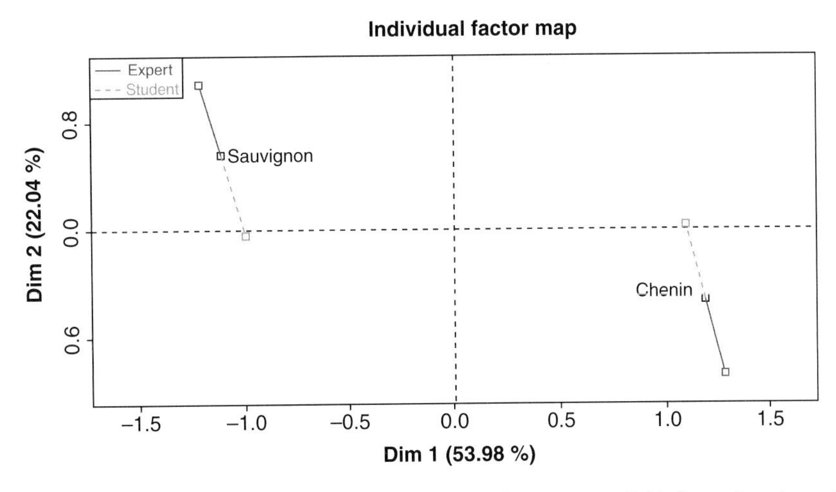


**Figure 6.5** Representation of mean and partial categories. (See colour version of this figure in colour plate section.)

## SUPPLEMENTARY CATEGORICAL VARIABLES

In this data set, one categorical variable plays an important role: the grape variety. Until now, the variety has been used empirically in the interpretations, which is possible partly due to the fact that we take into account the small number of individuals, but especially because there is only one categorical variable. To predict more complex cases (more individuals and more categorical variables), we need to formalize how such variables are accounted for.

In PCA, introducing a categorical variable as supplementary amounts to representing each category on the dimension of individuals. Rule: One category is placed at the barycentre of the individuals that carry it. In MFA, this rule is applied directly to the mean individuals. Furthermore, it is generalized to the partial individuals. Thus, a partial category ($m$) of a group ($j$) appears at the barycentre of the partial points of this group ($j$), which corresponds to the individuals carrying that category ($m$).

In the example, the result is the representation shown in Figure 6.5, in which:

- the deviation between Sauvignons and Chenins is clearly visible along dimension 1 and a little according to dimension 2;
- this deviation is greater for the experts than for the students, particularly according to dimension 2.

This last point is due to the greater number of olfactory descriptors in the experts' questionnaire, as the differences between the two wines are primarily olfactory.

## REPRESENTING THE DIMENSIONS OF SEPARATE ANALYSES

In MFA, the reference to separate analysis stems from two different viewpoints:

- We may wonder if the MFA dimensions are correlated with the dimensions of the separate PCAs. Question: are the principal dimensions of 'overall' variability (i.e. taking all the groups into account) close to the principal dimensions of 'local' variability (i.e. taking only one group into account)?

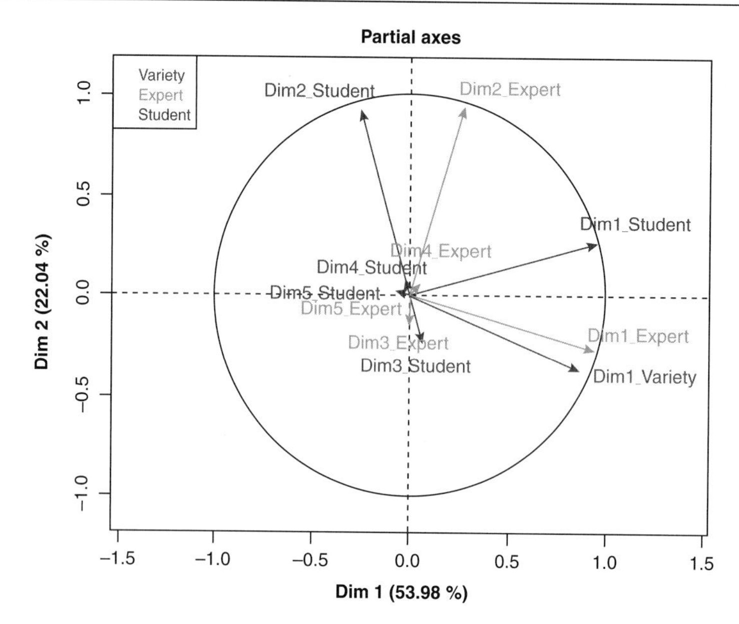


**Figure 6.6** Representations of the partial axes (principal components of the separate PCAs) on the first plane of the MFA. (See colour version of this figure in colour plate section.)

- In comparing groups of variables, we can examine the correlations between the principal components of the separate PCAs. In the example: are the principal dimensions of variability of the wines identical for the two panels?

In MFA, in order to do this, we represent the principal components of the separate PCAs, also known as 'partial axes', as supplementary variables. The result is a 'correlation circle' representation (the coordinates are correlation coefficients).

In the example, this representation (Figure 6.6) shows that:

- The first dimension of the MFA is closely linked to the first dimension of each of the separate PCAs; in addition, it is something of a compromise, which is very common for both groups.
- The second dimension of the MFA is also closely linked to the second dimension of the separate PCAs of which it is also a compromise.
- The first two (partial) axes of each separate analysis are well represented; the first plane of the MFA coincides (almost) with the first plane for each separate analysis, discarding a rotation.

The first two correlation coefficients between partial axes of the same rank are high but they are slightly different from 1 (Table 6.3). They give a 'pessimistic' view of the relationships between the separate PCAs: indeed, these PCAs lead more or less to the same first plane (see previous), but discarding a rotation.

The first dimension of the 'grape variety' group is very specific as this group has only one variable with two categories: it corresponds to a single dimension of the variables' space (thus equal to its first principal component). This direction is next to the first expert dimension. This seems to suggest that the experts structured their descriptions essentially from the variety (which they probably recognized).

**Table 6.3** Correlation between the principal components of the separate PCAs. EF1 (and SF1 respectively): first principal component of the 'expert' panel (and 'student' panel respectively).

| | EF1 | EF2 | EF3 |
|---|---|---|---|
| SF1 | 0.84 | 0.51 | –0.03 |
| SF2 | –0.50 | 0.76 | –0.32 |
| SF3 | 0.15 | –0.35 | –0.51 |

# REPRESENTING GROUPS OF VARIABLES

## Measuring the relationship between a variable and a group of variables

Especially when there are many groups, it can be interesting to directly measure the relationship between the principal components of the MFA and the groups themselves. As an example of a question: Is it possible to say that a given dimension is closely related to a given group of variables? By denoting $F_s$ the principal component of rank $s$ and $K_j$ the set of variables of group $j$, the objective is to measure the relationship between $F_s$ and $K_j$.

The measurement $Lg(F_s, K_j)$ is defined as the projected inertia of the variables of $K_j$ along $F_s$. We recognize the contribution of group $K_j$ in the construction of $F_s$. This indicator appears as a relationship measurement which can be interpreted for both active and supplementary groups.

In addition we have $0 \leq Lg(F_s, K_j) \leq 1$. The indicator $Lg$ is worth 0 if each of the variables of $K_j$ *is* uncorrelated with $F_s$. It is worth 1 if $F_s$ coincides with the group's first principal component, which represents the 'maximum' relationship situation between a variable and a group (the first principal component of $K_j$ can be defined as the linear combination of the variables the most closely linked to the variables in $K_j$).

## Relationship square

The $Lg$ measurements can be represented in the 'relationship square' graph (Figure 6.7). We thus obtain a representation of the groups of variables themselves (one group = one point). This representation is more useful as the number of groups increases. This is not the case in the example, but it illustrates how to interpret the relationship square. According to Figure 6.7:

- the first dimension is closely linked to each of the groups, that is to say it corresponds to an important direction of inertia in both of them;
- the fact that the groups are almost exactly superimposed shows that, for the most part (i.e. the two most important directions of inertia) the images of wines resulting from the experts' and students' description are identical.

If group $K_j$ is reduced to a single variable, the $Lg$ measurement coincides with the square of the correlation coefficient. It is, therefore, possible to represent an isolated variable in the relationship square (which is not the case here).

Without getting into the technical details, the Lg measurement is applied to the categorical variables. To define $Lg\ (F_s,K_j)$ when $K_j$ is composed of categorical variables, we replace the first eigenvalue $\lambda_1^j$ of the PCA of $K_j$ with that of the MCA of $K_j$ and we replace the squared correlation coefficient $r^2(F_s, K_j)$ with the squared correlation ratio $\eta^2(F_s, K_j)$.

The squared correlation ratio is an indicator used in one-way analysis of variance: the percentage of variance explained by the categorical variable (classically denoted $R^2$). Here, we will simply say that the

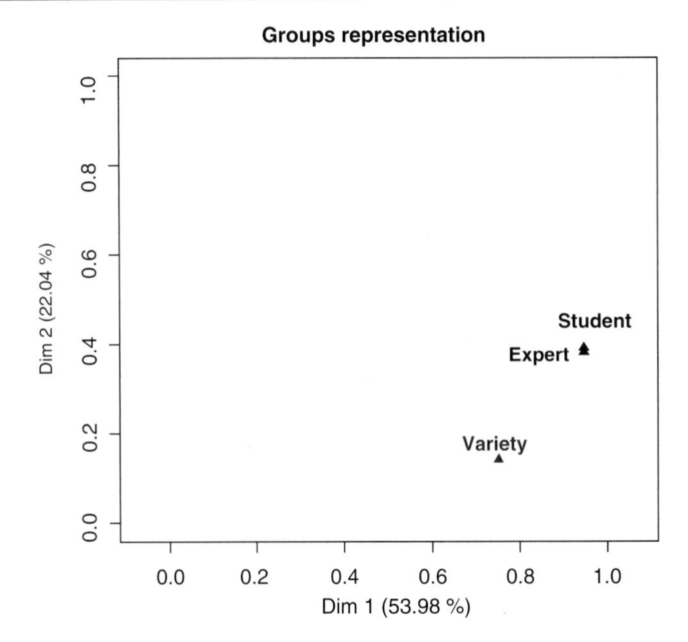


**Figure 6.7** Relationship square (representation of groups of variables).

square of the correlation ratio (between a quantitative variable $X$ and a categorical variable $V$) varies between 0 and 1 (in the same way as the squared correlation coefficient). Its value is 0 when there is no relationship and 1 for the maximum relationship (the belonging of an individual to a class of $V$ perfectly determines its value for $X$).

In the example (Figure 6.7), the grape variety is very closely related to the first dimension and not to the second. In this simple case (one single categorical variable with two categories), the contribution of this graph is weak compared to the explanations given previously. However, this contribution will be much greater in cases where the data include a lot of variables with a great number of categories.

## Relationship square and projection

The representation in Figure 6.7 can also be seen as the projection of a cloud in which each point represents a group. In this cloud, the position of a group, $j$ is defined from the inter-individual distances brought about by group $j$ alone (i.e. the distances in the cloud of partial individuals associated with group $j$).

The proximity of two groups, therefore, corresponds to a similarity of shape between the corresponding clouds of partial individuals. Or, from another point of view, the shapes of the clouds are identical between the wines described by each of the panels.

That said, interpreting the representation of the groups as projections suggests that the representation quality of each of the points should be calculated. Indeed, the figure highlights a strong similarity of shape between the two clouds of wines but other dimensions may highlight this resemblance to a greater or lesser extent. Table 6.4 brings together the representation qualities associated with the group representation; qualities which are measured in the usual way [(projected inertia)/(total inertia)] (a ratio which can also be interpreted as a squared cosine). It shows good representation quality for the 'expert' and 'student' groups: the similarity between the two clouds highlighted by the relationship square can be considered as an overall resemblance.

**Table 6.4** Representation quality of groups of variables in the relationship square.

| | F1 | F2 | Plane(1,2) |
|---|---|---|---|
| Expert | 0.770 | 0.132 | 0.902 |
| Student | 0.767 | 0.124 | 0.891 |
| Variety | 0.564 | 0.020 | 0.584 |

### Measuring the overall resemblance of shape between two similar clouds: the RV coefficient

The overall resemblance between the two clouds of wines mentioned in the previous section can be measured using the RV coefficient. Without going into the technical details, it must be noted that this coefficient is between 0 and 1. The RV is nil if each of the variables of a group is uncorrelated with each of the variables of the other group (which shows that this coefficient can also be seen as an indicator of the relationship between two groups of variables). The RV is at a maximum (equal to 1) if the two clouds (of individuals) are homothetic.

In this data, the RV coefficient has a value of 0.865. This would confirm, if indeed it were still necessary, the overall similarity of the shape of the two wine clouds. Naturally, this indicator is especially precious when there are a great number of groups. It must finally be noted that there is a test of significance associated with the RV coefficient (Josse *et al.*, 2008). In the example, the $p$ value associated to the RV is $7 \times 10^{-4}$, a value given here simply as an example, as there is no longer any doubt about the non-random nature of the similarity of the two wine clouds.

## CONSTRUCTING CONFIDENCE ELLIPSES AROUND THE WINES

When faced with principal component representations, users always wonder about the validity of what they see. Intuitively, they know that the positions of the points should not be taken as read and that they are tainted with uncertainty. This uncertainty remains to be defined.

The descriptive sensory data present particularities that we can exploit with this in mind. One of the reasons for uncertainty is the composition of the panel. Question: To what extent can we consider that the observations are 'robust' in terms of the composition of the panel? One way of answering this question is as follows (Husson *et al.* 2005a, 2005b; Pagès and Husson 2005):

(i) Construct a pair of virtual panels (one for the 'expert' panel and one for the 'student' panel) by drawing judges in each real panel with replacement.
(ii) Calculate the mean table associated with each of these panels (the table is thus constructed in the same format as the real table: the rows in this table are the products as 'seen' by the two virtual panels).
(iii) Project the products seen by this pair of virtual panels on the map.
(iv) Reiterate this procedure many times (500 times to be sure); for each product on the map we therefore obtain 500 representations 'around' the true representation.
(v) For each product construct an ellipse containing 95% of these representations.

This approach can also be applied both to mean points (Figure 6.8) and partial points (Figure 6.9).

In the example, the representation of ellipses around mean points validates the comments made until now. Overall, the wines are well separated. Furthermore the three most typical sauvignons

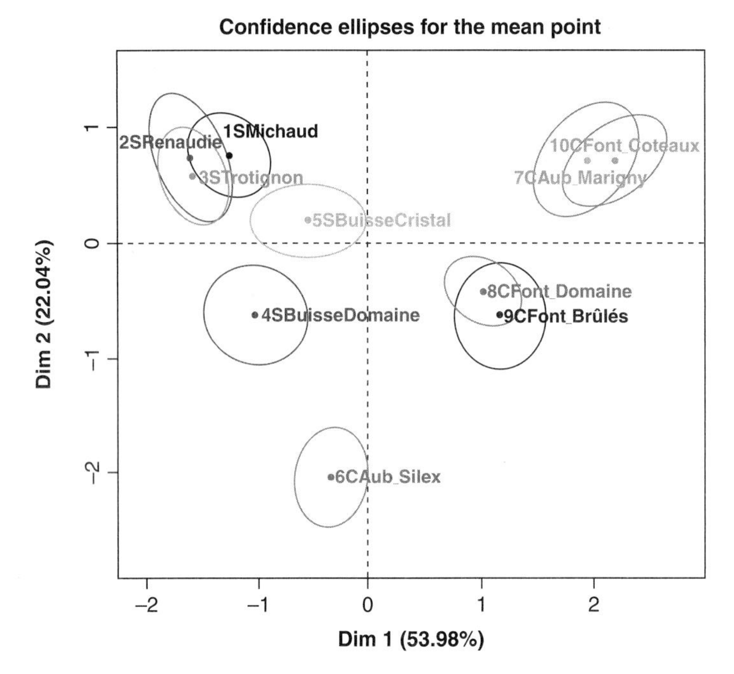


**Figure 6.8** Representation of confidence ellipses associated with mean points of Figure 6.2. (See colour version of this figure in colour plate section.)

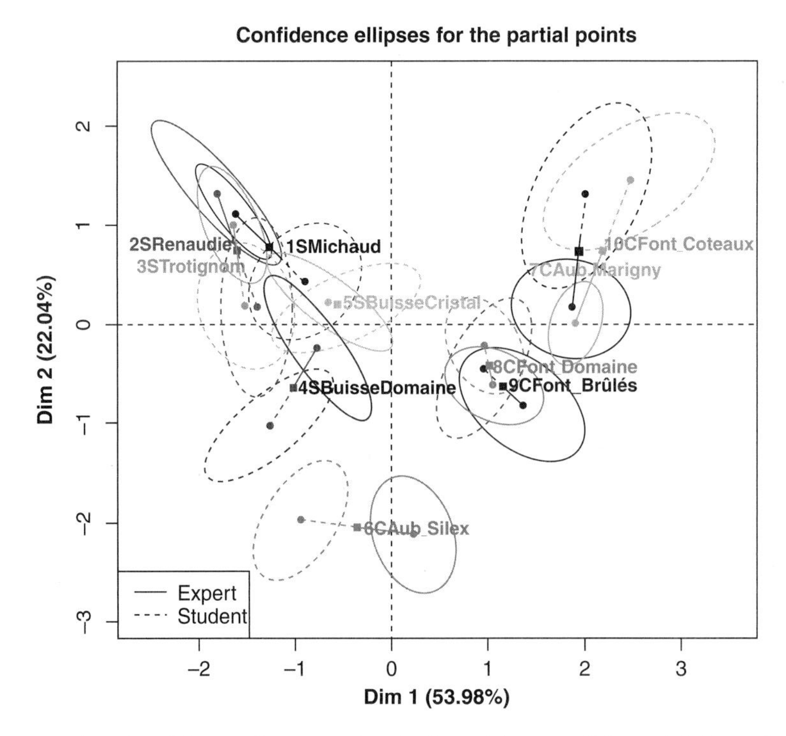


**Figure 6.9** Representation of confidence ellipses associated with partial points of Figure 6.4. (See colour version of this figure in colour plate section.)

(wines 1, 2 and 3) do indeed have the same profile; the same applies for the two woody wines (7 and 10).

The representation of the ellipses around partial points adds important elements to support the previous interpretations. Thus, for wine 6, perceived as sweet, there is a marked difference in profile between the experts and the students, with only the experts drawing a similarity between this wine and the other Chenins. In the same way, wines 7 and 10 were judged to be 'significantly' more unusual by the students.

## CONCLUSIONS AND IMPLICATIONS FOR OTHER APPLICATIONS

MFA takes into account all of the aspects of the principal component method in a table of individuals × variables in which the variables are organized into groups: balance between the groups, representation of overall individuals and variables, representation of partial clouds and their principal components, and group representations. All of these aspects are studied in a unique framework: in concrete terms, the rank $s$ dimensions of the different representations have the same interpretation. This is one of the most precious characteristics of the method: users no longer need to start all over again when passing from one aspect of the problem to another.

The case of categorical variables was discussed for a supplementary variable. In fact, the problem of multiple tables, mentioned again in the previous paragraph, also holds true for categorical variables. Therefore, MFA can also be applied to active categorical variables by replacing the framework of the PCA with that of multiple correspondence analysis (MCA). In particular, in the weighting of variables, the first eigenvalue of the separate PCAs becomes that of the separate MCAs. This weighting thus balances out the quantitative and categorical groups in the same way, and thus the two types of groups can be introduced simultaneously as active in one analysis. The result is, therefore, a method with a wide range of applications.

To illustrate this field of applications, three other examples of MFA in sensory analysis are:

*Sensory Description and Consumers' Hedonic Judgements*. Bringing together these two types of data is a major problem in sensory analysis. In a products × descriptors table, we can juxtapose the products × consumers table (which, at the intersections of row $i$ and column $k$, include the hedonic score given to product $i$ by consumer $k$). When analysing this data, we must balance the influence of the descriptive data on the one hand and the hedonic judgements on the other.

*Napping* (Pagès, 2005; Ares *et al.*, 2010). In this data gathering mode, each judge positions the products on a tablecloth as follows: the more similar two products are judged to be from a sensory point of view, the closer they are on the tablecloth. The data from one judge is gathered in a table with the products in the rows and the coordinates on the tablecloth in the columns. The data from all of the judges are juxtaposed in a table with the same number of blocks of two columns as there are judges. When analysing this data the influence of the different judges must be balanced out.

*Preference and Uses*. In a classic product test, the questionnaire is often made up of two parts that the user wants to connect: hedonic judgements and questions about consumer habits. These data are brought together in a table with the consumers in the rows and the questionnaire variables in the columns. These are sub-divided into two groups: hedonic scores and scores for consumer habits. It must be noted that the first group is made up of quantitative variables and the second is generally composed of categorical variables. This makes the issue of balancing the two groups of variables even more pertinent.

# APPENDIX 6.A SOFTWARE AND TECHNICAL POINT

## FactoMineR and SensoMineR

The FactoMineR package deals with multivariate exploratory data analysis. It is used to implement classic methods such as Principal Component Analysis, Correspondence Analysis, Multiple Correspondence Analysis, hierarchical clustering or Multiple Factor Analysis, as well as some advanced methods such as Hierarchical Multiple Factor Analysis, Mixed Data Analysis, Dual Multiple Factor Analysis. Many tools, indicators and graphical outputs are available to better interpret the results of each method. A website dedicated to this package presents the methods through examples: http://factominer.free.fr. A Graphical User Interface is also available.

The SensoMineR package of the free software R has been designed specifically for the statistical processing of sensory data. Sensometrics is a dynamic field of research, as proved by the great number of conferences focusing on the discipline (Pangborn, Sensometrics, Agrostat) where each year new methods are revealed and made available to practitioners. SensoMineR brings together many methods that are traditionally used to process sensory data with recently developed methods. It provides many graphical outputs which are easy to interpret, as well as summary tables resulting from different analysis of variance models or different principal component methods with confidence ellipses. SensoMineR is based on the main problems encountered by those practicing sensory analysis: characterizing products, relating sensory data with instrumental data, mapping preferences, evaluating a panel's performance, comparing the performance of multiple panels, characterizing products with holistic methods such as napping and categorization. The methods available in the package authorize different options but the most commonly used strategies are implemented by default; a description of these methods along with their outputs can be found at http://sensominer.free.fr.

The outputs presented in this chapter correspond to the MFA function of the FactoMineR package and the SensoMineR outputs for the confidence ellipses.

## Transition relations

*In PCA*

**Denotation**

$F_s(i)$: the score or the coordinate of individual $i$ along the principal component of rank $s$
$\lambda_s$: the inertia associated with the dimension of rank $s$
$G_s(k)$: the coordinate of variable $k$ along the dimension of rank $s$
$x_{ik}$: the value of individual $i$ for variable $k$. This value is always centred and may be standardized in standardized PCA

The transition relation from the variables' space towards that of the individuals is expressed:

$$F_s(i) = \frac{1}{\sqrt{\lambda_s}} \sum_k x_{ik} G_s(k)$$

This specifies that an individual is on the side of variables for which it has high values and opposite variables for which it has low values.

*In MFA*

We retain the denotations of the PCA. In addition we denote:

$K_j$: set of variables of group $K_j$
$J$: number of groups
$\lambda_1^j$: first eigenvalue of the separate PCA for group $j$

The PCA's relationship is directly transcribed to the MFA:

$$F_s(i) = \frac{1}{\sqrt{\lambda_s}} \sum_j \sum_{k \in K_j} \frac{1}{\lambda_1^j} x_{ik} G_s(k)$$

This relationship is interpreted in the same way as that of the PCA. The only difference is that the variables in group $j$ are weighted by $1/\lambda_1^j$, thus ensuring that the groups are balanced.

By restricting this relationship to the variables in group $j$, we obtain the coordinate (on the dimension of rank $s$) of individual $i$ as seen by group. Thus:

$$F_s\left(i^j\right) = \frac{1}{\sqrt{\lambda_s}} \frac{1}{\lambda_1^j} \sum_{k \in K_j} x_{ik} G_s(k)$$

The position of a partial individual with regards to the variables of group $j$ is thus interpreted in the usual PCA of group $j$.

In addition:

$$F_s(i) = \sum_j F_s\left(i^j\right) = \frac{1}{J} \sum_j J F_s\left(i^j\right)$$

A 'total' (or 'mean') individual is at the barycentre of its partial images (as long as the partial clouds are dilated with coefficient $J$: this is done by the software).

## REFERENCES


Ares, G., Deliza, R., Barreiro, C. *et al.* (2010) Comparison of two sensory profiling techniques based on consumer perception. *Food Quality and Preference* **21**, 417–426.

Escofier, B. and Pagès, J. (2008) Analyses factorielles simples et multiples; objectifs, méthodes et interprétation, 4th edn. Dunod, Paris, France.

Husson, F., Lê, S. and Pagès J. (2010) *Exploratory Multivariate Analysis by Example Using R*. CRC Press, Boca Raton, FL.

Husson, F., Le Dien, S. and Pagès, J. (2005a) Confidence ellipse for the sensory profiles obtained by Principal Component Analysis. *Food Quality and Preference* **16**, 245–250.

Husson, F., Bocquet, V. and Pagès, J. (2005b) Use of confidence ellipses in a PCA applied to sensory analysis: Application to the comparison of monovarietal ciders. *Journal of Sensory Studies* **19**, 510–518.

Josse, J., Pagès, J. and Husson, F. (2008) Testing the significance of the RV coefficient *Computational Statistics and Data Analysis* **53**, 82–91.

Pagès, J. (2005) Collection and analysis of perceived product inter-distances using multiple factor analysis; application to the study of ten white wines from the Loire Valley. *Food Quality and Preference* **16**, 642–649.

Pagès, J. and Husson, F. (2005) Multiple factor analysis with confidence ellipses: a methodology to study the relationships between sensory and instrumental data. *Journal of Chemometrics* **19**, 1–7.

Pagès, J., Bertrand, C., Ali, R., Husson, F. and Lê, S. (2007) Compared sensory analysis of eight biscuits by French and Pakistani panels. *Journal of Sensory Studies* **22**, 665–686.

# 7 Cluster analysis: Application in food science and technology

**Gastón Ares**
*Department of Food Science and Technology, Facultad de Química, Universidad de la República, Montevideo, Uruguay*

## ABSTRACT

Cluster analysis is a generic name for a wide range of exploratory multivariate techniques. It is used for classifying a data set in homogeneous groups in an unsupervised situation where little or no a priori information about group structure is available. This chapter focuses on the three most commonly used techniques in food science and technology (hierarchical cluster analysis, k-means and fuzzy clustering), discusses the theory behind the methods and presents examples of application and implementation in R free statistical software.

## INTRODUCTION

Cluster analysis is a generic name for a wide range of exploratory multivariate statistical procedures which aim at identifying homogeneous groups within a data set. Clustering is basically a procedure that enables the classification of objects or variables, according to their similarities and differences, taking into consideration previously assigned characteristics. Cluster analysis is usually referred to as unsupervised classification (Mardia *et al.*, 1979) due to the fact that objects or variables are classified without the use of prior information. This type of procedure is one of the most common operations in thinking and plays a relevant role in our understanding and description of the world (Coxon, 1999), being a fundamental practice in science.

Cluster analysis has several applications in food science and technology, which include the following:

- *Classification of products*: Cluster analysis could be used to classify a group of samples according to a wide range of characteristics. For example Cruz *et al.* (2013) applied cluster analysis to discriminate low and full fat yogurts using instrumental measurements (pH, instrumental firmness and instrumental colour). Lee *et al.* (2012) used this technique to classify different rice cultivars based on their hydration and pasting properties, whereas da Silva Torres *et al.* (2006) categorized foods according to their composition (moisture, ashes, lipids, proteins, dietary fibre, carbohydrates and calories).
- *Analysis of instrumental measurements*: Cluster analysis can be used to better analyse large data sets from instrumental measurements. De Luca *et al.* (2011) applied cluster analysis on Fourier transformed infrared spectra for differentiation and classification of olive oils from several producing regions of Morocco.

*Mathematical and Statistical Methods in Food Science and Technology*, First Edition.
Edited by Daniel Granato and Gastón Ares.

- *Characterization of bacteria strains*: Genetic information could be used to classify bacteria strains via cluster analysis, as performed by Oomes *et al.* (2007) when studying *Bacillus* spores occurring in the manufacturing of canned products.
- *Identification of consumer segments*: Cluster analysis is commonly used to identify groups of consumers with different preference patterns based on their liking of a set of samples, which enables food products to be developed for specific consumer segments (Punj and Stewart, 1983; Yenket *et al.*, 2011)
- *Identification of dietary patterns*: Wirfalt and Jeffery (1997) applied cluster analysis on food frequency consumption data of 38 food groups to identify consumers with different dietary patterns.

### Types of cluster analysis

There are several methodologies for cluster analysis, which can be divided according to different criteria (Jacobsen and Gunderson, 1986):

- *Hierarchical versus partitional*: The most frequent distinction between clustering techniques is whether the clusters come from a nested structure (hierarchical) or not (partitional).
- *Exclusive versus overlapping versus fuzzy*: Exclusive clustering techniques assign each object to a single cluster, whereas in overlapping or nonexclusive clustering an object can simultaneously belong to more than one group or cluster. Finally, in fuzzy clustering every object belongs to every cluster with a membership weight which ranges from 0 to 1.
- *Complete versus partial clustering*: Complete clustering assigns every object to a cluster, whereas in partial clustering some objects may not belong to any group.

This chapter does not cover all types of cluster analysis and focuses on the most commonly used methodologies in food science and technology: hierarchical cluster analysis, k-means and fuzzy clustering.

## HIERARCHICAL CLUSTER ANALYSIS

The most popular type of cluster analysis in food science and technology is agglomerative hierarchical cluster analysis, which aims at identifying a series of clusters within a nested structure (Jacobsen and Gunderson, 1986).

This technique assumes a hierarchical structure in the data set. It starts with each object as a separate cluster and then merges the two closest clusters sequentially until only one cluster is left (Burns and Burns, 2009). The basic steps of an agglomerative hierarchical cluster analysis are the following (Tan *et al.*, 2005):

(i) Calculate a distance matrix which includes the distance between all the clusters.
(ii) Merge the two closest clusters.
(iii) Update the distance matrix to include the distance between the new cluster and the original ones, considering a clustering procedure.
(iv) Repeat the procedures above until only one cluster is left.

### Calculating the distance between objects and clusters

The first step of the analysis is to calculate the degree of similarity or dissimilarity between each pair of objects. Objects are considered as points in a coordinate space such as that their degree of difference is estimated by calculating the distance between them (Beebe *et al.*, 1998). Distances can be calculated by different mathematical approaches; being Euclidean and Manhattan distances the most frequently used.

The Euclidean distance between two objects is calculated as the geometric distance, using an extension of Pythagoras' theorem:

$$d_{ij} = \sqrt{\sum_{k=1}^{K} (x_{ik} - x_{jk})^2} \tag{7.1}$$

where $d_{ij}$ is the Euclidean distance between clusters i and j, $x_{ik}$ is the value of variable $x_k$ for cluster i, $x_{jk}$ is the value of variable $x_k$ for cluster j and K is the total number of variables in the data set. This type of distance is by far the most popular (Tan *et al.*, 2005).

On the other hand, Manhattan distance is commonly used when working with qualitative or binary data and is calculated as the sum of the lengths of the projections of the line segment between the points onto the coordinate axes:

$$d_{ij} = \sum_{k=1}^{K} |x_{ik} - x_{jk}| \tag{7.2}$$

where $d_{ij}$ is the Manhattan distance between clusters i and j, $x_{ik}$ is the value of variable $x_k$ for cluster i, $x_{jk}$ is the value of variable $x_k$ for cluster j and K is the total number of variables in the data set.

Figure 7.1 shows the difference between Euclidean and Manhattan distances between two points in a bi-dimensional space. In this approach the largest the distance, the most different the objects are.

The distance between objects can also be estimated using similarity measures, such as Pearson or Spearman correlation coefficient. In this case, the distance between objects increases with their degree of similarity.

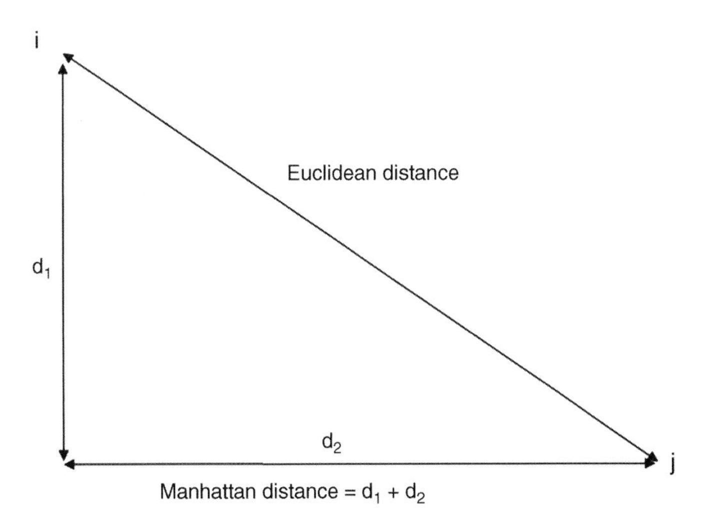


**Figure 7.1** Euclidean and Manhattan distances between two points (i and j) in a bi-dimensional space.

## Clustering procedure or amalgamation rule

Agglomerative hierarchical clustering starts with every object as a cluster. Then, the two clusters with the smallest distance are merged into a new cluster. Once the clusters have been merged, the next step is to calculate the distance between the new and the original clusters using a clustering procedure or amalgamation rule (Beebe *et al.*, 1998). This can be carried out in different ways; the most common amalgamation rules are: single linkage, complete linkage, group average and Ward's method (Tan *et al.*, 2005).

In the single linkage or nearest neighbour approach the distance between two clusters is calculated as the smallest distance between two objects that belong to different clusters. This approach is useful for identifying outliers but has the main disadvantage of being sensitive to a chaining effect. This occurs when clusters are merged due to single objects being close, even if the majority of the objects in each cluster may be very distant to each other (Næs *et al.*, 2010).

On the contrary, when complete linkage or furthest neighbour criterion is used the distance between two clusters is calculated as the distance between the furthest two objects in different clusters. It usually produces compact clusters but it has problems when the cluster structure is linear or when no clearly distinct clusters exist (Mardia *et al.*, 1979).

A third alternative, the average linkage, calculates the distance between two clusters as the average distance of all pairs of objects from different clusters. It is one of the most popular clustering procedures in food science and tends to create ball-shaped clusters with similar variance (Beebe *et al.*, 1998).

Figure 7.2 shows an example of the three distances between two clusters in a simple bi-dimensional space.

The fourth method, Ward's aggregation criterion, assumes that each cluster is represented by its centroid, that is the 'average' of all objects in the clusters. Cluster membership is determined using the total sum of squared distances of objects from centroid of the cluster (Burns and Burns, 2009). The criterion for merging two clusters is that the increase in the error sum of squares is minimized. This method provides small and even sized clusters and is not sensitive to chaining effects; it is a popular alternative in food science and technology (Jacobsen and Gunderson, 1986).

Clusters are sequentially merged according to their similarity until only one cluster is left, that is all the objects belong to the same group.

## Construction of the dendrogram

Results of agglomerative hierarchical clustering are represented graphically in a tree-like structure, which is called dendrogram (Næs *et al.*, 2010). The dendrogram provides a simple way of visualizing the hierarchical structure of the clustering and the level at which each cluster is formed, as well as cluster

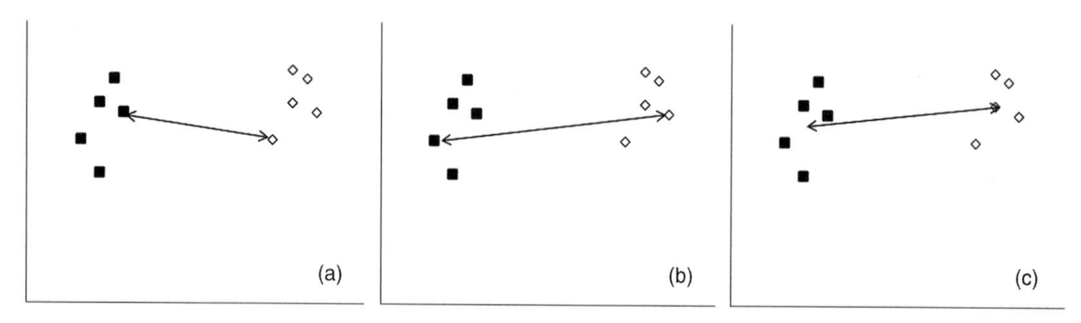


**Figure 7.2** Distance between two clusters in a bi-dimensional space according to three criteria: (a) single linkage, (b) complete linkage, and (c) average distance.

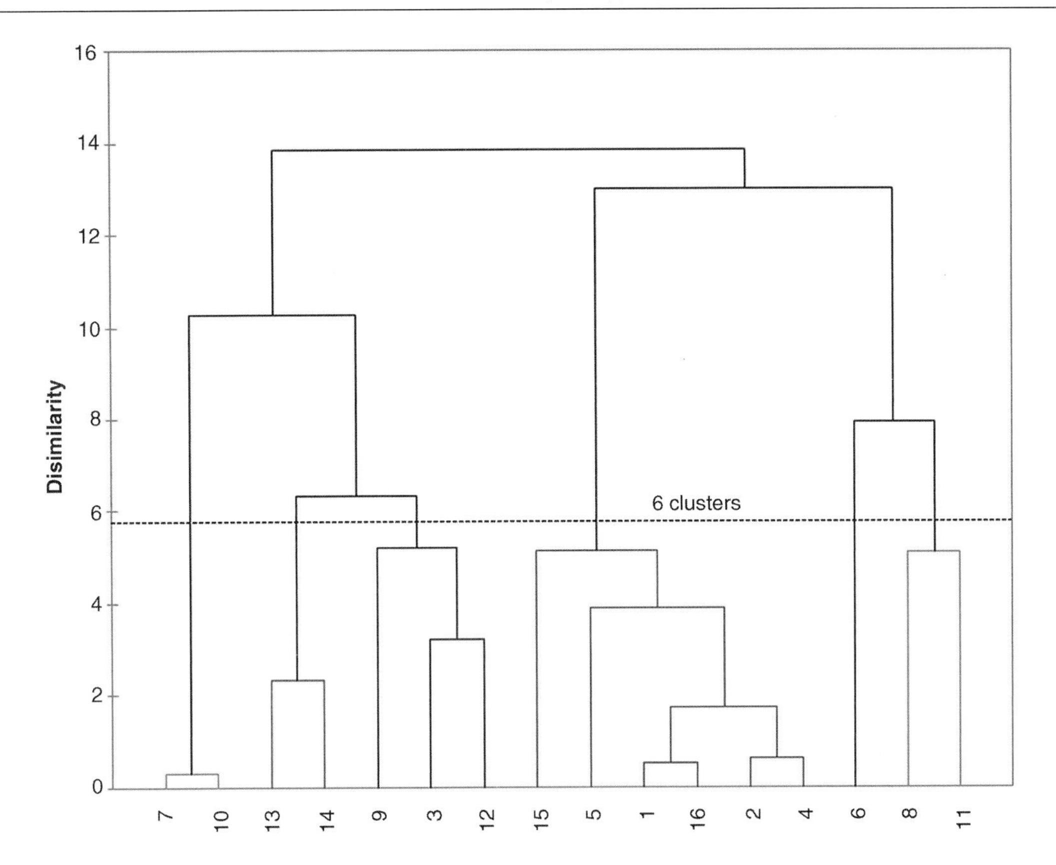


**Figure 7.3** Example of a dendrogram of 16 objects.

membership. A typical dendrogram of 16 objects is shown in Figure 7.3. Objects are represented as equally spaced points in the horizontal axe, in an order that enables a clear representation of the hierarchical structure of the clusters. The vertical axe represents the distance at which clusters are merged.

As shown in Figure 7.3, at the beginning of the procedure all objects are considered separate clusters. Then, objects 7 and 10 were identified as the most similar objects in the data set and were merged at a dissimilarity distance of 0.3, forming the first cluster. Then, objects 1 and 16 were merged at a distance of 0.5 and objects 2 and 4 at a distance of 0.6. These last clusters were similar and were merged in the fourth step of the procedure at a distance of 1.7. The process continues until all objects are merged in a single cluster, at a distance of 13.9. It is important to highlight that clearly different objects are identified by long vertical line segments before being merged. In the example depicted in Figure 7.3, objects 7, 10, 13, 14, 9, 3 and 12 are markedly different from objects 15, 5, 1, 16, 2, 4, 6, 8 and 11.

In many cases it is interesting to stop the procedure at a fixed number of clusters (Næs *et al.*, 2010). In the example of Figure 7.3, clustering was stopped to identify six separate clusters: Cluster 1 (objects 7 and 10), Cluster 2 (objects 13 and 14), Cluster 3 (objects 9, 3 and 12), Cluster 4 (objects 15, 1, 16, 2 and 4), Cluster 5 (object 6) and Cluster 6 (objects 8 and 11).

Hierarchical clustering methods are useful for identifying outliers in the data set. Outliers are objects that are very different from the rest of the data set and, therefore, will have long vertical line segments before being merged to the rest of the objects (Næs *et al.*, 2010). As shown in Figure 7.4, an outlier (object 4) is identified in a dendrogram as a separate cluster which is merged to the rest of the objects at the top level.

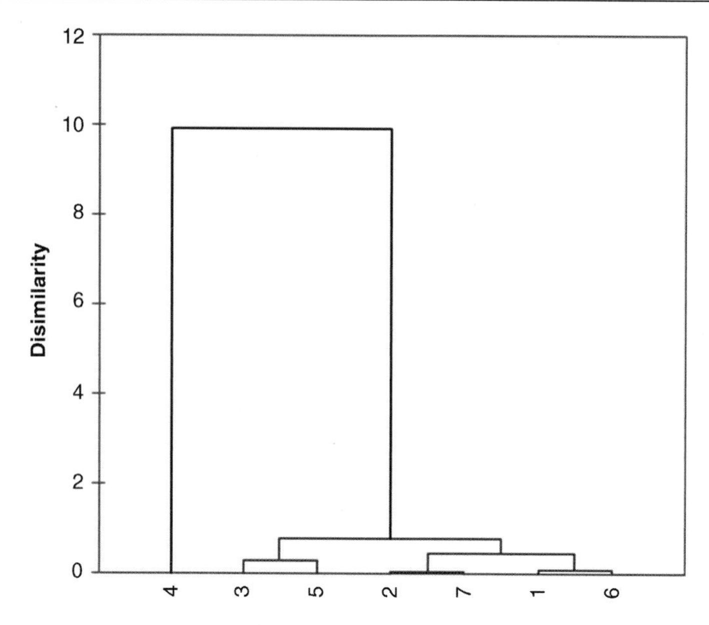


**Figure 7.4** Example of a dendrogram in which an outlier (object 4) can be identified.

## Examples

*Classification of milk desserts based on their rheological properties using hierarchical cluster analysis*

Table 7.1 shows results from a rheological characterization of 14 milk desserts using two different tests. Three variables were obtained from a dynamic stress sweep: stress at the end of the linear viscoelastic region ($\sigma_{LVE}$), critical stress ($\sigma_c$) and storage module ($G'_C$) at the crossover-point between storage ($G'$)

**Table 7.1** Results from a rheological characterization of fourteen milk desserts using dynamic stress sweeps and flow curves.

| Sample | $G'_c$ (Pa) | $\sigma_c$ (Pa) | $\sigma_{LVE}$ (Pa) | K | n |
|---|---|---|---|---|---|
| 1 | 42.6 | 227.2 | 12.3 | 60.2 | 0.319 |
| 2 | 85.1 | 608.6 | 30.3 | 103.8 | 0.323 |
| 3 | 83.3 | 660.0 | 35.2 | 138.9 | 0.321 |
| 4 | 19.4 | 49.7 | 3.0 | 21.5 | 0.343 |
| 5 | 113.5 | 721.8 | 38.8 | 154.3 | 0.303 |
| 6 | 93.2 | 257.0 | 11.6 | 132.7 | 0.312 |
| 7 | 59.0 | 350.9 | 16.0 | 97.2 | 0.395 |
| 8 | 77.3 | 518.6 | 23.3 | 146.1 | 0.337 |
| 9 | 24.4 | 52.1 | 6.4 | 23.5 | 0.373 |
| 10 | 45.2 | 177.9 | 8.4 | 72.6 | 0.378 |
| 11 | 137.4 | 602.5 | 26.0 | 200.3 | 0.332 |
| 12 | 56.2 | 137.0 | 2.8 | 54.4 | 0.364 |
| 13 | 36.7 | 132.5 | 3.2 | 48.6 | 0.307 |
| 14 | 13.4 | 52.7 | 1.0 | 15.8 | 0.421 |

Description of the variables: stress at the end of the linear viscoelastic region ($\sigma_{LVE}$); critical stress ($\sigma_c$) and storage module ($G'_C$) at crossover-point between storage ($G'$) and loss ($G''$) modulus; consistency index (K) and flow behaviour index (n) from the flow curve adjustment to the power-law model.

**Table 7.2** Distance matrix calculated from standardized data of Table 7.1 considering Euclidean distances.

| **Sample** | **Sample** | | | | | | | | | | | | | |
|---|---|---|---|---|---|---|---|---|---|---|---|---|---|---|
| | **1** | **2** | **3** | **4** | **5** | **6** | **7** | **8** | **9** | **10** | **11** | **12** | **13** | **14** |
| 1 | 0 | 0.199 | 0.151 | 0.498 | 0.113 | 0.600 | 0.056 | 0.097 | 0.613 | 0.320 | 0.165 | 0.538 | 0.292 | 0.191 |
| 2 | 0.199 | 0 | 0.098 | 0.675 | 0.088 | 0.793 | 0.226 | 0.240 | 0.771 | 0.518 | 0.362 | 0.696 | 0.481 | 0.354 |
| 3 | 0.151 | 0.098 | 0 | 0.647 | 0.065 | 0.743 | 0.157 | 0.158 | 0.756 | 0.457 | 0.310 | 0.683 | 0.440 | 0.331 |
| 4 | 0.498 | 0.675 | 0.647 | 0 | 0.601 | 0.215 | 0.508 | 0.535 | 0.180 | 0.284 | 0.355 | 0.119 | 0.232 | 0.339 |
| 5 | 0.113 | 0.088 | 0.065 | 0.601 | 0 | 0.711 | 0.140 | 0.159 | 0.705 | 0.432 | 0.278 | 0.631 | 0.401 | 0.283 |
| 6 | 0.600 | 0.793 | 0.743 | 0.215 | 0.711 | 0 | 0.590 | 0.603 | 0.331 | 0.296 | 0.437 | 0.305 | 0.323 | 0.466 |
| 7 | 0.056 | 0.226 | 0.157 | 0.508 | 0.140 | 0.590 | 0 | 0.044 | 0.633 | 0.301 | 0.158 | 0.555 | 0.291 | 0.209 |
| 8 | 0.097 | 0.240 | 0.158 | 0.535 | 0.159 | 0.603 | **0.044** | 0 | 0.663 | 0.309 | 0.181 | 0.587 | 0.315 | 0.245 |
| 9 | 0.613 | 0.771 | 0.756 | 0.180 | 0.705 | 0.331 | 0.633 | 0.663 | 0 | 0.440 | 0.493 | 0.193 | 0.390 | 0.470 |
| 10 | 0.320 | 0.518 | 0.457 | 0.284 | 0.432 | 0.296 | 0.301 | 0.309 | 0.440 | 0 | 0.159 | 0.371 | 0.122 | 0.238 |
| 11 | 0.165 | 0.362 | 0.310 | 0.355 | 0.278 | 0.437 | 0.158 | 0.181 | 0.493 | 0.159 | 0 | 0.411 | 0.137 | 0.120 |
| 12 | 0.538 | 0.696 | 0.683 | 0.119 | 0.631 | 0.305 | 0.555 | 0.587 | 0.193 | 0.371 | 0.411 | 0 | 0.288 | 0.360 |
| 13 | 0.292 | 0.481 | 0.440 | 0.232 | 0.401 | 0.323 | 0.291 | 0.315 | 0.390 | 0.122 | 0.137 | 0.288 | 0 | 0.146 |
| 14 | 0.191 | 0.354 | 0.331 | 0.339 | 0.283 | 0.466 | 0.209 | 0.245 | 0.470 | 0.238 | 0.120 | 0.360 | 0.146 | 0 |

The minimum distance between two different objects is highlighted in bold.

and loss (G″) modulus. Flow curves were also obtained and a consistency index (K) and flow behaviour index (n) from the flow curve adjustment to the power-law model were calculated. Details from the measurements can be found in Ares *et al.* (2012).

The first step for performing hierarchical cluster analysis is selecting the type of distance and calculating the distance matrix between objects. This can be done with or without standardizing the data. A usual criterion is to standardize the data (centring the variables to mean zero and scaling to unit variance) when dealing with variables measured in different units (Jacobsen and Gunderson, 1986). Table 7.2 shows the distance matrix calculated using Euclidean distances on standardized data from Table 7.1. The distance between samples 7 and 8 is the smallest and, therefore, they should be merged in a single cluster in the first step of the clustering procedure.

Figure 7.5 shows dendrograms obtained using the four most common aggregation rules: single linkage, complete linkage, average distance and Ward's method. As shown, the hierarchical structure differed according to the clustering procedure selected. In this sense it is important to stress that the solution of a cluster analysis is not unique and that the selection of the type of distance and aggregation rule strongly influence the results. Therefore, it is important to carefully select the type of cluster and to rely on theoretical or a priori information to select the best solution.

## *Other applications*

Dziuba *et al.* (2007) applied Fourier-transformed infrared (FTIR) spectroscopy and hierarchical cluster analysis to identify lactic acid bacteria. The analysis was performed on FTIR spectra and their first derivatives for different bacterial species of the genus *Lactobacillus*. This procedure enabled the identification of homogeneous clusters formed by spectra of *L. hilgardii, L. helveticus, L. amylophorus, L. brevis, L. fermentum, L. casei* and *L. rhamnosus*.

Patras *et al.* (2011) used hierarchical cluster analysis to classify fruits and vegetables commonly consumed in Ireland based on *in vitro* antioxidant activity. Four main groups with different antioxidant activity were identified. The first group of samples was composed of berries (blueberry, cranberry, raspberry and strawberry), which were characterized by their high antioxidant activity, phenolic content and anthocyanins. The second group included red peppers, due to their high ascorbic acid content. The

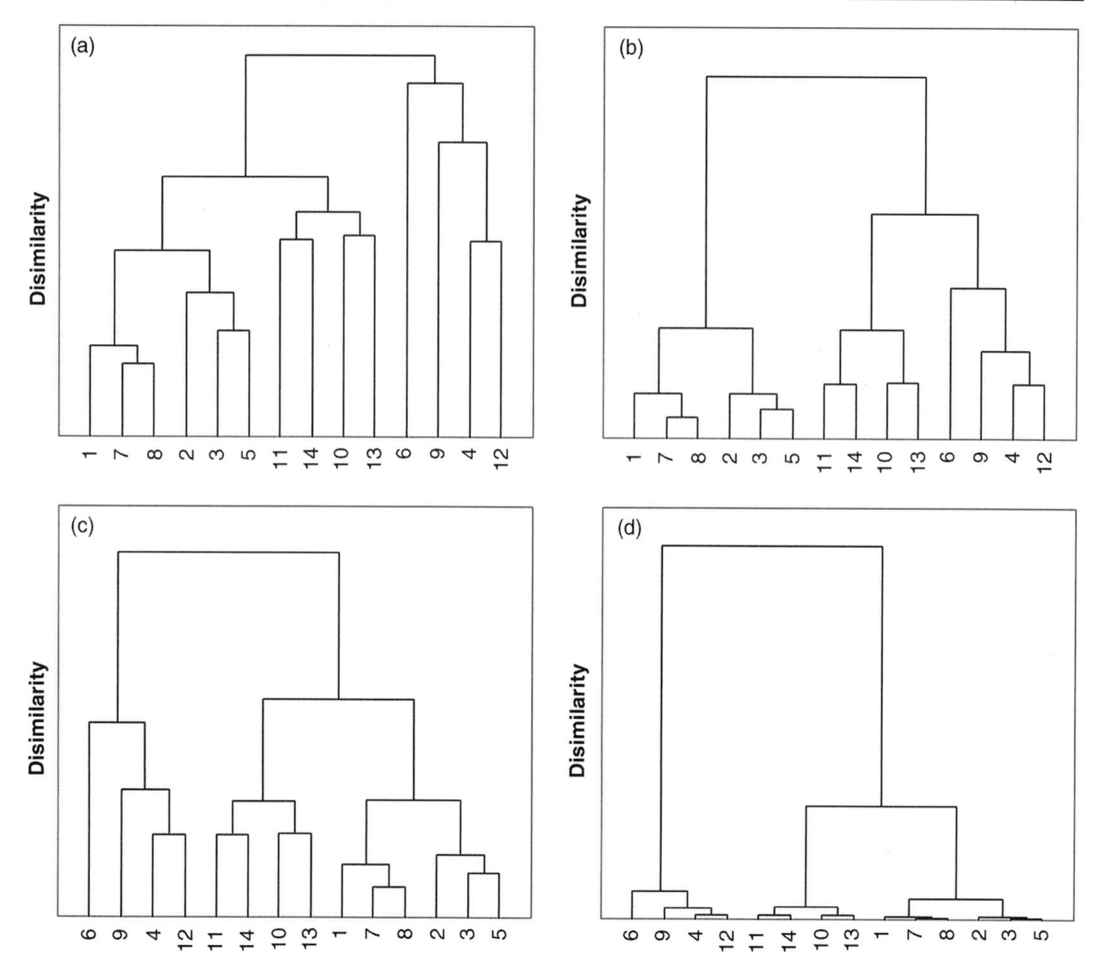


**Figure 7.5** Dendrograms obtained by applying different hierarchical aggregation rules to data from Table 7.2: (a) single linkage; (b) complete linkage; (c) average distance; (d) Ward's method.

third group consisted of broccoli, cherry tomato, lettuce and green pepper, while the last cluster was composed of carrots, yellow and red onion, and grapes.

da Silva Torres *et al.* (2006) applied hierarchical cluster analysis to study the composition of 54 food products sold in four different restaurants based on their nutritive values. Moisture, ash, lipid, protein, total carbohydrates and total dietary fibre content, as well as energy value, were determined for a total 18 main dishes, 20 salads, 10 side dishes and 5 desserts. Considering one of the restaurants the authors identified three main groups of food products based on their moisture content. The first group was composed of arugula, mixed salad, bean sprout salad, cooked carrot salad, chayote salad and heart-of-palm salad. This group was subdivided into two groups, one of which was composed of heart-of-palm salad alone because of its high ash content. The second group was composed of food products with intermediate moisture content (black bean and meat stew, beans and white bean salad), while the third group had the lowest moisture content and was composed of French fries, vegetable mayonnaise salad, fried zucchini milanese and rice.

## Implementation in R

Hierarchical cluster analysis can be easily performed in the free statistical software R (R Development Core Team, 2007) using the function *hclust*. The R-script shown in Box 7.1 performs hierarchical cluster analysis based on Euclidean distances on a given data set.

**Box 7.1 R-script performing hierarchical cluster analysis based on Euclidean distances**

```
#auxiliary routine to standardize the data (optional)

stand <- scale(Dataset)

#calculates a distance matrix (d) on the standardized dataset using Euclidean
distances

d <- dist(stand, method = "euclidean")

#performs hierarchical cluster analysis using Ward's method as aggregation
rule. Alternatively "single", "complete" or "average" could be used

Cluster <- hclust(d, method="ward")

#shows the dendrogram

plot(Cluster)

#If the objective is to identify a predetermined number of clusters, the
following commands could be used to visualize cluster membership. In this
example the number of clusters (k) is 2

groups <- cutree(Cluster, k=2)

rect.hclust(Cluster, k=2, border="red")

groups
```

The full description of the *hclust* function can be found at http://stat.ethz.ch/R-manual/R-patched/library/stats/html/hclust.html.

# K-MEANS CLUSTERING

K-means is the most popular partitioning clustering method. It aims at partitioning the data set into *k* different clusters of greatest possible difference between them (Burns and Burns, 2009). Initially, according to the number of desired clusters, *k* centroids (i.e. clusters' average points) are randomly selected, which define *k* initial clusters. The distance between each object and the *k* centroids is calculated, commonly using Euclidean or Manhattan distances. Considering these distances, each object is assigned to the closest centroid and all the objects associated to a centroid compose a cluster. Then, the centroid of each cluster is recalculated based on the objects assigned to it. The assignment of the objects to the centroids and an update of the centroids are repeated until they remain stable, according to a predetermined criterion (Tan *et al.*, 2005). This algorithm is illustrated in Figure 7.6 using a simple bi-dimensional example.

When Euclidean distances are used, the k-means algorithm uses the sum of the squared error (SSE) to select the best solution (Tan *et al.*, 2005), which is calculated as the total sum of the squared Euclidean

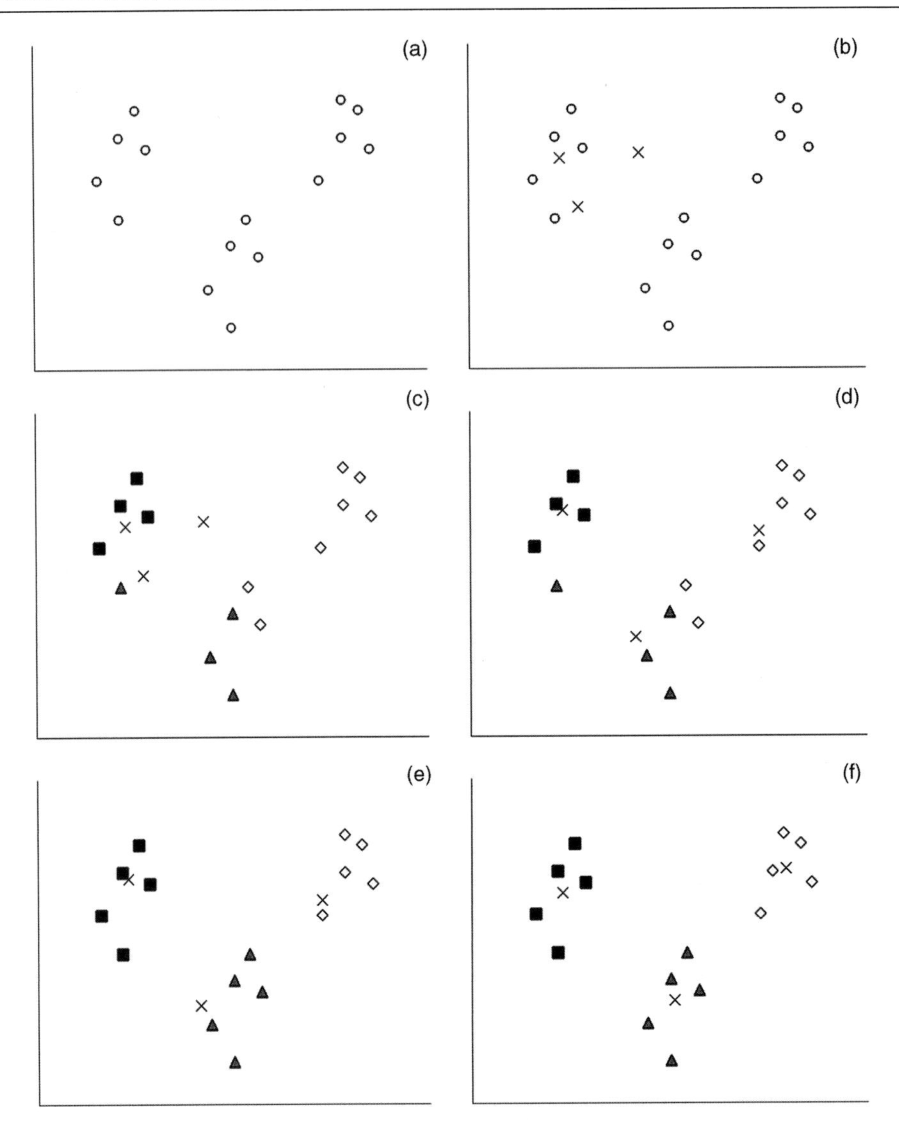


**Figure 7.6** Identification of three clusters using K-means clustering procedure in a simple bi-dimensional example: (a) original data set; (b) initial three randomly selected centroids; (c) assignments of the objects in the data set to the centroids to create three clusters; (d) update of the centroids; (e) reassignment of the objects to the updated centroids; (f) final clusters and centroids.

distance between each object and its centroid. Given two sets of clusters the one with the smallest squared error is selected. The SSE is calculated as follows:

$$SSE = \sum_{i=1}^{k} \sum_{x \in C_i} dist(c_i, x)^2 \tag{7.3}$$

where $x$ is an object, $C_i$ is the i-th cluster, $c_i$ is the centroid of cluster $C_i$, *dist* is the Euclidean distance between an object and the centroid of its cluster, and k is the total number of clusters.

One of the major drawbacks of k-means clustering is that the number of clusters to be identified should be decided before the analysis. However, in the majority of the cases the number of clusters is unknown to the researcher. A common way of solving this problem is to use different number of

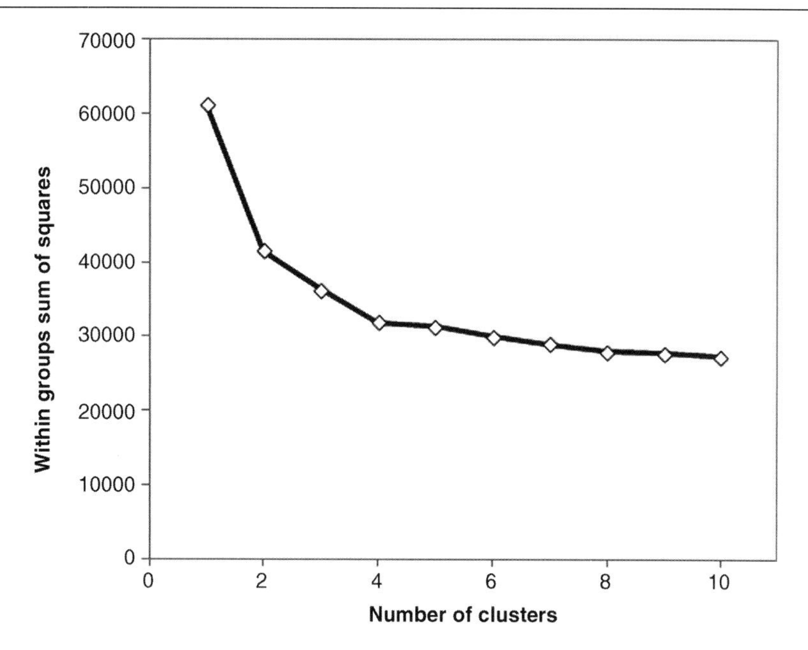


**Figure 7.7** Example of a plot of sum of squares as a function of the number of clusters for a k-means cluster procedure.

clusters and to visually evaluate the results considering the size and degree of difference of the clusters (Næs *et al.*, 2010). Sometimes hierarchical and k-means clustering techniques are used successively. Hierarchical cluster analysis is performed to get an idea of the number of clusters in the data set and then k-means clustering is applied to establish membership to those clusters (Burns and Burns, 2009).

There are other approaches that can be used to select the number of clusters in k-means based on statistical measures (Pham *et al.*, 2005). A common method is choosing the number of clusters based on the sum of squared error (SSE). As the number of clusters increases, the SSE should decrease due to the reduction in the size of the clusters. The optimum number of clusters could be estimated as the point at which the reduction in SSE markedly slows (Peeples, 2011). In the example shown in Figure 7.7 the reduction in SSE slows when the number of clusters is four, suggesting that solutions with more than four clusters do not have a large impact on SSE. However, there are many situations in which an obvious breaking point in the SSE plot does not exist.

## Examples

### *Identification of consumer segments with different preference patterns using k-means clustering*

Consumers do not have the same hedonic reaction to food products. However, groups of consumers usually like the same products and share the same preference pattern (MacFie, 2007). For this reason, cluster analysis is commonly applied in sensory consumer science to identify consumer segments with similar preference patters. In this context, it is usual to have a predefined idea of the number of clusters that should be identified for a given consumer sample. When working in industry, a usual criterion is to identify clusters with at least 40–50 consumers. Thus, the number of clusters is usually determined by dividing the total number of consumers by 50.

**Table 7.3** Example of a data set from a consumer study in which participants evaluated their overall liking of four cookie samples using a nine-point hedonic scale.

| Consumer | Overall liking score | | | |
|---|---|---|---|---|
| | **Sample A** | **Sample B** | **Sample C** | **Sample D** |
| 1 | 7 | 8 | 8 | 6 |
| 2 | 4 | 6 | 7 | 7 |
| 3 | 9 | 9 | 9 | 1 |
| 4 | 7 | 9 | 8 | 7 |
| ... | ... | ... | ... | ... |
| 110 | 7 | 7 | 8 | 4 |

Table 7.3 shows results from a consumer study with 110 consumers. Participants evaluated four cookies (Samples A to D) and rated their overall liking using a nine-point hedonic scale. By applying k-means clustering on standardized data from Table 7.3, two consumer segments could be identified: Cluster 1, composed of 66 consumers, and Cluster 2 with 44 consumers. Figure 7.8 shows average liking scores of the four cookies for the two clusters. As shown, the identified consumers segments clearly differed in their preference patterns: Cluster 1 preferred Samples B and C and rejected Sample D, whereas Cluster 2 liked samples A and D and clearly rejected Sample C. This information is important for food companies to identify target consumers of a specific product.

### *Identification of dietary patterns*

Stricker *et al.* (2013) applied k-means cluster analysis to identify dietary patterns based on food consumption data of 178 food items of 35 910 Dutch consumers. The authors identified two clusters:

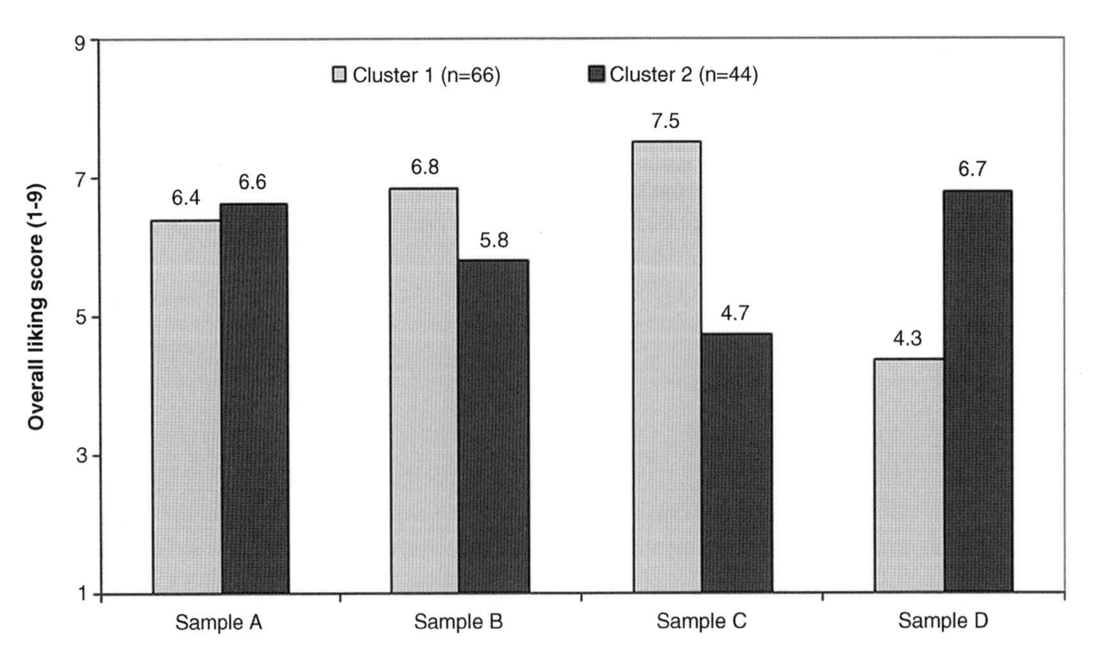


**Figure 7.8** Overall liking scores for four cookies of two consumer segments identified using k-means clustering on data from Table 7.3.

a prudent and a western cluster. The last cluster was characterized by high consumption of French fries, fast food, low-fibre products, soft drinks, sugar and also larger amounts of red meat, eggs, potatoes, fat, butter, sugar and sweets compared to the prudent cluster. Meanwhile, the prudent cluster had higher intakes of fish, high-fibre products, raw vegetables, wine, boiled vegetables and legumes, oils and margarine and low-fat dairy products and cheese than the other cluster.

## Implementation in R

K-means clustering can be performed in R (R Development Core Team, 2007) using the fuction *kmeans*. The R-script shown in Box 7.2 performs k-means clustering based on Euclidean distances on a given data set.

**Box 7.2 R-script performing k-means clustering based on Euclidean distances**

```
#auxiliary routine to standardize the data (optional)
Dataset.stand<-scale(Dataset)
#performs k-means clustering with k=2 clusters
cl <- kmeans(Dataset.stand, 2)
#shows the size of the clusters
cl$size
#shows the centroids of the clusters
cl$centers
#shows cluster membership of the original objects
cl$cluster
```

The full description of the *kmeans* function can be found at http://stat.ethz.ch/R-manual/R-patched/library/stats/html/kmeans.html.

# FUZZY CLUSTERING ALGORITHMS

Fuzzy clustering algorithms do not use the usual requirement that every object should belong to one and only one cluster. Instead, they consider that every object could simultaneously belong to more than one cluster with an associated degree of membership from 0 to 1 (0 = 'does not belong to' and 1 'belongs to'), with the condition that the membership values for an object for all clusters must sum up to 1 (Næs *et al*., 2010). According to Krishnapuram and Keller (1993) membership values can be interpreted as the probability of membership to the different clusters. Using the usual notation, the above requirements could be expressed as (Jacobsen and Gunderson, 1986):

$$0 \leq u_i(x_k) \leq 1 \tag{7.4}$$

$$\sum_{i=1}^{C} u_i(x_k) \leq 1 \tag{7.5}$$

where $x_k$ represents an object, $u_i(x_k)$ the degree of membership of object $x_k$ to cluster i, and C the total number of clusters.

As in any clustering technique, fuzzy clustering uses distances to determine cluster membership; Euclidean and Mahalanobis distances are the most common (Næs and Mevik, 1999). Using these distances fuzzy clustering algorithms partition the data set into clusters by an iterative solution of a system of simultaneous nonlinear equations (Jacobsen and Gunderson, 1986). There is a wide range of algorithms available (Rousseeuw *et al.*, 1996) but fuzzy k-means algorithm (also called c-means) is one of the most common in food science and technology (Næs and Mevik, 1999). This algorithm establishes cluster membership using the following criterion (Næs *et al.* 2010):

$$J = \sum_{i=1}^{C} \sum_{k=1}^{N} u_i(x_k)^m d_{ik}^2 \tag{7.6}$$

where $d_{ik}$ is the distance between object $x_k$ and the centroid of cluster i, and $m$ is called fuzzifier and takes a value larger or equal to 1 (Krishnapuram and Keller, 1993). The most usual value for $m$ is 2 (Næs *et al.*, 2010).

As in k-means clustering, fuzzy clustering requires deciding the number of clusters in the data set before the analysis. In the first iteration of the algorithm an initial guess of the position of the centroids is performed and the distance between each object in the data set and the centroids is calculated. The degree of membership of each object to each cluster is computed using the equation (Næs and Mevik, 1999):

$$u_i(x_k) = \frac{1}{\sum_{c=1}^{C} \left( \frac{d_{ik}}{d_{ck}} \right)^2} \tag{7.7}$$

After the degree of membership of each object to each cluster is determined, the new centroids are calculated using the following equation (Næs and Mevik, 1999):

$$c_i = \frac{\sum_{k=1}^{N} u_i(x_k)^2 . x_k}{\sum_{k=1}^{N} u_i(x_k)^2} \tag{7.8}$$

These new centroids are used to compute degree of membership using Equation 7.7, which are then used to calculate the centroids once again. This procedure is repeated until a stopping condition is reached, which usually is when the change in the position of the centroids between two consecutive iterations is lower than a predetermined threshold value (Jacobsen and Gunderson, 1986).

The main advantages of fuzzy clustering are that it provides membership values for the objects instead of forcing them to exclusively belong to only one cluster and that it is flexible with respect to the type of distance considered.

## Example of application: Classification of strawberry cultivars

Fuzzy clustering was used to classify 11 strawberry cultivars based on physicochemical data (firmness, weight, colour, soluble solids and titratable acidity), as shown in Table 7.4. Results of k-means fuzzy clustering for identifying two clusters based on Euclidean distances, considering m = 2, are shown in

**Table 7.4** Physicochemical data of 11 strawberry cultivars.

| Cultivar | Firmness ($g/cm^2$) | Weight of 10 fruits (g) | $L^*$ | $a^*$ | $b^*$ | Soluble solids (°Brix) | Titratable acidity (g/100mL) |
|---|---|---|---|---|---|---|---|
| A | 113.4 | 391.6 | 38.8 | 20.0 | 16.8 | 4.5 | 0.32 |
| B | 127.5 | 367.2 | 37.4 | 17.5 | 13.3 | 4.4 | 0.23 |
| C | 54.7 | 305.1 | 37.2 | 15.8 | 13.4 | 3.7 | 0.26 |
| D | 57.7 | 348.4 | 34.0 | 12.4 | 11.1 | 3.6 | 0.33 |
| E | 69.0 | 289.9 | 34.8 | 9.8 | 8.6 | 3.2 | 0.22 |
| F | 58.0 | 334.1 | 35.0 | 13.3 | 18.8 | 3.6 | 0.23 |
| G | 80.1 | 339.4 | 36.2 | 14.8 | 13.7 | 3.8 | 0.27 |
| H | 119.0 | 411.2 | 40.7 | 21.0 | 17.6 | 4.7 | 0.34 |
| I | 120.5 | 379.4 | 38.1 | 18.8 | 15.1 | 4.5 | 0.28 |
| J | 59.9 | 319.4 | 35.3 | 12.8 | 13.0 | 3.5 | 0.26 |
| K | 90.2 | 349.4 | 36.7 | 15.8 | 14.0 | 4.0 | 0.27 |

$L^*$, $a^*$ and $b^*$ are the parameters of the Hunter Lab system.

Table 7.5. In the example, all cultivars can be assigned to a unique cluster with a relatively high degree of membership. Cultivars A, B, H and I can be assigned to Cluster 1, whereas cultivars C, D, E, F, G, J and K can be assigned to Cluster 2. Cultivar K is the only one that belongs to both clusters to a certain extent.

A usual criterion is to assign a sample to a cluster when the degree of membership is higher than a certain predetermined value. When considering two clusters this value is usually 0.5. Westad *et al.* (2004) applied fuzzy clustering for consumer segmentation according to their preference of a set of cheese samples, following an approach similar to that described in the example section for k-means clustering. They used a conservative approach to assign consumers to two clusters: they considered that consumers with a degree of membership in the range 0.4–0.6 were not assigned to any cluster.

**Table 7.5** Results of k-means fuzzy clustering performed on data from Table 7.4. Degree of membership of the 11 strawberry cultivars to the two clusters considered in the analysis.

| Cultivar | Cluster | |
|---|---|---|
| | 1 | 2 |
| A | 0.99 | 0.01 |
| B | 0.93 | 0.07 |
| C | 0.04 | 0.96 |
| D | 0.11 | 0.89 |
| E | 0.10 | 0.90 |
| F | 0.03 | 0.97 |
| G | 0.11 | 0.89 |
| H | 0.94 | 0.06 |
| I | 0.99 | 0.01 |
| J | 0.01 | 0.99 |
| K | 0.38 | 0.62 |

## Implementation in R

Fuzzy clustering can be performed in R (R Development Core Team, 2007) using the fuction *cmeans* from package e1071. The R-script shown in Box 7.3 performs k-means (or c-means) fuzzy clustering based on Euclidean distances on a given data set.

**Box 7.3 R-script performing k-means (or c-means) fuzzy clustering based on Euclidean distances**

```
#auxiliary routine to standardize the data (optional)
Dataset.stand<-scale(Dataset)
#loads package e1071
library("e1071")
#performs k-means fuzzy clustering with c=2 clusters and m=2
cl<-cmeans(Dataset.stand,2,20,verbose=TRUE,method="cmeans",m=2)
#shows the degree of membership of each object to each cluster
print(cl)
```

The full description of the *cmeans* function can be found at http://rss.acs.unt.edu/Rdoc/library/e1071/html/cmeans.html.

## CONCLUSIONS

Cluster analysis is a very common and useful exploratory multivariate technique that enables the identification of groups of samples or variables with similar characteristics. It has a wide range of applications in almost every field. The most popular approach in the food science and technology literature is hierarchical cluster analysis. Despite the conceptual simplicity of the method, its main disadvantage is that it imposes a hierarchical structure to the data which might not exist. However, there are many situations in which the interest is not to uncover a hierarchical structure but to identify a certain predefined number of groups. In these situations, partitional clustering methods, such as k-means, may be more appropriate. Finally, fuzzy clustering methods present the clear advantage of enabling objects to simultaneously belong to more than one cluster with an associated probability, called degree of membership.

Clustering analysis consists of an unsupervised classification and, therefore, little a priori information is available about the number of clusters in the data set. One of the main drawbacks of the technique is that it is always able to identify clusters, even when working with random data, which raises the question about the validity of the results. The user has to be aware of the fact that the type of clustering technique strongly affects the results, and therefore the selection should be performed considering the characteristics of the methods and the type of data on which they are going to be applied. A classical review on the topic has been published by Dubes and Jain (1975). One of the authors' recommendation for studying cluster validity is to apply several cluster techniques and look for common clusters, as well as using available information about the samples and/or the variables for explaining the identified clusters, which is still good advice.

## REFERENCES


Ares, G., Budelli, E., Bruzzone, F. *et al.* (2012) Consumers' texture perception of milk desserts. I – Relationship with reological measurements. *Journal of Texture Studies* **43**, 203–213.

Beebe, K.R., Pell, R.J. and Seasholtz, M.R. (1998) *Chemometrics: A Practical Guide*. John Wiley and Sons, New York, pp. 64–80.

Burns, R., and Burns, R. (2009) Cluster analysis. In: *Business Research Methods and Statistics Using SPSS* (eds R. Burns and R. Burns). SAGE, London, pp. 552–567.

Coxon, A.P.M. (1999) *Sorting data: Collection and analysis*. Sage, Thousand Oaks, CA.

Cruz, A.G., Cadena, R.S., Alvaro, M.B.V.B. *et al.* (2013) Assessing the use of different chemometric techniques to discriminate low-fat and full-fat yogurts. *LWT – Food Science and Technology* **50**, 210–214.

da Silva Torres, E.A.F., Garbelotti, M.L. and Machado Moita Neto, J. (2006) The application of hierarchical clusters analysis to the study of the composition of foods. *Food Chemistry* **99**, 622–629.

De Luca, M., Terouzi, W., Ioele, G. *et al.* (2011) Derivative FTIR spectroscopy for cluster analysis and classification of morocco olive oils. *Food Chemistry* **124**, 1113–1118.

Dubes, R., and Jain, A.K. (1975) Clustering techniques: the user's dilemma. *Pattern Recognition* **8**, 247–260.

Dziuba, B., Babuchowski, A., Nalęcz, D. and Niklewicz, M. (2007) Identification of lactic acid bacteria using FTIR spectroscopy and cluster analysis. *International Dairy Journal* **17**, 183–189.

Jacobsen, T. and Gunderson, R.W. (1986) Applied cluster analysis. In: *Statistical Procedures in Food Research* (ed. J.R. Pigott). Elsevier Applied Science, New York, pp. 361–408.

Krishnapuram, R. and Keller, J.M. (1993) A possibilistic approach to clustering. *IEEE Transactions on Fuzzy Systems* **1**, 98–110.

Lee, I., We, G.J., Kim, D.E. *et al.* (2012) Classification of rice cultivars based on cluster analysis of hydration and pasting properties of their starches. *LWT – Food Science and Technology* **48**, 164–168.

MacFie, H. (2007) Preference mapping and food product development. In: *Consumer-Led Food Product Development* (ed. H. MacFie). Woodhead Publishing Ltd, Cambridge, UK, pp. 407–433.

Mardia, K.V., Kent, J.T. and Bibby, J.M. (1979). *Multivariate Analysis*. Academic Press, London.

Næs, T., Brockhoff, P.B. and Tomic, O. (2010) Cluster Analysis: Unsupervised classification. In: *Statistics for Sensory and Consumer Science* (eds T. Naes, P. B. Brockhoff and O. Tomic). John Wiley & Sons Ltd, Chichester, pp. 249–262.

Næs, T., and Mevik, B-H. (1999) The flexibility of fuzzy clustering illustrated by examples. *Journal of Chemometrics* **13**, 435–444.

Oomes, S.J.C.M., van Zuijlen, A.C.M., Hehenkamp, J. O. *et al.* (2007) The characterisation of Bacillus spores occurring in the manufacturing of (low acid) canned products. *International Journal of Food Microbiology* **120**, 85–94.

Patras, A., Brunton, N.P., Downey, G. *et al.* (2011) Application of principal component and hierarchical cluster analysis to classify fruits and vegetables commonly consumed in Ireland based on *in vitro* antioxidant activity. *Journal of Food Composition and Analysis* **24**, 250–256.

Peeples, M.A. (2011) *R Script for K-Means Cluster Analysis* [Online]. Available: http://www.mattpeeples.net/kmeans.html (last accessed 9 June 2013).

Pham, D.T., Dimov, S.S. and Nguyen, C.D. (2005) Selection of K in K-means clustering. *Proceedings of the Institution of Mechanical Engineers, Part C: Journal of Mechanical Engineering Science* **219**, 103–119.

Punj, G. and Stewart, D.W. (1983) Cluster analysis in marketing research: Review and suggestions for application. *Journal of Marketing Research* **10**, 134–148.

R Development Core Team (2007) *R: A Language and Environment for Statistical Computing*. R Foundation for Statistical Computing, Vienna, Austria, ISBN 3-900051-07-0.

Rousseeuw, P.J., Kaufman, L. and Trauwaert, E. (1996) Fuzzy clustering using scatter matrices. *Computational Statistics and Data Analysis* **23**, 135–151.

Stricker, M. D, Onland-Moret, N.C., Boer, J.M.A. *et al.* (2013) Dietary patterns derived from principal component and k-means cluster analysis: Long-term association with coronary heart disease and stroke. *Nutrition, Metabolism and Cardiovascular Diseases* **23**(3), 250–256. doi: 10.1016/j.numecd.2012.02.006.

Tan, P-N., Steinbach, M. and Kumar, V. (2005) Cluster analysis; basic concepts and algorithms. In: *Introduction to Data Mining* (eds P-N. Tan, M. Steinbach and V. Kumar). Addison-Wesley, Upper Saddle River, NJ, pp. 487–568.

Westad, F., Hersleth, M. and Lea, P. (2004) Strategies for consumer segmentation with applications on preference data. *Food Quality and Preference* **15**, 681–687.

Wirfalt, A.K.E. and Jeffery, R.W. (1997) Using cluster analysis to examine dietary patterns: Nutrient intakes, gender, and weight status differ across food pattern clusters. *Journal of the American Dietetic Association* **97**, 272–279.

Yenket, R., Chambers IV, E. and Johnson, D.E. (2011) Statistical package clustering may not be best for grouping consumers to understand their most liked products. *Journal of Sensory Studies* **26**, 209–225.

# 8 Principal component regression (PCR) and partial least squares regression (PLSR)

**Rolf Ergon**
*Telemark University College, Porsgrunn, Norway*

## ABSTRACT

Ordinary least squares regression is summarized, with emphasis on the variance problem caused by collinear predictor variables. A theoretical and optimal estimation solution to this problem is presented and the relations to Kalman filtering as well as to principal component regression (PCR) and partial least squares regression (PLSR) are pointed out. The theory behind PCR and PLSR is given, and alternative PLSR models and their different residuals are compared. The PLSR residual issue in process monitoring is discussed in some detail. PLSR score-loading correspondence is discussed, and PCR and PLSR model reduction methods are presented and discussed. Established PCR and PLSR practices are summarized, with reference to the literature; finally, some emerging methods in food science are mentioned.

## INTRODUCTION

In this chapter the focus is on theoretical aspects of principal component regression (PCR) and partial least squares regression (PLSR). An important reason for the use of PCR and PLSR, as compared with, for example, ridge regression, is the interpretational possibilities given by score and loading plots, as described for PCA (and thus for PCR) in Chapter 5. The advantage of PLSR is that in some cases it gives a lower number of components, and thus increased interpretability. A number of practical aspects of PCR and PLSR are merely listed, with references to literature where detailed discussions are given. Finally, some emerging PCR and PLSR methods in food science are mentioned.

## THEORY

In this section a close look is taken at the theory behind the biased multivariate calibration methods PCR and PLSR. Before doing that, however, we will recall the theory for the unbiased ordinary least squares regression (OLS) from Chapter 2, and point out the variance problems caused by collinearity between the predictor variables. We will then also establish the theory for optimal unbiased multivariate calibration, which at least in theoretical and simulation studies can be used as a reference for comparison with PCR and PLSR.

*Mathematical and Statistical Methods in Food Science and Technology*, First Edition.
Edited by Daniel Granato and Gastón Ares.

## Classical linear regression

For simplicity, assume a single response variable $y$ and multivariate predictor variables $\mathbf{x}$. Classical linear regression then assumes that $y$ is a weighted response on the predictor variables, with some random error $\varepsilon$ added, that is of $x$-variables

$$y = b_0 + x_1 b_1 + \cdots + x_p b_p + \varepsilon = b_0 + \mathbf{x}\mathbf{b} + \varepsilon \tag{8.1}$$

which after collection of $N$ independent observations or samples results in the OLS model

$$\mathbf{y} = 1b_0 + \mathbf{X}\mathbf{b} + \boldsymbol{\varepsilon} \tag{8.2}$$

where the expectation $E[\boldsymbol{\varepsilon}] = 0$, and $Cov(\boldsymbol{\varepsilon}) = E[\boldsymbol{\varepsilon}\boldsymbol{\varepsilon}^T] = \sigma^2\mathbf{I}$. Here, 1 is a $N \times 1$ vector of ones, while the regression parameters $b_0$ and $\mathbf{b}$, and the variance $\sigma^2$, are unknown.

If we, again for simplicity, assume centred data such that $b_0 = 0$, the unbiased ordinary least squares estimate of $\mathbf{b}$ is

$$\hat{\mathbf{b}}_{\text{OLS}} = \left(\mathbf{X}^T\mathbf{X}\right)^{-1}\mathbf{X}^T\mathbf{y} \tag{8.3}$$

which minimizes the sum of squared differences

$$\sum_{j=1}^{N} \hat{\varepsilon}_j^2 = \sum_{j=1}^{N} \left(y_j - x_1 b_{j1} - \cdots - x_{jp} b_p\right)^2 = (\mathbf{y} - \mathbf{X}\mathbf{b})^T(\mathbf{y} - \mathbf{X}\mathbf{b}) \tag{8.4}$$

It will be seen later that the corresponding PCR and PLSR estimates of $\mathbf{b}$ are modifications of Equation 8.3. If the data are not centred, Equation 8.3 must be replaced by $\left[\hat{b}_{0,\text{OLS}} \quad \hat{\mathbf{b}}_{\text{OLS}}\right]^T = \left([1 \quad \mathbf{X}]^T[1 \quad \mathbf{X}]\right)^{-1}[1 \quad \mathbf{X}]^T\mathbf{y}$.

The vector of residuals $\hat{\boldsymbol{\varepsilon}} = y - [1 \quad \mathbf{X}]\left[\hat{b}_{0,\text{OLS}} \quad \hat{\mathbf{b}}_{\text{OLS}}^T\right]^T$ contains the information about the remaining unknown parameter $\sigma^2$. Defining $s^2 = \hat{\boldsymbol{\varepsilon}}^T\hat{\boldsymbol{\varepsilon}}/(N - p - 1)$, we have $E[s^2] = \sigma^2$ (Johnson and Wichern, 1992).

The problem with OLS in industrial practice is that the inversion in Equation 8.3 causes a large variance in $\hat{\mathbf{b}}$ (Næs and Martens, 1988; Næs and Mevik, 2001). The fact that $\hat{\mathbf{b}}$ is unbiased means that the expectation of $\hat{\mathbf{b}}$ is $E[\hat{\mathbf{b}}] = \mathbf{b}$, but for a single data set the result may be very far from the true parameter vector. This problem is caused by collinearity in the $\mathbf{X}$ matrix, that is that the columns of $\mathbf{X}$ are similar except for a multiplication factor. In typical spectroscopic applications the number $p$ of predictor variables is very large, often much larger than the feasible number $N$ of independent samples, that is $p > N$, in which case collinearity must necessarily be present. In typical process monitoring applications the number of predictor variables are smaller, but still larger than the number of underlying phenomena that determine the value of the response variable; also, in such cases $\mathbf{X}$ will be collinear.

## Optimal estimation

### *Kalman filtering solution*

Under certain assumptions it is possible to obtain an expression for the optimal estimate of $\mathbf{b}$ in Equation 8.1, again assuming centred data and thus $b_0 = 0$ (Ergon and Esbensen, 2002). We must then assume an underlying model

$$\left.\begin{aligned} \boldsymbol{\tau}_{j+1} &= \mathbf{e}_j \\ y_j &= \mathbf{q}^T\boldsymbol{\tau}_j + v_{1,j} \\ \mathbf{x}_j &= \mathbf{W}\boldsymbol{\tau}_j + \mathbf{v}_{2,j} \end{aligned}\right\} \tag{8.5}$$

where $\boldsymbol{\tau}_j$ is a vector of latent variables and $\mathbf{e}_j$, $v_{1,j}$ and $\mathbf{v}_{2,j}$ are white noise sequences (i.e. random and not auto-correlated) with covariances $\mathbf{R}_e = E\left[\mathbf{e}_j\mathbf{e}_j^T\right]$, $r_{11} = E\left[v_{1,j}^2\right]$ and $\mathbf{R}_{22} = E\left[\mathbf{v}_{2,j}\mathbf{v}_{2,j}^T\right]$. We then assume a latent variables estimate

$$\hat{\boldsymbol{\tau}}_j = \mathbf{K}\mathbf{x}_j = \mathbf{K}\left(\mathbf{W}\boldsymbol{\tau}_j + \mathbf{v}_{2,j}\right) \quad (8.6)$$

where $\mathbf{K}$ is chosen such that the expected covariance

$$\begin{aligned}\mathbf{Z}_j &= E\left[\left(\boldsymbol{\tau}_j - \mathbf{K}\left(\mathbf{W}\boldsymbol{\tau}_j + \mathbf{v}_{2,j}\right)\right)\left(\boldsymbol{\tau}_j - \mathbf{K}\left(\mathbf{W}\boldsymbol{\tau}_j + \mathbf{v}_{2,j}\right)\right)^T\right] \\ &= (\mathbf{I} - \mathbf{K}\mathbf{W})E\left[\boldsymbol{\tau}_j\boldsymbol{\tau}_j^T\right](\mathbf{I} - \mathbf{K}\mathbf{W})^T + \mathbf{K}E\left[\mathbf{v}_{2,j}\mathbf{v}_{2,j}^T\right]\mathbf{K}^T\end{aligned} \quad (8.7)$$

is minimized. Using $E\left[\boldsymbol{\tau}_j\boldsymbol{\tau}_j^T\right] = E\left[\boldsymbol{\tau}_{j+1}\boldsymbol{\tau}_{j+1}^T\right] = E\left[\mathbf{e}_j\mathbf{e}_j^T\right] = \mathbf{R}_e$ and $E\left[\mathbf{v}_{2,j}\mathbf{v}_{2,j}^T\right] = \mathbf{R}_{22}$, we find (Gelb, 1974)

$$\frac{\partial}{\partial \mathbf{K}} trace\left(\mathbf{Z}_j\right) = -2(\mathbf{I} - \mathbf{K}\mathbf{W})\mathbf{R}_e\mathbf{W}^T + 2\mathbf{K}\mathbf{R}_{22} \quad (8.8)$$

that is $(\partial/\partial\mathbf{K})trace\left(\mathbf{Z}_j\right) = 0$ gives the optimal solution

$$\mathbf{K} = \mathbf{R}_e\mathbf{W}^T\left(\mathbf{W}\mathbf{R}_e\mathbf{W}^T + \mathbf{R}_{22}\right)^{-1} \quad (8.9)$$

Here, $\mathbf{K}$ is a special case of the gain matrix in Kalman filtering (Kalman, 1960), used in optimal estimation of state variables in dynamical systems, with numerous applications in signal processing and engineering control.

The resulting optimal response estimate follows from Equations 8.5 and 8.6 as

$$\hat{y}_j = \mathbf{q}^T\mathbf{K}\mathbf{x}_j = \mathbf{x}_j^T\mathbf{K}^T\mathbf{q} \quad (8.10)$$

corresponding to the model

$$y_j = \mathbf{x}_j^T\mathbf{K}^T\mathbf{q} + \eta_j \quad (8.11)$$

where it can be shown that $\eta_j$ is white noise (Ergon, 1999). Optimality here means that Equation 8.10 gives the best linear unbiased estimate (BLUE) and the best possible estimate whatsoever assuming Gaussian noise distributions (Grewal and Andrews, 1993).

In order to compare with the OLS regression vector (Equation 8.3), we may apply Equation 8.11 on all modelling samples, and use

$$\mathbf{y} = \mathbf{X}\mathbf{K}^T\mathbf{q} + \eta \quad (8.12)$$

from which we obtain the OLS solution

$$\hat{\mathbf{q}} = \left(\mathbf{K}\mathbf{X}^T\mathbf{X}\mathbf{K}^T\right)^{-1}\mathbf{K}\mathbf{X}^T\mathbf{y} \quad (8.13)$$

and thus the optimal prediction vector

$$\hat{\mathbf{b}}_{\text{OPTIMAL}} = \mathbf{K}^T\hat{\mathbf{q}} = \mathbf{K}^T\left(\mathbf{K}\mathbf{X}^T\mathbf{X}\mathbf{K}^T\right)^{-1}\mathbf{K}\mathbf{X}^T\mathbf{y} \quad (8.14)$$

Note that this is a modified version of the OLS regression vector $\hat{\mathbf{b}}_{\text{OLS}}$ in Equation 8.3. We will find similar modified versions of $\hat{\mathbf{b}}_{\text{OLS}}$ when we come to PCR and PLSR later. Also note that instead of $\mathbf{K}^T$ here we may use $\mathbf{K}^T\mathbf{M}$, where $\mathbf{M}$ is an invertible matrix that gives an oblique rotation and scaling of the column vectors in $\mathbf{K}^T$; what matters is the subspace spanned by $\mathbf{K}^T$.

The optimal regression vector $\hat{\mathbf{b}}_{\text{OPTIMAL}}$ requires knowledge of $\mathbf{K}$, that is according to Equation 8.9 of $\mathbf{W}$, $\mathbf{R}_e$ and $\mathbf{R}_{22}$. In practice we may find estimates of $\mathbf{W}$ from PCR (there called $\mathbf{P}_{\text{PCA}}$) or PLSR as described later, while $\mathbf{R}_e$ and $\mathbf{R}_{22}$ may be estimated from the data (Ergon and Esbensen, 2002; Gujral *et al.*, 2011). These covariance estimates may then be based also on easily available $x$ data that do not correspond to any of the $y$ samples used in the PCR or PLSR algorithms, so called unlabelled data, which may give significant improvements in the prediction results.

### *PCR and PLSR approximations*

From what we will see below, PCR is characterized by an orthonormal loading matrix $\mathbf{P}_{\text{PCA}}$, with $\mathbf{P}_{\text{PCA}}^T\mathbf{P}_{\text{PCA}} = \mathbf{I}$, while PLSR is characterized by an orthonormal loading weights matrix $\mathbf{W}$, with $\mathbf{W}^T\mathbf{W} = \mathbf{I}$, in both cases giving the connection between latent variables and $x$ variables, as shown in model (Equation 8.5). Using PLSR as an example we find from Equation 8.9 that

$$\mathbf{M}^T\mathbf{K} = \mathbf{W}^T\mathbf{W}\mathbf{R}_e\mathbf{W}^T\left(\mathbf{W}\mathbf{R}_e\mathbf{W}^T + \mathbf{R}_{22}\right)^{-1} \rightarrow \mathbf{W}^T \text{ when } \mathbf{R}_e \rightarrow \infty \tag{8.15}$$

where $\mathbf{M}$ is an oblique rotation and scaling matrix, as discussed above. With sufficient excitation of the latent variables, and provided that the number of components used equals the number of true latent variables, the matrices $\mathbf{P}_{\text{PCA}}$ and $\mathbf{W}$ will thus approach the transposed optimal gain matrix $\mathbf{K}^T\mathbf{M}$, and the predictor vectors $\hat{\mathbf{b}}_{\text{PCR}}$ and $\hat{\mathbf{b}}_{\text{PLSR}}$ that we will find later will then approach $\hat{\mathbf{b}}_{\text{OPTIMAL}}$ according to Equation 8.14. Simulation examples illustrating this are given in Ergon (2002a); not surprisingly, the need for excitation is found to be larger for small sample sizes.

Since the theory above shows that there exist optimal loading weights matrices that span the column space of $\mathbf{K}^T$, at least in cases with well defined numbers of latent variables, it is tempting to to try to find such a matrix by a numerical search. This was done in Ergon (2003a), with considerable improvement of predictions for a laboratory data set based on metal ion mixtures.

## Principal component regression

Principal component regression (PCR) is entirely based on principal component analysis (PCA) of the $\mathbf{X}$ matrix, as described in Chapter 5. The $\mathbf{X}$ matrix is thus factorized by, for example, singular value decomposition, into

$$\mathbf{X} = \mathbf{U}\mathbf{S}\mathbf{V}^T = \begin{bmatrix}\mathbf{U}_1 & \mathbf{U}_2\end{bmatrix}\begin{bmatrix}\mathbf{S}_1 & 0 \\ 0 & \mathbf{S}_2\end{bmatrix}\begin{bmatrix}\mathbf{V}_1 \\ \mathbf{V}_2\end{bmatrix} = \mathbf{U}_1\mathbf{S}_1\mathbf{V}_1^T + \mathbf{U}_2\mathbf{S}_2\mathbf{V}_2^T \tag{8.16}$$

which with the notation $\mathbf{T}_{\text{PCA}} = \mathbf{U}_1\mathbf{S}_1$, $\mathbf{P}_{\text{PCA}} = \mathbf{V}_1$, and $\mathbf{E} = \mathbf{U}_2\mathbf{S}_2\mathbf{V}_2^T$ gives

$$\mathbf{X} = \mathbf{T}_{\text{PCA}}\mathbf{P}_{\text{PCA}}^T + \mathbf{E} \tag{8.17}$$

where $\mathbf{T}_{\text{PCA}}$ has orthogonal column vectors and $\mathbf{P}_{\text{PCA}}$ is orthonormal. The number of components, that is of columns in the score matrix $\mathbf{T}_{\text{PCA}}$ and the loading matrix $\mathbf{P}_{\text{PCA}}$, must here be determined by validation, either cross-validation based on the modelling data (Martens and Næs, 1989) or preferably by use of an independent test set (as shown in Example 8.1 later).

Since the loading matrix $\mathbf{P}_{\text{PCA}}$ is orthonormal, the score matrix can be expressed as

$$\mathbf{T}_{\text{PCA}} = \mathbf{X}\mathbf{P}_{\text{PCA}} \tag{8.18}$$

such that the model of $\mathbf{X}$ becomes

$$\mathbf{X} = \mathbf{X}\mathbf{P}_{\text{PCA}}\mathbf{P}_{\text{PCA}}^T + \mathbf{X}\left(\mathbf{I} - \mathbf{P}_{\text{PCA}}\mathbf{P}_{\text{PCA}}^T\right) \tag{8.19}$$

This shows that the modelled part $\hat{\mathbf{X}}_{\text{PCA}} = \mathbf{X}\mathbf{P}_{\text{PCA}}\mathbf{P}_{\text{PCA}}^T$ is found by projection of $\mathbf{X}$ onto the column space of $\mathbf{P}_{\text{PCA}}$, while the row vectors of the residual $\mathbf{E}_{\text{PCA}} = \mathbf{X}\left(\mathbf{I} - \mathbf{P}_{\text{PCA}}\mathbf{P}_{\text{PCA}}^T\right)$ are contained in the left nullspace or orthogonal complement of the column space of $\mathbf{P}_{\text{PCA}}$ (Strang, 2006).

The score variables are now used as prediction variables in an ordinary least squares model

$$\mathbf{y} = \mathbf{T}_{\text{PCA}}\mathbf{q}_{\text{PCA}} + \boldsymbol{\varepsilon} \tag{8.20}$$

leading to the regression vector

$$\hat{\mathbf{q}}_{\text{PCA}} = \left(\mathbf{T}_{\text{PCA}}^T\mathbf{T}_{\text{PCA}}\right)^{-1}\mathbf{T}_{\text{PCA}}^T\mathbf{y} \tag{8.21}$$

Predictions $\hat{\mathbf{y}}_{\text{new}}$ based on new data $\mathbf{X}_{\text{new}}$ may be found by first forming $\mathbf{T}_{\text{PCA,new}} = \mathbf{X}_{\text{new}}\mathbf{P}_{\text{PCA}}$ and then finding $\hat{\mathbf{y}}_{\text{new}} = \mathbf{T}_{\text{PCA,new}}\hat{\mathbf{q}}_{\text{PCA}}$. They are thus found as

$$\hat{\mathbf{y}}_{\text{new}} = \mathbf{X}_{\text{new}}\mathbf{P}_{\text{PCA}}\left(\mathbf{P}_{\text{PCA}}^T\mathbf{X}^T\mathbf{X}\mathbf{P}_{\text{PCA}}\right)^{-1}\mathbf{P}_{\text{PCA}}^T\mathbf{X}^T\mathbf{y} \tag{8.22}$$

and the prediction vector corresponding to to $\mathbf{y} = \mathbf{X}\mathbf{b} + \boldsymbol{\varepsilon}$ is, therefore

$$\hat{\mathbf{b}}_{\text{PCR}} = \mathbf{P}_{\text{PCA}}\left(\mathbf{P}_{\text{PCA}}^T\mathbf{X}^T\mathbf{X}\mathbf{P}_{\text{PCA}}\right)^{-1}\mathbf{P}_{\text{PCA}}^T\mathbf{X}^T\mathbf{y} \tag{8.23}$$

This can be seen as a modified version of the OLS regression vector (Equation 8.3). Also note the similarity with the optimal prediction vector (Equation 8.14).

## Partial least squares regression

### *Basic theory*

The loading matrix $\mathbf{P}_{\text{PCA}}$ in PCA and thus in PCR is found by maximizing the variance in $\mathbf{X}$ in orthogonal directions, with the resulting prediction vector (Equation 8.23). The drawback with this is that some of the score vectors in $\mathbf{T}_{\text{PCA}}$ may have very little in common with the response vector $\mathbf{y}$. The basic feature in PLSR is that it finds an alternative matrix $\mathbf{W}$, the loading weights matrix, that also takes the covariance with $\mathbf{y}$ into account. This may lead to a reduction of the number of components, and thus increased interpretability (Example 8.1 later in this chapter). As shown later in this chapter there are many algorithms that do that, but the result is in any case the common prediction vector:

$$\hat{\mathbf{b}}_{\text{PLSR}} = \mathbf{W}\left(\mathbf{W}^T\mathbf{X}^T\mathbf{X}\mathbf{W}\right)^{-1}\mathbf{W}^T\mathbf{X}^T\mathbf{y} \tag{8.24}$$

Just as $\mathbf{P}_{\text{PCA}}$ in PCR, the $\mathbf{W}$ matrix is orthonormal. And in line with PCR we can model $\mathbf{X}$ by projection onto the column space of $\mathbf{W}$, resulting in the Martens factorization (Martens and Næs, 1987)

$$\mathbf{X} = \mathbf{X}\mathbf{W}\mathbf{W}^T + \mathbf{X}\left(\mathbf{I} - \mathbf{W}\mathbf{W}^T\right) = \hat{\mathbf{X}}_{\text{M}} + \mathbf{E}_{\text{B}} \tag{8.25}$$

where the row vectors of the residual $\mathbf{E}_\mathrm{B} = \mathbf{X}(\mathbf{I} - \mathbf{W}\mathbf{W}^T)$ are contained in the left nullspace or orthogonal complement of the column space of $\mathbf{W}$ (the subscript B will be explained below). And in line with PCR we find a score matrix $\mathbf{T}_\mathrm{M} = \mathbf{X}\mathbf{W}$, with score vectors that span the model space. There is, however, an important difference, in that the score vectors in $\mathbf{T}_\mathrm{M}$ are *not* orthogonal. In the original PLSR algorithm of Wold *et al.* (1983), this is corrected by an orthogonalization step, resulting in a score matrix $\mathbf{T}$ with orthogonal score vectors. This also results in a different residual matrix, as discussed below.

### A didactically motivated non-orthogonalized PLSR algorithm

A comparison of some of the common algorithms for finding $\mathbf{W}$, and the resulting properties of $\hat{\mathbf{X}}$ and $\mathbf{E}$, is given later. But let us first take a look at an algorithm that in a simple way reveals the basic feature of PLSR (Ergon and Esbensen, 2001), namely covariance maximization. Start by trying to explain $\mathbf{y}$ by use of only one component, that is by modelling $\mathbf{X}$ as:

$$\mathbf{X} = \mathbf{X}\mathbf{w}_1\mathbf{w}_1^T + \mathbf{E}_1 = \mathbf{t}_1\mathbf{w}_1^T + \mathbf{E}_1 \tag{8.26}$$

Then find the score vector $\mathbf{t}_1$ that maximizes the sample covariance $cov(\mathbf{t}_1, \mathbf{y}) = \mathbf{t}_1^T\mathbf{y}/(N-1) = \mathbf{w}_1^T\mathbf{X}^T\mathbf{y}/(N-1)$ under the constraint that $\mathbf{w}_1$ is a unit vector with $||\mathbf{w}_1|| = 1$. This is obviously obtained when $\mathbf{w}_1$ has the same direction as $\mathbf{X}^T\mathbf{y}$, that is $\mathbf{w}_1 = \mathbf{X}^T\mathbf{y}/\sqrt{\mathbf{y}^T\mathbf{X}\mathbf{X}^T\mathbf{y}}$. From Equation 8.24 now follows the one component fitted response

$$\hat{\mathbf{y}}_1 = \mathbf{X}\mathbf{w}_1\left(\mathbf{w}_1^T\mathbf{X}^T\mathbf{X}\mathbf{w}_1\right)^{-1}\mathbf{w}_1^T\mathbf{X}^T\mathbf{y} \tag{8.27}$$

and the response residual $\boldsymbol{\varepsilon}_1 = \mathbf{y} - \hat{\mathbf{y}}_1$. Next, try to improve $\hat{\mathbf{y}}$ by using the original $\mathbf{X}$ matrix and the response residual $\boldsymbol{\varepsilon}_1$ to find a new score vector $\mathbf{t}_2$ that maximizes the sample covariance $cov(\mathbf{t}_2, \boldsymbol{\varepsilon}_1) = \mathbf{t}_2^T\boldsymbol{\varepsilon}_1/(N-1) = \mathbf{w}_2^T\mathbf{X}^T\boldsymbol{\varepsilon}_1/(N-1)$, under the constraint that also $\mathbf{w}_2$ is a unit vector with $||\mathbf{w}_2|| = 1$, resulting in $\mathbf{w}_2 = \mathbf{X}^T\boldsymbol{\varepsilon}_1/\sqrt{\boldsymbol{\varepsilon}_1^T\mathbf{X}\mathbf{X}^T\boldsymbol{\varepsilon}_1}$, and the improved fitted response

$$\hat{\mathbf{y}}_2 = \mathbf{X}\mathbf{W}_2\left(\mathbf{W}_2^T\mathbf{X}^T\mathbf{X}\mathbf{W}_2\right)^{-1}\mathbf{W}_2^T\mathbf{X}^T\mathbf{y} \tag{8.28}$$

where $\mathbf{W}_2 = [\mathbf{w}_1 \quad \mathbf{w}_2]$. The new response residual will now be $\boldsymbol{\varepsilon}_2 = \mathbf{y} - \hat{\mathbf{y}}_2$, giving $\mathbf{w}_3 = \mathbf{X}^T\boldsymbol{\varepsilon}_2/\sqrt{\boldsymbol{\varepsilon}_2^T\mathbf{X}\mathbf{X}^T\boldsymbol{\varepsilon}_2}$ and so on. Finally, it is easy to show that all the unit length loading weights vectors obtained in this way are orthogonal, that is that $\mathbf{W} = [\mathbf{w}_1 \ \mathbf{w}_2 \ \cdots \ \mathbf{w}_A]$ is orthonormal (Ergon and Esbensen, 2001). The final latent variables model is

$$\mathbf{X} = \mathbf{X}\mathbf{W}\mathbf{W}^T + \mathbf{E}_\mathrm{B} = \mathbf{T}_\mathrm{M}\mathbf{W}^T + \mathbf{E}_\mathrm{B} \tag{8.29}$$

$$\mathbf{y} = \mathbf{T}_\mathrm{M}\mathbf{q}_\mathrm{M} + \mathbf{f} \tag{8.30}$$

first developed by Martens (Martens and Næs, 1987). Here, $\mathbf{q}_\mathrm{M}$ is given by the OLS solution $\mathbf{q}_\mathrm{M} = \left(\mathbf{W}^T\mathbf{X}^T\mathbf{X}\mathbf{W}\right)^{-1}\mathbf{W}^T\mathbf{X}^T\mathbf{y}$. The residual $\mathbf{E}_\mathrm{B}$ is, however, the same as from the older Lanczos bidiagonalization algorithm (Bidiag2) of Golub and Kahan (1965), further discussed later in the chapter. Note that the prediction vector (Equation 8.24) follows directly from this model.

### Orthogonalized PLSR algorithm

The original and conventional PLSR algorithm (Wold *et al.*, 1983) was developed in order to obtain orthogonal score vectors. Assuming centred data and a single response variable, the algorithm can be summarized as follows:

1. Let $\mathbf{X}_0 = \mathbf{X}$ and $\mathbf{y}_0 = \mathbf{y}$.
For $a = 1, 2, \ldots, A$ perform steps 2 to 7 below.
2. $\mathbf{w}_a = \mathbf{X}_{a-1}^T\mathbf{y}_{a-1}/||\mathbf{X}_{a-1}^T\mathbf{y}_{a-1}||$ (with length 1)
3. $\mathbf{t}_a = \mathbf{X}_{a-1}\mathbf{w}_a$
4. $\mathbf{q}_a = \mathbf{y}_{a-1}^T\mathbf{t}_a(\mathbf{t}_a^T\mathbf{t}_a)^{-1}$
5. $\mathbf{p}_a = \mathbf{X}_{a-1}^T\mathbf{t}_a(\mathbf{t}_a^T\mathbf{t}_a)^{-1}$
6. $\mathbf{X}_a = \mathbf{X}_{a-1} - \mathbf{t}_a\mathbf{p}_a^T$
7. $\mathbf{y}_a = \mathbf{y}_{a-1} - \mathbf{t}_a\mathbf{q}_a$

The resulting latent variables model is:

$$\mathbf{X} = \mathbf{T}\mathbf{P}^T + \mathbf{E}_\mathrm{C} \tag{8.31}$$

$$\mathbf{y} = \mathbf{T}\mathbf{q}_\mathrm{C} + \mathbf{f}, \tag{8.32}$$

where $\mathbf{T} = [\mathbf{t}_1 \quad \mathbf{t}_2 \quad \cdots \quad \mathbf{t}_A]$ (with orthogonal score vectors), $\mathbf{P} = [\mathbf{p}_1 \quad \mathbf{p}_2 \quad \cdots \quad \mathbf{p}_A]$ (with nonorthogonal loading vectors), $\mathbf{q}_\mathrm{C} = [q_1 \quad q_2 \quad \cdots \quad q_A]^T$ and $\mathbf{E}_\mathrm{C} = \mathbf{X}_A$. The resulting matrix $\mathbf{W} = [\mathbf{w}_1 \quad \mathbf{w}_2 \quad \cdots \quad \mathbf{w}_A]$ is orthonormal, and the same matrix as introduced above. Especially note that the residual $\mathbf{E}_\mathrm{C}$ is different from $\mathbf{E}_\mathrm{B}$ in the nonorthogonalized PLSR model (Equations 8.29 and 8.30) above. Also note that the deflation of $\mathbf{y}$ in step 7 is not strictly necessary, although it may have some numerical advantages.

The prediction vector used in conventional PLSR is $\hat{\mathbf{b}}_\mathrm{PLSR} = \mathbf{W}(\mathbf{P}^T\mathbf{W})^{-1}\mathbf{q}_\mathrm{C}$ (Martens and Næs, 1989), which unfortunately is not consistent with the conventional model (Equations 8.31 and 8.32). This problem was discussed in Pell *et al.* (2007), after which followed a rather heated discussion on the proper choice of PLSR residuals (Bro and Eldén, 2009; Ergon, 2009; Manne *et al.*, 2009; Wold *et al.*, 2009). A solution to the inconsistence problem had been given already in Ergon (2002a), where the model $\mathbf{X} = \mathbf{T}\mathbf{P}^T\mathbf{W}\mathbf{W}^T + \mathbf{E}_\mathrm{B}$, and thus the OLS solution $\mathbf{T} = \mathbf{X}\mathbf{W}(\mathbf{P}^T\mathbf{W})^{-1}$, was introduced, from which directly follows $\hat{\mathbf{y}} = \mathbf{T}\mathbf{q}_\mathrm{C} = \mathbf{X}\mathbf{W}(\mathbf{P}^T\mathbf{W})^{-1}\mathbf{q}_\mathrm{C}$ and thus $\hat{\mathbf{b}}_\mathrm{PLSR} = \mathbf{W}(\mathbf{P}^T\mathbf{W})^{-1}\mathbf{q}_\mathrm{C}$. The solution is thus to project $\hat{\mathbf{X}}_\mathrm{C} = \mathbf{T}\mathbf{P}^T$ onto the column space of $\mathbf{W}$. Finally, since Equation 8.32 gives $\mathbf{q}_\mathrm{C} = (\mathbf{T}^T\mathbf{T})^{-1}\mathbf{T}^T\mathbf{y}$, this projection results in

$$\begin{aligned}\hat{\mathbf{b}}_\mathrm{PLSR} &= \mathbf{W}(\mathbf{P}^T\mathbf{W})^{-1}\mathbf{q}_\mathrm{C} = \mathbf{W}(\mathbf{P}^T\mathbf{W})^{-1}(\mathbf{T}^T\mathbf{T})^{-1}\mathbf{T}^T\mathbf{y} \\ &= \mathbf{W}(\mathbf{P}^T\mathbf{W})^{-1}\left((\mathbf{W}^T\mathbf{P})^{-1}\mathbf{W}^T\mathbf{X}^T\mathbf{X}\mathbf{W}(\mathbf{P}^T\mathbf{W})^{-1}\right)^{-1}(\mathbf{W}^T\mathbf{P})^{-1}\mathbf{W}^T\mathbf{X}^T\mathbf{y} \\ &= \mathbf{W}(\mathbf{W}^T\mathbf{X}^T\mathbf{X}\mathbf{W})^{-1}\mathbf{W}^T\mathbf{X}^T\mathbf{y}\end{aligned} \tag{8.33}$$

that is the prediction vector (Equation 8.24). It should be added that the relation $\mathbf{T} = \mathbf{X}\mathbf{W}(\mathbf{P}^T\mathbf{W})^{-1}$ has been used in conventional PLSR since the beginning, but then disregarding the inconsistency with the conventional model (Equations 8.31 and 8.32).

Finally, since we obtain the same loading weights matrix $\mathbf{W}$ in both orthogonalized and non-orthogonalized PLSR, the nonorthogonalized model (Equations 8.29 and 8.30) can easily be found by use of the conventional algorithm, simply by using $\mathbf{T}_\mathrm{M} = \mathbf{X}\mathbf{W}$ as score matrix, that is by a re-interpretation of the results from the algorithm (Ergon, 2009).

## Example 8.1

For a comparison of prediction ability of PCR and PLSR we employ a data set provided by the Cargill company and Eigenvector Research Inc. (2005). From these data 80 samples of corn measured on a NIR spectrometer labeled m5 are used. The wavelength range is 1100–2498 nm at 2 nm intervals (700 channels). Samples 1–40 are used for modelling, and samples 41–80 for validation and testing, using centred data with standardized $y$ values. Multiplicative scatter correction (MSC) was used to compensate for multiplicative and additive effects caused by light scattering in the spectra used for modelling (Næs *et al.*, 2004). This was done by the standard procedure in the Unscrambler software from Camo, and the same correction was then also used on the test spectra. The purpose of the model is to estimate the protein content in corn (response variable number three). Figure 8.1 shows the root mean square error of prediction

$$RMSEP = \sqrt{\sum_{1}^{N_{\text{pred}}} (\hat{y}_i - y_i)^2 / N_{\text{pred}}}, \tag{8.34}$$

where $N_{\text{pred}}$ is the number of samples in the test set.

Figure 8.1 shows that PCR needs nine components to obtain a close to optimal performance, while PLSR needs only five (the PLSR performance for eight components and the PCR performance for 16–17 components cannot be expected with a different test set). Also, five components is a high number for interpretational purposes by use of score and loading plots (as described for PCA in Chapter 5), which underlines the usefulness of model reduction methods as discussed later in the chapter.

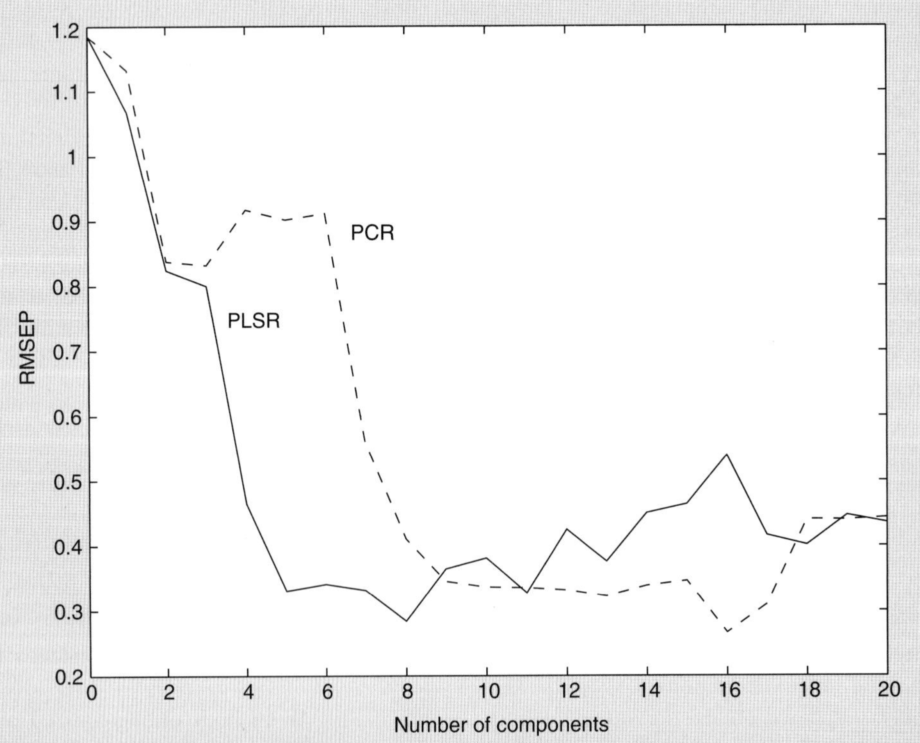


**Figure 8.1** Prediction ability of PCR and PLSR for NIR data in Example 8.1.

*Further discussion of different PLSR and related models*

In the data matrix $\mathbf{X} = [\mathbf{x}_1 \quad \mathbf{x}_2 \quad \cdots \quad \mathbf{x}_p] = [\mathbf{z}_1 \quad \mathbf{z}_2 \quad \cdots \quad \mathbf{z}_N]^T$, each column $\mathbf{x}_j$ is a vector in an $N$-dimensional vector space, while each sample or row $\mathbf{z}_i^T$ is a vector in a $p$-dimensional vector space. Using $\mathbf{X}$ and the response vector $\mathbf{y}$ as inputs, the original PLSR algorithm above produces the orthonormal loading weights matrix $\mathbf{W} \in \mathbb{R}^{p \times A}$ and the orthogonalized score matrix $\mathbf{T} \in \mathbb{R}^{N \times A}$, and corresponding low-dimensional subspaces, where $A$ is the number of PLS components used. In addition the algorithm produces the nonorthogonal loading matrix $\mathbf{P} = \mathbf{X}^T\mathbf{T}(\mathbf{T}^T\mathbf{T})^{-1} \in \mathbb{R}^{p \times A}$. The matrix $\mathbf{W}$ is also found by the nonorthogonalized PLSR, Bidiag2 and RE-PLSR algorithms as discussed later, and the RE-PLSR algorithm also produces $\mathbf{T}$. Both $\mathbf{W}$ and $\mathbf{T}$ can alternatively be found from Krylov subspaces (Manne, 1987; Di Ruscio, 2000; Bro and Eldén, 2009).

Using standard linear algebra (Strang, 2006), we may define the fitted matrix $\hat{\mathbf{X}}$ and the residual matrix $\mathbf{E} = \mathbf{X} - \hat{\mathbf{X}}$ in essentially two different ways, and with that follow essentially two different models of $\mathbf{X}$. One central question in a comparison of these models is which projections we perform when scores for individual samples are found, both for the modelling data and for new data, that is how do we find the individual row vectors $\boldsymbol{\tau}^T$ and $\boldsymbol{\tau}_{\mathrm{M}}^T$ both in $\mathbf{T}$ and $\mathbf{T}_{\mathrm{M}}$, and for new samples. Another central question is which sample residuals we should use.

Projection of all row vectors of $\mathbf{X}$ onto the column space of $\mathbf{W}$ results in the nonorthogonalized PLSR model of Martens (Martens and Næs, 1987)

$$\mathbf{X} = \mathbf{X}\mathbf{W}\mathbf{W}^T + \mathbf{X}(\mathbf{I} - \mathbf{W}\mathbf{W}^T) = \mathbf{T}_{\mathrm{M}}\mathbf{W}^T + \mathbf{E}_{\mathrm{B}} \tag{8.35}$$

where $\mathbf{T}_{\mathrm{M}}$ is the nonorthogonalized score matrix. As pointed out, Equation 8.35 shows that the row vectors of $\mathbf{E}_{\mathrm{B}} = \mathbf{X}(\mathbf{I} - \mathbf{W}\mathbf{W}^T)$ are contained in the left nullspace or orthogonal complement of the column space of $\mathbf{W}$. From this follows a fundamental theorem, also illustrated in Figure 8.2 (Ergon *et al.*, 2011):

**Theorem 8.1** *In non-orthogonalized PLSR, each sample $z$ (row in $X$ or new sample) is split into two orthogonal components $\hat{z}_B = WW^T z$ and $\varepsilon_B = (I - WW^T)z$, with $\hat{z}_B$ contained in the sample projection space (column space of $W$), and $\varepsilon_B$ contained in the sample residual space (orthogonal complement of the column space of $W$).*

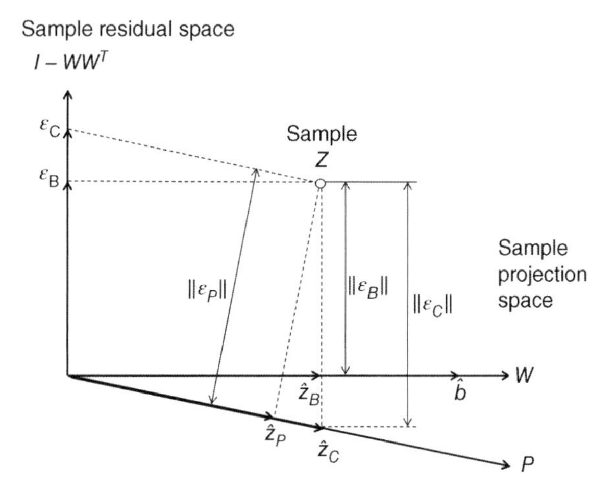


**Figure 8.2** Orthogonal splitting of sample $\mathbf{z}$ into $\hat{\mathbf{z}}_{\mathrm{B}}$ and $\boldsymbol{\varepsilon}_{\mathrm{B}}$, and nonorthogonal splitting into $\hat{\mathbf{z}}_{\mathrm{C}}$ and $\boldsymbol{\varepsilon}_{\mathrm{C}}$. The result is the projection $\hat{\mathbf{z}}_{\mathrm{B}}$ and the nonorthogonal mapping $\hat{\mathbf{z}}_{\mathrm{C}}$. Note that the orthogonality relations between fitted samples $\hat{\mathbf{z}}$ and residuals $\boldsymbol{\varepsilon}$ are valid for both the modelling data and new data. Ergon *et al.*, 2011. Reproduced with permission from John Wiley & Sons.

Note that a similar splitting into fitted samples $\hat{\mathbf{z}}_B$ and residuals $\boldsymbol{\varepsilon}_B$ in orthogonal subspaces also occurs in PCA and thus in PCR.

With $\mathbf{XW} = \mathbf{UR}$ model Equation 8.35 gives the Bidiag2 model (Golub and Kahan, 1965; Manne, 1987)

$$\mathbf{X} = \mathbf{URW}^T + \mathbf{E}_B \tag{8.36}$$

where **U** and **V** are orthonormal and **R** upper right bidiagonal. From model Equation 8.36 it is a short step to the RE-PLSR model (Ergon, 2009)

$$\mathbf{X} = \mathbf{UZ}^{-1}\mathbf{ZRW}^T + \mathbf{E}_B = \mathbf{TP}^T\mathbf{WW}^T + \mathbf{E}_B \tag{8.37}$$

where **Z** is a diagonal scaling matrix such that $\mathbf{P}^T\mathbf{W} = \mathbf{ZR}$ is upper right bidiagonal with ones along the main diagonal, and where the score vectors in $\mathbf{T} = \mathbf{UZ}^{-1}$ are orthogonal but no longer normalized. This model is also found by simply using $\mathbf{w}_A$ instead of $\mathbf{p}_A$ in the results from the original algorithm above (Ergon, 2002a, 2009). A comparison of Equations 8.35 and 8.37 shows that:

$$\mathbf{T} = \mathbf{XW}\left(\mathbf{P}^T\mathbf{W}\right)^{-1} \tag{8.38}$$

The scores $\boldsymbol{\tau}$ for an individual sample **z** are thus found by projection onto axes given by the columns of $\mathbf{W}\left(\mathbf{P}^T\mathbf{W}\right)^{-1}$, which are obtained through an oblique rotation of the **W** axes. Note that these scores are used also in conventional PLSR as discussed next, and that it follows that **T** and $\mathbf{T}_M = \mathbf{XW}$ span the same subspace. For a new set of samples Equation 8.38 gives $\mathbf{T}_{new} = \mathbf{X}_{new}\mathbf{W}\left(\mathbf{P}^T\mathbf{W}\right)^{-1}$, but it is essential to realize that $\mathbf{T}_{new}$ will not be orthogonal, that is $\left(\mathbf{W}^T\mathbf{P}\right)^{-1}\mathbf{W}^T\mathbf{X}_{new}^T\mathbf{X}_{new}\mathbf{W}\left(\mathbf{P}^T\mathbf{W}\right)^{-1}$ will not be a diagonal matrix (a discussion of asymptotic properties is given in the section *The PLSR residual issue in process monitoring*).

The models Equations 8.35, 8.36 and 8.37 are essentially the same, in that they are found by projection of rows in **X** onto the column space of **W**, as shown in Figure 8.2, and they therefore result in the same residuals $\mathbf{E}_B$. The conventional PLSR model, on the other hand, is found by projection of all column vectors of **X** onto the column space of **T**, as shown in Figure 8.3 (Ergon *et al.*, 2011). The result is the model

$$\mathbf{X} = \mathbf{T}\left(\mathbf{T}^T\mathbf{T}\right)^{-1}\mathbf{T}^T\mathbf{X} + \left(\mathbf{I} - \mathbf{T}\left(\mathbf{T}^T\mathbf{T}\right)^{-1}\mathbf{T}^T\right)\mathbf{X} = \mathbf{TP}^T + \mathbf{E}_C \tag{8.39}$$

that is as in model Equation 8.31, where both **T** and **P** follow from the conventional PLSR algorithm above. The column vectors of $\mathbf{E}_C$ are contained in the orthogonal complement of the column space of **T**, that is $\mathbf{T}^T\mathbf{E}_C = 0$, which is the conventional argument for use of model Equation 8.31 (Bro and Eldén, 2009; Wold *et al.*, 2009). Note, however, that this in practice is true for the modelling data only (see the discussion in the section *The PLSR residual issue in process monitoring*). We underline some other important facts in a theorem:

**Theorem 8.2** *In the conventional PLSR model (*Equation 8.31*) with* $\hat{X}_C = \boldsymbol{TP}^T$, *the fitted samples* $\hat{z}_C = \boldsymbol{P\tau} = \boldsymbol{P}\left(\boldsymbol{W}^T\boldsymbol{P}\right)^{-1}\boldsymbol{W}^T z$ *are contained in the column space of* $\boldsymbol{P}$, *while the residuals* $\varepsilon_C$ *are contained in the orthogonal complement of the column space of* $\boldsymbol{W}$. *Each sample is thus split into two in general non-orthogonal components* $\hat{z}_C$ *and* $\varepsilon_C$, *one in the column space of* $\boldsymbol{P}$, *and the other in the common sample residual space. (See Ergon et al. (2011) for a proof.)*

Since $\hat{\mathbf{X}}_M = \mathbf{TP}^T\mathbf{WW}^T$ is the projection of $\hat{\mathbf{X}}_C$ onto the column space of **W**, the location of a specific $\hat{\mathbf{z}}_C$ will in principle be as shown in Figure 8.2. The fact that $\hat{\mathbf{z}}_C$ and $\boldsymbol{\varepsilon}_C$ are *nonorthogonal* explains the ambiguities found in the section *The PLSR residual issue in process monitoring*, where so-called squared prediction error (SPE) plots used in process monitoring systems are discussed.

**Remark 8.1** *When* $\mathbf{W}$ *has been established as described above, the orthogonalization of the non-orthogonal score matrix* $\mathbf{T}_{\mathrm{M}} = \mathbf{XW}$ *can be performed by QR factorization. This will give both* $\mathbf{T}$ *and the upper bidiagonal matrix* $\mathbf{ZR} = \mathbf{P}^T\mathbf{W}$ *in model* Equation 8.37*(Ergon,2009), and if needed we can then also find* $\mathbf{P} = \mathbf{X}^T\mathbf{T}\left(\mathbf{T}^T\mathbf{T}\right)^{-1}$.

**Remark 8.2** *The axes* $\mathbf{W}$ *and* $\mathbf{P}$ *in* Figure 8.2 *indicate the different column spaces for these matrices. However, these spaces are overlapping in that* $\mathbf{p}_1, \mathbf{p}_2, \ldots, \mathbf{p}_{A-1}$ *are contained in the column space of* $\mathbf{W}$, *that is it is* $\mathbf{p}_A \neq \mathbf{w}_A$ *that makes* $\hat{\mathbf{z}}_{\mathrm{C}} \neq \hat{\mathbf{z}}_{\mathrm{B}}$ *and thus* $\varepsilon_{\mathrm{C}} \neq \varepsilon_{\mathrm{B}}$ (Ergon *et al.*, 2011).

As already mentioned the RE-PLSR model (Equation 8.37) can also be found by use of the original algorithm for conventional PLSR, simply by using $\mathbf{w}_A$ instead of $\mathbf{p}_A$ in the last component (Ergon, 2002a, 2009). From this follows

$$\mathbf{E}_{\mathrm{C}} = \mathbf{E}_{\mathrm{B}} - \mathbf{t}_A\left(\mathbf{p}_A^T - \mathbf{w}_A^T\right) = \mathbf{E}_{\mathrm{B}} + \sqrt{\mathbf{p}_A^T\mathbf{p}_A - 1}\,\mathbf{t}_A\mathbf{w}_{A+1}^T, \tag{8.40}$$

where we make use of the fact that $\mathbf{p}_A - \mathbf{w}_A = -\sqrt{\mathbf{p}_A^T\mathbf{p}_A - 1}\,\mathbf{w}_{A+1}$ (Ergon *et al.*, 2011). It is essential to note that this relation between the residuals is valid also for new samples. Also note that the term $\sqrt{\mathbf{p}_A^T\mathbf{p}_A - 1}\,\mathbf{t}_A\mathbf{w}_{A+1}^T$ in Equation 8.40 is contained in the common sample residual space, where it accounts for the difference between $\varepsilon_{\mathrm{C}}$ and $\varepsilon_{\mathrm{B}}$ in Figure 8.2.

**Remark 8.3** *Note that the residuals* $\varepsilon_{\mathrm{B}}$ *for both the modelling and new data can be found without use of the Bidiag2 algorithm with its numerical shortcomings (Andersson,* 2009*).*

**Remark 8.4** *The reason for the term* $\sqrt{\mathbf{p}_A^T\mathbf{p}_A - 1}\,\mathbf{t}_A\mathbf{w}_{A+1}^T$ *in* Equation 8.40 *is that* $\mathbf{E}_{\mathrm{C}}$ *is formed by projection of the columns of* $\mathbf{X}$ *onto the column space of* $\mathbf{T}$, *and not by projection of samples. A corresponding term does not occur in PCA, where projections onto loading and score spaces give the same residuals.*

*Residual properties*

Residual properties for new samples as well as for samples in the modelling data follow from Figure 8.2; these are especially relevant for the squared prediction error (SPE) discussion in the section *The PLSR residual issue in process monitoring* below. Note especially that the residuals $\varepsilon_{\mathrm{B}}^T$ and $\varepsilon_{\mathrm{C}}^T$ are both contained in the common sample residual space, that is in the orthogonal complement of the column space of $\mathbf{W}$, which means that once $\mathbf{W}$ is established none of them will have any influence on the predictions.

Some other residual properties that are valid for the modelling data only are shown in Figure 8.3, where $\mathbf{e}_{\mathrm{C}}$ and $\mathbf{e}_{\mathrm{B}}$ are column vectors of $\mathbf{E}_{\mathrm{C}}$ and $\mathbf{E}_{\mathrm{B}}$, respectively. An essential detail in Figure 8.3 may be underlined in a theorem:

**Theorem 8.3** *The column vectors in the residual matrix* $\boldsymbol{E}_B$ *are orthogonal to the response vector* $\boldsymbol{y}$, *that is* $\boldsymbol{y}^T\boldsymbol{E}_B = 0$. *(For proof see Ergon et al.,* 2011.*)*

Note especially that the orthogonality between $\mathbf{T}$ and $\mathbf{e}_{\mathrm{C}}$ shown in Figure 8.3 is valid only for the modelling data (see the subsection *Covariance structures* for a discussion of the corresponding property for a set of new samples).

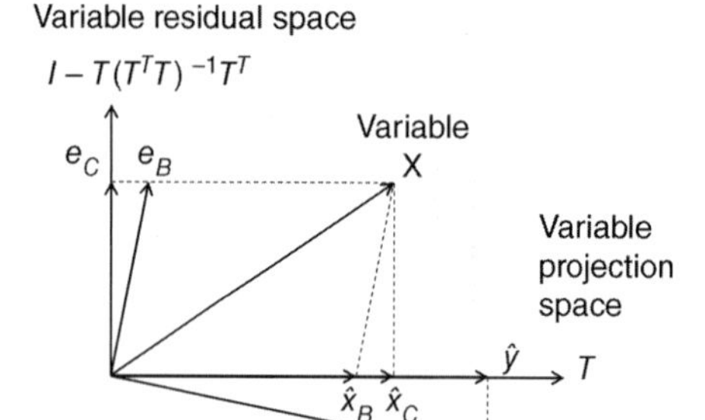


**Figure 8.3** Residual column properties for the modelling data. Note that the orthogonality relations between fitted variable vectors $\hat{\mathbf{x}}$ and residuals $\mathbf{e}$ are valid for the modelling data only. Ergon *et al.*, 2011. Reproduced with permission from John Wiley & Sons.

# THE PLSR RESIDUAL ISSUE IN PROCESS MONITORING

## Problem formulation

Industrial applications of PLSR are found in, for example, process monitoring systems. The typical situation is then that a primary product quality that is difficult or impossible to measure online is estimated by use of secondary measurements around the process, and PLSR is here a useful alternative to PCA (Kourti and MacGregor, 1996; Wise and Gallagher, 1996). In such cases it is necessary to keep an eye on the squared perpendicular distance, or squared prediction error (SPE) in $\mathbf{X}$, from a sample $\mathbf{z}$ to the projection space where the scores are found and the regression solution lives. Using PCA we thus find:

$$\mathrm{SPE}_{\mathrm{PCA}} = \boldsymbol{\varepsilon}_{\mathrm{PCA}}^T \boldsymbol{\varepsilon}_{\mathrm{PCA}} = \mathbf{z}^T(\mathbf{I} - \mathbf{P}_{\mathrm{PCA}}\mathbf{P}_{\mathrm{PCA}}^T)\mathbf{z} \tag{8.41}$$

For the nonorthogonalized PLSR, Bidiag2 and RE-PLSR models (Equations 8.35, 8.36 and 8.37), the sample projection space is the column space of $\mathbf{W}$, and as shown in Figure 8.2 the distance of interest is thus found from the projection onto the orthogonal complement of this space (the sample residual space). For both the modelling data and new data, the distance from a sample $\mathbf{z}$ to the projection space is $\boldsymbol{\varepsilon}_{\mathrm{B}} = (\mathbf{I} - \mathbf{W}\mathbf{W}^T)\mathbf{z}$, from which follows:

$$\mathrm{SPE}_{\mathrm{B}} = \boldsymbol{\varepsilon}_{\mathrm{B}}^T \boldsymbol{\varepsilon}_{\mathrm{B}} = \mathbf{z}^T(\mathbf{I} - \mathbf{W}\mathbf{W}^T)\mathbf{z} \tag{8.42}$$

The situation for the conventional PLSR model (Equation 8.31) is ambiguous, however. As shown in *Theory* section, the scores are also found here by projection onto the column space of $\mathbf{W}$, or more precisely onto the axes defined by $\mathbf{W}(\mathbf{P}^T\mathbf{W})^{-1}$, and this gives $\mathrm{SPE}_{\mathrm{B}}$, as in Equation 8.42. From model Equation 8.31 and Equation 8.38, however, it follows that $\boldsymbol{\varepsilon}_{\mathrm{C}}^T = \mathbf{z}^T(\mathbf{I} - \mathbf{W}(\mathbf{P}^T\mathbf{W})^{-1}\mathbf{P}^T)$ or, as found from Equation 8.40, $\boldsymbol{\varepsilon}_{\mathrm{C}}^T = \boldsymbol{\varepsilon}_{\mathrm{B}}^T + \sqrt{\mathbf{p}_A^T\mathbf{p}_A - 1}\,\tau_A \mathbf{w}_{A+1}^T$, where $\tau_A$ is computed from $\boldsymbol{\tau}^T = [\tau_1 \quad \tau_2 \quad \cdots \quad \tau_A] = \mathbf{z}^T\mathbf{W}(\mathbf{P}^T\mathbf{W})^{-1}$. As shown in Ergon *et al.* (2011) the result of this for both the modelling and new data is that

$$\mathrm{SPE}_{\mathrm{C}} = \boldsymbol{\varepsilon}_{\mathrm{C}}^T\boldsymbol{\varepsilon}_{\mathrm{C}} = \mathrm{SPE}_{\mathrm{B}} + 2\sqrt{\mathbf{p}_A^T\mathbf{p}_A - 1}\,\tau_A \mathbf{z}^T\mathbf{w}_{A+1} + (\mathbf{p}_A^T\mathbf{p}_A - 1)\tau_A^2 \tag{8.43}$$

$\mathrm{SPE}_{\mathrm{C}}$ can thus be both over- and underestimating the $\mathrm{SPE}_{\mathrm{B}}$ magnitude, depending on the sign and value of $\tau_A \mathbf{z}^T\mathbf{w}_{A+1}$.

From a geometrical point of view, model Equation 8.31 and thus $\hat{\mathbf{z}}_C^T = \boldsymbol{\tau}^T\mathbf{P}^T$ gives fitted samples $\hat{\mathbf{z}}_C$ that are contained in the the column space of $\mathbf{P}$, but then through *nonorthogonal mappings* and not orthogonal projections. The position of $\hat{\mathbf{z}}_C$ is thus determined by scores found from projection of $\mathbf{z}$ onto the column space of $\mathbf{W}$, but these scores are then applied in the column space of $\mathbf{P}$. As shown in Figure 8.2, the mapping of $\hat{\mathbf{z}}_C$ is given by the projection $\hat{\mathbf{z}}_B = \mathbf{W}\mathbf{W}^T\hat{\mathbf{z}}_C$. Note that a different position of $\mathbf{z}$ may give $\|\boldsymbol{\varepsilon}_C\| < \|\boldsymbol{\varepsilon}_B\|$.

The perpendicular distance from a sample to the column space of $\mathbf{P}$ is not $\boldsymbol{\varepsilon}_C$ but $\boldsymbol{\varepsilon}_P = (\mathbf{I} - \mathbf{P}(\mathbf{P}^T\mathbf{P})^{-1}\mathbf{P}^T)\mathbf{z}$, which corresponds to $\hat{\mathbf{z}}_P = \mathbf{P}(\mathbf{P}^T\mathbf{P})^{-1}\mathbf{P}^T\mathbf{z}$, with $\hat{\mathbf{z}}_P$ located as shown in Figure 8.2. This solution was discussed by Pell *et al.* (2007).

## Covariance structures

When comparing the different PLSR models and the residuals $\mathbf{E}_C$ and $\mathbf{E}_B$, it is of importance to discuss possible covariation between sample scores $\boldsymbol{\tau}$ and sample residuals $\boldsymbol{\varepsilon}_C$ and $\boldsymbol{\varepsilon}_B$; we will focus here on this problem in the process monitoring context. In conventional PLSR, according to model Equation 8.31, this covariation in the modelling data is given by the expectation

$$E\{\boldsymbol{\tau}\boldsymbol{\varepsilon}_C^T\} = \lim_{N\to\infty}\left\{\frac{1}{N-1}\sum_{i=1}^{N}\boldsymbol{\tau}_i\boldsymbol{\varepsilon}_{C,i}^T\right\} = \lim_{N\to\infty}\left\{\frac{1}{N-1}\mathbf{T}^T\mathbf{E}_C\right\} = 0 \tag{8.44}$$

Although $\mathbf{T}^T\mathbf{E}_C = 0$ by design, the matrices $\mathbf{T}$ and $\mathbf{E}_C$ will depend on the number of samples and stabilize only when $N \to \infty$. Since scores and residuals for new samples are based on the model according to $\boldsymbol{\tau}_{\text{new}}^T = \mathbf{x}_{\text{new}}^T\mathbf{W}(\mathbf{P}^T\mathbf{W})^{-1}$ (Equation 8.38) and $\boldsymbol{\varepsilon}_{C,\text{new}}^T = \mathbf{x}_{\text{new}}^T - \boldsymbol{\tau}_{\text{new}}^T\mathbf{P}^T$ (Equation 8.31), the expected covariance for new data approaches zero only when the number of modelling samples is large, that is when $\mathbf{W}$ and $\mathbf{P}$ have stabilized. This means that $E\{\boldsymbol{\tau}_{\text{new}}\boldsymbol{\varepsilon}_{C,\text{new}}^T\} \to 0$ only when $N \to \infty$, which for a new set of data would give $\mathbf{T}_{\text{new}}^T\mathbf{E}_{C,\text{new}} \to 0$ for $N_{\text{new}} \to \infty$ (assuming no drift). Correspondingly, the matrix $\mathbf{T}_{\text{new}}^T\mathbf{T}_{\text{new}}$ will be diagonal when and only when both $N \to \infty$ and $N_{\text{new}} \to \infty$.

From the alternative PLSR models (Equations 8.35, 8.36 and 8.37) follows

$$E\{\boldsymbol{\tau}\boldsymbol{\varepsilon}_B^T\} = \lim_{N\to\infty}\left\{\frac{1}{N-1}\sum_{i=1}^{N}\boldsymbol{\tau}_i\boldsymbol{\varepsilon}_{B,i}^T\right\} = \lim_{N\to\infty}\left\{\frac{1}{N-1}\mathbf{T}^T\mathbf{E}_B\right\} \neq 0 \tag{8.45}$$

which shows that the residuals $\boldsymbol{\varepsilon}_B$ will be correlated with the scores also for the modelling data. This is the conventional argument against the alternative PLSR models (Bro and Eldén, 2009).

### Example 8.2

As an illustration of the covariance theory above a simple simulation example is included here. We assumed a three component mixture with Gaussian frequency spectra $s_i(f) = 100(2\pi\sigma^2)^{-0.5}\exp(-(f-\mu_i)^2/2\sigma^2)$ over 500 frequencies, with $\sigma = 25$ and $\mu_i = 200$, 250 and 300 respectively. We varied the three concentrations randomly over $N$ modelling samples and $N_{\text{test}}$ test samples, using normal distribution $\mathcal{N}([10\quad 10\quad 10]^T, \mathbf{I})$, and added random noise with distribution $\mathcal{N}([0\quad 0\quad 0]^T, 0.5\mathbf{I})$ to the spectra. We used the concentration of the component with $\mu = 200$ as response variable, with added normal noise with variance $\sigma_y^2 = 0.04$. Models were found by use of various values of $N$ and centred data, and tested against data with the fixed value $N_{\text{test}} = 16000$. Mean results for RMSEP (root mean square error of prediction) and Frobenius norms from $M = 100$ Monte Carlo runs are shown in Figure 8.4. In all cases we found $\|\frac{1}{1-N}\mathbf{T}^T\mathbf{E}_C\|_F = 0$. The main result in the present context is that $\|\frac{1}{1-N_{\text{test}}}\mathbf{T}_{\text{test}}^T\mathbf{E}_{C,\text{test}}\|_F$ depends on the number of modelling

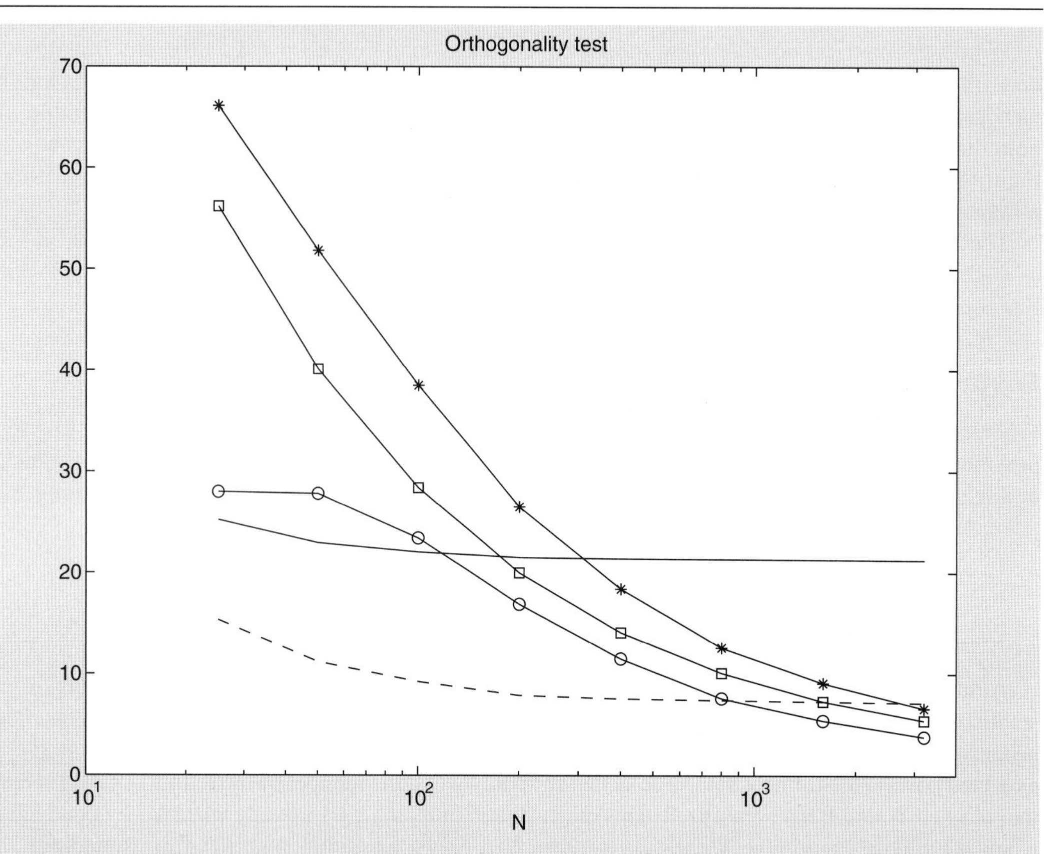


**Figure 8.4** $\|\frac{1}{1-N}\mathbf{T}^T\mathbf{E}_\mathrm{B}\|_F$ (circles), $\|\frac{1}{1-N_\mathrm{test}}\mathbf{T}_\mathrm{test}^T\mathbf{E}_\mathrm{C,test}\|_F$ (squares) and $\|\frac{1}{1-N_\mathrm{test}}\mathbf{T}_\mathrm{test}^T\mathbf{E}_\mathrm{B,test}\|_F$ (stars) as functions of $N$. 100*RMSEP values with (solid line) and without (dashed line) random noise on the test responses are also shown. The test set had $N_\mathrm{test} = 16000$. Ergon *et al.*, 2011. Reproduced with permission from John Wiley & Sons.

samples, and that it is far from zero for normal values of $N$. The reason why $\|\frac{1}{1-N}\mathbf{T}^T\mathbf{E}_\mathrm{B}\|_F$ does not increase for low values of $N$ is that RMSEC (root mean square error of calibration) $\rightarrow 0$ when $N \rightarrow A+1$, such that $\hat{\mathbf{y}} \rightarrow \mathbf{y}$ and $\mathbf{E}_\mathrm{B} \rightarrow \mathbf{E}_\mathrm{C}$ (Figure 8.4). Although it cannot be seen in Figure 8.4, we must expect that $\|\frac{1}{1-N_\mathrm{test}}\mathbf{T}_\mathrm{test}^T\mathbf{E}_\mathrm{C,test}\|_F \rightarrow 0$ when both $N \rightarrow \infty$ and $N_\mathrm{test} \rightarrow \infty$, while $\|\frac{1}{1-N_\mathrm{test}}\mathbf{T}_\mathrm{test}^T\mathbf{E}_\mathrm{B,test}\|_F \rightarrow \|\frac{1}{1-N}\mathbf{T}^T\mathbf{E}_\mathrm{B}\|_F \neq 0$. These asymptotic properties were more clearly seen in a separate test with $\sigma_y^2 = 0.5$, which as must be expected gave RMSEP $> \sqrt{0.5}$ with noise also on the test responses (but RMSEP $> 0.0763$ without test response noise; Næs *et al.* (2004) provides a general discussion of how errors in $y$ average out).

The discussion above shows that there is a theoretical covariance difference between the two alternative sample residuals $\varepsilon_\mathrm{C}$ and $\varepsilon_\mathrm{B}$. However, as shown in the simulation example and also as found in the real data examples in Ergon *et al.* (2011), $\mathbf{T}_\mathrm{test}^T\mathbf{E}_\mathrm{C,test}$ is far from zero in practical applications with limited modelling data, also when $N_\mathrm{test} \rightarrow \infty$. Rather, in all the examples we find that $\mathbf{T}_\mathrm{test}^T\mathbf{E}_\mathrm{C,test} \approx \mathbf{T}_\mathrm{test}^T\mathbf{E}_\mathrm{B,test}$, that is $\varepsilon_\mathrm{C}$ and $\varepsilon_\mathrm{B}$ will be correlated with the scores to approximately the same extent, such that $\mathrm{SPE}_\mathrm{C}$ and $\mathrm{SPE}_\mathrm{B}$ are quite similar.

What is even more important is that the discussion so far has been limited to normal process operation, where the few SPE values above the defined statistical upper limit must be ignored by the fault detection system, in order to avoid false alarms. However, the main motivation behind the use of SPE plots in process monitoring systems is not to detect extreme values from normal process operation, but the need to detect the occurrence of new types of special events, which were not present in the modelling data used to develop the in-control model (Kourti and MacGregor, 1996). Such events cannot be expected to follow the covariance structure of the modelling data and the score-residual orthogonality argument for use of conventional PLSR is, therefore, not valid for process fault data, not even in the unrealistic case of a very large number of modelling samples. Process fault data are, in other words, not representative for normal process operation (Esbensen and Geladi, 2010).

## Combination of SPE and $T^2$ plots

When SPE plots from PLSR are used in process monitoring systems, they are often used together with Hotelling's $T^2$ plots (Kourti and MacGregor, 1996; Wise and Gallagher, 1996), based on the definition

$$T^2 = \boldsymbol{\tau}^T \left( \frac{1}{N-1} \mathbf{T}^T \mathbf{T} \right)^{-1} \boldsymbol{\tau} = (N-1)\mathbf{z}^T \mathbf{W} \left( \mathbf{W}^T \mathbf{X}^T \mathbf{X} \mathbf{W} \right)^{-1} \mathbf{W}^T \mathbf{z} \tag{8.46}$$

where $\boldsymbol{\tau}^T$ follows from Equation 8.38 as $\boldsymbol{\tau}^T = \mathbf{z}^T \mathbf{W}(\mathbf{P}^T \mathbf{W})^{-1}$. The value of $T^2$ for a specific sample is thus the sum of squares of the scores, adjusted to equal variance, that is a statistical measure of the distance from the origin within the sample projection space. From this follows that $T^2$ is the squared length of the vector $\mathbf{v} = \mathbf{WMW}^T\mathbf{z} = \mathbf{WMW}^T\hat{\mathbf{z}}_\text{B}$, where $\mathbf{M} = \sqrt{N-1}\left(\mathbf{W}^\mathbf{T}\mathbf{X}^\mathbf{T}\mathbf{XW}\right)^{-\frac{1}{2}}$. The vector $\mathbf{v}$ is contained in the column space of $\mathbf{W}$, that is the $T^2$ value for each sample is defined in this space. From this and the definitions of $\text{SPE}_\text{B}$ and $\text{SPE}_\text{C}$ in Equations 8.42 and 8.43 follow two important propositions (Ergon *et al.*, 2011):

**Proposition 8.1** Under the assumption of a process fault that is not governed by a specific covariance structure, $\text{SPE}_\text{B}$ and $T^2$ are uncorrelated. (For proof see Ergon *et al.*, 2011.)

**Proposition 8.2** Under the same assumption as in Proposition 1, $\text{SPE}_\text{C}$ and $T^2$ are correlated, except for samples where $\tau_A = 0$. (For mathematical and geometrical proofs, see Ergon *et al.*, 2011.)

**Remark 8.5** *In the case of a specific fault covariance structure, it is highly unlikely that it is the same as for normal process data, such that $SPE_\text{C}$ should be used.*

**Remark 8.6** *The situation is further complicated by the fact that the SPE values may be the combined results of normal process data for some variables and abnormal fault data for other variables, such that both $SPE_\text{B}$ and $SPE_\text{C}$ to some extent are correlated with $T^2$.*

**Remark 8.7** *Owing to the very limited number of modelling samples often used in practice, both $SPE_\text{B}$ and $SPE_\text{C}$ will to some extent be correlated with $T^2$ also in normal process operation (as discussed above).*

## Conclusion on the PLSR residual issue

From the discussion above, the conclusion on the PLSR residual issue in the process monitoring setting appears to be that the conventional residuals $\varepsilon_\text{C}$ should be used to detect extreme values of samples from

**X** data that are representative. However, once one decides that a sample is not representative, the alternative residual $\varepsilon_{\mathrm{B}}$ should be used for further fault investigation. Industrial data examples with combined use of $\mathrm{SPE_B}$ and $T^2$ plots can be found elsewhere (Ergon *et al.*, 2011). These examples show that there are very small differences between $\mathrm{SPE_C}$ and $\mathrm{SPE_B}$ plots for normal numbers of modelling samples and small values of $T^2$, which is an argument for use of $\mathrm{SPE_B}$ plots exclusively.

## PLSR SCORE LOADING CORRESPONDENCE

PCA has the nice property that there is a correspondence between loading and score plots. This follows from the fact that a sample $\mathbf{z}_i^T$ according to Equation 8.18 gives a score

$$\boldsymbol{\tau}_{i,PCA}^T = \mathbf{z}_i^T \mathbf{P}_{PCA} = \mathbf{z}_i^T [\mathbf{p}_1 \quad \mathbf{p}_2 \quad \cdots \quad \mathbf{p}_A] = \mathbf{z}_i^T [\pi_1 \quad \pi_2 \quad \cdots \quad \pi_K]^T \tag{8.47}$$

such that a given sample $\mathbf{z}_i^T = [0 \quad \cdots \quad 0 \quad z_{ij} \quad 0 \quad \cdots \quad 0]$ gives a score

$$\boldsymbol{\tau}_{i,PCA}^T = z_{ij}\pi_j \tag{8.48}$$

The position of such a sample in a score plot, will thus be in the same direction as the corresponding loading in the loading plot, and for $z_{ij} = 1$ the distance from the origin will be the same. When more than one element in $\mathbf{z}_i^T$ is different from zero, the position of the score will be given by the vector sum for the elements involved. This correspondence can be used by showing both scores and loadings in the same bi-plot (Thielemans *et al.*, 1988).

The same type of correspondence does not exist for orthogonalized PLSR, for the reason that **T** is given by Equation 8.38, where $\mathbf{W}(\mathbf{P}^T\mathbf{W})^{-1}$ is not orthogonal. It does, however, exist for nonorthogonalized PLSR according to model Equation 8.35, for the reason that the loading weights matrix here is orthogonal, such that the score matrix is $\mathbf{T}_{\mathrm{M}} = \mathbf{XW}$ and thus $\boldsymbol{\tau}_{i,\mathrm{M}}^T = \mathbf{z}_i^T\mathbf{W}$. The interpretational drawback may here be that $\mathbf{T}_{\mathrm{M}}$ has nonorthogonal column vectors. There is a solution to this problem, based on the following reformulation of the nonorthogonalized PLSR model, making use of singular value decomposition of the score matrix:

$$\begin{aligned}\mathbf{X} &= \mathbf{T}_{\mathrm{M}}\mathbf{W}^T + \mathbf{E}_{\mathrm{B}} = \mathbf{USV}^T\mathbf{W}^T + \mathbf{E}_{\mathrm{B}} = [\mathbf{U}_1 \quad \mathbf{U}_2]\begin{bmatrix}\mathbf{S}_1\\0\end{bmatrix}\mathbf{V}^T\mathbf{W}^T + \mathbf{E}_{\mathrm{B}}\\ &= (\mathbf{U}_1\mathbf{S}_1)(\mathbf{WV})^T + \mathbf{E}_{\mathrm{B}} = \mathbf{T}_{\mathrm{B}}\mathbf{V}_{\mathrm{B}}^T + \mathbf{E}_{\mathrm{B}}\end{aligned} \tag{8.49}$$

In this way we obtain a so-called bi-orthogonal factorization of **X**, with orthogonal score vectors in $\mathbf{T}_{\mathrm{B}}$ and an orthonormal loading matrix $\mathbf{V}_{\mathrm{B}}$, with $\mathbf{V}_{\mathrm{B}}^T\mathbf{V}_{\mathrm{B}} = \mathbf{I}$. After this factorization, the order of components according to explanatory power may be lost, although the total explanatory power for all $A$ components will not be altered. The ordering may be restored by use of a permutation matrix (Ergon, 2002b).

Score-loading bi-plots are especially informative for models with only two components, that is in combination with model reduction methods, as described next. Industrial data examples using bi-plots for reduced models are given elsewhere (Ergon, 2004, 2007).

## MODEL REDUCTION METHODS

The interpretation of score and loading plots from PCR and PLSR is difficult when many components are involved and several ways to obtain reduced models have, therefore, been developed. Short summaries of such methods are given here.

## OPLS

The OPLS method of Trygg and Wold (2002) has gained quite some interest. The OPLS algorithm is, unfortunately, difficult to understand, but the essential point is that it results in the factorization (assuming only one $\mathbf{y}$ relevant component, and thus $A-1$ $\mathbf{y}$ orthogonal components)

$$\mathbf{X} = \mathbf{T}_{\text{ortho}}\mathbf{P}_{\text{ortho}}^T + \mathbf{t}_A^{\text{OPLS}}\left(\mathbf{p}_A^{\text{OPLS}}\right)^T + \mathbf{E}_{\text{C}} \tag{8.50}$$

where the score vectors in $T_{\text{ortho}}$ are orthogonal and also orthogonal to $\mathbf{y}$, $\hat{\mathbf{y}}$ and $\mathbf{t}_A^{\text{OPLS}}$. The algorithm also finds the first ordinary loading weights vector $\mathbf{w}_1$, and an OPLS loading weights matrix $\mathbf{W}_{\text{ortho}}$, and the key for understanding the algorithm is that $\mathbf{W}_{\text{ortho}} = -[\mathbf{w}_2 \quad \mathbf{w}_3 \quad \cdots \quad \mathbf{w}_A]$. OPLS has been presented as a preprocessing method, but that is not true. Exactly the same result will be obtained by first determining $\mathbf{W}$ by use of the conventional PLSR algorithm and then repeating the algorithm with predetermined $\mathbf{w}$ vectors in the order $\mathbf{w}_2, \quad \mathbf{w}_3, \quad \cdots, \quad \mathbf{w}_A, \quad \mathbf{w}_1$ (Ergon, 2005). The reason is that the vectors in $\mathbf{W}_{\text{ortho}}$, and thus also $\mathbf{w}_2$ to $\mathbf{w}_A$, are found by the algorithm, and then always one step ahead. As a consequence of that, OPLS gives exactly the same predictions as the original PLSR algorithm. Other aspects of the practical utility of OPLS are given in Tapp and Kemsley (2009).

## PLS + ST

A simple alternative to OPLS was presented in Ergon (2005). It is based on the RE-PLSR model Equation 8.37 and uses a straightforward similarity transformation to obtain

$$\mathbf{X} = \mathbf{T}_{\text{ortho}}\mathbf{P}_{\text{ortho}}^T\mathbf{W}\mathbf{W}^T + \mathbf{t}_A^{\text{OPLS}}\mathbf{w}_1^T + \mathbf{E}_{\text{B}} \tag{8.51}$$

This approach has been further developed by Kemsley and Tapp (2009) and by Kvalheim *et al.* (2008). Kemsley *et al.* (2010) at the Institute of Food Research in Norwich, UK, have developed it into a Statistical Toolkit for Metabolomics. A similar approach, called X-tended target projection (XTP), was developed by Kvalheim *et al.* (2008).

Similarity transformations have also been used to find the smallest possible part of $\mathbf{X}$ that can be used for explanation of multivariate response variables $\mathbf{Y}$ (Ergon, 2007).

## 2PLS

Ergon (2003b) has developed a method called 2PLS, where a nonorthogonalized PLSR model of $\mathbf{X}$ with a single response variable and many components is compressed into a model with only two components, that is into a form that can be visualized by 2-D plots. This is done with retained predictive power, that is with the same prediction vector $\hat{\mathbf{b}}_{\text{PLSR}}$. 2PLS exploits three useful properties of the nonorthogonalized factorization Equation 8.35. Firstly, only the first score vector is covariant with the response vector $\mathbf{y}$, while the second and subsequent score vectors are all orthogonal to $\mathbf{y}$. Secondly, the associated loading vectors are orthonormal. Thirdly, the first loading vector corresponds to a least-squares estimate of the 'true spectral profile'. Tapp *et al.* (2011) have developed the 2PLS method further into the so-called TinyLVR method.

## PCP

Langsrud and Næs (2003) have developed a multiresponse method called PCP (principal components of prediction), aimed at finding which directions in $\mathbf{Y}$ space can be predicted by which directions in $\mathbf{X}$ space.

### Model reduction in multiresponse cases

Ergon (2006) has shown that a multiresponse PCR model with a common number of components for all responses, as well as a PLSR model for several responses (PLS2), without loss of predictive power can be compressed into a model with as many components as responses. This is done by projection onto the subspace spanned by the prediction vectors $\hat{\mathbf{b}}_1$, $\hat{\mathbf{b}}_2$ and so on, and the obvious advantage is easier interpretation. This type of model reduction can also be done with only one response vector, resulting in a one-component model consisting of $\hat{\mathbf{b}}$ itself. That will, however, make score and loading plots impossible, and an alternative is then the 2PLS approach.

## ESTABLISHED PCR AND PLSR PRACTICES

Many aspects in relation to PCR and PLSR practice are well established. Accordingly, they are well discussed in the literature (Martens and Næs, 1989; Höskuldsson, 1996; Bereton, 2003; Næs *et al.*, 2004; Otto, 2007; Esbensen, 2010) and included in commercial programs. Therefore, only short comments on some of the main aspects are included here.

### Pre-treatment of data

#### *Centring and scaling*

In many cases the main focus is on variations around a fixed operating point, and centring of the data is then the natural choice. In spectral data the measurement units for the variables (wavelengths or frequencies) are often comparable across the whole spectrum and no form of scaling is then normally performed. In process monitoring cases, however, the **X** variables may be measured on very different scales and some form of scaling is then necessary. One may, for example, adjust the **X** variables to similar noise levels, or to similar variance, or use some other experience based scaling.

#### *Outlier detection*

Outliers in calibration data are unavoidable in practice and will be found in both **X** and **y** data. The most important tool for detection of **X** outliers in PCR and PLSR is score plots, where the Mahalanobis (statistical) distance from $\bar{\mathbf{t}}$ (the origin with centred data) will tell about the leverage a given sample has on the result. The most common tool for detecting **y** outliers is the residual $y_i - \hat{y}_i$. Obvious outliers due to mistakes should if possible be corrected, or otherwise removed, but excessive removal of outliers must be avoided.

#### *Orthogonal signal correction*

The idea behind orthogonal signal correction is to remove all information in **X** that is not related to **y**. The most widely used method for this is probably OPLS, but as discussed previously, this is in fact a disguised post-processing method.

#### *Variable selection*

Variable selection in PCR and PLSR may improve predictions, give better interpretations, or give lower measurement costs, and there are several approaches available (Andersen and Bro, 2010).

#### *Smoothing*

The quality of measurements that are contaminated by high frequency noise may be improved by, for example, moving average filters, in the simplest form just the mean values from several consecutive measurements. When used in prediction, this will obviously impair the ability to detect rapid changes. Also note that the term *smoothing* in signal processing literature refer to a specific form of Kalman filtering, where time series data are run in both directions.

#### *Fourier transformations*

In, for example, acoustic chemometrics (Halstensen and Esbensen, 2010), the **X** variables are found by Fourier analysis of a recorded analogue signal over a specific period of time, such that the frequency content of the signal is found.

#### *Scatter correction*

Scatter in spectroscopic data are irrelevant variations caused by, for example, light and particle surface interactions. Scatter correction is in some cases absolutely necessary and there exist several methods for doing this.

#### *Validation*

Model validation is essential in PCR and PLSR, both for finding the appropriate number of components to use and for evaluation of the prediction performance. Statistical methods for comparison of models must penalize the number of variables; this is done by use of the residual sum of squares (RSS) in the Mallows $C_p$ criterion (Mallows, 1973), which is a special case of Aikaikes Information Criterion (AIC) (Aikaike, 1974). When the number of variables is given or determined, a common approach is to use cross-validation, especially in the many practical cases where the number of modelling samples is low. There are, however, strong arguments for use of separate test sets (Esbensen and Geladi, 2010).

### Uncertainty estimation

The uncertainty of multivariate prediction coefficients are obliviously of great interest. There are basically two ways of doing this, namely, error propagation and various resampling strategies (Geladi, 2002). A detailed discussion is given in Olivieri *et al.* (2006). More recently, Faber (2009) has pointed out that neither of the alternative residuals in the alternative PLSR models Equations 8.31 and 8.35 discussed previously provide the right input for uncertainty estimation, and that rather the residuals from PCA optimized for fitting the **X** data should be used.

### Calibration transfer

Calibration transfer is useful when collecting data for similar samples on two or more different instruments (or on the same instrument at two or more different points in time). The goal is to create a model to compensate for differences in the instruments that can then be used to eliminate or reduce variation caused by the change in instrument. A review of methods for this has been given in Feudale *et al.* (2002) and a more recent comparative study in Pereira *et al.* (2008).

### Three-way methods

Batch process data can be arranged in a three-way matrix (batch $\times$ variable $\times$ time) as discussed in Nomikos and MacGregor (1994, 1995a, 1995b), and analysed by use of multiway PCA and PLSR. This is quite useful, especially since there are not many other alternatives for such data. Alternative approaches for such analyses are discussed in Westerhuis *et al.* (1999).

## SOME EMERGING METHODS IN FOOD SCIENCE

Analysis of relations between large blocks of data is becoming more and more important in many fields of science. In food science such data blocks may include, for example, raw material information, controlled process variables, secondary process measurements, spectral product measurements, sensory attributes like visual appearance, smell and taste, and consumer preferences. New multivariate methods for this are, for example, sequential and orthogonalized partial least squares regression (SO-PLS) (Næs *et al.*, 2011) and parallel orthogonalized partial least squares regression (PO-PLS) (Måge *et al.*, 2012).

## REFERENCES


Aikaike, H. (1974) A new look at the statistical model identification. *IEEE Transactions on Automatic Control* **AC-19**, 716–723.

Andersen, C.M. and Bro, R. (2010) Variable selection in regression – a tutorial. *Journal of Chemometrics* **24**, 728–737.

Andersson, M. (2009) A comparison of nine PLS1 algorithms. *Journal of Chemometrics* **23**, 518–529.

Brereton, R.G. (2003) *Chemometrics*. John Wiley & Sons Ltd, Chichester, UK.

Bro, R. and Eldén, L. (2009) PLS works. *Journal of Chemometrics* **23**, 69–71.

Di Ruscio, D. (2000) A weighted view on the partial least-squares algorithm. *Automatica* **36**, 831–850.

Eigenvector Research, Inc . (2005) *Data set*. [Online] Available: http://www.eigenvector.com/data/Corn/index.html (last accessed 9 June 2013).

Ergon, R. (1999) *Dynamic System Multivariate Calibration for Optimal Primary Output Estimation*, PhD thesis, Norwegian University of Science and Technology, Norway.

Ergon, R. (2002a) Noise handling capabilities of multivariate calibration methods. *Modeling, Identification and Control* **23**, 259–273.

Ergon, R. (2002b) PLS score-loading correspondence and a bi-diagonal factorization. *Journal of Chemometrics* **16**, 368–373.

Ergon R. (2003a) Constrained numerical optimization of PCR/PLSR predictors. *Chemometrics and Intelligent Laboratory Systems* **65**, 293–303.

Ergon, R. (2003b) Compression into two-component PLS factorizations. *Journal of Chemometrics* **17**, 303–312.

Ergon, R. (2004) Informative PLS score-loading plots for process understanding and monitoring. *Journal of Process Control* **14**, 889–897.

Ergon, R. (2005) PLS post-processing by similarity transformation (PLS + ST): a simple alternative to OPLS. *Journal of Chemometrics* **19**, 1–4.

Ergon R. (2006) Reduced RCR/PLSR models by subspace projections. *Chemometrics and Intelligent Laboratory Systems* **81**, 68–73.

Ergon, R. (2007) Finding Y-relevant part of X by use of PCR and PLSR model reduction methods. *Journal of Chemometrics* **21**, 537–546.

Ergon, R. (2009) Re-interpretation of NIPALS results solves PLSR inconsistency problem. *Journal of Chemometrics* **23**, 72–75.

Ergon, R. and Esbensen K.H. (2001) A didactically motivated PLS prediction algorithm. *Modeling, Identification and Control* **22**, 131–139.

Ergon, R. and Esbensen K.H. (2002) PCR/PLSR optimization based on noise covariance estimation and Kalman filtering theory. *Journal of Chemometrics* **16**, 401–407.

Ergon, R., Halstensen, M. and Esbensen, K.H. (2011) Model choice and squared prediction errors in PLS regression. *Journal of Chemometrics* **25**, 301–312.

Esbensen, K.H. (2010) *Multivariate Data Analysis in Practice*. Camo, Oslo, Norway.

Esbensen, K.H. and Geladi, P. (2010) Principles of proper validation: use and abuse of re-sampling for validation. *Journal of Chemometrics* **24**, 168–187.

Faber, N.M. (2009) The X-residuals calculated by partial least squares are problematic for uncertainty estimation. *Chemometrics and Intelligent Laboratory Systems* **96**, 264–265.

Feudale, R.N., Woody, N.A., Tan, H. *et al.* (2002) Transfer of multivariate calibration models: a review. *Chemometrics and Intelligent Laboratory Systems* **64**, 181–192.

Geladi, P. (2002) Some trends in the calibration literature. *Chemometrics and Intelligent Laboratory Systems* **60**, 211–224.

Gelb, A. (1974) *Applied Optimal Estimation*. MIT Press, Cambridge, MA.

Golub, G.H. and Kahan, W. (1965) Calculating the singular values and pseudoinverse of a matrix. *SIAM Journal on Numerical Analysis, Series B* **2**, 205–224.

Grewal, M.S. and Andrews A.P. (1993) *Kalman filtering: Theory and Practice*. Prentice-Hall, Englewood Cliffs, NJ.

Gujral, P., Amrhein, M., Ergon, R. *et al.* (2011) On multivariate calibration with unlabeled data. *Journal of Chemometrics* **25**, 456–465.

Halstensen, M. and Esbensen, K.H. (2010) Acoustic chemometric monitoring of industrial production processes. In: *Process Analytical Technology* (ed. K. A. Bakeev), 2nd edn. John Wiley & Sons Ltd, Chichester, UK.

Høskuldsson, A. (1996) *Prediction Methods in Science and Technology, Vol. 1 Basic Theory*. Thor Publishing, Copenhagen, Denmark.

Johnson, R.A. and Wichern, D.W. (1992) *Applied Multivariate Statistical Analysis*. Prentice-Hall, Englewood Cliffs, NJ.

Kalman, R.E. (1960) A new approach to linear filtering and prediction problems *Journal of Basic Engineering* **82**, 35–45. [Online] Available: http://www.elo.utfsm.cl/~ipd481/Papers%20varios/kalman1960.pdf (last accessed 9 June 2013).

Kemsley, E.K. and Tapp, H.S. (2009) OPLS filtered data can be obtained directly from non-orthogonalized PLS1. *Journal of Chemometrics* **23**, 263–264.

Kemsley, E.K., Tapp, H.S., Le Gall, G. and Colquhou, I.J. (2010) A *Statistical Toolkit for Metabolomics*. [Online] Available: http://www.ifr.ac.uk/Bioinformatics/BSResources/Metabolomics_Boston08_Kemsley_et_al.pdf (last accessed 9 June 2013).

Kourti, T. and MacGregor, J.F. (1996) Multivariate SPC methods for process and product monitoring. *Journal of Quality Technology* **28**, 409–428.

Kvalheim, O.M., Rajalahti, T. and Arneberg, R. (2008) X-tended target projection (XTP) – comparison with orthogonal partial least squares (OPLS) and PLS post-processing by similarity transformation (PLS+ST). *Journal of Chemometrics* **23**, 49–55.

Langsrud, Ø. and Næs, T. (2003) Optimised score plot by principal components of predictions. *Journal of Chemometrics* **24**, 168–187.

Måge, I., Menichelli, E. and Næs, T. (2012) Preference mapping by PO-PLS: Separating common and unique information in several data block. *Food Quality and Preference* **24**, 8–16.

Mallows, C.L. (1973) Some comments on $C_p$. *Technometrics* **15**, 661–675.

Manne, R. (1987) Analysis of two partial-least-squares algorithms for multivariate calibration. *Chemometrics and Intelligent Laboratory Systems* **81**, 68–73.

Manne, R., Pell R.J. and Ramos, L. S. (2009) The PLS model space: the inconsistency persists. *Journal of Chemometrics* **23**, 76–77.

Martens, H. and Næs, T. (1987) Multivariate calibration by data compression. In *Near Infra-red Technology in Agricultural and Food Industries* (eds P. C. Williams and K. Norris). American Association of Cereal Chemists, St Paul, MN, pp. 57–87.

Martens, H. and Næs, T. (1989) *Multivariate calibration*. John Wiley & Sons, Inc., New York.

Næs, T. and Martens, H. (1988) Principal component regression in NIR analysis: Viewpoints, background details and selection of components. *Journal of Chemometrics* **2**(2), 155–167.

Næs, T. and Mevik, B-H. (2001) Understanding the collinearity problem in regression and discriminant analysis. *Journal of Chemometrics* **15**, 413–426.

Næs, T., Isaksson, T., Fearn, T. and Davies, T.A. (2004) *A User-friendly Guide to Multivariate Calibration and Classification*. NIR Publications, Chichester, UK.

Næs, T., Måge, I. and Segtnan, V.H (2011) Incorporating interactions in multiblock sequential and orthogonalised partial least squares regression. *Journal of Chemometrics* **25**, 601–609.

Nomikos, P. and MacGregor, J.F. (1994) Monitoring of batch processes using multiway principal component analysis. *AIChE Journal* **40**, 1361–1375.

Nomikos, P. and MacGregor, J.F. (1995a) Multi-way partial least squares in monitoring batch processes. *Chemometrics and Intelligent Laboratory Systems* **30**, 97–108.

Nomikos, P. and MacGregor, J.F. (1995b) Multivariate SPC charts for monitoring batch processes. *Technometrics* **37**, 41–59.

Olivieri, A.C., Faber, N.M., Ferré, J. *et al.* (2006) Uncertainty estimation and figures of merit for multivariate calibration. *Pure and Applied Chemistry* **78**, 633–661.

Otto, M. (2007) *Chemometrics*. Wiley-VCH, Weinheim, Germany.

Pell, R.J., Ramos, L.S. and Manne, R. (2007). The model space in partial least squares regression. *Journal of Chemometrics* **21**, 165–172.

Pereira, C.F., Pimentel, M.F., Galavo, R.K.H. *et al.* (2008) A comparative study of calibration transfer methods for determination of gasoline quality parameters in three different near infrared spectrometers. *Analytica Chimica Acta* **611**, 41–47.

Strang, G. (2006) *Linear Algebra and Its Application*, 2nd edn. Cengage Learning, Boston, MA.

Tapp, H.S. and Kemsley, E.K. (2009) Notes on the practical utility of OPLS. *Trends in Analytical Chemistry* **28**, 1322–1327.

Tapp, H.S., Penfold, R. and Kemsley, E.K. (2011) TinyLVR: A utility for viewing single predictor multivariate models in terms of a two factor latent vector model. *Chemometrics and Intelligent Laboratory Systems* **105**, 19–26.

Thielemans, A., Lewi, P.J. and Massart, D.L. (1988) Similarities and differences among multivariate display techniques illustrated by Belgian cancer mortality distribution data. *Chemometrics and Intelligent Laboratory Systems* **3**, 277–300.

Trygg, J. and Wold, S. (2002) Orthogonal projections to latent structures, O-PLS. *Journal of Chemometrics* **16**, 119–128.

Westerhuis J.A., Kourti, T. and MacGregor, J.F. (1999) Comparing alternative approaches for multivariate statistical analysis of batch process data. *Journal of Chemometrics* **13**, 397–413.

Wise, B.M. and Gallagher, N.B. (1996) The process chemometrics approach to process monitoring and fault detection. *Journal of Process Control* **6**, 329–348.

Wold, S., Martens, H. and Wold, H. (1983) The multivariate calibration problem in chemistry solved by PLS method. *Lecture Notes in Mathematics* **973**, 286–293.

Wold, S., Høy, M., Martens, H. *et al.* (2009) The PLS model space revisited. *Journal of Chemometrics* **23**, 67–68.

# 9 Multiway methods in food science

**Åsmund Rinnan, José Manuel Amigo and Thomas Skov**
*Department of Food Science, Faculty of Science, University of Copenhagen, Frederiksberg, Denmark*

## ABSTRACT

This chapter gives an introduction to the world of multiway chemometrics, as well as showing how it can be applied to different types of multidimensional data related to food science and technology. It describes how information can be extracted from multiway data in a meaningful way by keeping the multiway dimensionality, or even by adding a new dimensionality to traditional two-dimensional data (data tables). The chapter mainly focuses on fluorescence and gas chromatography mass spectrometry (GC-MS) data. However, the methods discussed are not limited to the few cases shown, but rather to any multiway data which contain (or can be transformed to contain) a trilinear data structure.

## INTRODUCTION

The amount of multiway data produced in the framework of food research has rapidly increased, especially for two main reasons:

(i) The well-known ability of excitation-emission fluorescence spectroscopy for giving fast and reliable information from food samples even without the need of any sample pre-treatment (*in situ* measurements) (Christensen *et al.*, 2006).
(ii) The explosion of hyphenated and hypernated chromatographic systems, such as GC-MS (Amigo *et al.*, 2010a).

In order to analyse such data it is necessary to use specialized chemometric tools. During this chapter it will be shown that by keeping the natural data structure (three- or even multiway structure), better information will be extracted, compared to unfolding all this good multiway information into two-dimensional data. The chapter is divided into three parts. In the first section the different methods and concepts currently being used, and developed, in chemometrics are discussed. The subsequent section focuses on the different types of sources for multiway data. We have here chosen to focus on fluorescence landscapes and hyphenated chromatographic systems, as these are the most common sources for multiway data. Finally, we have added a section with necessary problem solving in this field in order to spread the word to the community as a whole.

## METHODS AND CONCEPTS

In this section multiway chemometric methods are explained. The main emphasis is put on the model structure and how these models can be used to analyse multiway data.

*Mathematical and Statistical Methods in Food Science and Technology*, First Edition.
Edited by Daniel Granato and Gastón Ares.

**Table 9.1** Parameters in different chemometric models when evaluating a $50 \times 200 \times 100$ data array using two factors/components (R = 2).

| Model | Parameters | Number of parameters |
|---|---|---|
| PCA (unfold) | $R(I + JK)$ | 40 100 |
| PARAFAC2 | $R(I + IJ + K)$ | 20 300 |
| Tucker3 (2,2,2) | $R(I + J + K) + R^3$ | 708 |
| PARAFAC | $R(I + J + K)$ | 700 |

[a]In the PARAFAC2 model the shift takes place in the second mode, but is located across the first mode. As will be shown later the first and third mode is still very interpretable with the same number of components, as for the Tucker3 and PARAFAC model.

According to Bro (Bro, 1997) the following order of the chemometric methods exists regarding complexity and how well they will describe multiway data:

$$\text{Two-way PCA (unfold)} \geq \text{PARAFAC2} \geq \text{Tucker3} \geq \text{PARAFAC} \tag{9.1}$$

This shows that a two-way PCA model always will fit the data better than a PARAFAC2, a Tucker3, and a PARAFAC model, except in situation where the models may fit equally well. The models to the left are the most complex and flexible, whereas models to the right are simpler and restricted (Bro, 1997). This fit refers to the mathematical description of data, meaning that a higher fraction of the variance in the data has been explained. However, this might not coincide with the chemical variance (i.e. chemical information); the solution may be more difficult to interpret. The numbers of parameters estimated for each of the models are shown in Table 9.1. The numbers are based on data set of 50 samples measured with GC-MS (200 scans and 100 *m/z* measured) using two factors.

The different number of parameters to estimate makes some models more affected by low signal-to-noise ratio and also on how consistent the chemical features are across samples. Another major difference is that the parameters extracted would be chemically meaningful (chemically as well as mathematically unique) for the PARAFAC and PARAFAC2 model when the correct number of factors is chosen. PCA and Tucker3, on the other hand, extracts parameters based on the orthogonality principle (mathematical unique) under normal conditions.

## Structure of the data

Data collected from samples are increasing in complexity. It is generally thought that the more data that are collected for one sample, the more knowledge it will be possible to obtain afterwards (Harris, 2000; Caggiano *et al.*, 2011). Thus, knowing the structure of the data collected plays a fundamental role when choosing the proper data analysis technique for the target pursued. As a brief introduction to the latter, Table 9.2 compiles the different orders of instruments that are used nowadays. The focus is on multiway data, nevertheless readers are encouraged to read the article by Booksh and Kowalski (Booksh and Kowalski, 1994) in which more details can be found.

Two of the most popular instruments used in food science today that readily produce multiway data are fluorescence spectroscopy and hyphenated chromatographic systems (i.e. gas chromatography

**Table 9.2** Orders of instruments.

| Data | Instruments | Example | One sample | Many samples |
|---|---|---|---|---|
| Univariate | Zero order | pH, °C | Scalar | Vector |
| Multivariate | First order | UV-Vis spectrum | Vector | 2D matrix |
| | Second order | Spectrum × elution time | 2D matrix | 3D array |
| Multiway | | Excitation × emission matrix | | |
| | Higher order | Elution time 1 × elution time 2 × spectrum | 3D array | N-way array |

Adapted from Booksh, K. S. & Kowalski, B. R., 1994.

connected to mass spectrometry – GC-MS). These two techniques are used as examples. However, the inner structure of this tensor is totally different depending on the measurement technique used. There are other instruments that give rise to multiway arrays; these can all be analysed by the same techniques mentioned.

Fluorescence spectroscopy is a function of two variables: excitation and emission. In many fields of science and research, only a single excitation and/or emission wavelength is measured. However, by measuring emission spectra for several excitation wavelengths the basic structure of the fluorescence signal will become as shown in Equation 9.2. One sample measured by fluorescence can thus conveniently be presented as a matrix of fluorescence intensities as a function of excitation and emission wavelengths. The fluorescence landscape $\mathbf{X}(I,J)$ can, therefore, be described as a function of a concentration dependent factor, $\alpha$, and its excitation, $\mathbf{b}(\lambda_{\mathrm{Em}})$, and emission, $\mathbf{c}(\lambda_{\mathrm{Ex}})$, characteristics; establishing the following linear relationship for each member of $\mathbf{X}$:

$$\mathbf{X} = \sum_{f=1}^{F} \alpha_f \times \mathbf{b}_f(\lambda_{Em}) \times \boldsymbol{c}_f(\lambda_{Ex}) \tag{9.2}$$

where $F$ is the total number of fluorescent species present in the sample. Having $F$ independent fluorophores with different concentration, Equation 9.2 can easily be extended with an additional dimension to become trilinear (Christensen *et al.*, 2006).

Chromatography for three-way analysis is linked to hyphenated and hypernated methods. As observed in Figure 9.1, the mathematical structure of $\underline{\mathbf{X}}$ obtained for chromatography is the same as the one for fluorescence. Nevertheless, the chemistry behind it is totally different, a fact that must be

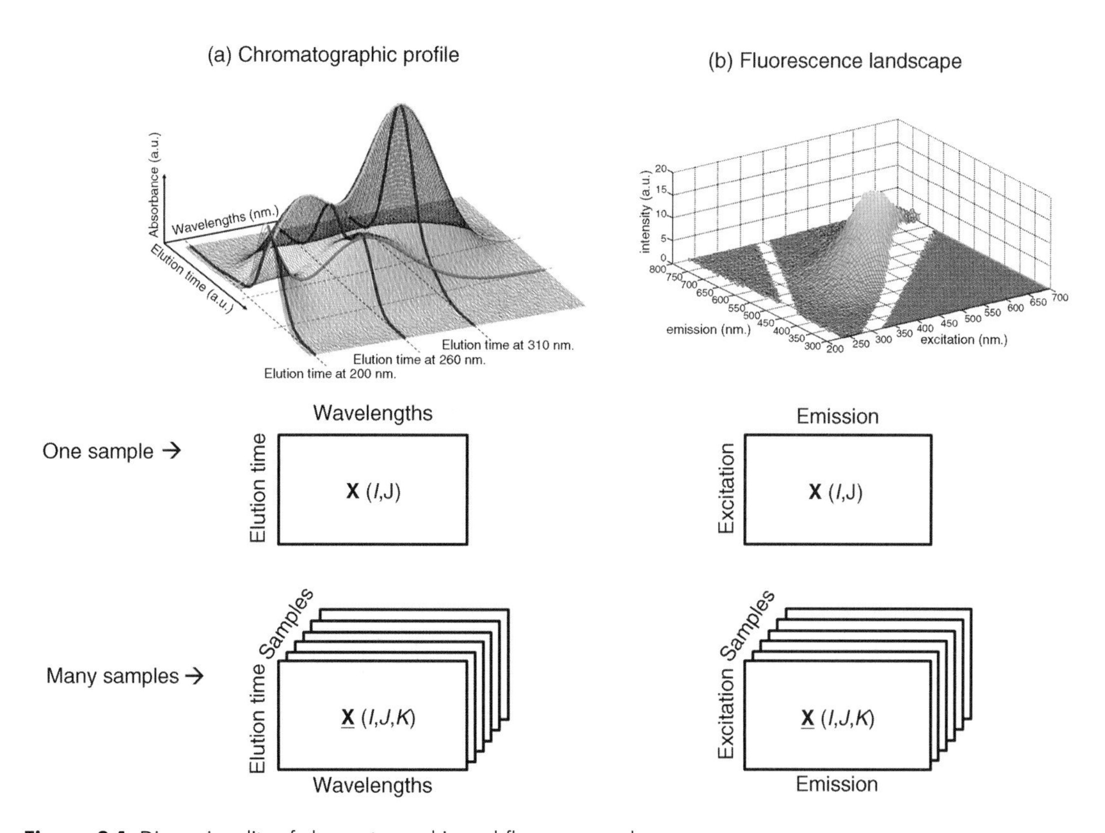


**Figure 9.1** Dimensionality of chromatographic and fluorescence data sets.

considered in the application of three- or multiway methods. A comprehensive description about this data structure is given by Amigo *et al.* (Amigo *et al.*, 2010). Despite the difference to fluorescence Equation 9.2 can be adapted for chromatography. In this case, the signal **X**(*I,J*) is proportional to the concentration α of each analyte, having a specific elution time **b**(et) and a spectral signal **c**(λ) (if, for instance, the detector is a spectral detector) as indicated in Equation 9.3:

$$\mathbf{X} = \sum_{f=1}^{F} \alpha_f \times \mathbf{b}_f(et) \times \mathbf{c}_f(\lambda) \tag{9.3}$$

## Second-order advantage

The major advantage of second-order instruments is the so-called second-order advantage. That is, the capability of giving accurate predictions even in the presence of new and un-calibrated interferents in future samples (Booksh and Kowalski, 1994; Bro, 1998; Rinnan *et al.*, 2007). This property is especially relevant in food science, where seasonal and species variation may lead to new uncalibrated interferents in future samples. In two-way methods (such as PCA and PLS), these interferents can only be detected but cannot be modelled. However, as Rinnan *et al.* (Rinnan *et al.*, 2007) showed, these uncalibrated interferents can easily be handled by PARAFAC, and thus also PARAFAC2. This is not the case for Tucker3 though, due to the rotational ambiguity of Tucker3 models (i.e. no uniqueness as is the case for PARAFAC and PARAFAC2).

## Tucker3

The Tucker3 model decomposes the three-way data array $\underline{\mathbf{X}}$ into three orthonormal loading matrices, denoted as **A** ($I \times$ P), **B** ($J \times Q$), **C** ($K \times R$), and the core matrix $\underline{\mathbf{G}}$ ($P \times Q \times R$), which describes the interactions between the components of **A**, **B**, and **C**. The core matrix consists of core elements and the largest squared elements indicate the most important components that describe $\underline{\mathbf{X}}$, that is explain most variations in the data (Henrion, 1994). The graphical description of the Tucker3 model is depicted in Figure 9.2, where the residual matrix $\underline{\mathbf{E}}$ is included.

Another way of writing the Tucker3 model is to use the individual elements of the shown matrices (Kiers, 2000). This is shown below, where $a_{ip}$ is an element of the **A** matrix, $b_{jq}$ an element in the **B** matrix, $c_{kr}$ is an element of the **C** matrix, $g_{pqr}$ is the core array, and where the indices (*ijk*) are running from $i = 1$ . . . . . . ,$I$, $j = 1$, . . . . . . . , $J$, $k = 1$, . . . . . . ,$K$, respectively.

$$x_{ijk} = \sum_{p=1}^{P} \sum_{q=1}^{Q} \sum_{r=1}^{R} a_{ip} b_{jq} c_{kr} g_{pqr} + e_{ijk} \tag{9.4}$$

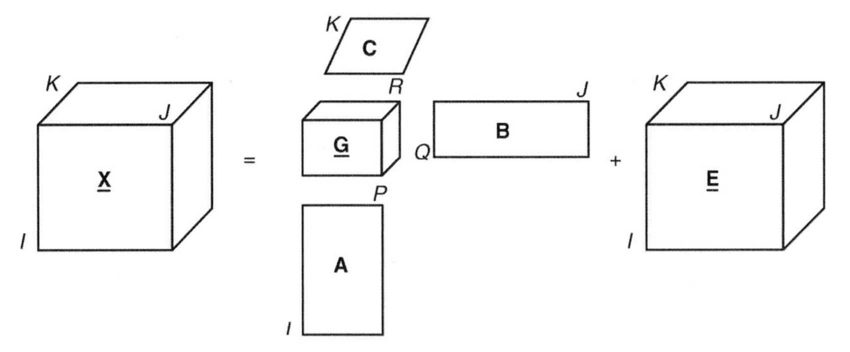


**Figure 9.2** Graphical description of the Tucker3 model.

**Box 9.1 Description of the core array**

Let the unfolded core array equal the following:

$$\begin{pmatrix} 20 & -1 & 6 & \vdots & 2 & 16 & -9 \\ 12 & 10 & 4 & \vdots & 9 & -8 & 26 \end{pmatrix}$$

The element (2, 2, 2) links the second factor of **A** with the second factor of **B** and the second factor of **C**. Let the 1st mode be the samples, the 2nd mode the time and the 3rd mode the sensors.

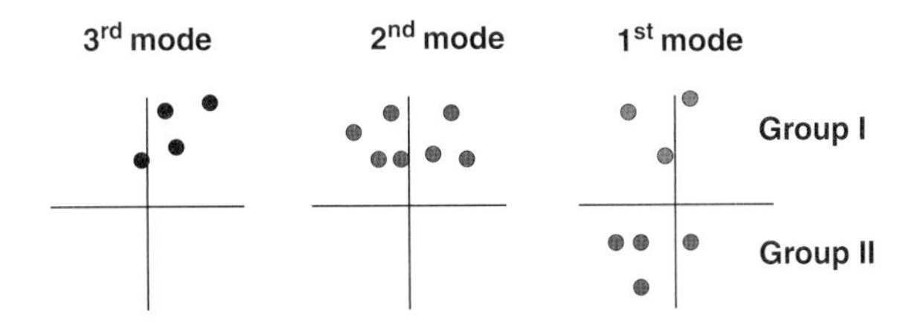


**Interpretation of the core array:**

Let **b** and **c** have positive loading values for the second factor, and let the variables of a be divided into two groups with different signs for the loadings for the second factor. The negative sign of the core element (2, 2, 2) can then be used to interpret the correlation between the two groups found in a as follows:

Loading sign of **a** · **b** · **c** · g = global sign
Group I: + · + · + · − = −
Group II: − · + · + · − = +

This means that over all variables of the 2nd mode (all times), group I samples will have lower sensor signals than group II sample for all variables of the 3rd mode (all sensors).

The number of factors ($P$, $Q$ and $R$) in each of the three modes are allowed to be different, which makes the model more flexible compared to, for example PARAFAC. This means that the Tucker3 model is able to handle nontrilinear data, which sometimes fits for chemical data. The size of the squared core element $g^2_{pqr}$ is proportional to the variation explained by the combination of component $p$ from mode **A**, component $q$ from mode **B** and component $r$ from mode **C** (Henrion, 1994). This is especially useful when interpreting the core array. The largest core element will, therefore, explain most of the variation in the data. This special feature is used when analysing the core array and is illustrated in Box 9.1.

The interpretation of the core array can often be simplified by rotating the core array to maximize the variance of the squared core elements or to enlarge the elements in the diagonal of the core (Andersson and Henrion, 1999). The latter method can only be conducted on quadratic core arrays. By rotating the core array to maximize the variance of the squared core elements a simplified core interpretation can most often be achieved (Henrion and Andersson, 1999).

## Parallel factor analysis – PARAFAC model

PARAFAC was developed independently in 1970 by Harshman (Harshman, 1970) and by Carroll and Chang under the name CANDECOMP (Carroll and Chang, 1970). Both algorithms were based on a principle of parallel proportional profiles suggested in 1944 by Cattell (Cattell, 1944). PARAFAC

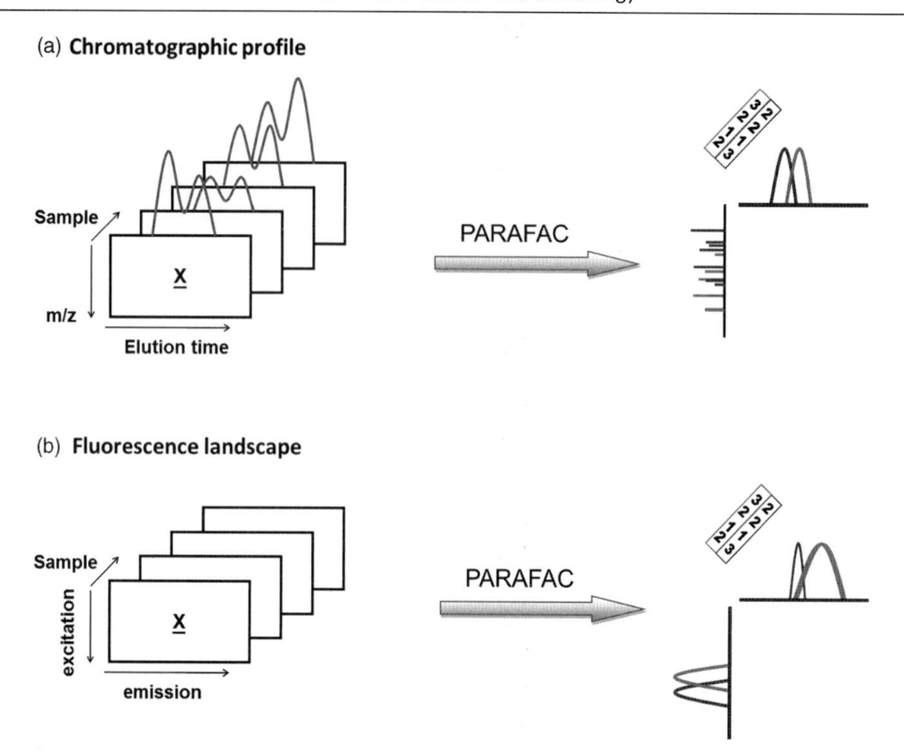


**Figure 9.3** PARAFAC application to (a) chromatographic and (b) fluorescence data sets with two components.

decomposes the data cube into three loading matrices, each one corresponding to the modes/directions of the data cube. One of the most used algorithms to obtain the pure signals in two of the modes and concentrations in the third mode is alternating least squares (ALS) (Bro, 2006). To explain PARAFAC in terms of data arrays, the model has been depicted in Figure 9.3 and Equation 9.5:

$$x_{ijk} = \sum_{f=1}^{F} a_{if} b_{jf} c_{kf} + e_{ijk} \tag{9.5}$$

where $x_{ijk}$ denotes each element of the $\underline{\mathbf{X}}$ array; $a_{if}$, $b_{jf}$ and $c_{kf}$ are the index of the first, second and third mode for each $f$ factor. $e_{ijk}$ represents the elements of the residual matrix $\underline{\mathbf{E}}$.

The two most important steps for PARAFAC can be summarized as follows (Garcia *et al.*, 2004):

(i) *Determining the number of factors/analytes:* In general, a factor (also called component) is any effect that causes variations in the signal in a higher level than the signal-to-noise ratio expected from the device and/or the samples. This definition encompasses the signal of the analytes and also the different artifacts that affect the signal. As an example, the baseline drift between samples in chromatography can be considered as an additional factor, since its effect is usually higher than the general signal-to-noise ratio. Choosing the proper number of factors is the most crucial step. Several dedicated methods to estimate the number of components have been developed for PARAFAC. Harshman and Lundy (Harshman and Lundy, 1984) came up with a method named 'split-half analysis' to estimate the number of factors. It uses the intrinsic properties of PARAFAC and the samples, stating that the same **B** and **C** loadings should be found in different subsets of the data. Another common method is the core consistency (Bro and Kiers, 2003) which estimates the appropriateness of the PARAFAC solution. A third method was presented by Hoggard and Synovec (Hoggard and Synovec, 2007), where they evaluated the so-called degenerate solution that can be observed for PARAFAC models with too many factors.

(ii) *Imposing constraints to the model:* A constraint is a chemical or mathematical property that the profiles should fulfil (Garcia *et al.*, 2004). For this question, the chemical structure of the data is taken into consideration in the selection of the proper constraints. The most common constraints are non-negativity and unimodality. Non-negativity forces the profiles to only contain nonzero values. This is especially useful for spectral and chromatographic profiles. Unimodality constraint can help to preserve the presence of only one peak in each profile extracted.

Furthermore, PARAFAC has two important features: uniqueness in the solution and factors are estimated simultaneously. Firstly, the PARAFAC solution is unique, that is there is no rotational freedom as there is in PCA or even Tucker3. Secondly, all factors are found simultaneously, which means that they are not independent of each other. This implies that, for example a three-factor PARAFAC solution is not the same as a one-factor PARAFAC model based on the residual of a two-factor PARAFAC model. In addition, PARAFAC concedes to the concept of trilinear data (de Juan and Tauler, 2001); trilinearity can be viewed as an extension of Lambert Beer's law to second-order data. This concept assumes that the measured signal is the sum of the individual peaks of each analyte and that the profiles in each mode for the analytes are proportional in all the samples (Comas *et al.*, 2004). The feature of uniqueness and trilinearity are closely related (Bro, 1997). If the data are trilinear, the true underlying signal will be found if the right number of factors is estimated and the signal-to-noise ratio is appropriate (Harshman, 1972).

## Parallel factor analysis 2 – PARAFAC2 model

One of the assumptions of trilinear data is that two of the loading matrices must be linearly independent. This can be explained by an example: for chromatography, the elution profile for one component must always appear in the same place and with the same shape in all the samples and only differs from one another in the intensity of the signal (Ortiz and Sarabia, 2007). Unfortunately, this is not very usual in chromatography. Peak shape changes occur when one of the elution mechanisms in the column are overloaded (e.g. peak tailing due to too high a concentration of a specific analyte (Dolan, 2002, 2003)). Moreover, a large concentration range across samples may result in different peak shapes. These drawbacks in the data collection will cause the data to deviate from trilinearity and provide a suboptimal PARAFAC solution.

To handle slightly nontrilinear data a less constrained version of PARAFAC can be applied, named PARAFAC2. PARAFAC2 (Bro *et al.*, 1999; Kiers *et al.*, 1999) decomposes three-way data arrays into two-way loading matrices (Figure 9.3), as PARAFAC does. The main difference to PARAFAC is that PARAFAC2 does not impose as strong restrictions on the data structure (de Juan and Tauler, 2001). The PARAFAC2 model allows every sample to have its own distinct set of loadings in one of the modes (Bro *et al.*, 1999), whereas the other independent mode is restricted to be linear. In this case, each sample (each slab of $\underline{\mathbf{X}}$) $\mathbf{X}_k$ $(I \times J)$ is modelled as is indicated in Figure 9.4 and Equation 9.6:

$$\mathbf{X}_k = \mathbf{A}\mathbf{D}_k(\mathbf{B}_k)^T + \mathbf{E}_k \tag{9.6}$$

where $\mathbf{X}_k$ $(I \times J)$ represents the sample related to the $k$-th sample. $\mathbf{A}$ $(I \times F)$ holds the independent mode that holds the linearity. $\mathbf{D}_k$ $(F \times F)$ is a diagonal matrix that holds the $k'$-th row of the sample mode loading matrix $\mathbf{C}$ (the dependent loading matrix). The matrix $\mathbf{B}_k$ $(I \times F)$ ideally holds an estimate of the individual loadings for each $F$ factor. The matrix $\mathbf{E}_k$ represents the residuals and $k = 1, \ldots, K$.

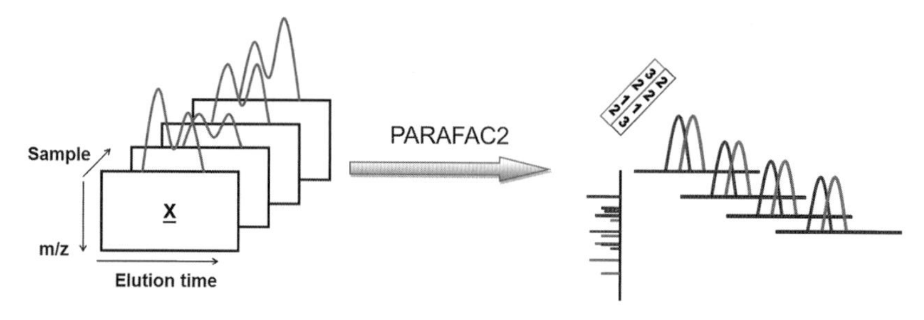


**Figure 9.4** PARAFAC2 application in a chromatographic profile with two analytes. The residuals are not included in the visualization.

Even though PARAFAC2 allows one of the loading profiles to differ within them in different samples, it still possesses uniqueness properties that are very similar to PARAFAC (Berge and Kiers, 1996). This means that PARAFAC2 can separate mixture data into the main contributions of the underlying analytes directly. The uniqueness property is achieved due to an important constraint: the cross-product of $\mathbf{B}_k$ has to be constant over all the samples (Equation 9.7). The implication of this cross-product constraint is that the loading profiles of **B** in different samples may differ, but their cross-product should stay constant.

$$(\mathbf{B}_1)^T(\mathbf{B}_1) = (\mathbf{B}_2)^T(\mathbf{B}_2)\ldots. = (\mathbf{B}_k)^T(\mathbf{B}_k) \tag{9.7}$$

## PARAFAC and PARAFAC2 versus Beer's law

It should be noted that as long as the right number of factors and the correct constraints are applied to PARAFAC and PARAFAC2, there is a good correlation between the scores found by these two methods and the actual concentration of the different (chemical) constituents. This can easily be verified by taking a look at the different equations of Beer's law (Equation 9.8), compared to PCA (Equation 9.9) and PARAFAC (Equation 9.10) (PARAFAC2 can be seen as an extension to PARAFAC and is therefore not shown here):

$$A_{i,\lambda} = \sum_{n=1}^{N} \varepsilon_{n,\lambda} \cdot c_{i,n} + e_{i,\lambda} \tag{9.8}$$

$$x_{i,j} = \sum_{f}^{F} t_{i,f} \cdot p_{j,f} + e_{i,j} \tag{9.9}$$

$$x_{i,j,k} = \sum_{f}^{F} a_{i,f} \cdot b_{j,f} \cdot c_{k,f} + e_{i,j,k} \tag{9.10}$$

In Beer's law there are no constraints on the different constituents (chemical components) in the system. However, in PCA, there is such a constraint and, thus, the loadings are probably not correlated to the concentration of the pure constituents. On the other hand, PARAFAC does not apply any orthogonality constrain to the system, and thus the scores **a** will be directly correlated to the product of $\varepsilon$ and c, in the perfect case. This also means that if the real concentration of one analyte (e.g. in g $ml^{-1}$) is known then this, together with the PARAFAC (and PARAFAC2) scores, can provide real concentrations for all analytes in all samples.

### Multilinear partial least squares regression. N-PLS

There exist several methods to solve multivariate regression problems. One of the most common, though, is the so-called partial least squares regression (PLS) (Martens and Næs, 1989). In three- and multiway systems, the natural extension of PLS is the so called N-PLS method (Bro, 1996). A comparison between PLS and N-PLS was performed by Amigo *et al.* (Amigo *et al.*, 2005).

In N-PLS the three way array of independent variables is decomposed into a trilinear model similar to the PARAFAC model. However, the inner relationship between the scores of $\underline{\mathbf{X}}$ and $\mathbf{Y}$ continues being like in the PLS model:

$$\mathbf{U} = \mathbf{T}\mathbf{B}_{PLS} + \mathbf{E} \tag{9.11}$$

where $\mathbf{U}$ and $\mathbf{T}$ are the scores of $\underline{\mathbf{X}}$ and $\mathbf{Y}$, respectively, both having the maximal covariance; and $\mathbf{B}_{PLS}$ is the matrix containing the regression vectors. Note that, despite not being common, the independent matrix $\mathbf{Y}$ can also be an n-way array, depending of the nature of the measurements. The main feature of the algorithm is that it produces score vectors that, in a trilinear sense, have maximum covariance with the explained part of the dependent variable. Therefore, the model is unique, like PARAFAC. Mathematically speaking, there is no difference between scores and loadings, because both are calculated in the same way. When there are a large number of variables, it is important to note that the solution is normally easier to interpret compared to unfolding methods. However, the main drawback of N-PLS is the loss of fit when compared with a bilinear model due to the more severe restrictions (Bro, 1998). It should also be noted that due to the dependence of Y during the decomposition step N-PLS does not hold the second order advantage (Rinnan *et al.*, 2007).

## SOURCES FOR MULTIWAY DATA

In this section four different types of multiway data are examined. The two first types are commonly used in food science and have also been referred to in the previous section, namely fluorescence and chromatography. In addition to these two methods, two additional sources of multiway data are discussed: low-field NMR and sensory data.

### Fluorescence

Fluorescence spectroscopy has a low detection range compared to other spectroscopic techniques. In addition, if fluorescence landscapes are measured, it gives a good selectivity. Samples give rise to fluorescence signals if the sample contains fluorophores. Fluorophores absorb energy from an incoming light source and shortly after will emit light at a slightly different wavelength; light which is recorded by a detector.

In the past it was common to measure the fluorescence by a certain excitation and emission pair (i.e. excitation 230 nm and emission 360 nm). This way of using the signal from fluorescence makes it highly prone to unwanted interferents. The expansion to measuring either a range of emission wavelengths for one specific excitation wavelength or a range of excitation wavelength at one specific emission wavelength is straight forward, and increases the amount of informational input (Lakowicz, 1999, p. 25; Karoui *et al.*, 2003). By recording a range of wavelengths it becomes easier to visually inspect a group of samples, but by doing so all the available information is still not recorded. Another way of collecting data from a fluorescence instrument is to collect several emission spectra at different excitation wavelengths (Matthews *et al.*, 1996). From this procedure the emission spectra can be set side-by-side, thus creating a fluorescence landscape, with the excitation wavelength along the x-axis, the

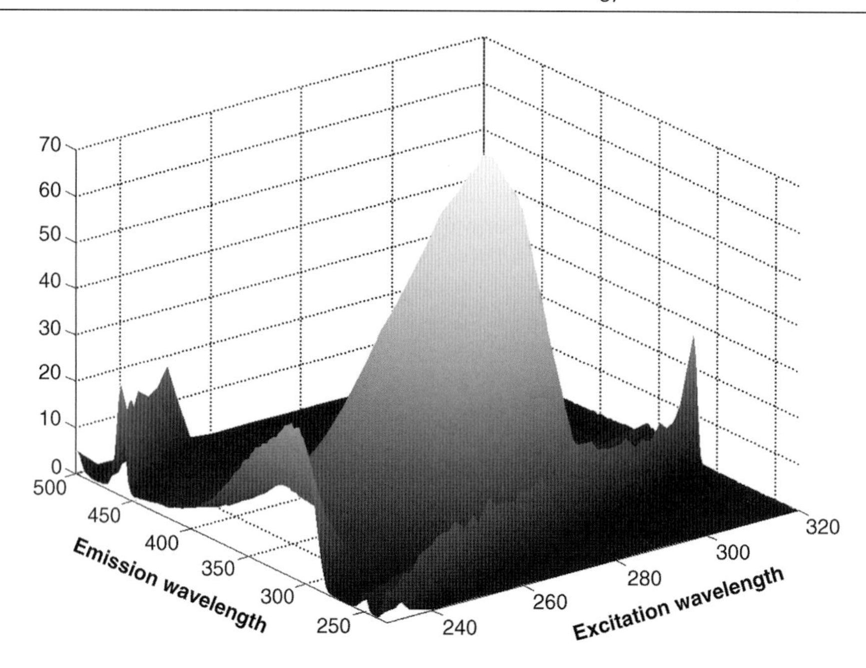


**Figure 9.5** An example of an excitation-emission matrix (EEM) of a sample containing a mixture of fluorophores in deionized water.

emission along the y-axis and the intensity of the signal along the z-axis. This landscape is also known as the excitation-emission matrix, abbreviated EEM (Figure 9.5).

The emission spectrum from a given fluorophore measured at different excitation wavelengths will only vary in intensity, not in shape. This is also true for the excitation spectrum; it will only vary in intensity depending on the emission wavelength (Lawaetz, 2011, pp. 19–20). Because the fluorescence spectrum is recorded as a function of two independent factors, excitation and emission, it is natural to take advantage of this bilinear nature of the data. Thus, using a trilinear curve resolution technique (such as PARAFAC) to decompose a set of EEM samples is recommended. PARAFAC has proven to be a valuable tool for the analysis of fluorescence spectra as a curve resolution method (Booksh *et al.*, 1996; Bro, 1997, 1999; Jiji *et al.*, 2000; McKnight *et al.*, 2001; Moberg *et al.*, 2001).

*Organizing the data*

Throughout this work, the fluorescence data have been arranged in a specific manner, with samples along the first mode, emissions along the second mode and excitations along the third mode. Therefore, the scores (**A** in Equation 9.10) from PARAFAC are (ideally) proportional to the concentrations of the fluorophores in the samples, the second mode loadings (**B**'s) are the estimated emission spectra, while the third mode loadings (**C**'s) are the corresponding excitation spectra.

*Effects causing the EEM to deviate from bilinearity*

Even though the excitation and emission spectrum are independent of each other, there are some effects which may cause a set of EEMs to deviate from the optimal trilinearity. Effects which can have a detrimental effect to the subsequent decomposition of the data are: pH, polarity of solvent, temperature, quenching and light scattering effects. It is possible to control the first three of these factors quite well, and in this chapter it is assumed that the amount of quenching is negligible. The scatter, however, is a factor that cannot easily be neglected and needs to be taken into account.

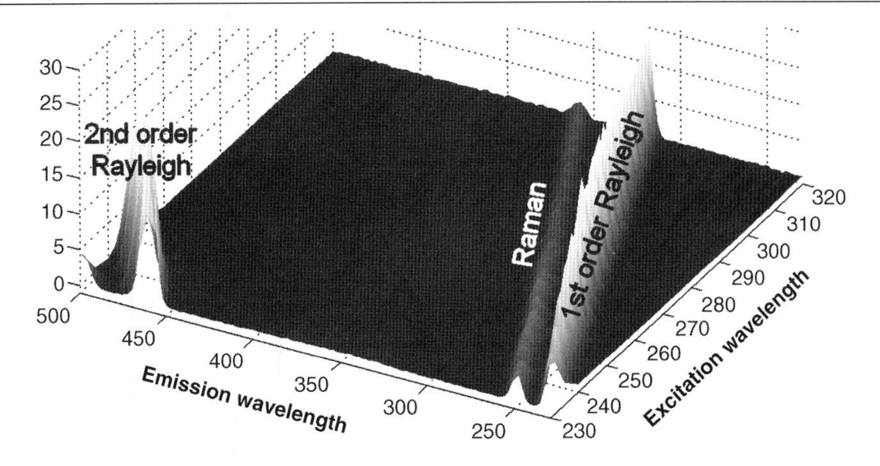


**Figure 9.6** Fluorescence landscape of a water sample showing the normal scatter effects.

*The nature of scatter*

An EEM will typically have areas with Rayleigh and Raman scatter (Ingle and Crouch, 1988, pp. 462–463; Lakowicz, 1999, pp. 39–40). These two types of scatter can readily be seen in an EEM of a (nonfluorescing) water sample, as shown in Figure 9.6. It is important, and comforting it might be said, that the scatter effects mainly originate from the solute, and thus do not hold any information regarding the fluorophores. Scatter cannot be described by a few PARAFAC factors, because it does not conform to the trilinear structure that is inherent for for PARAFAC. The Rayleigh and Raman scatter lines, as seen in Figure 9.6, are diagonal and thus cannot be explained by one vector in the emission mode and one in the excitation mode. The Rayleigh scatter line is of most concern since the intensity of this is higher than for Raman scatter (Skoog and Leary, 1992, pp. 298–299), thus affecting the decomposition to a larger extent. Sometimes the signal of the fluorophores lies away from the first-order Rayleigh peak and cutting away the area containing the scatter is sufficient (Beltrán *et al.*, 1998; Moberg *et al.*, 2001). However, in several natural samples, the first-order Rayleigh peak is partly overlapping the signal from one or more of the fluorophores. Removing this area with the scatter, would also remove some of the information of the fluorophore(s). Therefore, it is of interest to use a method to either remove exclusively the Rayleigh scatter, or to not let the Rayleigh scatter influence the decomposition.

*Rayleigh scatter*

Rayleigh scatter is predominantly caused by the solute, but may also originate from the fluorophores themselves (Ingle and Crouch, 1988, pp. 495–497). Since there is no loss of energy in Rayleigh scattering it is a type of elastic scatter, and occurs in integer multiples (i.e. first-order Rayleigh, second-order Rayleigh, etc.). Thus the maxima of the Rayleigh scatter will be found at emission wavelength equal an integer multiple of the excitation wavelength.

*Raman scatter*

While Rayleigh scatter is perfectly elastic, Raman is inelastic (Ingle and Crouch, 1988, pp. 497–499). It is caused by the molecules of the solute absorbing some of the incident light, followed by the emission of a photon. However, the energy in this photon is less than the energy absorbed. This energy difference, though, is constant and the Raman scatter line will be at a constant energy loss from the elastic Rayleigh scatter line. It is important to notice that a constant energy loss means a constant wave number shift – increasing the wavelength shift by increasing excitation wavelengths. The energy loss is dependent on the solute; for example for water it is $3600\,cm^{-1}$.

### *Handling scatter effects*

There are several ways of removing the Rayleigh scatter or enforcing PARAFAC not to take the Rayleigh scatter into account. The easiest is to omit the problematic area completely (Guimet *et al.*, 2005a). The next option is to replace the Rayleigh scatter and the area below and above these scatters by missing values (Munck *et al.*, 1998), possibly using additional constraints (Andersen and Bro, 2003) (Bro, 1999). This may lead to the loss of information and as such is not generally recommended. Thygesen *et al.* (Thygesen *et al.*, 2004) suggested replacing the area below and above the scatter lines by a mixture of missing values and zeros instead. Another solution is to subtract the spectrum of a standard from all the samples (Ho *et al.*, 1978, 1980; McKnight *et al.*, 2001), which at the same time will reduce or remove the Raman scatter. The problems of these scatter areas can also be solved in a way which takes into account the shape and the position of the scatter. Two different methods taking into account the shape and the position of the Rayleigh scatter are the use of weights (Jiji and Booksh, 2000; Bro *et al.*, 2002), which also can take into account the Raman scatter, or modelling the Rayleigh scatter line separately (Rinnan *et al.*, 2005; Bahram *et al.*, 2006). Furthermore, Engelen *et al.* (Engelen *et al.*, 2007) suggested a method using ROBPCA (Hubert *et al.*, 2005).

*Omit the scatter areas*

The easiest way to handle the Rayleigh scatter is to not measure the area in which the Rayleigh scatter is present. For example by measuring excitation from 230 to 320 nm and emission from 340 to 440 nm will omit any area with scatter (both first- and second-order Rayleigh). There are several papers in the literature which use this technique (Table 9.3). However, upon inspecting their solutions it can be seen that they probably would have achieved a better curve resolution by extending the emission and or excitation range into the scatter areas, and handling the scatter in one of the following methods.

**Table 9.3** Applications of fluorescence and PARAFAC in food science and how scatter correction was performed.

| **Food products** | **Scatter correction** | | | |
|---|---|---|---|---|
| | **Avoid** | **NaN** | **0 and NaN** | **Standard** |
| Oils | Guimet *et al.*, 2004, 2005a, 2005b; Tena *et al.*, 2012 | | | |
| Vinegar | | | Callejon *et al.*, 2012 | |
| Diary | Wold *et al.*, 2006; Canada-Canada *et al.*, 2009; Rodriguez *et al.*, 2009 | Christensen *et al.*, 2003; Andersen *et al.*, 2005; Diez *et al.*, 2007; Diez Azofra *et al.*, 2010; Morales *et al.*, 2011 | Christensen *et al.*, 2005 | |
| Sugar | | Baunsgaard *et al.*, 2000a, 2000b | | |
| Fish | | | Bassompierre *et al.*, 2007; Christensen *et al.*, 2009; Eaton *et al.*, 2012 | Christensen *et al.*, 2009; Eaton *et al.*, 2012 |
| Cereals | Hashemi *et al.*, 2008 | | | |
| Nuts | | | Yaacoub *et al.*, 2009 | |
| Wine | | | Airado-Rodriguez *et al.*, 2009, 2011 | |
| Fruit | Zhu *et al.*, 2007 | | | |
| Meat | | Møller *et al.*, 2003 | | |

*Subtracting a standard*

This requires that a standard of the solvent is available. However, such a standard is not always possible, especially for food samples. Subtracting a standard would normally only reduce – not eliminate – the Rayleigh scatter and may introduce negative values to the EEM. Introducing negative values is not a problem as such in the decomposition step but it indicates that more than the Rayleigh scatter in the sample has been removed, indicating that some information may have been lost.

*Inserting missing values and zeros*

In some practical applications of fluorescence spectroscopy, the emission is only recorded at wavelengths higher than the incident excitation light. The non-recorded part of the spectra is then filled with missing values (Christensen *et al.*, 2003; Møller *et al.*, 2003). This, however, may cause the PARAFAC solution to include unwanted artefacts. Inserting zeros far below the first-order Rayleigh scatter line (excitation wavelength larger than emission wavelength) to avoid these artefacts was found to be valuable (Matthews *et al.*, 1996; Stedmon *et al.*, 2003). It should be noted that on most data sets inserting zeros all up to the Rayleigh scatter line is not a good idea, since this can destroy the trilinearity of the data (Andersen and Bro, 2003).

Inserting a combination of zeros and missing values will stabilize the PARAFAC decomposition as fewer missing values need to be estimated. Thygesen *et al.* (Thygesen *et al.*, 2004) investigated how the insertion of a mixture of missing values and zeros affected the resolved spectra. They showed that by inserting zeros outside of the data and scatter areas will speed up the PARAFAC algorithm, as the number of data points to be estimated by the algorithm is reduced.

*Weights*

Weights are used during the decomposition to focus the modelling on the fluorophores and not on the areas with the Rayleigh scatter (and possibly Raman). The areas containing the scatter will be weighted down, either decreasing steadily towards the peak of the scatter or plainly setting the weight to a fixed value as long as the scatter is present (Jiji and Booksh, 2000; Bro *et al.*, 2002). Jiji and Booksh (Jiji and Booksh, 2000) came up with two different ways of assigning weights: either to hard-weight the areas with scatter to zero (the remaining areas were given a weight of one), or using the measurement uncertainty of the instrument as the basis of the weights. Rinnan and Andersen (Rinnan and Andersen, 2005), however, showed that using soft-weights of this type only slowed down the algorithm and gave poorer results than by using hard-weights as suggested by Jiji and Booksh (Jiji and Booksh, 2000) or, even better, the smoother weights suggested by Bro *et al.* (Bro *et al.*, 2002). Even though this method has been shown to give good results, the use of this method is limited in the literature.

*Modelling*

A different approach to the problem altogether is to model the scatter part. This can either be done as suggested by Bahram *et al.* (Bahram *et al.*, 2006), who simply interpolated the fluorescence signal by fitting a second-order polynomial to measurement points right before and right after the Rayleigh scatter lines. Rinnan *et al.* (Rinnan *et al.*, 2005) came up with a different suggestion, where they actively modelled the Rayleigh scatter line through shifting the EEM, so that the scatter line forms a straight line in the data. One of the advantages of modelling the Rayleigh scatter versus using weights is that the modelling is more automatic than the weighting scheme. However, Rinnan *et al.*'s (Rinnan *et al.*, 2005) method has not been made available for Raman scatter nor for higher order Rayleigh scatter. Thus, the method proposed by Bahram *et al.* (Bahram *et al.*, 2006) is currently more flexible, as it in theory also should work on Raman scatter. The challenge is just to detect the Raman scatter areas. The use of robust PCA to detect these areas (Engelen *et al.*, 2007) is at least a good start to solving this problem.

*Summary on handling scatter effects*
Inserting missing values or a mix of missing values and zeros and avoiding the scatter area are the most common techniques for handling Rayleigh found in the food literature (Table 9.3). There is some work, where first a combination of missing values and zeros were used. Subsequently a standard sample was subtracted to remove most of the Raman signal. The resulting decomposition seems to have been improved by this treatment. There are several papers that discuss the use of a mixture of missing values and zeros, and although Andersen and Bro (Andersen and Bro, 2003) argue that one should be very careful with inserting zeros, as there might be ghost peaks in the area outside the scatter, all the others point to the fact that inserting zeros greatly speeds up the PARAFAC algorithm, and reduces the amount of spurious peaks. By visual inspection of the raw data and the resulting deconvolution it should be possible to detect the data that may lead to ghost peaks, and this can thus be taken into account during the pre-processing of the EEM.

However, as very few applications have been performed with the use of weighting and modelling of the scatter, it is unknown if this would lead to better PARAFAC models.

### *Constraints in PARAFAC*

Andersen and Bro (Andersen and Bro, 2003) and Rinnan and Andersen (Rinnan and Andersen, 2005) both discuss the use of constraints during the deconvolution step. They both conclude that applying non-negativity constraints on both emission and excitation loadings in general will lead to better estimates of the pure fluorphores. Adding the unimodality constraint to this is a bit more 'tricky', as there are several fluorphores that especially have several peaks in the excitation mode, but this is also possible for the emission mode. Thus, the appropriateness of applying uni-modality depends on how the results from the first curve resolution behaves (i.e. are there more than one peak in any of the factors).

### *Applications of EEM and PARAFAC in food science*

The number of applications of EEM and PARAFAC in food science is increasing in number, showing that fluorescence spectroscopy coupled with PARAFAC is becoming an important tool in analysing food products. This is probably due to its higher sensitivity than other spectroscopic techniques (ppb, or even ppt, results have been showed in literature (Li *et al.*, 2003)). Furthermore, since only few chemical components fluoresce, many common interferents in other methods are invisible in fluorescence EEM.

## Chromatographic data

The evolution of hyphenated and hypernated chromatographic systems for the detection, quantification and/or identification of compounds has become one of the most important developments in different fields of chemistry in the last 30 years. A small number of the possibilities with regards to these systems is shown in Figure 9.7. Reliability of the coupled detectors joined with separation power of different high resolution chromatographic methods has changed many routines in analytical chemistry (Amigo *et al.*, 2010a).

### *Challenges – Increased capability, more problems?*

There are a great number of sources of variability in chromatographic runs, that is pumps, temperature gradients, stability of stationary phases and detectors. Most of these sources are reflected in the signal and may cause problems in the chromatographic profile: loss of separation and individual detection of the different analytes. This can either occur due to the lack of separation power of the analytical method or be due to low sample-to-sample reproducibility.

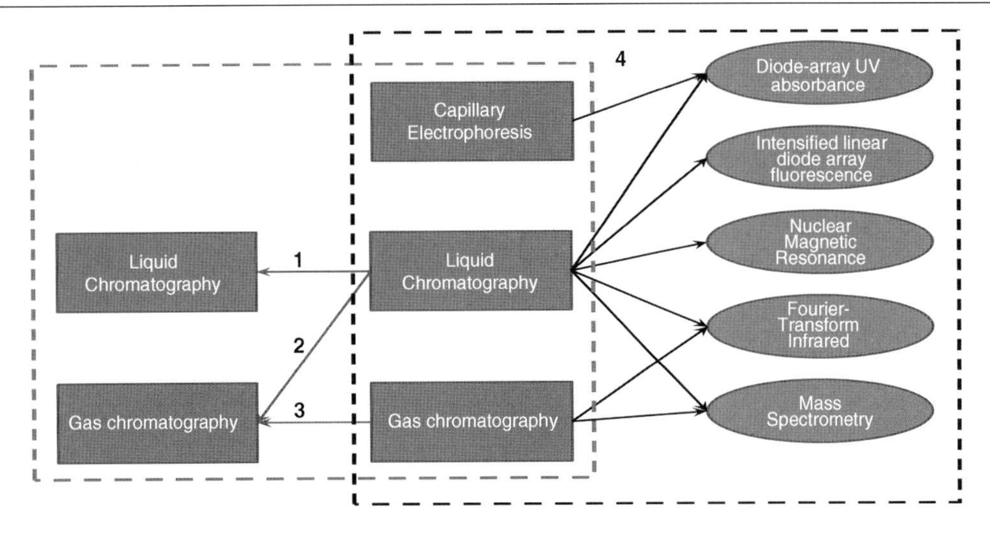


**Figure 9.7** An overview of hyphenated and hypernated systems in chromatography. 1 – Stoll *et al.*, 2006, 2007. 2 – Hyotylainen and Riekkola, 2003; Quinto Tranchida *et al.*, 2007. 3 – Xu *et al.*, 2003; Mohler *et al.*, 2007. 4 – Brinkman, 1999; Wilson and Brinkman, 2003, 2007.

One of the first challenges the chromatographer has to solve upon analysing a new set of samples is to re-parameterize the chromatographic method (new column, gradient, eluents, etc.). Optimizing some of the parameters is time consuming, while it is not possible to change others. Moreover, tuning one parameter to solve one problem may generate new problems (e.g. changing the temperature gradient in GC may improve the separation of two analytes but may lead to worse separation for other analytes). Sometimes it is not possible to achieve perfect separation, either because of the complexity of the sample or because faster chromatographic runs are preferred. Also, problems with drifts in the baseline, changes of shapes of the peaks and shifts in the elution times may decrease the quality of the final result of the analysis (Ortiz and Sarabia, 2007). Traditionally, the use of one or more standards helps to control and locate some of these problems. However, the capability of the internal standards is limited in more severe cases of baseline drifts and change of shapes of the peaks. The variability in the chromatographic system is reflected in the chromatographic signal, reducing the quality of the expected results.

*Low signal-to-noise/noisy peaks*

The problems of low signal-to-noise ratio and noisy peaks are correlated. It is evident that low concentration of an analyte gives a low peak and sometimes these peaks are below the detection limit of the system (cannot separate signal from noise), see Figure 9.8b. The noise is directly linked to the capability of the detector, that is UV-DAD has a better signal-to-noise ratio than a MS detector.

*Baseline offset*

The baseline offset (Figure 9.8b) may generate problems in quantification and is directly linked to the chromatographic conditions. For example gas chromatography gradients are achieved by programming temperature gradients in the column, generating the expansion of the mobile phase, thus promoting the displacement of the baseline.

*Co-eluting and embedded peaks*

Another problem derived from the chromatographic conditions is the co-elution of peaks. This is probably one of the major issues, as the separation of peaks is the main goal of chromatography.

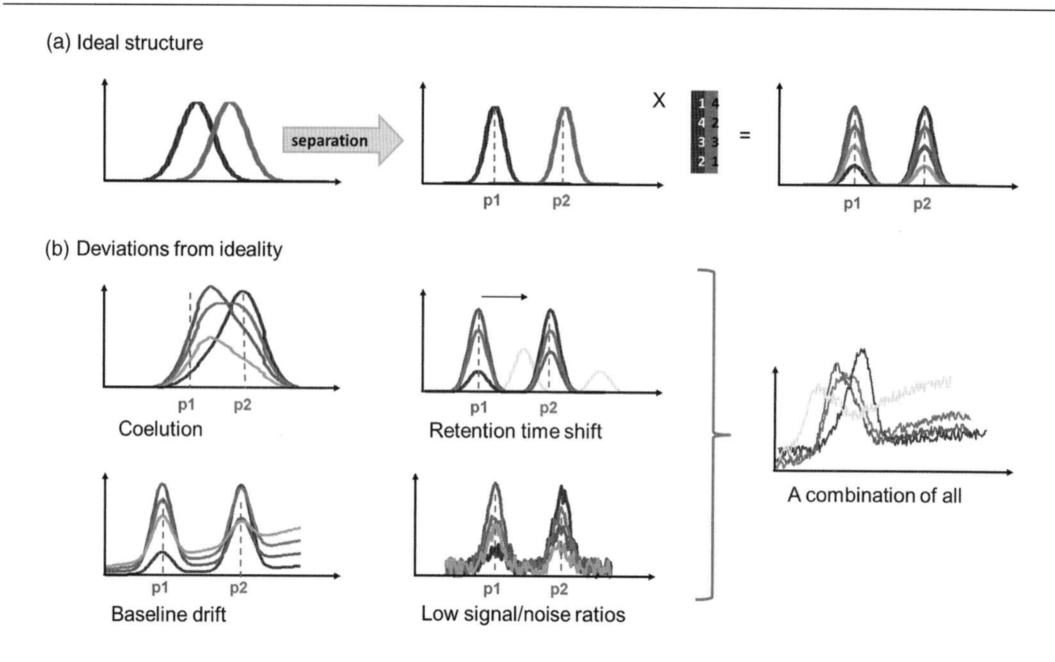


**Figure 9.8** Chromatographic issues and deviations from the ideal data structure.

Embedded peaks are the cause of a large number of derived troubles. Sometimes, the presence of a shoulder offers enough information to detect the overlapping peaks. However, the cases where one peak is totally embedded in a major peak are more complex to handle and, therefore, to solve. The problem of embedded peaks is nontrivial and may give rise to difficulties in both quantification as well as classification.

*Problems derived from the analysis of some samples at the same time – peak shifts*

So far, only problems related to one single sample have been discussed. However, one of the main interests in analysing chromatographic profiles is the analysis of several samples at the same time. With more information about the system, more possibilities of extracting information from the data set can be achieved. In addition, the problem of embedded peaks may be solved if one sample contains only one of the peaks. Nevertheless, more samples also may mean additional problems, mainly related to the sample-to-sample reproducibility. One of the classical problems derived from the low sample-to-sample reproducibility is the shift between peaks (Figure 9.8b). This can be so severe that some peaks might be mixed up between different samples in the same chromatographic run (Figure 9.8).

### Good news!

Many of these problems may be solved using proper mathematics – by baseline fitting, alignment and curve resolution techniques. The goal is to give the user a satisfactory solution without involving any change in the analytical methodology. The mathematical solutions should thus be able to solve these problems (e.g. overlapping peaks, peak shifts, baseline drift, etc.) and extract pure resolved peaks and their relative concentration in the sample. Curve resolution techniques, baseline removal methods as well as peak alignments algorithms have been applied in chromatography for the last 30 years, offering robust and reliable alternatives to time consuming re-parameterization.

*Alignment techniques*

Hyphenated separation techniques (e.g. GC-MS landscapes) include additional information from the spectral dimension (e.g. UV, MS and fluorescence), which can be used during the alignment process. This extra dimension can be used actively or passively in the alignment procedure. Passive means that the spectral dimension is summed/collapsed and vector-based alignment techniques used, for example Correlation Optimized Warping (Bylund *et al.*, 2002; Tomasi *et al.*, 2004, 2011; Skov, 2008). Active means that the spectral dimension is used when aligning, either as an additional tool in vector-based alignment or included in the alignment quality measure (e.g. matrix correlation) (Nielsen *et al.*, 1998; Gong *et al.*, 2004; Krebs *et al.*, 2006; Xu *et al.*, 2006; Szymanska *et al.*, 2007; Christin *et al.*, 2008, 2010). The use of the mass spectral information is an advantage if many peaks are overlapping and if peaks are missing in some samples to avoid aligning peaks that originate from different chemical compounds. The good thing about aligning data is that tri linear methods can be used afterwards, for example PARAFAC. However, for complex data with many co-eluting peaks, alignment can be difficult, it can introduce even worse artifacts. In this case, applying PARAFAC2 can result in a better solution (Skov *et al.*, 2009).

*Baseline methods for multidimensional chromatographic data*

For vector-based chromatographic data, fitting a baseline to a certain number data points and subtracting this from the chromatographic profile can often remove an unwanted baseline contribution. The fitting is often done using polynomials, either on a set of found baseline points in the whole chromatogram or in smaller sections allowing for larger flexibility. The pros and cons of local and global methods are discussed in Skov (2008). Removing a baseline from multiway data has been attempted a few times (Liang *et al.*, 1993; Amigo *et al.*, 2008). The methods work by finding and extracting parts of the data that originates from the baseline e.g. as a unique factor in a PARAFAC model.

### Curve resolution by PARAFAC and PARAFAC2

When peaks are co-eluting a traditional chromatographic software will use built in methods to extract the proper peak area such as division of peaks into individual contributions, for example by a vertical line and the use of a single mass fragment when this exists in only one of the co-eluting peaks. However, the methods are prone to errors due to different in peak height ratios of the co-eluting peaks and when no unique mass fragments are present. In this cases the optimal method would be to use the full mass spectral fingerprint as this often differs even between two very similar chemical compounds. This is exactly what PARAFAC for well-aligned peaks and PARAFAC2 for shifted peaks can do. Not only do these methods provide a much better peak division with respect to quantification but the mass spectral loading obtained hold information that can be directly transferred to mass spectral libraries (e.g. John Wiley & Sons and NIST) where the chemical compounds may be identified.

### Applications of multiway methods in chromatography

As stated elsewhere (Amigo *et al.*, 2010b) one of the main drawbacks of all factor analysis methods is that they can only be applied in local areas of the chromatographic profiles where the sources of variability (i.e. analytes and also physical variations like baseline drifts) do not exceed a determined number. This number depends on the quality of the data set in the selected interval as well as on the levels of co-elution of the different analytes (Figure 9.9). Nevertheless, the scores of the samples obtained after three-way analysis can be used for quantification or classification purposes (Pravdova *et al.*, 2002; Ortiz and Sarabia, 2007; Zhang *et al.*, 2007; Gomez and Callao, 2008; Schmidt *et al.*, 2008; Skov and Bro, 2008; Amigo *et al.*, 2010b; Morales *et al.*, 2011; Bosque-Sendra *et al.*, 2012; Murphy *et al.*, 2012; Varming *et al.*, 2013).

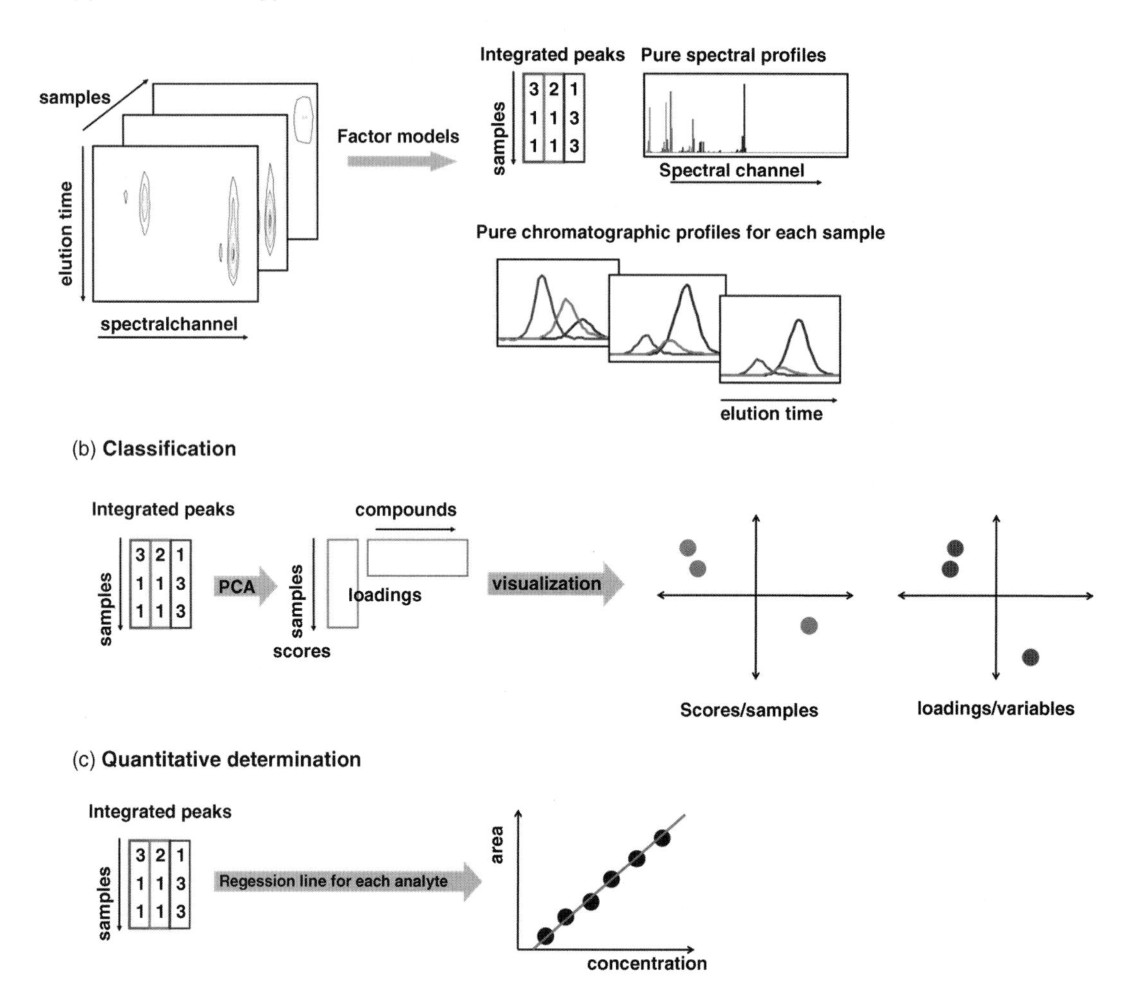


**Figure 9.9** Possibilities of PARAFAC2 model in selective areas of the chromatographic profile and the usefulness of the obtained scores for (b) classification or (c) quantification purposes. Amigo, *et. al*, 2010. Reproduced with permission from Elsevier.

The main interest though, is the complete quantification and/or understanding-classification of the peaks appearing in a data set (Bosque-Sendra *et al.*, 2012). Despite this drawback, three-way methods are very useful in the specific areas in which more focused knowledge is needed. As an example, Figure 9.10 shows a full GC-MS profile (TIC chromatograms) from different apples being ripened (detailed information in (Amigo *et al.*, 2010b)).

Here, the main interest is the fingerprinting of apples to assess the proper ripening period. It can be argued that this classification could have been done by just using the TIC profile and PCA. This is partially right. Nevertheless, by doing this, detailed chemical information is totally lost if, for instance, the interval highlighted in Figure 9.10 is considered. Here it can be appreciated that the interval has problems of baseline drift, elution time shifts and, especially, overlapping between different analytes. In this case, PARAFAC2 clearly demonstrated that it is possible to fully understand the chemical behaviour of that interval. In another recent study PARAFAC and PARAFAC2 were used

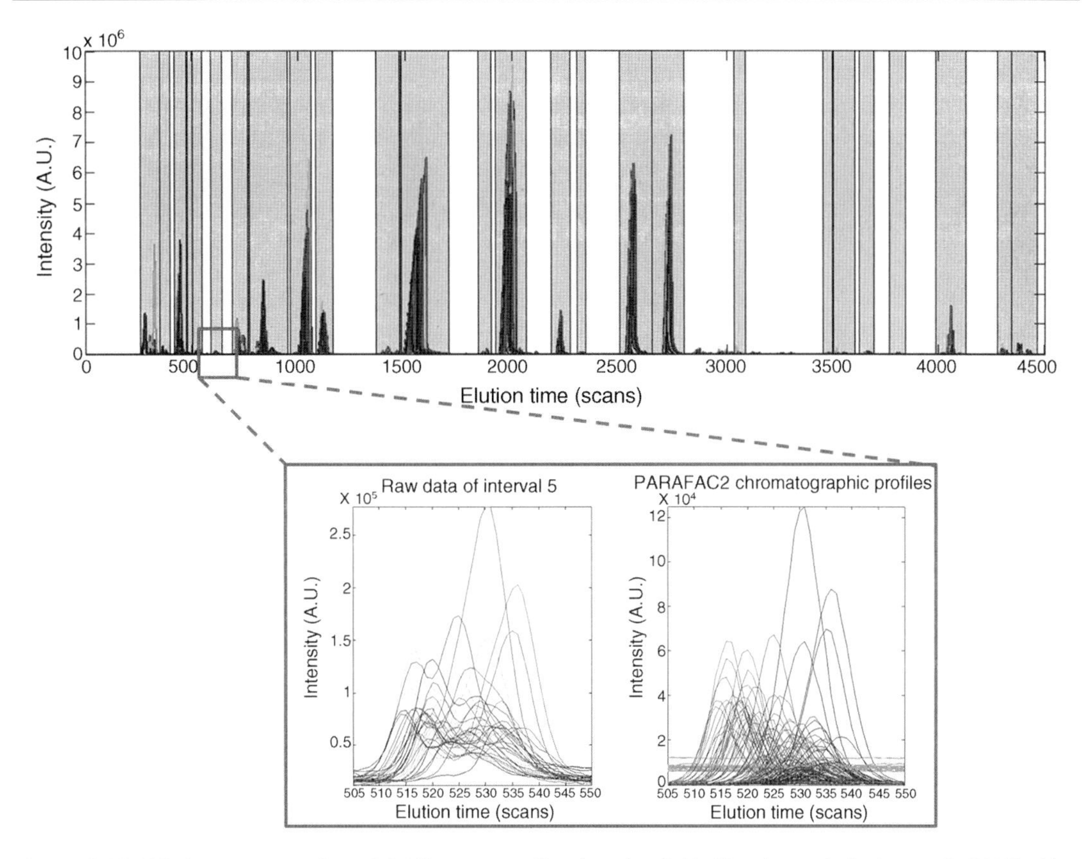


**Figure 9.10** TIC chromatogram from GC-MS aroma profile of apples divided into intervals (Amigo *et al.*, 2010) with a detail in one of the intervals and the PARAFAC2 solution with four chemical components.

on non-methane volatile organic compound emissions sampled on an Australian poultry farm (Murphy *et al.*, 2012). The study showed that it was possible to assign (identify and quantify) more peaks in a highly complex GC-MS data set using multiway methods compared to using classical chromatographic software.

As a general rule, we can say that the combination of individual three-way methods in selected areas of the chromatographic profile and other classification/regression methods is a very useful strategy in chromatography to obtain rich information to define the addressed problem. In that sense, problems like co-elution, as well as retention time shifts and baseline drifts have been overcome by multiway analysis (Amigo *et al.*, 2008, 2010a, 2010b).

## LF-NMR analysis of food

### Basic theory of LF-NMR

Nuclear magnetic resonance (NMR) spectroscopy and relaxometry are widely used as analytical techniques in research and industry (Yan *et al.*, 1996; Rutledge, 2001). In the beginning, the focus was aimed towards obtaining higher resolution spectra. The only way to reach this goal was to build bigger and stronger superconductor magnetic fields. Strong magnetic fields require a lot of space, making it unsuitable for anything but laboratory work. However, it was realized that the high resolution of these instruments was not essential for all scientific and industrial applications. The development of smaller, lighter bench-top NMR instruments was a result. A bench-top NMR typically has a magnetic field-strength of 0.23–0.70 Tesla equal to 10–30 MHz for protons (Rutledge, 1992), whereas high-field NMR has magnetic fields from 4.7 Tesla up to 21 Tesla equal to 900 MHz. Bench-top NMR instruments are typically used for obtaining relaxation curves, which is sample magnetization as a function of time. This is in contrast to high-field NMR where a spectrum in the frequency domain is obtained.

Relaxation curves primarily give information about water or fat content in the sample. They can further reveal whether the water is in a tightly bound, compartmentalized state or in free state. Carr and Purcell, and Meiboom and Gill introduced a technique (Carr and Purcell, 1954; Meiboom and Gill, 1958) that gives relaxation curves with a higher signal-to-noise ratio than previously achieved. The signal from a CPMG measurement is given as a sum of exponential decays:

$$m(t) = \sum_{n=1}^{N} M_{0,n} \cdot \exp\left(-\frac{t}{T_{2,n}}\right) + E \tag{9.12}$$

where $m(t)$ is the total relaxation signal, $N$ is the number of underlying pure mono-exponential relaxation curves present in the raw data, $M_0$ is the magnitude of the relaxation curve, $t$ is the time, $T_2$ is the characteristic transverse relaxation time, and $E$ is the unmodelled part of the data (the noise).

### *Multiway methods on LF-NMR signals*

The application of multiway chemometric techniques for exponentially decaying contribution profiles was initiated by the work of Windig and Antalek in 1997 (Windig and Antalek, 1997), who showed the use of Direct Exponential Curve Resolution Algorithm (DECRA) applied to pulsed gradient spin echo (PGSE) nuclear magnetic resonance (NMR) curves. Although their sample was of a chemical mixture created in the laboratory, it worked as a starter for the later use of this technique on LF-NMR data. DECRA takes into account the nature of the relaxation data. Since the data are made up of one or several mono-exponential curves, the contribution from each of these exponentials change along the relaxation curve. If a part of the relaxation curve is copied and put behind the other, the same group of relaxation curves will be present in both slabs, but the ratio between the different $T_2$-times will differ in the two slabs. The faster relaxing components will have a larger part of the signal in the first slab, while the latter slab(s) will be dominated by the slower relaxing components.

Windig and Antalek elaborated on their technique through two articles in 1998 (Antalek *et al.*, 1998; Windig *et al.*, 1998), where they showed how to apply the DECRA on PGSE NMR images of brains. In following years, they showed that DECRA also could be used to resolve kinetic profiles based on UV-VIS (Windig *et al.*, 1999) and mid-IR (Windig *et al.*, 2000) spectra.

The technique was introduced to low-field NMR by Pedersen *et al.* in 2002 (Pedersen *et al.*, 2002). In the original algorithm of Windig and Antalek the number of slabs was limited to two, as they used the general rank annihilation method (GRAM) (Sanchez and Kowalski, 1986) to extract the pure exponential curves. However, in the paper by Pedersen *et al.* (Pedersen *et al.*, 2002), they suggested the use of direct trilinear decomposition (DTLD) (Sanchez and Kowalski, 1990) instead of GRAM in the decomposition step, thus allowing a larger number of slabs to be added in the pre-processing step. They named this procedure SLICING. The advantage of adding more slabs to the matrix is to get a better representation of all relaxations curves in the subsequent three-way matrix. In addition to the term slab, it was necessary to add another term defining how the pre-processing was performed. This term is lag, indicating how many variables in the time domain have been removed in the specified slab. In other words, if a total of three copies are made, starting from time points 1, 2 and 5, there are three

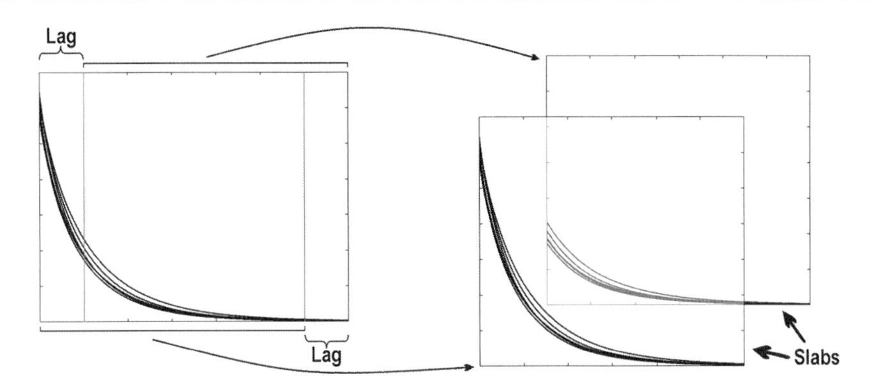


**Figure 9.11** The idea behind SLICING. Only two slabs shown, but the number of slabs is normally larger.

slabs, with lags 0, 1 and 4. The dimensionality of the data is thus decreased by four in the time domain and increased by two in the new slab direction. This is because in the first slab the four last time points are removed; in the second, the first and the three last ones are removed; and in the third slab, the four first time points are removed. An example of a simple SLICING with only two slabs is shown in Figure 9.11.

The dimensionality of the matrix will then increase from $I$ (samples)x$L$ (measurement points) to $I$x($L$-maximum lag)xnumber of slabs ($K$) – simply shortened to $I$x$J$x$K$. The three mode array (**X**) has the size $I$x$J$x$K$ and it contains the elements $x_{ijk}$, where the first index ($i$) refers to the samples, the second ($j$) refers to the time and the third ($k$) refers to the slab number. The rearranged three mode data follow a three-way model (DTLD or PARAFAC). In the case of LF-NMR, the scores in the PARAFAC model (**A**'s) are proportional to the $M_0$-value in Equation 9.12, while the second mode loadings (**B**'s) are the estimated decay curves. The third mode loadings (**C**'s) hold the same information as the second mode loadings but with a smaller dimension and, as such, are not of any interest. If the model is adequate, each second mode loading (**B**) should be uni-exponential because the three-way model can be shown to uniquely recover the underlying model when correctly specified (Windig and Antalek, 1997).

Andersen and Rinnan (Andersen and Rinnan, 2002) and Povlsen *et al.* (Povlsen *et al.*, 2003) suggested using PARAFAC rather than DTLD as the decomposition method of the three-way matrix. Often the DTLD solution was the optimal solution but in some cases refinement of the model was obtained through the use of PARAFAC. This is especially the case where one or more of the factors are small compared to the others, as was the case in both these articles.

As mentioned, the SLICING algorithm allowed for more than two slabs. However, the problem with this new feature was the subsequent optimization of the two terms lags and slabs. The number of possibilities is vast; not only do the number of slabs have to be decided, but so too is how the lags should be defined. That is, if three slabs are chosen, the first lag would be 0 but what of the two others, should they be 1 and 2, or some other combination of numbers, like 1 and 4, 2 and 10, etc.? Optimizing the SLICING proved to be rather time consuming. This problem was realized by Engelsen and Bro (Engelsen and Bro, 2003), who introduced power-slicing, a specific manner to define the lags to be used. The size of the lag was defined as $2^{(s-1)}$, where $s$ is the slab number. Through this method, the need for optimizing the number of lags is removed and the number of slabs is a more trivial variable to optimize. The last addition to the technique was made by Andrade *et al.* (Andrade *et al.*, 2007) who showed that it is even possible to perform the deconvolution on only one sample, by slicing the spectra not only in one direction but subsequently on the already sliced relaxation curves. They named this extension to the algorithm DOUBLESLICING. This means that each of the slabs is sliced again (Figure 9.11). In this way, each sample will create a three-dimensional matrix per sample, which thus could be analysed by DTLD or PARAFAC, thus leading to the same information to be in both **A**, **B** and **C** in the solution. Only the loading with the longest dimension will be used in the subsequent analysis.

The continuous challenge of estimating the right number of factors is more trivial using SLICING, as if too many components are extracted, the curves will reflect this by one or more of them being non-exponential. The number of factors can further be validated by the use of bootstrapping (Wehrens *et al.*, 2000), jack-knifing (Martens and Martens, 2001) and/or split-half analyses (Harshman and de Sarboe, 1994). In addition, the shape and distribution of the residuals of a model indicate whether there is information left in the data to model or not. The residuals should, in the perfect case, be randomly distributed with a zero mean. For an adequate model, the **B**'s, being uni-exponentials, can be fitted to Equation 9.12, and thus the $T_2$-values can be estimated. The $M_0$-value found in this fitting is dependent on the $T_2$-value of the curve, because in DTLD and PARAFAC the **B**-loadings are normalized. The sample specific $M_0$ values are found by multiplying the $M_0$ values found in the exponential fitting with the corresponding **A** scores.

### *A few applications in food science*

Andersen and Rinnan (Andersen and Rinnan, 2002) show an application of SLICING to fish samples measured in LF-NMR. They also introduced a correction of the LF-NMR curves, as it could be seen on the raw spectra that, even though the curves had stabilized, they did not vary around zero but rather around a slightly positive value. By subtracting the average value of the last points in the curve, a better deconvolution was ensured. This is an artifact that is inherent in the LF-NMR curves and which had not earlier been directed.

Povlsen *et al.* (Povlsen *et al.*, 2003) used the technique in the evaluation of LF-NMR curves from potato cultivars. Furthermore, they used the results from SLICING to predict the sensory evaluation of the potato cultivars with $r^2$ ranging from 0.57 to 0.79, very high numbers when it comes to the prediction of sensory data by other analytical techniques.

Hansen *et al.* (Hansen *et al.*, 2010) applied DOUBLESLICING on a set of rennet coagulation and syneresis samples. By performing DOUBLESLICING they showed that during gel formation two relaxation components were present. However, during syneresis, initiated by cutting the gel, an additional relaxation component was produced.

## Sensory data

Sensory data analysis is not one specific method but cover many types – time series, descriptive tests, napping, triangle tests and so on. However, sensory data have many things in common. Regardless of the chosen technique, often several attributes are evaluated by many assessors for several samples/products. By having more assessors, more attributes and several samples means that data are arranged in a data cube (a three-way data matrix) as exemplified in Figure 9.12(1) and, as such, would be suitable for multiway data analysis.

For traditional analysis of sensory cubed data the data cube structure is usually simplified by unfolding the cube or by finding the average of assessors. This is done in order to be able to use standard statistical and mathematical data descriptive methods. But can one be certain that such approaches do not hide important information, are too simple or even complicate the subsequent data analysis?

In the following, multiway methods for handling sensory data of different structures are discussed and guidelines for when to do what are provided. A regular sensory data set can be presented as shown in Figure 9.12(1). Such descriptive data are arranged in a cube but are easily

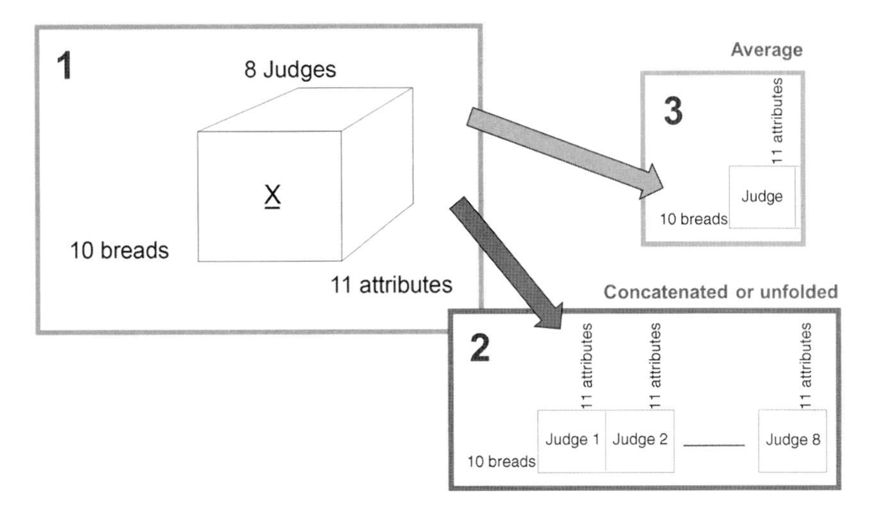


**Figure 9.12** Example of bread data arranged in three different structures: (1) data cube with intact data information; (2) data matrix with average of assessor score; and (3) data matrix with concatenated/unfolded assessor/attribute mode (Bro, 1998).

transformed into two simplified versions by either concatenating attributes and assessors or averaging across assessors shown in Figure 9.12.

Taking the average of assessors is the traditional way of treating descriptive sensory data. However, before taking the average it is necessary to be sure that all assessors perform in the same way and, if not, correct for this prior to any modelling (Dijksterhuis *et al.*, 2002). Often a well-trained panel can be looked upon as replicates in the same way as replicates in an analytical instrument, but it is recommended always to check the assessor performances, for example using PanelCheck (Tomic *et al.*, 2007) or as will be shown later a multiway model. By calculating the average, the data can be analysed with a simple PCA or other two-way methods. If it is preferred to include the assessors in the model, a PCA on concatenated data provides information about the assessor performances as well. However, the loading plot becomes very complex, as both attributes and assessors are now included in the variable mode. Another disadvantage is that the original data structure is lost. The use of PCA on averaged data and concatenated data is illustrated in Figure 9.13 for a sensory data set on different breads (see figure legend for information of the bread experiment).

An alternative is to use PARAFAC for sensory data. PARAFAC will split up attributes and assessors into two model terms and from these it is possible to evaluate relationships between attributes and assessors in separate plots, as visualized in Figure 9.13. In the example it can be seen that PARAFAC makes the replicates of the breads more similar as seen from the score plot. This suggests that concatenating removes some information about the replicates that was present in the original data. Also taking the average across assessors seems to be a too harsh simplification.

From the PARAFAC model it can be seen that assessor 8 deviates in the first factor, which indicates that this assessor has a different evaluation of the attributes salt, tough and sweet. The PARAFAC model and PCA on averaged data provide similar loading plots for the attribute mode, which also indicates that PARAFAC is able to find the right information and that such sensory data can be treated with trilinear methods such as PARAFAC. This small example (elaborated in Bro, 1998) shows that, through the use of multiway methods, additional information about the system is gained.

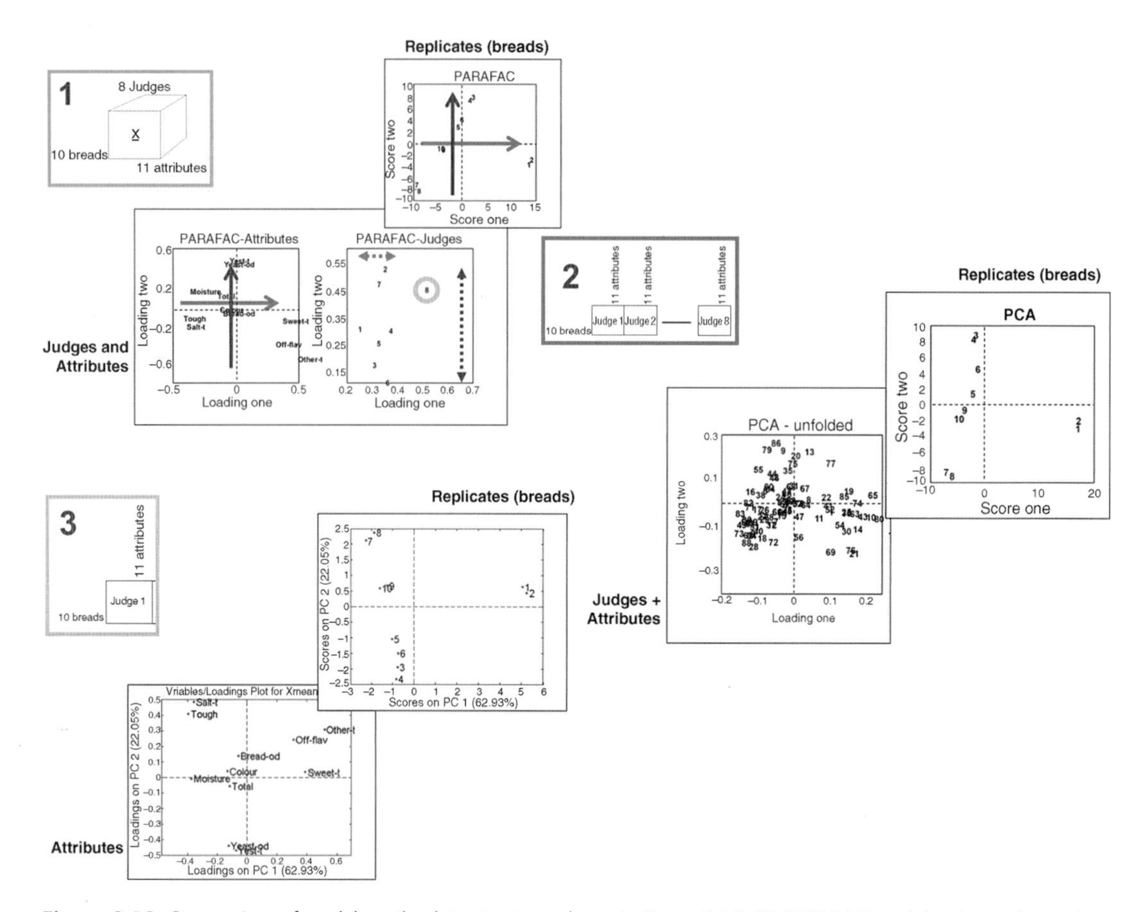


**Figure 9.13** Comparison of models on the data structures shown in Figure 9.13: (1) PARAFAC model on intact data cube; (2) PCA model on averaged data across assessors; and (3) PCA on concatenated/unfolded data. Bread data – 5 types of bread in duplicates (1 and 2, 3 and 4, 5 and 6, 7 and 8 and 9 and 10) with 8 assessors evaluating 11 attributes (Bro, 1998) providing a data cube of $10 \times 11 \times 8$.

## *Applications of sensory data in food science*

Other examples of multiway models applied to different types of sensory data can be found in the literature; these are summarized in Table 9.4.

For time intensity studies – for example development of chilli burn in the mouth – each assessor will evaluate different attributes over time. This makes data more complex, as the time profiles for different assessors will be different and, thus, conflicts with principles of trilinearity of the PARAFAC model. In such cases PARAFAC2 has proven to be applicable and provide a sensory meaningful description (Ovejero-Lopez *et al.*, 2005; Reinbach *et al.*, 2007).

## *Concluding remarks*

For profiling data it is suggested to use PARAFAC in combination with PCA to get a comprehensive overview of the data. The trilinearity constraint of PARAFAC allows complementary results to the initial PCA to be extracted, which can add to the understanding of product (dis)similarities, attribute correlations and assessor performances.

**Table 9.4** Sensory studies using multiway methods for handling the data.

| Data type | Food | Data modes | Data structure | Multiway model used | Reference |
|---|---|---|---|---|---|
| Consumer study | Yogurt | Samples × attributes × consumers | 6 × 5 × 120 | PARAFAC | Cruz *et al.*, 2012 |
| | Chewing gum | Samples × time × attributes | 10 × 231 × 5 | PARAFAC2 | Ovejero-Lopez *et al.*, 2005 |
| Flavour release – time intensity | Chilli burn | Samples × time × attributes | 8 × 150 × 8 | PARAFAC2 | Reinbach *et al.*, 2007 |
| | Meat flavour | Samples × time × attributes | 12 × 150 × 8 | PARAFAC2 | |
| | Bread | Samples × attributes × assessors | 10 × 8 × 11 | PARAFAC | Wu *et al.*, 2003 |
| | Balsamic vinegar | Samples × attributes × assessors | 142 × 8 × 16 | PARAFAC | Cocchi *et al.*, 2006 |
| Sensory profiling | Noodles | Samples × replicates × sessions × assessors × attributes | 8 × 4 × 2 × 12 × 8 | TUCKER | Cordella *et al.*, 2011 |
| | Maillard reactions | Samples × assessors × attributes | 66 × 5 × 9 | TUCKER | Pravdova *et al.*, 2002 |

For time intensity studies where the time profile for different assessors will be different, PARAFAC2 is the most optimal multiway method. However, by extracting features from the time profiles, such as maximum, steepness, plateau and range of taste/smell impression, it is possible to get well defined parameters that in many cases also hold valuable information of the products – for example by using PCA.

## FUTURE PERSPECTIVES

During this chapter the main features of multiway methods and their use in classification, quantitation, quality control and assessment in food science have been highlighted. With this perspective there are two main ideas to be said:

(i) A lot of work has been done to demonstrate the strength of multiway analysis but a lot of work must still be done to improve the algorithms and the applications. There are many questions to be solved. The answers to those depend on many things (quality of the data, target pursued with the analysis, etc.) and inherently no explicit answers can be given.

(ii) So far, the methods are mostly used in academia. Huge efforts must be done to transfer this knowledge to instrument vendors and industry.

Herein, some of the improvements are highlighted that can make multiway methods, first of all, more robust and, afterwards, more accessible to any researcher and company interested in them. They are based on the need to implement multiway methods with devices able to offer a fast and robust answer to either sample quantitation or process monitoring.

### Direct measurements in the raw samples – Process data

Many processes need to be monitored using robust and rapid methodologies capable of giving a response in real time. This need encompasses several characteristics:

- The measurement must be done in the raw/process sample without the need of any chemical treatment. EEM, among other techniques, has demonstrated its feasibility for this.
- The models must be able to give a fast response to take a fast decision.

Fermentation processes are clear examples in which the combination of EEM with PARAFAC gives a perfect framework for a real time and fast decision setup (Mortensen and Bro, 2006; Lillhonga and Geladi, 2011).

## Automation of three-way methods

Murphy *et al.* (Murphy *et al.*, 2012) demonstrated that through the use of open-source software it is possible to get superior identification and quantification of peaks from a highly complex GC-MS data set. The only thing missing is how to automate all this – several attempts have been made for this but still none of these are publicly available or open source (Hoggard and Synovec, 2007; Hoggard *et al.*, 2009).

### *Automating the curve resolution in EEM*

There are currently two suggested algorithms to make the whole curve resolution step of fluorescence more automatic. Engelen *et al.* (Engelen *et al.*, 2007) proposed a method that optimally should both handle scatter and remove potential outliers from the data, as long as the correct number of factors is selected. Thus it reduces the amount of parameters that must be estimated to one single! Bro and Vidal (Bro and Vidal, 2011) aim even higher, stating that they are capable of handling scatter, removing noisy wavelengths, remove outliers AND estimate the correct number of factors. If Bro and Vidal (Bro and Vidal, 2011) indeed have managed to make such a versatile tool, it still needs testing but it is a very appealing idea for the food scientist if such a method exist!

## How many samples are really needed?

The number of samples must also be considered when discussing low-rank aspects. However, due to the properties of second-order data and multiway models it is possible to extract the unique contribution from the analytes having rather few samples compared to ordinary two-way multivariate models, where more samples are often needed (Bro, 2003). Rinnan *et al.* (Rinnan *et al.*, 2007) showed that including as many samples as there are fluorophores, plus one, gives good predictions. This does need to be verified, but if it is the case, it is good news for calibration development with multiway methods.

## REFERENCES

Airado-Rodriguez, D., Galeano-Diaz, T., Duran-Meras, I. and Wold, J.P. (2009) Usefulness Of fluorescence excitation-emission matrices in combination with PARAFAC, as fingerprints of red wines. *Journal of Agricultural and Food Chemistry*, **57**, 1711–1720.

Airado-Rodriguez, D., Duran-Meras, I., Galeano-Diaz, T. and Wold, J.P. (2011) Front-face fluorescence spectroscopy: A new tool for control in the wine industry. *Journal of Food Compoisition and Analysis*, **24**, 257–264.

Amigo, J., Coello, J. and Maspoch, S. (2005) Three-way partial least-squares regression for the simultaneous kinetic-enzymatic determination of xanthine and hypoxanthine in human urine. *Analytical and Bioanalytical Chemistry*, **382**, 1380–1388.

Amigo, J.M., Skov, T., Coello, J. *et al.* (2008) Solving GC-MS problems with PARAFAC2. *TRAC-Trends in Analytical Chemistry*, **27**, 714–725.

Amigo, J.M., Skov, T. and Bro, R. (2010a) Chromathography: solving chromatographic issues with mathematical models and intuitive graphics. *Chemical Reviews*, **110**, 4582–4605.

Amigo, J.M., Popielarz, M. J., Callejón, R. M. *et al.* (2010b) Comprehensive analysis of chromatographic data by using PARAFAC2 and principal components analysis. *Journal of Chromatography A*, **1217**(26), 4422–4429.

Andersen, C.M. and Bro, R. (2003) Practical aspects of PARAFAC modeling of fluorescence excitation-emission data. *Journal of Chemometrics*, **17**, 200–215.

Andersen, C.M. and Rinnan, Å. (2002) Distribution of water in fresh cod. *Lebensmittel-Wissenschaft und-Technologie*, **35**, 687–696.

Andersen, C.M., Vishart, M. and Holm, V. (2005) Application of fluorescence spectroscopy in the evaluation of light-induced oxidation in cheese. *Journal of Agricultural and Food Chemistry*, **53**, 9985–9992.

Andersson, C.A. and Henrion, R. (1999) A general algorithm for obtaining simple structure of core arrays in N-way PCA with application to fluorometric data. *Computational Statistics and Data Analysis*, **31**(3), 255–278.

Andrade, L., Micklander, E., Farhat, I. *et al.* (2007) Doubleslicing: A non-iterative single profile multi-exponential curve resolution procedure – Application to time-domain NMR transverse relaxation data. *Journal of Magnetic Resonance*, **189**, 286–292.

Antalek, B., Hornak, J. and Windig, W. (1998) Multivariate image analysis of magnetic resonance images with the direct exponential curve resolution algorithm (DECRA) – Part 2: Application to human brain images. *Journal of Magnetic Resonance*, **132**, 307–315.

Bahram, M., Bro, R., Stedmon, C. and Afkhami, A. (2006) Handling of Rayleigh and Raman scatter for PARAFAC modeling of fluorescence data using interpolation. *Journal of Chemometrics*, **20**, 99–105.

Bassompierre, M., Tomasi G., Munck, L. *et al.* (2007) Dioxin screening in fish product by pattern recognition of biomarkers. *Chemosphere*, **67**, S28–S35.

Baunsgaard, D., Andersson, C. A., Arndal, A. and Munck, L. (2000a) Multiway chemometrics for mathematical separation of fluorescent colorants and colour precursors from spectrofluorimetry of beet sugar and beet sugar thick juice as validated by HPLC analysis. *Food chemistry*, **70**, 113–121.

Baunsgaard, D., Norgaard, L. and Godshall, M. (2000b) Fluorescence of raw cane sugars evaluated by chemometrics. *Journal of Agricultural and Food Chemistry*, **48**, 4955–4962.

Beltrán, J. L., Ferrer, R. and Guiteras, J. (1998) Multivariate calibratioh of polycyclic aromatic hyrdocarbon mixtures from excitation-emission fluorescence spectra. *Analytica Chimica Acta*, **373**, 311–319.

Berge, J. and Kiers, H. (1996) Some uniqueness results for PARAFAC2. *Psychometrika*, **61**, 123–132.

Booksh, K.S. and Kowalski, B. R. (1994) Theory of analytical chemistry. *Analytical Chemistry*, **66**(15), 782A–791A.

Booksh, K.S., Muroski, A. R. and Myrick, M. L. (1996) Singe-measurement excitation-emission matrix spectrofluorometer for determination of hydrocarbons in ocean water. 2. Calibration and quantitation of napthalene and styrene. *Analytical Chemistry*, **68**(20), 3539–3544.

Bosque-Sendra, J. M., Cuadros-Rodríguez, L., Ruiz-Samblás, C. and Mata, A. P. d. l. (2012) Combining chromatography and chemometrics for the characterization and authentication of fats and oils from triacylglycerol compositional data – A review. *Analytica Chimica Acta*, **724**(0), 1–11.

Brinkman, U. A.T. (1999) Preface. *Journal of Chromatography A*, **856**, 1–2.

Bro, R. (1996) Multiway Calibration. Multilinear PLS. *Journal of Chemometrics*, **10**, 47–61.

Bro, R. (1997) PARAFAC. Tutorial and application. *Chemometrics and Intelligent Laboratory Systems*, **38**, 149–171.

Bro, R. (1998) *Multiway Analysis in the Food Industry – Models, Algorithms and Applications*. PhD Thesis, University of Amsterdam, The Netherlands.

Bro, R. (1999) Exploratory study of sugar production using fluorescence spectroscopy and multiway analysis. *Chemometrics and Intelligent Laboratory Systems*, **46**, 133–147.

Bro, R. (2003) Multivariate calibration – What is in chemometrics for the analytical chemist. *Analytica Chimica Acta*, **500**, 185–194.

Bro, R. (2006) Review on multiway analysis in chemistry – 2000–2005. *Critical Reviews in Analytical Chemistry*, **36**, 279–293.

Bro, R. and Kiers, H. A. (2003) A new efficient method for determining the number of components in PARAFAC models. *Journal of Chemometrics*, **17**, 274–286.

Bro, R. and Vidal, M. (2011) Eemizer: Automated modeling of fluorescence EEM data. *Chemometrics and Intelligent Laboratory Systems*, **106**(1), 86–92.

Bro, R., Andersson, C. A. and Kiers, H. A. (1999) PARAFAC2 – Part II. Modeling chromatographic data with retention time shifts. *Journal of Chemometrics*, **13**, 295–309.

Bro, R., Sidiropoulos, N. D. and Smilde, A. K. (2002) Maximum likelihood fitting using ordinary least squares algorithms. *Journal of Chemometrics*, **16**, 387–400.

Bylund, D., Danielsson, R., Malmquist, G. and Markides, K. (2002) Chromatographic alignment by warping and dynamic programming as a pre-processing tool for PARAFAC modelling of liquid chromatography-mass spectrometry data. *Journal of Chromatography A*, **961**, 237–244.

Caggiano, G., Kapetanios, G. and Labhard, V. (2011) Are more data always better for factor analysis? Results for the euro area, the six largest euro area countries and the UK. *Journal of Forecasting*, **30**, 736–752.

Callejon, R. M., Amigo, J. M., Pairo, E. *et al.* (2012) Classification of sherry vinegars by combining multidimensional fluorescence, PARAFAC and different classification approaches. *Talanta*, **88**, 456–462.

Canada-Canada, F., Espinosa-Mansilla, A., de la Pena, A. M. *et al.* (2009) Determination of danofloxacin in milk combining second-order calibration and standard addition method using excitation-emission fluorescence data. *Food Chemistry*, **113**, 1260–1265.

Carr, H. and Purcell, E. (1954) Effects of diffusion on free precession in nuclear magnetic resonance experiments. *Physical Review*, **94**, 630–638.

Carroll, J. and Chang, J.-J. (1970) Analysis of individual differences in multidimensional scaling via an n-way generalization of 'Eckart–Young' decomposition. *Psychometrika*, **35**, 283–319.

Cattell, R. (1944) 'Parallel proportional profiles' and other principles for determining the choice of factors by rotation. *Psychometrika*, **9**, 267–283.

Christensen, J., Povlsen, V. T. and Sørensen, J. (2003) Application of fluorescence spectroscopy and chemometrics in the evaluation of processed cheese during storage. *Journal of Diary Science*, **86**, 1101–1107.

Christensen, J., Becker, E. and Frederiksen, C. (2005) Fluorescence spectroscopy and PARAFAC in the analysis of yogurt. *Chemometrics and Intelligent Laboratory Systems*, **75**, 201–208.

Christensen, J., Norgaard, L., Bro, R. and Engelsen, S. (2006) Multivariate autofluorescence of intact food systems. *Chemical Reviews*, **106**, 1979–1994.

Christensen, J., Tomasi, G., Strand, J. and Andersen, O. (2009) PARAFAC modeling of fluorescence excitation-emission spectra of fish bile for rapid en route screening of PAC exposure. *Environmental Science and Technology*, **43**, 4439–4445.

Christin, C., Smilde, A. K., Hoefsloot, H. C. J. *et al.* (2008) Optimized time alignment algorithm for LC-MS data: Correlation optimized warping using component detection algorithm-selected mass chromatograms. *Analytical Chemistry*, **80**, 7012–7021.

Christin, C., Hoefsloot, H. C. J., Smilde, A. K. *et al.* (2010) Time alignment algorithms based on selected mass traces for complex LC-MS data. *Journal of Proteome Research*, **9**, 1483–1495.

Cocchi, M., Bro, R., Durante, C. *et al.* (2006) Analysis of sensory data of Aceto Balsamico Tradizionale di Modena (ABTM) of different ageing by application of PARAFAC models. *Food Quality and Preference*, **17**, 419–428.

Comas, E., Gimeno, R.A., Ferre, J. *et al.* (2004) Quantification from highly drifted and overlapped chromatographic peaks using second-order calibration methods. *Journal of Chromatography A*, **1035**, 195–202.

Cordella, C. B. Y., Leardi, R. and Rutledge, D. N. (2011) Three-way principal component analysis applied to noodles sensory data analysis. *Chemometrics and Intelligent Laboratory Systems*, **106**, 125–130.

Cruz, A. G., Cadena, R. S., Faria, J. A. F. *et al.* (2012) PARAFAC: Adjustment for modeling consumer study covering probiotic and conventional yogurt. *Food Research International*, **45**, 211–215.

de Juan, A. and Tauler, R. (2001) Comparison of three-way resolution methods for non-trilinear chemical data sets. *Journal of Chemometrics*, **15**, 749–772.

Diez, R., Sarabia, L. and Ortiz, M. C. (2007) Rapid determination of sulfonamides in milk samples using fluorescence spectroscopy and class modeling with n-way partial least squares. *Analytica Chimica Acta*, **585**, 350–360.

Diez Azofra, R., Sarabia, L. A. and Cruz Ortiz, M. (2010) Optimization of a solid-phase extraction procedure in the fluorimetric determination of sulfonamides in milk using the second-order advantage of PARAFAC and D-optimal design. *Analytical and Bioanalytical Chemistry*, **396**, 923–935.

Dijksterhuis, G., Frost, M. and Byrne, D. (2002) Selection of a subset of variables: minimisation of Procrustes loss between a subset and the full set. *Food Quality and Preference*, **13**, 89–97.

Dolan, J. W. (2002) Peak tailing and resolution. *LC GC Europe 2002*, **15**(6), 334–337.

Dolan, J. W. (2003) Why do peaks tail?. *LC GC Europe 2002*, **16**(9), 610.

Eaton, J. K., Alcivar-Warren, A. and Kenny, J. E. (2012) Multidimensional fluorescence fingerprinting for classification of shrimp by location and species. *Environmental Science and Technology*, **46**, 2276–2282.

Engelen, S., Frosch Møller, S. and Hubert, M. (2007) Automatically identifying scatter in fluorescence data using robust techniques. *Chemometrics and Intelligent Laboratory Systems*, **86**, 35–51.

Engelsen, S. B. and Bro, R. (2003) PowerSlicing. *Journal of Magnetic Resonance*, **163**, 192–197.

Garcia, I., Sarabia, L., Ortiz, M. and Aldama, J. (2004) Building robust calibration models for the analysis of estrogens by gas chromatography with mass spectrometry detection. *Analytica Chimica Acta*, **526**, 139–146.

Gomez, V. and Callao, M. P. (2008) Analytical applications of second-order calibration methods. *Analytica Chimica Acta*, **627**, 169–183.

Gong, F., Liang, Y., Fung, Y. and Chau, F. (2004) Correction of retention time shifts for chromatographic fingerprints of herbal medicines. *Journal of Chromatography A*, **1029**, 173–183.

Guimet, F., Ferré, J., Boqué, R. and Rius, X.F. (2004) Application of unfold principal component analysis and parallel factor analysis to the exploratory analysis of olive oils by means of excitation-emission matrix fluorescence spectroscopy. *Analytica Chimica Acta*, **515**, 75–85.

Guimet, F., Ferré, J., Boqué, R. *et al.* (2005a) Excitation-emission fluorescence spectroscopy combined with three-way methods of analysis as a complementary technique for olive oil characterization. *Journal of Agricultural and Food Chemistry*, **53**, 9319–9328.

Guimet, F., Ferré, J. and Boqué, R. (2005b) Rapid detection of olive-pomace oil adulteration in extra virgin olive oils from the protected denomination of origin 'Siurana' using excitation–emission fluorescence spectroscopy and three-way methods of analysis. *Analytica Chimica Acta*, **544**, 143–152.

Hansen, C. L., Rinnan, Å., Engelsen, S. B. *et al.* (2010) Effect of gel firmness at cutting time, pH, and temperature on rennet coagulation and syneresis: An *in situ* H-1 NMR relaxation study. *Journal of Agricultural and Food Chemistry*, **58**, 513–519.

Harris, T. (2000) Think carefully, and take more data. *Analytical Chemistry*, **72**(21), 669 A.

Harshman, R. A. (1970) Foundations of the PARAFAC procedure – Models and conditions for an explanatory multi-modal factor analysis. *UCLA Working Papers in Phonetics*, **16**, 1–84.

Harshman, R. A. (1972) Determination and proof of minimum uniqueness conditions for PARAFAC1. *UCLA Working Papers in Phonetics*, **22**, 111–117.

Harshman, R. A. and de Sarboe, W. S. (1994). An application of PARAFAC to a small sample problem, demonstrating preprocessing, orthogonality constraints and split-half diagnostic techniques. In: *Research Methods for Multimode Data Analysis* (eds Law, H. G., Snyder, C. W., Hattie, J. A., and McDonald, R. P.) Praeger, New York, pp. 602–642.

Harshman, R. A. and Lundy, M. E. (1984). The PARAFAC model for three-way factor analysis and multidimensional scaling. In: *Research Methods for Multimode Data Analysis* (eds Law, H. G., Snyder, C. W., Hattie, J. A., and McDonald, R. P.). Praeger, New York, pp. 122–215.

Hashemi, J., Kram, G. A. and Alizadeh, N. (2008) Enhanced spectrofluorimetric determination of aflatoxin B1 in wheat by second-order standard addition method. *Talanta*, **75**, 1075–1081.

Henrion, R. (1994) N-way principal component analysis theory, algorithms and applications. *Chemometrics and Intelligent Laboratory Systems*, **25**, 1–23.

Henrion, R. and Andersson, C. (1999) A new criterion for simple-structure transformations of core arrays in N-way principal components analysis. *Chemometrics and Intelligent Laboratory Systems*, **47**, 189–204.

Ho, C.-N., Christian, G. D. and Davidson, E. R. (1978) Application of the method of rank annihilation to quantiative analyses of multicomponent fluorescence data from the video fluorometer. *Analytical Chemistry*, **50**(8), 1108–1113.

Ho, C.-N., Christian, G. D. and Davidson, E. R. (1980) Application of the method of rank annihilation to fluorescent multicomponent mixtures of polynuclear aromatic hydrocarbons. *Analytical Chemistry*, **50**, 1071–1079.

Hoggard, J.C. and Synovec, R. E. (2007) Parallel factor analysis (PARAFAC) of target analytes in GC × GC-TOFMS data – Automated selection of a model with an appropriate number of factors. *Analytical Chemistry*, **79**, 1611–1619.

Hoggard, J.C., Siegler, W. C. and Synovec, R. E. (2009) Toward automated peak resolution in complete GC × GC-TOFMS chromatograms by PARAFAC. *Journal of Chemometrics*, **23**, 421–431.

Hubert, M., Rousseeuw, P. and Verdonck, T. (2005) ROBPCA: A new approach to robust principal component analysis. *Technometrics*, **47**, 64–79.

Hyotylainen, T. and Riekkola, M. (2003) On-line coupled liquid chromatography-gas chromatography. *Journal of Chromatography A*, **1000**, 357–384.

Ingle, J. D. J. and Crouch, S. R. (1988). *Spectrochemical Analysis*. Prentice Hall, NJ.

Jiji, R.D. and Booksh, K. S. (2000) Mitigation of Rayleig and Raman spectra interferences in multivariate calibration of excitation-emission matrix fluorescence spectra. *Analytical Chemistry*, **72**, 718–725.

Jiji, R.D., Andersson, G. G. and Booksh, K. S. (2000) Application of PARAFAC for calibration with excitation-emission matrix fluorescence spectra of three classes of environmental pollutants. *Journal of Chemometrics*, **14**, 171–185.

Karoui, R., Mazerolles, G. and Dufour, E. (2003) Spectroscopic techniques coupled with chemometric tools for structure and texture determinations in dairy products. *International Dairy Journal*, **13**, 607–620.

Kiers, H. A. (2000) Towards a standardized notation and terminology in multiway analysis. *Journal of Chemometrics*, **14**, 105–122.

Kiers, H., Ten Berge, J. and Bro, R. (1999) PARAFAC2 Part I. A direct fitting algorithm for the PARAFAC2 model. *Journal of Chemometrics*, **13**, 275–294.

Krebs, M., Tingley, R. D., Zeskind, J. E. *et al.* (2006) Alignment of gas chromatography-mass spectrometry data by landmark selection from complex chemical mixtures. *Chemometrics and Intelligent Laboratory Systems*, **81**, 74–81.

Lakowicz, J. R. (1999). *Principles of Fluorescence Spectroscopy.*, 2 edn. Kluwer Academic/Plenum Publishers, New York.

Lawaetz, A. J. (2011). Fluorescence spectroscopy and chemometrics applied in cancer diagnostics and metabonomics. PhD Thesis, University of Copenhagen (Faculty of Life Sciences), Denmark.

Li, J.-S., Wang, H., Zhang, X. and Zhang, H.-S. (2003) Spectrofluorimetric determination of total amount of nitrite and nitrate in biological sample with a new fluorescent probe 1,3,5,7-tetramethyl-8-(3′,4′-diaminophenyl)-difluoroboradiaza-s-indacence. *Talanta*, **61**, 797–802.

Liang, Y., Kvalheim, O., Rahmani, A. and Brereton, R. (1993) A two-way procedure for background correction of chromatographic spectroscopic data by congruence analysis and least-squares fit of the zero-component regions comparison with double-centering. *Chemometrics and Intelligent Laboratory Systems*, **18**, 265–279.

Lillhonga, T. and Geladi, P. (2011) Three-way analysis of a designed compost experiment using near-infrared spectroscopy and laboratory measurements. *Journal of Chemometrics*, **25**(4), 193–200.

Martens, H. and Martens, M. (2001). *Multivariate analysis of quality – An introduction*. John Wiley & Sons Ltd, Chichester.

Martens, H. and Næs, T. (1989). *Multivariate Calibration*. John Wiley & Sons, Inc., New York.

Matthews, B., Jones, A., Theodorou, N. and Tudhope, A. (1996) Excitation-emission-matrix fluorescence spectroscopy applied to humic acid bands in coral reefs. *Marine Chemistry*, **55**, 317–332.

McKnight, D. M., M., Boyer, E. W., Westerhoff, P. K. *et al.* (2001) Spectrofluorometric characterization of dissolved organic matter for indication of precursor organic material and aromaticity. *Limnology and Oceanography*, **46**(1), 38–48.

Meiboom, S. and Gill, D. (1958) Modified spin-echo method for measuring nuclear relaxation times. *Review of Scientific Instruments*, **29**, 688–691.

Moberg, L., Robertsson, G. and Karlberg, B. (2001) Spectrofluorometric determination of chlorophylls and pheopigments using parallel factor analysis. *Talanta*, **54**, 161–170.

Mohler, R. E., Dombek, K. M., Hoggard, J. C. *et al.* (2007) Comprehensive analysis of yeast metabolite GC × GC-TOFMS data: combining discovery-mode and deconvolution chemometric software. *Analyst*, **132**, 756–767.

Møller, J.K.S., Parolari, G., Gabba, L. *et al.* (2003) Monitoring chemical changes of dry-cured Parma ham during processing by surface autofluorescence spectroscopy. *Journal of Agricultural and Food Chemistry*, **51**, 1224–1230.

Morales, R., Cruz Ortiz, M., Sarabia, L. A. and Sagrario Sanchez, M. (2011) D-optimal designs and N-way techniques to determine sulfathiazole in milk by molecular fluorescence spectroscopy. *Analytica Chimica Acta*, **707**, 38–46.

Mortensen, P. P. and Bro, R. (2006) Real-time monitoring and chemical profiling of a cultivation process. *Chemometrics and Intelligent Laboratory Systems*, **84**, 106–113.

Munck, L., Nørgaard, L., Engelsen, S. B. *et al.* (1998) Chemometrics in food science – a demonstration of the feasability of a highly exploratory, inductive evaluation strategy of fundamental scientific significance. *Chemometrics and Intelligent Laboratory Systems*, **44**, 31–60.

Murphy, K. R., Wenig, P., Parcsi, G. *et al.* (2012) Characterizing odorous emissions using new software for identifying peaks in chemometric models of gas chromatography – mass spectrometry datasets. *Chemometrics and Intelligent Laboratory Systems*, **118**, 41–50.

Nielsen, N., Carstensen, J. and Smedsgaard, J. (1998) Aligning of single and multiple wavelength chromatographic profiles for chemometric data analysis using correlation optimised warping. *Journal of Chromatography A*, **805**, 17–35.

Ortiz, M. C. and Sarabia, L. (2007) Quantitative determination in chromatographic analysis based on n-way calibration strategies. *Journal of Chromatography A*, **1158**, 94–110.

Ovejero-Lopez, I., Bro, R. and Bredie, W. (2005) Univariate and multivariate modelling of flavour release in chewing gum using time-intensity: a comparison of data analytical methods. *Food Quality and Preference*, **16**, 327–343.

Pedersen, H. T., Bro, R. and Engelsen, S. B. (2002) Towards rapid and unique curve resolution of low-field NMR relaxation data – trilinear SLICING versus two-dimensional curve fitting. *Journal of Magnetic Resonance*, **157**, 141–155.

Povlsen, V. T., Rinnan, Å., van den Berg, F. *et al.* (2003) Direct decomposition of NMR relaxation profiles and prediction of sensory attributes of potato samples. *Lebensmittel-Wissenschaft und-Technologie*, **36**(4), 423–432.

Pravdova, V., Boucon, C., de Jong, S. *et al.* (2002) Three-way principal component analysis applied to food analysis: an example. *Analytica Chimica Acta*, **462**(2), 133–148.

Quinto Tranchida, P., Donato, P., Dugo, P. *et al.* (2007) Comprehensive chromatographic methods for the analysis of lipids. *TRAC-Trends in Analytical Chemistry*, **26**, 191–205.

Reinbach, H. C., Meinert, L., Ballabio, D. *et al.* (2007) Interactions between oral burn, meat flavor and texture in chili spiced pork patties evaluated by time-intensity. *Food Quality and Preference*, **18**, 909–919.

Rinnan, Å. and Andersen, C. M. (2005) Handling of first-order Rayleigh scatter in PARAFAC modelling of fluorescence excitation-emission data. *Chemometrics and Intelligent Laboratory Systems*, **76**, 91–99.

Rinnan, Å., Booksh, K. S. and Bro, R. (2005) First order Rayleigh scatter as a separate component in the decomposition of fluorescence landscapes. *Analytica Chimica Acta*, **537**, 349–358.

Rinnan, Å., Riu, J. and Bro, R, (2007) Multiway prediction in the presence of uncalibrated interferents. *Journal of Chemometrics*, **21**, 76–86.

Rodriguez, N., Cruz Ortiz, M. and Sarabi, L. A. (2009) Study of robustness based on n-way models in the spectrofluorimetric determination of tetracyclines in milk when quenching exists. *Analytica Chimica Acta*, **651**, 149–158.

Rutledge, D. N. (1992) Low resolution pulse nuclear magnetic resonance (LPR-NMR). *Analusis*, **20**, 58–62.

Rutledge, D. N. (2001) Characterisation of water in agro-food products by time domain-NMR. *Food Control*, **12**, 437–445.

Sanchez, E. and Kowalski, B. R. (1986) Generalized rank annihilation factor analysis. *American Chemical Society*, **58**, 496–499.

Sanchez, E. and Kowalski, B. R. (1990) Tensorial resolution – A direct trilinear decomposition. *Journal of Chemometrics*, **4**, 29–45.

Schmidt, B., Jaroszewski, J. W., Bro, R. *et al.* (2008) Combining PARAFAC analysis of HPLC-PDA Profiles and structural characterization using HPLC-PDA-SPE-NMR-MS experiments: Commercial preparations of St. John's Wort. *Analytical Chemistry*, **80**(6), 1978–1987.

Skoog, D. A. and Leary, J. J. (1992). *Principles of Instrumental Analysis*, 4th edn. Saunders College Publishing, Orlando, FL.

Skov, T. (2008). *Mathematical Resolution of Complex Chromatographic Measurements*. PhD Thesis, University of Copenhagen (Faculty of Life Sciences), Denmark.

Skov, T. and Bro, R. (2008) Solving fundamental problems in chromatographic analysis. *Analytical and Bioanalytical Chemistry*, **390**, 281–285.

Skov, T., Hoggard, J. C., Bro, R. and Synovec, R. E. (2009) Handling within run retention time shifts in two-dimensional chromatography data using shift correction and modeling. *Journal of Chromatography A*, **1216**, 4020–4029.

Stedmon, C.A., Markager, S. and Bro, R. (2003) Tracing dissolved organic matter in aquatic environments using a new approach to fluorescence spectroscopy. *Marine Chemistry*, **82**, 239–254.

Stoll, D.R., Cohen, J. D. and Carr, P. W. (2006) Fast, comprehensive online two-dimensional high performance liquid chromatography through the use of high temperature ultra-fast gradient elution reversed-phase liquid chromatography. *Journal of Chromatography A*, **1122**, 123–137.

Stoll, D.R., Li, X., Wang, X. *et al.* (2007) Fast, comprehensive two-dimensional liquid chromatography. *Journal of Chromatography A*, **1168**, 3–43.

Szymanska, E., Markuszewski, M. J., Capron, X. *et al.* (2007) Evaluation of different warping methods for the analysis of CE profiles of urinary nucleosides. *Electrophoresis*, **28**, 2861–2873.

Tena, N., Aparicio, R. and Garcia-Gonzalez, D. L. (2012) Chemical changes of thermoxidized virgin olive oil determined by excitation-emission fluorescence spectroscopy (EEFS). *Food Research International*, **45**, 103–108.

Thygesen, L. G., Rinnan, Å., Barsberg, S. and Møller, J. K. S. (2004) Stabilizing the PARAFAC decomposition of fluorescence spectra by insertion of zeros outside the data area. *Chemometrics and Intelligent Laboratory Systems*, **71**, 97–106.

Tomasi, G., van den Berg, F. and Andersson, C. A. (2004) Correlation optimized warping and dynamic time warping as preprocessing methods for chromatographic data. *Journal of Chemometrics*, **18**, 231–241.

Tomasi, G., Savorani, F. and Engelsen, S. B. (2011) icoshift: An effective tool for the alignment of chromatographic data. *Journal of Chromatography*, **1218**, 7832–7840.

Tomic, O., Nilsen, A., Martens, M. and Naes, T. (2007) Visualization of sensory profiling data for performance monitoring. *LWT-Food Science and Technology*, **40**, 262–269.

Varming, C., Petersen, M. A., Skov, T. and Ardö, Y. (2013) Challenges in quantitative analysis of aroma compounds in cheeses with different fat content and maturity level. *International Dairy Journal*, **29**, **15**20.

Wehrens, R., Putter, H. and Buydens, L. M. (2000) The bootstrap – a tutorial. *Chemometrics and Intelligent Laboratory Systems*, **54**, 35–52.

Wilson, I. and Brinkman, U. (2003) Hyphenation and hypernation – The practice and prospects of multiple hyphenation. *Journal of Chromatography A*, **1000**, 325–356.

Wilson, I. D. and Brinkman, U. A. T. (2007) Hype and hypernation: multiple hyphenation of column liquid chromatography and spectroscopy. *TRAC-Trends in Analytical Chemistry*, **26**, 847–854.

Windig, W. and Antalek, B. (1997) Direct exponential curve resolution algorithm (DECRA): A novel application of the generalized rank annihilation method for a single spectral mixture data set with exponentially decaying contribution profiles. *Chemometrics and Intelligent Laboratory Systems*, **37**, 241–254.

Windig, W., Hornak, J. P. and Antalek, B. (1998) Multivariate image analysis of magnetic resonance images with the direct exponential curve resolution algorithm (DECRA). I – Algorithm and model study. *Journal of Magnetic Resonance*, **132**, 298–306.

Windig, W., Antalek, B., Sorriero, L. J. *et al.* (1999) Applications and new developments of the direct exponential curve resolution algorithm (DECRA). Examples of spectra and magnetic resonance images. *Journal of Chemometrics*, **13**, 95–110.

Windig, W., Antalek, B., Robbins, M. J. *et al.* (2000) Applications of the direct exponential curve resolution algorithm (DECRA) to solid state nuclear magnetic resonance and mid-infrared spectra. *Journal of Chemometrics*, **14**, 213–227.

Wold, J.P., Bro, R., Veberg, A. *et al.* (2006) Active photosensitizers in butter detected by fluorescence spectroscopy and multivariate curve resolution. *Journal of Agricultural and Food Chemistry*, **54**, 10197–10204.

Wu, W., Guo, Q., Massart, D. L. *et al.* (2003) Structure preserving feature selection in PARAFAC using a genetic algorithm and Procrustes analysis. *Chemometrics and Intelligent Laboratory Systems*, **65**, 83–95.

Xu, X., van Stee, L. L. P., Williams, J. *et al.* (2003) Comprehensive two-dimensional gas chromatography (GC x GC) measurements of volatile organic compounds in the atmosphere. *Atmospheric Chemistry and Physics*, **3**, 665–682.

Xu, C.-J., Liang, Y.-Z., Chau, F.-T. and van der Heyden, Y. (2006) Pretreatments of chromatographic fingerprints for quality control of herbal medicines. *Journal of Chromatography A*, **1134**, 253–259.

Yan, Z.-Y., McCarthy, M. J., Klemann, L. *et al.* (1996) NMR applications in complex food systems. *Magnetic Resonance Imaging*, **14**, 979–981.

Yaacoub, R., Saliba, R., Nsouli, B. *et al.* (2009) Rapid assessment of neoformed compounds in nuts and sesame seeds by front-face fluorescence. *Food Chemistry*, **115**, 304–312.

Zhang, Y., Wu, H. L, Xia, A.-L. *et al.* (2007) Interference-free determination of Sudan dyes in chilli foods using second-order calibration algorithms coupled with HPLC-DAD. *Talanta*, **72**(3), 926–931.

Zhu, S.-H., Wu, H.-L., Xia, A.-L. *et al.* (2007) Determination of carbendazim in bananas by excitation emission matrix fluorescence with three second-order calibration methods. *Analytical Sciences*, **23**, 1173–1177.

# 10 Multidimensional scaling (MDS)

**Eva Derndorfer[1] and Andreas Baierl[2]**
[1] *University of Applied Sciences Burgenland, University of Salzburg, UMIT Hall in Tirol, and independent sensory consultant, Vienna, Austria*
[2] *Department of Statistics and Operations Research, University of Vienna, Vienna, Austria*

## ABSTRACT

Multidimensional scaling (MDS) is a multivariate statistical method, which aims to uncover structures within a given data set. The method is becoming more and more popular in sensory science and product development, and may also be applied in nutritional science. An applied introduction is given based on recent examples in food sciences. Aspects of data collection, a methodological overview, and software implementations are covered, including practical guidelines to successfully apply MDS in food sciences.

## INTRODUCTION – WHAT IS MDS?

Multidimensional scaling (MDS) is a *multivariate* statistical method, an *explorative* technique that aims to uncover structures within a given data set. The method is becoming more and more popular in sensory science and product development, and may also be applied in nutritional science.

Let us begin with an example, the identification of perceived sensory similarities and dissimilarities within a set of products. Imagine the following scenario: You wish to obtain a rapid overview on sensory similarities of 20 chocolate samples, without needing descriptions of their odour or flavour. You do not have a group of already trained assessors but you can easily find 30 untrained people for a sensory test. Time is limited. What could you do to obtain meaningful results?

(A) You may let each of the 30 assessors rate the difference perceived between each product pair on a five-point scale. However, this procedure requires a large number of pairwise comparisons (in case of 20 products: $20 \times 19/2 = 190$), and thus produces sensory fatigue, carry-over and adaptation. Several testing sessions are necessary. This is time and cost consuming, and it is difficult to recruit and motivate assessors to complete the task. With many comparisons, assessors might change their criteria for judging similarity, which will cause inconsistent results.

(B) Another option is to let assessors sort the 20 chocolate samples in groups based on their perceived similarity. This faster procedure produces less fatigue and may be done within a single session. In sorting tasks, assessors may choose their own criteria for forming groups; thus, it is fairly free from investigatory bias (Lawless, 1993). Some assessors may sort samples in three groups, whereas others make up to ten groups.

*Mathematical and Statistical Methods in Food Science and Technology*, First Edition.
Edited by Daniel Granato and Gastón Ares.

Now, which of the chocolate samples are, overall, considered closer, which are more different? To answer this question, you first have to process your data into a matrix made up of 20 rows and columns that describes the proximities between all pairs of chocolates. Just looking at the data matrix does not provide you with a good overview on similarities between samples. Multidimensional scaling is an answer to your problem in both cases (A) and (B). It reduces complex data and visualizes similarities of the chocolates in few dimensions by means of a product plot. Clusters of similar products can be visually distinguished from dissimilar ones.

Our chocolate example in Figure 10.1 demonstrates which samples lie closest together and which lie furthest apart, thus are more different.

When do we need such a rapid overview? There are several applications.

- One of them is *Benchmarking*. Every company wants to compare its own products with those from competitors on a regular basis.
- In *New product development*, a food technologist may want to know which of a series of prototypes, made with varying processing parameters, are perceived as most similar and which are perceived as most different.
- In *Consumer testing* of food, the number of samples is practically limited, as the cost of a large sample set is high. Thus, if a larger number of products exists only a subset is selected for testing. The subset of samples has to be selected carefully, as it should represent the diversity of the whole sample set. Sorting samples and analysing data by MDS can be used in a *pre-test*, a tool for samples selection for consumer testing. You could select a few samples from Figure 10.1 that cover the whole sensory space, by choosing chocolates from different locations of the map.
- A MDS similarity map may also be linked with *Consumer preference* data.
- In *Nutritional sciences*, comparison of products within a food category may be based on quantitative variables, such as *nutritional values* (amount of carbohydrates, fat, protein, fibre, sodium, etc.).

As an explorative method, MDS does not include hypothesis testing, and therefore no significance level applies. Most statistical software offer MDS procedures in their standard packages. As with other

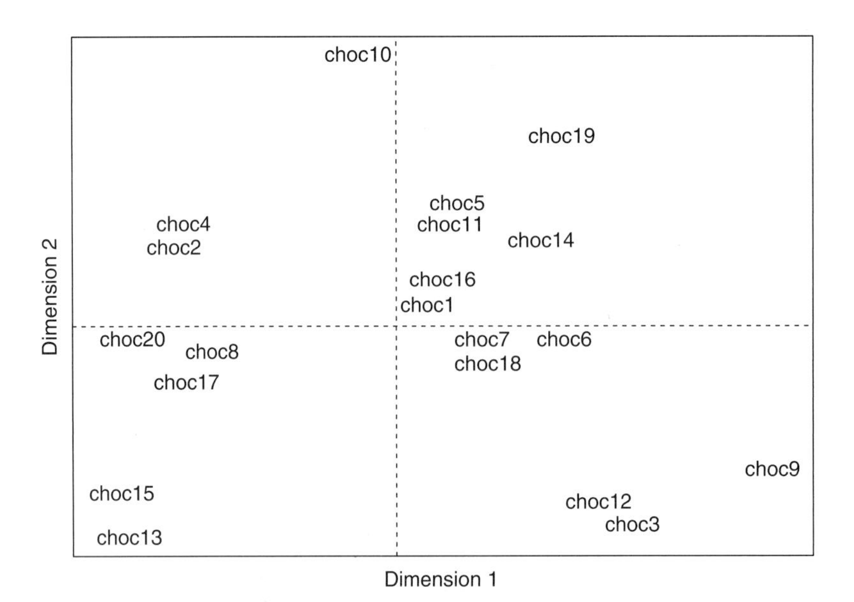


**Figure 10.1** MDS map of 20 chocolates.

explorative methods, interpretation of the MDS map is not always clear-cut. A final plot may well show relative similarities but the underlying dimensions, thus the interpretations of the axis of a two- or three-dimensional plot, are somewhat subjective. This issue is discussed in more detail.

This chapter first presents applications of MDS in food science, with an emphasis on sensory analysis, and gives an overview on current literature. Secondly, the statistical procedure is illuminated in detail. Finally, an example is discussed in detail, beginning with the questionnaire, covering raw data, data analysis and then interpretation of the plot.

## APPLICATION OF MDS IN FOOD SCIENCE

In sensory analysis, most studies use the sorting method (B) rather than similarity scaling of all product pairs (A). Sorting samples may be done based on a single sensory modality, such as flavour or odour, colour or sound, or based on overall sensory impression.

Falahee and MacRae (1995) analysed the taste of 13 drinking waters by means of a sorting task. They let assessors taste the waters and allowed them to form as many groups as they wished. Results were compared with preference data. In a second experiment, the same authors (Falahee and MacRae, 1997) let a group of assessors sort distilled, tap and bottled water, and blends thereof, into groups.

Lim and Lawless (2005) investigated the taste of divalent salts with iron, calcium, magnesium or zinc. They provided assessors with 15 taste solutions, among them a standard sweet, sour, salty and bitter taste solution, a mixture of salty and bitter, an astringent sample consisting of aluminium ammonium sulfate in water, and solutions of divalent salts. Sorting was performed once with an open nose and once with a nose-clip. The divalent salts fell outside the area of the basic tastes. The biggest difference between the two nose conditions arose from ferrous salts.

Sinesio *et al.* (2010) examined discrimination of umami taste in asparagus soup with varying levels of umami-rich ingredients. A group of tasters trained in recognizing and quantifying umami taste, as well as an untrained group of assessors performed a sorting experiment. Both groups distinguished soups according to their umami ingredients but the MDS maps obtained were not identical. The two groups apparently did not focus on exactly the same criteria when sorting.

In several studies, MDS has been used to study similarity of odours. Chrea *et al.* (2004) studied agreement between French, Vietnamese and Americans in terms of odour categorization. They let students from the three countries sort 40 odorants into groups on the basis of similarity. Although the three groups differed in their perceptual judgements, they categorized odours into four clusters, which the authors labelled sweet, floral, natural and bad. A fifth cluster called medicine was found by the French assessors. Chrea *et al.* (2005) compared odour sorting results from the above cultures; they let one group sort odorants based on odour similarity, a second group sorted odour names based on name similarity, and a third group sorted odours based on imagined odour similarity without perceiving the odours to smell. Results showed that odour sorting was conceptual and partly independent of word and imagined categories. Higuchi *et al.* (2004) let 50 participants sort fragrances in groups based on their perceptual similarity. Participants were allowed to form as many groups of fragrances as they felt necessary. Results were compared to those of an earlier experiment where 50 volunteers rated the same odour stimuli in their intensity of nine attributes on a seven-point scale. The correlation between the two MDS configurations was significant, thus similar spatial locations of fragrances were obtained by both methods.

Derndorfer and Baierl (2006) let two groups of assessors sort spices in groups based on odour similarity. One group performed the task under normal light, the second one under red light conditions. Spice mixtures were also included. The red light map can be considered as an objective aroma map of spices. It differed from the daylight map.

Heymann (1994) focused on vanilla types and compared MDS with results from *free choice profiling* (FCP), a sensory method where assessors use individual descriptors to differentiate between samples. As the MDS space was similar to the FCP space from sensory-savvy assessors, but does not provide attributes, Heymann recommended MDS as a useful technique for screening samples for further analysis.

Apart from developing olfactory perceptual maps, MDS was used to study the perceptual structure of everyday sounds (Bonebright, 2001) or to create perceptual maps of cheeses as well as cheese names (Lawless *et al*., 1995). Lawan *et al*. (2003) used the method to generate ideas for new product development with rice. Perceived similarity of brands may also be visualized (Schiffman and Knecht, 1993).

Sensory sorting tasks have also been performed with a nutritional focus. Schiffman *et al*. (1978) let 16 obese and 27 normal weight individuals taste and smell different foods while blindfolded. Then the assessors rated each food on 51 semantic differential adjective scales relating to aroma, taste and trigeminal aspects of the foods. Proximity measures were developed from the ratings. Then, two multidimensional scaling procedures were applied. Obese subjects found the hedonic and flavour dimension more important, and were better at identifying blended foods compared to normal weight subjects. In a second study with young and elderly subjects, Schiffman and Pasternak (1979) let assessors rate odour similarities of 91 combinations of pairs of 14 food flavours and collected hedonic ratings. Two multidimensional scaling procedures were applied. Results suggested that the ability to judge odour differences may decrease with age.

## QUALITY OF RESULTS

As sorting is becoming more popular, a fundamental question is: How robust is the sorting task? Studies have shown fairly good repeatability. Chollet *et al*. (2011) performed duplicated sorting tasks with beer using the same group of assessors, as well as sorting tasks with selected duplicated beers. They found that configurations from two repetitions were similar. However, when only selected duplicated samples were used, they were not satisfactorily grouped together. Bazala (2008) included one duplicate sample in a blind wine tasting, and the two identical samples were neighbours on the visual map.

As a reality check, we thus advise to include a duplicate sample in every sorting task. This allows you to roughly estimate the quality of your MDS map.

## MDS PROCEDURE

Let us now focus on the statistical aspect of MDS. Input data for MDS describes – in a direct or indirect way – relations between a number of objects (products etc.), usually based on assessor judgments. We outline the MDS procedure in the following section divided into four steps:

- Data
- Model selection and goodness of fit
- MDS methods and software
- Interpretation of MDS maps.

### Data

Input data for MDS can be collected in numerous ways; two options are mentioned in the introduction: pairwise comparisons (A) and sorting (B). Independently of the data collection procedure, input data for MDS have to be organized in square matrices. The number of rows (and columns) is equal to the number

**Table 10.1** Number of pairs of products. The number of pairs is derived from the number of products $n$ by the following formula: $n(n - 1)/2$.

| Number of products | Number of pairs |
|---|---|
| 5 | 10 |
| 6 | 15 |
| 7 | 21 |
| 8 | 28 |
| 9 | 36 |
| 10 | 45 |
| 15 | 105 |
| 20 | 190 |
| 25 | 300 |
| 30 | 435 |
| 50 | 1225 |

of assessed products. An example of an input matrix is shown in Table 10.4. Each cell represents the dissimilarity of a pair of products. Diagonal elements refer to distances between products and themselves, and are therefore zero-valued. The matrix is symmetric if an entry in row $i$ (e.g. 2) and column $j$ (e.g. 3) is equal to the entry in row $j$ and column $i$. Asymmetric distances can arise, for example ratings of closeness between two persons, but rarely occur in food science.

To summarize, a MDS analysis of n products needs proximity measures of all possible pairs of different products. The number of pairs of products increases rapidly with the number of included products and, therefore, constitutes a major challenge for the data collection process. Table 10.1 provides exact numbers for the amount of pairwise proximities.

A straightforward approach is to measure the similarity or dissimilarity between all pairs of products (Option (A) in the Introduction). The sampling of products and replication numbers largely depend on the nature of investigated products and the measuring process. If replicates of the same product exhibit large variation, sampling size has to be increased to achieve meaningful results. The same is true for the measuring process: if human assessors are involved, higher replication numbers will be necessary compared to situations where measuring is done through a technical process. Standard deviations for measurements of each pair of products can be calculated to estimate the amount of variation. In the case of human assessors, the choice of the sampling population and experimental conditions should correspond to the settings to which the mapping results will be applied. Although MDS is an explorative method and no statistical tests, confidence intervals and so on will be derived, sampling design still plays an important role.

The measuring process depends on a particular problem. In the case of food science, mainly human assessors will be engaged and scales are employed for measuring. Two common scales are discrete and line scales. Figure 10.2 shows examples for a five-point discrete scale and a line scale with the anchors 'same' and 'different'.

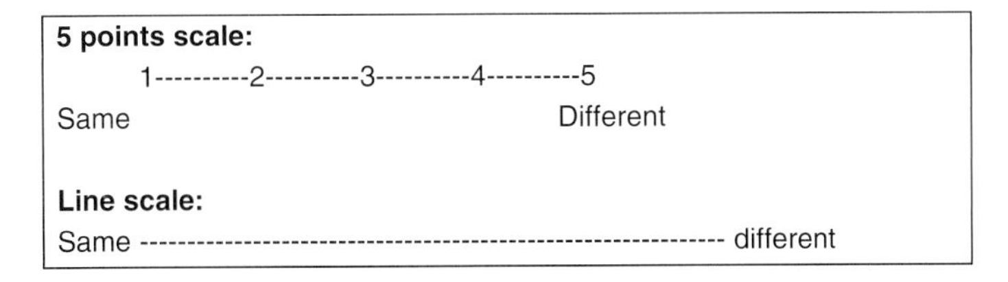


**Figure 10.2** Scales to rate the similarity of sample pairs.

The measurement of proximities of all pairs of products results in one matrix for each assessor. A major limitation of the described procedure constitutes the rapidly increasing number of comparisons with the number of assessed products (Table 10.1). Therefore, this procedure is usually limited to small mapping tasks with a maximum of 10–11 products.

There are modifications of explicitly measuring proximities between products to overcome the size limitation. One option applies incomplete designs, which involves that each assessor only scores a subset of pairs. A number of studies (Popper and Heymann, 1996) successfully applied incomplete designs.

A very common modification that can greatly reduce the amount of comparisons is sorting (Option (B) in the Introduction): assessors are asked to divide products into groups and are told to use a minimum (e.g. two) and a maximum (e.g. one minus the number of products) number of groups. Sorting may be done with as many as 40 products (Jamieson and Griffiths, 2009) but for certain food groups, for example alcoholic drinks, fewer samples may be the maximum. Kuhnt (2010) let tasters sort 22 red wine samples. Chollet *et al.* (2011) found that sorting can be carried out with 20 beers without losing efficiency compared to 15 beers, whereas 24 beers were too many for a sorting task.

Explicit proximity data can be directly applied to MDS by calculating the average dissimilarity between each pair over all assessments. Some MDS procedures can only process dissimilarities and no similarities. Usually, similarities can be easily converted into dissimilarities: if they are measured on a scale between 0 and 5, dissimilarities are just 5 minus the similarity value. With sorting tasks, first similarity values are calculated by counting the number of times each pair of products is sorted into the same group. Then the number of assessor minus the similarity value leads to dissimilarity measures.

All data collection procedures so far measure in one way or another direct proximities between products. Hence, they do not require a priori knowledge of the attributes or stimuli to be scaled, either by assessors or by experimenters. They provide a space that reveals dimensions relevant to the subjects – a major advantage of MDS over related methods.

It is also possible to use direct comparisons to investigate proximities in pre-specified attributes: assessors can be advised to judge similarities/dissimilarities in a certain dimension (e.g. creamy) instead of overall similarity. This of course assumes that assessors have a common understanding of this dimension.

A completely different approach to generate dissimilarity matrices is using attribute data, for example from sensory descriptive analysis, where trained assessors rate products in their intensity of a number of sensory attributes on a scale, or from nutritional data as described in the Introduction. Pairwise dissimilarities are derived by calculating distances, for example Euclidean distances, over all attributes. In the case of attribute measures, the amount of required data increases only linearly with the number of assessed products, that is the number of required measurements equals the number of attributes times the number of subjects times the number of products. Of course, we already have to specify relevant attributes in advance and we assume that assessors understand these attributes and can scale them in the same way.

In general, if the number of products becomes too large and you still want to collect explicit comparison data, you can consider dividing products into groups and do separate MDS maps. Representatives of each group can finally be analysed in a second step MDS analysis.

## Model selection and goodness of fit

Considering possible locations of three products they can either be arranged along a line (in one dimension) or span a two-dimensional plane. A relevant number of samples, for example 10, will hence span a nine-dimensional plane, which is not truly nice to look at. MDS procedures try to find the best representation of distances between products in a low-dimensional space. It depends on the nature of the data whether a reasonably good representation can be achieved. Geographical distances between cities can be displayed in two dimensions without major loss of information. In general, the goodness of a representation increases if assessors use only few underlying dimensions to judge distances between products, or if some dimensions are dominating the perception of similarity.

The number of dimensions used for mapping is limited by visual comprehension. A 3D representation of products is already much more difficult to assess visually compared to a 2D map. There exist further options to represent additional dimensions, for example by colouring points in varying hues as well as colour intensity, by varying the shape of points or by labelling points with values. Often, 2D maps with varying colours or labels are easier to read than 3D maps.

A low goodness of representation results in vague interpretation of distances or complex visual presentation. The common goodness of fit measure used in MDS analysis is *Stress*. It sums up the deviations of distances in the original data and distances displayed in the produced map. Stress values vary between 0 and 1. A low Stress value indicates a good fit. Increasing the number of dimensions in the map improves the fit and decreases the Stress value while Stress always increases with the number of products in the analysis.

Guidelines exist for classifying MDS solutions based on Stress: Kruskal (1964) describes values below 0.025 as excellent, below 0.05 as good, below 0.1 as fair and up to 0.2 as acceptable. This categorization is really only meant as a rough guideline. There are various situations where simple Stress values are misleading in judging the goodness of representing initial proximities in a low-dimensional space (Borg and Groenen, 2005, Chapter 3). The problem can be compared to interpreting correlation coefficients without looking at the data by using, for example a scatterplot. In MDS a so-called *Shepard diagram* (Figure 10.4 gives an example) plots all observed proximities from the observed proximity matrix on the horizontal axis against the corresponding MDS distances. Because MDS methods can apply various forms of transformations to the data (see the section on MDS methods), we usually do not expect the points to lie on a straight line, even in an ideal scenario of perfect fit. Therefore, a line is drawn in the Shepard diagram representing the transformed input proximities. Now, the vertical distances between the line and the points give the error of representation for each pair of products. The Shepard diagram lets us judge the goodness of fit and helps us detecting outliers and systematic deviations.

In case of outliers we might identify data errors or observations that do not fit and should be excluded. Observed systematic deviations can be eliminated by applying a different MDS method that uses an alternative transformation. If the goodness of fit is still unsatisfactory, there exists no low-dimensional space into which the observed proximities can be mapped without losing a substantial amount of information. Measurements might be too erroneous or assessors implicitly use too many underlying dimensions to judge similarities.

## MDS methods and software

A number of different approaches exist to find the optimal representations of a distance matrix, that is the one with the best fit and the smallest Stress value. Methods differ by their criterion of what they mean by optimal, their computationally efficiency and presentation of results. In the following, we give a brief overview of common methods. For a comprehensive presentation of MDS methodology see, for example Borg and Groenen (2005), Cox and Cox (1994) or Schiffman *et al.* (1981).

A major distinction can be made between methods that preserve the actual or algebraically transformed proximities in the data matrix, so-called *metric MDS* methods, and methods that only aim to preserve the order of the proximities, so-called *nonmetric MDS* methods. If the measurements of proximities were only done on an ordinal scale, metric MDS methods are not appropriate. On the other hand, if measurements are available on a metric scale, metric as well as nonmetric MDS can be applied.

A computationally fast, metric MDS method is classical MDS, also known as principal coordinate analysis (PCO). Nonmetric methods are computationally more intensive, especially for large numbers of products. Implementations of nonmetric MDS include Kruskal's nonmetric MDS.

When distinct proximity matrices for a number of assessors were collected (e.g. in the case where direct proximities between pairs of products were measured but not when sorting was applied), individual difference models, also called *weighted MDS* models, can be applied. The output of individual difference models contains an overall (group) map as well as individual plots for each assessor. INDSCAL (Individual Difference SCAling) by Carroll and Chang (1970) is a popular individual difference model.

A complete framework covering metric/nonmetric MDS, individual scaling and also allowing for symmetric and asymmetric proximity matrices is ALSCAL (Alternating Least Square Scaling). ALSCAL is available as stand-alone software as well as being implemented in the statistical package IBM SPSS. Software SAS includes `PROC MDS`, which shares many features of ALSCAL. The statistical package Stata supports metric and nonmetric MDS.

The free statistical software R contains various implementations of MDS: function `cmdscale` (package `stats`) for classical MDS, package `MASS` includes two versions of nonmetric MDS: `isoMDS` and `sammon`. A wrapper function called `metaMDS` for nonmetric MDS covering post-processing of results is part of package `vegan`. Further metric and nonmetric as well as individual scaling methods are implemented in packages `smacof` and `SensoMineR` (function `indscal`).

## Interpretation of MDS maps

As previously mentioned, the outcome of a MDS analysis is usually a two- or three-dimensional plot that is used to find structures within a group of assessed products. Several issues have to be considered when interpreting MDS maps:

- Axes are, in themselves, meaningless and the orientation of the map is arbitrary (except for results from individual scaling models).
- Usually, the final Stress value is larger than zero. Hence, you must keep in mind that the distances among items are imperfect, distorted, representations of the relationships given by your data. The greater the Stress, the greater the distortion.
- Generally, distinct patterns, such as clusters of products that are clearly separated or products with large distances to clusters, are more reliably represented than minor patterns, such as relative locations within a cluster.
- Dominant dimensions that were used by assessors in producing similarities/dissimilarities can cause products to be arranged along a straight line. The orientation of the line does not have to be vertical or horizontal but can be in any direction.

Additional information on the products may support the interpretation of MDS maps. These can be either the amount of ingredients, or physical and chemical measurements. In sensory science, assessors are sometimes asked after completion of their sorting task to write down a few attributes that characterize each group. Such information may facilitate interpretation of the final plot. However, it has to be kept in mind that additional attributes of untrained assessors are not of the same quality as a description from a trained panel. If sensory descriptions are of major interest, descriptive sensory methods with trained assessors are better suited than sorting tasks with additional descriptions, but they are much more time consuming. The interested reader is referred to books on sensory methods.

Product specific information can be visualized within the MDS map using labels or colour codes. Another option involves univariate or multivariate correlation analysis applied to MDS dimensions. Bivariate correlation coefficients can be calculated between each MDS dimension and all additional attributes. Alternatively, each MDS dimension can be defined as dependent variable in a multiple regression model and all additional attributes used as potential explanatory variables.

# EXAMPLE – A SORTING TASK OF WINE GLASS SHAPES

15 students from the masters degree 'international wine marketing' at the University of Applied Sciences Burgenland were asked to sort 14 wine glasses based on their visual appearance into groups. They were provided with pictures of the glasses that were given three-digit codes; every student received them in an individual order. The following written instructions were given to them:

> "With this test, we want to analyse the similarity of wine glasses. Please sort the provided pictures in groups. Glasses which are similar should be grouped together; glasses which are more different should be sorted into separate groups. Each group must consist of at least two glasses. You may form as many groups as you wish, with a minimum of two groups. Please list your groups on the ballot. Thank you."

A data matrix was established, with one column per student and one row per wine glass. Samples that were grouped together by a person receive the same number. It does not matter which group is called 1, 2, 3, as data analysis only considers how often two glasses are sorted into the same group. Table 10.2 shows the raw data matrix from this task. For example Person 1 put glasses 227, 579 and 918 into the same group, glasses 499, 737 and 902 into another group.

From this raw data matrix, firstly a similarity matrix is built by counting the number of times each glass *i* was grouped together with any glass *j*. Table 10.3 shows that some glasses were sorted into the same group by all assessors: g499, g737 and g902 and g655 and g790. Hence, one triplet and one pair of glasses were not differentiated by any assessor. Although the duplicate/triplicate glasses are excluded from subsequent MDS analysis, it is still very valuable information that some glasses were obviously perceived as very similar.

Our new data set consists of 11 instead of 14 data rows. Similarities were transformed into dissimilarities by first subtracting the number of times two glasses were grouped together from the number of assessors and by subsequently dividing by the number of assessors (Table 10.4). Applying this transformation leads to values that range from 0 (no dissimilarity) to 1 (maximum dissimilarity).

After processing the data, nonmetric MDS was carried out using the statistical package R (R Development Core Team, 2012) with function `isoMDS`. The Stress value for a two-dimensional solution

**Table 10.2** Raw data matrix from sorting task. Rows refer to glasses and columns to assessors (person 1 – person 15). Glasses that are grouped together by one assessor are indicated by identical numbers per row.

| | p1 | p2 | p3 | p4 | p5 | p6 | p7 | p8 | p9 | p10 | p11 | p12 | p13 | p14 | p15 |
|---|---|---|---|---|---|---|---|---|---|---|---|---|---|---|---|
| g118 | 4 | 2 | 3 | 3 | 6 | 4 | 1 | 2 | 3 | 2 | 1 | 3 | 2 | 3 | 5 |
| g227 | 1 | 3 | 3 | 4 | 5 | 4 | 3 | 3 | 1 | 5 | 4 | 3 | 1 | 5 | 2 |
| g258 | 3 | 2 | 4 | 3 | 1 | 2 | 3 | 2 | 3 | 1 | 1 | 1 | 3 | 3 | 4 |
| g342 | 5 | 3 | 1 | 4 | 5 | 4 | 1 | 4 | 2 | 3 | 3 | 2 | 4 | 1 | 2 |
| g499 | 2 | 2 | 2 | 2 | 3 | 3 | 2 | 1 | 3 | 4 | 2 | 4 | 2 | 4 | 3 |
| g543 | 3 | 2 | 4 | 4 | 1 | 2 | 3 | 2 | 3 | 1 | 1 | 1 | 3 | 2 | 4 |
| g579 | 1 | 3 | 2 | 2 | 2 | 3 | 2 | 3 | 1 | 5 | 4 | 3 | 1 | 5 | 2 |
| g601 | 4 | 2 | 4 | 3 | 1 | 2 | 3 | 2 | 3 | 2 | 1 | 1 | 3 | 3 | 5 |
| g655 | 5 | 1 | 1 | 1 | 4 | 1 | 1 | 4 | 2 | 3 | 3 | 2 | 4 | 1 | 1 |
| g691 | 4 | 2 | 3 | 3 | 6 | 2 | 3 | 2 | 3 | 2 | 1 | 3 | 1 | 3 | 4 |
| g737 | 2 | 2 | 2 | 2 | 3 | 3 | 2 | 1 | 3 | 4 | 2 | 4 | 2 | 4 | 3 |
| g790 | 5 | 1 | 1 | 1 | 4 | 1 | 1 | 4 | 2 | 3 | 3 | 2 | 4 | 1 | 1 |
| g902 | 2 | 2 | 2 | 2 | 3 | 3 | 2 | 1 | 3 | 4 | 2 | 4 | 2 | 4 | 3 |
| g918 | 1 | 3 | 3 | 4 | 2 | 3 | 3 | 3 | 1 | 4 | 4 | 3 | 3 | 2 | 2 |

**Table 10.3** Similarity matrix. Each cell represents the number of times the glass – indicated by the row label and the column label – were grouped together. Rows and columns in grey colour are excluded from further analysis as they could not be separated from glasses g499 and g655, respectively.

| | g118 | g227 | g258 | g342 | g499 | g543 | g579 | g601 | g655 | g691 | g737 | g790 | g902 | g918 |
|---|---|---|---|---|---|---|---|---|---|---|---|---|---|---|
| g118 | 15 | 3 | 6 | 2 | 3 | 4 | 1 | 9 | 1 | 11 | 3 | 1 | 3 | 2 |
| g227 | 3 | 15 | 1 | 5 | 0 | 2 | 10 | 1 | 0 | 4 | 0 | 0 | 0 | 10 |
| g258 | 6 | 1 | 15 | 0 | 2 | 13 | 0 | 12 | 0 | 9 | 2 | 0 | 2 | 2 |
| g342 | 2 | 5 | 0 | 15 | 0 | 1 | 2 | 0 | 10 | 0 | 0 | 10 | 0 | 3 |
| g499 | 3 | 0 | 2 | 0 | 15 | 2 | 4 | 2 | 0 | 2 | 15 | 0 | 15 | 2 |
| g543 | 4 | 2 | 13 | 1 | 2 | 15 | 0 | 10 | 0 | 7 | 2 | 0 | 2 | 4 |
| g579 | 1 | 10 | 0 | 2 | 4 | 0 | 15 | 0 | 0 | 2 | 4 | 0 | 4 | 9 |
| g601 | 9 | 1 | 12 | 0 | 2 | 10 | 0 | 15 | 0 | 10 | 2 | 0 | 2 | 2 |
| g655 | 1 | 0 | 0 | 10 | 0 | 0 | 0 | 0 | 15 | 0 | 0 | 15 | 0 | 0 |
| g691 | 11 | 4 | 9 | 0 | 2 | 7 | 2 | 10 | 0 | 15 | 2 | 0 | 2 | 3 |
| g737 | 3 | 0 | 2 | 0 | 15 | 2 | 4 | 2 | 0 | 2 | 15 | 0 | 15 | 2 |
| g790 | 1 | 0 | 0 | 10 | 0 | 0 | 0 | 0 | 15 | 0 | 0 | 15 | 0 | 0 |
| g902 | 3 | 0 | 2 | 0 | 15 | 2 | 4 | 2 | 0 | 2 | 15 | 0 | 15 | 2 |
| g918 | 2 | 10 | 2 | 3 | 2 | 4 | 9 | 2 | 0 | 3 | 2 | 0 | 2 | 15 |

amounts to 0.102. The resulting MDS map is displayed in Figure 10.3. Glasses that could not be distinguished are marked with grey labels in close proximity to the label of the related glass.

In order to judge the goodness of representation, a Shepard diagram was produced (Figure 10.4). The dots in the diagram, that is the dissimilarity values, are evenly spread above and below the line. Neither obvious outliers nor systematic deviations are visible.

For analysing the results, we look at the MDS map in Figure 10.3: Glasses 499, 737 and 902 in the top right quadrant were grouped together by all assessors. On the right side close to the x-axes, samples g118, g543, g258, g601 and g691 are also fairly close and form a group, although they were not sorted together by all assessors. These five glasses seem to be closer together than the triplet g737/g902/g499 on the top right.

On the left side of the map, relative distance between samples is obvious, but glasses do not seem to separate into clear groups as is the case on the right side of the map. Only glasses g655 and g790 were not distinguished. Glass g342 is slightly closer to the double g655/g790 than to glass g227 in the top left quadrant.

**Table 10.4** Normalized dissimilarity matrix. Values range between 1 (maximum dissimilarity) and 0 (no dissimilarity).

| | g118 | g227 | g258 | g342 | g499 | g543 | g579 | g601 | g655 | g691 | g918 |
|---|---|---|---|---|---|---|---|---|---|---|---|
| g118 | 0.00 | 0.80 | 0.60 | 0.87 | 0.80 | 0.73 | 0.93 | 0.40 | 0.93 | 0.27 | 0.87 |
| g227 | 0.80 | 0.00 | 0.93 | 0.67 | 1.00 | 0.87 | 0.33 | 0.93 | 1.00 | 0.73 | 0.33 |
| g258 | 0.60 | 0.93 | 0.00 | 1.00 | 0.87 | 0.13 | 1.00 | 0.20 | 1.00 | 0.40 | 0.87 |
| g342 | 0.87 | 0.67 | 1.00 | 0.00 | 1.00 | 0.93 | 0.87 | 1.00 | 0.33 | 1.00 | 0.80 |
| g499 | 0.80 | 1.00 | 0.87 | 1.00 | 0.00 | 0.87 | 0.73 | 0.87 | 1.00 | 0.87 | 0.87 |
| g543 | 0.73 | 0.87 | 0.13 | 0.93 | 0.87 | 0.00 | 1.00 | 0.33 | 1.00 | 0.53 | 0.73 |
| g579 | 0.93 | 0.33 | 1.00 | 0.87 | 0.73 | 1.00 | 0.00 | 1.00 | 1.00 | 0.87 | 0.40 |
| g601 | 0.40 | 0.93 | 0.20 | 1.00 | 0.87 | 0.33 | 1.00 | 0.00 | 1.00 | 0.33 | 0.87 |
| g655 | 0.93 | 1.00 | 1.00 | 0.33 | 1.00 | 1.00 | 1.00 | 1.00 | 0.00 | 1.00 | 1.00 |
| g691 | 0.27 | 0.73 | 0.40 | 1.00 | 0.87 | 0.53 | 0.87 | 0.33 | 1.00 | 0.00 | 0.80 |
| g918 | 0.87 | 0.33 | 0.87 | 0.80 | 0.87 | 0.73 | 0.40 | 0.87 | 1.00 | 0.80 | 0.00 |

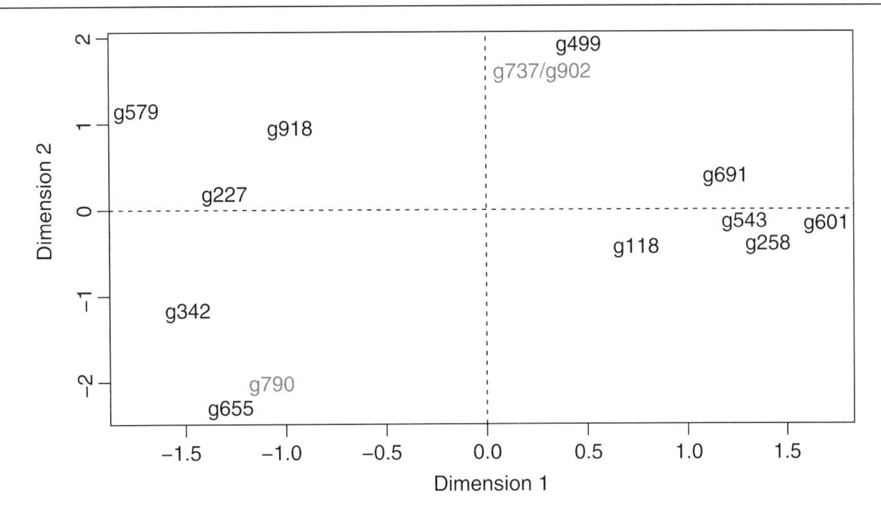


**Figure 10.3** MDS map of wine glasses. Grey labels indicate glasses that were not distinguished by any assessor from the closest glass with a black label.

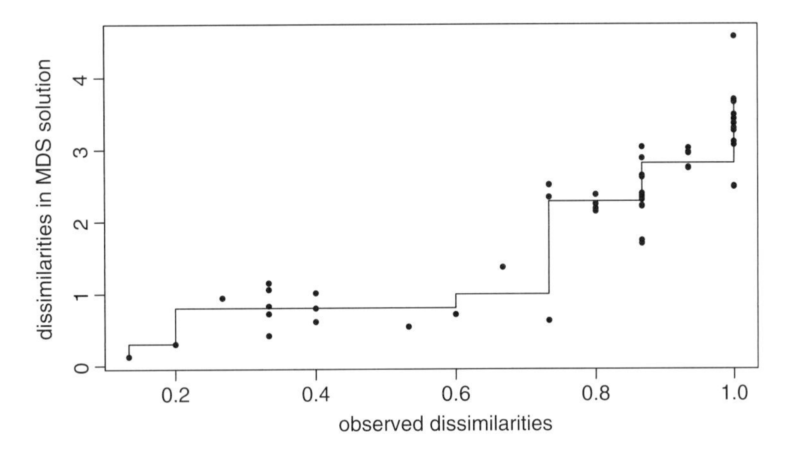


**Figure 10.4** Shepard diagram. The line represents transformed proximities.

Without additional information on the products, further interpretation is subjective. Additional information, such as glass height, shape or volume, can help with interpretation of the axis. This way, one might be able to explain x and y-axes.

As in our chocolate example in the introduction of this chapter, if one was to select four different glasses out of the 14, the map would help. One glass could be selected from the top left and one from the bottom left quadrant, and two glasses could be chosen from the right, one from each group.

## REFERENCES

Bazala, B. (2008) *Die empfundene sensorische Ähnlichkeit von Weinproben: blind versus branded.* Erste Bachelorarbeit, Fachhochschulstudiengänge Burgenland, Eisenstadt, Austria.

Bonebright, T. (2001) Perceptual structure of everyday sounds: a multidimensional scaling approach. In: *Proceedings of the 2001 International Conference on Auditory Display* (eds J. Hiipakka, N. Zacharov and T. Takala). Laboratory of

Acoustics and Audio Signal Processing and the Telecommunications Software and Multimedia Laboratory, Helsinki University of Technology, Espoo, Finland, pp. 73–78.

Borg, I. and Groenen P.J.F. (2005) *Modern Multidimensional Scaling: Theory and Applications*. Springer, New York.

Caroll, J.D. and Chang, J.J. (1970) Analysis of individual differences in multidimensional scaling via an n-way generalization of 'Eckart-Young' decomposition. *Psychometrika*, **35**, 283–319.

Chollet, S., Lelièvre, M., Abdi, H. and Valentin, D. (2011) Sort and beer: Everything you wanted to know about the sorting task but did not dare to ask. *Food Quality and Preference* **22**, 507–520.

Chrea, C., Valentin, D., Sulmont-Rossé, C. *et al.* (2004) Culture and odor categorization: agreement between cultures depends upon the odors. *Food Quality and Preference* **15**, 669–679.

Chrea C., Valentin D., Sulmont-Rossé C. *et al.* (2005) Semantic, typicality and odor representation: A cross-cultural study. *Chemical Senses* **30**, 37–49.

Cox, T.F. and Cox, M.A.A. (1994) *Multidimensional Scaling*. Chapman and Hall, London.

Derndorfer, E. and Baierl, A. (2006) Development of an aroma map of spices by multidimensional scaling. *Journal of Herbs, Spices and Medicinal Plants* **12**, 39–50.

Falahee, M. and MacRae, A.W. (1995) Consumer appraisal of drinking water: Multidimensional Scaling Analysis. *Food Quality and Preference* **6**, 327–332.

Falahee, M. and MacRae, A.W. (1997) Perceptual variation among drinking waters: The reliability of sorting and ranking data for multidimensional scaling. *Food Quality and Preference* **8**, 389–394.

Heymann H. (1994): A comparison of free choice profiling and multidimensional scaling of vanilla samples. *Journal of Sensory Studies* **9**, 445–453.

Higuchi, T., Shoji, K. and Hatayama, T. (2004) Multidimensional scaling of fragrances: A comparison between the verbal and non-verbal methods of classifying fragrances. *Japanese Psychological Research* **46**, 10–19.

Jamieson L. and Griffiths A. (2009) Application of the sorting methodology – strengths and limitations in sensory evaluation of food products. *Poster presentation at the Eighth Pangborn Sensory Science Symposium*, 26–30 July, Florence, Italy.

Kruskal, J.B. (1964) Multidimensional scaling by optimizing goodness of fit to a nonmetric hypothesis. *Psychometrika* **29**, 1–27.

Kuhnt, M. (2010) *Welche Assoziationen hat der Konsument mit Wein und stimmt er den von Experten verfassten Weinbeschreibungen zu?* Masters' thesis, Fachhochschulstudiengänge Burgenland, Eisenstadt, Austria.

Lawan, S., Nanthachai, K. and Tangwongchai, R. (2003) Application of consumer perceptual measurement for generating new rice product ideas. *Poster presentation at the Fifth Pangborn Sensory Science Symposium*, 20–24 July, Boston, MA.

Lawless, H.T. (1993) Characterization of odor quality through sorting and multidimensional scaling, In: *Flavor Measurement* (eds C. T. Ho and C. H. Manley). Marcel Dekker Inc., New York, pp. 159–183.

Lawless, H.T., Sheng, N. and Knoops, S.S.C.P. (1995) Multidimensional scaling of sorting data applied to cheese perception. *Food Quality and Preference* **6**, 91–98.

Lim, J. and Lawless, H.T. (2005) Qualitative difference of divalent salts: Multidimensional scaling and Cluster analysis. *Chemical Senses* **30**, 719–726.

Popper, R. and Heymann, H. (1996) Analyzing differences among products and panelists by multidimensional scaling. In: *Multivariate Analysis of Data in Sensory Science* (eds T. Naes and E. Risvik). Elsevier Science, London, pp. 159–184.

R Development Core Team (2012) *R: A language and environment for statistical computing*. R Foundation for Statistical Computing, Vienna, Austria.

Schiffman, S.S. and Knecht, T.W. (1993) Basic concept and programs for multidimensional scaling. In: *Flavor Measurement* (eds C. T. Ho and C. H. Manley). Marcel Dekker Inc., New York, pp. 133–157.

Schiffman, S.S. and Pasternak M. (1979) Decreased discrimination of food odors in the elderly. *Journal of Gerontology* **34**, 73–79.

Schiffman, S.S., Musante, G. and Conger, J. (1978) Application of multidimensional scaling to ratings of foods for obese and normal weight individuals. *Physiology and Behavior* **21**, 417–422.

Schiffman, S.S., Reynolds, M.L. and Young, F.W. (1981) *Introduction to Multidimensional Scaling*. Academic Press, New York.

Sinesio, F., Peparaio, M. and Comendador, F.J. (2010) Perceptive maps of dishes varying in glutamate content with professional and naïve subjects. *Food Quality and Preference* **21**, 1034–1041.

# 11 Application of multivariate statistical methods during new product development – Case study: Application of principal component analysis and hierarchical cluster analysis on consumer liking data of orange juices

**Paula Varela**

*Propiedades físicas y sensoriales de alimentos y ciencia del consumidor, Instituto de Agroquímica y Tecnología de Alimentos (CSIC), Paterna (Valencia), Spain*

## ABSTRACT

This chapter aims at showing, in a real application setting, the importance of applying principal component analysis (PCA) and hierarchical cluster analysis (HCA) to study consumers' liking segmentation during new product development.

The case study presented is an industry-focused consumer test to guide product development at a fruit juices manufacturer. The objectives of the study was to understand consumers' liking towards natural orange flavour in industrial orange juices and the potential segmentation of the market, to identify the size of the market segments and to verify whether any of the prototypes met consumers' expectations.

Hierarchical clustering analysis and internal preference mapping were used for successfully unveiling the distinct preference patterns of the population. Analysis of overall liking data is presented in this chapter, resulting in an interesting discussion that is allowed by a deep understanding of the raw data and the use of multivariate statistical tools.

## INTRODUCTION

In the last decade, the importance of developing or improving products through listening to the consumer has become undeniable (van Kleef *et al.*, 2006). Measuring consumer liking through hedonic scaling is a widespread practice. In general, 'liking data' are expressed as mean scores, assuming that consumers' preferences are homogeneous. However, preference responses are usually heterogeneous and mean scores are not always representative of real preference patterns (MacFie, 2007; Felberg *et al.*, 2010). One of the most important steps of new product development process is product optimization, that is identifying products that maximize consumers' acceptance ('ideals'). For this aim, preference mapping techniques have been widely used as tools for product improvement (Ares *et al.*, 2011). Preference mapping (PM) comprises a group of methods for investigating consumer hedonic response towards a set of products by using multivariate statistical mapping methods, such as principal component analysis (PCA), principal component regression (PCR) or

*Mathematical and Statistical Methods in Food Science and Technology*, First Edition.
Edited by Daniel Granato and Gastón Ares.

partial least squares regression (PLS) (Næs *et al.*, 2010). In PM applications, the technique helps in identifying product opportunities in the market, so they are particularly well suited as a tool for product development feedback (Jaeger *et al.*, 2003).

Internal or external preference mapping approaches could be applied to understand consumer preference patterns; and together with sensory data, obtained with a trained panel, both preference analyses can be applied to look for underlying dimensions that drive consumer preferences and to provide an understanding of the competitive positioning of products in the marketplace. In particular, PM allows marketing and managerial counterparts to understand where competitors are moving and where to position their own products. Choosing one or the other has been a hot topic in the sensory and consumer science community for the last few years and has 'a philosophical component to it' (van Kleef *et al.*, 2006; MacFie, 2007; Ares *et al.*, 2011). However, internal preference analysis has been identified as advantageous for marketing actionability and new product creativity as compared to external preference analysis; its most important advantage is that the preference space is created based upon consumers responses only (van Kleef *et al.*, 2006; MacFie, 2007).

Internal preference mapping (IPM) can be conducted on preference data alone (van Kleef *et al.*, 2006). IPM approaches allow examination of individual ratings by consumers and also include alternative techniques such as Landscape Segmentation Analysis (LSA). IPM and LSA do not make assumptions about the underlying product space structure and create spatial representations that provide insights into consumers' preferences based solely on consumer' liking scores. The major difference is that IPM is based on a vector model while LSA involves an ideal point model (Felberg *et al.*, 2010; Rousseau *et al.*, 2012).

The multidimensional representation of products and consumers achieved via IPM is most commonly obtained via PCA of a matrix of products × consumers, the data being the hedonic scores derived from a scaling exercise. The analysis is then based on individual consumer judgments (Nunes *et al.*, 2011). IPM can be regarded as an application of the biplot display of matrices developed by Gabriel (1971). The analysis results in a multidimensional space but generally the focus is on the first two components, which define a plane on which products are represented as points and consumers as vectors. The obtained map allows the visualization of the samples that received the highest hedonic scores together with the consumers that preferred them: the vectors indicate liking directions for each consumer (Nunes *et al.*, 2011; Rousseau *et al.*, 2012). Also, apart from the information on the variability of consumer individual opinions, it may be of interest to identify segments of consumers with similar preference patterns. Internal preference approaches allow consumers to be segmented into groups of similar preference criteria; this can be done visually by observing the groups of vectors pointing in different directions (IPM) or the densities of ideal points in various areas of the map (LSA). Further to this, techniques based on cluster analysis of overall liking scores can be applied in order to highlight the groups and identify the consumers belonging to each cluster. Eventually, subsequent IPMs could be performed on the resulting subgroups (Felberg *et al.*, 2010; Johansen *et al.*, 2010; Rousseau *et al.*, 2012).

From a practical point of view, 8–12 products are the maximum used in preference mapping studies; less than five samples is not recommended because of statistical power issues when estimating individual regression models. In terms of the number of consumers, general use is between 100 and 150. In some situations a lower number can be used successfully but not less than 50 consumers should be considered (Næs *et al.*, 2010).

This chapter aims to show in a real industrial application setting, why it is important to study consumers' preference patterns and liking segmentation, and what information can be unveiled by the use of internal preference mapping via principal component analysis (PCA) and hierarchical clustering analysis (HCA).

# CASE STUDY: CONSUMER RESEARCH TO GUIDE THE DEVELOPMENT PROCESS OF A 'MORE NATURAL' TASTING PROCESSED ORANGE JUICE

## Background

Orange juice is the most popular fruit beverage in the world, accounting for over 60% of juice and juice-based drinks sales in North America and Western Europe. Added to this, Eastern Europe has been experiencing increasing orange juice consumption in recent years. This trend is greatly affected by the growing economies, since it used to be considered as a luxury item in many of these markets. Processed orange consumption will continue to expand in emerging economies such as Eastern Europe countries. Moreover, the market segment for freshly-squeezed, processed juice has increased rapidly in the past decade. (Fry *et al.*, 1995; Spreen, 2001; Hodgins *et al.*, 2002). Commercial processed orange juice products have been reported to lack some of the much appreciated natural fruity flavour and aroma of fresh orange juice (Foley *et al.*, 2002). Furthermore, increased consumer awareness of diet and nutrition has also contributed to the increasing demand for orange juice that is as natural as possible and without extensive processing (Lee and Coates, 1999). This means orange juice manufacturers are looking for ways that allow the production of a more natural tasting product.

## Materials and methods

This consumer study is an example of a typical industry-focused consumer research test to guide product development. It was performed by a big global company that wanted to move its product to a more naturally tasting processed orange juice. It was the second player in market share in the target market. The objectives of the study were to understand consumers' liking towards natural orange flavour in orange juices and the segmentation of the market, to identify the potential size of the market segments and to verify whether any of the new prototypes met the expectations of the consumers towards natural orange juice flavour.

### *Samples*

The study included 11 orange juice samples: the company's own product (OWN), a competitor that at the moment of the study was the market leader (COMP), and eight prototypes of different processing parameters (P1–P8), trying to achieve a more natural orange flavour. Also, a natural, recently squeezed, filtered orange juice was added to the sample set as the 'golden reference' (NAT). Twenty-five millilitres of each juice were served cold to consumers (5–7°C), in plastic transparent containers, coded with three-digit random numbers. Samples were served following a balanced rotation design (multiple orthogonal Latin square). Mineral water was available for rinsing between samples.

### *Consumers*

300 consumers participated in the study, from 19–40 years old, primary shoppers of fruit juice category, 70% were female and 30% male, according to consumption and buying structure of the product under study in the target market (eastern Europe country). The quotas were taken as per market share, depending on the most used brand, as follows: 30% usual consumers of the market leader (COMP), 20% usual consumers of the company's product (OWN), 20% consumers of a private label and the rest were consumers of smaller brands.

### *Consumer test*

The test was performed in a central location setting and the recruitment was managed by a market research agency. Consumers were recruited from its database in order to fit in the sampling quotas. The study was a multiproduct sequential monadic test, where samples were tasted on three consecutive days.

Data were collected in paper ballots, through self-administered questionnaires previously explained to each consumer on a 1:1 basis. Consumers were asked to try each of the orange juices and to evaluate their overall liking (OL) using a nine-point hedonic box scale. They also answered other questions regarding particular attributes, but this chapter only discusses OL data. In all cases, the OL question was the first to be asked in the questionnaire.

### *Data analysis*

Analysis of variance (ANOVA) was performed on consumer overall liking scores considering consumer and sample as sources of variation. Mean ratings were calculated and significant differences were checked using Fisher's LSD test ($p < 0.05$).

Hierarchical cluster analysis (HCA) was performed to identify groups of consumers with different preference patterns. This analysis was performed on standardized liking scores, considering Euclidean distances and Ward's method as agglomeration criterion.

Internal preference mapping was carried out using a principal component analysis (PCA) on the correlation matrix of consumer individual liking data.

## Results and discussion

### *Overall liking*

Results obtained from the OL question showed that consumers reacted differently to the orange juices samples. ANOVA and the LSD test separated samples with significantly different OL mean scores (Figure 11.1). Prototype P4 was the most liked sample, while prototype P8 the least liked sample. There were some interesting facts in these results. For example the recently squeezed orange juice (NAT) did not top the chart as expected, but was the third least liked. Also, the competitor sample, the market leader (COMP), was not a good performer and appeared as the second least liked product. Both facts were surprising as both products were expected to be found in the first positions of OL scores. The company's product (OWN) had a relatively good performance in the overall results, appearing between the top liked

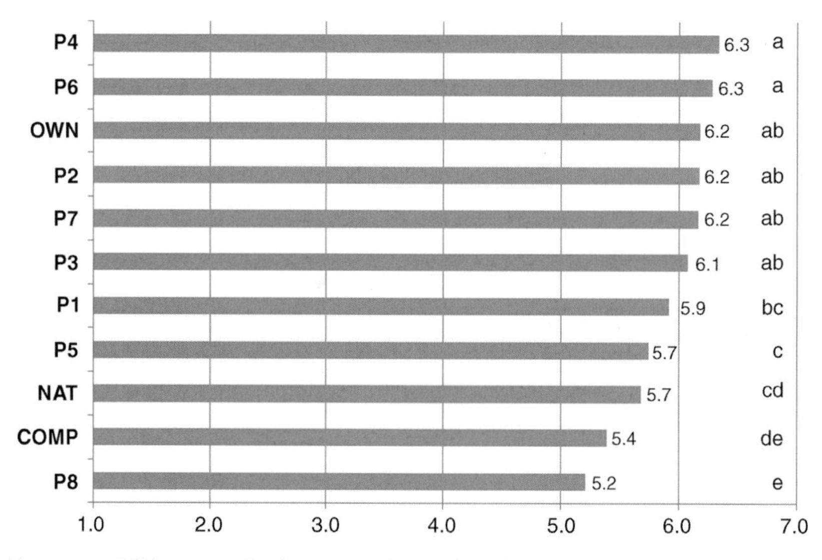


**Figure 11.1** Mean overall liking scores for the 11 samples, evaluated in a structured nine-point hedonic box scale. All consumers ($n = 300$). (Different letters indicate significant differences according to Fisher's LSD test ($p \leq 0.05$).)

products, in the third place in preference, with no significant difference to the top liked products. So, if one was to make inferences by only seeing the averages and ANOVA results based on the general population, conclusions could have been:

- Although the company's product is not going well in the target market, it is doing well in blind tasting, so the problem may be something else (pack, brand, marketing campaign, etc.).
- The competitor's product is not well liked in blind tasting, so its success may be the result of external cues and marketing strategies.
- The natural product is not really liked, so no improvement could be attained by going towards that kind of flavour.

### Segmentation by subgroups

In these days and times it is not common practice to only look at averaged results. Instead, it is usual to go deeper into the test responses and to look at *subgroups of consumers*, classifying them into groups that have some set of similar characteristics, separating them by demographics (gender, age, family size, income, occupation, education, etc.), by geographics (city, country) or by brand consumption, among others. In this particular case, the segmentation by most used brand was one of particular interest. It is noteworthy that consumers of the company's own product (OWN) 'recognized' their product, which means they preferred the product they most often bought (Figure 11.2a). Among those consumers OWN was the top liked, significantly more liked than the competitor, and with an OL rating (6.6) larger than the obtained considering the general population (6.2). When considering the consumers that most often bought the competitor's product (COMP) the case was completely different. These rated this sample with the lowest OL score (Figure 11.2b).

This example is a clear case where external cues played an important role in food choice decisions. Consumers of the market leader orange juice are most probably buying the product based on marketing strategies, and not switching to other products because of brand loyalty. Brand has been reported to be one of the most important nonsensory factors affecting consumers' decisions and has a key role in determining consumers' food choices, and even preference (Varela *et al.*, 2010). According to Keller (1998) brand consists of 'a promise', a symbolic sign of quality. Most importantly, brand has been reported to significantly affect consumers' affective reaction to food products. Guinard *et al.* (2001) reported that consumers' acceptability of beers significantly changed when consumers were aware of the brand. Gacula *et al.* (1996) stated that consumers were less critical when rating samples that were identified by brand; meaning that their perception was dependent on brand popularity. Apart from being able to influence consumers' liking and purchase intent, brand is also likely to affect consumers' sensory and hedonic expectations, since a well-recognized brand on the label can have a large positive influence in creating them (Di Monaco *et al.*, 2004). Interestingly enough, Varela *et al.* (2010) found that for powdered orange drinks the perception of the brands had a larger impact on liking scores than actual sensory characteristics. This suggests that blind tests may not predict consumers' affective reaction in this food category and advises that nonsensory characteristics should be taken into account during new product development, since they can significantly affect consumers' hedonic impression.

With regards to the recently squeezed orange juice (NAT), it had a medium performance in both groups, suggesting that consumers were not appreciating the natural taste, which is, at least, surprising.

### Frequency distribution

Another very valuable way of looking at OL scores is studying the *frequency distribution of the liking ratings*. Figure 11.3 shows, as an example, the frequency distributions for some of the products under study. Studying the frequency distribution is a first step towards detecting if liking segmentation existed

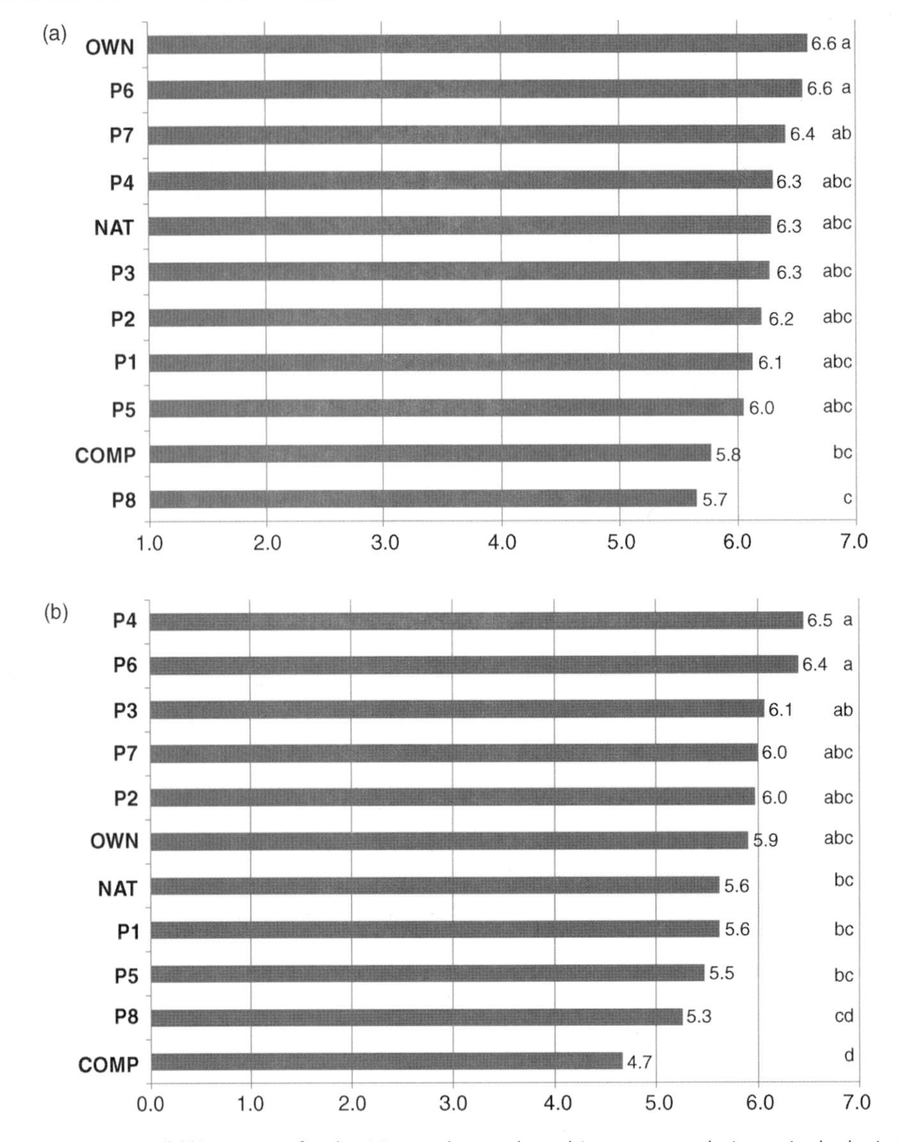


**Figure 11.2** Mean overall liking scores for the 11 samples, evaluated in a structured nine-point hedonic box scale. Segmented per brand: (a) consumers of OWN; (b) consumers of COMP. (Different letters indicate significant differences according to Fisher's LSD test ($p \leq 0.05$).)

in the interviewed population. Samples P6 and OWN are typical examples of well-liked products that had a non segmented response: a unimodal frequency distribution, negatively skewed (with a tail to the left), where the median is greater than the mean. In this case, frequent consumers of the category would typically like (in different degrees) the products, so the majority would rate the products with a five or more, with the majority of the responses of six, seven and eight, and few rejections, that is ratings below five ('neither like nor dislike'). Prototype P6 was particularly well liked, with more than 40 consumers giving it a nine. On the contrary, the market leader (COMP), presented a distribution where consumers were more homogeneously distributed over the nine points of the scale (flatter distribution), with a large number of consumers rejecting the sample (scoring less than five). The recently squeezed orange juice

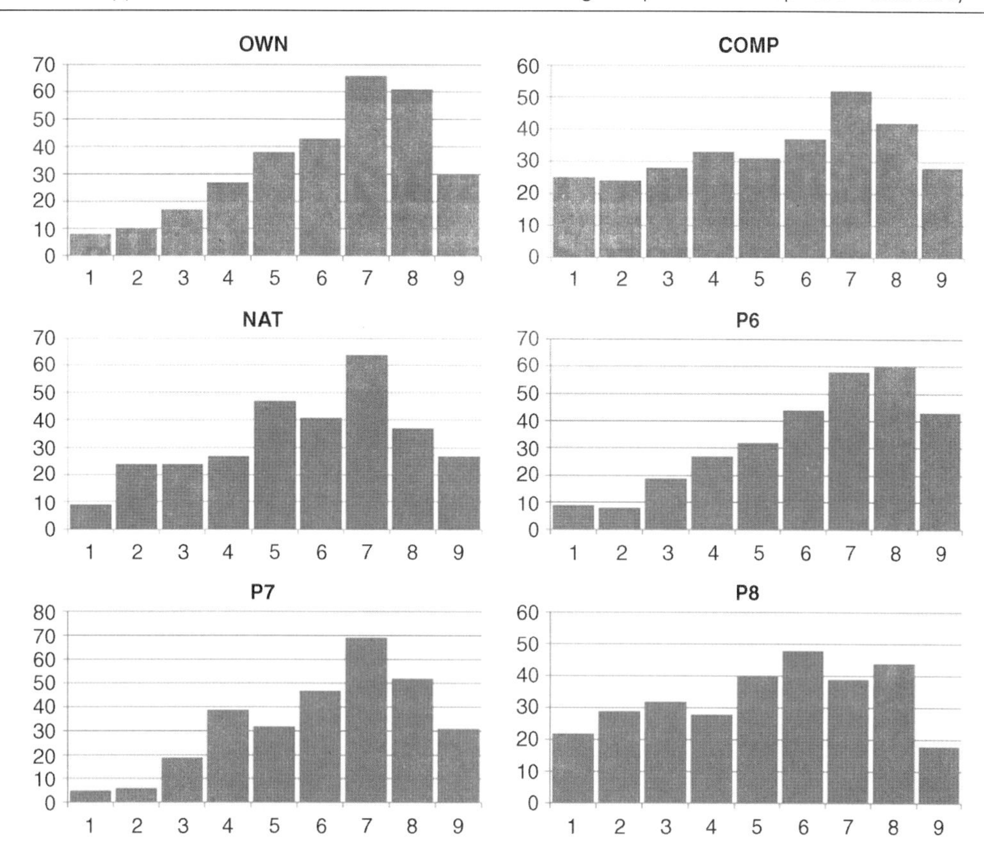


**Figure 11.3** Frequency distributions of the OL scores: number of consumers that rated the sample on each of the ratings (1–9). (Only the results of six products are displayed as examples.)

(NAT) and prototype P7 presented frequency distributions with two distinct peaks (bimodal), which is a typical sign of segmentation in the preferences towards a product, where two groups of consumers either like or reject it. In the case of prototype P8, it also had many rejections, which explains its bottom position in the OL ranking for the whole population. However, at a first glance it is not clear if there could be segmentation towards this product or not.

So far, with no 'complicated statistics', rich information has been obtained from the OL results, with the histograms of the frequency distributions already hinting that segmentation may be present towards some of the products in the set.

## *Agglomerative hierarchical clustering (AHC)*

MacFie (2007) described cluster analysis as 'the lowest level of preference mapping', in which it is assumed that not all people behave the same way and individuals or groups of consumers have a distinct liking pattern.

Agglomerative Hierarchical Clustering (AHC) is an iterative classification method that works from the dissimilarities between the objects to be grouped together; one of the results obtained is a dendrogram, which graphically shows the progressive grouping of the data. In a first step, it transforms liking data into a similarity or dissimilarity matrix, based on a chosen similarity index. A distance measure suited to the particular data has to be chosen. In consumer tests, the most common criteria is using *Euclidean distances* if the OL data are standardized (subtraction of the mean and division by the standard deviation) and *Pearson's correlation coefficient* if working with raw OL data. After that, a clustering algorithm must be selected in order to assign consumers to the different groups. The algorithm sequentially calculates the dissimilarity between two groups of objects. The most common practice in consumer hedonic data is to use either the *average linkage method* (also called *unweighted mean pair group*), which joins the clusters at the average similarity or dissimilarity level, or *Ward's method*, which tries to cluster consumers to form groups showing the maximum ratio of between to within variance. *Ward's* normally gives more compact clusters but it can be more affected by the presence of outliers.

The dendrogram is then used to decide how many clusters or groups of consumers with different liking patterns would be considered. This can be automatically determined by the statistical software in use (automatic truncation), but it is highly recommended to decide it by reasoned observation of the results. One criterion, particularly if working in the industry, is obtaining meaningful clusters with regards to size, as to being able to characterize them and to take decisions based on a sufficient number of consumers; the minimum being usually around 50. More importantly, the obtained clusters have to be consistent, in terms of having really distinct liking patterns, and to be understandable when looking at other product information to be actionable. An example would be contrasting the liking pattern of an obtained group with the sensory data (if available). The procedure would be observing common likes and dislikes: are these consumers liking a certain perceptual space, let us say more bitter or astringent products, while the other groups would prefer creamier or sweeter products? Other criteria may be used to study the composition of the clusters to see if the groups are representative of certain consumers, considering parameters like age, gender, education, being consumers of a brand, or even attitudinal characteristics. The idea behind it should always be that the obtained groups are 'actionable', permitting decisions to be taken based on the clustering. More detailed and very useful information on how to take decisions to apply AHC on consumer data can be found in MacFie (2007).

Going back to the case study, the best results were obtained for this data set using standardized OL scores, with Euclidean distances and Ward's agglomeration criterion. The resulting dendrogram is shown in Figure 11.4. The clustering solution gave three clusters, cluster 1 with 123 consumers, cluster 2

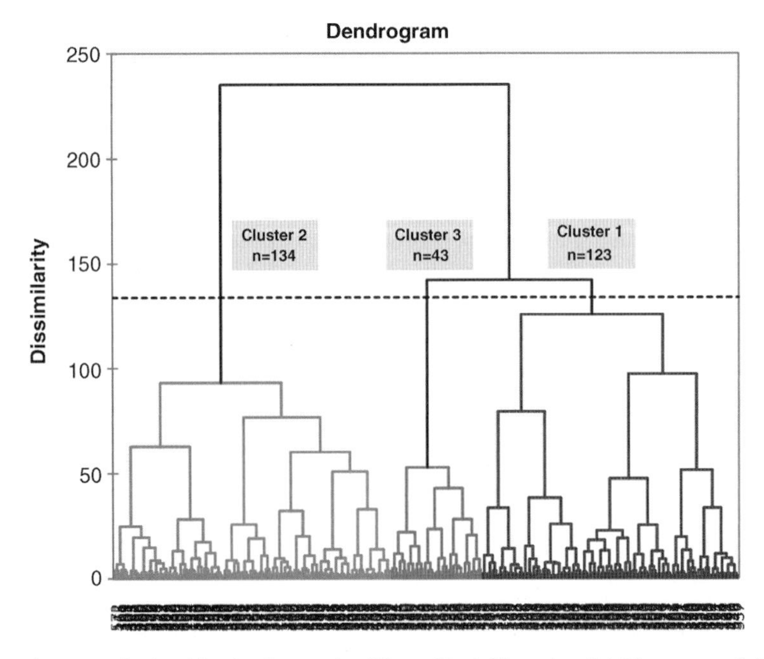


**Figure 11.4** Dendrogram obtained by Agglomerative Hierarchical Clustering (AHC) on overall liking data for all consumers (n = 300).

with 134 consumers and cluster 3 with 43. In this case, the decision of taking three clusters was done by observing the obtained liking scores for the three groups, in order to find meaningful liking patterns that discriminated among groups (Figure 11.5). Cluster 1, representing 41% of the market, showed a marked preference towards the natural product (NAT). Both the current product of the company (OWN) and the market leader (COMP) were less liked. In particular, these consumers widely rejected the market leader, giving it an average OL score of 4.3. Some of the prototypes had no significant difference in OL to the natural option for these consumers; in particular P8 and P5 were equally liked.

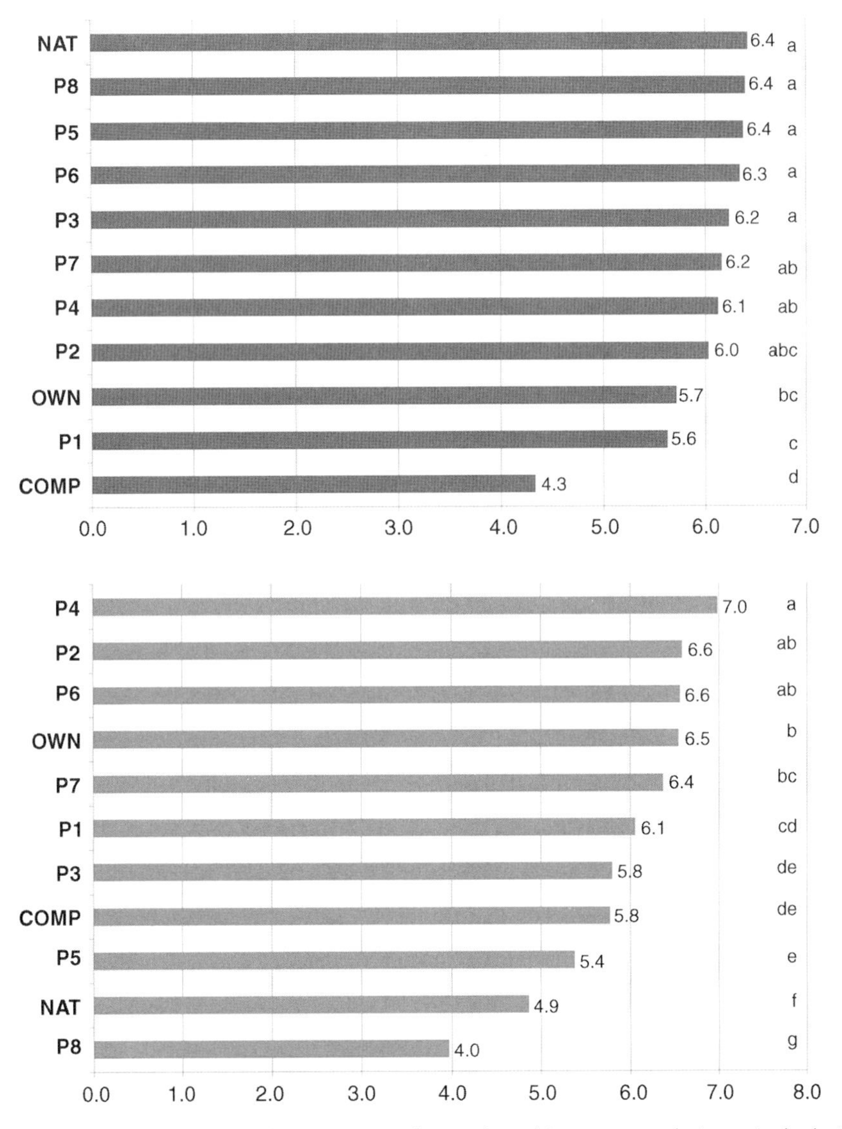


**Figure 11.5** Mean overall liking scores for the 11 samples, evaluated in a structured nine-point hedonic box scale. Segmented per liking pattern, by AHC: (a) cluster 1 (n = 123); (b) cluster 2 (n = 134); (c) cluster 3 (n = 43). (Different letters indicate significant differences according to Fisher's LSD test ($p \leq 0.05$).)

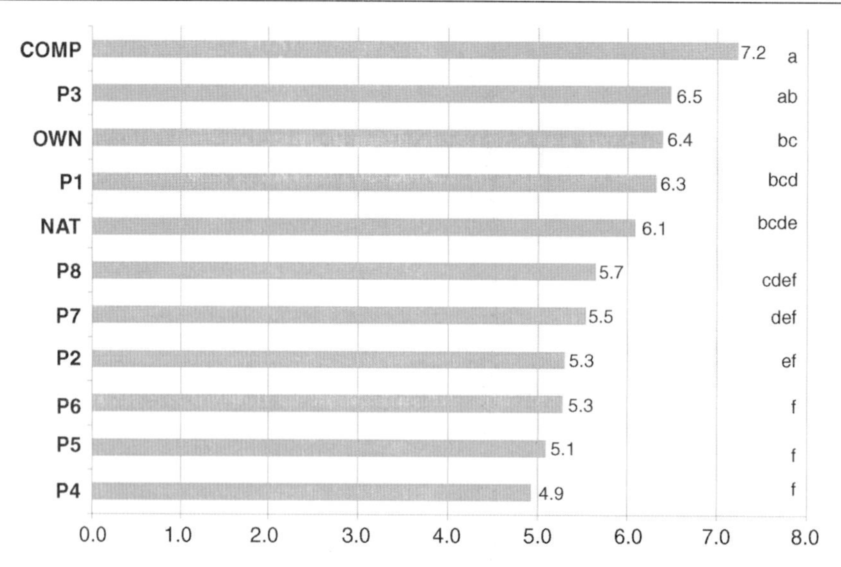


**Figure 11.5** *(Continued)*

Cluster 2, representing a 45% of the market, rejected the natural product, together with P8. This fact explains the average performance of NAT when taking the whole population. Both OWN and COMP had a medium performance for this cluster.

Cluster 3, with 14% of the market, represented consumers that liked the market leader, which topped the chart with a 7.2 OL score. The OWN product was significantly less liked for this group.

The AHC showed that there was indeed segmentation in the population. After identifying the groups and recalculating the averages for each cluster, it is noteworthy how OL scores were larger for the products topping the charts, or smaller for the bottom ones. A averaging OL scores taking the whole population 'dilutes' the preferences or rejections for the segmented products.

It looks like there probably is almost a half of the consumers of orange juice in the target market that do not like natural orange juice but prefer processed notes in their drink. The interesting point behind this observation is that if you asked them whether they like a natural tasting orange juice of a processed tasting one, they would for sure answer 'natural'. So this is a very interesting example where consumers do not consciously know what they like on a blind basis. The obtained segmented results showed that there is a big potential for the development of a naturally tasting orange juice as 45% of the population prefer this kind of taste. But, is the company going to undertake the risk of moving towards this flavour space? What would be the consequences? This is discussed later in the chapter; firstly, the outcomes of the internal preference mapping are examined.

### *Internal preference mapping (IPM)*

The advantage of preference mapping over clustering is that it generates a multidimensional graphical representation of consumers and products that can help communication between product developers and internal partners outside R&D, marketers or management. IP maps are generally easy to interpret. The direction of each vector represents the direction of increasing liking for each individual consumer, and those directions can be linked to the position of the prototypes and competitors in the product map, helping understanding where the ideal areas to target products can be located. Segmentation can be

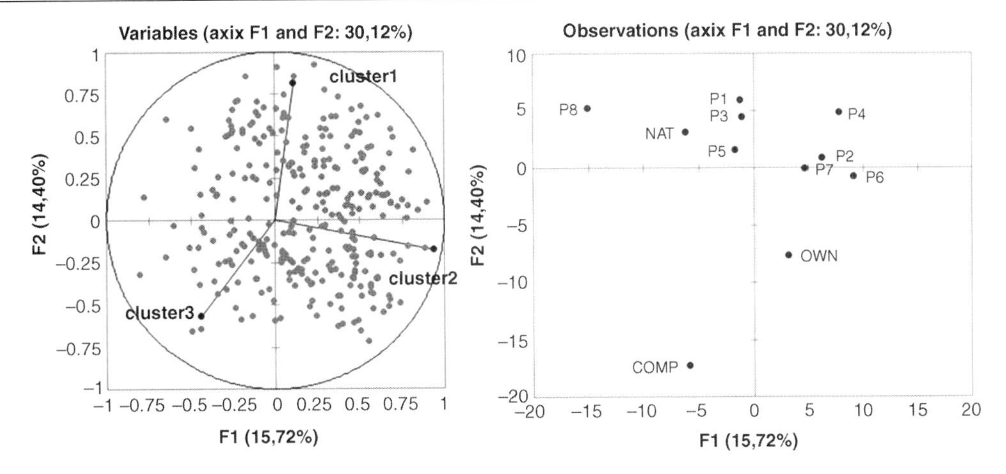


**Figure 11.6** Internal preference map of consumers' liking scores of the 11 evaluated samples: (left) consumers' individual representation and clusters projected as supplementary variables; (right) samples' representation. (Cluster 1 (n = 123); cluster 2 (n = 134); cluster 3 (n = 43).)

detected through this technique by visualizing groups of vectors (consumers) located pointing to different areas of the product space. IPM is usually considered as an approximation, since only two dimensions are being considered.

In this case study, though, the interpretation of the consumers's map is not that straightforward (Figure 11.6). It is clear that there was not a homogeneous liking pattern among the 300 consumers, as they appeared in the map distributed all over the space determined by the first two principal components. Also, the variance explained by the first two components was quite low (30.1%), which is not unusual in this method (MacFie, 2007).

Looking at consumers' vectors and the product map only it is difficult to identify groups of consumers preferring one product or another. One solution to clarify the results would be to combine the AHC and IPM tools, by projecting the average liking scores obtained for each of the clusters of the AHC onto the consumers map, as supplementary variables (Figure 11.6). Although conclusions would not be different to the clustering example, this would help understanding and communicating the results, by seeing together the map with the preference directions and the products location. However, in this case, the variance explained by the first two PC is still low.

Another option would be to apply the PCA on the clusters liking average scores, in other words doing an IPM after the clustering (Figure 11.7) (MacFie, 2007). With this method, the variability explained by the first two components in the case study went up to 93%, which would potentially be more convincing to the external partners to whom the researcher would be communicating the results. The outcomes are similar to the obtained by applying HCA, but the aid of the subsequent IPM, the graphical display enhanced the comprehension.

## General discussion and conclusions

The preliminary conclusion inferred, when looking at the averaged OL data is that the recently squeezed orange juice was not well liked proved to be wrong. Segmentation of the population determined that 41% preferred this kind of flavour. At the same time, the identification of liking patterns also unveiled the fact that 45% of the target market rejected the natural product in a blind test. This was a very difficult outcome to manage by a company. What should it do? Moving the current product (OWN) towards more natural flavour to target those consumers would be very tempting, particularly as it already had some

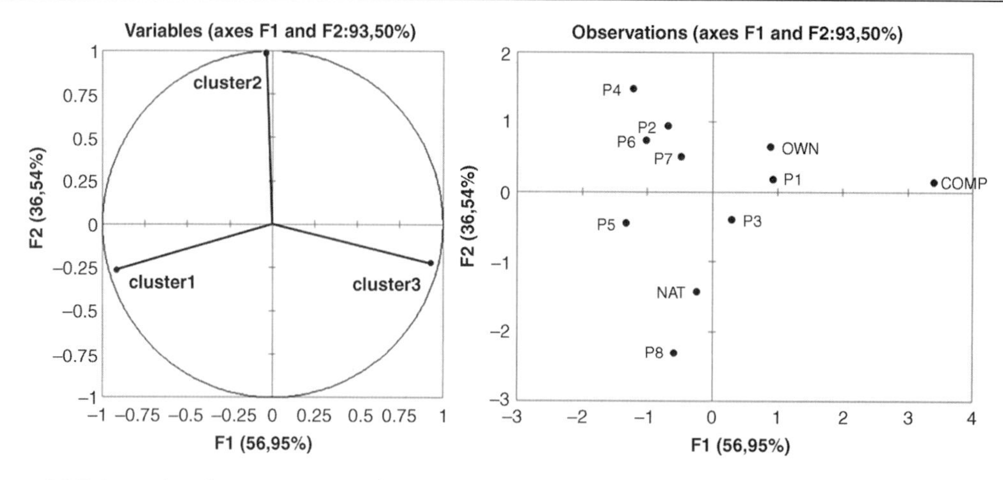


**Figure 11.7** Internal preference map using the average OL per cluster as active variables: (left) clusters representation; (right) samples' representation. (Cluster 1 (n = 123); cluster 2 (n = 134); cluster 3 (n = 43).)

prototypes that succeeded targeting this area (P8 and P5), but at the same time risky. The risk of new consumers believing that they like natural but they do not, trying and rejecting the new launch can be potentially harmful. Furthermore, by looking back at the company's current consumers (Figure 11.2) it can be observed that even though the NAT product performed somehow well among them, the prototypes P8 and P5 were rejected. Moving the current offering towards that sensory characteristic would not be a possibility then, as customers may be lost with the change. What about trying to design a specific product for the 'natural likers' while keeping the current product unchanged? That could be a good option, but a challenge to marketing. The idea of launching a 'naturally tasting' orange flavour, but leaving the other product as it is, can send the wrong message: 'Try our new naturally tasting orange juice! . . . if you do not like it, we still have the artificial old buddy'. Jokes apart, the outcome of this case study would keep the management of the company thinking for a while before taking a decision.

*Hierarchical clustering analysis* and *internal preference mapping* were used in this case study to unveil the distinct preference patterns of the population in a consumer test with great success. It has to be highlighted that only overall liking data were used in this exercise, and the discussion coming out of the data analysis was anyhow very rich, allowed by the use of multivariate statistical tools.

## ACKNOWLEDGEMENTS

The author is grateful to the Spanish Ministry of Science and Innovation for the awarded Juan de la Cierva contract.

## REFERENCES

Ares, G., Varela, P., Rado, G. and Giménez, A. (2011) Identifying ideal products using three different consumer profiling methodologies. Comparison with external preference mapping. *Food Quality and Preference* **22**, 581–591.

Di Monaco, R., Cavella, S., Di Marzo, S. and Masi, P. (2004) The effect of expectations generated by brand name on the acceptability of dried semolina pasta. *Food Quality and Preference* **15**, 429–437.

Felberg, I., Deliza, R., Farah, A. *et al.* (2010) Formulation of a soy–coffee beverage by response surface methodology and internal preference mapping. *Journal of Sensory Studies* **25**, 226–242.

Foley, D.M., Pickett, K., Varon, J. *et al.* (2002) Pasteurization of fresh orange juice using gamma irradiation: microbiological, flavor, and sensory analyses. *Journal of Food Science* **67**, 1495–1501.

Fry, J., Martin, G.G., and Lees, M. (1995) Authentication of orange juice. In: *Production and Packaging of Non-carbonated Fruit Juices and Fruit Beverages*. (ed. P.R. Ashurst). *Blackie* Academic and Professional, New York, pp. 1–52.

Gabriel, K.R. (1971). The biplot graphic display of matrices with application to principal component analysis. *Biometrika*, **58**, 453–467.

Gacula, M.C., Jr, Rutenbeck, S.K., Campbell, J.F. *et al.* (1996) Some sources of bias in consumer testing. *Journal of Sensory Studies* **1**, 175–182.

Guinard, J.X., Uotani, B. and Schlich, P. (2001) Internal and external mapping of preferences for commercial lager beers: Comparison of hedonic ratings by consumers blind versus with knowledge of brand and price. *Food Quality and Preference* **12**, 243–255.

Hodgins, A.M., Mittal, G.S. and Griffiths, M.W. (2002) Pasteurization of fresh orange juice using low-energy pulsed electrical field. *Journal of Food Science* **67**, 2294–2299.

Jaeger, S.R., Rossiter, K.L., Wismer, W.V. and Harker, F.R. (2003) Consumer-driven product development in the kiwifruit industry. *Food Quality and Preference* **14**, 187–198.

Johansen, S.B., Hersleth, M. and Næs, T. (2010) A new approach to product set selection and segmentation in preference mapping. *Food Quality and Preference* **21**, 188–196.

Keller, K.L. (1998). *Strategic brand management: Building, measuring and managing brand equity*. Prentice-Hall, Englewood Cliffs, NJ.

Lee, H.S. and Coates, G.A. (1999) Vitamin C in frozen, fresh squeezed, unpasteurized, polyethylene-bottled orange juice: a storage study. *Food Chemistry* **65**, 165–168.

MacFie, H. (2007). Preference mapping and food product development. In: *Consumer-Led Food Product Development* (ed. H. MacFie). Woodhead Publishing Ltd, Cambridge, UK, pp. 407–433.

Næs, T., Brockhoff, P.B. and Tomic, O. (2010) *Statistics for Sensory and Consumer Science*. John Wiley & Sons Ltd, Chichester, UK.

Nunes, C., Pinheiro, A. and Bastos, S. (2011) Evaluating consumer acceptance tests by three-way Internal preference mapping obtained by parallel Factor analysis (PARAFAC). *Journal of Sensory Studies* **26**, 167–174.

Rousseau, B., Ennis, D. and Rossi, F. (2012) Internal preference mapping and the issue of satiety. *Food Quality and Preference* **24**, 67–74.

Spreen, T. (2001) Projections of World Production and Consumption of Citrus to 2010. FAO Corporate Document Repository (Online). Avalilable: http://www.fao.org/docrep/003/X6732E/x6732e02.htm#2 (last accessed 21 June 2013).

van Kleef, E., van Trijp, H.C.M. and Pieternel Luning (2006) Internal versus external preference analysis: An exploratory study on end-user evaluation. *Food Quality and Preference* **17**, 387–399.

Varela, P., Ares, G., Giménez, A. and Gámbaro, A. (2010) Influence of brand information on consumers' expectations and liking of powdered drinks in central location tests. *Food Quality and Preference* **21**, 873–880.

# 12 Multivariate image analysis

**Marco S. Reis**
*CIEPQPF, Department of Chemical Engineering, University of Coimbra, Coimbra, Portugal*

## ABSTRACT

In this chapter, an overview is provided of the fundamental methodologies falling in the scope of multivariate image analysis (MIA). MIA is a body of knowledge dedicated to the extraction of useful information about products or processes from images, and their use on relevant tasks, such as diagnosis, monitoring and control. Particular attention is devoted to methodologies with potential relevancy for the food industry. The techniques are properly classified and described in some detail, in order to facilitate their future adoption by the interested readers. Furthermore, a variety of applications is also presented, involving the tasks of exploratory analysis and support to diagnosis, process monitoring and grade classification, process control and predictive modelling.

## INTRODUCTION

The automatic acquisition and analysis of images through so-called computer and machine vision systems, is already a current reality in a variety of tasks in the food industry from harvest supervision, product inspection, to process monitoring and control (Brosnan and Sun, 2004). As the main drivers, it is possible to point to their noninvasive nature, objectivity of the information collected and analysed, stability of the assessment process, speed and today's affordability of the necessary hardware components. With the progress in digital cameras and sensing technology, it is now possible to collect images containing a variety of layers of information. The most well known representatives of such images are colour images. Available globally since the middle of the 1980s, they are the simplest members of the class of spectral images. However, it was the emergence of hyperspectral images (which were initially collected and transmitted by satellites but now are being used on product and process related applications) and multiwavelength X-ray images, among others, that sparked the interest in the development of new tools for their proper processing and analysis. These new images, called in a general sense as multivariate images, carry much more information about the objects and scenes under analysis when compared to the classic grey-scale and colour formats. However, such potentially useful information is codified in rather complex and high-dimensional data structures, eventually immersed also within a large quantity of irrelevant data. Therefore, they require suitable and effective approaches that are able to find out and retrieve all the desired relevant information, but that also must be quite efficient and fast, in order to be applied in real world scenarios.

It was in this context that multivariate image analysis (MIA) emerged in the late 1980s, proposing a body of knowledge for handling the new higher dimensional images, especially multispectral and hyperspectral images. Noting that these images consist, in fact, of ordered arrays of spectra (each pixel corresponds to a spectroscopic measurement – the local spectrum of the corresponding narrow region of

*Mathematical and Statistical Methods in Food Science and Technology*, First Edition.
Edited by Daniel Granato and Gastón Ares.

the object under analysis), it is not surprising that its origins arose within a field familiar with the processing and analysis of spectroscopic data: *chemometrics*.[1]

Since its introduction, MIA has attracted a significant interest from academia and industry, and is currently being applied in a wide variety of tasks, such as exploratory analysis, support to diagnosis, process monitoring, process control and predictive modelling (classification and regression). Most of these tasks have direct counterparts in the food industry and, therefore, MIA bears considerable interest for all those with responsibilities in process management, control and improvement, as well as research, in this domain.

In this chapter, an overview is provided on the fundamentals of MIA and the way it addresses the main types of problems found in practice. The essential methodological aspects are presented with some detail, so that readers can replicate the analysis in their own contexts, for their own problems. More advanced applications are also briefly referred to and explained, along with suitable references where the complete treatments are available. It is not the aim of this chapter to provide a state of the art overview on this field, for which other texts have been written (Duchesne *et al.*, 2012; Prats-Montalbán *et al.*, 2011), but to offer an introductory perspective, of a tutorial nature, focused on the fundamental methodologies and principles of MIA and well supported by a number of applications, which hopefully guide potential users in their first steps on this exciting field.

In the next section, the basics and nomenclature regarding digital images are established, and the main types of MIA approaches described, according to a classification scheme based on the nature of image information explored and the level of resolution of the analysis. Then, the methods used more frequently to handle these types of problems are described. In the third section, several applications are referred to, illustrating the range of problems that can be addressed with MIA methodologies. The last section provides final remarks on the current status of MIA and some possible future developments in this field.

# METHODS

In this section, the essential methodologies employed in MIA are presented. It begins with the revision of some basic notions of digital images and sets the nomenclature that will be followed in the subsequent sections of this chapter, when referring to such data structures. Also introduced, in the second subsection, are the fundamental classes of MIA approaches.

## Digital images basics

A digital image is a data array depicting a scene or object in a certain environment. In two-dimensional imaging, images have two relevant spatial coordinates, $x$ (row index, $r = 1, \ldots, R$) and $y$ (column index, $c = 1, \ldots, C$). Each individual pair $I(c, r)$ is called a pixel (Figure 12.1) and the organized arrangement of all pixels, side by side, form the image. Pixels can be one dimensional, three dimensional, multidimensional or hyperdimensional entities. When pixels are one-dimensional quantities, they only contain a single intensity measurement from the original image, giving raise to grey-scale or intensity images. Three-dimensional pixels contain the intensities in three spectral bands, which are used to compose the colour for that element of the image. In the most frequent situation, these bands are usually referred as Red, Green and Blue (hence the name RGB for such images), even though other colour spaces can also be adopted (e.g. HSB CMYK, Lab, Luv, etc.). Therefore, in this case, each pixel corresponds to a three-dimensional vector and not a scalar as happens with grey-scale images. Multi- and hyperdimensional pixels generalize this concept to k-dimensional vectors, leading to multispectral

[1]According to the International Chemometrics Society (ICS), 'chemometrics is the science of relating measurements made on a chemical system or process to the state of the system via application of mathematical or statistical methods'.

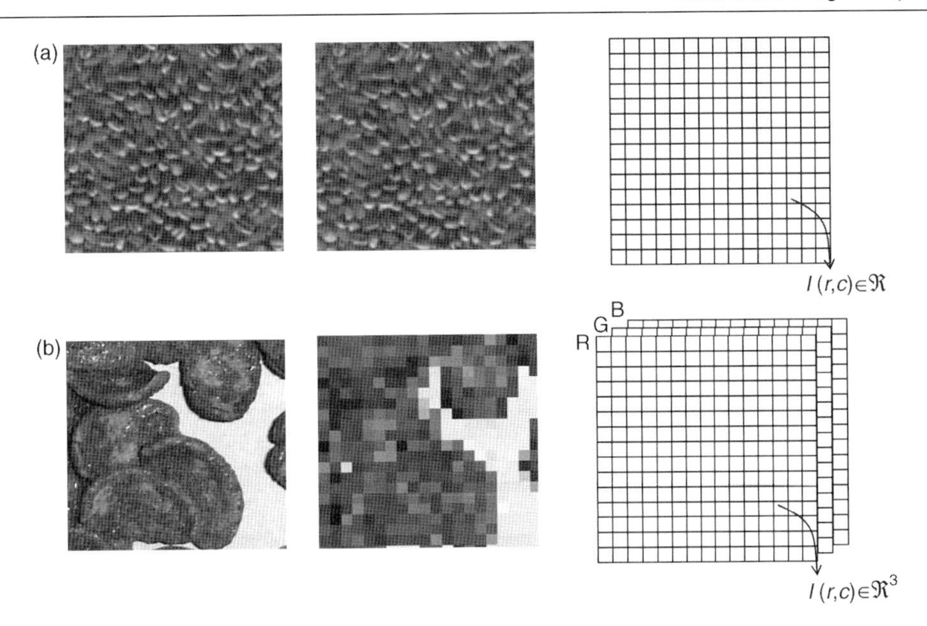


**Figure 12.1** Digital images and their underlying data structures: (a) a grey-scale image. Courtesy of Matti Pietikäinen and Outex Texture Database http://www.outex.oulu.fi/; (b) a colour image. (See colour version of this figure in colour plate section.)

($3 < k <= 10$) and hyperspectral images ($k \gg 10$), depending on the order of magnitude of $k$. Systematizing the data structures underlying each of the above mentioned types of images:

- Grey-level or intensity images: $I(r,c) \in \Re, r = 1,\ldots,R, c = 1,\ldots,C$
- Colour images: $I(r,c) \in \Re^3, r = 1,\ldots,R, c = 1,\ldots,C$
- Multispectral images: $I(r,c) \in \Re^K, 3 < K \leq 10, r = 1,\ldots,R, c = 1,\ldots,C$
- Hyperspectral images: $I(r,c) \in \Re^K, K \gg 10, r = 1,\ldots,R, c = 1,\ldots,C$

This chapter is dedicated to the analysis of multivariate images, which consist essentially of images of the type, $I(r,c) \in \Re^K, K \geq 3$ (usually, $k \gg 3$), where the different subimages, $k = 1,\ldots,K$, are congruent, that is, for all of them, each pixel contains information about the same spatial location. The $k$-th subimage contains the intensities for the $k$-th measurement quantity, which can be a different spectral band (such as the Red, Green and Blue bands) or even another type of measurement, collected for the same scene under analysis, such as optical, electron and ion measurements (as in NMR–X-ray tomography) (Geladi and Grahn, 1996). These different dimensions are frequently called, in a rather general way, *channels*.

In practice, not only is the spatial coordinate discretized into pixels but the spectral dimension must also undergo a similar quantization process. For instance, typically the intensities in the three channels of RGB images are codified in one byte (corresponding to intensity levels in the range 0–255), and the same applies to grey-level images. Therefore, the real nature of a digital image consists of an array of discrete $K$-dimensional vectors. However, computations involving intensities are usually approached using continuous algebra, in order to simplify programming and the analysis of results.

## A classification of MIA approaches

Having introduced the basic data structures underlying digital images, let us now address the main classes of approaches available in the MIA framework to analyse them. For this, we propose a classification scheme based on two distinct dimensions.

The first dimension regards the type of information used. On one hand, the analysis can be focused only on the information contained in the $K$-dimensional vectors (channels) that constitute the pixels, without contemplating any additional treatment regarding their spatial distribution. For example when performing segmentation based strictly on pixels colour or spectra, one is making use of this type of approaches. On the other hand, the analysis can be centred only in the spatial distribution of the characteristics. In this case, the spectral dimension is not relevant (images are usually converted to intensity images) and what really matters is the proper description of the spatial variation. Texture characterization and classification fall in this class of approaches. Of course, it can also happen that both the spectral and spatial domains contain relevant information regarding the problem under analysis. These three cases (channels/spatial/both channel and spatial) constitute the possible instances of the classification dimension 'type of relevant information in the MIA approach'.

The other dimension relevant to classify MIA methodologies concerns the central entity of the final analysis task. If the basic elements of analysis are the pixels, then the analysis is referred to as being conducted at the *pixel level* (Prats-Montalbán *et al.*, 2011) or using *distribution features* (Duchesne *et al.*, 2012). We will call this the *pixel-wise* approach, in order to emphasize the main entities of the analysis. The early developments of MIA fell in this category, which also attracted much interest into this area. In the next section one such methodology is covered: the use of principal component analysis in MIA problems.

On the other hand, the entire image may be characterized by a set of features, computed from all its pixels, concerning different aspects that are deemed to be relevant for the problem under analysis. This type of approach is known as being conducted at the *global image level* (Prats-Montalbán *et al.*, 2011), or using *overall* features (Duchesne *et al.*, 2012). Note that such features may even be a result of a preliminary pixel-wise analysis but, if what is fed to the final analysis stage is the set of global, image-wise features, then the problem is indeed of the *image-wise* type. Examples of image-wise features include the first-order statistics, directly extracted from the intensity histograms of the image channels, such as the mean, median, variance, skewness and kurtosis, which are useful for describing fairly spatially homogenous images, and the second-order statistics, computed from the grey-level co-occurrence matrix (GLCM), which already contain some information regarding the spatial distribution of features, being therefore more appropriate for describing images with more heterogeneous random structures (Haralick, 1979; Haralick *et al.*, 1973). Another relevant example of image-wise quantities, are the Wavelet Texture Analysis features, which will be addressed further ahead in this chapter. Thus, in this second classification dimension regarding the 'resolution level of the analysis', MIA approaches can be categorized as pixel-wise or image-wise.

Note that, in this two dimensional classification scheme, the type of analysis tasks (e.g. supervised or unsupervised analysis) and the final application (exploratory analysis and support to diagnostic, process/product monitoring, process control, property prediction) were not included. These two additional dimensions could also have been contemplated in the classification scheme, at the price of making it more complex (four dimensional). Instead, we opt here to treat these two extra dimensions rather explicitly in the text, in the next sections of this chapter.

## Principal components analysis (PCA) applied in MIA

When first introduced by Geladi and colleagues (Esbensen and Geladi, 1989; Geladi *et al.*, 1989), MIA consisted essentially of analysing a variety of possible outcomes obtained from the application of principal component analysis (PCA) over an unfolded multivariate image. This simple procedure grants access to a wealth of information regarding the spectral characteristics of the image and, after reconstruction, to some insight regarding the spatial locations of such features. The basic steps of this methodology are briefly reviewed here.

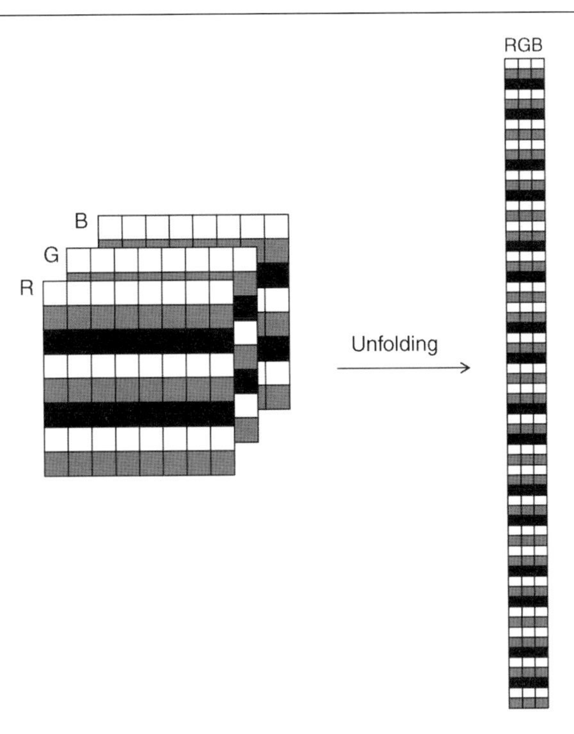


**Figure 12.2** Illustration of the unfolding operation usually conducted in a pixel-wise MIA analysis (in this case, the order of the pixels in the resulting matrix is obtained by slicing down the rows of the original image).

After a multivariate image is acquired, the first step is to convert it into a two-way data matrix, where each row represents a given pixel and each column a given channel of the image. Figure 12.2 illustrates such an operation for the case of an RGB image.

This operation is called unfolding and consists of the following mapping, $U$:

$$U : I \in \Re^{R \times C \times K} \rightarrow X \in \Re^{(R \times C) \times K} \tag{12.1}$$

This mapping can be conducted in two different ways, which only differ in the order in which pixels will appear distributed along the rows of the resulting matrix. In one case, pixels can be collected and stacked on the top of each other by slicing down rows (as depicted in Figure 12.2), whereas the other consists on slicing across columns. The only difference is the final position of the pixels in the new unfolded matrix, $X$. However, in pixel-wise computations, this will not affect the results of the analysis.

The inverse operation, $U^{-1}$, reconstructs the image, pixel by pixel, from the unfolded matrix:

$$U^{-1} : X \in \Re^{(R \times C) \times K} \rightarrow I \in \Re^{R \times C \times K} \tag{12.2}$$

This mapping (also called *backfolding*) is used quite often for analysing the location of a set of selected pixels in the original image, for instance during exploratory and diagnostic tasks.

After unfolding, the resulting matrix $(R \times C) \times K$ is usually pre-processed, by mean centring or autoscaling (i.e. by removing just the columns means, or by performing this operation followed by scaling columns to unit variance, respectively), and then principal components analysis (PCA) is applied on the resulting pre-processed matrix, $Z$. Matrix $X$ can also be treated without any pre-processing. PCA is a multivariate methodology that seeks to explain the overall variability of a data matrix (in the sense of the sum of the variances of all the variables or columns), using a reduced number of new variables, the principal components (PCs) (Martens and Naes, 1989; Jackson, 1991). PCs are just linear combinations of the original variables, with weights provided by the so-called loading vectors, which can be all gathered, column by column, in a single loading matrix, $L = [L_1 L_2 \cdots L_K]$, with dimension, $K \times K$. Applying such $K$ linear combinations to the $(R \times C)$ rows of values (regarding the different pixels) in the $Z$ matrix, will result in $K$ columns with $(R \times C)$ values, one column for each PC, $T = [T_1 T_2 \cdots T_K]$. These columns are called the PCA scores and can also be collected into a $(R \times C) \times K$ matrix of scores, $T$. The relationship between matrices, $Z$, $T$ and $L$ is simply:

$$Z = T_1 \cdot L_1^T + T_2 \cdot L_2^T + \cdots + T_K \cdot L_K^T = T \cdot L^T \tag{12.3}$$

It stems from the properties of PCA, that the first PC is the one explaining most of the variability present in the matrix $Z$. This means that the outer product of the first column of $T$ with the first column of $L$, $T_1 \cdot L_1^T$, is the best rank one approximation of $Z$. The next PCs will present a decreasing capability for explaining the variability in $Z$ and their rank one contributions, $T_i \cdot L_i^T$, will provide increasingly residual contributions to reconstruct $Z$. After a certain point, the additional PCs are just describing the nonstructured variability in the data set and may be gathered in a residual matrix, $E$:

$$E = \sum_{i=a+1}^{K} T_i L_i^T \tag{12.4}$$

The number of retained PCs, $a$, is usually referred as the pseudorank of $Z$. There are currently a large variety of techniques available to define the pseudorank, such as the Kaiser method, Scree test, parallel analysis, information theory – Akaike information criterion, minimum description length – cross-validation and cross-validation ratio, variance of the reconstruction error and F-test, among others (Wold, 1978; Jackson, 1991; Valle *et al.*, 1999; Vogt and Mizaikoff, 2003). The selected scores capture the variability in the PCA subspace defined by the corresponding loading vectors, whereas the residual matrix addresses the variability around this subspace, not captured by the PCA model. The final decomposition of $Z$, can then be written as:

$$Z = T_{1:a} \cdot L_{1:a}^T + E \tag{12.5}$$

where, $T_{1:a}$ and $L_{1:a}$, represent the first $a$ columns of the matrices of scores and loadings, respectively.

Having computed the PCA model for $Z$, the next step is to explore the potential of information available in the scores, loads and residuals. This is usually accomplished by plotting these quantities, or others derived from them, in an appropriate way. Among the derived quantities, the ones most often used are the $T^2$ and Q statistics (the last one also known as squared prediction error, *SPE*). These statistics summarize the information for each pixel (row in $Z$) regarding its statistical distance to the centroid of the pixels distribution in the PCA subspace, and the squared Euclidian distance of each pixel to the PCA subspace, respectively.

As an example of such an analysis, consider the original RGB image presented in Figure 12.3a, regarding cereal flakes. Unfolding the corresponding structure as described above and performing the PCA decomposition directly on this matrix (without any pre-processing), gives a model where the first two PCs concentrate more than 99% of the original variability. Therefore, they will be retained for describing the structural part of the data set ($a = 2$). In fact, the first PC would be almost sufficient for conducting the analysis, as it explains more than 98% of the original variability, a fact that could also be verified visually by observing the so called *score image* for PC1 (not shown), an intensity image that is

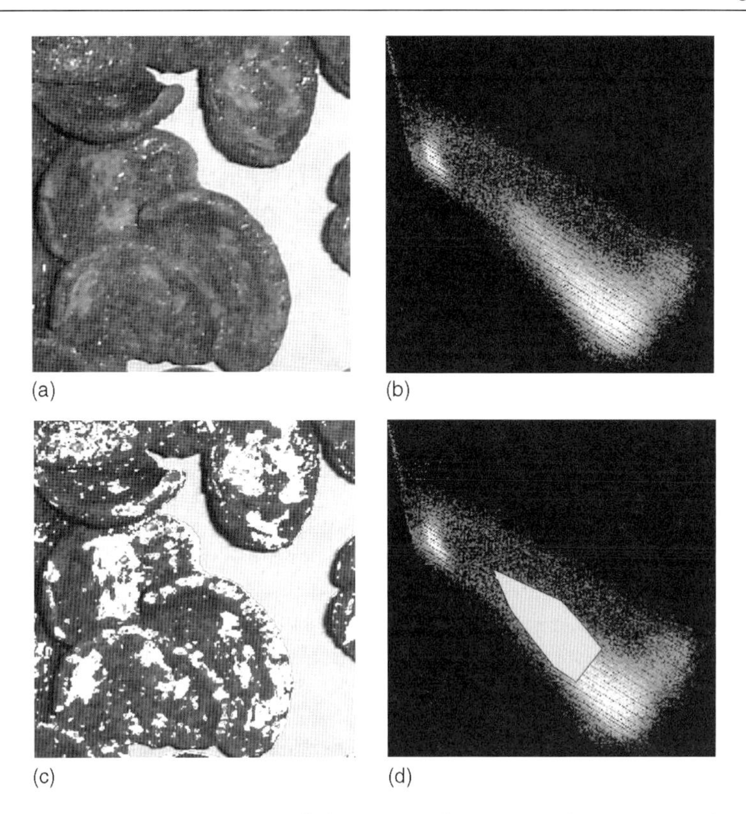


**Figure 12.3** An original image regarding cereal flakes (a) and the corresponding score plot (b). In (d) a mask was constructed in the score plot, by trial-and-error, for selecting the pixels regarding badly coated areas. The corresponding positions of such pixels in the original image are presented in (c). (This analysis was conducted with the MACCMIA software, developed by the McMaster University Advanced Control Consortium, freely available at http://macc.mcmaster.ca/maccmia.php). (See colour version of this figure in colour plate section.)

obtained by applying the operator $U^{-1}$ to $T_1$, instead of $X$ (this encompasses some abuse of notation, as this operator was defined for the whole unfolded matrix, and not for just one column; however it would be still applicable, without change, by redundantly considering three replicates of the same column). Figure 12.3b presents the corresponding scores plot, where each pixel appears as a point (in fact these plots are *scores density histograms*, as the colour represents the local density of points). Points clustering together exhibit similar characteristics in the channels under analysis. In this case, this means that they have similar colours. In this context, it is very easy to perform colour-based segmentation in the scores space (or in general, spectral-based segmentation) by just constructing envelops or masks in the regions where the pixels exhibiting the colour or spectra of interest lie. For instance, in Figure 12.3d one presents a mask constructed by trial-end-error, with the purpose of selecting the badly coated areas in the cereals (without much chocolate, having therefore a light brown colour). The positions of these pixels in the original images are represented in Figure 12.3c. This information paves the way to the computation of areas with bad coating and then to the development of monitoring approaches for product quality, for instance (Pereira *et al.*, 2009).

The trial-and-error masking procedure referred in the example above, can be sped up by first sampling several pixels representing the variety of colours in the structures of interest on the original image and marking their position in the scores plot. This may require a preliminary composite image, built from several representative images of the product under analysis, which will also be useful for estimating a better overall PCA model. Then, the masks can begin to be constructed by involving such points and adjusted by overlaying the selected pixels in the original image. Other alternative methodologies to such trial-and-error masking procedures have been proposed for addressing this task in an automated fashion, such as the method developed by Liu *et al.* based on Support Vector Machines (Liu *et al.*, 2005), which also allows operating with more than two scores, contrary to the manual trial-and-error method, and the so-called Feedback Multivariate Model Selection (FEMOS) proposed by Nordam *et al.* (Noordam *et al.*, 2005).

The methodology described above, is an example of an unsupervised pixel-wise image analysis, where no spatial information was incorporated in the modelling stage (spatial information is lost in the unfolding operation), even though some qualitative spatial inferences can still be made by overlaying the selected pixels in the original image. In the next subsection, a supervised MIA methodology is addressed.

## Partial least squares (PLS) applied in MIA: Multivariate image regression (MIR)

Multivariate images can also be used to predict a quantity of interest, $Y$. If the values of $Y$ are represented in a continuous scale, then we are in the presence of a multivariate image regression problem. The features extracted to predict $Y$ can have different resolutions. They may consist of image-wise features, such as those referred in an earlier section (A classification of MIA approaches), that will act as regressors corresponding to the values of $Y$ collected for the same scene depicted in the image (each image will give rise to a row in the data set under analysis, containing the associated image-wise features and value of $Y$). Alternatively, they can consist of pixel-wise features, in which case, the procedure usually consists of: (i) developing an overall PCA model for all the images in the training set (i.e. considering the whole set of images available to estimate the model; the value of $Y$ is known for each of one of such images); (ii) building the scores density histograms for each image (by projecting the rows of the unfolded image onto the global PCA model, see also Figure 12.3b and 12.3d for an example of such density histogram) and make all the histograms congruent by using the same range and binning for all of them; (iii) selecting the features to extract from the histograms; (iv) using the extracted features to develop a model for predicting $Y$.

Step (iv) is very similar in both image-wise and pixel-wise approaches. In both classes, a $(N \times M)$ feature matrix is obtained, $X_F$, where $N$ stands for the number of images and $M$ represents the number of features. This matrix will then be regressed onto the response $Y$, in order to derive the corresponding predictive model. It is the nature of the features extracted in step (iii) that makes the two approaches different. In the pixel-wise approaches, these features arise from the analysis of the scores density histogram and represent local information in the score space (at the pixel scale). These features can be selected in a variety of ways. They can simply result from a preliminary pixel-wise MIA masking and consist of counting the number of pixels falling inside the mask or other simple operations on the segmented pixels (Yu and MacGregor, 2004), or the regions in the scores space can be selected by regressing $Y$ onto the scores histograms and selecting the more predictive bins (features consist of counts in the selected bins in the score space) (Yu and MacGregor, 2003). More advanced and robust approaches for generating these pixel-wise features involve the use of covariance between the bins of the score histograms and $Y$, from which a covariance map is build. This covariance map having information about the regions with the highest covariance with the target response, is then used to develop curved bins, is more adequate to describe the regions with higher prediction power for $Y$ (Yu and MacGregor, 2003; Yu *et al.*, 2003).

A more direct way to conduct pixel-wise MIR is presented by Lied and Esbensen (2001); it consists of building a composite of multivariate images regarding different responses levels and another Y-image, where such response levels are placed in the corresponding pixels (an example is given in Figure 12.7).

Then, after applying the unfolding operator, $U$, to both the composite image and the Y-image, a regression model can be derived that predicts the response for each pixel in a given image. As the final result should be only one response level for each image, some averaging procedure should finally be made over the responses obtained for all the pixels, in order to come up with a single final quantitative prediction for the image.

The regression approach typically used in MIR is partial least squares (PLS) (Geladi and Kowalski, 1986; Höskuldsson, 1988; Martens and Naes, 1989; Wold *et al.*, 2001). As in PCA, PLS computes a set of linear combinations of the variables, which are now given by the columns of $X_F$. However, now the criterion used to compute the linear combinations is not focused on explaining the variability of $X_F$ (as in PCA), but to maximize the covariance with $Y$. Therefore, after proper pre-processing the matrix of regressors, $X_F$, thus obtaining $Z_F$, PLS will extract successively the linear combinations showing maximum covariance with $Y$, and updates the regression vector accordingly, $\beta_{PLS}$, for the model $Y = Z_F \cdot \beta_{PLS}$. For establishing the final model, the number of linear combinations (latent variables) to use has to be selected, for which different criteria are available, usually consisting of a minimization of prediction errors estimated by some form of cross-validation (Wold, 1978; Martens and Naes, 1989).

PLS can also be adapted for handling classification problems, that is situations where the response is not a continuous variable but a finite set of discrete entities (Barker and Rayens, 2003), being then called partial least squares for discriminant analysis (PLS-DA). The basic procedure consists of including a binary response for each class under analysis in the $Y$ matrix (PLS can handle multiple responses), with 1 signalling the presence of the class, and 0 its absence (for the 2-class problem only one response is necessary, of course). The resulting binary $Y$ matrix, with the binary responses, codifies the class membership of each object in a row (e.g. a pixel or image). Then a PLS model is fitted. Classes of future samples are assigned according to some rule, such as the maximum value in the predicted response vector, but more often thresholds for each response are used and tuned, in order to assign classes only if the predicted response is above a certain minimum level.

## Incorporating spatial information in MIA

MIA methodologies referred so far are strictly based on colour or spectral oriented features. However, frequently, spatial information also plays a relevant role in the characterization of products and processes depicted in images. A wide variety of approaches were developed for addressing this issue, which can also be distinguished primarily as pixel-wise or image-wise approaches. As the focus now is just on the spatial information, let us assume that images are now composed by just a single intensity channel ($K = 1$).

An example of an image-wise MIA method for incorporating spatial information consists of aggregating to each pixel, information about the intensity levels of pixels belonging to some neighbourhood. This can be made by upgrading an intensity image to a multivariate image through the addition of new artificial channels corresponding to displaced versions of the same original image (Bharati *et al.*, 2004). Figure 12.4 illustrates this process for the case where only one shift is incorporated in the eight directions mentioned (0°, 45°, 90°, 135°, 180°, 225°, 270° and 315°), besides the original image, which will result, after unfolding such artificial multivariate image, in a $(R \times C) \times 9$ matrix. Therefore, in each row corresponding to a given pixel of the original image, there also appears information about the intensities in neighbourhood pixels, whose correlation can be explored using pixel-wise techniques, such as those referred to above.

Regarding image-wise methodologies for incorporating spatial information, a wide variety of methodologies were proposed in the literature falling in this category. In the class of texture analysis problems, for instance, methods were proposed that try to model the variation of pixel intensities using several model structures (Markov Random Fields, Autoregressive and Fractal models), while others are

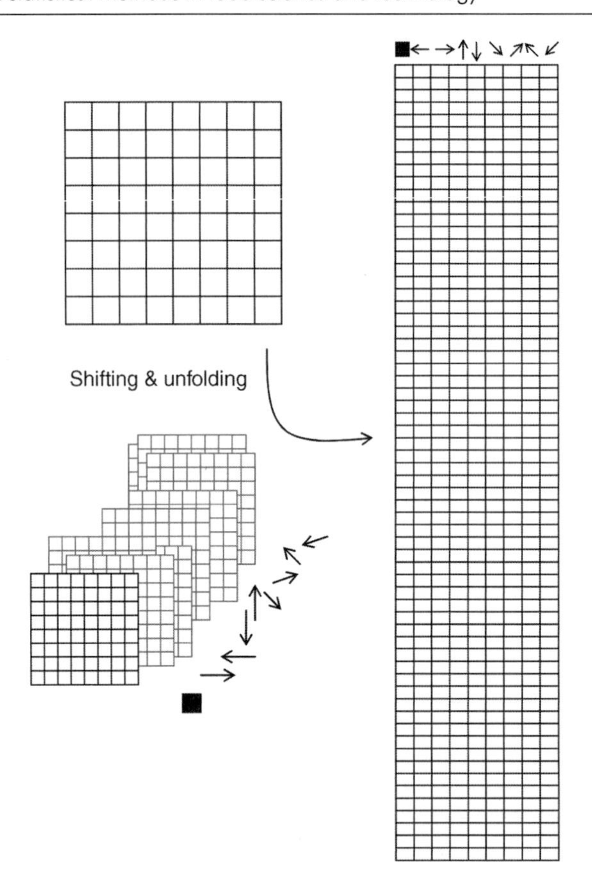


**Figure 12.4** Incorporation of spatial information by creating an artificial multivariate image whose channels results from applying shifting operations on the original intensity image. After unfolding, the rows corresponding to all pixels in the image also contain information about the intensities in a certain neighbourhood.

based on global statistical descriptions of the image (such as those referred in the section 'A classification of MIA approaches', involving the use of first-order statistics, such as mean, median, variance, skewness, kurtosis etc., and second-order statistics, computed from the grey-level co-occurrence matrix), and still others based on some data transformation, from which spatial features can be computed (Prats-Montalbán *et al.*, 2011). Such methods are compared elsewhere (Bharati *et al.*, 2004). Here, one particular methodology is addressed; it is representative of the class of transformed-based methods, called Wavelet Texture Analysis (WTA).

WTA is based on the multiresolution decomposition of an image performed with a given wavelet transform (Vetterli and Kovačević, 1995; Strang and Nguyen, 1997; Mallat, 1999). This multiresolution decomposition decomposes the original image into several subimages regarding different frequency bands along three different directions (horizontal, vertical and diagonal), as depicted in Figure 12.5. Intrinsic to each frequency band is a characteristic spatial scale. The transform coefficients, called wavelet coefficients, for each scale (or frequency band) contain information about the spatial features at that particular scale. This fact, along with the way frequency bands are generated, makes this framework very interesting for extracting images containing relevant features at different scales (multiscale features) or for addressing problems where the characteristic scale is not known beforehand.

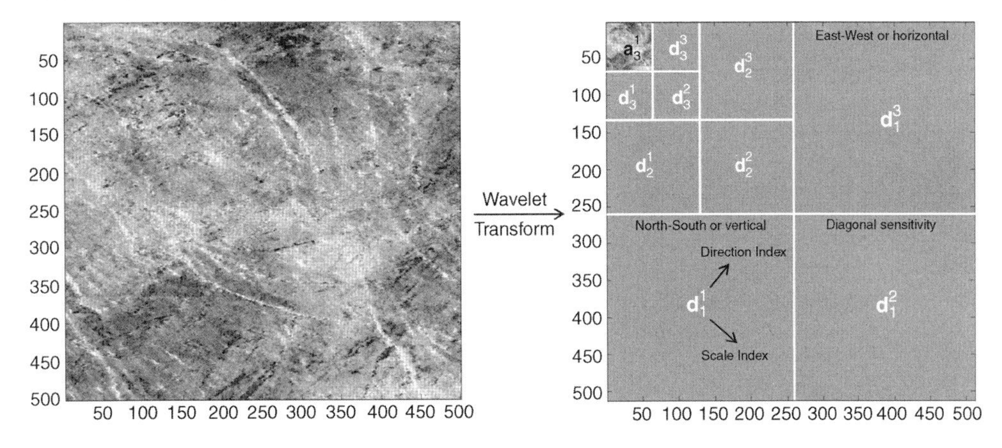


**Figure 12.5** Wavelet transform of a grey-scale image using three levels of decomposition.

These wavelet coefficients are organized in the following blocks: $a^{Jdec}, \{d_i^j\}\Big|_{\substack{i=1,3 \\ j=1,J_{dec}}}$, where $J_{dec}$ stands for the number of scales used in the transform (also known as the decomposition level), $a^{Jdec}$ represents the approximation coefficients block and $d_i^j$ the detail coefficients blocks. The number of scales contemplated in the transform is set by the user. As this number is increased, more bands appear in the low frequency region of the spatial spectra.

After computing the wavelet transform of the grey-scale image, WTA proceeds by summarizing the information contained in the various blocks of wavelet coefficients, in a suitable way. Examples of summary features that can be computed from wavelet coefficients in each block include: energy ($E_{jk} = \left|\left|\mathbf{d}_j^k\right|\right|_F^2$, where $||\bullet||_F$ stands for the Frobenius norm of a matrix), entropy, averaged $l_1$-norm and the standard deviation. These are global features that will be used to compose matrix $X_F$, in an image-wise MIA analysis.

## Integration of spectral and spatial information in MIA

When both colour/spectral features and spatial characteristics are relevant for the problem under analysis, then these different dimensions must be brought together in an integrated way in the image analysis methodology. Quite often, the solutions proposed combine, in an appropriate way, the techniques that were developed for individually taking into account each of the two sources of information. This is justifiable by the fact that, usually, there is no strong correlation between spatial and spectral features. This means that they can be addressed simultaneously in this way, without compromising useful information arising from potential interactions between the two sources of information.

For example an integrated image-wise approach for combining spectral and spatial information consists in replicating the image-shifting procedure depicted in Figure 12.4 for each spectral channel. Then, such shifted versions of the intensities images for each channel are stacked together, side by side, after unfolding (Figure 12.6). In this context, each row, relative to a single pixel, contains information regarding the various spectral channels and neighbourhood intensities, which can be described simultaneously using, for instance, PCA (Prats-Montalbán and Ferrer, 2007).

Another integrated approach, proposed by Liu and MacGregor (2007), combines the wavelet transform multiresolution decomposition with PCA. The authors suggested two ways for performing this combination: (i) multiresolution decomposition of the intensity image at each channel of the

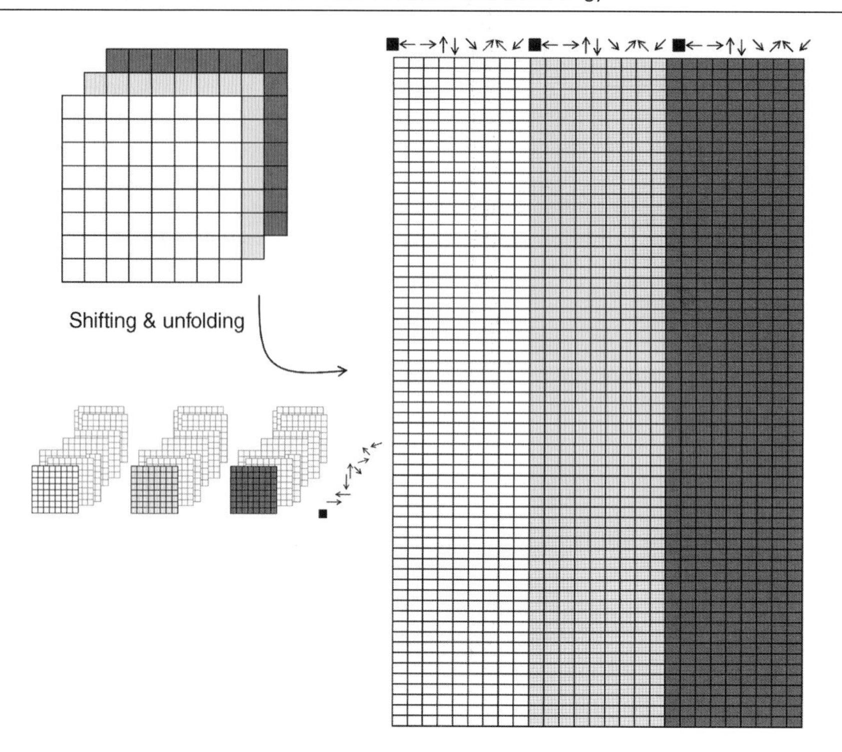


**Figure 12.6** Image shifting and unfolding of all the channels, for the simultaneous pixel-wise analysis of spectral and spatial characteristics.

multivariate image followed by the implementation of PCA at each scale (this is the straightforward generalization of MIA to a multiresolution context); (ii) firstly, apply PCA to the multivariate image and then implement the multiresolution decomposition of the scores images thus obtained (i.e. of the images resulting from the application of the backfolding operator, $U^{-1}$, to each column of PCA scores). This approach provides the possibility of having score plots at each resolution, because all the scores are orthogonal among themselves, and the same applies to the wavelet coefficients at different scales. Methodology (i) is referred as MR-MIA (I) and, according to the authors, is advisable for situations in which spatial and spectral features are expected to be correlated. Methodology (ii), called MR-MIA (II), is preferable for situations where the two information sources do not interact too much with each other and where the number of channels is large (Liu and MacGregor, 2007).

As to image-wise integrated methods, a simple way for combining spectral and spatial information would be to gather all the spatial features (e.g. first- and second-order statistical features, WTA features) computed independently at the different channels, side by side, in the features matrix, $X_F$ – if the total number of images is $N$, the number of features $M$, and the number of channels, $K$, then the dimensions of this matrix will be, $N \times (M \times K)$. Then, this matrix can be explored through multivariate supervised or unsupervised techniques, according to the specific goal of the analysis.

## APPLICATIONS

In this chapter, the application of MIA approaches in different application contexts is illustrated with concrete examples from the food sector, in order to consolidate the knowledge on the techniques presented in the previous sections. In this context, and following what has been already stated, this section does not aim to provide an extensive state of the art review of MIA in the food industry. Instead it aims to complement the conceptual introduction to the fundamental MIA approaches and to show the variety of scenarios where they can be useful and bring added value to processes and organizations.

## Exploratory and diagnostic analysis

MIA techniques can be used to detect and isolate (or segment) specific structures in images from objects or scenes that are relevant for the quality of products or processes. Such structures may consist, for instance, of specific types of imperfections or other quality related features, whose analysis usually requires the intervention of trained operators, being, therefore, quite labour intensive in skilled personnel and suffering from the usual problems of sensorial grading systems, namely limited objectivity in the measurements and stability of criteria, as well as reduced processing throughput. MIA methodologies provide ways to improve this type of approach, either by assisting the analyst in the diagnostic tasks – for instance by overlaying the pixels corresponding to identified structures of interest in the original image, which can also be accompanied with quantitative figures, to provide further support to decision making and thus work out the objectivity, stability and speed limitations – or even by conducting the whole process in an autonomous fashion, as will be addressed in a later section.

Geladi and Grahn presented an example where small vegetation particles are identified by applying suitable masks in the score space from an image taken with a microscope, regarding a sample obtained after boiling a small amount of raw peat in diluted sodium hydroxide (to remove colloidal particles and humic acids) (Geladi and Grahn, 1996). PCA was also applied to the detection of bruise damage on white mushrooms (*Agaricus bisporus*), where the score images obtained from the second PC led to the best contrast between damaged and undamaged regions (the pre-processing method used was mean normalization). Then, a thresholding operation implemented on the scores of this second PC segments the image and isolates the damaged regions. Misclassifications due to border effects were also taken into account in the proposed methodology (Gowen *et al.*, 2008).

Prats-Montalbán *et al.* (2011) exemplified the masking procedure in the scores space to segment and extract cocoa beans from the image background. The authors have also applied such methodology to isolate the affected areas in the diagnostic of diseases in oranges (such as the *Rodolia cardenalis*) and tested the use of a supervised methodology, instead of PCA, to improve the separation between sound and affected areas in the latent variables space and to provide a classification of the image pixels as sound or affected. The methodology used in this case was PLS for discriminant analysis (PLS-DA).

PLS-DA was also used in the pixel-wise discrimination: of normal and flawed (broken, perforated and burnt) Scandinavian crispbread (Lied and Esbensen, 2001), powdered food products (maize, pea, soya bean meal and wheat) (Chevallier *et al.*, 2006), grains and other materials (Wallays *et al.*, 2009).

## Process monitoring and grade classification

Process monitoring regards the activity of supervising the process continuously in order to decide whether, at each instant, it is operating under normal conditions or if something abnormal has happened that must be localized and accommodated or fixed as soon as possible. This is also one of the main goals of statistical process control (Shewhart, 1931; Montgomery, 2001), within which a variety of methodologies were proposed for monitoring univariate (Shewhart, 1931; Page, 1954; Hunter, 1986), multivariate (Hotelling, 1931; Crosier, 1988; Lowry *et al.*, 1992) and megavariate (Jackson, 1959; Jackson and Mudholkar, 1979; Kresta *et al.*, 1994; MacGregor and Kourti, 1995; Nomikos and MacGregor, 1995) processes using data collected from typical process sensors and also from images (Bharati and MacGregor, 1998).

Prats-Montalbán and Ferrer used the pixel-wise procedure for incorporating the spectral and spatial information presented in Figure 12.6 in the development of classifiers for predicting the disease type in oranges. Pereira *et al.* (2009) presented a pixel-wise approach conducted in the original colour space (and not in the scores space, as usual in MIA applications) for the assessment of coating quality in cereals. In this approach, two masking procedures were implemented using the Mahalanobis distance: one for removing the background and selecting the image regions corresponding to cereal flakes, and the other for identifying the regions with coating defects within the flakes. With this tool, it was possible to accurately reproduce the quality assessments performed by a panel of experts (Pereira *et al.*, 2009).

## Process control

In the section 'Exploratory and diagnostic analysis', the use of MIA for assisting analysts in their tasks of diagnosis, where they have to decide about the current state of the product or process and act accordingly, was addressed. In the section 'Process monitoring and grade classification', MIA was applied to perform the two first tasks (analysis and decision) and it was up to the operator to know what actions to take next once a problem was detected. In this section the situation where the whole sequence of tasks analysis/decision/actuation is carried out by an image based system, in an autonomous way is address.

A good example of this application scenario, and perhaps the first successful application of MIA in the automatic control of product quality during its manufacturing, is the one described by Yu and colleagues in 2003. After testing several feature extraction approaches for obtaining more predictive information regarding the response variable to control (coating concentration in snack food), an advanced and robust approach generating pixel-wise features was selected, involving the use of the covariance between the score density histograms bins and the response (Yu *et al.*, 2003). This methodology proved to be able to track down changes in the product on-line; it was applied, therefore, in a feedback control loop for automatically controlling the coating concentration. The system has been implemented for more than a decade now and has led to a significant reduction in coating variability.

## Predicting product quality and process variables

Most frequently, MIA is applied to address classification or grading problems, where products or processes are labelled according to a finite set of categories, for instance regarding their quality in some relevant aspect from the end user perspective. These situations were already covered in the sections 'Exploratory and diagnostic analysis' and 'Process monitoring and grade classification'. Addressed now are applications where the property to infer is of a quantitative continuous nature.

Yu and MacGregor (2003) presented an approach for predicting the coating content and distribution in snack foods. Several feature extraction methods were tested to build the feature matrix, $X_F$, and derive the predictive model, including image-wise features (average colour and loading vector of the first PCA component) and pixel-wise features (scores density histogram, one-dimensional histogram and cumulative histogram obtained using projections in the scores space and the cumulative histogram based on correlation-oriented segmentation) (Yu and MacGregor, 2003). Pixel-wise features led to more robust predictions, especially those obtained with the correlation-oriented segmentation method. The coefficients of determination ($R^2$) obtained in the study were of the order of 0.99 for the test sets.

Another example concerns the monitoring of the storage stability of fruit products (Lied and Esbensen, 2001). A sequence of images was collected for the same items (bananas) in typical storage conditions, with the purpose to provide information regarding their deterioration. The response variable is the effective storage time, which should be inferred by a set of image features. The sequence of images corresponding to 20 days of ageing was gathered into a single composite image and the age of the corresponding pixels recorded in the associated Y-image (Figure 12.7). After unfolding both images,

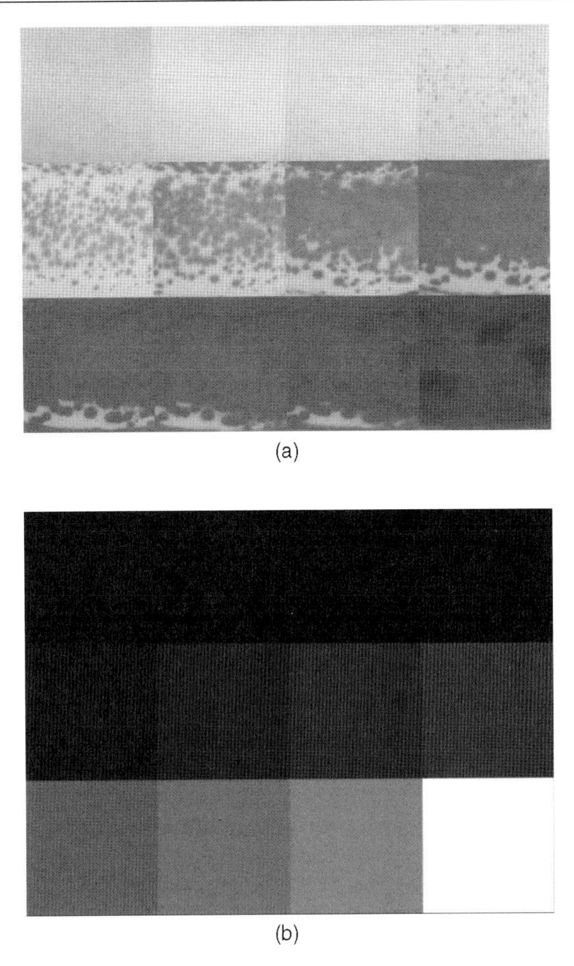


**Figure 12.7** (a) Composite image of fruit (banana) with different storage times in the interval 1–20 days; (b) the associated Y-image with the corresponding storage times for each pixel of the composite image. Lied, T. T. & Esbensen, K. H. (2001) Reproduced with permission from Elsevier. (See colour version of this figure in colour plate section.)

using the *U* operator, a PLS model was estimated that allowed for the proper prediction of storage time of the bananas. The authors also underline that such applications are not restricted at all to perishable fruits but can equally well be implemented in products such as cereals, bread, meat, fish, beverages and milk, among others.

Two methodologies for predicting the rheological properties of cheeses (fracture work) were proposed, one based on PLS and the other on N-PLS, which is the three-way counterpart of PLS whose inputs consist of a third-order tensor (Huang *et al.*, 2003). Each cheese is characterized by four transmission electron micrographs taken at different positions, which are therefore not congruent. To make the whole set of images congruent, a transform approach can be implemented, such as the two-dimensional Fast Fourier Transform (FFT) or the wavelet transform. In this case, the authors opted for FFT, obtaining a two-dimensional transform image for each original image. This operation makes all the transformed images coherent, as now, in each position, information relates to the same FFT wavelength under analysis. These transformed images were then used to build predictive models, along with the measured values for the property of interest (fracture work; one value per image), after unfolding, using PLS, or as they are, using N-PLS. In this case, the N-PLS model showed better prediction accuracy.

Based on hyperspectral images, Trong *et al.* (2011) used PLS-DA for the detection of the cooking front in potatoes, from which the cooked and raw areas can be segmented and employed in the prediction of the optimal cooking time.

## CONCLUSIONS

Measurement devices are currently available that are able to collect highly accurate information either in the spatial domain (high spatial resolution) or in the wavelength domain (high spectral resolution). In fact, the detectors' sensitivity and density has been steadily increasing over time, as well as the availability of suitable optical equipment that allows for the extraction of detailed information on the objects of analysis. Multispectral and hyperspectral cameras are, therefore, able to collect images with a notably high potential for providing useful information about products and processes. Multivariate image analysis (MIA) emerged in the late 1980s within the chemometrics community as a body of knowledge for addressing applications involving these types of images.

In this chapter an overview has been provided about some of the most fundamental and important methodologies used in MIA, covering both pixel-wise and image-wise techniques for exploring spectral/spatial/spectral and spatial information in multivariate images. Furthermore, a variety of applications has been presented regarding the gradient of integration of MIA approaches in processes: from their use as a supporting diagnosis tool, through their adoption as an analysis and decision tool, to their implementation as an autonomous analysis, decision and actuation tool. Applications in the quantitative prediction of properties were also referred to.

So far, MIA has only addressed situations where pixels are characterized by first-order tensors of a variety of sizes, such as one (grey-level images), three (RGB images) or more (multivariate and hyperspectral images). The issue of considering higher-order tensors has not been addressed yet. Furthermore, the variable 'time' has been absent from most of the methodologies presented so far, all of them assuming processes without any relevant dynamical behaviour. The analysis of *image time series* can also be an interesting topic to pursue in the near future, in this exciting and fast evolving field.

## ACKNOWLEDGEMENTS

The author acknowledges financial support through project PTDC/EQU-ESI/108597/2008 cofinanced by the Portuguese FCT and European Union's FEDER through 'Eixo I do Programa Operacional Factores de Competitividade (POFC)' of QREN (with ref. FCOMP-01-0124-FEDER-010398).

## REFERENCES

Barker, M. and Rayens, W. (2003) Partial least squares for descrimination. *Journal of Chemometrics*, **17**, 166–173.

Bharati, M. H. and MacGregor, J. F. (1998) Multivariate image analysis for real-time process monitoring and control. *Industrial and Engineering Chemistry Research*, **37**, 4715–4724.

Bharati, M. H., Liu, J. J. and MacGregor, J. F. (2004) Image texture analysis: methods and comparisons. *Chemometrics and Intelligent Laboratory Systems*, **72**, 57–71.

Brosnan, T. and Sun, D.-W. (2004) Improving quality inspection of food products by computer vision – a review. *Journal of Food Engineering*, **61**, 3–16.

Chevallier, S., Bertrand, D., Kohler, A. and Courcoux, P. (2006) Application of PLS-DA in multivariate image analysis. *Journal of Chemometrics*, **20**, 221–229.

Crosier, R. B. (1988) Multivariate generalizations of cumulative sum quality-control schemes. *Technometrics*, **30**, 291–303.
Duchesne, C., Liu, J. J. and Macgregor, J. F. (2012) Multivariate image analysis in the process industries: a review. *Chemometrics and Intelligent Laboratory Systems*, **117**, 116–128.
Esbensen, K. and Geladi, P. (1989) Strategy of multivariate image analysis (MIA). *Chemometrics and Intelligent Laboratory Systems*, **7**, 67–86.
Geladi, P. and Grahn, H. (1996) *Multivariate Image Analysis*. John Wiley & Sons Ltd, Chichester, UK.
Geladi, P. and Kowalski, B. R. (1986) Partial least-squares regression: a tutorial. *Analytica Chimica Acta*, **185**, 1–17.
Geladi, P., Isaksson, H., Lindqvist, L. *et al.* (1989) Principal component analysis of multivariate images. *Chemometrics and Intelligent Laboratory Systems*, **5**, 209–220.
Gowen, A. A., O'Donnell, C. P., Taghizadeh, M. *et al.* (2008) Hyperspectral imaging combined with principal component analysis for bruise damage detection on white mushrooms (*Agarius bisporus*). *Journal of Chemometrics*, **22**, 259–267.
Haralick, R. M. (1979) Statistical and structural approaches to texture. *Proceedings of the IEEE*, **67**, 780–803.
Haralick, R. M., Shanmugam, K. and Dinstein, I. (1973) Textural features for image classification. *IEEE Transactions on Systems, MAn, and Cibernetics*, **3**, 610–621.
Höskuldsson, A. (1988) PLS regression methods. *Journal of Chemometrics*, **2**, 211–228.
Hotelling, H. (1931) The generalization of student's ratio. *The Annals of Mathematical Statistics*, **2**, 360–378.
Huang, J., Wium, H., Qvist, K. B. and Esbensen, K. H. (2003) Multi-way methods in image analysis – relationships and applications. *Chemometrics and Intelligent Laboratory Systems*, **66**, 141–158.
Hunter, J. S. (1986) The exponentially weighted moving average. *Journal of Quality Technology*, **18**, 203–210.
Jackson, J. E. (1959) Quality control methods for several related variables. *Technometrics*, **1**, 359–377.
Jackson, J. E. (1991) *A User's Guide to Principal Components*. John Wiley & Sons, Inc., New York.
Jackson, J. E. and Mudholkar, G. S. (1979) Control procedures for residuals associated with principal component analysis. *Technometrics*, **21**, 341–349.
Kresta, J. V., Marlin, T. E. and MacGregor, J. F. (1994) Development of inferential process models using PLS. *Computers and Chemical Engineering*, **18**, 597–611.
Lied, T. T. and Esbensen, K. H. (2001) Principles of MIR, Multivariate Image Regression I: Regression typology and representative application studies. *Chemometrics and Intelligent Laboratory Systems*, **58**, 213–226.
Liu, J. J. and MacGregor, J. F. (2007) On the extraction of spectral and spatial information from images. *Chemometrics and Intelligent Laboratory Systems*, **85**, 119–130.
Liu, J. J., Bharati, M. H., Dunn, K. G. and MacGregor, J. F. (2005) Automatic masking in multivariate image analysis using support vector machines. *Chemometrics and Intelligent Laboratory Systems*, **79**, 42–54.
Lowry, C. A., Woodall, W. H., Champ, C. W. and Rigdon, C. E. (1992) A multivariate exponentially weighted moving average control chart. *Technometrics*, **34**, 46–53.
MacGregor, J. F. and Kourti, T. (1995) Statistical process control of multivariate processes. *Control Engineering Practice*, **3**, 403–414.
Mallat, S. (1999) *A Wavelet Tour of Signal Processing*. Academic Press, San Diego, CA.
Martens, H. and Naes, T. (1989) *Multivariate Calibration*. John Wiley & Sons Ltd, Chichester, UK.
Montgomery, D. C. (2001) *Introduction to Statistical Quality Control*. John Wiley & Sons, Inc., New York.
Nomikos, P. and MacGregor, J. F. (1995) Multivariate SPC charts for monitoring batch processes. *Technometrics*, **37**, 41–59.
Noordam, J. C., Van Den Broek, W. H. A. M., Geladi, P. and Buydens, L. M. C. (2005) A new procedure for the modelling and representation of classes in multivariate images. *Chemometrics and Intelligent Laboratory Systems*, **75**, 115–126.
Page, E. S. (1954) Continuous inspection schemes. *Biometrics*, **41**, 100–115.
Pereira, A. C., Reis, M. S. and Saraiva, P. M. (2009) Quality control of food products using image analysis and multivariate statistical tools. *Industrial and Engineering Chemistry Research*, **48**, 988–998.
Prats-Montalbán, J. M. and Ferrer, A. (2007) Integration of colour and textural information in multivariate image analysis: defect detection and classification issues. *Journal of Chemometrics*, **21**(1) 10–23.
Prats-Montalbán, J. M., De Juan, A. and Ferrer, A. (2011) Multivariate image analysis: a review with applications. *Chemometrics and Intelligent Laboratory Systems*, **107**, 1–23.
Shewhart, W. A. (1931) *Economic Control of Quality of Manufactured Product*. D. Van Nostrand Company, Inc., New York.
Strang, G. and Nguyen, T. (1997) *Wavelets and Filter Banks*. Wellesley-Cambridge Press, Wellesley, MA.
Trong, N. N. D., Tsuta, M., Nicolaï, B. M. *et al.* (2011) Prediction of optimal cooking time for boiled potatoes by hyperspectral imaging. *Journal of Food Engineering*, **105**, 617–624.
Valle, S., Li, W. and Qin, S. J. (1999) Selection of the number of principal components: the variance of the reconstruction error criterion with a comparison to other methods. *Industrial and Engineering Chemistry Research*, **38**, 4389–4401.
Vetterli, M. and Kovačević, J. (1995) *Wavelets and Subband Coding*. Prentice Hall, Upper Saddle River, NJ.

Vogt, F. and Mizaikoff, B. (2003) Dynamic determination of the dimension of pca calibration models using F-statistics. *Journal of Chemometrics*, **17**, 346–357.

Wallays, C., Missotten, B., De Baerdemaeker, J. and Saeys, W. (2009) Hyperspectral waveband selection for on-line measurement of grain cleanness Biosystems Engineering, **107**, 1–7.

Wold, S. (1978) Cross-validatory estimation of the number of components in factor and principal components models. *Technometrics*, **20**, 397–405.

Wold, S., Sjöström, M. and Eriksson, L. (2001) PLS-Regression: a basic tool of chemometrics. *Chemometrics and Intelligent Laboratory Systems*, **58**, 109–130.

Yu, H. and MacGregor, J. F. (2003) Multivariate image analysis and regression for prediction of coating content and distribution in the production of snack foods. *Chemometrics and Intelligent Laboratory Systems*, **67**, 125–144.

Yu, H. and MacGregor, J. F. (2004) Monitoring flames in an industrial boiler using multivariate image analysis. *AIChE Journal*, **50**, 1474–1483.

Yu, H., MacGregor, J. F., Haarsma, G. and Bourg, W. (2003) Digital imaging for online monitoring and control of industrial snack food processes. *Industrial and Engineering Chemistry Research*, **42**, 3036–3044.

# 13 Case Study: Quality control of *Camellia sinensis* and *Ilex paraguariensis* teas marketed in Brazil based on total phenolics, flavonoids and free-radical scavenging activity using chemometrics

**Débora Cristiane Bassani[1], Domingos Sávio Nunes[2] and Daniel Granato[3]**

[1] *Department of Biomedicine, Centro Educacional das Faculdades Metropolitanas Unidas, São Paulo, SP, Brazil*

[2] *Department of Chemistry, Universidade Estadual de Ponta Grossa, Ponta Grossa, PR, Brazil*

[3] *Food Science and Technology Graduate Programme, State University of Ponta Grossa, Ponta Grossa, Brazil*

## ABSTRACT

This study aimed at monitoring the quality of Brazilian ready-to-drink teas and teas marketed in bags from *Camellia sinensis* and *Ilex paraguariensis* based on chemical composition and antioxidant activity by using chemometrics. Tea samples could be clustered into three distinct groups: Cluster 1, composed of 50% green tea samples, presented the highest content of antioxidant polyphenols, while Cluster 3, composed of 62.5% yerba mate teas, presented intermediate content of flavonoids, total phenolics and antioxidant capacity, and Cluster 2, which contained a mixture of black, red and white teas, presented the lowest antioxidant capacity and contents of phenolics and flavonoids. The total content of phenolic compounds, flavonoids, and antioxidant activity of teas marketed in bags were up to 1.5, 3.85, and 3.21 times higher than ready-to-drink samples. Data indicated that the use of chemometrics, especially cluster analysis, is suitable to monitor the quality of teas based on the content of bioactive compounds.

## INTRODUCTION

Hot water infused brews of *Camellia sinensis* (yellow, red, white, green and black teas) and *Ilex paraguariensis* (yerba mate) leaf infusions are the types of teas most consumed by Brazilians. Based on ethnomedical claims, clinical, epidemiological, *in vitro* and other experimental evidence, aqueous extracts (teas) from these plants can be regarded as functional foods (Gomes *et al.*, 1995; Wheeler and Wheeler, 2004; Chen *et al.*, 2009; Mao *et al.*, 2010; Abeywickrama *et al.*, 2011). Some beneficial effects of regular consumption of yerba mate and *Camelia sinensis* on humans are: weight reduction (Auvichayapat *et al.*, 2008), increased total serum antioxidant potential (Carmargo *et al.*, 2006), reduced inflammation biomarkers (Basu *et al.*, 2011), reduced oxidative stress biomarkers (Boaventura *et al.*, 2012; Leite *et al.*, 2012), increased expression of endogenous antioxidant enzymes and decreased lipid peroxidation products (Matsumoto *et al.*, 2009), decreased frequency of cancer development (Beltz

*Mathematical and Statistical Methods in Food Science and Technology*, First Edition.
Edited by Daniel Granato and Gastón Ares.

*et al.*, 2006), reduced risk of cardiovascular diseases (Davies *et al.*, 2003) and improvement of hypertriacylglycerolemia (Leite *et al.*, 2012), among others.

The antioxidant activity towards chemically reactive species, such as those that come from oxygen, nitrogen and carbon, has been linked with some of the above-mentioned health effects of teas (Valko *et al.*, 2007). These reactive species take an electron from neighbouring molecules/atoms to become stable; however, this process generates other free radicals. This chain reaction has been recognized as contributing to increasing the risk of pathogenesis of several human diseases, such as atherosclerosis and other cardiovascular diseases, diabetes mellitus, chronic inflammation, neurodegenerative diseases and some types of tumours (Zhang *et al.*, 2006). In this sense, the phytochemicals present in teas, especially flavonoids, may represent suitable alternatives to provide an exogenous protection to the body.

Teas, such as green/black/red/yellow and yerba mate, contain a considerable amount of catechins, which are flavonoids that present considerable antioxidant-related effects. There are several catechins present in *Camellia sinensis* and *Ilex paraguariensis* leaves; being the most relevant (−)–catechin, (+)–epicatechin, (−)–epigallocatechin, (−)–epigallocatechin gallate, (−)–epicatechin, and (−)–epicatechin gallate (Song *et al.*, 2012). These phytochemicals are the main components responsible for the *in vitro* and *in vivo* biological activities of teas.

The quality control of foods by using multivariate statistical techniques (chemometrics) is well established and increased in many fields of food science and technology once these tools are able to extract the maximum amount of information from chemical data, including chemical composition and antioxidant activity of beverages (Granato *et al.*, 2010, 2011). In relation to commercial teas, it is important to study their chemical composition and bioactivity. Considering that leaf infusions from yerba mate and *Camellia sinensis* leaves are able to increase the antioxidant capacity in a human's plasma and multiple organs, the evaluation of such products is important to provide information about their possible health benefits. Additionally, to the best of our knowledge, this is the first report in the literature regarding the assessment of the phenolic composition and antioxidant capacity of Brazilian commercial teas from *Camellia sinensis* and *Ilex paraguariensis*. Therefore, this study aimed at monitoring the quality of Brazilian ready-to-drink teas and teas marketed in bags from *Camellia sinensis* and *Ilex paraguariensis* based on total phenolics, flavonoids and antioxidant activity by using unsupervised statistical techniques.

## MATERIAL AND METHODS

### Reagents

Folin-Ciocalteau and 1,1-diphenyl-2-picrylhydrazyl (DPPH) were purchased from Sigma-Aldrich (St Louis, MO, USA) and the other reagents used in the experiments were of analytical grade.

### Tea samples and extraction procedure

A total of 25 commercial samples (*Camellia sinensis* and roasted *Ilex paraguariensis*) were purchased in São Paulo, Brazil. Eight corresponded to ready-to-drink products and the remaining samples ($n = 17$) were marketed in tea bags. It is noteworthy that the ready-to-drink samples evaluated in this study correspond to brands marketed in the State of São Paulo, Brazil. All these samples have a detailed quality control report about their botanical authenticity (performed by each industry).

To standardize the extraction procedure of samples marketed in bags, a total of 2.0 grams of each dried leaves sample was used in the experiment. Initially, water was heated to 80°C and 100 ml were added to a flask containing the leaves; the flask was covered with a lid to avoid evaporation. The extraction procedure was carried out under magnetic stirring for a period of 7.5 minutes. Then, the

**Table 13.1** Total flavonoids, phenolics and antioxidant activity of Brazilian teas.

| Coded samples | Type of tea | Type | Total flavonoids[a] | Total phenolic compounds[b] | Antioxidant capacity[c] |
|---|---|---|---|---|---|
| 1 | Green tea | Ready-to-drink | 190.86 | 430.97 | 51.10 |
| 2 | Green tea with ginger + orange | Ready-to-drink | 258.77 | 423.74 | 58.42 |
| 3 | Black tea | Ready-to-drink | 46.46 | 464.65 | 4.66 |
| 4 | Black tea | Ready-to-drink | 43.60 | 139.47 | 7.43 |
| 5 | White tea with lychee | Ready-to-drink | 94.36 | 402.81 | 27.04 |
| 6 | Red tea with red fruits | Ready-to-drink | 124.38 | 416.32 | 2.78 |
| 7 | Red tea with mulberry | Ready-to-drink | 167.98 | 417.84 | 13.85 |
| 8 | White tea with balm | Ready-to-drink | 56.47 | 429.83 | 20.26 |
| 9 | Yellow tea | Sachet | 528.26 | 690.95 | 74.43 |
| 10 | White tea | Sachet | 507.53 | 556.36 | 82.56 |
| 11 | Red tea | Sachet | 285.93 | 355.05 | 22.75 |
| 12 | Green tea | Sachet | 656.92 | 840.36 | 73.53 |
| 13 | White tea | Sachet | 443.91 | 556.39 | 93.98 |
| 14 | Green tea | Sachet | 667.65 | 775.68 | 94.48 |
| 15 | Green tea | Sachet | 600.45 | 702.48 | 93.90 |
| 16 | Yerba-mate tea | Sachet | 913.55 | 966.57 | 92.06 |
| 17 | Green tea | Sachet | 686.95 | 796.35 | 93.90 |
| 18 | Black tea | Sachet | 437.47 | 592.41 | 92.98 |
| 19 | Yellow tea with orange | Sachet | 208.01 | 441.84 | 72.16 |
| 20 | Green tea | Sachet | 564.00 | 603.28 | 94.15 |
| 21 | Yerba-mate tea | Sachet | 280.93 | 394.44 | 59.36 |
| 22 | Yerba-mate tea | Sachet | 313.81 | 420.13 | 53.61 |
| 23 | Yerba-mate tea | Sachet | 317.38 | 417.84 | 55.11 |
| 24 | Yerba-mate tea | Sachet | 321.54 | 419.36 | 54.31 |
| 25 | Yerba-mate tea | Sachet | 315.24 | 417.84 | 62.15 |
| Pooled Standard Deviation | | | 232.13 | 184.74 | 31.85 |

Note: Data presented as mean (n = 3) ± PSD.
[a]Expressed as mg CTE/l.
[b]Expressed as mg GAE/l.
[c]Expressed as % of inhibition of the DPPH radical.

mixture was filtered and the tea transferred to Falcon tubes and immediately frozen at −20°C until further analysis. The ready-to-drink samples (1 litre) were homogenized and transferred to Falcon tubes and then frozen. All the samples of teas are described in Table 13.1.

## Determination of total phenolic content

The total phenolic content (TPC) of tea samples was determined in triplicate according to the Folin–Ciocalteu spectrophotometric method (Singleton and Rossi, 1965). Briefly, 250 μl of diluted tea (1:3) was mixed with 250 μl of twofold diluted Folin–Ciocalteu's phenol reagent and 2 ml of distilled water. The mixture was allowed to react for 5 minutes and then 250 μl of a 10% sodium carbonate solution was added. After 45 minutes of reaction at room temperature, the absorbance at 725 nm was read using a spectrophotometer (Model 432, Femto Ltda, São Paulo, Brazil). The measurement was compared to a calibration curve of a gallic acid (GA) solution (total phenolic concentration = 126.85 × absorbance; $r = 0.9869$; $p < 0.001$) and results were expressed as milligrams of gallic acid equivalents (GAE) per litre of tea (mg GAE/l).

## Determination of total flavonoid content

The total flavonoid content (TFC) was estimated in triplicate by an aluminium chloride colorimetric assay (Jia *et al.*, 1999). Briefly, 250 μl of diluted tea sample (1:3) were mixed with 2720 μl of a 30% ethanolic solution and 120 μl of sodium nitrite. This solution was mixed well and allowed to react for 5 minutes, then 120 μl of a 10% aluminium chloride solution were added to the test tubes and the solution was allowed to react for a further 5 minutes. 800 μl of a 1 mol/l s solution were added to the tubes and the absorbance was read against a reagent blank (ethanol) at a wavelength of 510 nm using a spectrophotometer (Model 432, Femto Ltda, São Paulo, Brazil). The measurement was compared to a calibration curve of a catechin (CT) solution (flavonoid concentration = 476.55 × absorbance; $r = 0.9996$; $p < 0.001$) and the results were expressed as milligrams of catechin equivalents (CTE) per litre of tea (mg CTE/l).

## Free-radical scavenging assay (DPPH)

The DPPH radical scavenging activity of tea was determined in triplicate according to the method proposed by Brand-Williams *et al.* (1995), with minor changes. This method determines the hydrogen donating capacity of antioxidants such as phenolic compounds and does not produce oxidative chain reactions or react with free radical intermediates. Every diluted (1:7) tea sample (100 μl) was added to 3.9 ml of a 125 μmol/l methanolic DPPH solution. The absorbance at 517 nm was measured after the solution had been allowed to stand in the dark for 30 minutes. Methanol was used as a negative control (blank). The DPPH free-radical scavenging activity of each tea sample was calculated using Equation 13.1:

$$\text{Free-radical scavenging activity } (\%) = [1 - (A_{517}\,\text{sample}/A_{517}\,\text{blank})] \times 100 \qquad (13.1)$$

## Statistical evaluation

Results were expressed as mean ± pooled standard deviation. Initially, all results were checked for normality by using the Kolmogorov–Smirnov test. In this study, a data matrix containing 25 rows (samples) and three columns (variables) was built, and pattern recognition methods were applied to the experimental data; these were principal component analysis (PCA) and hierarchical cluster analysis (HCA) as unsupervised statistical methods.

PCA is a procedure that allows identifying relationships between objects and variables and also the overall correlation of the variables (Alezandro *et al.*, 2011). In this study, PCA was used to classify the tea samples ($n = 25$) according to their chemical composition and bioactivity. The data matrix was prepared by including the total phenolic and flavonoid contents as well as the antioxidant activity as columns and tea samples as rows. Analyses were based on correlations and variances were computed as $SS/(n - 1)$. Data were auto-scaled prior to analysis (Cruz *et al.*, 2011).

In this study, tree clustering was used to group the samples with a higher degree of similarity and also to assess the association between total phenolic compounds, total flavonoids and the *in vitro* antioxidant activity. HCA was applied to the auto-scaled data using Ward's method and the Euclidean distances generated a dendrogram for the samples (Granato *et al.*, 2010). To compare the results among the formed clusters, Levene's test was carried out to check for homogeneity of variances, while one-way analysis of variance (ANOVA) and Fisher's LSD *post hoc* tests were then applied to identify noted differences among clusters. For the total phenolic compounds, which presented nonhomogenous variances ($p < 0.05$), the Welch test was used as nonparametric analysis of variance, followed by the multiple comparison Kruscall–Wallis test. Statistical differences ($p < 0.05$) between ready-to-drink products and samples marketed in sachets were assessed by the Student-$t$ test after assessing the homogeneity of variances by the $F$-test.

Correlation coefficients (r) were calculated and expressed by the Pearson product. Linear regression analysis was used to evaluate how much variability could be explained by each independent variable (total phenolic and flavonoid compounds) for the dependent variable (antioxidant activity). For all analyses, p-values below 0.05 were regarded as significant. The statistical analyses were performed using the software Statistica v. 11 (Statsoft, USA) and Action v. 2.4 (Estatcamp, Brazil).

## RESULTS AND DISCUSSION

The results of the chemical and antioxidant assays are presented in Table 13.1. It is possible to observe that the phenolic compound content ranged from 139.47 to 966.57 mg GAE/l and the flavonoid content ranged from 43.60 to 913.55 mg CTE/l, while the percentage of inhibition of the DPPH radical ranged from 2.78 to 94.48, indicating that there is a remarkable difference among tea samples marketed in Brazil. These data are in accordance with other studies published elsewhere where *Camellia sinensis* (red, black, yellow, white and green teas) and *Ilex paraguariensis* (yerba mate) were evaluated (Bravo *et al.*, 2007; Alarcón *et al.*, 2008; Seeram *et al.*, 2008; Fukushima *et al.*, 2009). These differences in phenolic composition and antioxidant capacity among teas can be explained by several factors, such as tea variety, brewing techniques, blending, chopping grade, addition of other ingredients such as ascorbic acid and fruit juices, and specific unit operation processes. All these factors affect the composition and, consequently, the *in vitro* antioxidant capacity of teas.

To the best of the authors' knowledge, this research is the first report of application of chemometric statistical techniques (principal component analysis and hierarchical cluster analysis) to identify the simultaneous association between the main phenolic compounds that, in a more pronounced way, exerted the *in vitro* antioxidant activity measured by the DPPH assay in different types of *Camellia sinensis* and yerba mate teas marketed in Brazil. By using a scatter plot built with PCA (Figure 13.1), it was possible to observe differences among samples in terms of phenolic composition and antioxidant activity, and this projection was able to explain up to 98% of variability in data. Samples located in the second quartile (upper left) presented a higher antioxidant capacity as well as higher contents of phenolics and flavonoids, while samples located in the right side of the scatter plot presented the lowest means for all response variables. By using PCA it was not possible to clearly observe groups of samples with similar characteristics; therefore, PCA was only used to visualize all teas and response variables simultaneously.

In order to overcome this limitation, HCA was applied to group samples that presented the highest degree of similarity within the same group, and the highest dissimilarity in comparison to the samples contained in other clusters (Granato *et al.*, 2012). In this regard, HCA applied to samples (Figure 13.2) was used to explain more suitably the experimental data and allowed checking for association between tea variety and chemical/antioxidant properties. Cluster 1 ($n = 10$) contained a total of five green, two white, one black, one yerba mate and one yellow tea, while Cluster 2 ($n = 7$) contained a total of two black, three red and two white teas, and Cluster 3 ($n = 8$) presented five yerba mate, two green and one

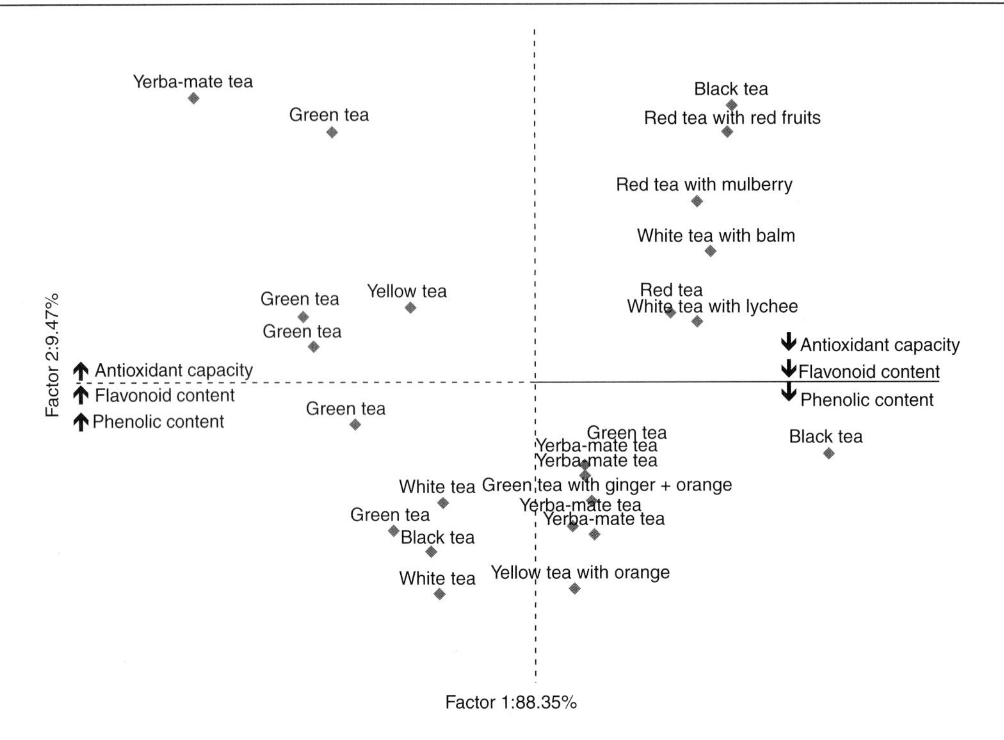


**Figure 13.1** A scatter plot of Principal Component 1 versus Principal Component 2 of the main sources of variability between the commercial tea samples.

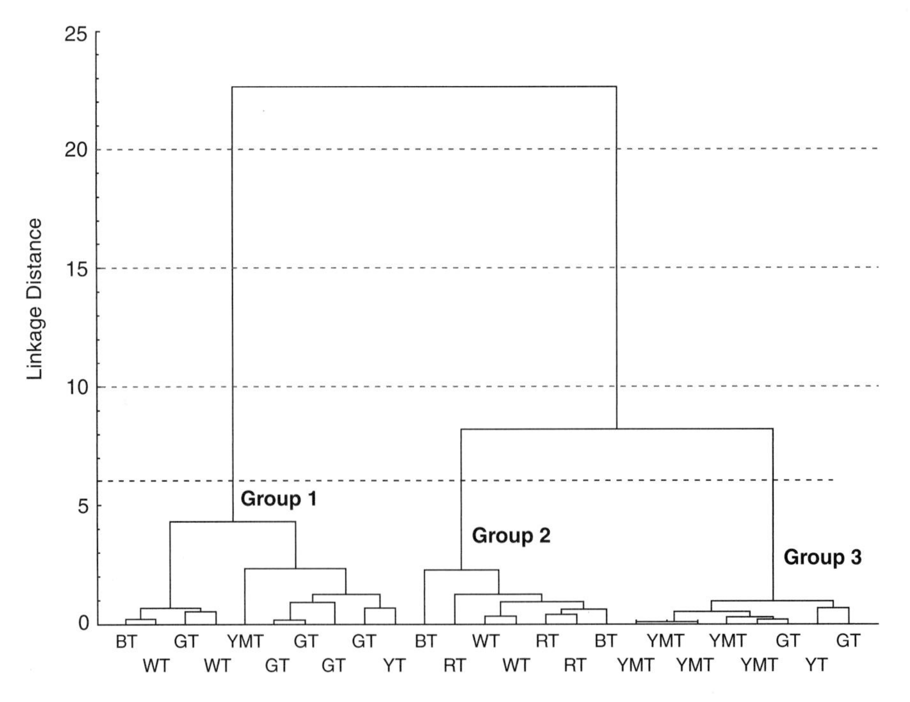


**Figure 13.2** HCA applied for the samples and the clusters. BT= black tea, GT= green tea, WT= white tea, RT= red tea, YT= yellow tea, YMT= yerba mate tea.

**Table 13.2** Statistical comparison among clusters in relation to the total contents of flavonoids and phenolic compounds and antioxidant activity towards the DPPH radical.

| Response variables | Group 1 (n = 10) | Group 2 (n = 7) | Group 3 (n = 8) | PSD[a] | p-value[b] | p-value[c] |
|---|---|---|---|---|---|---|
| Total flavonoids | 600.67[a] | 117.03[c] | 275.82[b] | 232.13 | 0.14 | <0.001 |
| Total phenolic compounds | 708.08[a] | 375.14[b] | 420.77[b] | 184.74 | 0.01 | <0.001 |
| DPPH | 88.59[a] | 14.11[c] | 58.04[b] | 31.85 | 0.32 | <0.001 |

[a]PSD = pooled standard deviation.
[b]Probability value obtained by Levene's test for homogeneity of variances.
[c]Probability value obtained by one-factor analysis of variances or by the Welch test. Different letters in the same line represent statistically different ($p < 0.05$) results.

yellow tea. The statistical comparison among clusters is presented in Table 13.2. Cluster 1, which was composed of 50% green tea samples, presented the highest ($p < 0.001$) means for all response variables, while cluster 3, which was composed of 62.5% yerba mate teas, presented intermediate mean values for the responses, and Cluster 2, which contained a mixture of black, red and white teas, presented the lowest mean values of antioxidant capacity as well as of phenolic and flavonoid compounds.

The comparison between ready-to-drink teas and samples marketed in tea bags is presented in Figure 13.3. It is noteworthy that the total contents of phenolic compounds and flavonoids and the antioxidant activity of teas marketed in bags were up to 1.5, 3.85, and 3.21 times higher ($p < 0.05$) than ready-to-drink samples, demonstrating that it is preferable to prepare teas marketed in sachets rather than consuming commercial ready-to-drink teas if a higher intake of antioxidant phytochemicals is being sought. Indeed, in this study a high and significant correlation between DPPH and the flavonoid content (Figure 13.4) was verified, corroborating the results obtained by Deetae *et al.* (2012).

The use of unsupervised statistical methods has increased recently. For example Deetae *et al.* (2012) evaluated the relationship between total phenolic compounds and total nonflavonoid compounds and the antioxidant capacity measured by the ABTS (2,2′-azino-bis(3-ethylbenzothiazoline-6-sulfonic acid) and FRAP (Ferric Reducing Antioxidant Power) assays of different types of teas, including green, black and oolong teas, using multivariate statistical methods (PCA and HCA). The authors verified that the model proposed by using PCA was able to explain up to 96% of the variance in results, while HCA was able to characterize stevia tea in the same cluster of green/black tea, while sappan tea was grouped with oolong tea, corroborating that chemometrics is a suitable and effective tool to assess the relationship between phenolic composition and the antioxidant capacity of teas. Likewise, Nakamura *et al.* (2009) used HCA to evaluate the content of triterpenes, saponins, methylxanthines and chlorogenic acid, as well as the antioxidant activity (DPPH) of yerba mate teas. The authors verified that cluster analysis separated samples with similar characteristics while the levels of caffeine ($r = 0.23$; $n = 8$; $p > 0.05$) and theobromine ($r = -0.01$; $n = 8$; $p > 0.05$) were not significantly correlated with DPPH.

Overall, the experimental data obtained in our study support the hypothesis that the antioxidant properties of green tea are higher than those of red, black, yellow, yerba mate and white teas. More commercial samples of *Camellia sinensis* and yerba mate teas from different countries should be evaluated in order to corroborate these results and to create a more robust statistical method to classify samples in relation to the content of phenolic compounds and antioxidant capacity.

Although our study has shown that white, red, black and yellow teas marketed in Brazil present a less pronounced antioxidant activity and lower contents of total phenolics and flavonoids in comparison with green or yerba mate teas, many studies worldwide have shown that all the aqueous extracts (teas) from *Camellia sinensis* and also *Ilex paraguariensis* show health-promoting effects in humans, such as increased fat oxidation at rest, prevent obesity and improve insulin sensitivity (Venables *et al.*, 2008), antioxidant and potential anti-inflammatory activity on primary human dermal fibroblast cells (Thring *et al.*, 2011), weight and plasma lipid control (Richard *et al.*, 2009), protection against oxidative stress

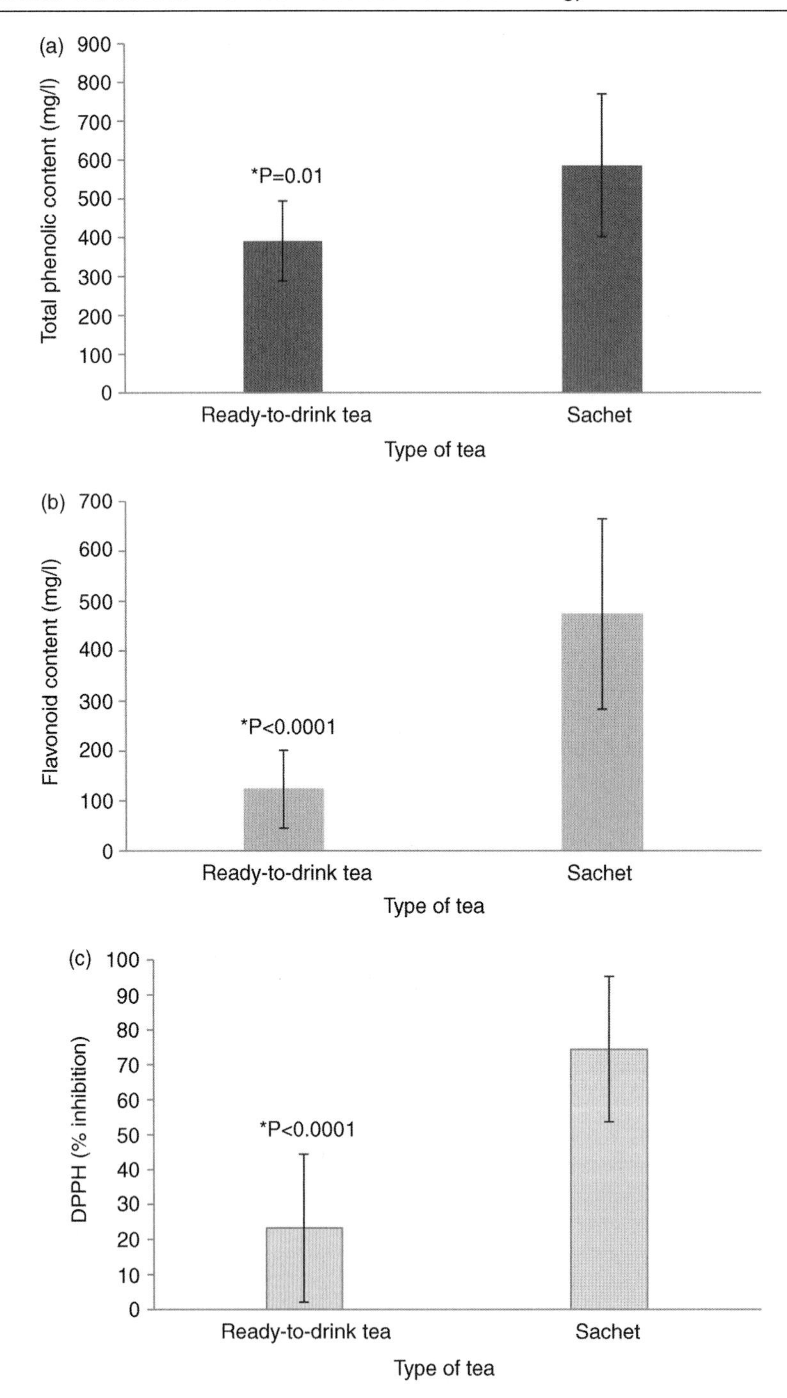


**Figure 13.3** Comparison between ready-to-drink teas (n=8) and teas marketed in bags (n=17) based on phenolic compounds (a), flavonoids (b), and antioxidant capacity (c). Probability values obtained by the Student-*t* test for independent samples are shown.

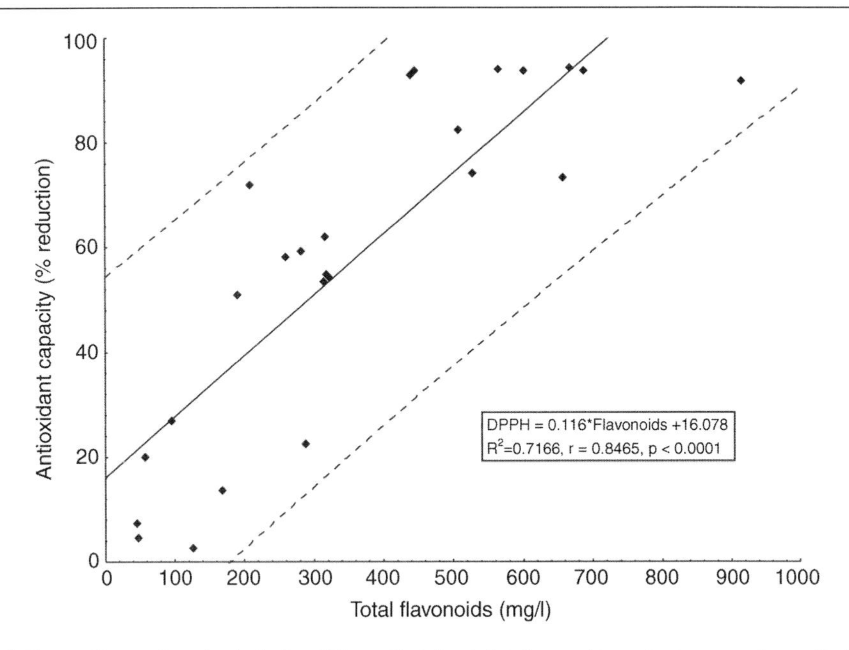


**Figure 13.4** Linear Regression Analysis (n=25) applied for total flavonoid content versus the antioxidant activity measured by the DPPH method with 95% of confidence interval.

and DNA damage (Kumar *et al.*, 2012), and so many others. All these studies corroborate the fact that yerba mate as well as white, green, red, black and yellow teas have functionality and can be consumed daily in order to improve health.

## CONCLUSIONS

A multivariate approach was used to establish associations among the phenolic composition and the *in vitro* antioxidant capacity of Brazilian commercial teas from *Camellia sinensis* and *Ilex paraguariensis* species either marketed in bags or as teas ready for consumption. Hierarchical cluster analysis showed that three distinct groups were formed: one group composed predominantly by green tea that presented the highest antioxidant capacity and the higher contents of phenolic compounds and flavonoids, while yerba mate teas presented intermediate values of antioxidant capacity and phenolic compounds. On the other hand, red, black and white teas presented the lowest content of antioxidant phenolic compounds. More samples should be evaluated in order to confirm the results obtained in this work. However, it is important to note that the use of chemometrics, especially cluster analysis, is a suitable tool to monitor the quality of different types of teas based on phenolic composition and *in vitro* antioxidant activity.

## REFERENCES


Abeywickrama K.R.W., Ratnasooriya W.D. and Amarakoon A.M.T. (2011) Oral hypoglycaemic, antihyperglycaemic and antidiabetic activities of Sri Lankan Broken Orange Pekoe Fannings (BOPF) grade black tea (*Camellia sinensis* L.) in rats. *J Ethnopharmacol.* **135**, 278–286.

Alarcón E., Campos A.M., Edwards A.M. *et al.* (2008) Antioxidant capacity of herbal infusions and tea extracts: A comparison of ORAC-fluorescein and ORAC-pyrogallol red methodologies. *Food Chem* **107**, 1114–1119.

Alezandro, M.R., Granato, D., Lajolo, F.M., and Genovese, M.I. (2011) Nutritional aspects of second generation soy foods. *J Agri Food Chem*, **59**, 5490–5497.

Auvichayapat, P., Prapochanung, M., Tunkamnerdthai, O. *et al.* (2008) Effectiveness of green tea on weight reduction in obese Thais: A randomized, controlled trial. *Physiol Behav* **93**, 486–491.

Basu, A., Du, M., Sanchez, K. *et al.* (2011) Green tea minimally affects biomarkers of inflammation in obese subjects with metabolic syndrome. *Nutr* **27**, 206–213.

Beltz, L.A., Bayer, D.K., Moss, A.L. and Simet, I.M. (2006) Mechanisms of cancer prevention by green and black tea polyphenols. *Anticancer Agents in Med Chem* **6**, 389–406.

Boaventura, B.C.B., Di Pietro, P.F., Stefanuto, A. *et al.* (2012) Association of mate tea (*Ilex paraguariensis*) intake and dietary intervention and effects on oxidative stress biomarkers of dyslipidemic subjects. *Nutr* **28**, 657–664.

Brand-Wiliams, W., Cuvelier, M.E. and Berset, C. (1995) Use of a free radical method to evaluate antioxidant activity. *LWT – Food Sci Technol* **28**, 25–30.

Bravo, L., Goyaa, L. and Lecumberria, E. (2007) LC/MS characterization of phenolic constituents of mate (*Ilex paraguariensis*, St. Hil.) and its antioxidant activity compared to commonly consumed beverages. *Food Res Int* **40**, 393–405.

Camargo, A.E.I., Daguer, D.A.E. and Barbosa, D.S. (2006) Green tea exerts antioxidant action *in vitro* and its consumption increases total serum antioxidant potential in normal and dyslipidemic subjects. *Nutr Res* **26**, 626–631.

Chen, H., Qu, Z., Fu, L. *et al.* (2009) Physicochemical properties and antioxidant capacity of 3 polysaccharides from green tea, oolong tea and black tea. *J Food Sci* **74**, 469–474.

Cruz, A.G., Granato, D., Faria, J.A.F. and Branco, G.F. (2011) Characterization of Brazilian lager and brown ale beers based on color, phenolic compounds, and antioxidant activity using chemometrics. *J Sci Food Agri* **91**, 563–571.

Davies, M.J., Judd, J.T., Baer, D.J. *et al.* (2003) Black tea consumption reduces total and LDL cholesterol in mildly hypercholesterolemic adults. *J Nutr* **133**, 3298–3302.

Deetae, A., Parichanon, P., Trakunleewatthana, P. *et al.* (2012) Antioxidant and anti-glycation properties of Thai herbal teas in comparison with conventional teas. *Food Chem* **133**, 953–959.

Fukushima, Y., Ohie, T., Yonekawa, Y. *et al.* (2009) Coffee and green tea as a large source of antioxidant polyphenols in the Japanese population. *J Agri Food Chem* **57**, 1253–1259.

Gomes, A., Vedasiromoni, J.R., Das, M. *et al.* (1995) Anti-hyperglycemic effect of black tea (*Camellia sinensis*) in rat. *J Ethnopharmacol* **45**, 223–226.

Granato, D., Katayama, F.C.U. and Castro, I.A. (2010) Assessing the association between phenolic compounds and the antioxidant activity of Brazilian red wines using chemometrics. *LWT – Food Sci Technol* **43**, 1542–1549.

Granato, D., Castro, I.A. and Katayama, F. (2011) Phenolic composition of South-American red wines classified according to their antioxidant activity, retail price, and sensory quality. *Food Chem* **129**, 366–373.

Granato, D., Katayama, F.C.U. and Castro, I.A. (2012) Characterization of red wines from South America based on sensory properties and antioxidant activity. *J Sci Food Agri* **92**, 526–533.

Jia, Z., Tang, M. and Wu, J. (1999) The determination of flavonoid contents in mulberry and their scavenging effects on superoxide radicals. *Food Chem* **64**, 555–559.

Kumar, M., Sharma, V.L., Sehgal, A. and Jain, M. (2012) Protective effects of green and white tea against benzo(a)pyrene induced oxidative stress and DNA damage in murine model. *Nutr Cancer* **64**, 300–306.

Leite, J.I.A., Teixeira, L.G., Lages, P.C. *et al.* (2012) White tea (*Camellia sinensis*) extract reduces oxidative stress and triacylglycerols in obese mice. *Ciên Tecnol Alim* **32**(4), 733–741. doi: 10.1590/S0101-20612012005000099.

Mao, J.T., Nie, W.X., Tsu, I.H. *et al.* (2010) White tea extract induces apoptosis in non-small cell lung cancer cells the role of perixisome proliferator-activated receptor-{gamma} and 15- lipoxygenases. *Cancer Prev Res* **3**, 1132–1140.

Matsumoto, R.L.T., Bastos, D.H.M., Mendonça, S. *et al.* (2009) Effects of mate tea (*Ilex paraguariensis*) ingestion on mRNA expression of antioxidant enzymes, lipid peroxidation, and total antioxidant status in healthy young women. *J Agri Food Chem* **57**, 1775–1780.

Nakamura, K.L., Cardozo Junior, E.L. Donaduzzi, C.M. and Schuster, I. (2009) Genetic variation of phytochemical compounds in progenies of *Ilex paraguariensis* St. Hil. *Crop Breeding Applied Biotech* **9**, 116–123.

Richard, D., Kefi, K., Barbe, U. *et al.* (2009) Weight and plasma lipid control by decaffeinated green tea. *Pharmaceutical Res* **59**, 351–354.

Seeram, N.P., Aviram, M., Zhang, Y. *et al.* (2008) Comparison of antioxidant potency of commonly consumed polyphenol-rich beverages in the United States. *J Agri Food Chem* **56**, 1415–1422.

Singleton, V.L. and Rossi, J.A. Jr. (1965) Colorimetry of total phenolics with phosphomolybdic-phosphotungstic acid reagents. *Am J Enol Viticult* **16**, 144–158.

Song, R., Kelman, D., Johns, K.L. and Wright, A.D. (2012) Correlation between leaf age, shade levels, and characteristic beneficial natural constituents of tea (*Camellia sinensis*) grown in Hawaii. *Food Chem* **133**, 707–714.

Thring, T.S., Hili, P., and Naughton, D.P. (2011) Antioxidant and potential anti-inflammatory activity of extracts and formulations of white tea, rose, and witch hazel on primary human dermal fibroblast cells. *J Inflam* **8**, 1–7.

Valko, M., Leibfritz, D., Moncol, J. *et al.* (2007) Free radicals and antioxidants in normal physiological functions and human disease. *Int J Biochem Cell Biol* **39**, 44–84.

Venables, M.C., Hulston, C., Cox, H. and Jeukendrup, A.E. (2008) Green tea extract ingestion, fat oxidation, and glucose tolerance in healthy humans. *Am J Clin Nutr* **87**, 778–84.

Zhang, J.L., Stanley, R.A., Adaim, A. *et al.* (2006) Free radical scavenging and cytoprotective activities of phenolic antioxidants. *Mol Nutr Food Res* **50**, 996–1005.

Wheeler, D. and Wheeler, W. (2004) The medicinal chemistry of tea. *Drug Dev Res* **61**, 45–65.

# Section 3

# 14 Statistical approaches to develop and validate microbiological analytical methods

**Anthony D. Hitchins**
*Rockville, MD, USA [Center for Food Safety and Applied Nutrition, United States Food and Drug Administration (retired)]*

## ABSTRACT

Recent developments in the statistical validation of qualitative analyses for the detection of pathogenic and nonpathogenic microorganisms in foods are reviewed in the context of similarities and differences between chemical and microbial analyses. Two calculations for determining the limit of detection (LOD) of a method in single and multilaboratory studies are considered. Two calculations for determining the reproducibility of methods among laboratories are described. Although not qualitative methods, colony count and most probable number enumerations are described because they form the crucial foundation upon which the qualitative method calculations depend. Any differences between results of alternative calculations are liable to be insignificant relative to enumeration uncertainty. Application of the new methods of calculating the LOD and reproducibility parameters to method validation is considered but acceptable performance criteria are not. These can vary among method validation bodies and are liable to flux due to current harmonization efforts.

## INTRODUCTION

In considering recent statistical approaches to developing and validating microbiological analytical methods it is helpful to begin by comparing and contrasting microbiological and chemical analytical methods. Conventionally, it has been convenient to consider chemical and microbiological analyses separately. However, with the advent of replicable macromolecule analysis, chemical analysis now overlaps with microbiological analysis, which solely involves replicable analytes. Thus, the conventional distinction is becoming blurred. Analysis of replicable entities involves, indeed requires, measuring very low numbers of replicable macromolecules or microorganisms. While most chemical analytes are nonreplicable and measured in mass concentration units, the masses do represent numbers, albeit tremendously large numbers, of chemical particles, so the conventional distinction was in any case somewhat artificial despite its convenience.

A pure chemical analyte is easily quantitated by measuring its mass but in practice, of course, there is always some degree of doubt about the purity of the analytical sample. Hence, it is necessary to measure out a quantity of the sample containing the analyte, the analytical portion, and determine indirectly the amount of the portion that is actually analyte. To do this requires the use of some specific chemical, physical or biological attribute of the analyte that can be quantitatively related to its mass in a standard curve relationship. The qualitative presence of an analyte in an analytical portion can be similarly

*Mathematical and Statistical Methods in Food Science and Technology*, First Edition.
Edited by Daniel Granato and Gastón Ares.

ascertained but it is not necessary to quantify the attribute detected by means of a standard assay curve. Most commonly, of course, it is the analyte in the presence of substantial amounts of matrix impurity that is being measured or detected.

A microbial analyte could in theory be analysed quantitatively or qualitatively just like a chemical analyte, that is in mass concentration units. In practice it is enumerated for two main reasons. Microbial cells in a pure culture differ in size and composition. Depending on culture age and conditions a proportion of a pure population may be nonviable. The cells, although functionally organized, are somewhat variable mixtures of cell components. Furthermore, separation of entities of interest in foods, such as microbial pathogens, microflora and food constituents, is not easy. It involves dilution to low numbers and, as a result, necessitates enumeration, since the masses involved would be below thresholds of chemical detection.

In chemical analyses the analyte is a particle such as an atom, molecule or macromolecule. In microbiological analyses the analyte particle may be a cell (microbe or microorganism) or a virion (virus particle); these are, essentially, more or less complex mixtures of atoms and molecules. Some kinds of macromolecules, that is the nucleic acids (DNA and RNA) and most microbiological particles, are replicable given the right conditions. This means that it is possible, in principle, to detect and enumerate single microbiological particles because they can be encouraged to replicate on a solid culture medium to yield visible enumerable entities, such as discrete populations of descendent cells (colonies), lysed populations of virus host cells (viral plaques in a lawn of host cells) or populations composed of replicates of a replicable macromolecule. The latter are detectable by the polymerase chain reaction (PCR) and enumerable by real-time PCR.

With chemical and microbial analyte particles, quantitation or enumeration over a very broad range of concentrations is possible given appropriate quantitative dilutions of the analytical portion. Theoretically, with microbial particles the concentration range can be from one per millilitre to $10^9$ or more per millilitre. With chemical particles, the upper limit is Avogadro's number/ml corrected for the molecular size of the particle. In contrast to microbial particles the lower limit, while at least $10^8$-fold lower than the upper limit, is still generally a very large number unless the particles are replicable. Very large numbers of nonreplicable particles are necessary to achieve the threshold of detection. In comparison, while large numbers of replicable particles are also needed to achieve the detection threshold, suitable spatial separation of individual particles permits an individual particle to achieve a detectable population threshold by replication. However, an important consideration is how the threshold level response is detected. Whether the detector is an instrument or the analysts' eye the numbers for detection will be still considerable.

## Variability

Analytical values inevitably vary about the true value, hence the need for replicate determination within laboratories and comparison of analyses between laboratories. It is well known that between laboratories variation is larger than intra-laboratory variation. Variability is the reason that validation is necessary. It cannot be overemphasized that even when a method has been properly validated a laboratory should nevertheless make sure it can run the method getting results within the parameters established by the validation.

Variability has several sources. A primary source of variability is due to sample inhomogeneity and a crucial part of sample preparation involves ensuring maximal homogeneity. Homogeneity is never perfect, being only an approachable ideal state. Typically, sufficient homogeneity is achieved in practice. Secondary sources of variation include pipetting and weighing variability and analytical instrument variability, including that due to instrument background noise. Secondary variability usually outweighs primary variability given effective homogenization when the threshold of detection involves large numbers of particles, as with nonreplicable analyte in a chemical analysis. In contrast, primary variability due to inhomogeneity looms larger when the threshold of detection involves small numbers of particles, as with replicable analyte in microbiological analyses and some chemical analyses where the limit of detection may approach one particle per test portion.

### Validation

Validation of an analytical method within or between laboratories involves determining the variability of the method under controlled conditions, particularly shared use of common characterized samples. Naturally, a method that has low variability is preferred to one with larger variability. However, other factors, such as cost and ease of performance, can come into play, so that not always is the least variable method preferred. As long as the variability is well characterized a more variable method may be chosen, as long as the method is fit for the purpose of its use. Purposes include official regulation, screening analyses and routine production analyses.

Validation protocols differ between the well-known international organizations like AFNOR, AOAC International, ISO and the NORDIC consortium. It is not the intention of this chapter to describe the various protocols but rather to devise a representative one for explanatory purposes. In any case, individual descriptions will tend to become redundant as internationally harmonized protocols are commonly accepted to be desirable.

Protocols for validation studies have to be carefully crafted so that they are clear to the investigators participating in the validation studies. When an unexpected value is obtained it is common for that value to be dropped from the analysis if it can be unequivocally ascribed to interpretative or technical investigator error. When the value is not ascribable to error validation authorities sometimes differ as to whether it should be used in the analysis of the study data. Authorities which permit discarding usually do a statistical test to see whether the value is an outlier relative to the study data in general.

Validation studies may be single laboratory (SLV) or multilaboratory (MLV) validations. MLV studies often depend on the acceptable performance of a method in a preliminary SLV study. In SLV studies it is the repeatability or intra-laboratory variability of the method that is validated. In MLV methods it is the reproducibility or inter-laboratory variability of the method in the hands of different laboratories or different investigators within a single laboratory that is of concern.

In validation of quantitative microbiological methods analytes are replicable. In quantitative chemical methods analytes are generally but not always nonreplicable. In any case, the analyte is diluted to a range where a segment of the original population is measured parametrically or by direct enumeration. The situation with regard to chemical and microbiological qualitative methods is similar. However, with replicable analytes very low numbers, approaching one per test aliquot, can be detected because they can be amplified to detectable threshold levels. This is not possible with chemical analytes that are nonreplicable. With all analytes it is possible to calculate limits of detection (LOD) by dilution toward extinction by measuring proportions of positive responses in a suitable series of dilutions.

## SAMPLE PREPARATION

In chemical and microbiological analyses of a food matrix, representative samples of the food lot are homogenized. Then the homogenate is analysed by one or more laboratories taking portions from it. Homogenization is particularly important in MLV studies, since in order to measure method reproducibility ideally identical portions must be provided to each laboratory. Ideality can only be conscientiously approached, since perfect homogeneity can never be achieved with certainty and, of course, there are always volumetric or weighing errors as well.

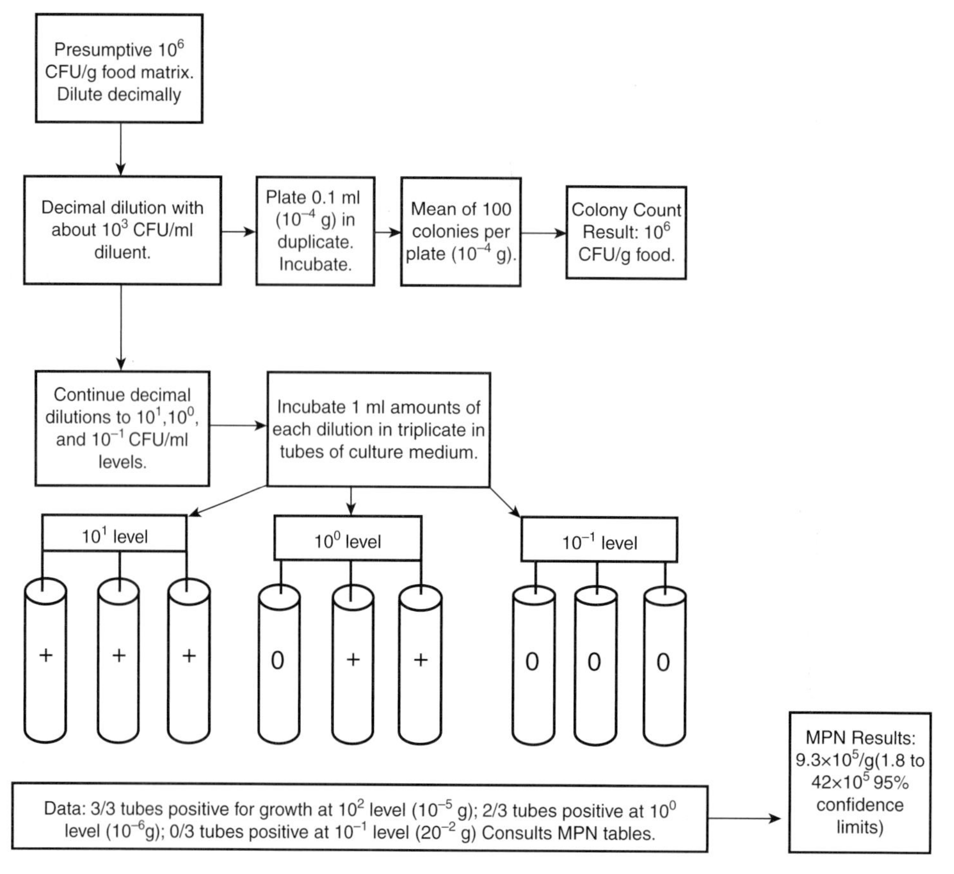


**Figure 14.1** Schemata for colony count and MPN enumerations of food microflora by a single or collaborating laboratories.

The nominally homogenous portion then has to be diluted, usually decimally in microbiology, to bring it into the linear range of the analytic method. In the case of standard microbial enumeration, that is the plate count, this is 25–250 colonies per test portion plated on solid medium. This usually involves a series of decimal dilution steps (Figure 14.1). To measure even lower concentrations dilution is extended to extinction, that is theoretically a minimum of about one cell per test sample and a most probable number is determined (see the section 'Most probable number enumeration' and Figure 14.1).

In qualitative microbiological analysis the test portion of food (typically 25 g/ml) is usually diluted decimally once and cultured in a suitably selective liquid culture medium for the microbe of interest, usually a pathogenic one. The limit of detection is thus, theoretically, one cell per 25 g/ml. How well a method approaches this minimal value is determined as described later in 'Qualitative methods'.

## QUANTITATIVE MICROBIOLOGY, ENUMERATION

Microbial biochemical products and cellular components are analysed according to chemical or physical–chemical principles, that is by parametric measurements, which can be treated by parametric descriptive statistics and tests. In contrast, whole viable microbes are analysed nonparametrically by enumerating the individual cells (as colonies) or viruses (as plaques). Individual cells can be enumerated microscopically but only at concentrations greater than about $10^7$ per millilitre and it is difficult to distinguish viable from nonviable cells and to distinguish a microbe of special interest from other microbes that may be present in a food microflora. Mass measurements can be done turbidometrically with pure cultures but in most cases of interest a particular food microbe, for instance, is not present on its own but is accompanied by other members of the natural food microflora, as well as food matrix particles. Furthermore, the measurement does not distinguish viable from dead cells.

Figure 14.1 (top part) outlines the colony count procedure for enumerating the food microflora, which is a mixed population of different kinds of microbes. Specific subpopulations or microbes would require the use of selective culture media. Having pointed out that microbe enumeration is a nonparametric process it remains that the enumeration data, in logarithmic transform, are analysed parametrically. The numbers actually counted are typically in the range 25–250 colonies representing 25–250 cells per analytical portion plated on microbial culture media (selective for a specific microbe or a related group of microbes). When corrected for dilution, numbers orders of magnitude greater may be involved, further justifying the use of parametric statistical descriptives. Thus, microbial enumeration results are presented as mean concentrations per g or ml $\pm 2$ standard deviations: $s_r$ for replicative variation in a single laboratory and $S_R$ for reproducibility variation in a multilaboratory study (definitions are shown in Table 14.1). Nevertheless, because the numbers actually counted are relatively low the relative standard deviations are large. Table 14.2 shows an example of the experimental design of a hypothetical collaborative enumeration validation study (Steiner 1975; AOAC International, 2012a, 2012b). An actual collaborative study (Crowley *et al.*, 2009) illustrates the variances observed in counting the aerobic mesophilic flora of foods. Importantly replicate colony count data are not normally distributed, so are transformed to $\log_{10}$ counts. The transformed data are statistically analysed by the analysis of

**Table 14.1** Definitions of symbols.

| **Symbol** | **Meaning** |
|---|---|
| n | Number of laboratories providing useable results[a] |
| m | Mean analytical value |
| $s_r$ | Standard deviation of replication; $(s_r)^2$ is the variance |
| $RSD_r$ % | Relative standard deviation of replication |
| $S_R$ | Standard deviation of reproducibility; $(S_R)^2$ is the variance |
| $RSD_R$ % | Relative standard deviation of reproducibility |
| $S_L$ | Standard deviation of laboratory effect; $(S_L)^2$ is the variance (see *Reproducibility* section) |

[a]In quantitative validations special statistical tests are done to detect or confirm outliers (Steiner, 1975). Also, values may be eliminated for noncompliance with the experimental protocol.

**Table 14.2** Typical enumeration validation design: Microbial $\log_{10}$ count per g/ml of a matrix by test and reference method at a single concentration.[a]

| **Laboratory No.**[c] | **Type of method**[b] | | | |
|---|---|---|---|---|
| | ***Reference*** | | ***Test*** | |
| | ***I*** | ***II*** | ***I*** | ***II*** |
| 1 | 1.991 | 1.852 | 1.963 | 1.987 |
| n | 1.792 | 1.965 | 1.823 | 1.781 |

[a]Counts at medium and high concentration should be performed too, e.g. at about $\log_{10}$ 3.0 and 4.0. For each level the m, $s_r$ $RSD_r$, $S_R$, and $RSD_R$ parameters are calculated.
[b]I and II = replicate test portions.
[c]n = 10 or more laboratories is preferred.

variance method (Steiner, 1975). In this study the $RSD_R$ values typically range from 4 to 16% for the reference and test methods compared. Statistically the two methods were comparable. Colony enumeration has a lower limit of about 250 CFU (colony forming units) per g/ml, since the maximum sample matrix that can be plated on a single culture plate is about 0.1 g or ml. This can be improved somewhat by performing replicate plating but economically and pragmatically it is easier to perform a most probable number enumeration.

### Most probable number enumeration

For lower concentrations of microbes, such as food-borne pathogens, most probable number (MPN) enumeration is performed (Figure 14.1, bottom part). However, this incurs larger standard deviation values. In colony enumeration a single CFU self-replicates to form a visible enumerable colony that is discretely located on the surface of a solid culture medium (appropriately dried to remove any moisture condensate and so inhibit motile bacteria spreading, growing and preventing formation of distinct colonies). In MPN, one or more CFU grows to form a culture suspension in a separate container such as a tube or microplate well.

The test sample is decimally diluted to extinction. Then replicates (triplicates or pentuplicates) are cultured in tubes of selective medium at the extinction level and at several levels around the extinction level. The proportions of positive cultures at the various levels are converted to numbers of cells per g/ml using MPN tables (Blodgett, 2003). After 24–48 hours of incubation any resulting growth is streaked on selective solid culture medium. The identities of presumptive colonies of the target microbe are confirmed by suitable further tests.

In a collaborative validation of an MPN method (Crowley *et al.*, 2010) the $RSD_r$ values ranged from 2 to 12% and the $RSD_R$ values from 2 to 25% for the test method with a variety of food matrices in the range 2–4 $\log_{10}$ CFU/g or ml. Corresponding RSD ranges for the reference MPN method were 5–25% and 17–39%. It was concluded that the test method performed better than the reference method.

## QUALITATIVE MICROBIOLOGY METHODS

Qualitative methods are presence or absence (plus or minus) methods. A test portion of homogenized food is incubated in a suitably selective culture medium for 24–48 hours at a suitable growth temperature. Any resulting growth is detected, presumptively identified and confirmed by suitable further tests. Figure 14.2 depicts the general sort of experimental protocol that is involved.

Validation usually involves comparing the performance of a new method (the test method) with a standard method (the control method) whose performance is widely acceptable for the particular food matrices of interest. The validation begins with SLV-type studies that may be followed by an MLV usually involving at least 10 laboratories. Each laboratory receives centrally prepared replicates of the analytical portions. Also a low and a high concentration of analyte in the food matrix are tested. Figure 14.3 indicates the experimental design and the main stages in SLV and MLV studies.

If the methods have the same enrichment, a single enrichment culture can be used for both methods. This has the advantage of certainty of knowing if the analyte is detected in the test or control culture, when the concentration approaches the limit of one particle per test portion, that failure to detect in one culture was not due to absence of analyte. In this paired design it is possible to measure false positive and false negative rates and sensitivity. The significance of any difference between the test and control method results can be measured by a paired Chi-square test (Siegel, 1956). Table 14.3 summarizes the main features of the conventional and new tests for the statistical validity of reproducibility data.

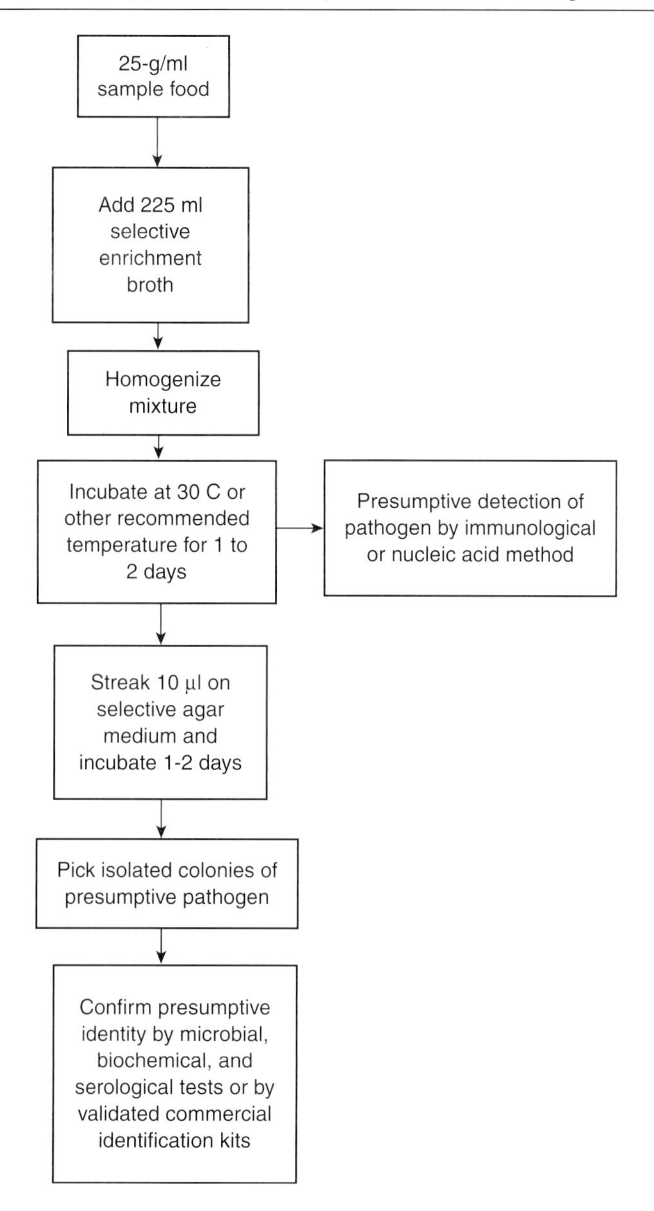


**Figure 14.2** Detection of a pathogenic microbe in a food by selective enrichment. If 1 CFU/25 g, i.e. 0.04 CFU/g, then about a 2500-fold enrichment (amplification) to 100 CFU per ml of the 250 ml of enrichment culture (contains 0.1 g food) is needed. This will ensure about 1 CFU per 10 μl enrichment culture to be streaked on a selective agar. Some methodologies use two different enrichments in succession for 1 day each.

If the methods have different enrichments there will be increasing uncertainty about the analyte's presence if the concentration approaches the limit of one particle per test portion, which is pertinent to levels of detection for food-borne pathogen methods. In this unpaired design it is not possible to measure false positive and false negative rates and sensitivity. The significance of any difference between the test and control method results can be measured by an unpaired Chi-square test (Siegel, 1956).

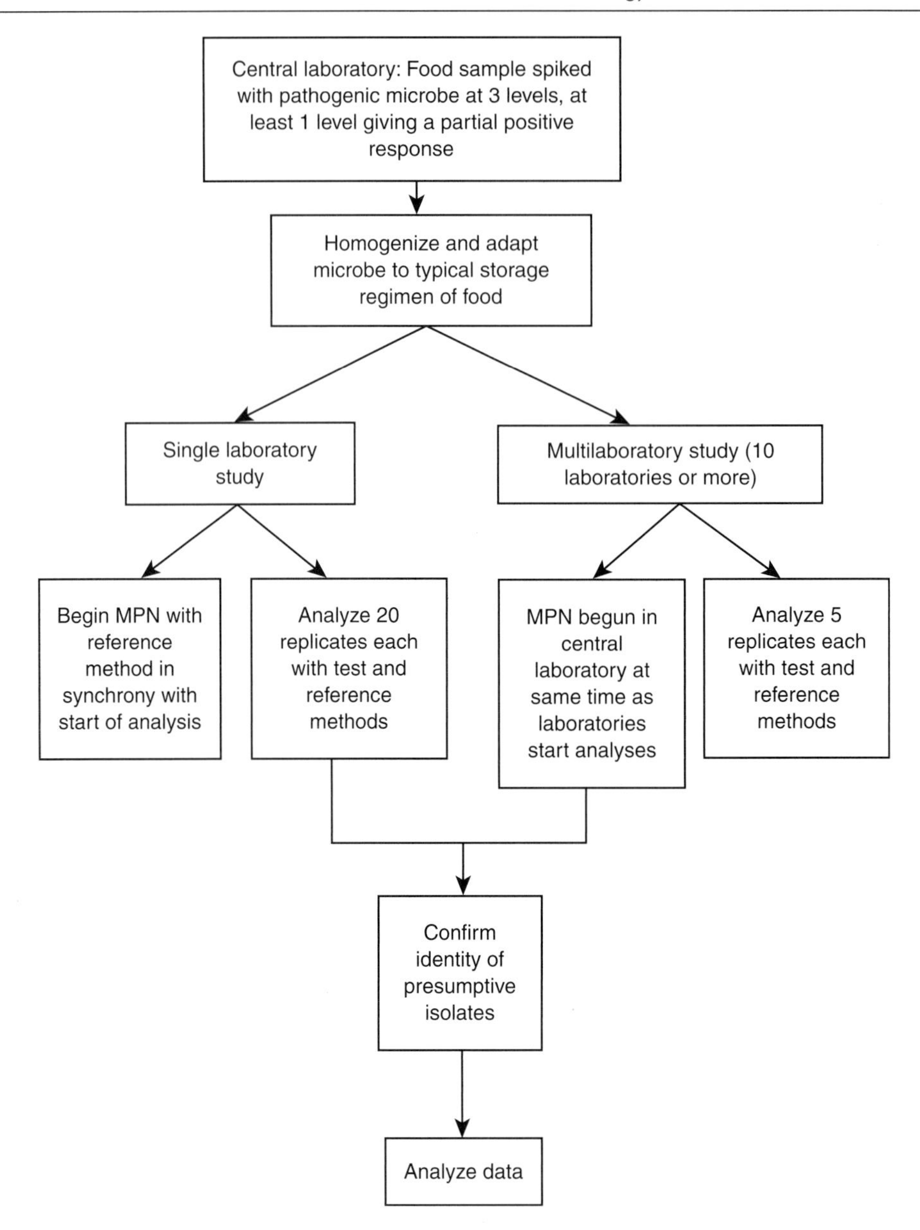


**Figure 14.3** Single and collaborative laboratory study formats for qualitative method validation.

## Reproducibility

The prime reason for MLV is to measure method reproducibility. The measured reproducibility is deemed acceptable if it is equivalent to or better than the standard method's reproducibility. The data in a qualitative method are nonparametric, that is presence or absence data. Nevertheless, parametric and nonparametric approaches have been developed for analysing MLV data.

In the parametric approach, data is evaluated as the number of positive results per laboratory replicate set, that is the probability of detection (POD) per laboratory (Wehling *et al.*, 2011; AOAC International,

**Table 14.3** Reproducibility tests for validation of qualitative microbiological methods at concentrations of a few CFU per g or ml.

| Reference and alternative method enrichments | Sensitivity and false positive and negative rates | Statistical test |
|---|---|---|
| Common enrichment | Determinable | McNemar for related samples |
| Different enrichments[a] | Problematical[b] | Chi-square for two independent samples |
| Common or different enrichments | Embedded[c] | POD model |

[a]Alternative methods usually have customized enrichments differing from reference method enrichments.
[b]Failure to spike, due to statistical variability at very low concentrations, cannot be distinguished from failure of method to detect.
[c]The POD parameter includes the information inherent in sensitivity rate, and false positive and negative rates. POD is dependent on concentration and allows for the statistical variability of spiking at very low concentrations.

2012b). These laboratory POD values are treated parametrically, with concentration as a continuous variable, to generate an $(S_R)^2$ value (AOAC International, 2012b). The number of positive results per laboratory replicate set is assumed to have a binomial distribution. The $(S_R)^2$ value is parsed into replicate and laboratory variability: $(S_R)^2 = (S_r)^2 + (S_L)^2$ where $(S_L)^2$ is the laboratory variance, $(S_r)^2$ is the replicate variance and $(S_R)^2$ is the reproducibility variance. Figure 14.4 shows the relationship of the three variances to the mean POD for the situation where there is no laboratory effect variance. Also shown is an example of a case, at a POD of 0.5, where there is a laboratory effect. The $(S_R)^2$ does not depart much from the no-laboratory effect value but the $(S_r)^2$ is markedly reduced to accommodate the observed laboratory effect.

The laboratory effect standard deviation, $S_L$ is calculated. A T-test based on the chi-square distribution indicates whether the $(S_L)^2$ value, and hence the $S_L$, is significant or not (Mendenhall, 1967). The confidence limits of parameters can be estimated. The method awaits application to new study data with the aim of developing a criterion for evaluating the acceptability or not of a given $S_L$ value.

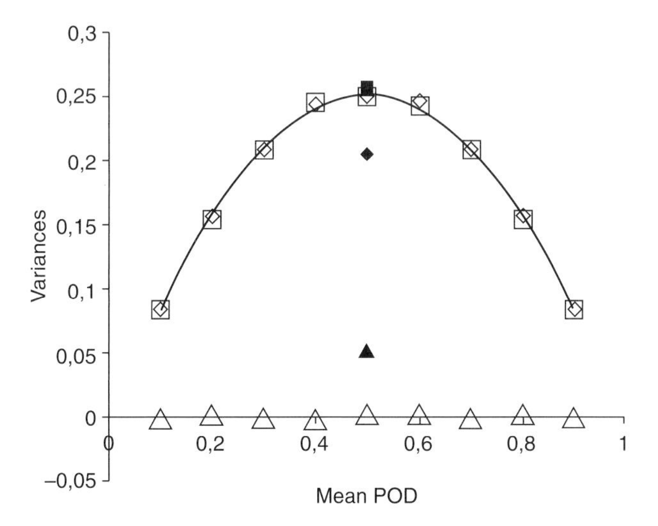


**Figure 14.4** Relationship of the replication, the laboratory effect and the reproducibility variances to the mean probability of detection in the POD model. Open symbols for no laboratory effect: □, reproducibility variance ($s_R{}^2$); , replicability variance ($s_r{}^2$); △, laboratory effect variance ($s_L{}^2$). Filled symbols for an example, at a mean POD of 0.5, of a laboratory effect. Best fit second-order polynomial curve describes both the replicability and reproducibility variances when there is no laboratory effect.

**Table 14.4** POD model for interlaboratory method reproducibility and laboratory effect value at concentrations of a few CFU per analytical portion.

| *Spike (MPN/25 g or ml)* | *Laboratory identity* | *Replicates/ laboratory* | **Reference method**[a] | | **Test method**[b] | |
|---|---|---|---|---|---|---|
| | | | *Positives detected* | *POD* | *Positives detected* | *POD* |
| 0.5 | 1 | 12 | 3 | 0.25 | 2 | 0.17 |
| 0.5 | 2 | 12 | 9 | 0.75 | 3 | 0.25 |
| 0.5 | 3 | 12 | 4 | 0.33 | 7 | 0.58 |
| 0.5 | 4 | 12 | 5 | 0.42 | 8 | 0.67 |
| 0.5 | 5 | 12 | 4 | 0.33 | 10 | 0.83 |
| 0.5 | 6 | 12 | 6 | 0.5 | 5 | 0.42 |
| 0.5 | 7 | 12 | 5 | 0.42 | 0 | 0.33 |
| 0.5 | 8 | 12 | 6 | 0.5 | 6 | 0.5 |
| 0.5 | 9 | 12 | 2 | 0.17 | 4 | 0.33 |
| 0.5 | 10 | 12 | 7 | 0.58 | 5 | 0.42 |
| 0.5 | All | 120 | 51 | 0.425 | 50 | 0.417 |

[a]Reference method: $(S_R)^2 = 0.247$; $(S_r)^2 = 0.239$; $(S_L)^2 = 0.008$.
[b]Test method: $(S_R)^2 = 0.249$; $(S_r)^2 = 0.206$; $(S_L)^2 = 0.043$. The observed value of the estimated test method laboratory variance is significantly different from that expected from the reference method laboratory variance value ($P_T < 0.0001$ by the $\chi^2$ test). Thus, the standard deviation values are significantly different too.

In any case, using the differential between their POD values, the POD method can be used to compare test methods with reference methods. POD values and confidence limits can also be calculated for SLV studies. The use of the POD approach harmonizes the statistical concepts and parameters of quantitative and qualitative method validations. The method awaits application to new study data with the aim of developing a criterion for evaluating the acceptability or not of a given $S_L$ value.

The experimental format, artificial data and test and reference method parameter variances based on a spike level of 0.5 MPN per 25 g/ml analytical portion are shown in Table 14.4. The laboratory effect standard deviations for these two examples with 12 replicates per laboratory per method are significantly different.

In the nonparametric approach, the data are evaluated as the number of laboratories detecting a given proportion of positives per replica-set that is a given POD. The distribution of the laboratories over the possible numbers of positives in a replicate set is assumed to be binomial. Analyses of data from past validations for *Listeria* and *Salmonella* detection methods are consistent with this assumption (Hitchins, 2011). The significance of departures, if any, from the binomially expected number of laboratories observing a given proportion of positives in a replicate set is determined using a one-sample binomial test (Siegel, 1956). Interestingly, the presence of some degree of non-reproducibility was found in some validations that had not been apparent with statistical tests originally applied to the data from the studies. Whether this approach really has more resolution will require further study. The experimental design (ten laboratories and six replicate portions per laboratory per method) and exemplary data for test and reference methods with a spike level of 0.5 MPM per 25 g/ml analytical portion are shown in Table 14.5. In this example the data of the reference method are not significantly different from that expected for a binomial distribution of the number of laboratories observing each possible number of positives (0–6) per six-replicate set. This implies that the $(S_R)^2$ was due essentially entirely to the replication effect, $(S_r)^2$ and that the laboratory effect $(S_L)^2$ was negligible. In contrast, with the test method some departure from the expected binomial distribution is apparent implying that there is a laboratory effect $(S_L)^2$ in addition to the replication effect, $(S_r)^2$.

**Table 14.5** Nonparametric model for interlaboratory method reproducibility at concentrations of a few CFU per analytical portion.

| Spike (MPN/25 g/or ml) | Positives detected/ 6-replicate category | Reference method | | Test method | |
|---|---|---|---|---|---|
| | | Number of laboratories/ category | Expected number of laboratories/ category[a] | Number of laboratories/ category | Expected number of laboratories/ category[a] |
| 0.5 | 0 | 1 | 0.5 | 3[b] | 0.5 |
| 0.5 | 1 | 2 | 1.94 | 0 | 1.94 |
| 0.5 | 2 | 3 | 3.14 | 3 | 3.14 |
| 0.5 | 3 | 3 | 2.72 | 3 | 2.72 |
| 0.5 | 4 | 1 | 1.32 | 1 | 1.32 |
| 0.5 | 5 | 0 | 0.34 | 0 | 0.34 |
| 0.5 | 6 | 0 | 0.04 | 0 | 0.04 |
| 0.5 | All | 10 | 10 | 10 | 10 |

[a]Expected binomially on the basis of a spike level of 0.5 MPN/25 g/ml food portion determined by the reference method.
[b]Significant departure from the expected binomial distribution value.

## LIMITS OF DETECTION OF QUALITATIVE MICROBIOLOGY METHODS

Limits of detection (LOD) have not been required in official qualitative microbiology method validations. However, it is in fact possible to calculate LOD values and their confidence limits using various methods. These methods are analogous to those used to calculate either lethal dose or infectious dose values in animal dose-response experiments (Meynell and Meynell, 1965). Results are usually presented as doses producing responses in 50% of the host animals. $LOD_{50}$ values calculated in qualitative microbiology method validations from growth of a microbe in spiked analytical portions in culture media are analogous to those calculated from infections in host animals, that is *in vivo* growth of a challenge microbe.

One advantage of calculating an $LOD_{50}$ value is that a single numerical value can be used to represent lots of unwieldy performance data. The generated values certainty is defined by confidence limits. Another important advantage is that it may enable a move toward using an absolute $LOD_{50}$ value as for validation of a method. Thus, a theoretically minimal LOD value of 1CFU/25 g analytical portion is the best that can be expected of a method. In principal, therefore, there would be no need to compare the test method with a standard method as described previously. This would cut the work load in half. In practice, however, the work load saving will be somewhat less, though still significantly lower. This is because the method depends on enumeration of the spiking levels used in the determination of the $LOD_{50}$. Such enumeration, generally MPN enumeration for reasons to be described, is done using the various standard methods. Thus, even though the collaborators need not run the standard method in parallel with the test method, the $LOD_{50}$ methods still require that the microbial analyte concentration be known. The way around this is for the central laboratory, which prepares the portions to be sent to the collaborating laboratories, to do the enumeration.

### Determining $LOD_{50}$ values by the Spearman–Kärber method

The Spearman–Kärber (SK) method is a nonparametric statistical method that can be used to calculate the limit of detection of a qualitative microbial detection method, specifically the 50% endpoint ($LOD_{50}$) (Spearman, 1908; Kärber, 1931; Cornell, 1982). Background information on its application to limits of

**Table 14.6** Estimating microbial $LOD_{50}$ and $LOD_{95}$ values by the Spearman–Kärber and log-log complementary methods.

| A. Experimental design and exemplary data | | |
|---|---|---|
| ***Spike level (CFU/ 25 g or ml)*[a]** | ***Number of replicates/level*** | ***Number of positive responses/level*** |
| 10 | 5 | 5 |
| 2 | 5 | 2 |
| 0.4 | 5 | 1 |
| 0.08 | 5 | 0 |
| 0.016 | 5 | 0 |

| B. Results | | |
|---|---|---|
| ***Analytical method*** | ***$LOD_{50}$ (CFU/g or ml) and 95% confidence interval*** | ***$LOD_{95}$ (CFU/g or ml) and 95% confidence interval*** |
| Spearman–Kärber | 0.068 (0.024–0.196) | 0.29 (0.103–0.843)[b] |
| Log-log complementary | 0.079 (0.033–0.190) | 0.34 (0.141–0.820) |

[a]Spiking culture is decimally diluted to about 10 CFU/25 g or ml followed by four fivefold dilutions to extinction.
[b]Spearman–Kärber $LOD_{95}$ value obtained by multiplying $LOD_{50}$ value by 4.3.

detection and the program for calculation have been published elsewhere (FDA, 2006, appendices K and L). A working version of the program accommodating up to five spiking levels is available at http://www.aoac.org/accreditation/DEMO.xls (last accessed 4 July 2013). Several studies on methods for the foodborne pathogens, *Listeria monocytogenes* and *Campylobacter* spp., have used this application of the SK method to determine $LOD_{50}$ values (Hitchins, 1989; Tran *et al.*, 1990; Twedt *et al.*, 1994; Thunberg *et al.*, 2000). In addition, the method has been applied retrospectively to data from published *Salmonella* method validation studies (Hitchins, 2012). The $LOD_{50}$ is expressible as CFU per analytical portion, which is usually 25 g or ml. This can be converted to the more scientific CFU per g or per ml if the test portion size is clearly indicated. The $LOD_{95}$ value can be calculated as $4.2 \times LOD_{50}$, since LOD values as a function of the spiking concentration follow a Poisson curve (see later).

Table 14.6 shows an idealized example of the kind of experimental design that is necessary and the sort of results obtained with the resulting SK 50% endpoint and confidence limits. (The complementary log-log analysis is discussed later.) Typically, a food is spiked at three, or more, known different concentration levels each replicated in triplicate or more. One of the levels should yield 0% recovery and another should yield 100% recovery. The calculation is based on the logarithms of the concentrations. The levels should preferably be spaced at a constant interval such as 10- or 2-fold. Since it is difficult to set up the spiking experiment to obtain the required range of recoveries, it is probably best to maximize the number of levels keeping the replicates to three per level. This range of levels should be about 1 CFU per test portion (e.g. 25 g/ml) because this is the maximal recovery level possible. This will be commonly attainable but it should be extended upward in case recovery is suboptimal for a particular matrix–method combination. A major reason that obtaining the right range of levels is somewhat chancy is that investigators often prefer to acclimatize the spiked inoculum to the particular food matrix properties, such as pH and its storage temperature. This may involve death and/or injury, for example with frozen storage. Thus, even though the spike concentration is nominally known it has to be measured synchronously with the performance of the SLV or MLV, since the degree of inurement will be uncertain, thus requiring more levels to allow for potential loss of spiked cells. If investigators prefer to spike and test without acclimation, culture turbidity can be used to estimate the spike levels if a turbidity-concentration curve is available. This makes it somewhat easier to choose the right spiking levels up front. The estimate should be confirmed by colony count.

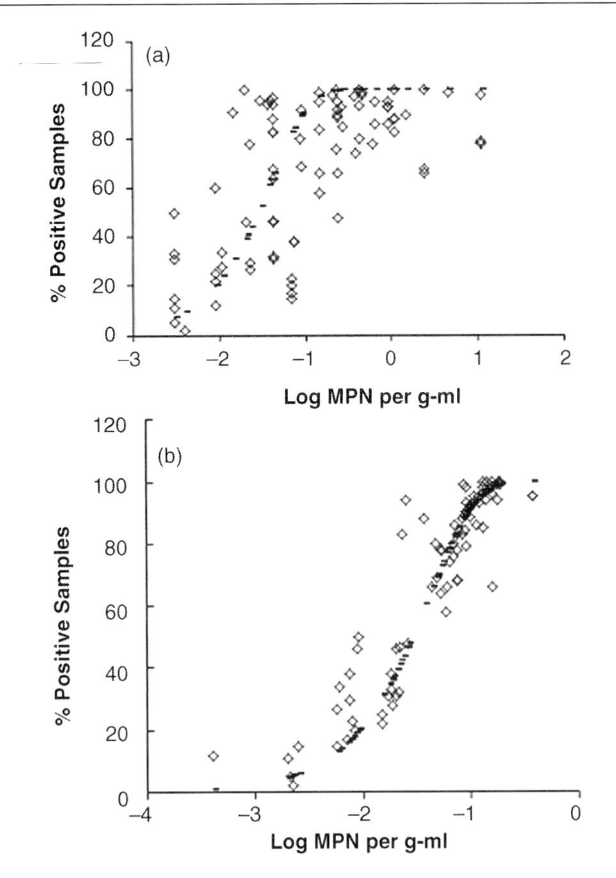


**Figure 14.5** Relationship of the *Listeria monocytogenes* detectability in foods by test and reference methods to its concentration. (a) spiking concentration determined by 3-tube MPN reference method; diamonds, reference and test method results (◇) broadly scattered about the Poisson curve; (b) spiking concentration determined by multitube MPN method based on the reference method's proportion of pooled positive responses; reference method results (◇) fall on the Poisson curve and test method data is less markedly scattered about it.

With retrospective analyses of published validation studies there is no control over the design. Thus, levels giving 0 and 100% recovery values may not be available. However, it is often possible to use dummy concentrations. If the recovery is near 100%, for example 90%, a concentration just above that for the 90% recovery is used for 100% recovery. A similar thing could be done with a recovery near 0% but usually a concentration of 0.004 CFU/g-ml (0.1 CFU/25g-ml) is chosen based on the proportion positive versus concentration curve. This is approximately described by the Poisson equation, as illustrated in Figure 14.5 using published results from MLV studies of rapid kits and standard methods for detecting the food pathogen *Listeria monocytogenes* in foods (Hitchins, 1998; Burney and Hitchins, 2004). The points generally follow the Poisson curve but there is considerable scatter due to the three-tube MPN enumeration variability. However, when the standard method validation results are used for a multitube (>50 tubes) one-level enumeration, the scatter for the test kit data is noticeably reduced. The use of standard method analytical data in this fashion is not inappropriate, since the original three-tube MPN enumerations in these validations were performed with the standard methods.

Measurement of the $LOD_{50}$ of a method can suffice by itself as a measure of method performance. There is no need for each collaborative laboratory to run the method in parallel with a reference method. However, the lead laboratory would still need to use one to enumerate the spike. A retrospective analysis of *Salmonella* method validation studies showed how the distribution curve of a historical collection of $LOD_{50}$ data could act as a reference standard for identifying outlier $LOD_{50}$ values (Hitchins, 2012). Such outlying values would imply the need for their possible rejection, especially if they were repeatable.

## Determining $LOD_{50}$ values by the complementary log-log model

Wilrich and Wilrich (2009) developed a model for the qualitative microbiology probability of detection (POD) as a function of a food's contamination level. The model is called a complementary log-log model because it relates the probability of detection to the log of the spike level. It is a special case of generalized linear models (McCullagh and Nelder, 1952; Wilrich and Margaritescu, 2013).

Table 14.6 shows that the designs for the Spearman–Kärber and the complementary log-log models can be identical if desired and that the two calculations give statistically indistinguishable results with the exemplary data set.

It was tested on *L. monocytogenes* food contamination data from an SLV study involving four different foods and food process water using four or five spiking level per matrix and six replicates per level. The $LOD_{50}$ values ranged from 0.017 to 0.033 CFU/g or ml that is 0.43 to 0.83 CFU/25 g or ml of analytical portion. This method yielded results that were comparable to those obtained by the SK method (Wilrich and Wilrich, 2009).

A useful additional feature of this model is a test for the existence of a food matrix effect. This test measures the deviation of the calculated POD versus concentration curve from the ideal curve, that is the one without measurement error. There was no matrix effect with any of the five matrices studied. The matrix test has some parameter boundary limitations. For instance, if the number of positives at a spike level is zero the matrix effect cannot be detected.

Wilrich has also devised a program to measure the relative LOD (RLOD) of two methods (Wilrich and Margaritescu, 2013).

# CONCLUSIONS

The foundation for all the analyses considered here is the microbial count. Microbial counts have considerable uncertainty bounds, most especially the MPN method. This limits in turn the degree of certainty of the results of analyses for the $LOD_{50}$ and the reproducibility, since they all involve knowing the concentration of the analyte as precisely and accurately as possible.

Two methods for calculating $LOD_{50}$ values of microbial detection methods have been presented. So far, direct comparison study data are limited but it appears that they give similar results. Nevertheless, in comparing detection methods it is probably sensible to compare the results using only the same calculation method. This does not preclude secondarily using the other method in the same fashion. Generally, differences between the results of the two calculations, unless they are unexpectedly large, will not matter. This is because of the large uncertainty inherent in microbial enumeration methods. This uncertainty can be ameliorated to some degree by increasing the replicate number. For example the one-level MPN with 50 or more tubes is recommended with adapted spikes, which requires use of the MPN enumeration. With unadapted spikes, count more colonies by using more replicates plates. Pragmatically, enumeration uncertainty can never be completely eliminated. One useful feature of the $LOD_{50}$ result is that it condenses the sizeable amount data from a validation study to one number per food matrix. It thus makes use of all the results from two or more spiking levels. Furthermore, the result has confidence limits, which counters exaggeration of differences between $LOD_{50}$ values. In the experimental design it is suggested that more levels be preferred to more replicates per spike. For example five levels and three replicates per level are preferable to three levels and five replicates per level. One reason is that it is tricky to find the range for partial positives, so not all levels may be useable. With the MPN method of spike enumeration a trial run will usually be necessary with a previously untested food matrix.

The caveat about enumeration uncertainty also applies to the two reproducibility validation methods discussed. Critical values of reproducibility for acceptable method performance are not stated here. They are more appropriately the province of official standards organizations.

## ACKNOWLEDGEMENTS

My thanks to Curtis N. Barton, James T. Peeler and Arshia A. Burney for help in my introductory ventures into the topic of detection limits.

## REFERENCES

AOAC International (2012a) Guidelines for collaborative study procedures to validate characteristics of a method of analysis, Appendix D. [Online] Available: http://www.coma.aoac.org/appendices.asp (last accessed 23 June 2013).

AOAC International (2012b) Food microbiology guidelines, Appendix X. [Online] Available: http://www.aoac.org/vmeth/guidelines (last accessed 23 June 2013).

Blodgett, R. (2003) Appendix 1. Most probable number determination from serial dilutions. *Bacteriological Analytical Manual*. [Online] Available: http://www.fda.gov/Food/FoodScienceResearch/LaboratoryMethods/ucm109656.htm (last accessed 4 July 2013).

Burney, A.A., and Hitchins, A.D. (2004) *Determination of the limits of detection of AOAC validated qualitative microbiology methods*. AOAC International Annual Meeting. Abstract. St Louis, MO.

Cornell, R.G. (1982) Kärber method. In: *Encyclopedia of Statistical Sciences*, volume 4 (eds S. Kotz and N. L. Johnson), John Wiley & Sons, Inc., New York, NY, pp. 354–357.

Crowley, E.S., Bird, P. M., Torontali, M. K. *et al.* (2009) TEMPO® TVC for the enumeration of aerobic mesophilic flora in foods: Collaborative study. *J AOAC Int* **92**, 165–174.

Crowley, E.S., Bird, P. M., Torontali, M. K. *et al.* (2010) TEMPO® EC for the enumeration of *Escherichia coli* in foods: Collaborative study. *J AOAC Int* **93**, 576–586.

FDA (2006) Final Report and Executive Summaries from the AOAC International Presidential Task Force on Best Practices in Microbiological Methodology. US Food and Drug Administration. [Online] Available: http://www.fda.gov/Food/FoodScienceResearch/LaboratoryMethods/ucm124900.htm (last accessed 23 June 2013).

Hitchins, A.D. (1989) Quantitative comparison of two enrichment methods for isolating *Listeria monocytogenes* from inoculated ice cream. *J. Food Protect* **52**, 898–900.

Hitchins A.D. (1998) *Retrospective interpretation of qualitative collaborative study results:* Listeria *methods*. AOAC International Annual Meeting, Abstract J-710

Hitchins, A.D. (2011) The determinacy of reproducibility assessments of qualitative microbial food borne pathogen methods for detecting a few microbes per analytical portion. *Food Microbiology* **28**, 1140–1144.

Hitchins, A.D. (2012) A meta-analytical estimation of the detection limits of methods for *Salmonella* in food. *Food Research International* **45**, 1065–1071.

Kärber, G. (1931) Beitrag zur kollektiven Behandlung pharmakologischer Reihenversuche [A contribution to the collective treatment of a pharmacological experimental series]. *Archiv für experimentelle Pathologie & Pharmakologie* **162**, 480–483.

McCullagh, P. and Nelder, J.A. (1952) *Generalized Linear Models*. Chapman and Hall, New York, NY.

Mendenhall, W. (1967) *Introduction to Probability and Statistics*. 1st edn. Wadsworth Publishing Company, Inc., Belmont, CA, pp. 205–209.

Meynell, G.G. and Meynell, E. (1965) *Theory and Practice in Experimental Bacteriology*, 1st edn. Cambridge University Press, London, pp. 179–182.

Siegel, S. (1956) *Nonparametric Statistics for the Behavioral Sciences*. McGraw-Hill Book Co., New York, NY.

Spearman, C. (1908) The method of "right and wrong cases" ("constant stimuli") without Gauss's formulae. *Br J Psychol* **2**, 227–242.

Steiner, E.H. (1975) Planning and analysis of results of collaborative tests. In: *Statistical Manual of the Association of Official Analytical Chemists*. AOAC International, Gaithersburg, MD, USA, pp. 65–83.

Thunberg, R.L., Tran, T.T. and Walderhaug, M. O. (2000) Detection of thermophilic *Campylobacter* spp. in blood-free enriched samples of inoculated foods by the polymerase chain reaction. *J Food Protect* **63**, 299–303.

Tran, T. T., Stephenson, P. and Hitchins, A.D. (1990) The effect of aerobic mesophilic microfloral levels on the isolation of inoculated *Listeria monocytogenes* strain LM82 from selected foods. *J. Food Safety* **10**, 267–275.

Twedt, R. M., Hitchins, A.D. and Prentice, G. (1994) Determination of the presence of *Listeria monocytogenes* in milk and dairy products: IDF collaborative study. *J AOAC Int* **77**, 395–402.

Wehling, P., LaBudde, R. A., Brunelle, S. L. and Nelson, M. T. (2011) Probability of detection (POD) as a statistical model for the validation of qualitative methods. *J AOAC Int* **94**, 335–347.

Wilrich, C. and Wilrich, P.-T. (2009) Estimation of the POD function and the LOD of a qualitative microbiological measurement method. *J AOAC Int* **92**, 1763–1772.

Wilrich, P. and Margaritescu, I. (2013) Determination of the RLOD of a qualitative microbiological measurement method with respect to a reference measurement method. *J AOAC Int* **96** (In Press).

# 15 Statistical approaches to the analysis of microbiological data

**Basil Jarvis**

*Department of Food and Nutrition Sciences, School of Chemistry, Food and Pharmacy, The University of Reading, Whiteknights, Reading, Berkshire, UK*

## ABSTRACT

Statistics are important in the planning and execution of experiments in food microbiology and in the subsequent analysis of data. To use statistics effectively it is essential to understand the basis of population distributions and their relevance to experimental techniques and statistical tests. Various statistical tests with particular relevance to food microbiology are discussed with worked examples that illustrate diverse applications. Modelling of microbiological data is important in extending statistical considerations of experimental data that may be limited due to practical considerations; and, in a predictive role, to illustrate the effects of factors that affect the growth, survival and death of micro-organisms in foods.

## INTRODUCTION

Statistics is essential to the interpretation of much scientific data, especially in biological sciences. Population statistics cannot usually be determined directly but estimates of the parameters of independent samples enable conclusions to be drawn about the population from which the random samples were taken. Tests for statistical significance are used to assess the probability of wrongly rejecting a null hypothesis of 'no observed difference' between results from related sample sets. It is important to distinguish between the size of an effect, which may be statistically significant, and its importance, which may not be critical. Data sets may include, for instance: estimates of prevalence and/or numbers of microorganisms in samples of foods and the environment; experimental studies of population change, i.e. growth or death of microorganisms in processing or storage conditions; evaluation of the suitability of a culture medium (Jarvis, 2011); and the validation and verification of microbiological methods for particular purposes (Chapter 17). Estimates of parameters can be used also as input data for modelling microbial distributions.

## MICROBIAL POPULATION DISTRIBUTIONS

The spatial distribution of microorganisms describes the manner in which they occur in a natural substrate. Spatial distributions comprise both individual cells and colonies of organisms distributed randomly but in some circumstances clusters of organisms may occur in isolated foci due to localized

*Mathematical and Statistical Methods in Food Science and Technology*, First Edition.
Edited by Daniel Granato and Gastón Ares.

contamination. When samples are taken for analysis, spatial distributions become disrupted, so that suspensions in a diluent will contain both individual cells and clumps. Theoretically, such cell suspensions will be distributed randomly but, in many cases, cell and colony counts do not reflect randomness because of the presence of cell clumps. Hence, counts of the number of colony forming units (CFU) in dilutions of replicate food samples will estimate the apparent level of contamination, but will not reflect the original spatial distribution of organisms in the samples. The accuracy and precision of the tests undertaken will also affect the apparent numbers of organisms that are detected. Such differences are reflected in the stochastic (or probability) distribution of colony counts.

## Spatial distributions

Electron microscopy of frozen sections can reveal the inter-relationship between organisms of different types, for example yeasts and lactic acid bacteria, as a consequence of colonization of a substrate. If a freshly cut piece of meat were exposed to aerial contamination a random distribution of cells would be expected; some grow to form colonies, whilst others die or remain dormant, depending on environmental conditions. Hence, the distribution of organisms after storage will differ from that on a freshly cut surface; Figure 15.1 provides a schematic illustration of random surface contamination and a 'contagious' (i.e. over-dispersed) distribution of organisms after cell growth and death (Jarvis, 2008). On mixing and dilution some, but not all, chains and clumps will become disrupted and both individual cells and clumps will form colonies when plated onto a suitable culture medium.

## Statistical distributions

A frequency distribution represents the frequency of occurrence of results that occur within a particular range of values. For instance, in an examination of chilled raw milks it might be found that the majority of samples tested contain less than $10^4$ CFU/ml but some contain more. The probability distribution of such data can be described by a mathematical expression and modelled as a curve. If the area under the

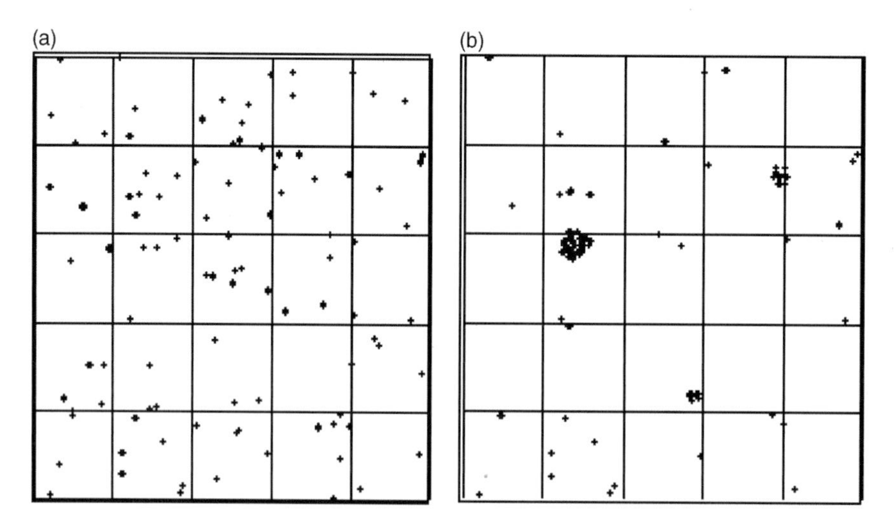


**Figure 15.1** Schematic illustration of the spatial distribution of organisms on a surface (a) after initial random contamination; (b) after growth and death of some organisms, showing individual cells and micro-colonies to illustrate development of a 'contagious' or 'over-dispersed' distribution.

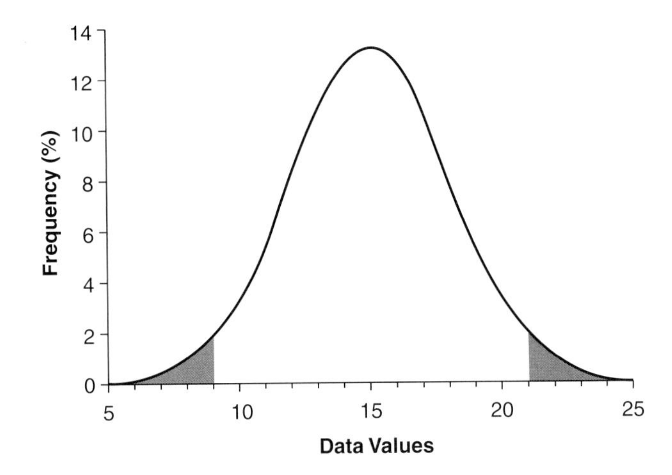


**Figure 15.2** The Normal Distribution Curve for $\mu = 15$ and $\sigma = 3$; the shaded areas in the tails of the curves each cover 2.5% of the distribution and show that 95% of the central area of the curve is covered by $\mu \pm 2\sigma$.

curve is defined as one, the area under specific parts of the curve will represent the probability for occurrence of fractions of the whole.

### *The normal distribution*

The effect of random influences causes an event to converge on a central value and the distribution of data follows a symmetrical 'bell shape', generally referred to as a normal (Gaussian) distribution (ND). Values close to the centre of the distribution are more likely to occur than are values towards either extreme (Figure 15.2). The shape of the ND is determined by the population mean value ($\mu$) and its standard deviation ($\sigma$); but since it is not possible to measure these parameters directly they are estimated from the mean ($\bar{x}$) and standard deviation ($s_x$) of randomly drawn samples. Approximately 95% of all values lie within the range $\mu \pm 2\sigma$ and 99% within the range $\mu \pm 3\sigma$.

Many statistical procedures are based on the ND with inferences based on the extent of agreement to normality, so the ND is critical to many statistical analyses. Although measurements of levels of products of microbial metabolism are likely to conform to ND, other population-related parameters, especially cell numbers and colony counts, do not do so, until they have been mathematically transformed.

### *The lognormal distribution*

About 90% of colony counts approximate to a lognormal distribution. Suppose that samples of raw minced meat from a production plant are examined each day for aerobic colony counts. Even though the ingredients and the process are 'standardized', it would be unrealistic to expect that colony counts (as CFU/g) would be identical for replicate samples tested on a single day and certainly not for samples of product manufactured over a period of time. Figure 15.3a shows some daily average colony counts that provide an estimate of the microbial load of the product on different days. Sometimes colony counts are higher than at other times, which may reflect higher microbial levels in the ingredients and/or environmental contamination on those days.

To assess whether the colony counts are generally within an acceptable range, the frequency with which particular counts occur is determined and these are plotted as a frequency bar chart (Figure 15.3b). The data distribution is skewed (biased) towards the lower end of the CFU scale and does not conform to a ND; the locations of the overall mean and the median counts are also offset towards the lower end of the

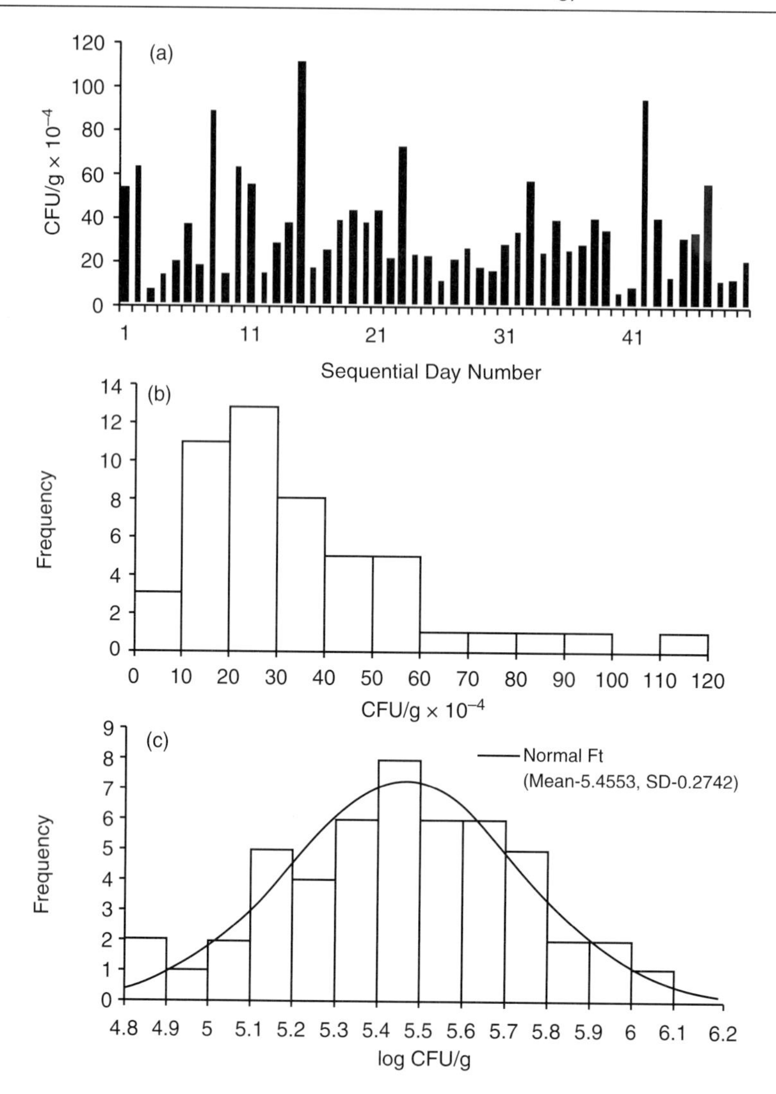


**Figure 15.3** Distribution of colony counts from raw minced meat (a) as daily mean CFU/g over a period of 49 days; (b) frequency plot of CFU/g and (c) frequency plot of $\log_{10}$ CFU/g, overlaid with a ND plot.

scale. However, the colony counts can be 'normalized' by deriving $\log_{10}$ values of the counts. A plot of the log-transformed data (Figure 15.3c) more closely approximates a ND. If the overall mean colony count (5.455 log CFU/g) is back-transformed it provides an estimate of the geometric mean ($28.5 \times 10^4$ CFU/g), which for a ND also approximates the median of the original data; the geometric mean is smaller than the arithmetic mean ($34.6 \times 10^4$ CFU/g). We can 'adjust' the geometric mean count using Equation 15.1 (Rahman, 1968):

$$\log m = \hat{m} + \ln 10 \cdot s^2/2 \tag{15.1}$$

where $\log m = \log_{10}$ of the arithmetic mean, $\hat{m}$ = estimate of the mean $\log_{10}$ colony count and $s^2$ = the variance of the $\log_{10}$ colony count. For the data cited, $\hat{m} = 5.455$ and $s^2 = 0.075$, so $\log m = 5.530$; hence $m = 33.9 \times 10^4$ CFU/g, which approximates the arithmetic mean. Although the lognormal distribution adequately describes the heterogeneous distribution of most colony counts on foods more extreme distributions (e.g. the Negative Binomial) also occur.

*The Poisson distribution*

A perfect suspension of microbial cells would be expected to be distributed randomly and to conform to a Poisson distribution (Jarvis, 2008). The Poisson distribution has a single parameter ($\lambda$ = mean value) that defines its shape; its variance also equals its mean value. At low values of $\lambda$ the distribution is highly skewed but as the value of $\lambda$ increases to 20 or more, the shape of the distribution approaches that of a ND. Suppose that 100 colony counts were carried out using 1 ml of a suspension of cells with an expected mean count of 2 CFU/ml. Table 15.1 shows the frequency of occurrence of 0, 1, 2, . . . 7 colonies/plate; the total frequency (derived by multiplying the number of colonies/plate by their frequency) is 197 colonies on 100 plates, giving an average inoculum level of $1.97 \approx 2$ CFU/ml inoculum. The frequency distribution (Figure 15.4a) shows the actual data and that for a theoretical Poisson distribution with $\lambda = 1.97$. Note that the distributions are skewed with a long tail to the right.

**Table 15.1** Distribution of colony counts from a dilute suspension of organisms.

| Number of colonies/plate (*n*) | Number of results for | | | | | | | | Total |
|---|---|---|---|---|---|---|---|---|---|
| | 0 | 1 | 2 | 3 | 4 | 5 | 6 | 7 | |
| Number of plates (frequency; *f*) | 14 | 27 | 27 | 18 | 9 | 4 | 1 | 0 | 100 |
| Cumulative frequency (*nf*) | 0 | 27 | 54 | 54 | 36 | 20 | 6 | 0 | 197 |

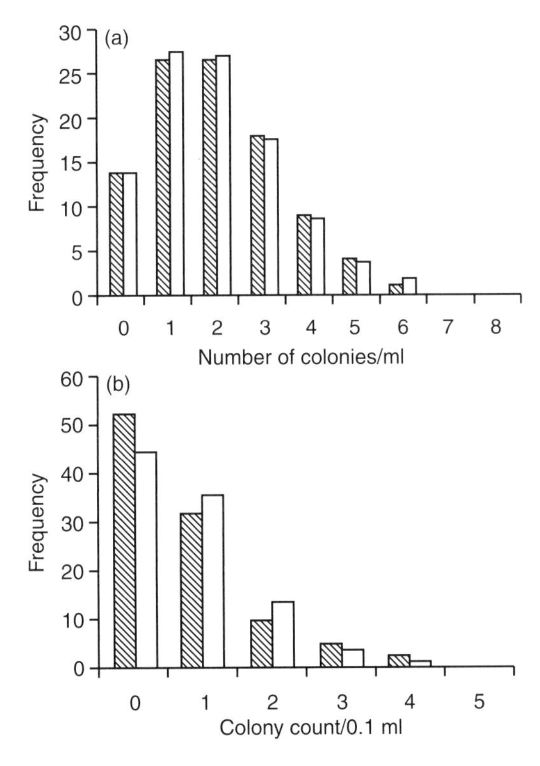


**Figure 15.4** Frequency plot for a Poisson distribution: (a) an almost perfect random distribution – actual counts (hatched) and theoretical Poisson distribution (open) for $\lambda = 1.97$; (b) plot of counts from an 'over-dispersed' suspension of *Listeria* cells with mean = 0.83 cells/0.1 ml (hatched) and a Poisson distribution (open) for $\lambda = 0.83$.

Data obtained by plating 60 replicate aliquots, each of 0.1 ml of a suspension of *Listeria* cells expected to contain about 10 CFU/ml, did not conform well to Poisson (Figure 15.4b). For this suspension, the calculated mean count was 0.83 cells/0.1 ml and its variance was 1.30. The frequency of zero counts and of counts $\geq$4 cells/0.1 ml were both higher than would be expected from a suspension that was truly random indicating divergence from Poisson.

A simple test for randomness (Fisher's Index of Distribution; $I$) is the ratio of the estimated variance ($s^2$) to the mean value ($\bar{x}$): $I = s^2/\bar{x}$; for a perfectly random distribution, $\lambda = \bar{x} = s^2$ so $I = 1$. For the data in Figure 15.4a, $I = 1.91/1.97 = 0.97$, which is consistent with the expectation; but the *Listeria* data (Figure 15.4b) has a greater variance than expected, so $I = 1.30/0.83 = 1.57$. This indicates over-dispersion, i.e. the distribution does not conform to Poisson. Further examination shows the distribution to conform to a negative binomial (qv). Other tests to assess conformance to a particular distribution include the use of Fisher's $\chi^2$-test of the observed distribution frequency against the expected frequency derived for the same mean value (Jarvis, 2008).

Since the Poisson distribution has only one defining parameter the probability for the occurrence of a count of $x$ is given by:

$$P_x = e^{-\lambda}(\lambda^x/x!) \tag{15.2}$$

where $P_x$ = probability of occurrence of $x$ cells, $\lambda$ = the expected mean value, $x!$ = factorial $x$ and e is the exponential factor (2.718). The probabilities of 0, 1, 2, 3 and so on individuals per sampling unit are given by the individual terms of the expansion of Equation 15.2, thus:

$$\text{for } x = 0 \quad P_{(x=0)} = e^{-\lambda} \tag{15.2.1}$$

$$\text{for } x = 1 \quad P_{(x=1)} = P_{(x-1)}\frac{\lambda}{x!} = P_{(x=0)}\lambda \tag{15.2.2}$$

$$\text{for } x = 2 \quad P_{(x=2)} = P_{(x-1)}\frac{\lambda}{x!} = P_{(x=1)}\frac{\lambda}{2!} \tag{15.2.3}$$

and so on.

The probability of a positive result is given by $P_{(x\geq 1)} = (1 - P_{(x=0)})$. This is useful for estimating the relative frequency of positive and negative results knowing the expected mean value of the cell density of a suspension of organisms. For example, if $\lambda = 2$, $P_{(x=0)}$ is 0.135, thus the probable frequency of one or more positive results ($P_{(x\geq 1)}$) will be 0.865. Conversely, to ensure that at least one viable organism is present in 99% of subcultures inoculated with 1 ml of diluted culture, then $P_{(x=0)}$ should be $<0.01$, for which the mean inoculum level ($\lambda$) would be about 5 CFU/ml ($P_{(x=0)} = 0.007$).

A 'General Homogeneity ($G^2$) Test' can be used to assess randomness in a data set, for example the distribution of colonies derived from sequential serial dilutions (Jarvis, 2008).

### *The negative binomial (NB) or Gamma-Poisson distribution*

The NB is a group of discrete probability distributions that provide appropriate models for over-dispersed microbiological data. Since the Poisson distribution has only a single defining parameter it cannot provide an appropriate model for over-dispersion. In the NB distribution, the Poisson parameter ($\lambda$) is itself distributed according to a Gamma ($\Gamma$) or other distribution.

Extreme spatial distributions are sometimes found in dried foods examined for pathogens, for example the prevalence and level of contamination by organisms such as *Cronobacter sakazachii* in dried infant feeds. Clusters of high level contamination occur in localized parts of batches of food that

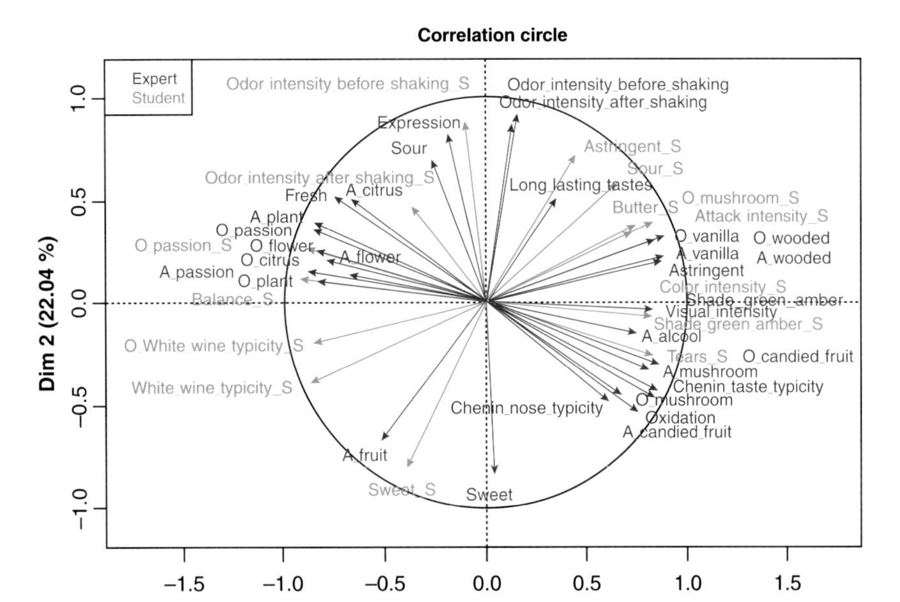


**Plate 6.3** Representation of descriptors on the first plane of the MFA.

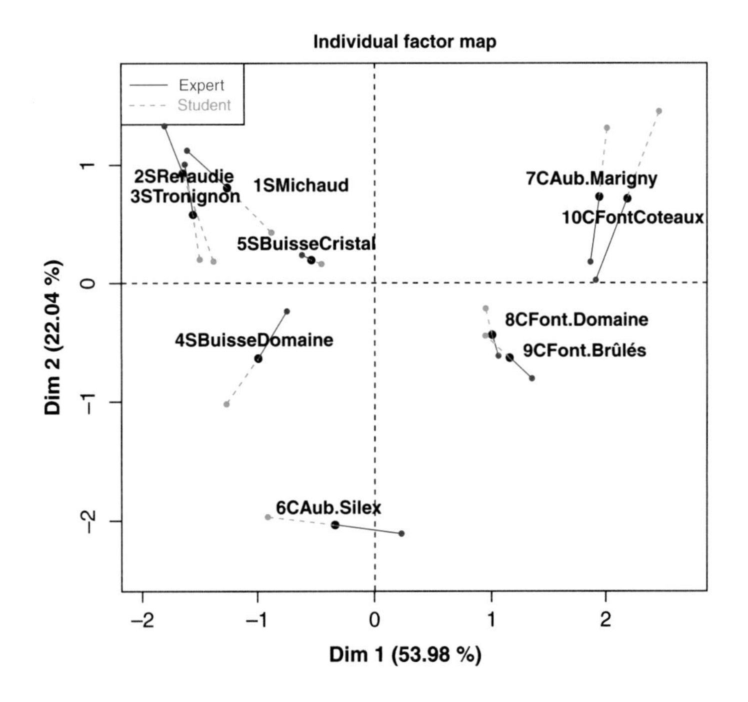


**Plate 6.4** Representation of the mean and partial individuals on the first plane of the MFA.

*Mathematical and Statistical Methods in Food Science and Technology*, First Edition.
Edited by Daniel Granato and Gastón Ares.

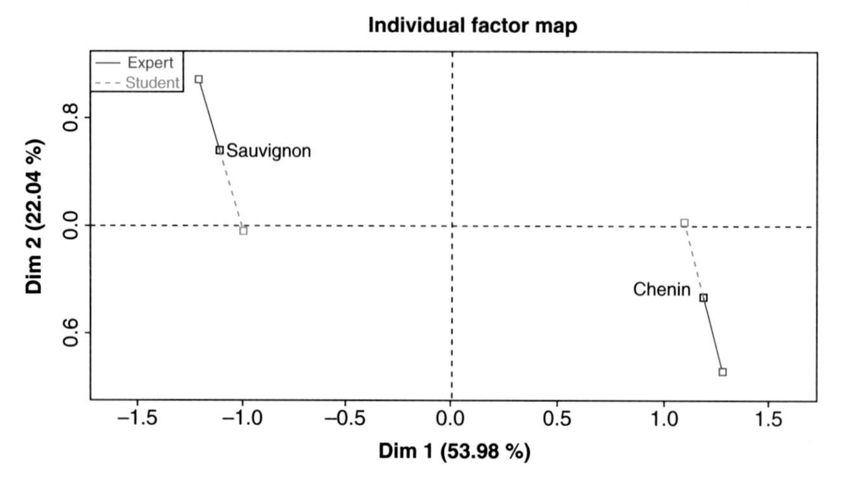


**Plate 6.5** Representation of mean and partial categories.

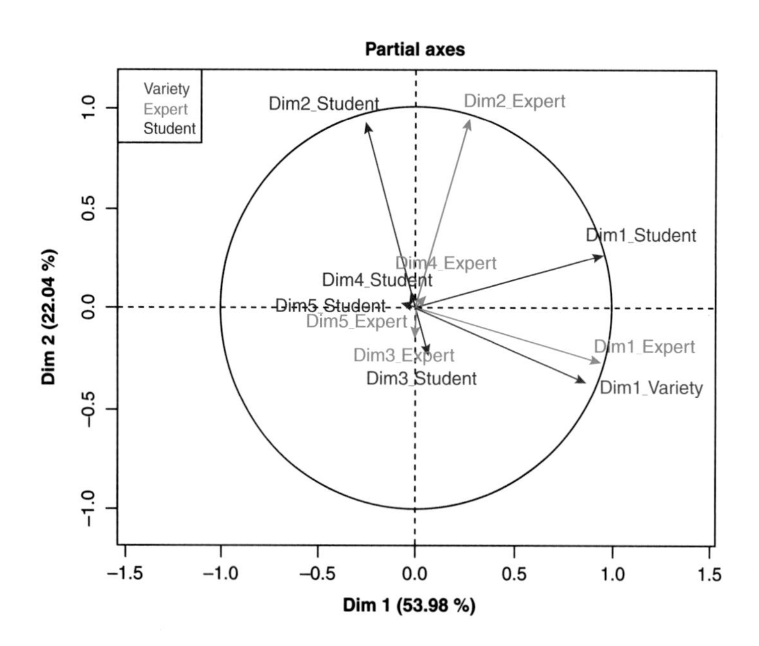


**Plate 6.6** Representations of the partial axes (principal components of the separate PCAs) on the first plane of the MFA.

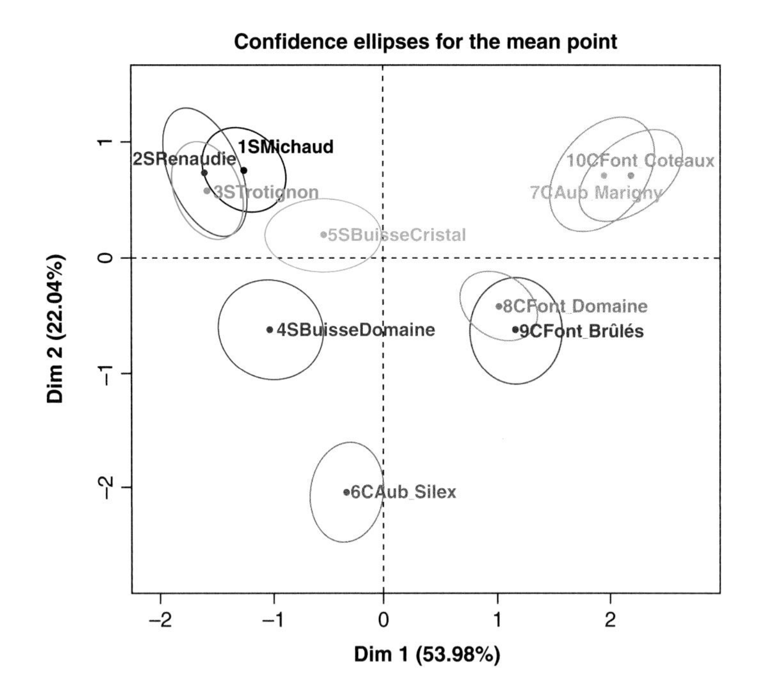


**Plate 6.8** Representation of confidence ellipses associated with mean points of Figure 6.2.

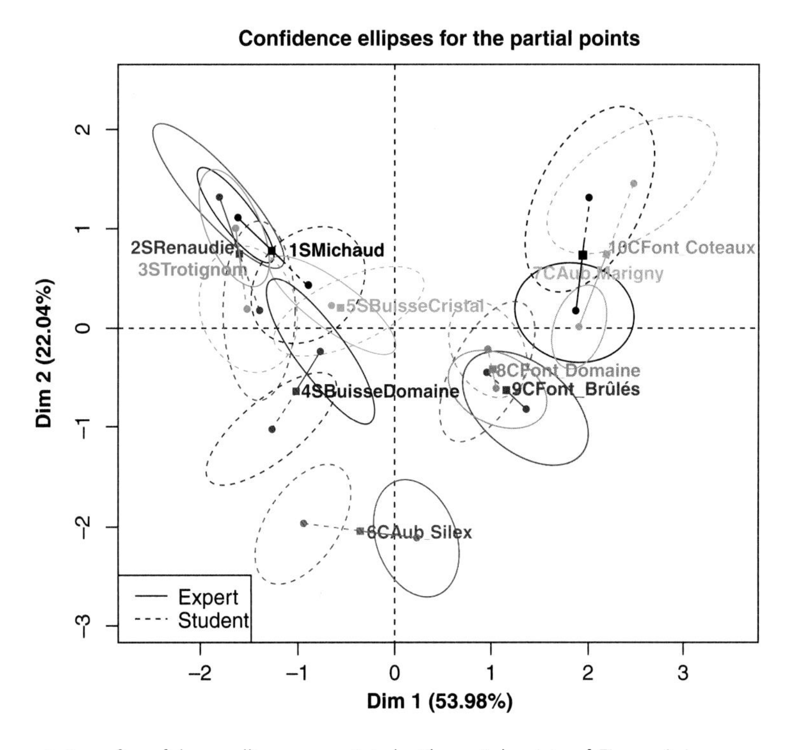


**Plate 6.9** Representation of confidence ellipses associated with partial points of Figure 6.4.

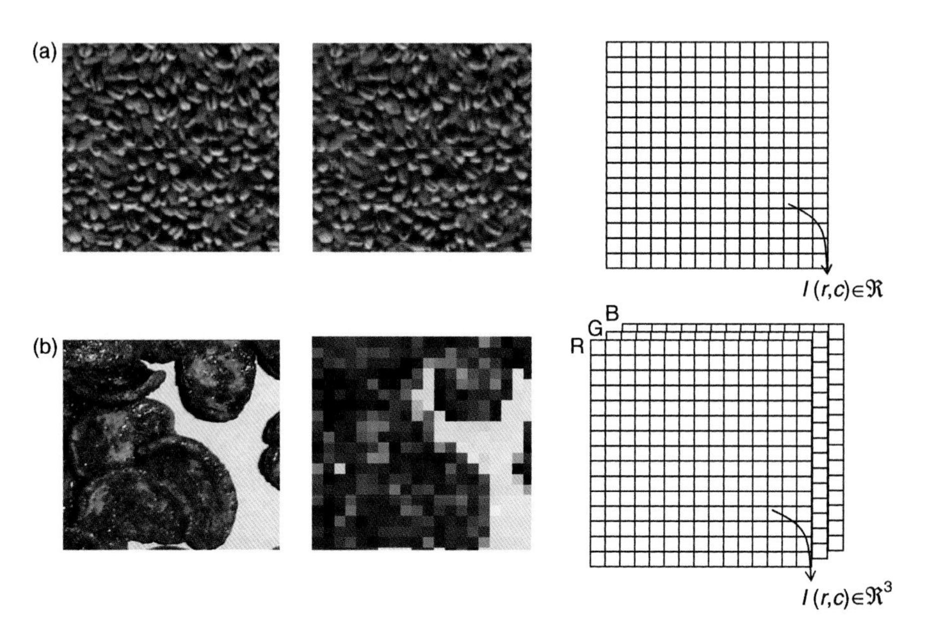


**Plate 12.1** Digital images and their underlying data structures: (a) a grey-scale image. Courtesy of Matti Pietikäinen and Outex Texture Database http://www.outex.oulu.fi/; (b) a colour image.

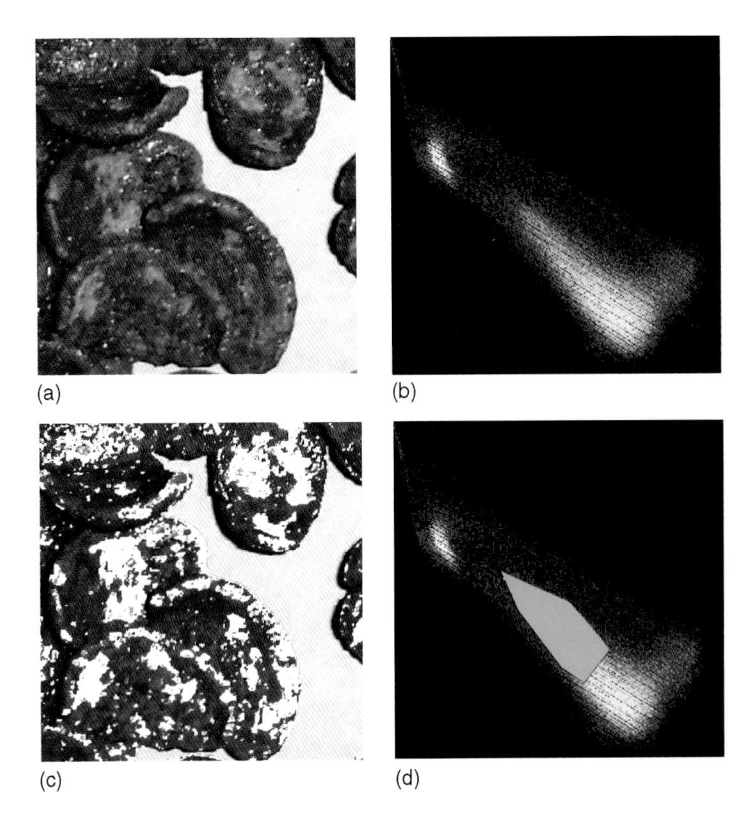


**Plate 12.3** An original image regarding cereal flakes (a) and the corresponding score plot (b). In (d) a mask was constructed in the score plot, by trial-and-error, for selecting the pixels regarding badly coated areas. The corresponding positions of such pixels in the original image are presented in (c). (This analysis was conducted with the MACCMIA software, developed by the McMaster University Advanced Control Consortium, freely available at http://macc.mcmaster.ca/maccmia.php).

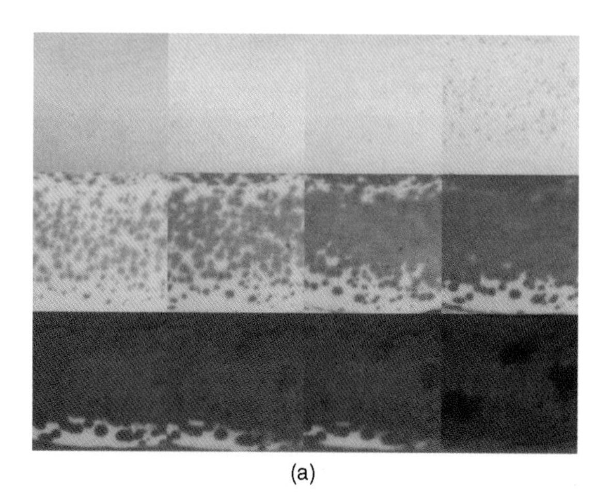

(a)

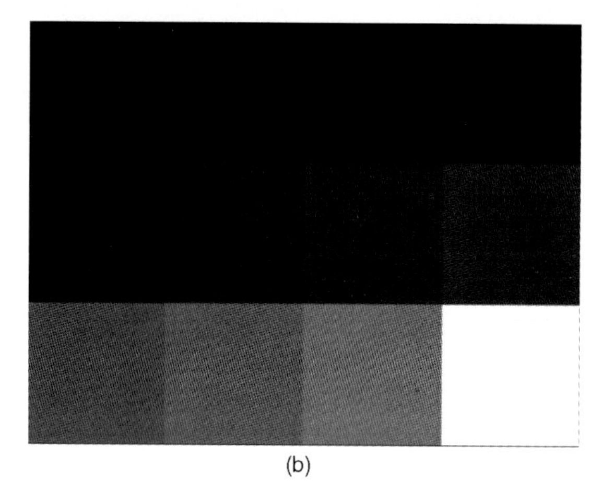

(b)

**Plate 12.7** (a) Composite image of fruit (banana) with different storage times in the interval 1–20 days; (b) the associated Y-image with the corresponding storage times for each pixel of the composite image. Lied, T. T. & Esbensen, K. H. (2001) Reproduced with permission from Elsevier.

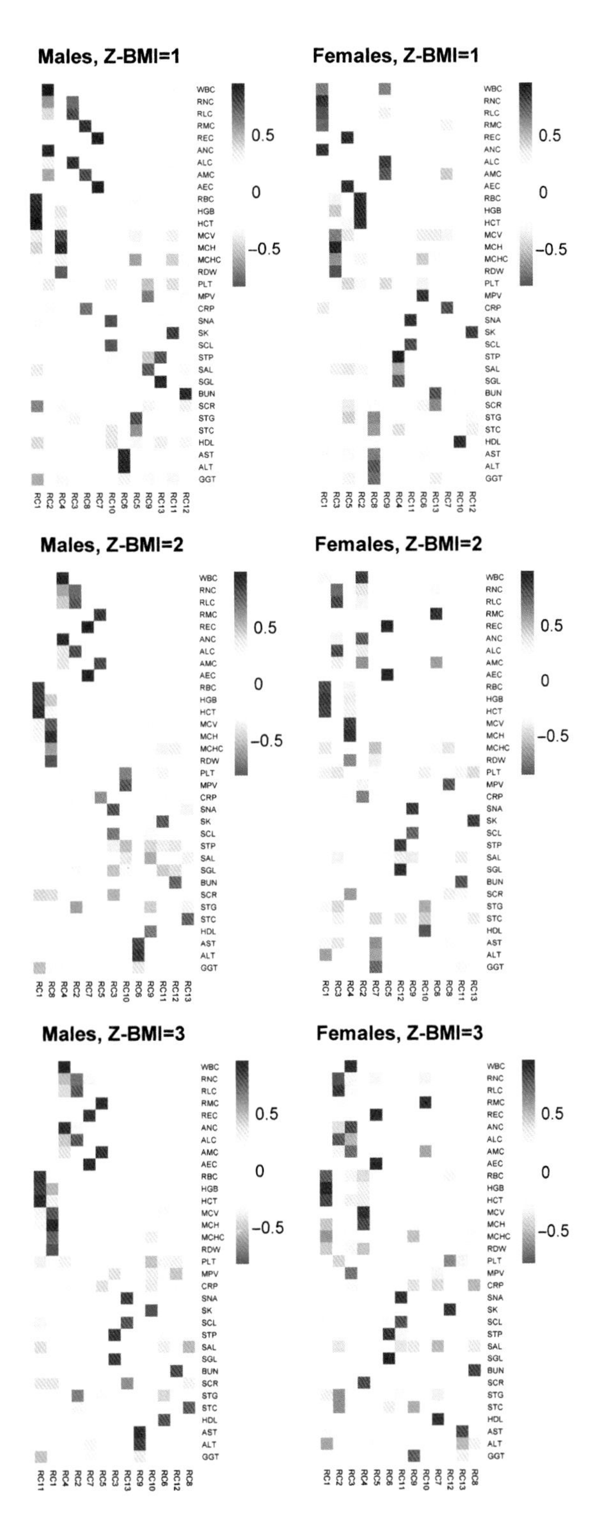


**Plate 17.3** Results of the PCA (structure matrices visualized with heat maps) for different Z-BMI levels, segregated according to sex. White colour indicates zero correlation, red means +1, blue means −1.

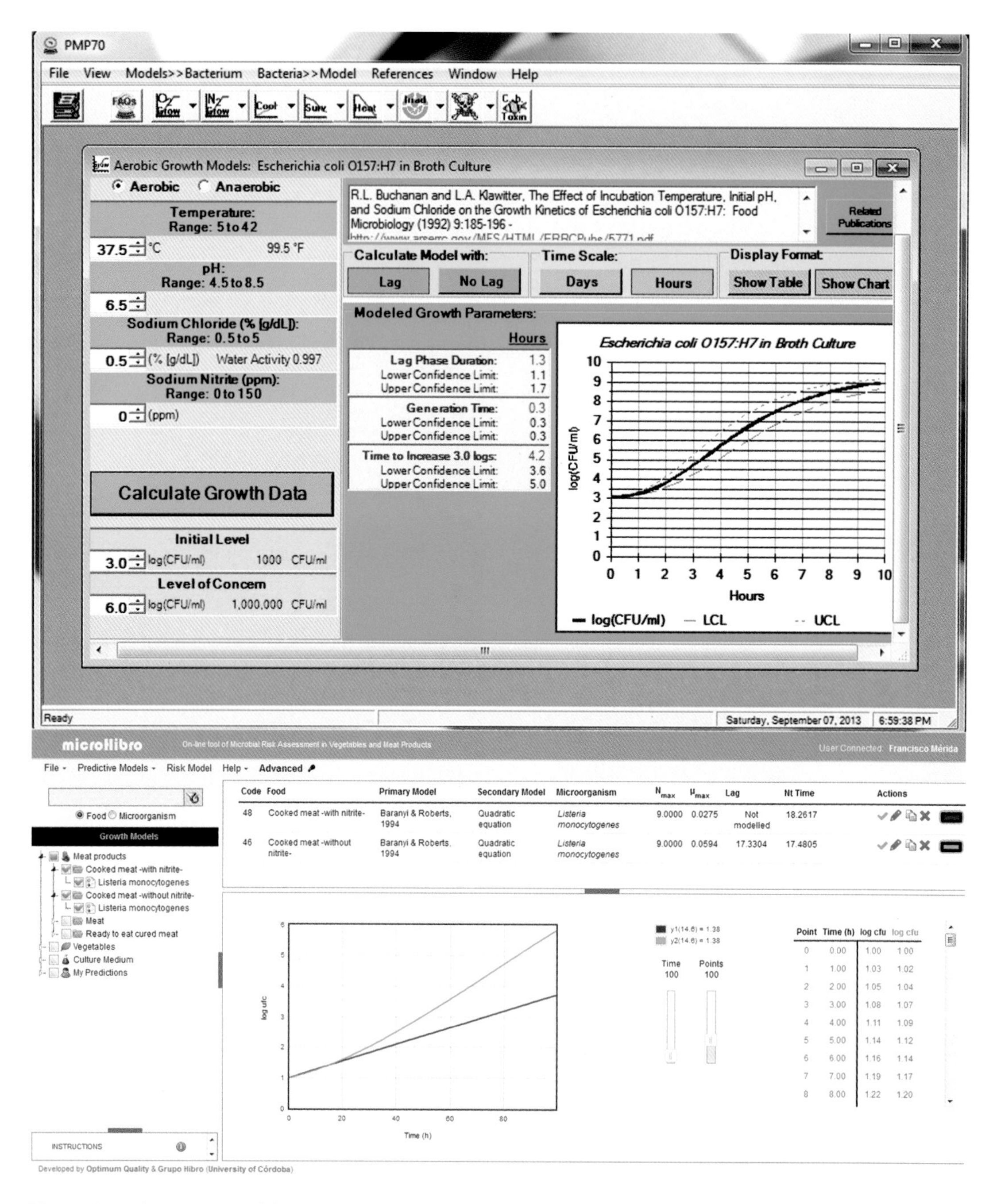


**Plate 18.10** Screen prints of the tertiary models Pathogen Modeling Program (PMP) (on the top). Courtesy Pathogen Modeling Program (PMP), USDA Agricultural Research Service, http://portal.arserrc.gov/PMP.aspx and Microhibro (www.microhibro.com) (on the bottom). Courtesy of Microhibro www.microhibro.com.

appear otherwise to be uncontaminated (Jongenburger, 2012). Thus, examination of such foods requires a totally different approach to sampling surveillance. The data generally conform to a NB or one of the other related distributions such as the Beta-Poisson distribution.

The general form of the NB probability distribution is given by:

$$\begin{aligned} P_x &= \left(1+\frac{\mu}{k}\right)^{-k}\left(\frac{(k+x-1)!}{x!(k-1)!}\right)\left(\frac{\mu}{\mu+k}\right)^x \\ &= \left(1+\frac{\mu}{k}\right)^{-k}\frac{\Gamma(k+x)}{x!\Gamma(k)}\left(\frac{\mu}{\mu+k}\right)^x \end{aligned} \tag{15.3}$$

where $P_x$ is the probability that $x$ organisms occur in a sample unit, $\mu$ is the population mean, with variance $\sigma^2$, $k$ is the negative binomial parameter and $\Gamma$ is the gamma function. Estimates of $\mu$ and $\sigma^2$ are obtained from the experimentally determined values of the mean ($\bar{x}$) and variance ($s^2$) and an approximate estimate of the parameter $k$ can be obtained from:

$$k = \frac{\mu^2}{(\sigma^2-\mu)} \approx \frac{\bar{x}^2}{(s^2-\bar{x})} \tag{15.4}$$

In practice, it is necessary to refine the estimated approximate value for $k$ (Jarvis, 2008).

For a series of values, the zero term is found by setting $x=0$ and simplifying Equation 15.3 to:

$$P_{(x=0)} = \left(1+\frac{\mu}{k}\right)^{-k} \tag{15.5}$$

Successive terms are computed iteratively:

$$P_x = P_{(x-1)}\left(\frac{k+x-1}{x}\right)\left(\frac{\mu}{\mu+k}\right) \tag{15.5.1}$$

So, for example, for $x=1$:

$$P_{(x=1)} = P_{(x=0)}\left(\frac{k+1-1}{1}\right)\left(\frac{\mu}{\mu+k}\right) = P_{(x=0)}\,k\left(\frac{\mu}{\mu+k}\right) \tag{15.5.2}$$

and for $x=2$:

$$P_{(x=2)} = P_{(x=1)}\left(\frac{k+2-1}{2}\right)\left(\frac{\mu}{\mu+k}\right) = P_{(x=1)}\left(\frac{k+1}{2}\right)\left(\frac{\mu}{\mu+k}\right) \tag{15.5.3}$$

and so on.

Figure 15.5 shows that the Listeria data shown previously (Figure 15.4b) conform well to a NB distribution with $k=1.45$. Since $k$ cannot be a negative value, the NB distribution cannot be fitted to a data set where the variance is less than the mean; here a binomial distribution is more appropriate.

### *The binomial distribution*

The binomial distribution has two parameters $p$ and $n$ that describe the proportion of successes in a series of $n$ Bernoulli trials, where the outcomes are random and can only be positive or negative. A common example of a Bernoulli trial in microbiology is a test for the presence or absence of a microorganism in a sample; only one of two possible outcomes can occur, i.e. growth or no growth.

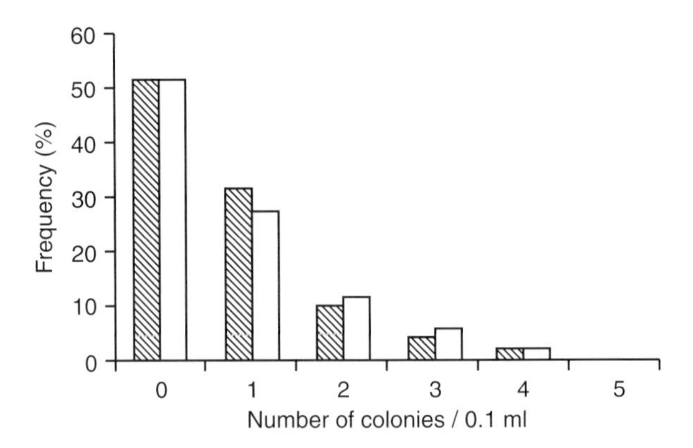


**Figure 15.5** Frequency plot for counts (as Figure 15.4b) from an 'over-dispersed' suspension of *Listeria* cells with mean = 0.83 cells/0.1 ml (hatched) and a negative binomial distribution (open) for mean = 0.83 and *k* (amended) = 1.45.

The proportion of positive results (successes) in a trial is referred to as $p$ and that of negative results (failures) as $q$. Since only a positive or a negative result is possible, $p+q=1$; or, $p=(1-q)$. The probability ($P_{(x)}$) of getting $x$ successes in $n$ parallel trials is given by:

$$P_{(x)} = \binom{n}{x} p^x q^{n-x} = \binom{n}{x} p^x (1-p)^{n-x} \tag{15.6}$$

where $\binom{n}{x}$, the binomial coefficient (Equation 15.7), calculates the number of ways of choosing $x$ from $n$,

$$\binom{n}{x} = \frac{n!}{x!(n-x)!} \tag{15.7}$$

and ! means 'factorial'. So if $n=6$ and $x=4$, there are 15 ways of selecting one positive:

$$\binom{n}{x} = \binom{6}{4} = \frac{6!}{4!(6-4)!} = \frac{6 \times 5 \times 4 \times 3 \times 2}{(4 \times 3 \times 2)(2)} = \frac{6 \times 5}{2} = 15 \tag{15.7.1}$$

The mean of the binomial is $np$ and its variance is $npq$. Since $q = <1$, the variance of the distribution will be less than the mean. It can, therefore, be used as a model for any data that conform to a 'regular' distribution.

### *Summary*

Statistical distributions, and the relationships between distribution models, are important for understanding the analysis of data. Although equations provide mathematical descriptions of the different distributions, most critical values for deriving probability can be obtained from published tables of statistical values (Fisher and Yates, 1963; Murdoch and Barnes, 1986). But such tables do not always provide an explanation of how they are derived or their relevance. In addition to the distributions discussed above, others, for example the $t$, $\chi^2$ and $F$ distributions, are also important in various tests.

## EXPERIMENTAL DESIGN AND PLANNING

Much has been written about experimental design (Fisher, 1971; Mead *et al.* 2002; Crawley, 2005; Casella, 2008). It is axiomatic that the purpose of an experiment is to draw conclusions about specific attributes of a population. Such attributes may be descriptive observations (e.g. aroma of a product) or chemical, physical and microbiological measurements (e.g. acidity, pH value, presence or absence of specific microbes, numbers of CFU/g etc.). The 'population' may relate to a set of food items (e.g. a production batch) or, in a laboratory experiment, to replicate cultures of defined microbes.

The overriding requirements for any experiment are to decide the objectives, define null and alternative hypotheses and plan the work to ensure that the objectives can be achieved, including identifying the statistical methods to be used. Developing and testing a set of hypothetical data can be useful to ensure that proposed statistical tests are appropriate and that the experimental plan should work, thereby avoiding frustration (Dytham, 2011). It is essential to avoid bias by using a sampling plan that ensures samples are drawn randomly: the selection of any one sample must be independent of the selection of all other samples. Randomization can be achieved by effective use of random numbers.

An experimental objective may be to determine whether or not specific organisms are present in one or more batches of food or to estimate levels of microbial contamination of a product, knowing the processing and packaging conditions. In the laboratory, the objective may be to compare treatments applied to a population of microbes to assess resistance to thermal or chemical treatment or to determine growth characteristics under defined conditions. Experimental planning requires identification of dependent and independent variables, with some prior knowledge of errors (variances) associated both with distribution of the target analyte and the method used for its determination. Also important is the power of any statistical test. The probability that a statistical test will wrongly reject a null hypothesis ($H_0$) when it is true is defined as $\alpha$ and referred to as a Type I error. Similarly, the probability for accepting a null hypothesis ($H_0$) when it is false is defined as $\beta$ and called a Type II error. Statistical power ($=1-\beta$) is the probability of rejecting $H_0$ when it is false, so the greater the power the greater the chance of correctly rejecting $H_0$. Statistical power is related to the size of sample required to detect an effect of a given size: the larger the sample size the greater the power of the test to detect small differences in effects.

### Sampling for surveillance

Before carrying out the work many questions need to be answered, including how best to ensure that samples are both representative and obtained randomly. If it is intended to take samples from a production line or from a warehouse, can the entire product be accessed or must sampling be stratified, i.e. from parts easily accessed? How many replicate samples should be taken? How should the samples be transported to minimize changes in microbial types and levels before testing? When the primary samples arrive in the laboratory, how representative is each test sample (i.e. the amount that is actually tested) of each primary sample? Is replication of laboratory sampling and testing necessary? What tests are to be done? And so on.

Suppose that the requirement is to establish the levels of organisms in 500 g packs of a product sampled from a production line during processing. If the total mass of product is say, one tonne, then the manufacturing process should yield 2000 packs of product. How many packs should be tested? Many microbiological criteria require five replicate random samples to be tested (ICMSF, 1986; EU, 2005) but this presumes, amongst other things, that products are produced under steady state conditions. For a study on a new process, or for a comparative study of product manufactured in several process plants, such sample sizes would be too small.

Firstly, it is necessary to define limits for acceptable numbers of 'defective' units and a statistical probability for acceptance of defective product. A 'defective' sample might be one containing a pathogen or an unacceptable level of microbes. The number of samples to be tested can be determined using an equation based on binomial theory. For detection of an agreed maximum prevalence of defectives (say, $d = 5\%$), with a probability of $P = 0.90$, the number of sample packs ($n$) to be tested would be derived as:

$$n = \frac{\log_{10}(1 - P)}{\log_{10}(1 - d)} = \frac{\log_{10} 0.10}{\log_{10} 0.95} = \frac{-1.00}{-0.022} \approx 45 \quad (15.8)$$

Similarly, to be 99% certain of detecting 5% defectives then from (15.8) $n = 90$ samples; but if $d = 0.1\%$, and $P = 0.99$, $n = 4603$ samples!

One solution is to examine a smaller number of samples than might be considered ideal and then to mathematically model the expectation based on the available data (see the section 'Modelling procedures' and Example 9). Start by defining a realistic sample size by reference to the precision of the test. For a set of data that is expected to conform to ND, precision is measured as the standard error of the mean (SEM) $= \sqrt{(s^2/n)}$, where $s^2$ = variance and $n$ = number of samples. Figure 15.6 shows that as $n$ increases $1/\sqrt{n}$ decreases: the initial decrease is rapid but the rate of change slows at $n = 8$. For $n = 2$, $1/\sqrt{n} = 0.707$ and for $n = 8$, $1/\sqrt{n} = 0.354$; hence, the precision of the mean is doubled by testing eight rather than two samples; above $n = 8$, the increased benefit gained by increasing sample numbers is reduced and a further doubling of the precision requires analysis of 32 samples ($1/\sqrt{32} = 0.177$). If the data do not conform to lognormal then it would be essential to analyse at least 12 replicate samples (Jarvis and Hedges, 2011). More detail on sampling for microbiological examination of foods can be found elswhere (Jarvis, 2000, 2008).

## Sampling and testing in the laboratory

When a set of samples arrives in the laboratory it is often assumed that each individual sample is homogeneous; but in the same way that microbes are distributed throughout a batch of product, they will also be distributed within the sample. Ideally the total mass of each sample should be tested but this is not normally feasible. In order to reduce within-sample variation chemists often combine many subsamples

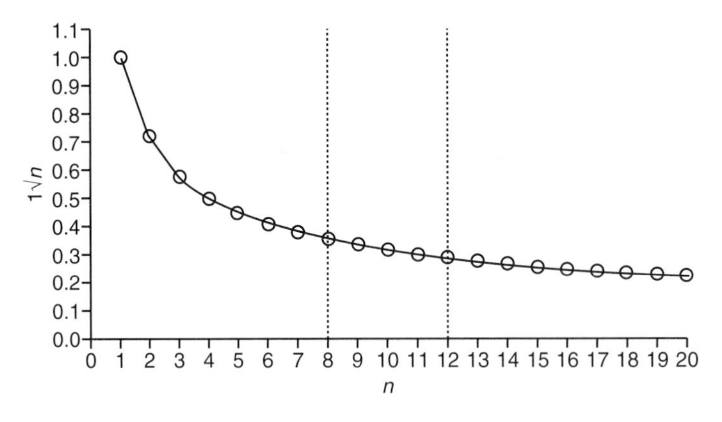


**Figure 15.6** Relationship between $1/\sqrt{n}$ and $n$, where $n$ = number of sample units tested. Note that the slope of the curve flattens out at about $n = 8$ where $1/\sqrt{n}$ is half the value obtained when $n = 2$ but still 22% greater than when $n = 12$. Reproduced with permission of the copyright holder from Jarvis & Hedges, 2011.

to derive a mixed test sample for analysis. Traditionally, microbiologists have argued that this is not possible because of the risks of cross-contamination but this is a fallacy. Provided that appropriate aseptic techniques are used to blend and extract organisms from a sample, subsamples can be composited. It is also essential to ensure that the mass of sample is sufficiently large to be representative of the whole, for example 100 g is better than 10 g. Corry *et al.* (2010) demonstrated how to minimize intrasample variation for a diverse range of foods by use of aseptic blending and sampling procedures.

### Setting up laboratory experiments

Laboratory experiments are more controllable than those based on surveillance studies and require two key concepts: replication and randomization. Replication is essential to increase reliability, randomization to reduce bias. Recognition of the difference between experimental and observational data is also essential. Other important issues include the principles of parsimony, the importance of properly designed controls and the power of statistical tests. The term 'parsimony' implies that models should have as few parameters as possible; linear models are preferred to nonlinear models and experiments relying on a few assumptions are preferred to those relying on many. But there is a limit to simplification. Einstein is reputed to have said: 'everything should be as simple as it is, but not simpler'. More information on these and other criteria for experimental design can be found in the references cited previously. If the experiment is not properly designed, not thoroughly randomized, or lacks adequate controls then no matter how capable an operator is at statistical analysis, the experimental efforts will be wasted!

## STATISTICAL METHODS FOR ANALYSIS OF DATA

### Descriptive statistics

For any set of quantitative data, it is useful to estimate descriptive statistics of the data, for example the mean, median, mode, variance, standard deviation, standard error of the mean (SEM) and coefficient of variation. In addition, tests can be included for outlier data values (i.e. values that would not be likely to occur in a set that complies with the ND) and distribution plots for a ND. If the data do not conform to ND, as with microbial cell and colony counts, it will be necessary to transform the data before doing these analyses. For low numbers of cells and colony numbers that conform to Poisson distribution a square root transformation is used, whilst colony counts are usually log transformed. Most analyses can be done using standard statistical calculations included in a spreadsheet, such as Excel, or using one of the many Excel statistical add-on packages, such as Analyse-it® or Minitab®, or using high-power statistical packages, such as SPSS.

### Comparing independent data sets

#### *Use of 'Student's' t-test to compare quantitative data*

Student's *t*-test is used to test the null hypothesis that two sets of data are from the same population (i.e. they are equal). The approach depends on whether the data are paired or are independent and sometimes it is difficult to choose the correct version of the test. To satisfy the requirement that data are paired it is essential to be able to identify the basis of the pairing. For instance, Paulson (2008) provides an example relating to recovery of bacterial spores on two media after treatment with a sporicidal agent. Because both sets of colony counts were derived from the same set of test dilutions the counts are 'paired'. However, if two parallel series of tests had been done with spores in one group being treated with a different compound to the other group then a nonpaired $t$ test would be used.

*The* t-*test for unpaired samples*

Student's $t$-test measures the ratio (Equation 15.9) of the difference between two mean results (i.e. $\bar{x}_1 - \bar{x}_2$) divided by the combined standard error ($SE_{com}$) of the two data sets, which is determined using Equation 15.10 or Equation 15.10.1:

$$t = (\bar{x}_1 - \bar{x}_2)/SE_{com} \quad (15.9)$$

$$SE_{com} = \sqrt{s_1^2/n_1 + s_2^2/n_2} \quad (15.10)$$

where $s_1^2$ and $s_2^2$ are the variances of two data sets with $n_1$ and $n_2$ replicates, respectively for $\nu = (n_1 + n_2 - 2)$ degrees of freedom (df). If sample numbers are equal (i.e. $n_1 = n_2 = n$) the equation simplifies to:

$$SE_{com} = \sqrt{(s_1^2 + s_2^2)/n} \quad (15.10.1)$$

for $\nu = (n - 1)$ df.

The critical value of $t$ is obtained from tables for $\nu$ df and the relevant probability. Note that depending on the definition of the hypotheses, a choice must be made between a one-sided and a two-sided $t$-value. For instance, if $H_0$: C = T and $H_1$: C ≠ T, then a two-sided $t$-test is appropriate, but if $H_0$: C = T and $H_1$: C < T or $H_1$: C > T, then a one-sided $t$-value should be used.

A basic requirement of the $t$-test is that the data should be normally distributed, after transformation if appropriate, and that the variances of the (transformed) data sets should be roughly equal. If the variances are unequal, it is necessary to use Welch's approximation for unequal variances (Equation 15.11) (Welch, 1947). This test does not alter the calculation of $t$ but calculates a lesser df ($\nu$) for the test:

$$\nu = \frac{\left(\frac{s_1^2}{n_1} + \frac{s_2^2}{n_2}\right)^2}{\frac{s_1^4}{n_1^2 \cdot (n_1 - 1)} + \frac{s_2^4}{n_2^2 \cdot (n_2 - 1)}} \quad (15.11)$$

If the variances of the two data sets differ markedly, use of Welch's approximation can avoid errors in interpretation of the significance of the null hypothesis. Most software packages apply Welch's approximation and, as shown in Example 15.1, the df may change considerably.

## Example 15.1 Analysis of data from a fermentation trial

You carry out an experiment to test the effect of two supplements (S1 and S2) on cell growth and alcohol production by *Saccharomyces cerevisiae* and set up three temperature controlled fermentation vessels, each containing 10 litres of apple juice. You add a defined quantity of a sterile solution of S1 to one vessel, an equal quantity of S2 to the second and water to the third (control; C). Each vessel is inoculated with $10^6$ viable yeast cells/ml and immediately after inoculation a sample is withdrawn aseptically for estimation of yeast numbers and alcohol content. During incubation, further samples are withdrawn after defined time periods using a pre-determined sampling schedule (Table Ex 15.1.1) that minimizes the risk of bias in sampling. Each sample is immediately chilled to prevent further

**Table Ex 15.1.1** Example of a sampling schedule for three fermentation vessels each of which is sampled on six occasions; the purpose of changing the sequence is to avoid bias in the timing of sampling.

| | Sampling sequence at time (h) | | | | | |
|---|---|---|---|---|---|---|
| | **0** | **6** | **12** | **24** | **48** | **72** |
| Control | 1 | 2 | 3 | 2 | 1 | 3 |
| Test S1 | 2 | 1 | 1 | 3 | 3 | 2 |
| Test S2 | 3 | 3 | 2 | 1 | 2 | 1 |

growth and metabolism and, after taking a subsample to estimate yeast cell numbers, the remainder is centrifuged to provide a cell-free extract for alcohol analysis (and a pellet of biomass to estimate changes in cell mass). Results are recorded for each independent treatment and incubation time and graphed using an *x-y* plot (Figure Ex15.1.1). Compared to the control (C), alcohol production is stimulated by supplement S1 but repressed by S2; this suggests that the effects require confirmation by repeating the experiment.

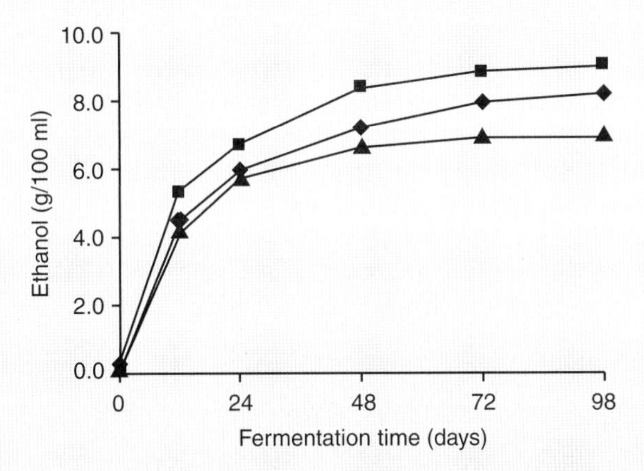


**Figure Ex 15.1.1** Ethyl alcohol production by *S. cerevisiae* in control (◆) and supplemented (S1 ■, S2 ▲) apple juice in the first trial.

The trial is repeated three times and the data for each trial are combined (Table Ex 15.1.2) to show the overall mean, variance and standard deviation for alcohol (g/100 ml) for each sampling time and each independent trial, with $n = 8$.

Inspection of the data confirms that supplement S1 increased the yield of alcohol but S2 reduced it. Are the differences statistically significant? For this example we will consider only the total alcohol content after 96 h incubation. We define the null hypothesis as $H_0$: C = S and the alternative hypothesis as $H_1$: C ≠ S.

We assume that the data conform to ND and for comparison of C and S1 we use Student's *t*- test for unpaired samples with equal variances. The test compares the difference between the mean values of each set of results against the combined standard error. As $n_C$ and $n_T$ are equal we use the combined version of the equation, with $\nu = n - 1$

$$t = \frac{\bar{C} - \bar{T}}{\sqrt{\left(s_C^2 + s_T^2\right)/n}}$$

**Table Ex 15.1.2** Summary data for alcohol levels (g/100 ml) for the three treatments over four replicate trials.

| Sample time (h) | C | | | S1 | | | S2 | | |
|---|---|---|---|---|---|---|---|---|---|
| | Mean | VAR | SD | Mean | VAR | SD | Mean | VAR | SD |
| 0 | 0.111 | 0.000 | 0.022 | 0.11 | 0.001 | 0.024 | 0.12 | 0.001 | 0.024 |
| 12 | 4.613 | 0.006 | 0.076 | 5.41 | 0.018 | 0.134 | 4.51 | 0.018 | 0.133 |
| 24 | 6.043 | 0.016 | 0.126 | 6.72 | 0.018 | 0.133 | 5.75 | 0.007 | 0.082 |
| 48 | 7.276 | 0.002 | 0.050 | 8.64 | 0.025 | 0.157 | 6.68 | 0.007 | 0.083 |
| 72 | 8.108 | 0.009 | 0.096 | 8.95 | 0.003 | 0.056 | 7.15 | 0.039 | 0.198 |
| 96 | 8.433 | 0.002 | 0.044 | 9.07 | 0.002 | 0.041 | 7.78 | 0.021 | 0.146 |

where $\bar{C}$ and $\bar{T}$ are the mean values of the control and test, respectively, for which $s_C^2$ and $s_T^2$ are the respective variances, with $n=8$ replicate tests and $\nu=n-1=7\,df$

$$t=\frac{\bar{C}-\bar{T}}{\sqrt{\left(s_C^2+s_T^2\right)/n}}=\frac{8.433-9.068}{\sqrt{(0.00194+0.00168)/8}}=\frac{-0.635}{\sqrt{0.0004525}}=\frac{-0.635}{0.02127}=-29.9$$

The absolute value of $t=|t|=29.9$ is much greater than the two-sided critical value of $t_{0.05,\nu=7}=2.262$, from tables of the $t$ distribution, so we reject the null hypothesis and accept the alternative hypothesis that $C\neq S_1$.

We repeat this calculation for data from C and S2, but with the alternative hypothesis that $H_1$: C > S2 (C − S2 = 0.655). Because the variances of these data are unequal (0.002 and 0.021) we must use the full version of the unpaired $t$-test together with Welch's approximation for unequal variances (Equation 15.10) (Welch, 1947), which recalculates the $df$. We insert the values into the equation:

$$t=\frac{\bar{C}-\bar{T}}{\sqrt{\left(s_C^2/n_C+s_T^2/n_T\right)}}=\frac{8.43-7.78}{\sqrt{(0.044^2/8+0.146^2/8)}}=\frac{0.65}{\sqrt{0.00291}}=\frac{0.65}{0.0539}=12.06$$

If the variances had been equal the degrees of freedom would have been $\nu=8+8-2=14$ but Welch's correction gives $\nu=10.2$ for which the one-sided tabulated $t$ is $t_{0.05,\nu=10.2}=1.82$; since the calculated value of $t=12.066>1.82$, we again reject $H_0$ and accept $H_1$. The product yield for the S2 supplemented media is significantly smaller than that for the control. An alternative approach to analysis of these data uses analysis of variance (Example 15.6).

## Example 15.2 A paired $t$-test analysis of bacterial colony counts (data from Paulsen, 2008)

This example uses a 'paired' $t$-test to examine data for a suspension (about $10^5$/ml) of spores of *B subtilis* treated with a solution of sodium hypochlorite and plated in duplicate onto two media to determine the level of residual spores. The culture medium is a soil extract nutrient agar prepared either with reagent grade deionized water (Control = C) or with a locally produced deionized water (Test = T). Samples taken after treatment of the spores for 5 min were immediately neutralized

**Table Ex 15.2.1** Colony counts of *B subtilis* spores on two different media after treatment for 5 min with sodium hypochlorite at 300 μg/l (reproduced and modified with permission from Table 3.2 of Paulson (2008); Bioscience Laboratories Inc., Bozeman, MT, USA). Paulson, 2008.

| Block pair (*n*) | $Log_{10}$ CFU/ml | | | $d^2$ |
|---|---|---|---|---|
| | **Medium C** | **Medium T** | **Difference *d* = (C − T)** | |
| 1 | 3.8 | 4.1 | −0.3 | 0.09 |
| 2 | 4.7 | 4.3 | +0.4 | 0.16 |
| 3 | 5.2 | 4.9 | +0.3 | 0.09 |
| 4 | 4.7 | 4.7 | 0.0 | 0.00 |
| 5 | 3.9 | 3.8 | +0.1 | 0.01 |
| 6 | 3.3 | 3.2 | +0.1 | 0.01 |
| 7 | 3.6 | 3.4 | +0.2 | 0.04 |
| 8 | 3.2 | 3.0 | +0.2 | 0.04 |
| 9 | 3.3 | 3.1 | +0.2 | 0.04 |
| Total | | | +1.2 | 0.48 |
| Mean | | | +0.133 | |

to deactivate residual hypochlorite and dilutions were surface plated on both media. The colony counts are recorded as $\log_{10}$ CFU/ml. Each of several trials is regarded as a separate 'block'. The null hypothesis is $H_O$: C = T, i.e. there is no statistically significant difference between the media; the alternative hypothesis is $H_1$: C ≠ T, i.e. the media differ in their ability to recover treated spores. Data are shown in Table Ex 15.2.1.

The difference ($d$) for each pair of colony counts is calculated and the overall mean difference $\bar{d}$ and the square of the differences ($d^2$) determined. The variance of the differences is $s_d^2 = \frac{\sum d^2 - ((\sum d)^2/n)}{(n-1)}$, where $n = 9$. For these data, $\sum d^2 = 0.48$ and $(\sum d)^2/n = 1.2^2/9 = 0.16$ so $s_d^2 = (0.48 - 0.16)/8 = 0.040$ and $s_d = \sqrt{0.040} = 0.20$. The SE is $s_d/\sqrt{n} = 0.20/\sqrt{9} = 0.067$.

The value of $t$ is calculated as $t = \frac{\bar{d}}{s_d/\sqrt{n}} = 0.133/0.067 \approx 1.99$ with $\nu = n - 1 = 8$ *df*.

The calculated $t$-value of 1.99 is less than the critical two-sided estimate of $t_{(p=0.05,\nu=8)} = 2.306$ from tables, so the null hypothesis ($H_O$) is not rejected. We conclude that the difference in the recovery of treated spores on the two media is not statistically significant and reject the alternative hypothesis ($H_1$).

It is generally assumed that differences between data values will conform to ND, so use of the paired $t$-test is valid. In this example it was not actually necessary to log transform the colony counts before determining the differences, other than to simplify data handling. The overall difference between the original paired counts was almost 12 000 CFU/ml, the SD of the differences 27 200 and the SE was 9000. The $t$ statistic was 1.32 with $\nu = 8$ *df*, which again did not reject $H_O$.

*Paired samples*

The procedure (Example 15.2) differs from that for unpaired data. The approach is to assess differences between pairs of results using Equation 15.12:

$$s_d^2 = \frac{\sum d^2 - \left((\sum d)^2/n\right)}{(n-1)} \qquad (15.12)$$

where $d$ = the difference between each pair of values, $d^2$ = the square of the differences and $n$ = number of data pairs. The $SE_d$ is determined as $SE_d = s_d/\sqrt{n}$; it is used (Equation 15.13), with $\bar{d}$ = overall mean difference, to determine $t$:

$$t = \frac{\bar{d}}{SE_d} \text{ for } \nu = (n-1)\text{df} \tag{15.13}$$

The critical value of $t$ is obtained from tables for the appropriate level of probability and df.

Only pairs of data can be tested using this procedure – if one result is missing the other must be excluded; introducing a hypothetical replacement value totally undermines the credibility of the test! Note also that difference values usually conform to ND so that transformation of data is not required – but high numbers involved in comparing, for example nontransformed colony counts, make calculation unwieldy.

### *Nonparametric tests for comparison of quantitative data*

Nonparametric tests can be useful when data distributions do not conform to ND and when data consist of ordinal values, for example scored responses. Example 15.3 provides illustrations for unpaired and paired nonparametric data comparisons. The procedures require the data values to be ranked and the ranked values contrasted – hence these are considered to be 'distribution-free' analyses. Although ranking methods do not depend on any particular distribution, they do assume that both distributions have the same shape and, if the variances are markedly unequal, the tests should not be used. For a two-sample test with disparate variances it is preferable to use 'Welch's $t$-test' that is based upon means (which tend to ND by the central limit theorem). Parametric and nonparametric tests can also be used to compare data from experiments with more than two independent variables. Nonparametric analysis of variance is discussed in the section 'Analysis of Variance (ANOVA)'.

## Example 15.3 Nonparametric methods for comparison of non-normal data

### Ex 15.3.1 The Mann–Whitney test for unpaired data

An experiment was done to compare the release of Enterobacteriaceae from meat. Two subsamples, each taken from one of 12 separate samples, were 'extracted' in either a Stomacher® (S) or a Pulsifier® (P) and serial tenfold dilutions were plated onto VRBG Agar. The null hypothesis is that the counts derived from the two methods will be the same, $H_0$: P = S; the alternative hypothesis is that the two results will differ significantly, $H_1$: P ≠ S.

These data are interesting both to illustrate calculation of a nonparametric statistic and because they are tested as unpaired samples. Each individual value was obtained by examination of a separate subsample, so it is **not** appropriate to consider them to be paired! However, if in each case the two subsamples had been carefully mixed before testing the individual subsamples the differences in the counts could have been tested using the Wilcoxon signed-rank test for paired data (see below), in which case the overall conclusion would have been that there is no statistically significant difference in the two methods of extraction. This illustrates the importance of proper experimental planning before carrying out practical work!

The colony counts (as CFU × $10^{-3}$/g) are tabulated and ranks are assigned in increasing order to the entire set of observations (note that identical counts each receive the average of their ranks).

| Sample # | Stomacher® | | Pulsifier® | |
|---|---|---|---|---|
| | CFU × $10^{-3}$/g | Rank ($R_S$) | CFU × $10^{-3}$/g | Rank ($R_P$) |
| 1 | 870 | 2 | 820 | 1 |
| 2 | 1900 | 21 | 1360 | 11 |
| 3 | 1600 | 15.5 | 1020 | 3 |
| 4 | 1700 | 19 | 1260 | 8 |
| 5 | 1030 | 4 | 1430 | 12 |
| 6 | 1280 | 9 | 1610 | 17 |
| 7 | 1990 | 22 | 2170 | 23 |
| 8 | 1590 | 14 | 1600 | 15.5 |
| 9 | 1040 | 5 | 1080 | 6 |
| 10 | 1090 | 7 | 1460 | 13 |
| 11 | 2560 | 24 | 1740 | 20 |
| 12 | 1350 | 10 | 1670 | 18 |
| Total Rank | 152.5 | | 147.5 | |

The totals of the ranks are: $\sum R_S = 152.5$; $\sum R_P = 147.5$

Calculate:

$$U_S = [n_S(n_S + 1)/2 + n_S n_P] - \sum R_S = (12 \times 13)/2 + (12 \times 12) - 152.5 = 69.5$$

$$U_P = [n_P(n_P + 1)/2 + n_S n_P] - \sum R_P = (12 \times 13)/2 + (12 \times 12) - 147.5 = 74.5,$$

where $n_S$ and $n_P$ are the numbers of observations in each group. The smaller value of $U_P$ or $U_S$ is used as the test statistic and must be $\leq$ Wilcoxon's tabulated critical value for $U$ to be of statistical significance (Snedecor and Cochran, 1980).

In this example, the smaller value, $U_S = 69.5$, is less than the critical value (=115) at $P = 0.05$ for $n_S = n_P = 12$ so $H_0$ is rejected: the difference between the two data sets is statistically significant. We conclude that in this experiment the average recovery of Enterobacteria is significantly lower using the Pulsifier than using the Stomacher but the difference is not of practical significance!

## Ex 15.3.2 The Wilcoxon signed-rank test for paired data

The data were produced in an experiment to assess the value of adding the surfactant Tween 20 to a culture medium for estimates of total viable count (TVC) on meats. A subsample of 10 g of each of 10 samples of bowl-chopped meats was 'stomached' in 90 ml of diluent. Serial tenfold dilutions of each initial homogenate were plated in duplicate on a nutrient agar containing 1.0% Tween 20 (A) and on standard nutrient agar (B). The colony counts on the two media are 'paired', because each was made from the same dilutions of the same subsample. Because each original sample was different, it represents a distinct population that may or may not be similar, so use of a *t*-test is inappropriate. A Wilcoxon signed-rank nonparametric test for paired samples is appropriate to assess the results. The null hypothesis is that the median difference between results on the two media is zero, $H_0$: A = B; the alternative hypothesis is that the median values of the two tests differ significantly, $H_1$: A ≠ B.

Set out the data in a table, determine the differences between the mean colony counts (A – B) and, ignoring the sign, rank each difference from 1 to 10. List the differences and restore the sign value to each of the ranks. Two (or more) identical differences each receive the average score. A zero difference is discarded and, for each zero value, the number of test pairs is reduced by one.

| Meat | Colony Count (CFU × $10^{-4}$/g) | | Difference | Rank |
|---|---|---|---|---|
| | Medium A | Medium B | (A – B) | (R) |
| 1 | 154 | 119 | 35 | +10 |
| 2 | 113 | 127 | –14 | –7 |
| 3 | 8 | 10 | –2 | –4 |
| 4 | 106 | 105 | 1 | +2 |
| 5 | 5 | 10 | –5 | –5.5 |
| 6 | 7 | 6 | 1 | +2 |
| 7 | 10 | 5 | 5 | +5.5 |
| 8 | 120 | 105 | 15 | +8 |
| 9 | 5 | 6 | –1 | –2 |
| 10 | 131 | 112 | 19 | +9 |
| Median | 58 | 57 | | |
| Sum of + ranks | | | | 36.5 |
| Sum of – ranks | | | | –18.5 |

Determine the sum of positive (+R) and negative (–R) ranks and compare the smaller number ($T_{-R}$) with Wilcoxon's tabulated critical values for T for $n$ pairs; the result is significant if $T_R \leq T_{critical}$. In this case, with $n = 10$ and $P = 0.05$, the two-sided value of $T_{critical} = 8 < T_{-R} = 18.5$, so the difference is not statistically significant and the null hypothesis is accepted. In other words, the test does not show that addition of Tween 20 to the agar increases the TVC.

### *Testing the significance of the difference between proportions*

Surveillance data for the prevalence of specific organisms in a food are not continuous distributions, such as occur in quantitative estimates of independent variables; nor are the numbers of positive results obtained in qualitative comparisons of culture media. Often such data sets may include categorical, nominal or ordinal data. Categorical data are those for which a descriptive label can be allocated, for instance a country or a factory. Nominal data are those that can be labelled with a code to define a class, for example a negative (0) or a positive (1) result; or some independent parameter, such as 'chilled' or 'ambient', that describes storage conditions. Other data may be ranked on an arbitrary (ordinal) scale that defines characteristics, such as the visual assessment of turbidity in a liquid culture, which might be graded as (0) none, (1) light, (2) moderate or (3) heavy. Before analysis it is essential to identify the format of the data, as different statistical approaches may be necessary.

Differences between frequencies (or proportions) can be assessed using nonparametric tests of association, such as Pearson's $\chi^2$–test on a $2 \times N$-way classification (a contingency table). Other nonparametric tests on proportions include Fisher's 'exact test', and Cochran's test for paired samples, but such tests should not be used for small data sets. Details of these and other tests can be found in statistical texts such as Snedecor and Cochran (1980), Hawkins (2005), Armstrong and Hilton (2011) and Sheskin (2011). In addition, computer-based 'unconditional exact tests' are available (for example http://www4.stat.ncsu.edu/~boos/exact/).

Suppose that two laboratories find differences in the incidence of positive results from samples inoculated at a low level; could this indicate a lack of competence in one of the laboratories or could the difference arise purely by chance? Example 15.4.1 illustrates how such differences can be analysed using McNemar's test of 'disagreement', which is also used in method validation (ISO, 2003). The general use of 2 × 2 contingency tables to assess whether two sets of surveillance frequency data differ is illustrated in Example 15.4.2 and tests based on population parameters in Example 15.4.3.

## Example 15.4 Analysing for differences in proportions

Proportions are derived in microbiological studies using Bernoulli trials and analysis of differences requires specific approaches.

### Ex 15.4.1 Proportion of positive responses in cultural tests (McNemar's Test)

The objective is to assess whether there is a difference in the performance of two laboratories using food samples inoculated at a level slightly greater than the limit of detection, such that about 50% of the test results should be positive. Suppose that each laboratory tests 10 inoculated food samples for salmonellae by preparing six replicate pre-enrichment cultures of each sample followed by inoculation of plates of a diagnostic medium. The presumptive presence of the target organism, scored as present (+) or absent (−), is based on defined reactions on the diagnostic media. The following results were obtained:

| Sample / Laboratory | Number of positive responses from 6 replicates/trial | | | | | | | | | | |
|---|---|---|---|---|---|---|---|---|---|---|---|
| | 1 | 2 | 3 | 4 | 5 | 6 | 7 | 8 | 9 | 10 | Total |
| A | 4 | 3 | 4 | 5 | 2 | 3 | 3 | 5 | 1 | 3 | 33/60 |
| B | 3 | 2 | 3 | 4 | 2 | 3 | 2 | 4 | 3 | 2 | 28/60 |
| Difference (A − B) | 1 | 1 | 1 | 1 | 0 | 0 | 1 | 1 | −2 | 1 | 7+; 2− |

The data show a lower proportion of positive tests from laboratory B than from laboratory A. Is the difference statistically significant and how can it be assessed? A contingency test cannot be used because the experiments were done using duplicate samples, so the results will be correlated. Instead the data are analysed by McNemar's test for agreement between results. The null hypothesis is $H_0$: Lab A = Lab B; the alternative hypothesis is $H_1$: Lab A ≠ Lab B.

For each sample, determine the number of positive and negative results in each laboratory and assess the results for agreements and disagreements. Note that Lab B found three positive results from sample 9 whilst laboratory A found only one positive; note also that seven positive results from Lab A were reported as negative by Lab B. Hence, 26 positive and 25 negative results agreed but there were two 'false' positives and seven 'false' negatives (false only in the sense that they disagree). Set out the data in a two-way table:

| | Lab A | | |
|---|---|---|---|
| | + | - | Total |
| **Lab B** + | 26 (a) | 2 (b) | 28 |
| **Lab B** − | 7 (c) | 25(d) | 32 |
| **Total** | 33 | 27 | 60 |

McNemar's test ignores results that agree (a = + +; d = − −) and uses the disparate results (b = − +; c = + −) to calculate the difference in proportions, which is assessed by:

$$\chi^2 = \frac{(b-c)^2}{(b+c)} = \frac{(2-7)^2}{(2+7)} = \frac{25}{9} = 2.78 \text{ with } \nu = 1 \text{ df.}$$

But since the value of $b < 5$, the calculation should be modified to include Yate's correction:

$$\chi^2 = \frac{(b-c-0.5)^2}{(b+c)} = \frac{(2-7-0.5)^2}{(2+7)} = \frac{20.25}{9} = 2.25$$

From tables, the two-tailed probability for $\chi^2 = 2.25$ with 1 df is $P = 0.18$, so the null hypothesis is not rejected. The same conclusion would have been reached if the alternative hypothesis been Lab A > Lab B, when a one-tailed test would have given $P = 0.09$. This outcome does not imply that the laboratories perform equally well (or badly) but only that it is not possible to demonstrate a difference based on the available results.

### Ex 15.4.2 Surveillance for campylobacter in chicken (contingency test)

Suppose that a survey showed that campylobacter were detected in 70 of 113 fresh chickens purchased from supermarkets (S) and in 76 of 122 fresh chickens purchased from local butcher shops (B). Does the incidence of contamination differ between the sources? The null hypothesis is $H_0$: S = B; and the alternative hypothesis as $H_1$: S ≠ B. Arrange the data in a contingency table:

**Observed (O):**

| Test Source | Campylobacter | | Total |
|---|---|---|---|
| | + | − | |
| Butchers (B) | 76 | 46 | 122 (=$R^B$) |
| Supermarkets (S) | 70 | 43 | 113 (=$R^S$) |
| **Total** | 146 (=$C^+$) | 89 (=$C^-$) | 235 (=N) |

Determine the expected number of positive and negative results by multiplying each row total (R) by the column total (C) and divide by N:

**Expected (E):**

| Test Source | Campylobacter | | Total |
|---|---|---|---|
| | + | − | **122** |
| Butchers (B) | 75.8 | 46.2 | 122 |
| Supermarkets (S) | 70.2 | 42.8 | 113 |
| **Total** | 146.0 (=$C^+$) | 89.0 (=$C^-$) | 235 |

Then for each value calculate $\chi^2$ as $\frac{(O-E)^2}{E}$ and add together:

$$\sum \chi^2 = \frac{(76-75.8)^2}{75.8} + \frac{(46-46.2)^2}{46.2} + \frac{(70-70.2)^2}{70.2} + \frac{(43-42.8)^2}{42.8}$$

$$= \frac{0.2^2}{75.8} + \frac{0.2^2}{46.2} + \frac{0.2^2}{70.2} + \frac{0.2^2}{42.8} = 0.0029$$

From statistical tables, the probability for $\chi^2$ with 1 df is $P = 0.96$. So the null hypothesis that the prevalence of contamination is the same in the two sources of chicken is not rejected.

### Ex 15.4.3 Comparison of surveillance data using ND assumptions

This example draws data from a very large database on the European prevalence of *Campylobacter* spp in broiler chicken carcasses (EFSA, 2010). Based on a total survey of 10 000 carcasses the prevalence of campylobacter across the European Union (EU) was 58.7%. Is a prevalence of 65% contamination in broilers in country X significantly different to the average EU prevalence, based on a sample size of 100 carcasses?

Let the proportions of positive carcases in country X $= p_X$ and that in the EU $= p_E$ and define the null hypothesis as $H_0$: $p_X = p_E$ and the alternative hypothesis as $H_1$: $p_X > p_E$. Note that this is a one-sided test so the null hypothesis is rejected if the prevalence in country X is significantly larger than in the EU. Carry out a $z$-test with a significance level of $P = 0.05$, on the realistic assumption that proportions conform to a ND provided that the population size is large compared to the sample size.

Determine the SD ($s$) of the prevalence ($p_X$) in the sample of 100 carcasses taken from the population in country X:

$$s = \sqrt{(p_X(1-p_X))/n} = \sqrt{(0.65(1-0.65))/100} = \sqrt{0.002275} = 0.0477$$

where $p_X$ is defined above and the test sample size $n = 100$.

Then determine $z$ as the difference in the proportions $p_X = 0.65$ and $p_E = 0.587$ divided by $s$:

$$z = (p_E - p_X)/s = (0.587 - 0.650)/0.0477 = -1.321$$

From tables of the area in the normal distribution for a one-tailed test, $z = -1.321$ would occur with a probability of $P = 0.0933$. Hence, the null hypothesis is not rejected at the 5% level and we conclude that prevalence of campylobacter in chicken from country X does not differ from that across the entire EU. But if the sample size had been 400 carcases, rather than 100, then $s = 0.0238$, so $z = -2.65$, and the observed difference for a one-sided test between the prevalence values would have been statistically significant at $P = 0.004$.

A $2 \times 2$ $\chi^2$ test based on the average prevalence value of 58.7% for the EU (with 10 000 samples) and a sample prevalence of 65% based on 100 carcasses in country X would also not reject the null hypothesis at $P = 0.05$ but the test is unbalanced, and therefore unreliable, because the total numbers of samples are significantly different. An alternative and very simple approach is to use a table of the 95% confidence limits (CLs) to a binomial distribution (Snedecor and Cochran,

1980, Table 1.4.1). Noting that a value of 65% positive results is the same as 35% negative results, we enter the table for 35 observed values from a sample 100 carcases and obtain 95% CLs of 26% and 45%; so for positive samples the 95% CLs range from 55 to 74%. Since the EU value of 58.7% falls within these CLs we again conclude that the prevalence in country X does not differ significantly at $P = 0.05$ from that in the EU.

*Regression, correlation and equivalence*

It is often necessary to assess changes in microbial numbers against an independent parameter, such as time, temperature or concentration of an antimicrobial agent using tests for regression and correlation (Chapter 2). In method validation (Chapter 17), data are tested for linearity and agreement but whilst Pearson's correlation coefficient provides an estimate of the strength of a relationship between two data sets it does not measure equivalence. A Bland–Altman plot (Bland and Altman, 1986, 1995) is used to determine the equivalence two methods based on estimation of bias (Example 15.5).

## Example 15.5 Test of equivalence using the Bland–Altman plot

The concept of equivalence is a better way to compare two methods than using correlation, which provides a measure of the strength of a relationship but not of its equivalence. When two methods are compared neither provides an unequivocally correct result, so we should assess the degree of agreement. The data for this example were obtained during validation of a microbiological method against a reference method, using naturally contaminated samples. The data set contains 22 valid pairs for which the correlation coefficient $r = 0.99$ ($p < 0.001$) indicates a very strong relationship.

Plot the data and draw a line of identity on which all points would lie if the two methods gave identical results for all samples. The line of identity is not a regression line; it provides visual assessment of the extent to which the two methods agree. Examination of Figure Ex15.5.1 shows that the data are offset against the line of identity.

Next, determine the average of, and the difference between, each pair of data values:

| Log CFU/g | | | |
|---|---|---|---|
| **Alternative (A)** | **Reference (R)** | **Mean (A + R)/2** | **Difference (A – R)** |
| 2.00 | 2.00 | 2.000 | 0.00 |
| 2.19 | 2.10 | 2.145 | +0.19 |
| 3.15 | 3.15 | 3.150 | 0.00 |
| 3.26 | 2.98 | 3.120 | +0.28 |
| 3.28 | 3.24 | 3.260 | +0.04 |
| 3.35 | 3.24 | 3.295 | +0.11 |
| 3.63 | 3.66 | 3.645 | −0.03 |
| 3.56 | 3.43 | 3.495 | +0.13 |
| 4.41 | 4.42 | 4.415 | −0.10 |
| etc. | etc. | etc. | etc. |

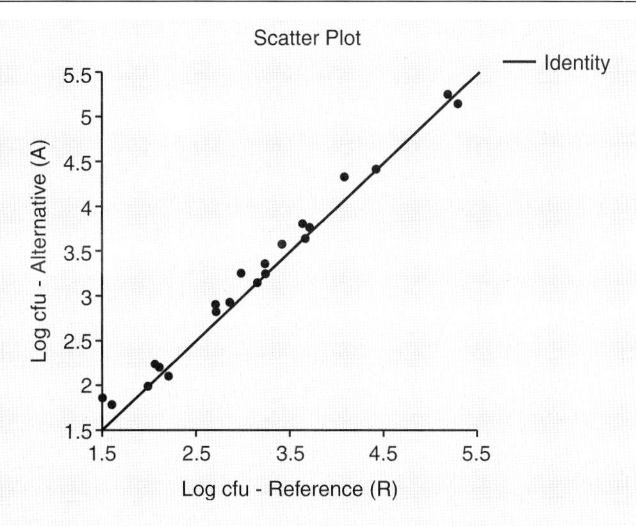


**Figure Ex 15.5.1** Identity Plot between the reference and alternative method.

Note that the procedure requires pairs of results so any value that lacks a 'partner' must be excluded. Plot the differences against the mean values and calculate the bias (mean difference) and the 95% CLs of the bias (i.e. mean $\pm$ $2s$), as shown in Figure Ex15.5.2. For these data, the overall bias is 0.087 $\log_{10}$ CFU/g. One of the 22 differences slightly exceeds the upper confidence limit (CL), but since difference values conform to ND, 1 in 20 values might be expected to lie outside the 95% CL.

We conclude that both methods give essentially identical results with a very small positive bias in favour of the alternative method, but the bias will not affect significantly any results obtained using the method. Hence, the alternative method can be considered to provide results equivalent to those from the reference method.

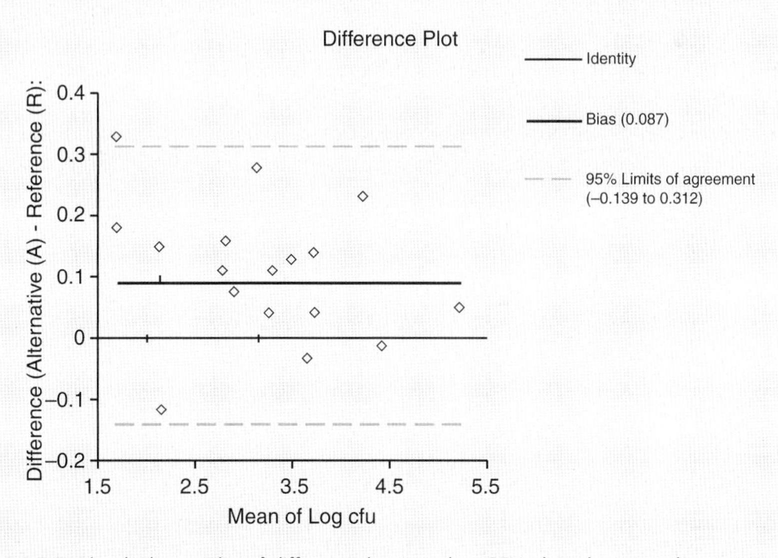


**Figure Ex 15.5.2** Bland-Altman Plot of difference between log CFU plotted against the average log CFU.

## Analysis of variance (ANOVA)

### The principles of ANOVA

ANOVA partitions the observed variance of a treatment into its component sources. In its simplest form, ANOVA provides a way to assess statistically whether the mean values of several groups are equal and, thereby, enables differences to be compared for more than two groups. This is important since to carry out multiple *t*-tests risks increasing the chances of committing a Type I error (i.e. incorrectly rejecting a null hypothesis).

ANOVA provides a form of statistical hypothesis testing that limits the rate of Type I and Type II errors. Typically the null hypothesis is that all test groups are random samples of the same population, which implies that all factors have the same effect, or none. Rejecting the null hypothesis requires acceptance of alternative hypotheses that one or more treatments cause significant effects. ANOVA can be used also without significance testing, for example to assess the relative contributions of components of variance in interlaboratory trials (Jarvis *et al.*, 2007).

The terminology of ANOVA derives largely from the statistical design of experiments, where the responses of selected factors are measured to assess an effect. To ensure validity, factors are assigned to experimental units by a combination of randomization and blocking (restricting randomization to isolate systematic effects that would otherwise obscure main effects). Responses show a variability that is partially caused by fixed effects and partially by random error. There are three main types of ANOVA models:

(i) A 'fixed effects' model, where a number of factors applied to the subject of the experiment cause different responses.
(ii) A 'random effects model' uses treatments that are themselves subject to random variation, for instance, even minor changes in test conditions may be the cause of random effects
(iii) A 'mixed effects model' tests for both fixed and random effects.

### Requirements of ANOVA tests

Assuming that the ANOVA test is based on a linear model, the following key assumptions refer to the probability distribution of the responses:

- Independence of observations, which assumes effective randomization of subjects.
- Compliance of observations with ND, after transformation if appropriate.
- Equality of variances: this is sometimes hard to achieve and is not always essential but if variances are markedly different it is important to assess whether the differences may affect the outcome of the analysis.

### The ANOVA approach

*Partitioning the sums of squares*

The basic approach is to determine the sums of the squares (SS) of the differences between the observations ($x_i$) and their mean value ($\bar{x}_i$). A one-way ANOVA, based on a single treatment factor, derives:

$$\mathrm{SS}_{\mathrm{Total}} = \mathrm{SS}_{\mathrm{Factor}} + \mathrm{SS}_{\mathrm{Error}} \tag{15.14}$$

Similarly, the degrees of freedom (df) can be apportioned into:

$$\mathrm{df}_{\mathrm{Total}} = \mathrm{df}_{\mathrm{Factor}} + \mathrm{df}_{\mathrm{Error}} \tag{15.15}$$

If the number of treatments $= i$, and the number of factors $= j$, then for a simple scheme $\mathrm{df}_{\mathrm{Factor}} = i - 1$, and the $\mathrm{df}_{\mathrm{Error}} = j - 1$.

The mean square is determined by dividing the SS by the df, so the mean square factor is:

$$\mathrm{MS}_{\mathrm{Factor}} = \mathrm{SS}_{\mathrm{Factor}}/(i - 1) \tag{15.16.1}$$

and the mean square error is:

$$MS_{Error} = SS_{Error}/(j-1) \quad (15.16.2)$$

*The* F *test*

Since the mean squares are the apportioned estimates of the variance, then the $F$ test is the ratio between the mean square of the factor and that of the error, i.e. $F = MS_{Factor}/MS_{Error}$. The significance of the estimate of $F$ is obtained from standard tables of the $F$ distribution for $i-1$, and $j-1$ df. If the observed value of $F$ is greater than the tabulated critical value at, for example, $P=0.05$, the observed value is statistically significant at that level of probability and the null hypothesis is rejected.

### *Types of ANOVA*

Different types of ANOVA are used depending on the experimental design, especially the protocol that specifies the random assignment of treatments to subjects; the description of the protocol should include a specification of the structure of the factors and of any blocking. A simple one-way ANOVA is used to test for differences among two or more independent group means (Example 15.6). Note that for comparison of two mean values, Student's $t$-test and the ANOVA $F$ test are equivalent, the relationship being given by $F=t^2$. A two-way, or factorial, ANOVA is an extension of the one-way ANOVA and is used to study interaction effects among the factors with one (or more) measurable variables and two, or more, nominal variables (Example 15.7). Note that a two-way ANOVA can be done with or without replication; in the absence of replication, any interactions are included in the residual variance. A repeated ANOVA is used when the same subjects are used for several sequential treatments, for example in a longitudinal study with total replication of test series. However, measurements repeated on the same unit are correlated and allowance for this correlation must be made in ANOVA. Often a better approach is through multilevel modelling (Goldstein *et al.*, 2002).

## Example 15.6 One-way analysis of variance and Tukey's *post hoc* test

Data for the levels of alcohol produced after 96 h in a control and two supplementation trials, analysed previously by the $t$-test (Example 15.1), are now reanalysed in a simple one-way ANOVA. The null hypothesis is that there is no difference in alcohol production in the three parallel trials, $H_0$: $C=S_1=S_2$; the alternative hypotheses, $H_{1A}$: $C \neq S_1 \neq S_2$.

The data are summarized together with key calculations:

| **Trial #** | **Ethyl alcohol (g/100 ml)** | | | **Total** |
|---|---|---|---|---|
| | **Control (C)** | **Supplement S1** | **Supplement S2** | |
| 1 | 8.30 | 9.05 | 7.54 | |
| 1 | 8.28 | 9.06 | 7.78 | |
| 2 | 8.40 | 8.99 | 7.90 | |
| 2 | 8.40 | 9.12 | 7.88 | |
| 3 | 8.48 | 9.11 | 7.58 | |
| 3 | 8.46 | 9.08 | 7.76 | |
| 4 | 8.38 | 9.05 | 7.86 | |
| 4 | 8.44 | 9.08 | 7.92 | |
| $\sum X=$ | 67.14 | 72.54 | 62.22 | 201.90 |
| $\bar{X}=$ | 8.39 | 9.07 | 7.78 | 25.24 |
| $\sum X^2=$ | 563.51 | 657.77 | 484.06 | 1705.34 |
| $(\sum X)^2/N=$ | 563.47 | 657.76 | 483.92 | 1705.15 |
| df | 7 | 7 | 7 | 21 |

We now determine:

Total Sum of Squares $(SS) = \sum X^2 - (\sum X)^2/N = 1705.34 - 201.9^2/24 = 6.856$
Between Groups $SS = \sum \left((\sum X)^2/N\right) - (\sum X)^2/N = 1705.15 - 1698.484 = 6.667$
Within Group (Error) $SS =$ Total $SS$ – Between Groups $SS = 6.856 - 6.667 = 0.190$

These values are entered into the ANOVA table (below) and the mean square (MS) and $F$ value are derived. The statistical probability is determined for $F_{(2,21)}$, which gives a critical value of 3.45 for $P = 0.05$. We therefore reject the null hypothesis that the mean values for alcohol production are equal in the three trials and accept the alternative hypothesis.

**ANOVA Table**

| Source | df | SS | MS | F | P |
|---|---|---|---|---|---|
| Between Groups | 2 | 6.856 | 3.433 | 381.4 | <0.0001 |
| Within Groups | 21 | 0.190 | 0.009 | | |
| Total | 23 | 7.00 | | | |

**NB.** For display the figures have been rounded to two (or three decimal places) but the calculations were done using the full decimal values.

### *Post hoc* tests

Tukey's 'wholly significant difference' (WSD) test (Tukey, 1955) provides a simple way to contrast the groups in a pairwise fashion. For the data tested above we have rejected the null hypothesis but we need to reconsider alternative hypotheses: $H_{1A}$: C < S1; $H_{1B}$: C > $S_2$; $H_{1C}$: $S_1 > S_2$.

For the three groups the mean levels of alcohol production were $\bar{C} = 8.39$, $\bar{S}1 = 9.07$ and $\bar{S}2 = 7.78$ g/100 ml; the $MS_{within} = 0.009$ with $df_{within} = 21$ and $df_{between} = 2$. The Standard Error ($SE_{\bar{x}}$) of the group sample means is $\sqrt{0.009/2} = 0.0474$.

From tables, for $k = 3$ groups with $n = 8$ replicates/group, Tukey's WSD = 4.8. To be statistically significant, the absolute difference between two means must be $> 4.8 \times SE_{\bar{x}} = 4.8 \times 0.0474 = 0.228$.

The observed differences in the means are: C – S1 = –0.64; C – S2 = 0.68; S1 – S2 = 1.29. So we conclude that all three differences are statistically significant at $P = 0.05$ and that none of the alternative hypotheses should be rejected.

***Note*:** Procedures for the one-way ANOVA are readily available in statistical packages, which include also various methods for comparisons of test groups.

## Example 15.7 Two-way ANOVA

In Example 15.6 we noted that, compared to the control, addition of supplement $S_2$ significantly depressed alcohol production whilst $S_1$ caused a significant increase. Colony counts for yeast were done at the same time intervals, so we now question whether changes in cell growth were affected by the supplements?

A two-way ANOVA provides a means of determining whether the colony counts (as $\log_{10}$ CFU/ml) are affected significantly by either treatment (presence of supplements and trials). A two-way ANOVA can be done on either original data or on summary data; for this example we will consider only the analysis of the average colony counts done after 96 h.

Set out the mean data in a two-way table; calculate row (r) and column totals (c) and the grand total (GT). Then calculate mean values for the columns ($\bar{c}$) and rows ($\bar{r}$), divide the sums of the squares of the row and column values by the number of replicates ($(\sum r)^2/n_c$ and $(\sum c)^2/n_r$) and determine the sums of the squares of the rows ($\sum r^2$) and column ($\sum c^2$) values. Finally, determine the 'Grand Total' (GT) = 85.270.

| Trial | Mean $log_{10}$ CFU/ml with Treatment Factor | | | | | | | |
|---|---|---|---|---|---|---|---|---|
| | C | S1 | S2 | $\sum r$ | $n_c$ | $\bar{r}$ | $(\sum r)^2/n_c$ | $\sum r^2$ |
| 96A | 7.05 | 6.77 | 7.50 | 21.320 | 3 | 7.107 | 151.5141 | 151.7854 |
| 96B | 7.08 | 6.79 | 7.49 | 21.350 | 3 | 7.117 | 151.9408 | 152.1879 |
| 96C | 7.07 | 6.80 | 7.46 | 21.325 | 3 | 7.108 | 151.5852 | 151.8058 |
| 96D | 7.07 | 6.78 | 7.43 | 21.275 | 3 | 7.092 | 150.8752 | 151.0875 |
| $\sum c$ | 28.260 | 27.135 | 29.875 | GT = 85.270 | | | C = 605.915 | |
| $n_r$ | 4 | 4 | 4 | | | | | |
| $\bar{c}$ | 7.065 | 6.784 | 7.469 | | | | | |
| $(\sum c)^2/n_r$ | 199.6569 | 184.0771 | 223.1289 | B = 606.8629 | | | | |
| $\sum c^2$ | 199.6574 | 184.0775 | 223.1317 | | | | | A = 606.8666 |

Determine the:

- Sum of all squared data values: $\sum r^2 = \sum c^2 = 606.8666$; call this A.
- Square of the sum of the column values divided by $n_r$: $(\sum c)^2/n_r = 606.8629$; call this B.
- Square of the sum of the row values divided by $n_c$: $(\sum r)^2/n_c = 605.9150$; call this C.
- Square of the GT divided by the product of the number of data entries in the columns and rows ($n_r n_c = 4 \times 3 = 12$) to obtain the overall mean square $= (85.27)^2/12 = 605.914$; call this D.

Set out the ANOVA table:

| Source of Variation | Sum of squares | df* | Mean square | F |
|---|---|---|---|---|
| Between columns (Treatments) | B − D = 0.949 | $(n_c - 1) = 2$ | 0.47445 | 1054 |
| Between rows (Trials) | C − D = 0.001 | $(n_r - 1) = 3$ | 0.00033 | 0.73 |
| Residual Error | 0.003 | $(n_c - 1)(n_r - 1) = 6$ | 0.00046 | |
| Total | A − D = 0.953 | $((n_c)(n_r) - 1) = 11$ | | |

* degrees of freedom

From tables, the critical value of $F_{(0.05,\ df\ 2,6)} = 5.14$; so we should reject the null hypothesis that all treatments are equal – for these data $P < 0.001$, so the difference between the treatments is very highly significant. By contrast, for the replicate trials the critical value of $F_{(0.05,\ df\ 3,6)} = 5.61$, so we accept the null hypothesis that the colony counts do not differ significantly between replicate trials.

Because only the mean colony count values were tested, the lack of replication data prevents assessment of any interaction between the factors. However, any variance due to interaction is included in the residual mean square so it is unlikely to be large. The following shows the results of analysis of the complete data set of duplicate log CFU/ml at 96 h, including the assessment for interaction between the factors:

| Source of variation | Sum squares | df | MS | F | P |
|---|---|---|---|---|---|
| Treatments | 1.897 | 2 | 0.9485 | 711.34 | <0.0001 |
| Trial | 0.002 | 3 | 0.0007 | 0.49 | 0.6973 |
| Treatment × Trial | 0.006 | 6 | 0.0009 | 0.69 | 0.6618 |
| Residual | 0.016 | 12 | 0.0013 | | |
| Total | 1.920 | 23 | | | |

No significant interaction (trial × treatment) was observed and the other conclusions are not changed. If a statistically significant interaction effect had been seen it would have been necessary to consider possible causes: to review the normality of the data, the equality of the variances and, possibly, the way in which the trials had been done.

A nested ANOVA for both balanced and unbalanced designs is used for one measurable variable with two, or more, nominal variables having subgroups 'nested' within a group variable. For instance, in examining the effects of sampling on overall measurement uncertainty there might be two levels of nesting. The primary group might comprise *i* target samples from each of which *j* subsamples are drawn for analysis. The test procedure is then a one-way ANOVA with two mean square calculations, $MS_{between\ samples}$ and $MS_{within\ samples}$ and two corresponding *F* values. The variance is partitioned between the residual error ($s_r^2$; repeatability) and the between-sample ($s_{between}^2$) and within-sample ($s_{within}^2$) error; reproducibility is then given by $s_{Reprod}^2 = s_{between}^2 + s_{within}^2 + s_r^2$ (Jarvis *et al.*, 2012). Multivariate analysis of variance (MANOVA) can be used when several response variables are assessed. Non-microbiological examples are described elsewhere (Chapters 7 and 10).

### Post hoc *tests*

*Post hoc* tests are used for further investigation of differences identified in an ANOVA, once planned comparisons have been made. Many procedures can be used to make such evaluations; of these Tukey's Wholly Significant Difference is possibly the most common (Example 15.6). Armstrong and Hilton (2011) compared the advantages and disadvantages of many *post hoc* tests, especially in relation to the risk of introducing Type I or Type II errors.

### *Nonparametric ANOVAs*

For circumstances where ND data are difficult to obtain even after transformation, various nonparametric forms of ANOVA are available. The procedures test the difference between median, rather than mean, values because the median is less affected by outlier data. Whether or not rejection of an outlier value, which could have occurred purely by chance, is justified is a matter for debate; there are arguments for and against outlier removal.

The UK Royal Society of Chemistry (RSC, 2001a) describes one approach for which 'robust' ANOVA software can be downloaded as an Excel Add-in (RSC, 2001b). The method is based on use of the median absolute difference (MAD) approach. An alternative method (Hedges and Jarvis, 2006), uses stepwise analysis based on recursive estimation of the median absolute paired deviation (MAPD) described by Rousseuux and Croux (1993). In a comparison using *in silico* generated data that conformed to ND, both robust procedures and the standard parametric ANOVAs generated similar estimates of key values; but when real data were used that did not conform to ND, the outputs from both robust methods were similar but differed considerably from those using the parametric ANOVA.

## USE OF MODELLING TECHNIQUES

Modelling can take many forms. For example, a set of data fitted to a probability model tests if the data conform to that particular distribution. Note that data must be tested against a model not *vice versa*! In a more general context, experimental data are used to develop mathematical expressions that describe and help to explain observations, such as the interactions between chemical preservatives that inhibit toxin production by *Clostridium botulinum* (Roberts and Jarvis, 1983). At a different level, modelling can be used to define a sampling programme for foods (Chapter 22) and to develop international recommendations for food safety (Bassett, 2010).

Generally, models provide only an approximation to a true situation since the 'real world' is too complex to provide an exact representation. The quotation 'all models are wrong but some are useful' is a truism attributed to George Box. In developing a model it is essential to apply the principles of parsimony and to achieve a balance between accuracy and manageability.

Modelling is used to help public and professional understanding of food safety issues associated with the presence of specific microbes in a food product, using models for 'Predictive Microbiology' (McMeekin *et al.*, 1993; McKellar and Lu, 2003). Such models are generally dynamic and identify inter-relationships between critical factors important in food processing. Possibly the most widely used microbiological model ever devised is that of the 'Botulinum Cook' based on the work of Esty and Meyer (1922).

## Approaches to mathematical modelling of processes for food safety

Investigations of food safety require multifactorial experiments to understand the intrinsic and extrinsic factors that impact survival and growth of organisms. Roberts and Jarvis (1983) identified a need for international collaboration to provide data for modelling key food safety needs. In the 1990s, microbial growth prediction studies led to the development of two pathogen-modelling programmes in the United Kingdom and the United States of America; Since then, international food safety prediction has developed significantly (McKellar and Lu, 2003). Combined databases are now freely available (Baranyi and Tamplin, 2004) together with predictive software (ComBase Predictor 3.0) that calculates growth and inactivation rates for various key organisms. The Food Risk Organization provides an online database of models for predictive microbiology. Koseki (2009) has developed a revised ComBase-derived predictor, the Microbial Responses Viewer. Provided that the physical and chemical composition of foods are sufficiently understood it is possible to use such databases to predict the potential fate of organisms in foods and the output from such predictions can be used as input for further modelling studies.

## Modelling procedures

Mathematical modelling of data can be used for many purposes; procedures such as bootstrapping (sampling with replacement), jackknifing (a modified bootstrap) and shuffling (sampling without replacement) permit the generation of computer-derived data (*in silico* samples) based on observational data. The outputs can be further modelled using Monte Carlo (MC) simulation and related procedures with applications in many fields (Davison and Hinckley, 1997; Manly, 2007). Example 15.8 provides two simple illustrations of some procedures involved.

## Example 15.8 Modelling

### Ex 15.8.1 Bootstrapping

A simple example of bootstrapping uses reanalysis of data to increase the number of values available for analysis. Suppose that you have carried out aerobic colony counts on 10 samples of a product – the results (as $\log_{10}$ CFU/g) range from 4.80 to 5.95 with $\bar{x} = 5.53$ and $s = 0.34$. The range of results is wide and the SD is large. You note that one count (4.80) differs considerably from the remaining data values. Is this count an outlier? Is it possible that the sample is from a different batch of product?

If the low-count value is excluded, $\bar{x}' = 5.61$ and $s' = 0.24$. What is the probability that the low count product does not come from the same population as the other samples? If it is assumed that the log counts conform to ND, the z-value can be determined for the low count sample: the difference between the actual count and the revised mean count ($\bar{x}'$) is given by $z = (5.61 - 4.80)/0.24 = 0.81/0.24 = 3.375$; so from tables we would expect that the low result could have occurred with a probability of $P = 0.0005$. Hence a null hypothesis that this result comes from the same population as the other nine samples would be rejected.

But was it realistic to assume that the data conformed to ND? Statistical tests, such as the Shapiro–Wilk, show that the hypothesis for ND cannot be rejected ($P = 0.74$) but the distribution is highly skewed. Further data need to be obtained to confirm the conclusion but it is not necessary to repeat the actual sampling and analysis; bootstrapping can be used to revalue the original data, since the distribution in the data set reflects the population from which it was drawn. For instance the 95% confidence limits (CLs) for the mean can be derived using an Excel Add-in such as "Resampling Stats for Excel" (2009) or other software. Resampling with replacement means that after a value has been selected for the new data set, that value is replaced in the original data set before another value is selected. This procedure can be repeated to develop a series of 1 000 000 or more iterations but generally 1000 iterations are adequate. The procedure involves the following steps:

1. Calculate the mean value of the original data set ($\bar{x} = 5.53$).
2. Resample the data values with replacement and calculate the 'resampled mean' value.
3. Repeat stage 2 for, for example 1000 iterations (using software this takes about 1.5 s) and record the output of 1000 mean values.
4. Sort the data using the standard Excel 'data sort' command.
5. Use the Excel 'paste function' to find the 2.5 and the 97.5 percentiles and the overall mean (50th percentile) values.

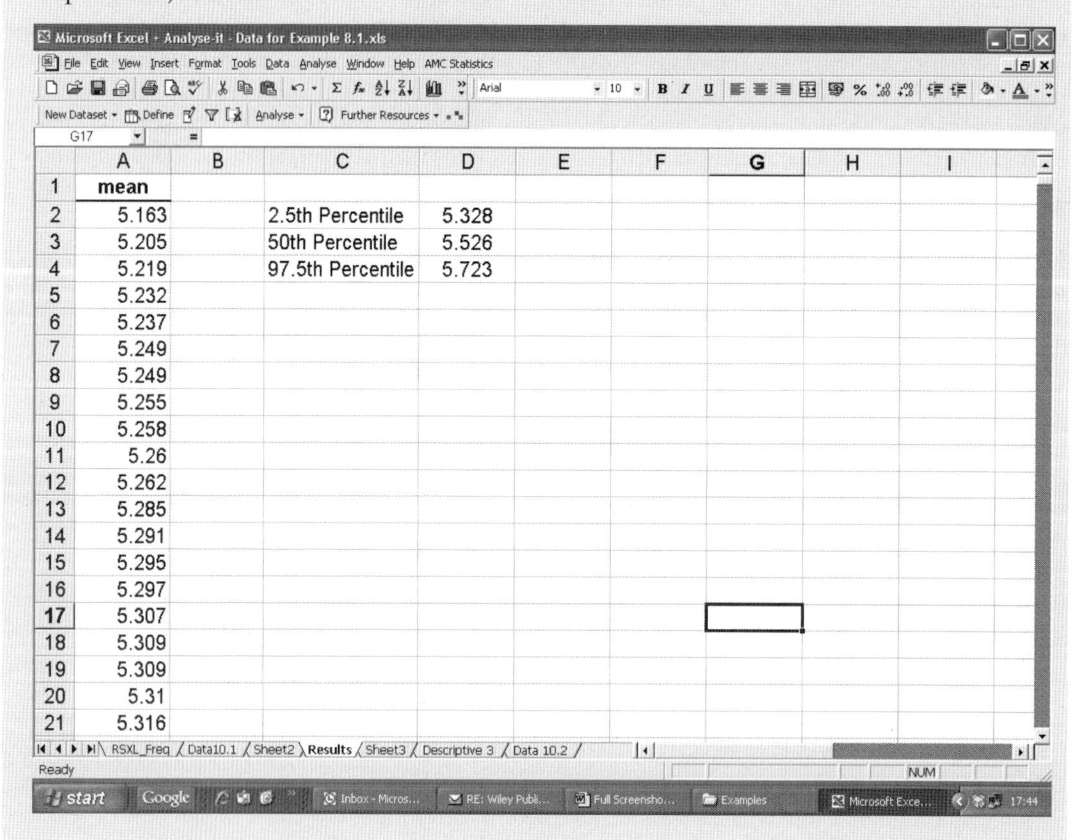

| | A | B | C | D |
|---|---|---|---|---|
| 1 | mean | | | |
| 2 | 5.163 | | 2.5th Percentile | 5.328 |
| 3 | 5.205 | | 50th Percentile | 5.526 |
| 4 | 5.219 | | 97.5th Percentile | 5.723 |
| 5 | 5.232 | | | |
| 6 | 5.237 | | | |
| 7 | 5.249 | | | |
| 8 | 5.249 | | | |
| 9 | 5.255 | | | |
| 10 | 5.258 | | | |
| 11 | 5.26 | | | |
| 12 | 5.262 | | | |
| 13 | 5.285 | | | |
| 14 | 5.291 | | | |
| 15 | 5.295 | | | |
| 16 | 5.297 | | | |
| 17 | 5.307 | | | |
| 18 | 5.309 | | | |
| 19 | 5.309 | | | |
| 20 | 5.31 | | | |
| 21 | 5.316 | | | |

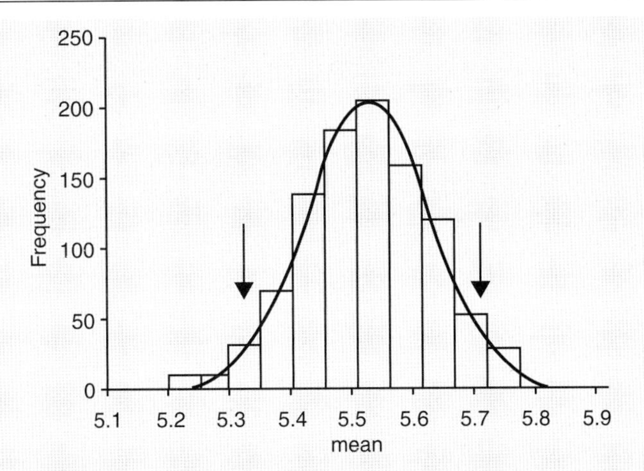


**Figure Ex 15.8.1** Distribution of 1,000 resampled mean values, overlaid with a ND curve for $\bar{x} = 5.53$ and $SE_{mean} = 0.100$; the arrows mark the upper and lower bounds of the 95% CLs.

The distribution of the mean values is shown in Figure Ex 15.8.1.

This procedure provides an overall mean value ($\bar{\bar{x}} = 5.53$) with upper and lower bounds to the 95% CLs of 5.72 and 5.33, i.e. $\bar{\bar{x}} \pm 0.20$. So the $SE_{mean}$ is $\pm$ 0.10. For $n = 10$, $s = SE_{mean} \times \sqrt{n} = 0.10 \times 3.16 = 0.316$. The $z$ value for the original outlier value of 4.80 is now determined as $z = (5.53 - 4.80)/0.316 = 2.3$, so the probability that it will occur by chance is $P = 0.01$. The original null hypothesis, that the sample comes from the same population as the other nine samples, is again rejected but this time on much stronger grounds.

**Ex 15.8.2 Sampling without replacement**

Jarvis and Hedges (2011) used 'resampling without replacement' (shuffling) to examine the effects of sample numbers on the precision of microbial colony counts. The procedure was used in preference to bootstrapping since it more closely mimics the effects of drawing samples for microbiological testing from a finite subpopulation as would be found in a retail outlet.

The example uses aerobic colony count data from replicate samples of sprouted mixed beans. Twenty colony count values are resampled sequentially without replacement for $n = 2$ to $n = 15$ samples. The procedure used is:

1. The values (column B) for $n$ samples (column C) are selected at random from the entire dataset (column A). The mean and variance (columns D & E) are calculated for the data in column B, together with 2.5 and 97.5 percentile limits based on a $\chi^2$ distribution (columns F & G; for detail see Jarvis and Hedges, 2011).
2. The resampling is repeated through 1000 iterations and mean values derived for each of the parameters (D to G). In addition, the actual 2.5 and 97.5 percentile values of the variances are determined from the 1000 estimates of the variance.

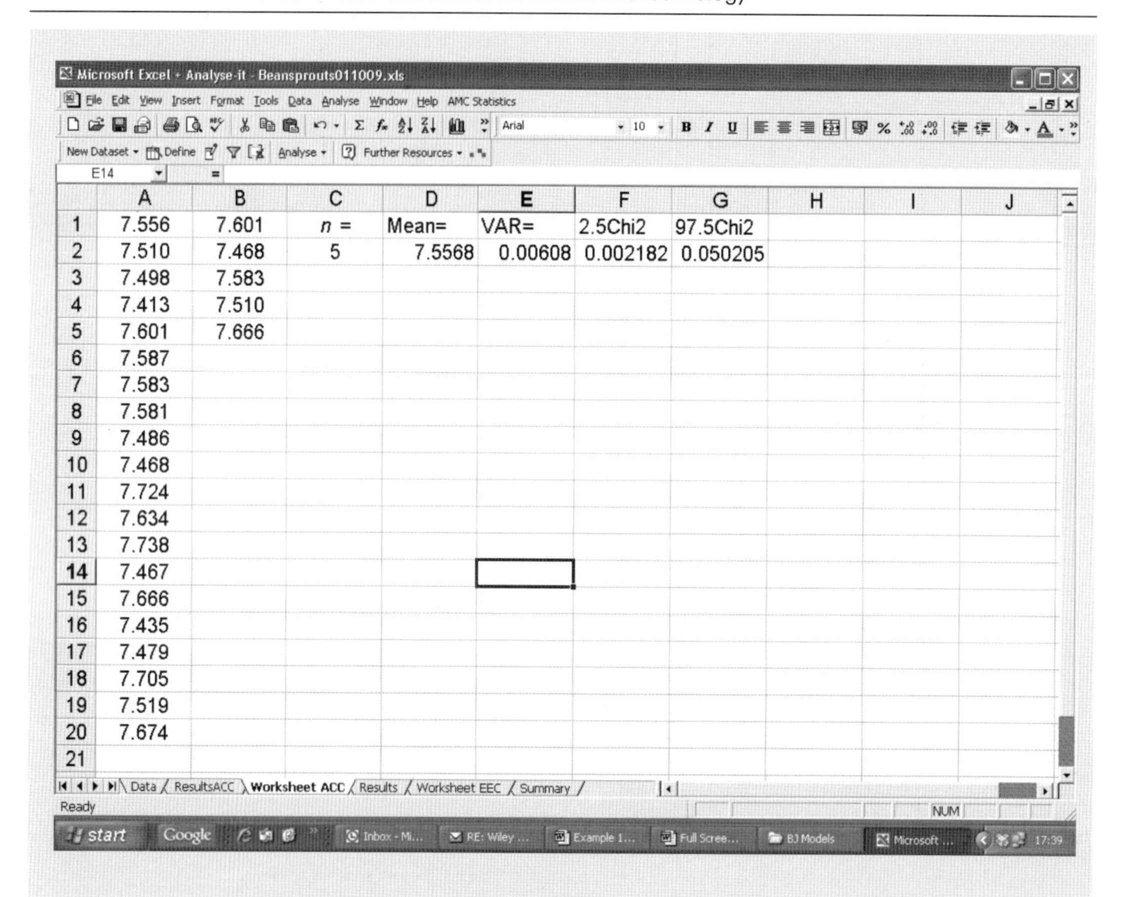

| | A | B | C | D | E | F | G | H | I | J |
|---|---|---|---|---|---|---|---|---|---|---|
| 1 | 7.556 | 7.601 | $n$ = | Mean= | VAR= | 2.5Chi2 | 97.5Chi2 | | | |
| 2 | 7.510 | 7.468 | 5 | 7.5568 | 0.00608 | 0.002182 | 0.050205 | | | |
| 3 | 7.498 | 7.583 | | | | | | | | |
| 4 | 7.413 | 7.510 | | | | | | | | |
| 5 | 7.601 | 7.666 | | | | | | | | |
| 6 | 7.587 | | | | | | | | | |
| 7 | 7.583 | | | | | | | | | |
| 8 | 7.581 | | | | | | | | | |
| 9 | 7.486 | | | | | | | | | |
| 10 | 7.468 | | | | | | | | | |
| 11 | 7.724 | | | | | | | | | |
| 12 | 7.634 | | | | | | | | | |
| 13 | 7.738 | | | | | | | | | |
| 14 | 7.467 | | | | | | | | | |
| 15 | 7.666 | | | | | | | | | |
| 16 | 7.435 | | | | | | | | | |
| 17 | 7.479 | | | | | | | | | |
| 18 | 7.705 | | | | | | | | | |
| 19 | 7.519 | | | | | | | | | |
| 20 | 7.674 | | | | | | | | | |
| 21 | | | | | | | | | | |

**Table Ex 15.8.1** Summary of the mean, variance and the 95% actual and $\chi^2$ percentiles of the variance distribution following resampling without replacement.

| Number of samples ($n$) | Mean colony count (log cfu/g) | Variance | | | | |
|---|---|---|---|---|---|---|
| | | | $\chi^2$ Percentile Percentile | | Actual Percentile | |
| | | Average | 2.5% | 97.5% | 2.5% | 97.5% |
| 2 | 7.57 | 0.010 | 0.0020 | 10.292 | 0.0000 | 0.0424 |
| 3 | 7.57 | 0.010 | 0.0027 | 0.4006 | 0.0004 | 0.0268 |
| 4 | 7.57 | 0.010 | 0.0032 | 0.1398 | 0.0014 | 0.0211 |
| 5 | 7.56 | 0.010 | 0.0035 | 0.0797 | 0.0020 | 0.0187 |
| 6 | 7.57 | 0.010 | 0.0039 | 0.0599 | 0.0029 | 0.0179 |
| 7 | 7.57 | 0.010 | 0.0041 | 0.0478 | 0.0037 | 0.0163 |
| 8 | 7.57 | 0.010 | 0.0043 | 0.0404 | 0.0043 | 0.0152 |
| 9 | 7.57 | 0.010 | 0.0045 | 0.0365 | 0.0047 | 0.0152 |
| 10 | 7.57 | 0.010 | 0.0046 | 0.0325 | 0.0056 | 0.0144 |
| 12 | 7.57 | 0.010 | 0.0050 | 0.0285 | 0.0061 | 0.0134 |
| 15 | 7.57 | 0.010 | 0.0052 | 0.0243 | 0.0070 | 0.0121 |

3. Stages 1 and 2 are repeated sequentially for $n = 2$ to $n = 15$.
4. Each data set is then summarized as shown in Table Ex 15.8.1.

The values of the mean and variance do not change for differing numbers of samples but the estimates of the actual and $\chi^2$-derived CL percentiles differ considerably and both change as $n$ increases from 2 to 15. The effect can be seen in Figure Ex15.8.2. As $n$ increases the 2.5% bound of the CL increases slowly and the 97.5% bound decreases rapidly. Hence, the precision of the variance increases with increasing numbers of samples but the rate of change becomes progressively smaller for values of $n > 8 - 12$.

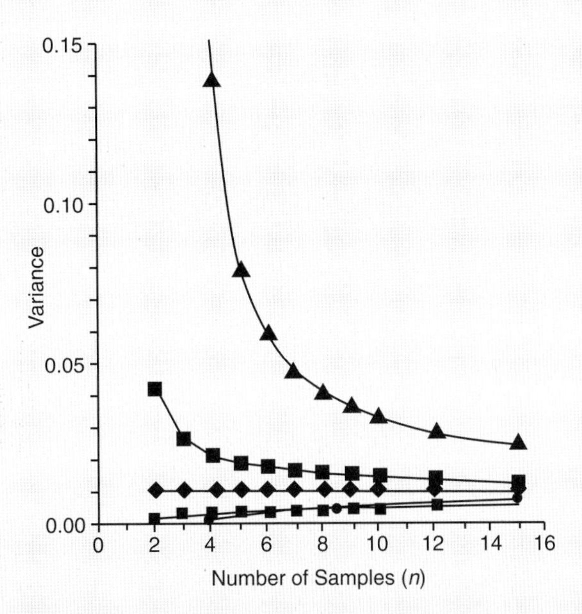


**Figure Ex 15.8.2** Changes in the variance parameters with increasing numbers of samples. Variance (◆), Upper (▲, ■) and lower (•, ×) bounds of the $\chi^2$ and actual distribution percentiles, respectively.

This example shows not only how data can be modelled but also reinforces the importance of ensuring sufficient sample numbers to optimize the precision of the variance, which is important in minimising microbiological uncertainty.

*Source:* Microsoft® Office Excel® 2003 and Resampling Stats, Simon (1997)

A MC simulation is essentially a computer experiment involving random sampling with high levels of replication from a probability distribution. MC simulations are used extensively for analysis of risk prediction (see www.foodrisk.com). There are constraints on the nature of the distributions that can be simulated and a number of issues must be considered before starting a simulation. For instance, assumptions must be relevant and risks for deviation from those assumptions determined. Is the 'estimator' based on ND? Is it consistent? Is it biased for finite samples? Is a hypothesis testing procedure achievable and, if so, what is the statistical power of the test against alternative hypotheses? In some cases, users may rush in without first considering not only what they want to achieve by a MC simulation but how they propose to communicate the outcome. MC simulation is often used in Bayesian statistics to provide models reputed to be of greater value than those obtained by standard statistical approaches.

## CONCLUSIONS

This chapter has sought to illustrate some of the more commonly used (and misused) statistical procedures that are relevant to analysis and interpretation of microbiological data. Many other statistical tests are available for special purposes and the reader is encouraged to consult appropriate standard texts on statistics for biological sciences.

Three key issues need to be faced. Firstly, it is important to involve a statistician in the planning stages of a study, afterwards is often too late! Secondly, microbiologists are often intimidated by the apparent complexity of statistical equations; but these are merely mathematical expressions that describe key parameters. Increasingly, low-cost computer software takes the tedium out of calculations but, before using such systems, it is essential to ensure that experiments have been properly planned according to defined objectives and correct hypotheses have been assigned. Thirdly, statisticians sometimes admit that they do not understand microbiological concepts. Time spent in explaining microbiological practices to statisticians, and statistical practices to microbiologists, increases the ability of both parties to work together effectively. It is really no different to the general international problem that people often do not understand the language and ethos of people in other countries. It is necessary to have confidence in our own skills to share them with others.

## ACKNOWLEDGEMENT

I am indebted to my friend and colleague Dr Alan J Hedges for helpful comments and criticisms. Data used for most examples comes from the work of past research students.

## REFERENCES

Analyse-it (2012) Analyse-it® 3.0 Software for Excel. Analyse-it Software Ltd. [Online] http://www.analyse-it.com (last accessed 23 June 2013).

Armstrong, R.A. and Hilton, A. (2011) *Statistical Analysis in Microbiology: Statnotes*. John Wiley & Sons, Inc., Hoboken, NJ, USA.

Baranyi J. and Tamplin M. (2004). ComBase: A Common Database on Microbial Responses to Food Environments. *Journal of Food Protection* **67**, 1834–1840.

Bassett, J. (2010) Impact of microbial distributions on food safety. *ILSI Europe Report Series*. [Online] http://www.ilsi.org/Europe/Publications/Microbial%20Distribution%202010.pdf (last accessed 24 June 2013).

Bland, J.M. and Altman, D.G. (1986) Statistical methods for assessing agreement between two methods of clinical measurement. *Lancet* **i**, 307–310.

Bland, J.M. and Altman D.G. (1995) Comparing methods of measurement: Why plotting difference against standard method is misleading. *Lancet* **346**, 1085–1087.

Casella, G. (2008) *Statistical Design*. Springer, New York.

ComBase Predictor 3.0 A web-based prediction model for food microbiology. http://modelling.combase.cc/membership/ComBaseLogin.aspx (last accessed 4 July 2013).

Corry, J.E.L., Jarvis, B. and Hedges, A.J. (2010) Minimising the between sample variance in colony counts on foods. *Food Microbiology* **27**, 598–603.

Crawley, M. (2005) *Statistics – An Introduction Using R*. John Wiley & Sons Ltd, Chichester, UK.

Davison, A.C. and Hinkley, D.V. (1997) *Bootstrap Methods and their Application*. Cambridge University Press, Cambridge, UK.

Dytham. C. (2011) *Choosing and Using Statistics: A Biologist's Guide*, 3rd edn. Blackwell Publishing Ltd, Oxford.

EFSA (2010) Analysis of the baseline survey on the prevalence of *Campylobacter* in broiler batches and of *Campylobacter* and *Salmonella* on broiler carcasses, in the EU, 2008 Part A: *Campylobacter* and *Salmonella* prevalence estimates. *EFSA Journal* **8**(8), 1522–1654.

Einstein quotation from http://www.alberteinsteinsite.com/quotes/einsteinquotes.html (last accessed 24 June 2013).

Esty, J.R. and Meyer, K.F. (1922) The heat resistance of the spores of *Bacillus botulinus* and allied anaerobes, XI. *Journal of Infections Disease* **31**, 650–663.

EU (2005) Commission Regulation No. 2073/2005 on microbiological criteria for foodstuffs. *Official Journal* **L338**, 1–26.

Fisher, R.A. (1971) *The Design of Experiments*, 8th edn. Hafner Pub Co., New York.

Fisher, R.A. and Yates, F. (1963) *Statistical Tables for Biological, Agricultural and Medical Research*, 6th edn. Oliver and Boyd, Edinburgh, UK.

Food Risk Organisation Predictive Microbiology Tools. [Online] http://foodrisk.org/tools/predictivemicro/ (last accessed 24 June 2013).

Goldstein, H., Browne, W. and Rashbash, J (2002) Multilevel modelling of medical data. *Statistics in Medicine* **21**, 3291–3315.

Hawkins, D (2005) *Biomeasurement: Understanding, Analysing and Communicating Data in the Biosciences*. Oxford University Press, Oxford.

Hedges, A.J. and Jarvis, B. (2006) Application of 'robust' methods to the analysis of collaborative trial data using bacterial colony counts. *Journal of Microbiological Methods* **66**, 504–511.

ICMSF (1986) *Microorganisms in Foods. 2. Sampling for microbiological analysis: Principles and specific applications*, 2nd edn. University of Toronto Press, Toronto, ON, Canada.

ISO (2003) *Microbiology of food and animal feeding stuffs – Protocol for the establishment of precision characteristics of quantitative methods by Interlaboratory studies*. ISO 16140:2003. International Standards Organization, Geneva, Switzerland.

Jarvis, B. (2000) Sampling for microbiological analysis. In *The Microbiological Safety and Quality of Foods*, Vol II (eds. B.M. Lund, A.C. Baird-Parker and G.W. Gould). Aspen Publishers, Gaithersburg, MD, USA, pp. 1691–1733.

Jarvis, B. (2008) *Statistical Aspects of the Microbiological Examination of Foods*, 2nd Edn. Academic Press, London.

Jarvis, B. (2011) Some practical and statistical aspects of the comparative evaluation of microbiological culture media. In *Handbook of Culture Media for Food and Water Microbiology*, 3rd Edn (eds J.E.L. Corry, G.D.W. Curtis and R.M. Baird). RSC Publishing, Cambridge, pp. 3–38.

Jarvis, B. and Hedges, A.J. (2011) The effect of the number of sample units tested on the precision of microbial colony counts. *Food Microbiology* **28**, 1211–1219.

Jarvis, B., Hedges, A.J. and Corry, J.E.L. (2007) Assessment of measurement uncertainty for quantitative methods of analysis: Comparative assessment of the precision (uncertainty) of bacterial colony counts. *International Journal of Food Microbiology* **116**, 44–51.

Jarvis, B., Hedges, A.J. and Corry, J.E.L. (2012) The contribution of sampling uncertainty to total measurement uncertainty in the enumeration of microorganisms in foods. *Food Microbiology* **30**, 362–371.

Jongenburger, I. (2012) *Distributions of microorganisms in foods and their impact on food safety*. PhD Thesis, Wageningen University, The Netherlands.

Koseki, S. (2009) Microbial Responses Viewer (MRV): a new ComBase-derived database of microbial responses to food environments. *International Journal of Food Microbiology* **134**, 75–82.

McKellar, R.C. and Lu, X. (eds) (2003) *Modelling Microbial Responses in Foods*. CRC Press, Boca Raton, FL, USA.

McMeekin, T.A., Olley, J.N., Ross, T. and Ratkowsky, D.A. (1993) *Predictive Microbiology*. John Wiley & Sons Ltd, Chichester, UK.

Manly, B.F.J. (2007) *Randomization, Bootstrap and Monte Carlo Methods in Biology*, 3rd edn. Chapman and Hall/CRC, Boca Raton, FL, USA.

Mead, R., Curnow, R.N. and Hasted, A.M. (2002) *Statistical Methods in Agriculture and Experimental Biology*, 3rd Edn. Chapman and Hall/CRC, Boca Raton, FL, USA.

Minitab 16 (2012) Software for quality improvement (Link to software). [Online] http://www.minitab.com/en-US/default.aspx (last accessed 24 June 2013).

Murdoch, J. and Barnes, J.A. (1986) *Statistical Tables for Science, Engineering, Management and Business Studies*, 3rd edn. MacMillan, London.

Paulson, D.S. (2008) *Biostatistics and Microbiology: A Survival Pack*. Springer, New York, USA.

Rahman, N.A. (1968) *A Course in Theoretical Statistics*. Griffin, London, pp. 298–299.

Roberts, T.A. and Jarvis, B. (1983) Predictive modelling of food safety with particular reference to *Clostridium botulinum* in model cured meat systems. In *Food Microbiology: Advances and Prospects* (eds T.A. Roberts and F.A. Skinner). Academic Press, London, pp. 85–95.

Rousseuuw, P.J. and Croux, C. (1993) Alternatives to the median absolute deviation. *Journal of the American Statistical Association* **88**, 1273–1283.

RSC (2001a) *Robust statistics: a method of coping with outliers*. AMC Technical Brief No. 6, Royal Society of Chemistry, Analytical Methods Committee. [Online] http://www.rsc.org/Membership/Networking/InterestGroups/Analytical/AMC/TechnicalBriefs.asp (last accessed 24 June 2013).

RSC (2001b) *MS Excel Add-in for Robust Statistics*. Royal Society of Chemistry, Analytical Methods Committee. [Online] http://www.rsc.org/Membership/Networking/InterestGroups/Analytical/AMC/Software/RobustStatistics.asp (last accessed 24 June 2013).

Sheskin, D.J. (2011) *Handbook of Parametric and Non-parametric Statistical Procedures*, 5th edn. Chapman Hall/CRC, Boca Raton, FL, USA.

Snedecor, G.W. and Cochran, W.D. (1980) *Statistical Methods*, 7th edn. Iowa State University Press, Ames, IA, USA.

Tukey, J.W. (1955) Answer to Queries. *Biometrics*, **11**, 111–113.

Welch, B.L. (1947) The generalisation of 'student's' problem when different population variances are involved. *Biometrika* **34**, 28–35.

# 16 Statistical modelling of anthropometric characteristics evaluated on nutritional status

**Zelimir Kurtanjek and Jasenka Gajdos Kljusuric**
*Faculty of Food Technology and Biotechnology, University of Zagreb, Zagreb, Croatia*

## ABSTRACT

The aim of this chapter is to illustrate to researchers in anthropometry and nutrition sciences how to apply chemometric methods for the extraction of data structures and their relations leading to development of mathematical models. Formally, this chapter gives a short review of basic methodologies in multivariate chemometric analysis from theoretical and practical view points and illustrates its high potential in analysis of complex life systems. As an example, chemometry is applied for the analysis of anthropometric parameters and nutritional status for secondary school girls and boys of age 14–18 situated in boarding schools in Croatia. In total 25 variables are measured; anthropometric, energy and macronutrients (fat, proteins and carbohydrates) and some life style factors. The data cover the period from 1997 to 2010 with a total number of 3440 individual recordings. The focus of the analysis is to determine the data structure of normal, overweight and obese subpopulations and relate their classification to principal latent variables and, consequently, to evaluate corresponding main important factors. Analysis shows clear appearance of body mass index (BMI) bimodal monovariable probability density function and its dynamics through the period of 15 years. Developed and compared are PLS models for body fat prediction based on Carnegie Mellon University data and boarding schools in Croatia. Generally, it is shown that chemometric analysis provides effective tools for data classification and extraction of important variables in analysis of complex phenomena such as nutritional status and obesity.

## INTRODUCTION

### Methodology

Chemometrics is a field of applied statistics and modelling which deals with the extraction of statistically significant information from multivariate experimental data (numerical and categorical) from complex systems, such as chemistry, life systems, agronomy, ecosystems, psychology, economy, bioinformatics, metabolomics, industrial process analytics and many others. In general, it is seen as a set of algorithms for information extraction from general multivariate complex systems. Despite its very broad definition, chemometrics' most popular profile is associated with chemical data analysis from spectrograms (IR, NIRS). Since living systems are not fully understood from the first principles (by equations of physics and chemistry), unlike human-designed technical systems, application of multivariate techniques is focused on extraction of a diversity of information from mostly uncontrolled experiments with large data sets. It is a common phrase that chemometrics is a data-driven methodology which is in contrast to major

*Mathematical and Statistical Methods in Food Science and Technology*, First Edition.
Edited by Daniel Granato and Gastón Ares.

scientific disciplines devoted to the application of exact principles to data from controlled experiments. Hence, some researchers do not view chemometrics as a 'hard' science, such as natural and technical sciences. However, chemometric algorithms are based on fundamental principles of statistics and numerical analysis, and are model oriented. It can be said that chemometric analysis 'inspires' modelling (as a preliminary stage proceeding hard modelling), provides model alternatives, applies numerous model validation algorithms and uses computer resources extensively. The applicative side of chemometrics is very broad in technical and biotechnical sciences, with limitless applications from analytical laboratory level to process control.

The basic ideas of multivariate analysis were introduced by Pearson (1901) and the fathers of the term chemometry, which was introduced at the beginning of 1970s, were Wold (1995) and Kowalski and Bender (1972). The numerical algorithms involved in chemometrics (eigenvalue problems) were developed by numerical mathematicians and physicists at the beginning of twentieth century. The number of manuscripts published in scientific journals is growing exponentially; a *ScienceDirect* search in 2012 on the key word 'chemometrics' produced about 20 000 articles. There are hundreds of books in the field, from introductory level to applications in specific fields. Some of 'well known' recent books are: Varmuza and Filmoser (2009), Brereton (2003, 2007), Wehrens (2011) and Venables and Ripley (2002).

In the last decade chemometric application of chemometric software tools became very popular due to the availability of numerous chemometric software packages with very important high-level graphical capabilities for 'quick' visual inspection of large multivariate data sets. Some of the most popular chemometric software systems are the SAS software (SAS®, 2012), SPSS software (IBM, 2012) Eigenvector Chemometric Tool Box (Eigenvector, 2012), The Unscrambler® (Camao, 2012), Statistica (StatSoft, 2011), R free statistical software (R Development Core Team, 2012) and the use of general scientific programming systems like Matlab (MathWorks®, 2012) and W.R. *Mathematica* (Wolfram, 2011). The importance of the software graphical capabilities is due to data space reduction by the use of data projections to the space of latent variables, mostly represented by two-dimensional plots. Formally, latent variables are not directly observable but can be expressed as linear combinations of the measured and observed variables, which in a given optimal sense provide most of the information. Deeper analysis of the nature of latent variables can lead to their semantic enrichment, that is to be associated with phenomena (for example with a specific metabolic activity) and initiate development of a formal theory. Vectors spanning the space of latent variables are called principal components and their analysis is the trademark of the whole field associated with acronym PCA (principal component analysis). Projections of experimental data on planes of principal components are called scatter plots. In popular jargon one could say that the main result of chemometric analysis is one or two scatter plots of multivariate data sets. Information is extracted from data structures identified in scatter plots and their relation to measured variables. As the basic concept of modelling is the mathematical relationship between input and output data (exogenous and endogenous in life systems), it is inferred from variable important analysis of PCA components. In cases of a relatively small number of variables (approximately less than 25) it is possible to use graphical joint plots of data and variables (bi-plots) from which by a visual inspection can be inferred linear multiple input–multiple output models (MIMO systems). The fact that linear chemometric models are successfully inferred from nonlinear complex (life) systems is due to the concept of latent variables reflected from numerous observed variables. The main characteristics of PCA analysis are:

- effective dimension reduction of complex systems with many (n) observed variables and objects (p), especially when $n > p$ (or $n \gg p$ as in bioinformatics);
- verbalization of key 'hidden' states inferred from latent variables;
- transformation of highly correlated observations that contribute to a smaller set of uncorrelated principal components;

- effective rejection of statistically unimportant information (for example errors in measured data) and irrelevant observations;
- application of robust statistical estimates by detection of multivariate outliers;
- simple visualization of multivariate data by two-dimensional scatter plots with principal components;
- application of methodologies for model selection and validation;
- provision of a basis for 'hard' modelling.

Although chemometrics is a broad collection of various algorithms and tools, the following are the three most common and are routinely applied:

(i) principal component analysis (PCA)
(ii) linear discriminant analysis (LDA)
(iii) partial least squares (PLS).

By the PCA method multivariate data are explored to extract similarities between data and variables. Based on score and latent variables projections clusters of multidimensional similarities (correlations) of objects and variables are recognized. The result is the extraction of data structures which imply their functional similarities. The result of PCA result is a reduction in the dimensionality of multivariate data, recognition of 'pooled' objects and variables, leading to interpretation of possible functional relationships.

LDA analysis is focused on exploring the main differences between multivariate data and variables. Usually, visual inspections reveal which objects and variables are the most dissimilar and provide interpretation for classification. It is also commonly used for classification of 'unknown' samples that need classification, like in inspection of product adulteration.

PLS is the most common multivariate calibration technique, applied for spectrometric methods and other analytical or process measurement techniques. It is applied for multiple input single output (MISO) and multiple input multiple output (MIMO) by the PLS2 method. It provides strictly validated and robust linear calibration models (multidimensional lines) with reliable estimations of errors of output predictions.

## Applications in food sciences

During the last two decades, numerous publications have reported applications of chemometric methodologies in food technology and life systems. One of the review articles on chemometric methodologies applied in monitoring food quality (Bro *et al.*, 2002) deals with the analysis of the food production chain in meat transport and storage, vegetable characterization, fish processing, sugar quality and adulteration. Presented are applications of PCA and PLS techniques applied to spectroscopic data and multivariate calibrations of process sensors. The important issue of food safety can be effectively addressed (without sample preparation and no chemical and time consuming procedure) based on chemometric determination of undesirable substances in food and feed. It includes rapid detection of impurities in vegetables and contamination of plants by various pathogens. Multi-dimensional image analysis techniques are combined with chemometric tools to get spatial information and link it (model) to physical and chemical characteristics (Pierna *et al.*, 2012). There are many publication focused on the detection of adulteration in food products, like wine, olive oil and fruit juices, among others. One of the recent publications reported on the application of linear discrimination analysis (LDA) on very similar NIR spectra of extra virgin olive oil and the product adulterated with rice bran oil. The very small deviations on the spectra were discriminated by LDA and provided calibration (Rohman and Man, 2012). Also reported are successful results of multivariate factor analysis of a 14-year study of fruit juice concentrate marked in Argentina (Oteiza *et al.*, 2011). The analysis provides insight into the relationship between industrial process factors and incidence of bacterium spoilage in various fruit and vegetable juices. Based on the chemometric results improvements in technology are considered to ensure better product quality and longer stable product shelf lives.

As an example of application of multivariate and chemometric reasoning (modelling) is the analysis of commercial beers based on phenolic composition and antioxidant activity (Granato *et al.*, 2011). Nowadays, the important issue of human health and food quality is related to the antioxidant potential of various food feedstocks and products. It is especially focused on 'the French paradox' and the antioxidant potential of red wine. Chemometric analysis of antioxidant Brazilian red wines has been reported (Granato *et al.*, 2010) in which hierarchical modelling based on principal component analysis is applied. Chemometric analysis of a sensory descriptive attribute of the Croatian dessert wine Prošek has been reported (Budić-Leto *et al.*, 2012). A complex chemometric analysis of wheat, as one of major foods, has been based on data integration of 30 major winter wheat brands determined by biochemical composition (RP-HPLC), high molecular analysis (HMW) by SDS-PAGE and numerous baking quality tests. The aim is to discriminate wheat cultivars based on their molecular structure and develop PLS calibration models for predicing 15 baking bread quality extensographic, farionographic and indirect quality parameters (Kurtanjek *et al.*, 2008, 2013). The important issue of perception of the food quality of functional food by the young population in Croatia has been analysed by chemometric analysis (Markovina *et al.*, 2011). Chemometric methodology can be also applied in the design stage of new functional foods, such as milk toffee with less sugar, more inulin and decreased energy content (Rumora *et al.*, 2013).

## Applications in anthropometry

For the last few decades the World Health Organization has recognized obesity as a common risk factor for numerous diseases. All over the world it has initiated numerous projects with application of multivariate statistical analysis and predictive chemometric models to elucidate complex relationship between human health, obesity, lifestyle, socio-economic factors, nutrition and anthropometric parameters. Most of these studies relate anthropometric data to population age and gender subgroups, socio-economic status, dietary profiles, education and life style factors, among others. In Brazil, Moraes *et al.* (2012) have reported on investigations conducted among adolescents (aged 14–18) on their dietary preferences, socio-economic status and life style. They applied PCA methodology to extract the main food groups and their association with the anthropometric parameters and life conditions. Results obtained are aimed to provide guideline for public health planners in Brazil, especially for planners in schools. In Mexico, Fernald (2007) conducted an extensive house-to-house survey among the poorest sections of the society. Data were collected on socio-economic status, anthropometric parameters and food intake profiles, especially consumption of sweetened drinks and obesity.

Multivariate analysis has been applied to establish predictive relationships for body mass index (BMI). From China is reported a study on trunk fat mass among the female population aged 20–40 years (Jiang *et al*, 2007). Fat mass was measured by X-ray technique in parallel with five anthropometric parameters (body mass index, waist circumference [WC], hip circumference [HC], waist to hip ratio [WHR], and conicity index [CI]). PCA analysis and multiple regression PCR was applied for prediction of fat mass on the parameters. Models were developed for subpopulations with regression coefficients $R^2 = 0.85–0.9$ depending on age range. The results show that BMI is the most important variable for prediction of fat mass. More complex methodologies, including data mining algorithms, were applied in a Greek project (Lazarou *et al.*, 2010, 2012). They studied a child population (9–13 years) in order to extract pattern profiles and their association with obesity. The analysis related to numerous life style variables, such as physical activity, time for TV viewing, computer time, socio-economic factors, food preferences, Mediterranean dietary habits and anthropometric parameters. The decision tree induction method, which is essentially nonlinear in contrast to linear chemometric models, was applied. The study showed the better predictivity of the decision tree but the results were less transparent to extract simple rules. Results obtained by chemometric models have similar performances but are transparent and easier for interpretation.

A US study (Waring *et al.*, 2009) investigated anthropometric data and some life style habits (alcohol consumption and tobacco smoking) and health risks for coronary dieses for a population from 28 to 62 years. PCA analysis was applied and proved to be insightful into extraction of multivariate patterns associated with heart diseases risks. Use of chemometric models for prediction of cardiovascular diseases and blood hypertension among Indian populations (Badaruddoza *et al.*, 2010; Bishnoi et al., 2010) were focused on younger generations, university going students and female populations age from 16 to 45. The populations cover several ethnic groups in India. Besides anthropometric parameters the model included standard biochemical tests. The analysis clearly showed that WHR is the most important variable for long term prediction of cardiovascular diseases. The ethnicity and gender of the populations also showed as important factors. Similar results from a study from Africa are available (Sanya *et al.*, 2009). This study covered Nigerians of a broad age range (from 15 to 85 years). Anthropometric data were supplemented with blood pressure measurements and life style variables, as well as history of hypertension, diabetes, cardiac and renal diseases. The multivariate analysis showed that BMI and WHR strongly correlate with hypertension. A Spanish study (Nieto *et al.*, 2011) focused on selection of the important anthropometric variables to characterize a healthy elderly generation from 65 to 95 age. They applied PCA analysis and dendrograms of the anthropometric variables and observations and found the same conclusion, that for the healthy elderly population regardless of gender and age only anthropometrical variables (height, weight, skinfold thickness and mid-upper arm circumferences) are sufficient for classification.

## EXPERIMENTAL DATA

Presented here are experimental data and results of chemometric analysis of Croatian high school male and female adolescents aged 14–18, situated in secondary boarding schools, from surveys conducted during 1997, 2000 and 2010 with a total of 3440 individual recordings. Included are boarding schools in the two main regions of the country, continental and Adriatic, and schools in the capital city Zagreb with mixed populations. A total of 25 variables were measured, covering anthropometric parameters, energy, macronutrients (fat, proteins, and carbohydrates) and some life style factors. The boarding school generations were selected with the aim to evaluate effects of common dietary profiles on anthropometric parameters and appearance of obesity. An overview of the data with calculated means and standard deviations is given in Table 16.1. Trends in the BMI for female and male populations during the period of the last 15 years are analysed by a nonparametric graphical 'box and whiskers' plot (Figure 16.1) and numerical approximation of BMI probability density functions by the 'pseudo' bimodal univariate normal distribution:

$$\begin{aligned}\rho(BMI) = N(x;\alpha,\mu_1,\mu_2,\sigma_1,\sigma_2) &= \alpha\cdot\frac{1}{\sigma_1\cdot\sqrt{2\cdot\pi}}\cdot Exp\left(-((x-\mu_1)/\sigma_1)^2\right)\\ &\quad+(1-\alpha)\cdot\frac{1}{\sigma_2\cdot\sqrt{2\cdot\pi}}\cdot Exp\left(-((x-\mu_2)/\sigma_2)^2\right)\\ &\quad 0\le\alpha\le 1\end{aligned} \tag{16.1}$$

where $\mu_1$ and $\mu_2$ are the expected values of the normal and obese populations, $\sigma_1$ and $\sigma_2$ are the corresponding standard deviations, and $\alpha$ accounts for relative contributions of the two populations.

**Table 16.1** Summary of anthropometric and nutritional status of female and male pupils situated in secondary boarding schools in Croatia during the years 1997, 2000 and 2010.

| Survey year | Student Age | Gender | Boarding school | Years in institution | Daily intake of energy and macronutrients | | | |
|---|---|---|---|---|---|---|---|---|
| 1997<br>N = 1094 | 15–19 | Female<br>N = 520<br>Male<br>N = 574 | Female<br>Male<br>Coed | 1–5 | Energy<br>(kJ) [kcal]<br>11 842<br>[2833] | Proteins<br>(g/day)<br>86.08 | Fat<br>(g/day)<br>133.25 | Carbohydrates<br>(g/day)<br>304.63 |

**Anthropometric data**

| Height (m) | Mass (kg) | Biceps (cm) | Chest (cm) | Waist (cm) | Hip (cm) | Upper arm (cm) | Forearm (cm) | BMI (kg/m²) | WHR |
|---|---|---|---|---|---|---|---|---|---|
| 1.73<br>±0.08 | 63.7<br>±10.5 | 25.2<br>±2.63 | 89.8<br>±7.06 | 73.9<br>±7.52 | 96.1<br>±6.48 | 11.59<br>±4.77 | 10.76<br>±4.07 | 21.2<br>±2.42 | 0.769<br>±0.058 |

| Survey year | Student Age | Gender | Boarding school | Years in institution | Daily intake of energy and macronutrients | | | |
|---|---|---|---|---|---|---|---|---|
| 2000<br>N = 1407 | 14–19 | Female<br>N = 718<br>Male<br>N = 689 | Female<br>Male<br>Coed | 1–5 | Energy<br>kJ(kcal)<br>11 714<br>(2802) | Proteins<br>(g/day)<br>96.74<br>±12.95 | Fat<br>(g/day)<br>112.8<br>±14.12 | Carbohydrates<br>(g/day)<br>356.01<br>±53.62 |

**Anthropometric data**

| Height m | Mass kg | BMI kg/m² | Height m | Mass kg | BMI kg/m² |
|---|---|---|---|---|---|
| **male** | **male** | **male** | **female** | **female** | **female** |
| 1.76<br>± 0.08 | 68.3<br>± 11.3 | 21.8<br>±3.0 | 1.65<br>±0.06 | 57.98<br>± 9.1 | 21.2<br>± 3.1 |

| Survey year | Gender | Age | L10: Regions | |
|---|---|---|---|---|
| 2010 | Female<br>N = 348 | 14–25 | Continental N = 80<br>Mediterranean N = 268 | |

| Mass (kg) | Height (m) | Body fat (%) | BMI (kg/m²) | Waist (cm) | Hip (cm) | WHR | L1 (year) | L2 (/week) |
|---|---|---|---|---|---|---|---|---|
| 61.41<br>±10.33 | 1.66<br>±0.063 | 23.9<br>±5.65 | 22.26<br>±3.29 | 72.91<br>±7.35 | 98.11<br>±6.86 | 0.742<br>±0.043 | 2.16<br>±1.37 | 4.32<br>±2.27 |

**Table 16.1** *(continued)*

| Survey year | Gender | Age | L10: Regions |
|---|---|---|---|
| 2010 | Male<br>N = 593 | 14–25 | Continental N = 199<br>Mediterranean N = 392 |

| Mass (kg) | Height (m) | Body fat (%) | BMI (kg/m$^2$) | Waist (cm) | Hip (cm) | WHR |
|---|---|---|---|---|---|---|
| 72.35<br>±12.97 | 1.78<br>±0.08 | 15.83<br>±6.67 | 22.56<br>±3.4 | 79.74<br>±8.8 | 98.41<br>±7.78 | 0.809<br>±0.04 |

| L1 (year) | L2 (/week) | L3 (/week) | L4 (/week) | L5 (/week) | L6 (/week) | L7 (/week) | L8 (/week) | L9 (/week) |
|---|---|---|---|---|---|---|---|---|
| 2.29<br>±1.38 | 4.47<br>±2.44 | 3.13<br>±2.07 | 6.69<br>±0.81 | 6.34<br>±1.15 | 4.51<br>±1.52 | 2.46<br>±2.14 | 17.51<br>±3.09 | 14.34<br>±3.13 |

| Energy (kJ) [kcal] | Fat (g/day) | Hydrocarbons (g/day) | Proteins (g/day) |
|---|---|---|---|
| 13 380<br>[3200] | 132.8<br>±14.49 | 419.5<br>±51.37 | 119.8<br>±10.17 |

'Life style' data: L1 = number of years spent in a boarding school, L2 = number of breakfasts, L3 = number of breakfasts at school, L4 = number of lunches at school, L5 = number of dinners, L6 = number of dinners at school, L7 number of consumed snacks, L8 = total number of meals, L9 = total number of meals consumed in a boarding school, L10 = Region (continental or Mediterranean)

The assumption is that there is overlapping of two populations, the primary or dominant normal population and the secondary increasing obese population. The parameters of the distributions (Equation 16.1) were estimated by the nonlinear least squares method using the Levenberg–Marquardt optimization method from corresponding approximated histograms; the results are presented in Figure 16.2. The nonparametric and parametric results show linearly increasing trends with time of the medians and

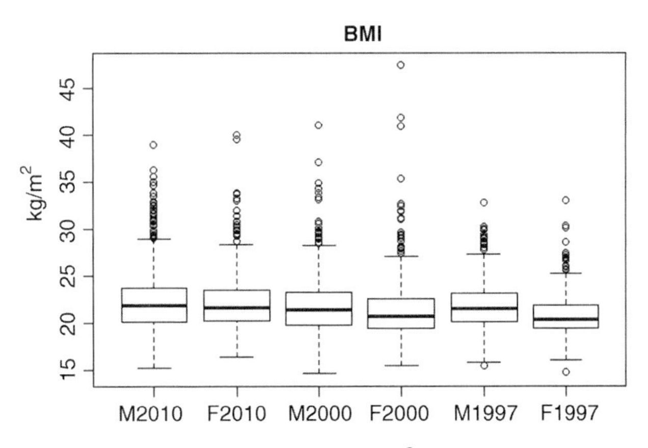


**Figure 16.1** 'Box and whiskers' plot of body mass index, BMI (kg/m$^2$), for male (M) and female (F) pupils of age 14–18 years situated in secondary boarding schools in Croatia surveyed in the years 1997, 2000 and 2010.

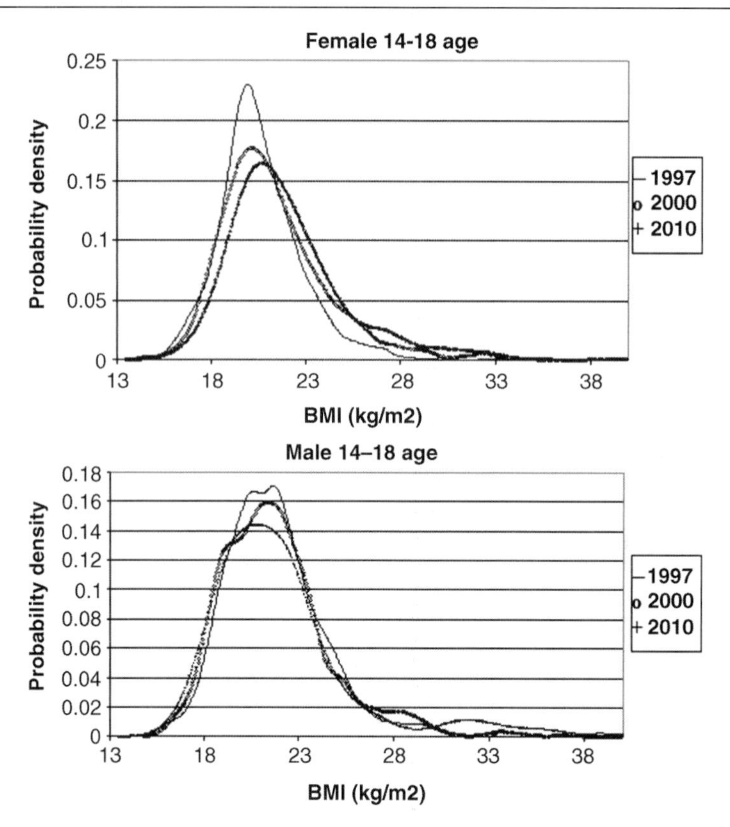


**Figure 16.2** Probability density functions of BMI for male (M) and female (F) students of age 14–18 years situated in secondary boarding schools in Croatia determined during the years 1997, 2000 and 2010.

quintiles range and parameters of the secondary population. Through the period from 1997 to 2010 for the male populations estimations of the primary distribution are approximately constant $\hat{\mu}_1 = 21$ and $\hat{\sigma}_1 = 3.1$ (kg/m$^2$) constant; however, the dispersion of the secondary distribution linearly increases with time from $\hat{\sigma}_2 = 1.4 - 3.6$ (kg/m$^2$) and the average contribution of the primary distribution is about $\alpha = 80\%$. For the female populations there is a linear increase in expected value of the primary distribution $\hat{\mu}_1 = 10.98 - 20.56$ (kg/m$^2$) but the secondary is almost constant $\hat{\mu}_2 = 22.5$ (kg/m$^2$), with average contribution $\alpha = 85\%$. The estimates obtained are most likely biased due to a significant presence of outliers. A much more objective interpretation of the data can be inferred from the 'box and whiskers' plot given in Figure 16.1. Although there is a noticeable difference between medians and estimated means of the subpopulations, the main conclusions are the same, that there is a trend of increasing BMI.

## PRINCIPAL COMPONENT ANALYSIS (PCA)

The method of principal components analysis (PCA) is the 'trade mark' of the whole field of chemometrics. The main objective is reduction of multivariate space to usually two or three dimensional principal subspaces. The dimension reduction is achieved by 'viewing' the data from a new coordinate system of latent variables (principal components or principal vectors). The term of latent variables has an important 'philosophical' interpretation, as being real variables but not directly observable (measurable) and usually expressible by linear (or sometimes nonlinear) combination of the observed variables. The effect of reduction is due to grouping of correlated variables into a smaller number of latent variables. In a new view the data present themselves in a more plausible manner, revealing data structures and their relations with observed variables. Graphical presentations of the data are scores of a scatter plot and are usually depicted in two-dimensional planes with principal components on the axes. The measured variables are also projected into the latent spaces, termed target vectors, and are also depicted as scores in two-dimensional plots. Due to the mathematical algorithm, the principal components are uncorrelated and scaled to one (i.e. orthonormal vectors).

The usual mathematical notion is to use $\boldsymbol{X}$ as the data matrix with variables associated with columns and individual sample or objects with rows. It has dimension $\boldsymbol{X}$ $(n,m)$ meaning that $m$ variables are measured for $n$ samples. The 'power' of chemometrics lies in the fact that usually $m > n$ (but not a necessary condition), which enables extraction of latent information from numerous observable variables. It is unlike the ordinary last squares method, for which it is required that $n > m$. When the data matrix contains variables of different nature and units of measure, it is advisable to autoscale data to relative values and zero averages. By this procedure the effects of units are removed but there is an adverse effect that increase of irrelevant data ('noise') may appear.

The projection is given in the terms of matrixes of principal components $\boldsymbol{P}$, targets $\boldsymbol{T}$ and the residual after projection named as an error $\boldsymbol{E}$:

$$\boldsymbol{X} = \boldsymbol{T} \cdot \boldsymbol{P}^{\mathrm{T}} + \boldsymbol{E} \tag{16.2}$$

As the projection itself is sometimes called a model, matrix $\boldsymbol{E}$ is interpreted as a model error. The key modelling issue is the determination of the dimension of the latent space (number of principal components). The difficulty lies in the fact that sometimes matrix $\boldsymbol{E}$ may contain essential information which is 'shadowed' by measurement errors and does not present itself as a part of the principal components. The simplest way to get a notion of the number of significant principal components is to use the graphical presentation of a 'scree' plot. It is a two-dimensional plot with number of principal components on the abscissa and partial variance of corresponding component. The graph is visually inspected to select a 'knee' point at which the 'deterministic' content of information levels off with 'stochastic' component. A better way is to use some metrics for statistical validation of the number of principal components.

Numerical evaluation of principal vectors is based on an optimization problem with orthonormality constraints:

$$\begin{aligned} &\boldsymbol{P} \rightarrow \max(\mathrm{var}(\boldsymbol{T})) \\ &\boldsymbol{P} \cdot \mathbf{P}^{\mathrm{T}} = \mathbf{I} \end{aligned} \tag{16.3}$$

where 'var' denotes variances of scores $\boldsymbol{X} \cdot \boldsymbol{P}$ and $\boldsymbol{I}$ is an identity matrix. The solution is determined by Langrangian multiplicator formalism, which yields eigenvalue problem:

$$(\boldsymbol{X}^{\mathrm{T}} \cdot \boldsymbol{X}) \cdot \boldsymbol{p}_i = \lambda_{\mathrm{i}} \cdot \boldsymbol{p}_i \tag{16.4}$$

where $\boldsymbol{X}^{\mathrm{T}} \cdot \boldsymbol{X}$ is a covariance matrix, $\boldsymbol{p}_{\mathrm{i}}$ is individual i-th principal component (eigenvector) and $\lambda_{\mathrm{i}}$ is corresponding eigenvalue. Due to the high sensitivity of the covariance matrix to outliers, a robust estimation of the covariance is advisable. The relative magnitude of eigenvalues corresponds to the fraction of variance accounted for by the corresponding vector:

$$\mathrm{var}(\boldsymbol{t}_i) = \lambda_i \quad \text{and} \quad \sum_{i=1}^{i=m} \lambda_i = m \tag{16.5}$$

The eigenvelues are sorted in a descending sequence and the first principal component corresponds to the first eigenvalue and so on. There are several numerical procedures available for the eigenvalue problem, such as singular value decomposition (SVD) and Jacobi rotation, which are standard methods in physics and numerical analysis, but due to the character of batch processing to is inefficient for large problems (many variables). In chemometrics, the most popular is the iterative method of nonlinear iterative partial least squares (NIPLS). In the first step the optimization is solved for the first component, the first scores obtained are subtracted, and the same procedure is repeated until eigenvalue becomes negligible. Usually only very few first components are calculated.

Application of chemometric software makes PCA evaluation easy, does not require mathematical knowledge, and is greatly intuitive due to graphical presentations. It is usually the first choice to view the multivariate data. However, validation of the dimension of latent space is always difficult and prone to errors.

In Figure 16.3 the results are given of PCA analysis of the male population of pupils situated age 14–18 years in the secondary boarding schools in Croatia during 2010. The schools are in continental and Mediterranean (Adriatic) regions. The data (Table 16.1) include anthropometric parameters, average daily intake of basic macronutrients (energy, proteins, fat, carbohydrates), life style parameters (L1–L9, Table 16.1) and regional location of a school as the main variable, which defines

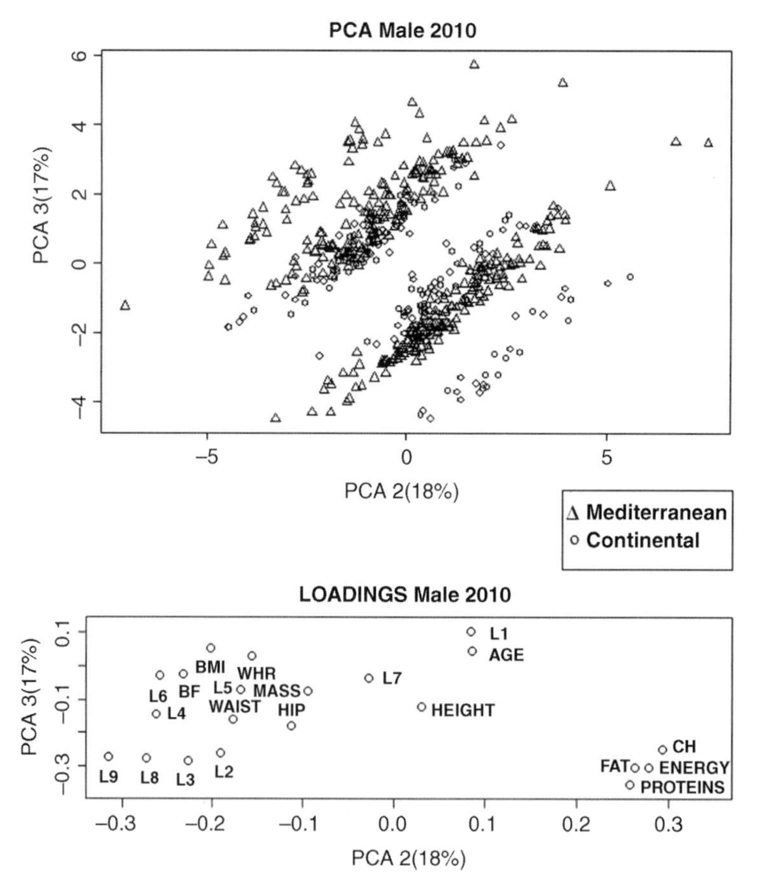


**Figure 16.3** Principal component analysis (PCA) of male pupil population of age 14–18 situated in boarding high schools in continental and Mediterranean Croatian regions in the year 2010.

regional way of life and locally specific dietary traditions. The objectives of PCA analysis is to see if there are distinctive data and variable patterns that account for the most of dispersion (variance) content. The contributions of the first three principal components obtained are 23%, 18% and 17%). From the scree plot it is concluded that higher components do not provide deterministic information. The first scatter plot (PC1 versus PC2) does not show any data structure, but on the next plot (PC2 versus PC3, Figure 16.3) the structures are clearly observed. To indicate which variable is dominant for the appearance of the structure, the scores are labelled according to their corresponding geographical regions. Observed are four clusters, the one at the left ($PC1 = -5$) has only pupils (samples) from Mediterranean and the one the right boundary ($PC1 = 5$) only pupils from the continental region. The two additional clusters contain scores from both regions. It clearly shows the dominant effect of regional way of life and nutritional preferences. The corresponding loading plot reveals how variables are interrelated and how they produce common effects. It is clearly seen that the dispersion along the PC2 is affected by increased total macronutrient intake in the continental schools. The vertical projection along PC2 is determined by the two groups of variables (L1 and AGE) and (L9, L8, L3 and L2). The variables L1 and AGE are correlated because number of years spent in a boarding school is proportional to age of a pupil. A separate cluster is formed by the anthropometric parameters and a collection of 'life style' variables (BMI, WHR, BF, WAIST, HIP; L4, L5, L6) which account for number of lunches and dinners consumed per week.

## LINEAR DISCRIMINATE ANALYSIS (LDA)

In contrast to PCA analysis, which determines latent variables responsible for similarity of data, linear discriminant analysis (LDA) determines latent variables which optimally explain differences between data belonging to predefined classes. This is also known as a supervised classification unlike unsupervised clustering algorithms. For example, in this research, based on BMI data (weight-for-age curves) published by the Centers for Disease Control and Prevention (CDC) the pupils are grouped into three classes: obese, overweight and 'normal' according to 95% and 75% percentiles. It is considered to be the modelling stage with the 'training' data, after which follows the model prediction phase for classification of new experimentally observed data or simulation of hypothetical cases. Importantly, analysis of the measured factors of the latent variables of a LDA discrimination model reveals which are the key parameters responsible or 'inputs' for a given classification.

Classical Fisher's LDA method is based on maximization of *t*-scores (Student variable) in the space of latent variables. Determination of the principle components is also an eigenvalue problem for the product of matrices, which account for the covariance ***C*** and variance ***V*** between groups:

$$\left(\mathbf{C}^{-1} \cdot \mathbf{V}\right) \cdot \mathbf{p}_i = \lambda_i \cdot \mathbf{p}_i \tag{16.6}$$

where the covariance is an average of covariances within each of $k$ classes weighted with observed 'a priori' probabilities for each class:

$$\begin{aligned} \mathbf{C} &= \sum_{j=1}^{j=k} p_j \cdot \mathbf{C}_j \\ \mathbf{C}_j &= \frac{1}{n-1} \cdot \left(\mathbf{X}_j - \mu_j \cdot \mathbf{I}\right)^T \cdot \left(\mathbf{X}_j - \mu_j \cdot \mathbf{I}\right) \end{aligned} \tag{16.7}$$

and the variance between classes:

$$\mathbf{V} = \sum_{j=1}^{j=k} p_j \cdot (\boldsymbol{\mu}_j - \mu \cdot \mathbf{1}) \cdot (\boldsymbol{\mu}_j - \mu \cdot \mathbf{1})^T \tag{16.8}$$

where $\mu$ is average over all classes. The eigenvectors $\boldsymbol{p}_i$ are the discrimination principal components, also noted as principal LDA components. The eigenvalues $\lambda_i$ account for successful classification by the corresponding LDA principal component.

In Figure 16.4 the results are presented of LDA analysis for the survey of male pupils of age 14–18 situated in secondary boarding schools for schools in the both regions. The first two principal LDA components account for 95% and 5% of classifications. Clearly the first principal LDA component is the main classification variable. Three clearly distinct clusters are obtained, the obese cases are associated with the maximum value and the 'normal' with the minimum projections on LDA1 vector. The overweight cases are in between the two but some cases are at the interface between overweight and obese. The importance of anthropometric and 'life style' parameters are separated and presented as relative percentages. The main 'life style' factors are regional position (55% influence), continental or Mediterranean region, and number of years (20% influence) of spent years in a boarding school. The main anthropometric variables discriminating the classes are WHR (50%), height (35%) and BMI (5%). In spite of the fact that CDC curves are functions of BMI, it is not the most influential variable for the classification. This can be explained by essential nonlinearity of CDC curves versus BMI.

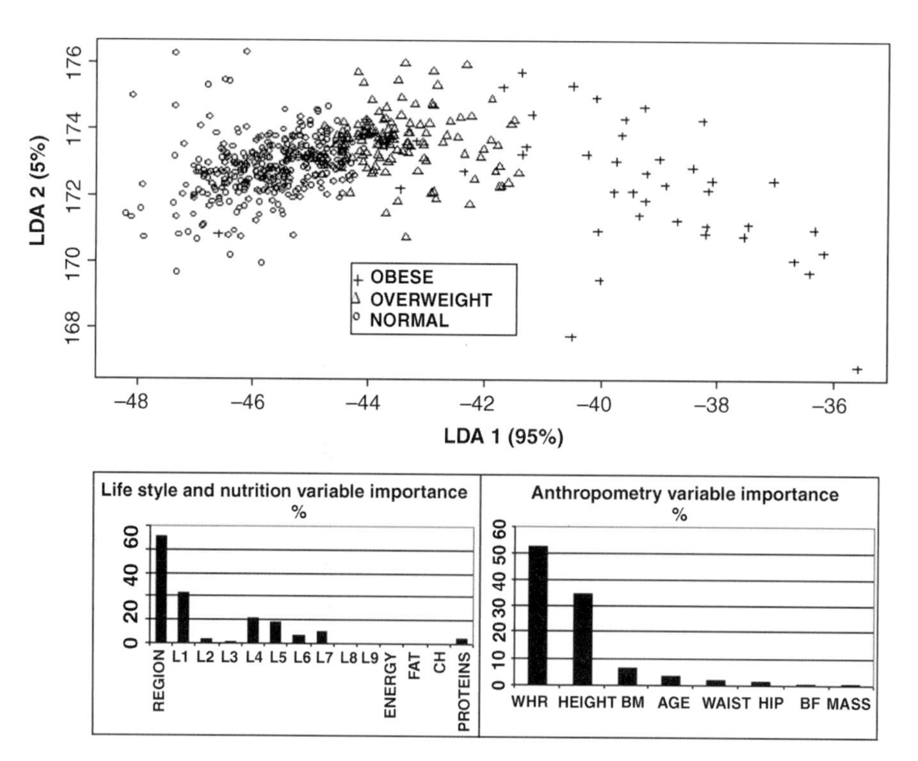


**Figure 16.4** Linear discriminant analysis (LDA) of obese, overweight and normal populations of male pupils of age 14–18 years during 2010 situated in boarding schools in all regions of Croatia.

Hence, a multivariate combination of anthropometric parameter is needed for relatively accurate classification.

## PARTIAL LEAST SQUARES PLS

The main objective of mathematical modelling of complex life systems is to relate by mathematical relations two sets of data, the input set $\boldsymbol{X}$ and output set $\boldsymbol{Y}$. The general (nonlinear) form of a mathematical model is:

$$Y = f(X) \tag{16.9}$$

Most chemometric models are focused on linear relationships between (multivariate form of a line):

$$\boldsymbol{Y} = \boldsymbol{L} + \boldsymbol{K} \cdot \boldsymbol{X} \tag{16.10}$$

Unlike the ordinary least squares method, the chemometric approach is based on use of latent variables, which are optimal in a statistical sense (maximum covariance, maximum correlation or minimum variance. There are numerous examples in all scientific fields, the most common are various multivariate calibration techniques based on spectrographic data. In this research models are developed between body fat (BF) and anthropometric parameters, which may also be interpreted as calibration. Determination of body fat is, in principle, a very complex measurement procedure and derived models can be of very practical importance for simple and quick assessment of BF based on simple anthropometric measurements.

The most common is the partial least squares methods (PLS) for single output (MISO) and PLS2 for multiple output (MIMO) models. Similar to ordinary least squares (OLS), data are regressed around a line in a space of latent variables. The criteria of maximum covariance was applied (although it may not be the optimal choice) as a compromise between model robustness from the PCA concept (maximum variance in X space) and high correlation in Y space. The main advantages of PLS models over OLS is better predictive performance and applicability to cases with few cases and many variables which may be partially correlated but are carriers of additional information, enabling the power of the latent space concept. Both input and output data spaces are projected into corresponding latent spaces by linear transformations (rotations):

$$\boldsymbol{X} = \boldsymbol{T} \cdot \boldsymbol{P}^{\mathrm{T}} + \boldsymbol{E}\mathrm{x} \tag{16.11}$$

$$\boldsymbol{Y} = \boldsymbol{U} \cdot \boldsymbol{Q}^{\mathrm{T}} + \boldsymbol{E}\mathrm{y} \tag{16.12}$$

where $\boldsymbol{P}$ and $\boldsymbol{Q}$ are principal components, $\boldsymbol{T}$ and $\boldsymbol{U}$ are targets, and $\boldsymbol{E}$x and $\boldsymbol{E}$y are matrices correspondingly belonging to input X and output Y spaces. It is assumed that the data are previously autoscaled to zero means and variances equal to one. Upon data projections is defined a linear input–output model between the targets:

$$\boldsymbol{U} = \boldsymbol{T} \cdot \boldsymbol{B} + \boldsymbol{E} \tag{16.13}$$

The model parameters are $\boldsymbol{B}$ and $\boldsymbol{E}$ is the model residual matrix. The principal components are found by nonlinear optimization of the covariance:

$$\max[\mathrm{Cov}(\boldsymbol{U}, \boldsymbol{T})] \tag{16.14}$$

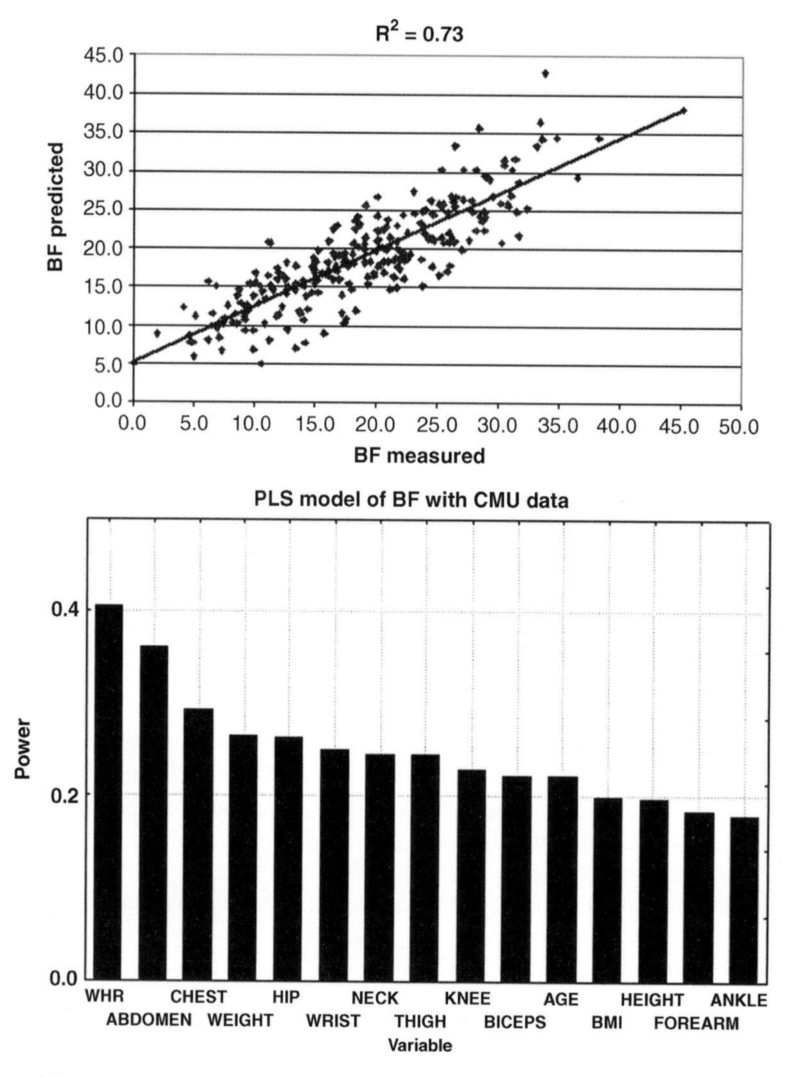


**Figure 16.5** Partial least squares model (PLS) of body fat (BF) based on Carnegie Mellon University data. Data from Johnson (1996). Adapted with permission from the author.

The orthonormality constraints are:

$$\begin{aligned} P \cdot P^{\mathrm{T}} &= I \\ Q \cdot Q^{\mathrm{T}} &= I \end{aligned} \tag{16.15}$$

The main problem in PLS modelling is determining the dimension of the latent space in order to provide required model predictivity. As in a PLC case, it is recommended to use numerical norms of statistical prediction criteria evaluated by separated training and validation subsets of input and output data, or one of the methods of cross-validation.

As an example for PLS, developed here are models for the estimation of body fat content based on two different methods of measurement. Used here are data reported (Jhonson, 1966) also available online from Carnegie Mellon University (CMU) at StatLib-Datasets Archive for a population of 252 men of age from 22 to 75 years and anthropometric measurements for 592 male pupils and students residing in Croatia boarding institutions aged 14–25 years surveyed in 2010. CMU data were obtained by the accurate and time consuming method of underwater weighing and body circumference measurements. In the Croatia study a quick method of measurements of electrical impendence using an Omron BF 306 instrument is applied. Derived are MISO PLS models with BF as the predicted ($y$ or output data) variable and anthropometric parameters as experimental predictor variables ($\boldsymbol{X}$ or input data). For the CMU model used there are 15 anthropometric parameters (given in Figure 16.5) and in the Croatia model 10 variables, of which six are anthropometric parameters and four nutritional accounting for average macronutrient intake in a boarding school (Figure 16.6). Decomposition of the both data sets resulted in only two statistically significant principal LDA components but for each

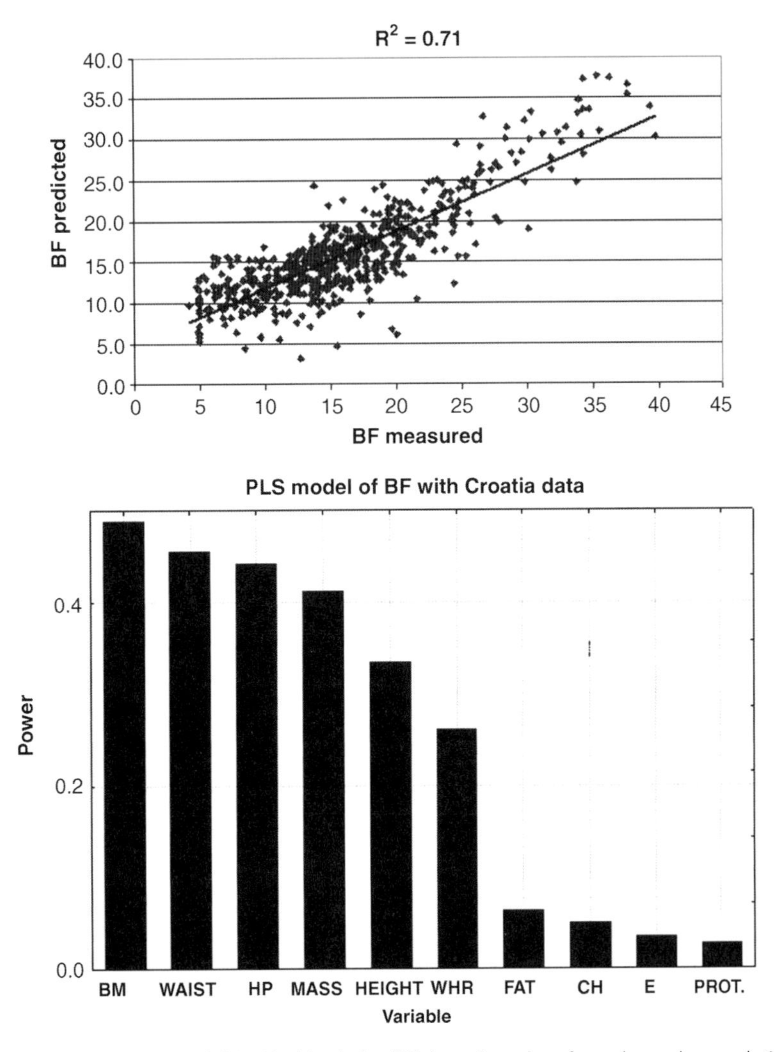


**Figure 16.6** Partial least squares model (PLS) of body fat (BF) based on data from the male population of pupils and students of age 14–21years in boarding schools during 2010 from all regions of Croatia.

prediction model the first three were applied. Both models have very similar 'prediction qualities' (for CMU $R^2 = 0.73$ and for Croatia $R^2 = 0.71$). An interesting observation is that the distribution of variable importance, given as 'power', for the CMU model is relatively flat with WHR being the most important predictor of BF. In the Croatia model the anthropometric parameters have high importance and the nutritional data have considerable lower importance. Unlike to the previous result here BMI is the most important predictor variable.

## CONCLUSIONS

This has presented a short review of statistical modelling techniques based on chemometrics with application in food engineering and nutrition. The main objective was to emphasize a systems view and multivariate concept in modelling complex living systems, such as relations between health, nutrition, 'life style' variables and socio-economic factors. Introduction of the concept of latent variables is the key in understanding high-dimensional data sets with numerous variables, of which some are relatively correlated but provide additional information. Reviewed were the three most common chemometric modelling approaches (PCA, LDA and PLS) and their algorithms with corresponding definitions of principal components. PCA and LDA are interpreted as tools valuable during the 'pre-modelling' stage of research, while PLS is the main linear input–output modelling algorithm for research of complex systems. The importance of PLS over the ordinary least squares method is in its robustness and superior predictive power due to the reduction of model dimensions by latent variables.

Chemometric algorithms are numerically demanding and sensitive to numerical errors, therefore application of chemometric software is essential. Most of the complex mathematical operations involved in chemometric algorithms are imbedded into software and do not represent an obstacle in application to a non-mathematical researcher. Emphasized was the visualization of multivariate data through various two-dimensional scatter plots as a main 'inductive' methodology for understanding data structures and their relationships with measured variables.

Although application of high-level chemometric software simplifies its application, the question of appropriate selection of model validation is still problem specific and open to research. Reliable validation of chemometric models is the key to their success in practice.

Application of the methods were illustrated by analysis of anthropometric and nutritional status of male and female pupils aged 14–18 years accommodated in secondary level boarding schools during the period 1997–2010. Two methods of raw data inspection were compared, the nonparametric 'box and whiskers' plot and the parametric pseudo two modal univariate probability distribution. Simplicity, transparency, observability of data skewness and insensitivity to outliers shows the important advantage of the nonparametric approach. The important potential of the PCA method is presented by analysis of anthropometric parameters, 'life style' variables and nutritional status (macronutrient intake) for a male population of students situated in the two main regions in Croatia. The analysis provides data structures and their correspondence to measured variables and clearly shows separation between data from continental and Mediterranean schools. Based on BMI data and CDC weight-for-age curves, the data are associated with three classes (obese, overweight and 'normal'). LDA analysis was applied as a tool for finding the most important variables responsible for classification. The results are in accordance with the PCA results, confirming the dominance of 'life style' parameters, including the importance of traditional Mediterranean food preference.

The chemometric input–output modelling concept was illustrated by PLS calibration of body fat predictions based on anthropometric measurement. This is an example where an 'expensive to measure' parameter is replaced by a multivariate model and easy to measure anthropometric parameters. Two models were compared using CMU and Croatia data sets. Both models showed very similar accuracy in prediction of BF.

In conclusion, systems view and chemometric algorithms based on high throughput data sets have become essential for modern scientific fundamental research such as nutrigenomics, medicine, food science, and also in practical fields of food engineering and management.

## REFERENCES

Badaruddoza, Kaur, N. and Barna, B. (2010) Inter-relationship of waist-to-hip ratio (WHR), body mass index (BMI) and subcutaneous fat with blood pressure among university-going Punjabi Sikh and Hindu females. *International Journal of Medicine and Medical Sciences* **2**(1), 5–11.

Bishnoi, D., Kaur, T. and Badaruddoza (2010) Predictor of cardiovascular disease with respect to BMI, WHR and lipid profile in females of three population groups. *Biology and Medicine* **2**(2), 32–41.

Brereton, R.G. (2003) *Chemometrics: Data Analysis for the Laboratory and Chemical Plant*. John Wiley & Sons Ltd., Chichester, UK.

Brereton, R.G. (2007) *Applied Chemometrics for Scientists*. John Wiley & Sons Ltd, Chichester, UK.

Bro, R., van den Berg, F., Thybo, A. *et al.* (2002) Multivariate data analysis as a tool in advanced quality monitoring in the food production chain. *Trends in Food Science and Technology* **13**, 235–244.

Budić-Leto, I., Gajdoš Kljusurić, J., Zdunić, G. *et al.* (2012) Comparison of the sensory descriptive attributes of taste and chemical parameters of Croatian dessert wine Prošek using multivariate analysis. *International Journal of Food, Agriculture & Environment – JFAE* **10**(1), 132–136.

Camao (2012) The Unscrambler. Camao International Co. Ltd. Available: http://www.camao.com (last accessed 24 June 2013).

CDC (Centers for Disease Control and Prevention) . About BMI for Children and Teens. Available: http://www.cdc.gov/healthyweight/assessing/bmi/childrens_bmi/about_childrens_bmi.html (last accessed 24 June 2013).

Eigenvector (2012) Chemometrics Software, Eigenvector Research Inc. Available: http://www.eigenvector.com (last accessed 24 June 2013).

Fernald, L.C.H. (2007) Socio-economic status and body mass index in low-income Mexican adults. *Social Science and Medicine* **64**, 2030–2042.

Granato, D., Katayama, F.C.U. and Castro, I.A. (2010) Assessing the association between phenolic compounds and the antioxidant activity of Brazilian red wines using chemometrics. *LWT – Food Science and Technology* **43**, 1542–1549.

Granato, D., Branco, G.F., Faria, J.A.F. and Cruz, A.G. (2011) Characterization of Brazilian lager and brown ale beers based on color, phenolic compounds, and antioxidant activity using chemometrics. *Journal of the Science of Food and Agriculture* **91**(3) 563–571.

IBM (2012) SPSS software. Available: http://www.ibm.com (last accessed 24 June 2013).

Jiang, C., Lei, S-F., Liu, M-Y. *et al.* (2007) Evaluating the correlation and prediction of trunk fat mass with five anthropometric indices in Chinese females aged 20-40 years. *Nutrition, Metabolism & Cardiovascular Diseases* **17**, 676–683.

Johnson, R. (1996) Fitting Percentage of Body Fat to Simple Body Measurements, *J. of Statistics Education*, **4**(1), http://www.amstat.org/publications/jse/v4n1/datasets.johnson.html

Kowalski, B.R. and Bender, C.F. (1972) Pattern recognition. A powerful approach to interpreting chemical data, *Journal of the American Chemical Society* **94**, 5632–5639.

Kurtanjek, Ž. Horvat, D., Magdić, D. and Drezner, G. (2008) Factor Analysis and Modelling for Rapid Quality Assessment of Croatian Wheat Cultivars with Different Gluten Characteristics. *Food Technology and Biotechnology* **46**, 270–277.

Kurtanjek, Ž. Horvat D., Drezner G. and Magdić, D. (2013) Prediction of wheat baking quality based on gliadin fractions and HMW-GS data by chemometric analysis (PLS modelling). *Acta Alimentaria*, **42**(4), (DOI: 10.1556/AAlim.2013.5555)

Lazarou, C., Karaolis, M., Matalas A.L. and Panagiotakos D.B. (2012), Dietary patterns analysis using data mining method. An application to data from the CYKIDS study. *Computer Methods and Programs in Biomedicine* **108**(2), 706–714.

Lazarou, C., Panagiotakos, D.B. and Matalas, A.L. (2010) Physical activity mediates the protective effect of the Mediterranean diet on children's obesity status: The CYKIDS study, *Nutrition* **26**, 61–67.

Markovina, J.,Čačić, J., Gajdoš Kljusurić, J. and Kovačić, D. (2011) Young Consumers' Perception of Functional foods in Croatia, *British Food Journal*, **113**(1), 7–16.

Mathworks® (2012) MATLAB, The MathWorks, Inc. Available: http://www.mathworks.com (last accessed 24 June 2013).

Moraes, A.C.F., Adami, F. and Falcão, M.C. (2012) Understanding the correlates of adolescents' dietary intake patterns. A multivariate analysis. *Appetite* **58**, 1057–1062.

Nieto, J.T., Esteban, E.B. and Martin, A.V. (2011) Selecting the best anthropometric variables to characterize a population of healthy elderly persons, *Nutrición Hospitalaria* **26**(2) 384–391.

Omron Corp. http://www.omron.com (last accessed 24 June 2013).

Oteiza, J.M., Ares, G., Sant'Ana, A.S. *et al.* (2011) Use of a multivariate approach to assess the incidence of *Alicyclobacillus* spp. in concentrate fruit juices marketed in Argentina: Results of a 14-year survey, *International Journal of Food Microbiology* **151**, 229–234.

Pearson, K. (1901) On lines and planes of closest fit to systems of points is space. *Philosophical Magazine Series 6* **2**(11), 559–572.

Pierna, J.A.F., Vermeulen P., Amand O. *et al.* (2012) NIR hyperspectral imaging spectroscopy and chemometrics for the detection of undesirable substances in food and feed. *Chemometrics and Intelligent Laboratory Systems* **117**, 233–239.

R Development Core Team (2012) The R Project for Statistical Computing. R Foundation for Statistical Computing, Vienna, Austria. Available: http://www.R-project.org (last accessed 24 June 2013).

Rohman, A. and Man, Y.B.C. (2012) The chemometrics approach applied to FTIR spectral data for the analysis of rice bran oil in extra virgin olive oil. *Chemometrics and Intelligent Laboratory Systems* **110**, 129–134.

Rumora, I., Kobrehel Pintarić, I., Gajdoš Kljusurić, J. *et al.* (2013) Efficient use of modelling in new food-product design and development. *Acta Alimentaria* (In press).

Sanya, A.O., Ogwumike, O.O., Ige A.P. and Ayanniyi, O.A. (2009) Relationship of waist–hip ratio and body mass index to blood pressure of individuals in Ibadan North local government. *African Journal of Physiotherapy and Rehabilitation Sciences* **1**(1) 7–11.

SAS (2012) SAS software. Available: http://www.sas.com (last accessed 24 June 2013).

StatSoft (2011) Statistica software version 10. StatSoft Inc. Available: http://www.statsoft.com (last accessed 24 June 2013).

Varmuza, K. and Filzmoser, P. (2009) *Introduction to Multivariate Statistical Analysis in Chemometrics*. CRC Press/Taylor & Francis Group, Boca Raton, FL, USA.

Venables, W.N. and Ripley, B.D. (2002) *Modern Applied Statistics with S*, 4th edn. Springer, New York, USA.

Waring, M.E., Eaton, C.B., Lasater, T.M. and Lapane, K.L. (2009) Correlates of Weight Patterns during Middle Age Characterized by Functional Principal Components Analysis, *Annals of Epidemiology* **20**(3) 201–209.

Wehrens R. (2011) *Chemometrics with R: Multivariate Data Analysis in the Natural Sciences and Life Sciences (Use R!)* Springer-Verlag, New York, USA.

Wold, S. (1995) Chemometrics; what do we mean with it and what do we want from it. *Chemometrics and Intelligent Laboratory Systems*, **30**, 109–115.

Wolfram (2011) Mathematica software. Wolfram Research Inc. Available: http://www.wolfram.com (last accessed 24 June 2013).

# 17 Effects of paediatric obesity: a multivariate analysis of laboratory parameters

**Tamas Ferenci and Levente Kovacs**

*Physiological Control Group, Institute of Information Systems, John von Neumann Faculty of Informatics, Obuda University, Budapest, Hungary*

## ABSTRACT

Obesity is a rapidly growing epidemic in almost every developed country. Because of this, and the comorbidities associated with obesity, it is in focus of public health, with paediatric obesity being especially of concern. The effects of obesity on the human body and homeostasis are very complex, involving several organ systems and mechanisms. One manifestation of these effects is the alteration of the results of blood tests – albeit mostly subclinical; obesity induces characteristic changes in laboratory parameters, which also shed light on the pathophysiological alterations caused by obesity. Several studies have addressed the association of obesity with a particular parameter but have not studied them comprehensively, much less their multivariate structure, especially for children. In this chapter, novel methods are employed to investigate the uni- and multivariate structure of laboratory parameters in a uniform way, involving 33 routinely used blood tests. A convenience sample of $n = 183$ Hungarian volunteer children was used.

## INTRODUCTION

Obesity is considered an epidemic in most parts of the developed world (Andersen, 2003). As an example, for some time overweight and obese people have been in the majority in the United States' population; according to the latest data, the prevalence of those who are classed as overweight is 34.2%, while the prevalence of those who are classed as obese and extremely obese is 39.5% among adults aged 20 and over (Ogden and Carroll, 2010a). The speed of progress is even more frightening, especially as far as obesity is concerned: the same prevalence was only 14.3% in 1960 (Ogden and Carroll, 2010a). The situation is similar in Hungary: the prevalence of overweight is 34.1%, the prevalence of obesity is 19.5% (OECD, 2012).

The same applies to paediatric obesity as well, although the available information is less detailed (Wang and Lobsten, 2006; Ogden *et al.*, 2007). In the United States, the prevalence of obesity among children and adolescents aged 2–19 is 16.9% (Ogden and Carroll, 2010b); in Hungary, the same prevalence is estimated to be about 5–10% (Kern, 2007; Antal *et al.*, 2009).

Obesity has been in the focus of public health for decades, as – in addition to its continuously increasing prevalence – it also increases all-cause morbidity and mortality (Visscher and Seidell, 2001; Pi-Sunyer, 2009). Type II diabetes, various cardiovascular diseases (including ischaemic heart disease), asthma, gallbladder disease and various malignant tumours are examples of diseases with increased occurrence causally linked to obesity (Guh *et al.*, 2009). These have been described in children too (Burke, 2006; Nyberg *et al.*, 2011).

*Mathematical and Statistical Methods in Food Science and Technology*, First Edition.
Edited by Daniel Granato and Gastón Ares.

It is well known that obesity, and even overweight, causes systematic changes in laboratory test results. The reasons for these changes are complex. On one hand, many changes are more or less a direct consequence of the manifestly altered homeostatic equilibrium induced by obesity, such as the elevated serum alanine aminotransferase (ALT) and aspartate aminotransferase (AST) levels found in obese adults (Ruhl and Everhart, 2003), and in children as well (Dubern *et al.*, 2006).

However, in some cases, the change in the laboratory parameters cannot be attributed to a single physiological alteration, or even to any well-defined alteration that causes a manifestly obesity-related finding at all when the laboratory parameter is already changed. A notable example is C-reactive protein (CRP), which is used predictively (Bo *et al.*, 2009; Juonala *et al.*, 2011; Ong *et al.*, 2011) because of this reason.

Note that these are all univariate, association-oriented findings, that is they describe changes in a certain laboratory result in obese subjects (as opposed to the healthy state). To our knowledge, no investigation has addressed the question of how obesity affects the laboratory results from a multivariate perspective (i.e. what is the effect of obesity if not only individual changes but also alterations in the correlation structure of the laboratory results are considered), especially not in children.

Therefore, our primary aim was to investigate how paediatric obesity influences the uni- and multivariate structure of common laboratory parameters in a precise, uniform way for all parameters. Traditional methods were employed for the univariate analyses, and novel approaches for the exploration of the multivariate structure.

## MATERIALS

### Data acquisition

We relied on a Hungarian cross-sectional, multicentre clinical observation that was arranged specifically for this investigation as the data source. For the study data were collected from subjects aged between 12 and 18 years, including both healthy volunteers and clinically obese ones. Sampling was done independently in the two groups up to a pre-specified quota, so the sample sizes are not representative for the prevalence of obesity. Clinically obese subjects were intentionally included to oversample the region of severe obesity.

The healthy control group consisted of volunteers from four Hungarian secondary schools, three of them being located in the capital city (Budapest) and one in a rural town (Mátészalka). Subjects were selected as a convenience sample, so results are not necessarily representative at national level. Each child participated with full written informed consent from its parents and the study was pre-authorized by the Hungarian Regional Bioethical Commission. The data were collected between April 2008 and May 2009. Examinations of healthy volunteers included anthropometrical measurements, body composition analysis (with an InBody 3.0 multifrequency bioelectric impedance analyser), fasting blood sample drawn for standard laboratory parameters and anamnestic data recording. Measurements were carried out by physicians of the Heim Pál Children's Hospital (Budapest) and results were manually recorded in electronic format (Ferenci, 2009).

The obese group consisted of children treated in the Heim Pál Children's Hospital, with no significant comorbidity. Data (including laboratory parameters) of the obese children were extracted from the hospital's electronic records with a custom application developed by the authors (Ferenci, 2009).

### Collected data

In this paper, the laboratory results from the database are analysed. These include 33 laboratory parameters (Table 17.1 shows a complete list).

Body Mass Index (BMI) is also included in the database and was used to assess the degree of overweight and obesity. Although the drawbacks of BMI for that use are well known (Romero-Corral

**Table 17.1** Laboratory parameters investigated in this chapter with name, abbreviation and unit of measurement. (Percentages of missing values in the study that were later imputed are also indicated in a male/female format; empty cell means no missing value.)

| Name | Abbr. | UoM | Missing (%) |
|---|---|---|---|
| White blood cell count | WBC | G/l | |
| Absolute neutrophil count | ANC | G/l | |
| Absolute lymphocyte count | ALC | G/l | |
| Absolute monocyte count | AMC | G/l | |
| Absolute eosinophil count | AEC | G/l | |
| Relative neutrophil count | RNC | % | |
| Relative lymphocyte count | RLC | % | |
| Relative monocyte count | RMC | % | |
| Relative eosinophil count | REC | % | |
| Red blood cell count | RBC | T/l | |
| Haemoglobin | HGB | g/l | |
| Haematocrit | HCT | % | |
| Mean corpuscular volume | MCV | fl | |
| Mean corpuscular haemoglobin | MCH | pg | |
| Mean corpuscular haemoglobin concentration | MCHC | g/l | |
| Red blood cell distribution width | RDW | % | |
| Platelet count | PLT | G/l | |
| Mean platelet volume | MPV | fl | 0.9/0.0 |
| C-reactive protein | CRP | mg/l | 5.3/8.6 |
| Serum sodium | SNA | mmol/l | 0.0/1.4 |
| Serum potassium | SK | mmol/l | |
| Serum chloride | SCL | mmol/l | |
| Serum globulin | SGL | g/l | 1.8/2.9 |
| Serum albumin | SAL | g/l | 1.8/2.9 |
| Serum total protein | STP | g/l | 1.8/1.4 |
| Serum creatinine | SCR | μmol/l | |
| Blood urea nitrogen | BUN | mmol/l | |
| Serum triglycerides | STG | mmol/l | 0.0/2.9 |
| Serum total cholesterol | STC | mmol/l | 0.0/2.9 |
| Serum HDL cholesterol | HDL | mmol/l | 9.7/17.1 |
| Aspartate transaminase | AST | IU/l | 0.9/1.4 |
| Alanine transaminase | ALT | IU/l | |
| Gamma glutamyl transpeptidase | GGT | IU/l | |

*et al.*, 2008; Okorodudu *et al.*, 2010), it is still the most widely used simple indicator of the degree of overweight/obesity (WHO, 2011). The growth of the subjects is obviously non-negligible in our setting, to handle this a metric that took the children's growth (i.e. age) into account was employed. We chose standardized BMI (BMI $z$-score, or Z-BMI (Cole *et al.*, 2005)), which is essentially the deviation of the child's BMI from the mean BMI of the child's age and sex, measured in standard deviation units. This can be calculated based on sex-specific growth charts; we employed the one from the Centers for Disease Control and Prevention (CDC, 2012), which also includes the necessary L-M-S parameters (Cole, 1990) to calculate Z-BMI. Note that extreme percentiles (i.e. Z-BMI values lower than $-2$ or higher than $+2$) should be handled with care (CDC, 2002) due to the extrapolation. (We used the CDC Growth Chart because the Hungarian Growth Charts (Joubert *et al.*, 2006) unfortunately do not include the necessary L-M-S parameters.) Note that we employed no hard threshold to specify overweight or obesity, instead we relied on Z-BMI as a continuous (scale) indicator (proxy) of the degree of obesity.

Our database consists of $n = 183$ subjects (113 males, 70 females). The distribution of the BMI $z$-scores of these subjects is shown in Figure 17.1. The oversampling of the obese population is obvious.

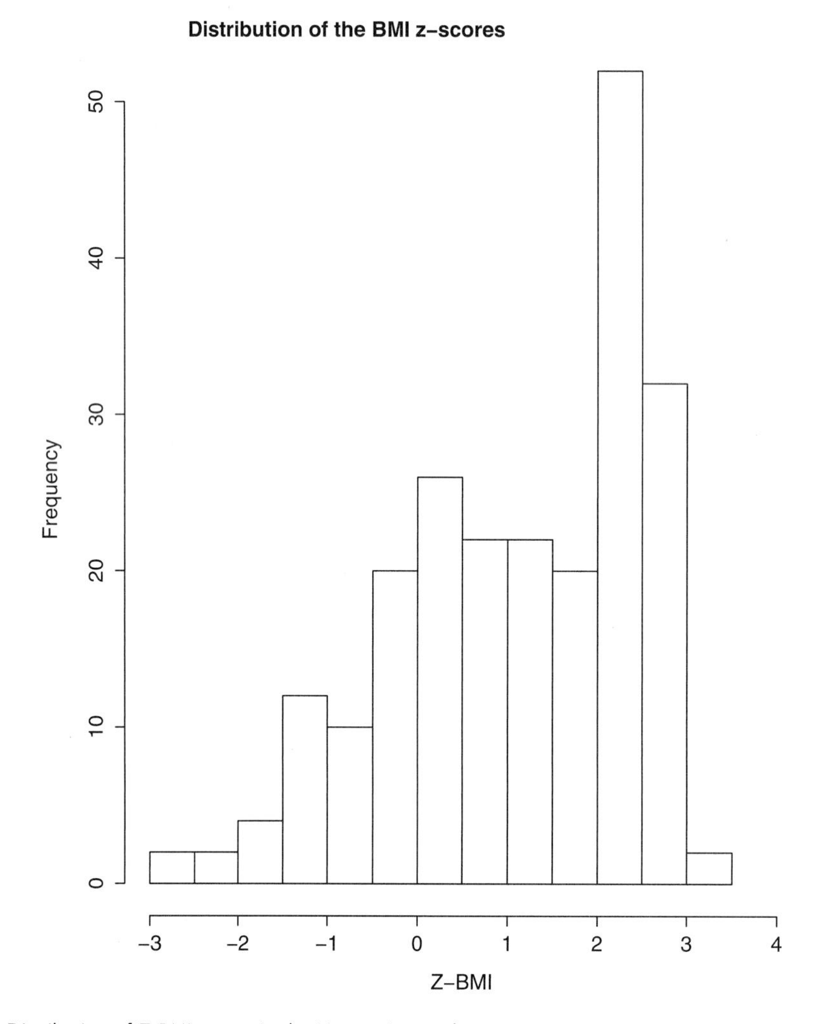


**Figure 17.1** Distribution of Z-BMI scores in the Hungarian study.

The sexes were analysed separately. Sex has a profound impact on many laboratory parameter (see (Kelly and Munan, 1977; Rodger *et al.*, 1987; Taylor *et al.*, 1997) for hematologic examples) and would introduce complex interactions that are hard to interpret from our point of view, so the most practical form of sex matching was simply the complete separation of the analysis according to sex.

# METHODS

## Statistical analysis

Statistical analysis was principally performed with the R statistical program package (R Core Team, 2012), version 2.15.1, with custom scripts. Libraries ks (Doung, 2012), psych (Revelle, 2012) and weights (Pasek, 2012) were used.

Every subject with more than five missing laboratory results or missing BMI, sex or age was dismissed from the analysis. Missing values for the retained subjects were univariately imputed with sample median value from the same sex. (The number of these missing values was 0% for most parameters of the study and never exceeded 10% with a single exception (Table 17.1), hence such a simple imputation is justified.)

## Univariate analysis

### *Univariate descriptive statistics*

Univariate reference values of the laboratory parameters for different severities of obesity, namely Z-BMI = +1, +2 and +3 were given, segregated according to sex. Classical descriptive statistics of mean and standard deviation, and more robust alternatives of median and interquartile range (IQR) were used (Armitage *et al.*, 2002). The usage of robust statistics is justified by the well-known fact that the distribution of many laboratory parameters is skewed, sometimes highly (Armitage *et al.*, 2002).

The question arises how to define these descriptors for a given Z-BMI value (e.g. for Z-BMI = +1). As Z-BMI is a continuous variable, there is no point in calculating an average value (or any other statistic) for subjects that have exactly Z-BMI = +1 (possibly there is not even a single subject in the database with a Z-BMI of exactly +1). The problem is obviously that we only have a finite sample drawn from an otherwise continuous distribution. One solution would be the binning of Z-BMI values, that is to give the average value for subjects having 0.5 < BMI < 1.5 (instead of Z-BMI= 1). While this method is quite robust, the drawback is that information is lost (by grouping everyone from Z-BMI = 0.5 to Z-BMI = 1.5 in the same category, regardless of the subject's actual Z-BMI), hence losing possible tendencies within the 0.5 < Z-BMI < 1.5 group. Therefore, we chose another alternative: we tried to reconstruct the – continuous – distribution based on the sample. What is needed is the joint distribution of the Z-BMI and the investigated laboratory parameter: from this, the (conditional) distribution of the laboratory parameter for any given Z-BMI value can be obtained. Given this conditional distribution, we can numerically calculate any statistic (mean, median, standard deviation etc.) of the investigated laboratory parameter for the exact Z-BMI value on which we conditioned (such as Z-BMI = +1), just as intended.

This is essentially a joint probability density function (pdf) estimation task, which we solved by employing kernel density estimation (KDE). KDE (Scott, 1992) can be considered as a method that replaces the data points in the sample with true, nondegenerate distributions and fits a 'smooth' function for the sample by mixing these distributions. For the bivariate KDE, we employed normal kernel function (as is almost universally done), with a bandwidth matrix obtained using smoothed cross-validation (Hall *et al.*, 1992).

Once the two-dimensional (BMI versus investigated variable) joint pdf is estimated, the distribution of the investigated parameter for any given BMI can be obtained by 'slicing' the two-dimensional surface perpendicular to the BMI axis at the point of interest (e.g. Z-BMI = +1). The 'slice' should be then normalized, so that the area under its curve equals one; that is, we obtain a conditional (one-dimensional) distribution from the two-dimensional joint pdf, conditioning on Z-BMI. The required statistic then can be directly computed from the obtained pdf of the conditional distribution (for example by numeric integration in case of mean). These are illustrated in Figure 17.2 for HDL of males.

What we have done is, essentially, a nonparametric estimation (Takezawa, 2005) of the statistical connection between Z-BMI and the investigated laboratory parameter, as we did not want to presume any function form (e.g. linear) for this regression.

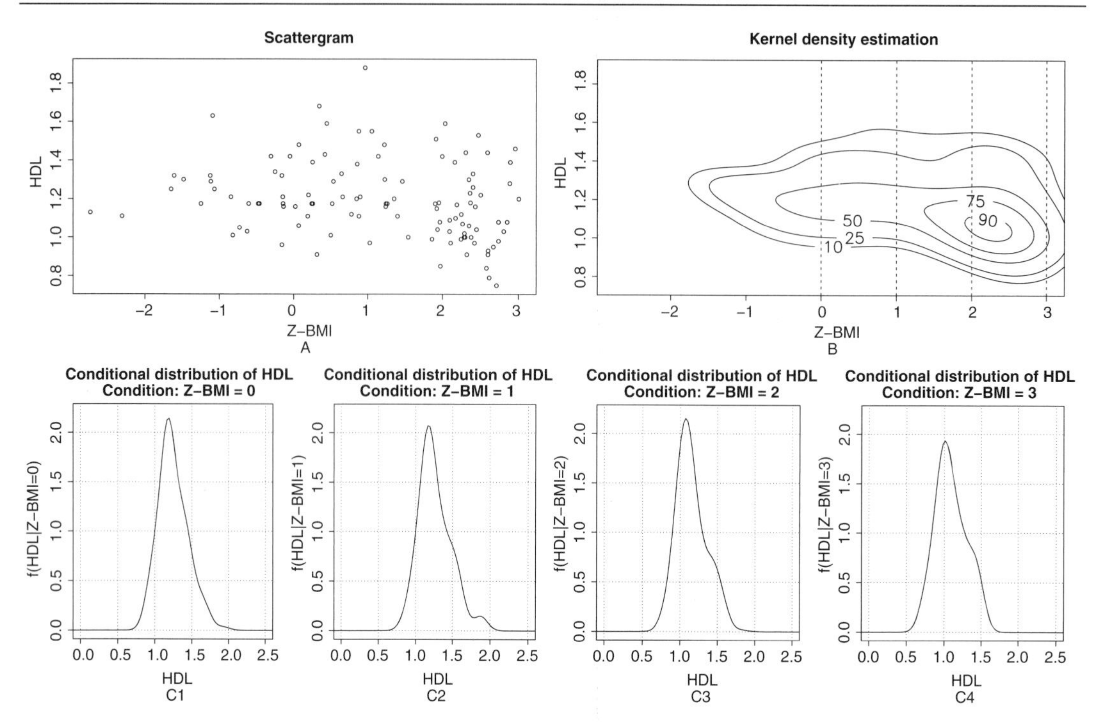


**Figure 17.2** Derivation of the conditional distribution of the HDL of males for a given Z-BMI. A shows the scatter plot of HDL and Z-BMI; B shows the estimated joint pdf of these variables (based on the same sample as shown in A) obtained with KDE, using a contour plot (the labels of the contours indicate the percentage of the distribution outside the encircled region). Dependence of HDL on the Z-BMI can be clearly seen. C1–C4 show how a conditional distribution for a given Z-BMI can be obtained by 'slicing' the surface of the joint pdf. Different statistics, such as mean, can be obtained from this univariate pdf.

### *Univariate association analysis*

We performed a classical (univariate) analysis on the association of obesity with systematic alteration of different laboratory parameters, parameter-by-parameter.

We had a continuous indicator of the degree of obesity (Z-BMI). Therefore, instead of binning the Z-BMI values (i.e. discretizing that variable) and then using either a *t*-test or an ANOVA-type statistical test, we retained every value of the Z-BMI variable unchanged and calculated its correlation coefficient with the investigated laboratory result. (The scatter plot of Figure 17.2 gives an overall impression on this correlation.) The advantage of binning would have been its ability to detect nonlinear relationships as well; to compensate for this we used Spearman's $\rho$ (Maritz, 1995) correlation coefficient (instead of the more classical Pearson (product-moment) coefficient), which detects monotone connections in general, not only linear connections. This is also justified by the already mentioned non-normality of the laboratory parameters.

For selecting the significant differences, we have to take the effects of the multiple comparisons situation (Miller, 1981) into account. As we are running several hypothesis testing in parallel (and consider their results disjunctively, i.e. we are looking for *any* difference) the usage of the pre-specified significance level (e.g. 5%) for every test would result in an experimentwise $\alpha$ far above the significance level. To protect against this, we used the Holm correction (Holm, 1979; Shaffer, 1995).

## Investigation of the multivariate structure

### *Traditional approaches' shortcomings, and the alternatives*

The correlation structure of a database is directly reflected in its correlation matrix (if we confine ourselves to linear connections) and can be visualized by (matrix) scatter plots. Both methods fail to facilitate to understanding of the correlation structure if the number of variables is too high: matrix scatter plots and correlation matrices can hardly be overseen for more than 10 variables (Venables and Ripley, 2002).

Having more than 30 variables, we employed other approaches of multivariate statistics: principal component analysis (PCA) and cluster analysis (CA). Both methods can be used to ease the understanding and interpretation of the correlation structure of large databases. Note that these methods do not require the database itself, only its correlation matrix.

### *Conditional correlation matrices*

The first question that arises is again the need to obtain the correlation matrix for a given Z-BMI. We again applied the KDE approach: we calculated the correlation matrix for a given Z-BMI ('conditional correlation matrix') element-by-element. That is, for every variable pair of the matrix, we reconstructed the three-dimensional pdf of the two involved laboratory parameters and Z-BMI (through KDE), and then marginalized for a given Z-BMI, the same way it has already been described. Covariance was determined by numerical integration and correlation was calculated.

The only problem with this approach is that the correlation matrix obtained this way – as opposed to a 'traditional' correlation matrix – might not be non-negative-semidefinite at all. This would prevent the performing of PCA (or any other method that expects a valid correlation matrix as input), so we applied smoothing (Wothke, 1993) to reconstruct a closely approximating non-negative-definite matrix from our correlation matrix by eliminating negative eigenvalues and rescaling positive ones. (This is acceptable in this case because we only had very few negative eigenvalues, with very small absolute values.)

We have used correlation (and not covariance) matrices because laboratory parameters have different measurement scales.

### *Principal components analysis (PCA)*

Principal components analysis (PCA) is one the most classic tools of multivariate data analysis (Flury, 1997). It can be employed (and interpreted) in many different ways; here we will use it as a tool to ease the understanding of a correlation matrix's structure.

This is based on one of the PCA's outputs, the so-called structure matrix, which shows the correlations between the principal components and the observed variables. This is useful, because those variables that are highly correlated (in absolute value) with the same principal component, are also correlated among themselves; hence, instead of searching for high correlations within a $k \times k$ matrix (where $k$ denotes the number of laboratory parameters), it is sufficient to consider $k$ values (a column of the structure matrix) at a time and repeat this $k$ times. This essentially means that the problem can be decomposed into subtasks that are much easier to solve. Furthermore, as principal components are in an order of decreasing importance (based on the error that occurs when the last principal components are omitted in the reconstruction of the observed variables), it is usually enough to consider the first few columns of the structure matrix in many practical cases. This way, we can obtain information even from large databases on what variables exhibit statistically connected behaviour, which might indicate causal (physiological, this time) connections between them.

Now we used PCA purely to transform the correlation matrix, that is we applied it in descriptive sense, with no inductive statistics performed. Because of this, neither the calculation of the KMO measure, nor any hypothesis testing was performed.

We performed PCA for the conditional correlation matrices for Z-BMI = +1, +2 and +3 (obtained as described above) consistently with what we have done in case of the univariate descriptors.

To ease the interpretation of the structure matrices, we applied varimax rotation (Kaiser, 1958) after performing PCA to achieve a well-interpretable component structure. The number of extracted components was set to 13, this was selected to support interpretability and also to ensure that we extract components that have an eigenvalue larger than one (Kaiser's criterion (Kaiser, 1960)).

#### *Cluster analysis (CA)*

Cluster analysis (CA) aims (Tan *et al.*, 2006) to form groups of data objects such that objects within a group are similar to each other, while object in different groups are dissimilar (according to some pre-specified metric of similarity). Such groups are called clusters in the context of CA. The result of CA depends not only on the metric used but also on the way the distance between the clusters is defined (in addition to the objects and the algorithm used, of course). We have already shown (Ferenci *et al.*, 2011) that the results in this task are not sensitive to the distance metric and the cluster definition that is used; hence, we employed the popular Ward method as cluster-distance definition (Everitt *et al.*, 2006) and 1 minus the absolute value of correlation (Glynn, 2005) a distance (dissimilarity) measure. (This distance measure is equivalent to the well-known Euclidean distance if standardized variables are used, save for a constant factor depending only on the number of coordinates.)

We performed CA for the conditional correlation matrices for Z-BMI = +1, +2 and +3 (obtained as described above), consistently with what we have done in case of the univariate descriptors.

There are several approaches to CA, we will now base it on the so-called agglomerative hierarchical clustering. We will cluster the variables of the data set (not the cases, as more usual), hence 'data objects' will be laboratory parameters, and 'similarity' will mean statistically connected behaviour. That is, we are aiming to identify groups of laboratory parameters that exhibit similar behaviour. (This task is essentially the same as the one we set forth with PCA but the approach to solve it is completely different.)

For our purpose, the most useful representation of the results of an agglomerative hierarchical CA is the dendrogram. Variables are connected with lines on a dendrogram, each connection having a so-called 'height' that is measured on the vertical axis. The smaller the height is, the more similar the variables are. Thus, the variables or clusters of variables that are more 'deeply' connected are more similar. This way, groups of similar variables can be formed for any minimum of similarity (that is reflected as a height at which connections are 'cut off'). This way, the dendrogram can be considered as a graphical interpretation of the correlation matrix.

## RESULTS

### Univariate descriptive statistics and association analysis

The most important univariate descriptors of the laboratory results for different levels of obesity (Z-BMI = +1, +2 and +3) and the results of the univariate association analysis are given in Table 17.2.

### Investigation of the multivariate structure

Results of PCA, that is the structure matrices (visualized with heat maps) for different levels of obesity (Z-BMI = +1, +2 and +3) segregated according to sex are shown in Figure 17.3.

**Table 17.2** Univariate descriptors of the laboratory parameters for different levels of obesity (Z-BMI = +1, +2 and +3), segregated according to sex in Mean (Median) ± SD (IQR) format and the result of the univariate association analysis (ρ correlation coefficient, and its *p*-value; ** marks association that is significant at 5%, * marks association that is significant at 10%, with Holm-correction in both cases).

| | | Z-BMI = +1 | Z-BMI = +2 | Z-BMI = +3 | ρ | p | |
|---|---|---|---|---|---|---|---|
| WBC | Male | 6.8 (6.7) ± 1.4 (1.8) | 7.5 (7.4) ± 1.5 (2) | 8 (7.9) ± 1.5 (1.9) | 0.53 | 0.0000 | ** |
| | Female | 7.4 (7.4) ± 1.3 (1.9) | 7.9 (7.7) ± 1.7 (2.3) | 8.5 (8.4) ± 1.9 (3.3) | 0.24 | 0.0431 | |
| RNC | Male | 50.5 (50.7) ± 7.3 (9) | 52.5 (53.2) ± 7.4 (9.1) | 54.8 (55.9) ± 8.4 (9.8) | 0.26 | 0.0061 | |
| | Female | 52.9 (52.2) ± 9.1 (11.8) | 54.7 (55) ± 9.6 (14.3) | 57.1 (57.9) ± 9.5 (14.5) | 0.11 | 0.3641 | |
| RLC | Male | 36 (35.5) ± 7.3 (9) | 35 (34.6) ± 7 (9.3) | 33.1 (32.5) ± 7.3 (9.2) | −0.21 | 0.0272 | |
| | Female | 34.5 (34.6) ± 8.2 (11.8) | 34.5 (34) ± 8.4 (12.3) | 32.8 (32.1) ± 8 (11.8) | −0.06 | 0.6196 | |
| RMC | Male | 9.6 (9.2) ± 2.2 (2.5) | 9.4 (8.7) ± 2.6 (2.8) | 9 (8.4) ± 2.4 (2.5) | −0.23 | 0.0123 | |
| | Female | 8.3 (8.2) ± 1.7 (2.6) | 8.2 (8) ± 1.6 (2.3) | 7.9 (7.8) ± 1.5 (2.1) | −0.22 | 0.0660 | |
| REC | Male | 3.2 (2.8) ± 1.8 (1.9) | 2.9 (2.6) ± 1.6 (1.7) | 3.1 (2.6) ± 2.1 (1.9) | −0.08 | 0.4155 | |
| | Female | 3.9 (2.7) ± 3.5 (3.1) | 2.9 (2.3) ± 2.2 (2.3) | 2.9 (2.4) ± 2.2 (2.4) | −0.10 | 0.4125 | |
| ANC | Male | 3.5 (3.4) ± 1.1 (1.3) | 4.1 (3.9) ± 1.2 (1.4) | 4.4 (4.4) ± 1.2 (1.5) | 0.48 | 0.0000 | ** |
| | Female | 4.1 (4) ± 1.2 (1.7) | 4.4 (4.2) ± 1.4 (1.9) | 5 (4.9) ± 1.5 (2.3) | 0.23 | 0.0600 | |
| ALC | Male | 2.4 (2.3) ± 0.6 (0.8) | 2.6 (2.6) ± 0.6 (0.9) | 2.7 (2.7) ± 0.6 (0.9) | 0.33 | 0.0004 | ** |
| | Female | 2.5 (2.4) ± 0.6 (0.8) | 2.7 (2.6) ± 0.7 (1) | 2.7 (2.6) ± 0.8 (1) | 0.16 | 0.1801 | |
| AMC | Male | 0.7 (0.6) ± 0.2 (0.2) | 0.7 (0.7) ± 0.2 (0.3) | 0.7 (0.7) ± 0.2 (0.3) | 0.31 | 0.0009 | ** |
| | Female | 0.6 (0.6) ± 0.1 (0.2) | 0.6 (0.6) ± 0.2 (0.2) | 0.7 (0.6) ± 0.2 (0.3) | 0.07 | 0.5755 | |
| AEC | Male | 0.2 (0.2) ± 0.1 (0.1) | 0.2 (0.2) ± 0.1 (0.1) | 0.2 (0.2) ± 0.2 (0.1) | 0.14 | 0.1495 | |
| | Female | 0.3 (0.2) ± 0.3 (0.2) | 0.2 (0.2) ± 0.2 (0.2) | 0.2 (0.2) ± 0.2 (0.2) | −0.02 | 0.8742 | |
| RBC | Male | 5.2 (5.2) ± 0.3 (0.4) | 5.3 (5.3) ± 0.3 (0.5) | 5.3 (5.4) ± 0.4 (0.5) | 0.13 | 0.1642 | |
| | Female | 4.8 (4.7) ± 0.3 (0.4) | 4.8 (4.8) ± 0.3 (0.4) | 4.9 (4.9) ± 0.2 (0.3) | 0.40 | 0.0006 | ** |
| HGB | Male | 151 (151) ± 11 (17) | 147 (146) ± 13 (20) | 147 (147) ± 14 (21) | −0.19 | 0.0422 | |
| | Female | 136 (136) ± 9 (12) | 135 (135) ± 10 (14) | 134 (134) ± 9 (14) | −0.07 | 0.5656 | |
| HCT | Male | 0.4 (0.4) ± 0 (0) | 0.4 (0.4) ± 0 (0.1) | 0.4 (0.4) ± 0 (0.1) | −0.23 | 0.0129 | |
| | Female | 0.4 (0.4) ± 0 (0) | 0.4 (0.4) ± 0 (0) | 0.4 (0.4) ± 0 (0) | −0.07 | 0.5900 | |
| MCV | Male | 85.4 (85.5) ± 4 (5.6) | 83.1 (83.2) ± 4.2 (5.8) | 82 (82) ± 4.3 (5.7) | −0.45 | 0.0000 | ** |
| | Female | 87.1 (87.3) ± 4.2 (6.2) | 85.6 (85.8) ± 4.1 (6) | 83.7 (83.8) ± 3.8 (5.4) | −0.47 | 0.0000 | ** |
| MCH | Male | 28.7 (28.8) ± 1.4 (2) | 28 (28.1) ± 1.7 (2.2) | 27.7 (27.7) ± 1.8 (2.4) | −0.36 | 0.0001 | ** |
| | Female | 28.6 (28.7) ± 1.6 (2.4) | 28.2 (28.2) ± 1.6 (2.2) | 27.2 (27.3) ± 1.6 (2.2) | −0.46 | 0.0001 | ** |
| MCHC | Male | 337 (336) ± 10 (13) | 337 (337) ± 11 (14) | 337 (337) ± 10 (15) | 0.03 | 0.7375 | |
| | Female | 329 (327) ± 11 (18) | 330 (329) ± 10 (15) | 326 (326) ± 9 (12) | −0.10 | 0.4071 | |
| RDW | Male | 13.5 (13.4) ± 0.7 (0.9) | 13.8 (13.7) ± 0.8 (1.1) | 13.9 (13.9) ± 0.8 (1.1) | 0.24 | 0.0104 | |
| | Female | 13.6 (13.6) ± 0.8 (1.1) | 13.6 (13.6) ± 0.9 (1.3) | 13.9 (13.8) ± 0.9 (1.2) | 0.26 | 0.0277 | |
| PLT | Male | 257 (254) ± 59 (80) | 277 (277) ± 54 (71) | 296 (296) ± 58 (77) | 0.40 | 0.0000 | ** |
| | Female | 277 (274) ± 45 (67) | 292 (288) ± 49 (68) | 304 (299) ± 50 (71) | 0.37 | 0.0018 | ** |

*(continued)*

**Table 17.2** *(continued)*

| | | Z-BMI = +1 | Z-BMI = +2 | Z-BMI = +3 | ρ | p | |
|---|---|---|---|---|---|---|---|
| MPV | Male | 10.7 (10.6) ± 0.8 (1.2) | 10.6 (10.6) ± 0.8 (1.2) | 10.7 (10.6) ± 0.8 (1.3) | −0.13 | 0.1683 | |
| | Female | 10.9 (11) ± 0.7 (1) | 10.7 (10.7) ± 0.6 (0.8) | 10.7 (10.7) ± 0.6 (0.9) | −0.12 | 0.3092 | |
| CRP | Male | 3.9 (2.1) ± 7.6 (2.6) | 5.6 (3.6) ± 9.7 (4.1) | 7 (4.9) ± 7.1 (5.7) | 0.64 | 0.0000 | ** |
| | Female | 2.2 (1.8) ± 2.4 (2.2) | 4.6 (2.8) ± 6.1 (4) | 5.8 (4.7) ± 4.8 (5.6) | 0.61 | 0.0000 | ** |
| SNA | Male | 139 (139) ± 2 (2) | 138 (138) ± 2 (3) | 139 (138) ± 2 (3) | −0.08 | 0.3756 | |
| | Female | 138 (138) ± 2 (2) | 138 (138) ± 2 (3) | 138 (138) ± 2 (3) | 0.02 | 0.8765 | |
| SK | Male | 4.2 (4.2) ± 0.3 (0.5) | 4.3 (4.3) ± 0.4 (0.5) | 4.4 (4.4) ± 0.3 (0.5) | 0.25 | 0.0082 | |
| | Female | 4.4 (4.4) ± 0.3 (0.5) | 4.3 (4.3) ± 0.3 (0.4) | 4.4 (4.4) ± 0.3 (0.4) | 0.12 | 0.3204 | |
| SCL | Male | 102 (102) ± 2 (3) | 103 (102) ± 2 (3) | 103 (103) ± 2 (3) | 0.07 | 0.4939 | |
| | Female | 103 (103) ± 2 (2) | 103 (103) ± 2 (3) | 104 (104) ± 2 (3) | 0.19 | 0.1073 | |
| STP | Male | 76.6 (76.5) ± 4.2 (5.9) | 76.5 (76.5) ± 3.9 (5.5) | 75.5 (75.1) ± 4.2 (6.1) | −0.14 | 0.1432 | |
| | Female | 76.4 (75.9) ± 4.4 (6.5) | 76.5 (76) ± 4.8 (7.3) | 76.4 (75.8) ± 4.5 (6.3) | 0.09 | 0.4768 | |
| SAL | Male | 50.2 (50.1) ± 2.6 (3.5) | 48.7 (48.5) ± 2.7 (3.8) | 47.5 (47.1) ± 2.4 (3.6) | −0.47 | 0.0000 | ** |
| | Female | 48.5 (48.6) ± 2.4 (3.4) | 47.5 (47.6) ± 2.8 (4) | 46.2 (46) ± 2.6 (3.5) | −0.35 | 0.0028 | * |
| SGL | Male | 26.6 (26.6) ± 3.7 (4.7) | 27.7 (27.5) ± 3.7 (4.6) | 28 (27.7) ± 3.7 (4.7) | 0.19 | 0.0458 | |
| | Female | 27.9 (27.3) ± 3.9 (6.3) | 29.3 (29) ± 3.8 (6.1) | 30.2 (30.1) ± 3.6 (5.2) | 0.39 | 0.0009 | ** |
| BUN | Male | 4.6 (4.6) ± 1.2 (1.9) | 4.5 (4.4) ± 1.1 (1.7) | 4.4 (4.4) ± 1 (1.6) | −0.11 | 0.2377 | |
| | Female | 4 (4) ± 0.8 (1.2) | 4.2 (4) ± 1 (1.4) | 4.2 (4) ± 1.1 (1.7) | −0.13 | 0.2739 | |
| SCR | Male | 73.2 (73.7) ± 9.5 (12.6) | 68.2 (67.7) ± 9.6 (14.4) | 65.1 (64.3) ± 8.7 (13.3) | −0.43 | 0.0000 | ** |
| | Female | 58.6 (58.6) ± 5.9 (8.4) | 60.6 (60.6) ± 7 (10.8) | 61.1 (61.4) ± 6.4 (8.9) | 0.05 | 0.6798 | |
| STG | Male | 1 (0.9) ± 0.5 (0.6) | 1.2 (1.1) ± 0.5 (0.6) | 1.2 (1.1) ± 0.6 (0.6) | 0.43 | 0.0000 | ** |
| | Female | 1.2 (1) ± 0.6 (0.9) | 1.3 (1.1) ± 0.7 (0.7) | 1.3 (1.2) ± 0.5 (0.6) | 0.34 | 0.0039 | |
| STC | Male | 3.9 (3.8) ± 1 (1.1) | 4.2 (4.1) ± 0.9 (1) | 4.2 (4.1) ± 0.8 (1) | 0.29 | 0.0017 | ** |
| | Female | 4.1 (4) ± 0.9 (0.9) | 4.3 (4.2) ± 0.8 (1.1) | 4.3 (4.2) ± 0.7 (1) | 0.23 | 0.0515 | |
| HDL | Male | 1.3 (1.2) ± 0.2 (0.3) | 1.2 (1.1) ± 0.2 (0.3) | 1.1 (1.1) ± 0.2 (0.3) | −0.27 | 0.0043 | * |
| | Female | 1.3 (1.3) ± 0.2 (0.3) | 1.3 (1.3) ± 0.2 (0.3) | 1.2 (1.2) ± 0.2 (0.3) | −0.26 | 0.0327 | |
| AST | Male | 22.7 (22.2) ± 7.3 (9.3) | 25 (23.9) ± 8.9 (10.6) | 26.2 (24.6) ± 9.3 (10.3) | 0.24 | 0.0119 | |
| | Female | 21 (18.7) ± 7.9 (8.7) | 20.4 (19) ± 7.4 (9.5) | 19.1 (17.9) ± 6.5 (8.3) | −0.16 | 0.1870 | |
| ALT | Male | 22 (19.3) ± 12.8 (11.2) | 30.4 (26.1) ± 18.5 (16.3) | 36.4 (30.8) ± 20.6 (21) | 0.56 | 0.0000 | ** |
| | Female | 17.4 (14) ± 10.1 (9.9) | 19.5 (16.7) ± 9.5 (12.1) | 22.2 (19.7) ± 9.8 (11.6) | 0.32 | 0.0063 | |
| GGT | Male | 23.5 (21.4) ± 9.3 (14) | 25.7 (22.8) ± 11.4 (12) | 32.8 (29.2) ± 13.9 (18.4) | 0.52 | 0.0000 | ** |
| | Female | 15.7 (14.2) ± 6.6 (8) | 18.5 (16.6) ± 8.7 (8.7) | 21.2 (19.4) ± 8.1 (9) | 0.25 | 0.0335 | |

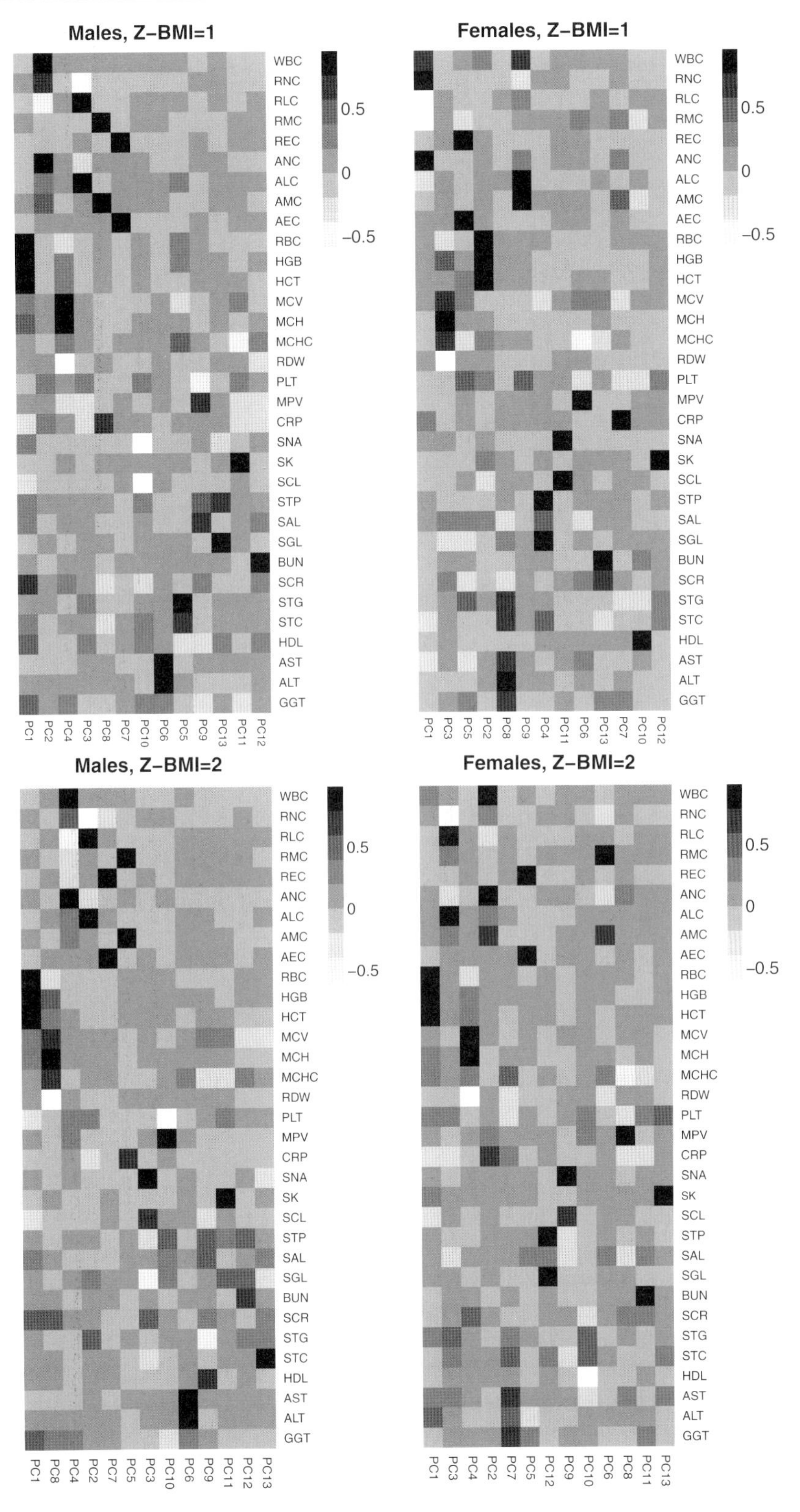


**Figure 17.3** Results of the PCA (structure matrices visualized with heat maps) for different Z-BMI levels, segregated according to sex. White colour indicates zero correlation, red means +1, blue means −1. (See colour version of this figure in colour plate section.)

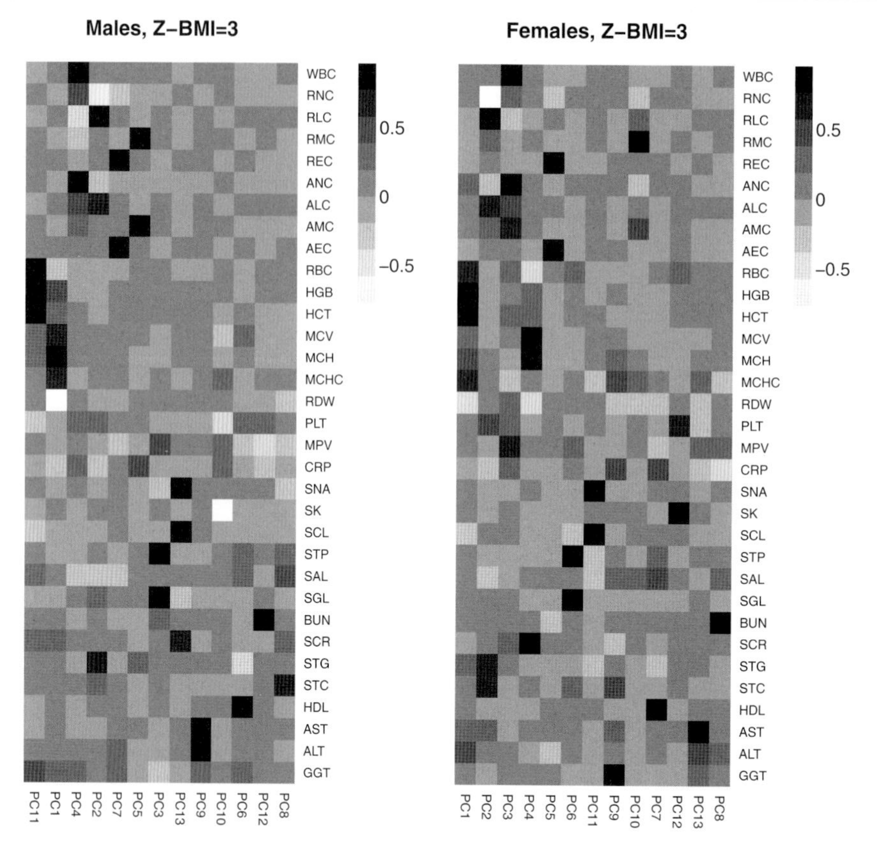


**Figure 17.3** *(Continued)*

Dendrograms, obtained with CA are shown in Figure 17.4, again for different levels of obesity (Z-BMI = +1, +2 and +3) segregated according to sex.

# DISCUSSION

## Univariate descriptive statistics and association analysis

The descriptive statistics are important as reference values and to facilitate further analyses and interpretation of the results.

According to our study, the laboratory parameters that are significantly altered by obesity in both sexes are mean corpuscular volume, mean corpuscular haemoglobin, platelet count, C-reactive protein and serum albumin. In addition, white blood cell count (with absolute neutrophil count, absolute lymphocyte count and absolute monocyte count), serum creatinine, serum triglycerides, serum total cholesterol, serum HDL cholesterol, Alanine transaminase and Gamma glutamyl transpeptidase only change significantly in the case of males. Red blood cell count and serum globulin only change significantly in the case of females. The high number of parameters changing significantly only for males is likely attributable to the fact that the sample size was much larger for males in our study – it is quite possible that these changes would have been significant for females as well were we able to use a larger female sample.

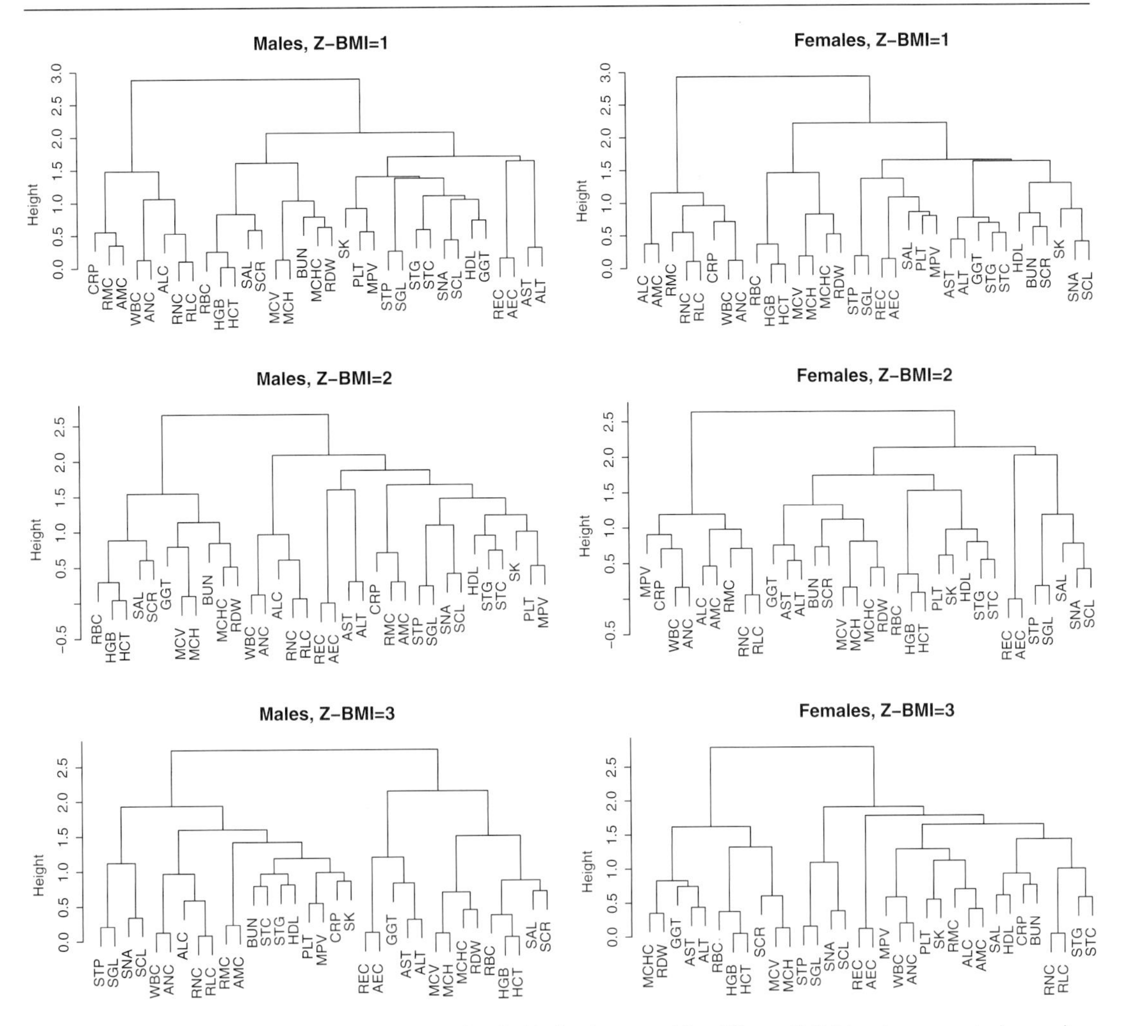


**Figure 17.4** Results of the cluster analysis (visualized with dendrograms) for different Z-BMI levels, segregated according to sex.

It is also worth noting that results from our study should be handled with more caution due to its nonrepresentative nature.

These results can be interpreted together with the descriptors for different Z-BMI values, which shed light on the medical content of differences by revealing the direction and the (clinical) size of the difference.

Most obvious is the presence of inflammation-related changes: elevated levels of the inflammation marker C-reactive protein, elevated platelet count and elevated white blood cell count and white blood cell count fractions (for males). These can be considered as an empirical confirmation that obesity is associated with inflammation. This was first noted decades ago, and is now hypothesized to be caused by the proinflammatory mediators released by the excessive white adipose tissue (Bastard *et al.*, 2006; Stienstra *et al.*, 2007). Our results are consistent with the literature: the idea that obesity can be considered as a systematic, low-level, chronic inflammation state is widely discussed (Ferroni *et al.*, 2004) and was described specifically in children, too (Sachek, 2008). In addition to theoretical results, the elevations of the above mentioned inflammation markers were clinically also observed (Oda and Kawai, 2010), specifically in children (Syrenicz *et al.*, 2006; Gilbert-Diamond *et al.*, 2010). Note that the relative counts remain unchanged (even in males), suggesting that the increase in the absolute counts (absolute neutrophil count, absolute lymphocyte count, absolute monocyte count) is caused by the overall increase in white blood cell count, not by the shifting of the proportions of the different white blood cell count fractions.

Another important finding is that mean corpuscular volume and mean corpuscular haemoglobin both have a significant decreasing tendency as the degree of obesity increases. Obesity is known to be associated with features of (low-level) anaemia (Ausk and Ioannou, 2008), which is logical if we consider that chronic inflammation is often associated with anaemia. However, clinically, we cannot speak of anaemia as the haemoglobin levels are not decreasing with the degree of obesity in our study. This is consistent with what Ausk and Ioannou (2008) have found but they have not reported results for mean corpuscular volume and mean corpuscular haemoglobin. Therefore, our findings not only confirm theirs but also extend it by pointing out that the nature of anaemia is hypochromic microcytic (which was to be expected for chronic inflammation).

The elevation of alanine transaminase and gamma glutamyl transpeptidase (males) can be considered as an indication of the effect of obesity, especially central obesity on liver function that is relatable to nonalcoholic fatty liver disease (Lam and Mobarhan, 2004; Colicchio *et al.*, 2005; Gholam *et al.*, 2007). It is worth noting that some even presume that this effect is mediating between obesity and type 2 diabetes (Lawlor *et al.*, 2005).

Changes in serum triglycerides, serum total cholesterol and serum HDL cholesterol (males) are almost trivial, as they are indicators of fat metabolism state (van Vliet *et al.*, 2011).

Decreasing tendency of serum albumin (seen in both sexes) with the degree of obesity seems to be contradictory to what has been reported in the literature, at least with respect to metabolic syndrome (Cho *et al.*, 2012; Ishizaka *et al.*, 2007). Note, however, that these results pertain to adult population.

The decreasing tendency of serum creatinine (males) with the degree of obesity is consistent with the fact that serum creatinine is a surrogate marker of muscle mass, with low muscle mass being associated in turn with obesity (Hjelmesæth *et al.*, 2010).

## Investigation of the multivariate structure

It is immediately obvious from the results of the PCA that the structure of the components was largely the same, irrespectively of the degree of obesity (Z-BMI) and sex (save for the order of components, which is sometimes varied).

The medical 'meaning' of a principal component can be given based on those variables that highly correlate with the component (which can be read from the structure matrix). Summing the information from the different structure matrices, the following common components can be identified:

- White blood cell components: Relative and absolute counts of the same fraction usually form separate components; white blood cell count is typically in the neutrophil component.
- Macroscopic red blood cell component: Consists of red blood cell count, haemoglobin and haematocrit, all being positively correlated with the component.
- Microscopic red blood cell component: Consists of mean corpuscular volume, mean corpuscular haemoglobin and mean corpuscular haemoglobin concentration (positively correlated with the component) and red blood cell distribution width (negatively correlated).
- Platelet component: Consists of platelet count (positively correlated) and mean platelet volume (negatively correlated).
- Liver enzymes components: Consists of alanine transaminase and aspartate transaminase (positively correlated). GGT's association is less pronounced.
- Inorganic constituents of the serum component: Consists of serum sodium and serum chloride (positively correlated). Serum potassium forms separate component.

- Organic constituents of the serum component: Consists of serum total protein and serum globulin (positively correlated).
- Blood lipids component: Consists of serum triglycerides (positively correlated) and serum HDL cholesterol (negatively correlated). Serum total cholesterol usually forms separate component.

These findings are consistent with the dendrograms of the CA: variables that belong to the same component are usually found to be closely connected on the dendrogram. For example the deepest connection is between haemoglobin and haematocrit, which is followed by red blood cell count almost invariably.

CA also confirms that the correlation structure is largely independent from both the sex and the degree of obesity. Although dendrograms are varied, structural blocks (like the already mentioned haemoglobin–haematocrit–red blood cell count trio) can be identified that are largely the same in every case.

## CONCLUSION

Univariate examination of laboratory results sheds light on the pathophysiological alterations that are associated with obesity. While these changes were mostly already well-known for particular parameters, we have now performed a comprehensive, uniform investigation for 33 routinely measured blood tests.

The analysis of the multivariate structure of the laboratory results reveals groups of variables that exhibit similar stochastical behaviour, pointing to a shared physiological background. On the other hand, this analysis also demonstrated that the correlation structure of the laboratory parameters is largely unaffected by the degree of obesity and sex.

The method we proposed for the analysis of the multivariate structure (obtaining conditional correlation matrices through KDE element-by-element with smoothing being applied afterwards, and the analysis of these matrices with PCA or CA) lived up to expectations and was demonstrated to be a useful tool in similar tasks.

## ACKNOWLEDGEMENTS

This work was supported in part by the Hungarian National Scientific Research Foundation grant OTKA K82066. It is connected to the scientific program of the 'Development of quality-oriented and harmonized R + D + I strategy and functional model at BME' project, supported by the New Széchenyi Plan (Project ID: TÁMOP-4.2.1/B-09/1/KMR-2010-0002).

Levente Kovács is supported by the János Bolyai Research Scholarship of the Hungarian Academy of Sciences.

The authors say thanks to the involved secondary schools: Fazekas Mihály (Budapest), Leövey Klára (Budapest), Puskás Tivadar (Budapest) and Esze Tamás (Mátészalka) Primary and Secondary Grammar Schools. We say special thanks to Bétéri Csabáné for her cooperation in organizing the study at Mátészalka.

## REFERENCES

Andersen, R.E. (2003) *Obesity: Etiology, Assessment, Treatment, and Prevention.* Human Kinetics Publishers, Champaign, IL.

Antal, M., Péter, S., Biró, L. *et al.* (2009) Prevalence of underweight, overweight and obesity on the basis of body mass index and body fat percentage in Hungarian schoolchildren: representative survey in metropolitan elementary schools. *Ann Nutr Metab* **54**(3), 171–176.

Armitage, P., Berry, G. and Matthews, J.N.S. (2002) *Statistical Methods in Medical Research*. John Wiley & Sons, Inc., New York.

Ausk, K.J. and Ioannou, G.N. (2008) Is obesity associated with anemia of chronic disease? A population-based study. *Obesity (Silver Spring)* **16**(10), 2356–2361.

Bastard, J.P., Maachi, M., Lagathu, C. *et al.* (2006) Recent advances in the relationship between obesity, inflammation, and insulin resistance. *Eur Cytokine Netw* **17**(1), 4–12.

Bo, S., Rosato, R., Ciccone, G. *et al.* (2009) What predicts the occurrence of the metabolic syndrome in a population-based cohort of adult healthy subjects? *Diabetes Metab Res Rev* **25**(1), 76–82.

Burke, V. (2006) Obesity in childhood and cardiovascular risk. *Clin Exp Pharmacol Physiol* **33**(9), 831–837.

CDC (2002) *2000 CDC Growth Charts for the United States: Methods and Development*. Vital and Health Statistics, Series 11, Number 246, Centers for Disease Control and Prevention, Atlanta, GA.

CDC (2012) *Growth Charts*. Centers for Disease Control and Prevention. [Online] Available: http://www.cdc.gov/growthcharts/cdc_charts.htm (last accessed 26 June 2013).

Cho, H.M., Kim, H.C., Lee, J.M. *et al.* (2012) The association between serum albumin levels and metabolic syndrome in a rural population of Korea. *J Prev Med Public Health* **45**(2), 98–104.

Cole, T.J. (1990) The LMS method for constructing normalized growth standards. *Eur J Clin Nutr* **44**(1), 45–60.

Cole, T.J., Faith, M.S., Pietrobelli, A. and Heo, M. (2005) What is the best measure of adiposity change in growing children: BMI, BMI %, BMI z-score or BMI centile? *Eur J Clin Nutr* **59**(3), 419–425.

Colicchio, P., Tarantino, G., del Genio, F. *et al.* (2005) Non-alcoholic fatty liver disease in young adult severely obese non-diabetic patients in South Italy. *Ann Nutr Metab* **49**(5), 289–295.

Doung, T. (2012) *ks: Kernel smoothing*. R package version 1.8.8.

Dubern, B., Girardet, J.P. and Tounian, P. (2006) Insulin resistance and ferritin as major determinants of abnormal serum aminotransferase in severely obese children. *Int J Pediatr Obes* **1**(2), 77–82.

Everitt, B., Landau, S., Leese, M. and Stahl, D. (2006) *Cluster Analysis*. John Wiley & Sons, Inc., Hoboken, NJ.

Ferenci T. (2009) *Biostatistical Analysis of Obesity-Related Parameters of Adolescent Hungarian Population* [in Hungarian, Kiskorú magyar populáció obesitassal összefüggo paramétereinek biostatisztikai elemzése]. MSc Thesis, Budapest University of Technology and Economics, Budapest, Hungary.

Ferenci, T., Almássy, Zs., Kovács, A. and Kovács, L. (2011) Effects of obesity: a multivariate analysis of laboratory parameters. *Buletinul Stiintific Al Universitatii Politehnica Din Timisoara-Seria Automatica Si Calculatoare* **56**(70), 145–152.

Ferroni, P., Basili, S., Falco, A. and Davi, G. (2004) Inflammation, insulin resistance, and obesity. *Current Atheroscler Rep* **6**(6), 424–431.

Flury, B. (1997) *A First Course in Multivariate Statistics*. Springer, New York.

Gilbert-Diamond, D., Baylin, A., Mora-Plazas, M. and Villamor E. (2012) Chronic inflammation is associated with overweight in Colombian school children. *Nutr Metab Cardiovasc Dis* **22**(3), 244–251.

Gholam, P.M., Flancbaum, L., Machan, J.T. *et al.* (2007) Nonalcoholic fatty liver disease in severely obese subjects. *Am J Gastroenterol* **102**(2), 399–408.

Glynn, E. F. (2005) *Correlation 'Distances' and Hierarchical Clustering*. Stowers Institute for Medical Research. [Online] Available: http://research.stowers-institute.org/efg/R/Visualization/cor-cluster/index.htm (last accessed 26 June 2013).

Guh, D.P., Zhang, W., Bansback, N. *et al.* (2009) The incidence of co-morbidities related to obesity and overweight: a systematic review and meta-analysis. *BMC Public Health* **9**(1), 88.

Hall, P, Maron, J. and Park, B. (1992) Smoothed cross-validation. *Probab Theory Related Fields* **92**(1), 1–20.

Hjelmesæth, J., Røislien, J., Nordstrand, N. *et al.* (2010) Low serum creatinine is associated with type 2 diabetes in morbidly obese women and men: a cross-sectional study. *BMC Endocr Disord* **10**, 6.

Holm, S. (1979) A simple sequentially rejective multiple test procedure. *Scand J Stat* **6**(2), 65–70.

Ishizaka, N., Ishizaka, Y., Nagai, R. *et al.* (2007) Association between serum albumin, carotid atherosclerosis, and metabolic syndrome in Japanese individuals. *Atherosclerosis* **193**(2), 373–379.

Joubert, K., Darvay, S., Gyenis, Gy. *et al.* (2006) *Results of the National Longitudinal Child Growth Survey from Birth to the Age of 18 Years* [in Hungarian, Az Országos Longitudinális Gyermeknövekedés-vizsgálat eredményei születéstől 18 éves korig]. KSH Népességtudományi Kutató Intézetének Kutatási jelentések 83, Budapest, Hungary.

Juonala, M., Juhola, J., Magnussen, C.G. *et al.* (2011) Childhood environmental and genetic predictors of adulthood obesity: The cardiovascular risk in young Finns study. *J Clin Endocrinol Metab* **96**(9), E1542–1549.

Kaiser, H.F. (1958) The varimax criterion for analytic rotation in factor analysis. *Psychometrika* **23**(3), 187–200.

Kaiser, H.F. (1960) The application of electronic computers to factor analysis. *Educ Psychol Meas* **20**(1), 141–151.

Kelly, A. and Munan, L. (1977) Haematologic profile of natural populations: Red cell parameters. *Br J Haematol* **35**(1), 153–160.

Kern, B (2007) *The Prevalence of Overweight and Obesity in Hungarian Children*. 1st Summer School of the European Anthropological Association, Prague, Czech Republic.

Lam, G.M. and Mobarhan, S. (2004) Central obesity and elevated liver enzymes. *Nutr Rev* **62**(10), 394–399.

Lawlor, D.A., Sattar, N., Smith, G.D. and Ebrahim, S. (2005) The associations of physical activity and adiposity with alanine aminotransferase and gamma-glutamyltransferase. *Am J Epidemiol* **161**(11), 1081–1088.

Maritz, J.S. (1995) *Distribution-Free Statistical Methods*. Chapman and Hall, New York.

Miller, R.G. (1981) *Simultaneous Statistical Inference*. Springer, New York.

Nyberg, G., Ekelund, U., Yucel-Lindberg, T.L. *et al.* (2011) Differences in metabolic risk factors between normal weight and overweight children. *Int J Pediatr Obes* **6**(3-4) 244–252.

Oda, E. and Kawai, R. (2010) Comparison between high-sensitivity C-reactive protein (hs-CRP) and white blood cell count (WBC) as an inflammatory component of metabolic syndrome in Japanese. *Intern Med* **49**(2), 117–124.

OECD (2012) *Factbook 2011–2012, Economic, Environmental and Social Statistics*. Organization for Economic Co-operation and Development, Paris, France.

Ogden, C.L. and Carroll, M.D. (2010a) *Prevalence of Overweight, Obesity, and Extreme Obesity Among Adults: United States, Trends 1960–1962 Through 2007–2008*. NCHS Health E-Stats, www.cdc.gov/nchs/data/hestat/obesity_adult_07_08/obesity_adult_07_08.pdf (last accessed 5 July 2013).

Ogden, C.L. and Carroll, M.D. (2010b) *Prevalence of Obesity Among Children and Adolescents: United States, Trends 1963–1965 Through 2007–2008*. NCHS E-Stats, www.cdc.gov/nchs/data/hestat/obesity_child_07_08/obesity_child_07_08.pdf (last accessed 5 July 2013).

Ogden, C. L., Yanovski, S. Z., Carroll, M. D. and Flegal, K. M. (2007) The epidemiology of obesity. *Gastroenterology* **132**(6), 2087–2102.

Okorodudu, D.O., Jumean, M.F., Montori, V.M. *et al.* (2010) Diagnostic performance of body mass index to identify obesity as defined by body adiposity: a systematic review and meta-analysis. *Int J Obes (Lond)* **34**(5), 791–799.

Ong, K.L., Tso, A.W., Xu, A. *et al.* (2011) Evaluation of the combined use of adiponectin and C-reactive protein levels as biomarkers for predicting the deterioration in glycaemia after a median of 5.4 years. *Diabetologia* **54**(10), 2552–2560.

Pasek, J. (2012) *weights: Weighting and Weighted Statistics*, R package version 0.72.

Pi-Sunyer, X. (2009) The medical risk of obesity. *Postgrad Med* **121**(6), 21–33.

R Core Team (2012) *R: A Language and Environment for Statistical Computing*. R Foundation for Statistical Computing, Vienna, Austria.

Revelle, W. (2012) *psych: Procedures for Psychological, Psychometric, and Personality Research*, R package version 1.2.1.

Rodger, R.S.C., Fletcher, K., Fail, B.J. *et al.* (1987) Factors influencing haematological measurements in healthy adults. *J Chronic Diseases* **40**(10), 943–947.

Romero-Corral, A., Somers. V.K., Sierra-Johnson, J. *et al.* (2008) Accuracy of body mass index in diagnosing obesity in the adult general population. *Int J Obes (Lond)* **32**(6), 959–966.

Ruhl, C.E. and Everhart, J.E. (2003) Determinants of the association of overweight with elevated serum alanine aminotransferase activity in the United States. *Gastroenterology* **124**(1), 71–79.

Sachek, J. (2008) Paediatric obesity: an inflammatory condition? *J Parenter Enteral Nutr* **32**(6): 633–637.

Scott, D.W. (1992) *Multivariate Density Estimation: Theory, Practice, and Visualization*. John Wiley & Sons, Inc., New York.

Shaffer, J.P. (1995) Multiple Hypothesis Testing. *Ann Rev of Psych* **46**(1), 561–584.

Stienstra, R., Duval, C., Muller, M. and Kersten, S. (2007) PPARs, Obesity, and Inflammation. *PPAR Res* 95974.

Syrenicz, A., Garanty-Bogacka, B., Syrenicz, M. *et al.* (2006) Low-grade systemic inflammation and the risk of type 2 diabetes in obese children and adolescents. *Neuro Endocrinol Lett* **27**(4), 453–458.

Takezawa, K. (2005) *Introduction to Nonparametric Regression*. John Wiley & Sons, Inc., New York.

Tan, P.M., Steinbach, M. and Kumar, V. (2006) *Introduction to Data Mining*. Addison-Wesley, New York.

Taylor, M.R., Holland, C.V., Spencer, R. *et al.* (1997) Haematological reference ranges for schoolchildren. *Clin Lab Haematol* **19**(1), 1–15.

Venables, V.N and Ripley, B.D. (2002) *Modern Applied Statistics with S*. Springer, New York.

Visscher, T.L. and Seidell, J.C. (2001) The public health impact of obesity. *Annu Rev Public Health* **22**, 355–375.

van Vliet, M., Heymans, M.W., von Rosenstiel, I.A. *et al.* (2011) Cardiometabolic risk variables in overweight and obese children: a worldwide comparison. *Cardiovasc Diabetol* **10**, 106.

Wang, Y. and Lobsten, T. (2006) Worldwide trends in childhood overweight and obesity. *Int J Pediatr Obes* **1**(1), 11–25.

WHO (World Health Organization) (2011) *Global Database on Body Mass Index – an interactive surveillance tool for monitoring nutrition transition*. [Online], Available: http://apps.who.int/bmi/index.jsp (last accessed 26 June 2013).

Wothke, W. (1993) Nonpositive definite matrices in structural modeling. In: *Testing structural equation models* (eds. Bollen, K.A. and Long, J.S.). Sage Publications, Newbury Park, UK, pp. 256–293.

# 18 Development and application of predictive microbiology models in foods

**Fernando Pérez-Rodríguez**

*Department of Food Science and Technology, University of Cordoba – International Campus of Excellence in the AgriFood Sector ceiA3, Campus Rabanales, Edificio Darwin – Córdoba, Spain*

## ABSTRACT

Predictive microbiology is a relatively novel scientific field belonging to food microbiology; it is aimed at developing mathematical models that account for the effect of intrinsic and extrinsic factors on microorganism responses in food. This area has experienced significant advances over the last decades, providing numerous and innovative predictive models to estimate the fate of microorganisms across the food chain. Predictive models can be classified, according to the type of modelled phenomenon, into growth models, inactivation and survival models and probability models. More recently, transfer models, single-cell-based models or genomic-scale models have been proposed as more mechanistic approaches to reflecting microbial responses in foods. Predictive microbiology is, in addition, the basis for developing quantitative microbial risk assessment by modelling different microbial process across the food chain, and thus supporting making decision processes within risk management. Likewise, the application of user-friendly software tools has enabled end-users to easily apply predictive models in different contexts (e.g. Pathogen Modeling Program – PMP, ComBase or Microhibro). In spite of these important achievements, predictive microbiology should still continue to improve model precision and accuracy and, based on more mechanistic approaches such as systems biology, to generate robust and reliable models for complex and real food systems.

## PREDICTIVE MICROBIOLOGY FRAMEWORK

### Historical background

The origin of predictive microbiology is usually associated with the works developed by Bigelow and Esty (1920), Bigelow (1921) and Esty and Meyer (1922) in which thermal inactivation of *Clostridium botulinum* is modelled by using a log-linear equation. However, until the 1980s modelling studies were scarce and limited. Nonetheless, the important role played by fermentation microbiology in the development of empirical models applied for controlling fermentation process should not be overlooked. The term Predictive Microbiology was first introduced by Roberts and Jarvis (1983), establishing the conceptual basis of the modern predictive microbiology. During 1980s, predictive microbiology showed a great growth thanks to the advances in computers and statistical software, which enabled complex and precise models to be developed. In these years, Ratkowsky *et al.* (1982) introduced Bêlehrádek-type models to describe bacteria growth rate as a function of temperature. Later, the cardinal temperature

*Mathematical and Statistical Methods in Food Science and Technology*, First Edition.
Edited by Daniel Granato and Gastón Ares.

model and the Gamma concept model were proposed as means of modelling growth rate based on model parameters with biological interpretation (Rosso *et al.*, 1995; Zwietering *et al.*, 1996). In parallel, several authors devoted efforts to derive mathematical functions able to describe the sigmoid growth patterns showed by microorganisms in closed systems (i.e. foods). Among the different models, we highlight for their relevance, the logistic model and the modified Gompertz equation introduced by Gibson *et al.* (1987) and the model of Baranyi and Roberts (1994), which has come to be one of the most used models together with the reparameterized Gompertz equation (Zwietering *et al.*, 1990). Other models and modelling approaches followed in subsequent years, such as the probability models (Genigeorgis, 1981; Roberts *et al.*, 1981), the effect of pre-culture conditions on kinetic parameters, single-cell based models (Dupont and Augustin, 2009), stochastic models, bacterial transfer models (Pérez-Rodríguez *et al.*, 2008) and, more recently, genomic-scale models (Brul *et al.*, 2008; Métris *et al.*, 2011). All these relevant advances in predictive microbiology are now the basis of the Quantitative Microbial Risk Assessment (QMRA), which is considered a fundamental tool to support decision-making processes for food risk management (Lammerding and Paoli, 1997).

## Definitions and concepts

Predictive microbiology is a scientific field belonging to food microbiology aimed at studying the effect of food-related factors on microorganisms deriving mathematical models to predict the fate of microorganisms in foods.

Models can be classified according the type of structure and variables (Figure 18.1). The primary models are those reflecting the microbial load change with respect to time. Secondary models relate the kinetic parameters from primary models to environmental factors. Finally, tertiary models are not models themselves but rather they refer to implementations of primary and secondary models in software tools (e.g. spreadsheet software) in order to provide estimates of microbial behaviour under specific conditions defined by users.

Other important aspect in predictive microbiology is the differentiation between mechanistic and empirical models. Mechanistic models are those whose development comes from the understanding of the underlying biochemical and biological processes governing microbial phenomena. In these cases, model parameters are supposed to have a biological meaning. On the contrary, empirical models are

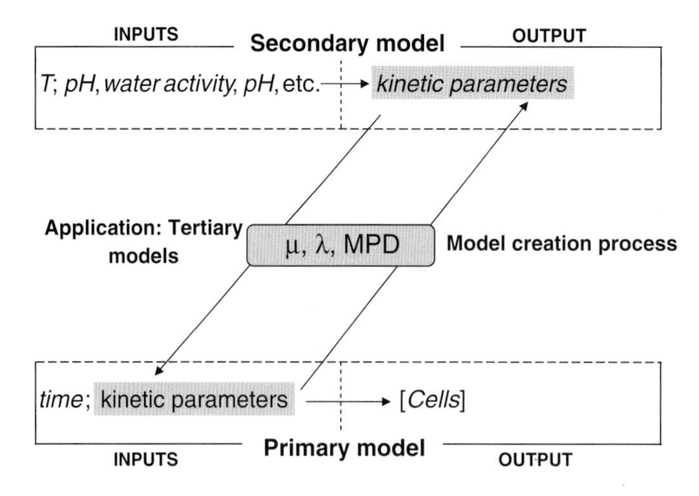


**Figure 18.1** Illustrative representation of predictive model classification: primary model, secondary model, and tertiary model.

mathematical functions simply describing observations on a certain phenomenon. Polynomial functions fitted to growth data by any regression method are a clear example of empirical models. These have important limitations concerning their capacity of extrapolation and generalization of results out of the experimental conditions where observations were obtained. Most models in predictive microbiology are empirical models although exceptions can be found, in which specific model parameters are claimed to have a biological meaning. They are not absolute mechanistic models, and hence they are so-called pseudo-mechanistic models. One example of this is the primary model by Baranyi and Roberts (1994), in which lag time is related to a parameter called $q_0$ that defines the initial physiological state of cells or readiness of cells for growth in a new environment. Furthermore, models can be distinguished according to the type of phenomenon modelled. Thus, there are kinetic models which describe growth and death processes. Probability models are a kind of model which estimate the probability of growth and/or define the growth/no growth boundaries as a function of certain environmental factors. More recently, transfer models have been proposed as models to account for bacterial transfer processes occurring in food-related environments (i.e. cross-contamination).

## FUNDAMENTAL STEPS FOR PREDICTIVE MODEL ELABORATION

Figure 18.2 shows a general scheme of the main steps needed to develop and apply predictive models in foods. Models can be built based on existing data or by generating new data in specific experiments. These data can be obtained on either artificial media or food matrices. The former enables a major number of experiments to be carried out saving human and material resources while keeping better control on experimental conditions. However, in these cases, the main factors impacting microbial behaviour should be selected carefully so as to obtain reliable and precise models. In contrast, experiments on food matrices often imply more complex inoculation procedures, high variability in outcomes and more human and material resources; in return, observations are usually closer to reality.

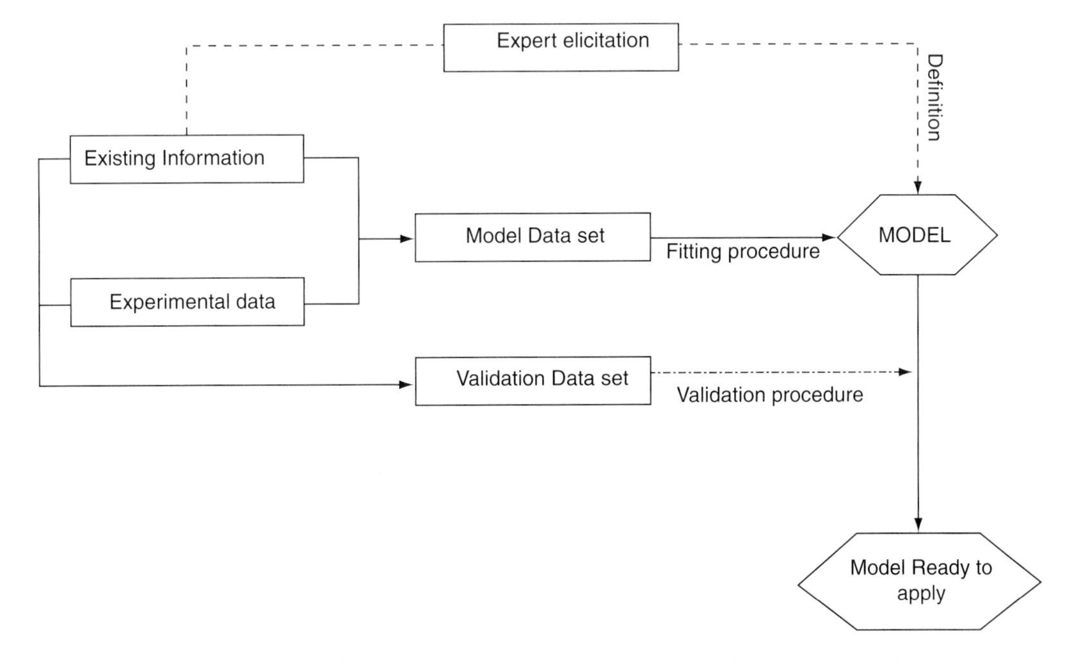


**Figure 18.2** General scheme including the main steps for predictive model elaboration and validation.

Microbial behaviour is usually measured by traditional microbiological techniques based on plate counts, even though other methods, such as direct observation by microscope, or use of indirect methods, such as impedance or optical density, are possible. Indirect methods based on optical density (i.e. turbidity) of microbial cultures have been widely used for growth models, as they allow the testing of multiple conditions at the same time in a cost-effective manner (Dalgaard *et al.*, 1994; Miles *et al.*, 1997; Augustin *et al.*, 1999). For example Bioscreen (Labsystems, Finland) is an automatic turbidimeter that is able to monitor growth in 100-well plates while controlling temperature conditions (Membré *et al.*, 2005; Lindqvist, 2006; Valero *et al.*, 2006, 2009).

When multiple factors are considered for modelling, an experimental design can be recommendable in order to optimize the experimental work and improve performance of the model. Nonetheless, a screen of the main factors affecting microbial behaviour should first be conducted. As the number and levels of involved factors rise, the experimental design becomes more complex. In this sense, different typologies of experimental design have been applied to predictive microbiology experiments, such as the Complete factorial design, Central composite design and Doehlert matrix.

Different methodologies are used to derive predictive models. Fitting models to data by using regression methods (linear and nonlinear regression) is the most used methodology. Cardinal-based models (secondary models), which are constructed on parameters with biological or graphical meaning, can be defined by using theoretical values derived from expert elicitation or literature data. In addition, these values can be used as starting values in nonlinear regression to optimize the fitting process. The neural networks approach has been also applied to derive predictive models with interesting results although their complexity makes this type of models less attractive to end-users and practitioners.

Once models are generated by one of the above procedures, they should be validated. Internal validation is performed to assess to the ability of the model to reflect observations in the same experimental conditions. For that, researchers usually generate two data sets, one for building the model and the other for validation (Figure 18.2). Likewise, models should be subject to an external validation, where the capacity of generalization of the model is challenged against observations obtained in food matrices. For that, specific indexes have been proposed, such as Accuracy factor and Bias factor which are explained in detail later.

## PRIMARY MODELS

### Growth curve: phases and parameters

Growth primary models are aimed at describing the growth curve shown by microorganisms in food environments. The growth curve is made of four stages, as shown in Figure 18.3: lag phase (or latency phase), exponential phase, stationary phase, and decline phase. These phases are defined in models through specific mathematical parameters; in concrete, lag time, growth rate and maximum population density (or asymptotic microbial level). The lag or latency phase ($\lambda$) is the initial stage in growth curves defining a certain period of time in which no increase of cell numbers is observed. This phase is given when microorganisms colonize a new habitat in which cells undergo adaptation to the new environment. Once adapted, microorganisms multiply (e.g. mitosis) showing an increase in number of cells. This phase corresponds to the exponential phase in the curve and is mathematically described by means of the growth rate (i.e. increase of cells per time unit), which from a value of zero in the lag phase accelerates to reach a maximum growth rate ($\mu_{max}$). The maximum growth maintains constant during a certain period of time and then decelerates reaching a final value of zero, coinciding with the Maximum Population Density (MPD). The stationary phase is a plateau in which the MPD is constant for a certain period of time (Zwietering *et al.*, 1990).

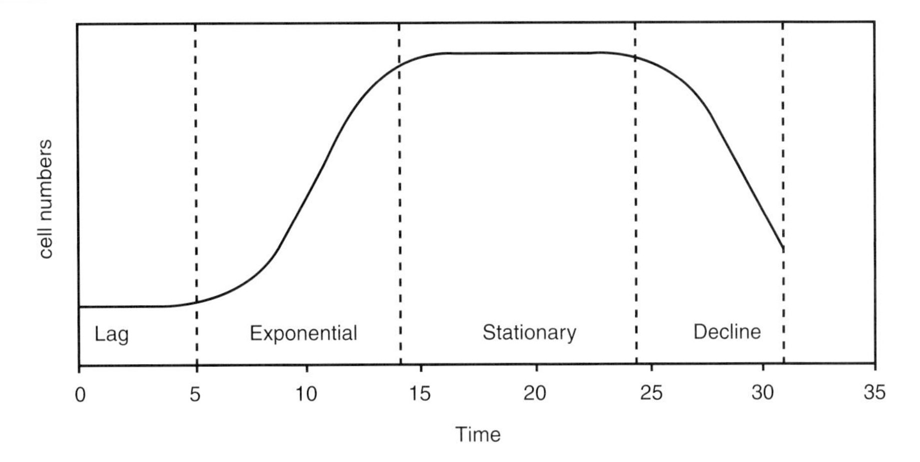


**Figure 18.3** Typical microbial growth curve comprising four phases: lag, exponential, stationary and decline.

The three first steps, when graphically represented appear as a sigmoid curve (Figure 18.4). Traditionally, growth rate is estimated as the tangent of the exponential phase, while maximum growth rate is the tangent at steepest part of it (Figure 18.4). The doubling time ($t_d$) can also be calculated using the growth rate according to the following equation when cell numbers are expressed in logarithm with base 10:

$$t_d = \frac{\mu_{\max}}{\log 2} \tag{18.1}$$

The generation time (i.e. mean of generation time of individual cells) is equal to the doubling time of microbial population when division cycles of individual cells occur synchronously. Therefore, in such cases, $t_d \approx t_g$.

If they do not grow synchronously, then the average cell cycle time of the individual cells is somewhat higher than the time necessary to double the whole population.

Finally, the lag time is estimated as the length of time between the initial level and the tangent to the steepest part of the exponential phases (Figure 18.4). On the other hand, the decline phase is not

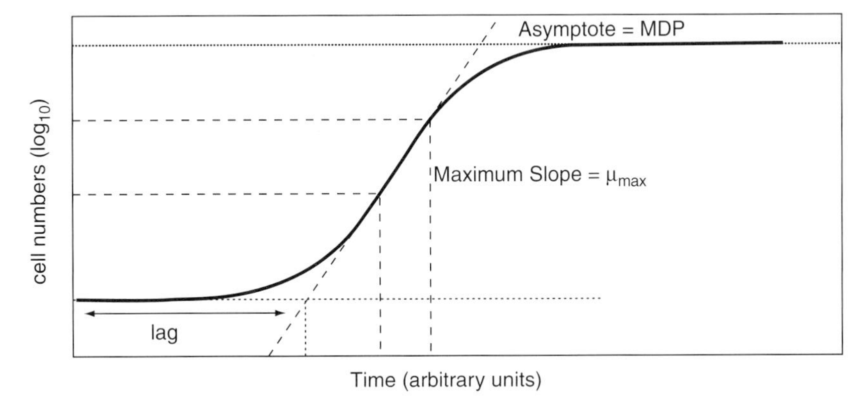


**Figure 18.4** Graphical meaning of kinetic parameters defining the typical bacterial growth curve: lag phase, maximum growth rate ($\mu_{max}$), and Maximum Population Density (MPD).

considered in predictive microbiology, since it is not expected to be reached during food shelf-life and, therefore, few models (e.g. a modified Fermi model) have been developed considering this fourth phase of the bacterial growth (Peleg, 2006a).

## Sigmoid functions

The sigmoid functions are a family of mathematical models that have been widely used to fit microbial growth data. The logistic function is one of these functions and was introduced by Gibson *et al.* (1987) to describe microbial growth with a symmetric growth pattern:

$$N(t) = A + \frac{C}{1 + \exp[-B(t - M)]} \tag{18.2}$$

where $N(t)$ and A stand for the logarithm of numbers of cells at time t and 0; $M$ is the time that determines the commence of growth and $B$ is the exponential growth are at the inflection point.

A modified logistic function, represented by Equation 18.3, was proposed to account for asymmetric growth curves by incorporating the parameter $m$ as exponent in the denominator of the equation (Peleg, 2006a). When $m$ is equal to 1, the function describes the symmetric pattern.

$$N(t) = A + \frac{C}{1 + \exp[-B(t - M)^{m}]} \tag{18.3}$$

This function has been successfully applied to describe growth from different microorganisms in different food matrices, including pathogenic bacteria such as *Listeria monocytognes* and *Salmonella spp.* and spoilage microorganisms (Dalgaard *et al.*, 1994; Watier *et al.*, 1996; Augustin *et al.*, 1999; Juneja and Marks, 2003; Mataragas *et al.*, 2006; Gurtler *et al.*, 2011).

Another widely applied primary model is the modified Gompertz equation, also introduced by Gibson *et al.* (1987). This equation is a modification of the Gompertz model to account for sigmoid growth pattern of logarithmic counts. Unlike the logistic function, the modified Gompertz equation, described by Equation 18.4, is able to describe the asymmetric growth curves.

$$N(t) = A + C \exp\{(-\exp[-B(t - M)]\} \tag{18.4}$$

here $C$ is the difference between $N_0$ and the asymptotic count; $B$ is the relative growth rate at time $M$ and $M$ is the time at which growth rate is maximum. These parameters are regression parameters although mathematical expressions have been derived relating these parameters to meaningful kinetic parameters:

$$\text{Lag time } (\lambda) = M - \frac{1}{B} \tag{18.5}$$

$$\text{Maximum growth rate } (\mu_{\max}) : \frac{BC}{e} (e = 2.718\ldots) \tag{18.6}$$

The Gompertz function was reparameterized by defining growth rate ($R_g$) and lag time ($\lambda$) as kinetic parameters (Equation 18.7), thus allowing a more interpretable model to be obtained (Zwietering *et al.*,

1990). With their clearer biological meaning, starting values for these parameters required by the fitting process can better defined.

$$N(t) = A + C\exp\left(-\exp\left(2.71\left(\frac{R_g}{C}\right)(\lambda - t) + 1\right)\right) \tag{18.7}$$

The primary model proposed by Baranyi and Roberts (1994) is a variation of the modified logistic function previously described (Equation 18.2). Both assume that the microbial kinetic in a nonlimiting environment follows a first order kinetic, which means:

$$\frac{\partial N}{\partial t} = k \cdot N \tag{18.8}$$

where $N$ is concentration of microorganism, $k$ is a constant and $t$ is time. In addition, it can be mathematically proved that the constant $k$ corresponds to the growth rate. Another commonly used concept is the specific growth rate, which refers to the rate of increase per the population density that has produced that increase (Baranyi and Roberts, 2000), that is:

$$\frac{\frac{\partial N}{\partial t}}{N} = k = \mu \tag{18.9}$$

In order to consider the limiting effect on growth derived from the $N$ increase, a new term (inhibition function) is added to Equation 18.9, yielding the following form:

$$\frac{\partial N}{\partial t} = k \cdot N \cdot \left(1 - \frac{N}{N_{MDP}}\right) \tag{18.10}$$

in which $N_{MDP}$ stands for the maximal concentration of microorganisms (i.e. Maximum Density Population). Equation 18.10 is equivalent to the Verhulst equation, which was proposed to explain the self-limiting growth of a population (Verhulst, 1838).

The peculiarity of the model developed by Baranyi and Roberts (1994) lies in that a term called the adjustment function is added to explain for the transition between the lag and exponential phases. The function can be written as:

$$\frac{\frac{\partial N}{\partial t}}{N} = \mu_{\max} \cdot u(N) \cdot \alpha(t) \tag{18.11}$$

The term $u(N)$ refers to the inhibition function coming from the Richards's family of growth models (Richards 1959; Baranyi *et al.*, 1993; Baranyi and Roberts, 1994, 1995), which includes a parameter $m$ to represent different curvatures in the transition from exponential to stationary phase, in other words:

$$u(N) = \left(1 - \frac{N}{N_{MDP}}\right)^m \tag{18.12}$$

The definition of the term $\alpha(t)$, the adjustment function, is based on the hypothesis that there is a bottleneck substance determining the commencement of the cell division cycles, and therefore the duration of the lag phase. The adjustment function can be defined as:

$$\alpha(t) = \frac{q_0}{q_0 + e^{vt}} \tag{18.13}$$

here, $q_0$ is the ratio between $P_0$ (substance at time 0) and $K_p$ (i.e. the Michaelis–Menten constant) and $v$ is the increase ratio of the substance. The value of $q_0$ can be used as a measure of the readiness of cells in the new environment. Similarly, $\alpha(0)$ can used to represent preparedness, in such cases:

$$\alpha(0) = \frac{q_0}{1 + q_0} \tag{18.14}$$

Additional transformations were introduced to facilitate the fitting procedure, for example assuming that $v = \mu_{max}$ and making the parameters defining curvature in stationary and lag phase dimensionless, i.e. $m = M$ and $v = h_0$. Therefore, the reparameterized form after integration remains as follows:

$$y(t) = y_0 + \mu_{\max} A(t) - \frac{y_{\max} - y_0}{M} \cdot \ln\left(1 - e^{-M} + e^{-M} \frac{y_{\max} - y_0 - \mu_{\max} A(t)}{y_{\max} - y_0}\right) \tag{18.15}$$

here $y(t)$ is ln $N(t)$ and $y_0$ is $\ln(N_0)$ and where

$$A(t) = t - \lambda\left(1 - \frac{1}{h_0}\ln\left(1 - e^{-h_0 \frac{t}{\lambda}} + e^{-h_0\left(\frac{t}{\lambda} - 1\right)}\right)\right) \tag{18.16}$$

This model (Equations 18.15 and 18.16) is widely used by predictive microbiology researchers, in part supported by the free distribution of a BASIC subroutine implemented (as add-in) in Excel (Microsoft $^{TM}$, Readmon), the so-called DmFit which enables Baranyi's model to be fitted to count data and optical density data.

## The three-phase model

In contrast to above models, Buchanan *et al.* (1997) proposed a more simple function to explain bacterial growth curves by applying straight lines to describe the three growth phases: lag, exponential and stationary. The authors also elaborated a biological basis for justifying the suitability of the linear model, theorizing on the variability of the behaviour of individual cells, cell energy balance and so on. The model, the so-called three-phase linear model, can be formulated as follows:

$$N(t)\begin{cases} \text{For } t \le t_{\text{lag}}, \mathrm{N_t} = \mathrm{N_0} & \text{-Lag phase-} \\ \text{For } t_{\text{lag}} < t < t_{\max}, N_t = N_0 + \mu_{\max}\left(t - t_{\text{lag}}\right) & \text{-Exponential growth phase-} \\ \text{For } t \ge t_{\max}, \mathrm{N_t} = N_{\max} & \text{-Stationary phase-} \end{cases} \tag{18.17}$$

where $N_t$ is the cell concentration at time $t$ in log units, $N_0$ is the initial cell concentration, $N_{max}$ is the maximum cell concentration (i.e. MPD), $t$ the elapsed time, $t_{lag}$, the time when the lag phase ends, $t_{max}$ is the time when the MPD is reached and $\mu_{max}$ is the maximum growth rate.

## Compartmental models

The Hill growth model is presented here because of its different underlying theoretical basis (Hills and Wright, 1994). This model focused on defining lag phase using a two-compartmental approach. The first compartment reflects the change of chromosomal material with respect to time ($\mu_{max}$) while the second compartment describes the evolution of all nonchromosomal material against time ($v$). This can be mathematically represented by the following function:

$$\begin{cases} \dfrac{\partial m}{\partial t} = \mu_{\max} \cdot m & \text{with } m(t=0) = x_0 \\ \dfrac{\partial x}{\partial t} = vx \cdot \left(\dfrac{m-x}{x}\right) & \text{with } x(t=0) = x_0 \end{cases} \tag{18.18}$$

where $m$ is the total biomass concentration of the bacterial culture measured in units of minimal biomass per cell, $x$ is the cell concentration, $x_0$ is the initial value of this cell concentration and $t$ is the time.

When cells colonize a new environment, at the beginning of the lag phase, $m = x_0$ in units of minimal biomass per cell (so all cells are at minimal biomass level). Then, cells start to incorporate nutrients from the surrounding environment by increasing their nonchromosomal material. When conditions are suitable for growth, the excess of biomass is used to initiate the chromosomal replication with $\partial x / x \partial t$, which is related to the excess of biomass $(m-x)/x$ through the constant rate $v$. The lag phase is defined as the elapsed time until an excess of biomass is reached, that is when chromosomal replication takes place and growth is possible. Thus, an explicit solution for the above function (Equation 18.18) is:

$$x(t) = \frac{x_0}{\mu_{\max} + v} \cdot \left(ve^{u_{\max} t} + \mu_{\max} e^{-vt}\right) \tag{18.19}$$

Therefore, the lag phase can be defined as:

$$\lambda = \frac{\ln\left(1 + \dfrac{\mu_{\max}}{v}\right)}{\mu_{\max}} \tag{18.20}$$

The model proposed by McKellar (1997) is another compartment model; it assumes two different bacterial populations in two different states (or compartments): growing and nongrowing. According to this author, growth is due to a small fraction of the total bacterial population. Hence, growth can be described by the following function:

$$\frac{\partial G}{\partial t} = G \cdot \mu_{\max} \cdot \left(1 - \frac{G}{N_{MDP}}\right) \tag{18.21}$$

in which $G$ is the concentration of cells in the growing compartment.

The models shows a great similarity to the growth model proposed by Baranyi and Roberts (1994), although the theoretical basis is different. In McKellar's growth model, a heterogeneous population is assumed, in which the transition from lag phase to exponential phase is accounted for by the sum of the two different populations (or compartments).

## Inactivation and survival models: log-linear model and the Weibull model

Inactivation models are intended to describe the inactivation patterns shown by microorganisms when they are exposed to a lethal process or agent. These can be either physical or chemical processes. The distinction between inactivation and survival models is not always clear but survival processes are usually associated with slowly declining patterns while inactivation refers to lethal process showing a rapid decrease of microbial population. In the literature, different inactivation patterns have been reported, although they could be summarized into two basic types of curve corresponding to linear and nonlinear inactivation curves. The former type is represented by the traditional log-linear model, in which the decline pattern followed by microbial counts resembles a straight line. The log-linear models are more adequate to reflect thermal inactivation. The latter type emerges as a consequence of observations departing from the linear approach. The nonlinear inactivation patterns basically include upward and downward concavity. Some hypothesis has been put forward to explain the nonlinear pattern shown by some inactivation curves. These could be related to the variability in heating procedure, use of mixed cultures and clumping or protective effect of the food matrix or dead cells (Stringer *et al.*, 2000).

### *Bigelow model (log-linear model)*

This model was first proposed to quantify microbial inactivation in the canning industry, assuming first-order kinetics (Bigelow and Esty, 1920; Bigelow, 1921). The theoretical basis consists of assuming inactivation of a critical enzyme following a first-order kinetics (although there might be some exceptions to this rule). The model has the following form:

$$\frac{\partial N_t}{\partial t} = -k'(T) \cdot N(t) \tag{18.22}$$

with $N_t$ defining the cells surviving at time $t$ and $k$ as the first-order rate constant dependent on temperature ($T$). After integrating the Equation 18.22 and arranging to be expressed in logarithms with base 10, the equation becomes:

$$\log S_t = -k(T) \cdot t \tag{18.23}$$

where $S_t$ defines the ratio $N_t/N_0$ ($N_0$, the initial number of cells) and $k = k'/\ln(10)$.

In a more recognized form, using the well-known D-value which corresponds to the reciprocal of the rate constant (Figure 18.5), that is $k(T)$, Equation 18.22 can be rewritten as:

$$\log S_t = -\frac{t}{D} \tag{18.24}$$

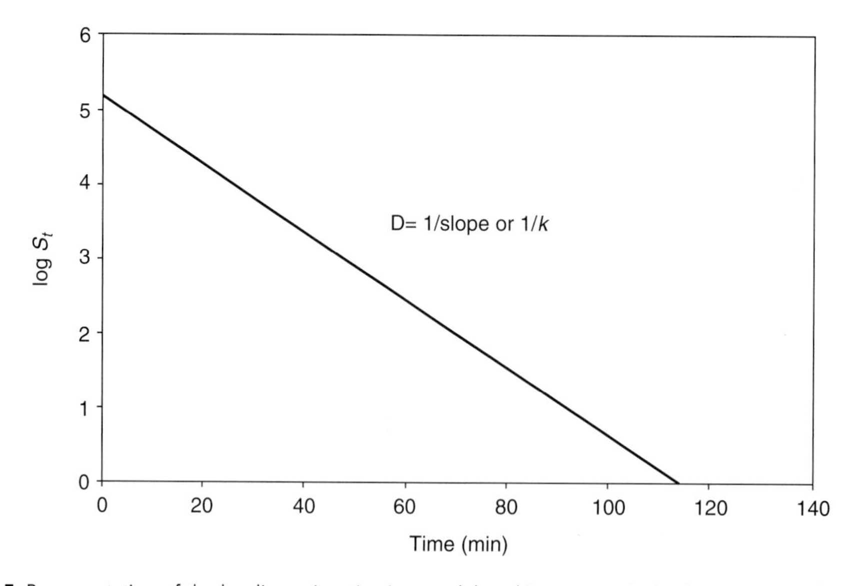


**Figure 18.5** Representation of the log-linear inactivation model and its geometrical relations to D-value.

### Biphasic and multiexponential decay models

This type of models has been used to explain enzymatic activity and can be used also to reflect the different inactivation patterns shown by microorganism when exposed to lethal or toxic agents. The typical expression corresponds to:

$$S_t = f \cdot e^{-k_1 t} + (1 - f) \cdot e^{-k_2 t} \tag{18.25}$$

In this case, $k_1$ and $k_2$ define inactivation rates of two microbial population, while $f$ is used to define the relative fraction of each population on the mixed population. This equation can be expanded to consider more than two populations by increasing the number of terms and constants. In such cases, there would be as many parameters $f$ and $k$ as types of microbial population. One of the weaknesses of this model is the great dependency of model parameter estimates on the duration of experiments, indicating that multiexponential models are purely empirical models without any valid biological interpretation. Therefore, special precaution should be taken when theoretical conclusions regarding the type of population are drawn from multiexponential decay models.

### Weibull model

The Weibull distribution is often used to model the time interval between successive, random, independent events that occur at a variable rate (Cullen and Frey, 1999). The Weibull model is built on the basis of the Weibull distribution function (Equation 18.26) and has been proposed as an alternative to the log-linear model in order to reflect curvilinear inactivation or survival patterns observed when logarithm of cell numbers versus time are plotted (Peleg, 2006b). The model is defined by two parameters, the scale parameter $\alpha$ and the dimensionless shape parameter $\beta$. The shape parameter accounts for upward concavity of a survival curve ($\beta < 1$), a linear survival curve ($\beta = 1$), and downward concavity ($\beta > 1$) (Figure 18.6). The Weibull model assumes that there are different microbial populations with different sensitivity to the lethal or toxic agent. When $\beta$ is $<1$ it implies that there is a more sensitive microbial population that dies rapidly leaving a residual population more resistant and resulting in a tail region in the inactivation or survival curve. In contrast, $\beta > 1$ indicates that as the survival ratio decreases, cells become more susceptible, dying more rapidly. An especial case is when $\beta = 1$, for which no difference between microbial populations is assumed, meaning that all cells have the same probability to die off when exposed to the lethal agent.

In terms of a survival curve, the cumulative function is:

$$\log S_t = -\frac{1}{2.303}\left(\frac{t}{\alpha}\right)^{\beta} \tag{18.26}$$

### Shoulder/tail models

The great variability of nonlinear inactivation patterns has led to authors taking several modelling approaches. The *anatomy* of a nonlinear inactivation curve can be described on the basis of the existence of shoulder (or lag time) and tail region. The shoulder represents the initial period of time in which microbial population remains in similar levels. On the other hand, the tail region corresponds to the final segment of the inactivation curve, which is a consequence of a residual population showing a more resistance to the lethal agent.

The work by Geeraerd *et al.* (2005) includes a complete review of existing inactivation models describing the different inactivation patterns by combining linear and nonlinear modelling approaches: (i) linear curves, (ii) shoulder + linear decrease, (iii) curves displaying a so-called tail after a log-linear decrease, (iv) survival curves displaying both shoulder and tailing behaviour, (v) concave curves,

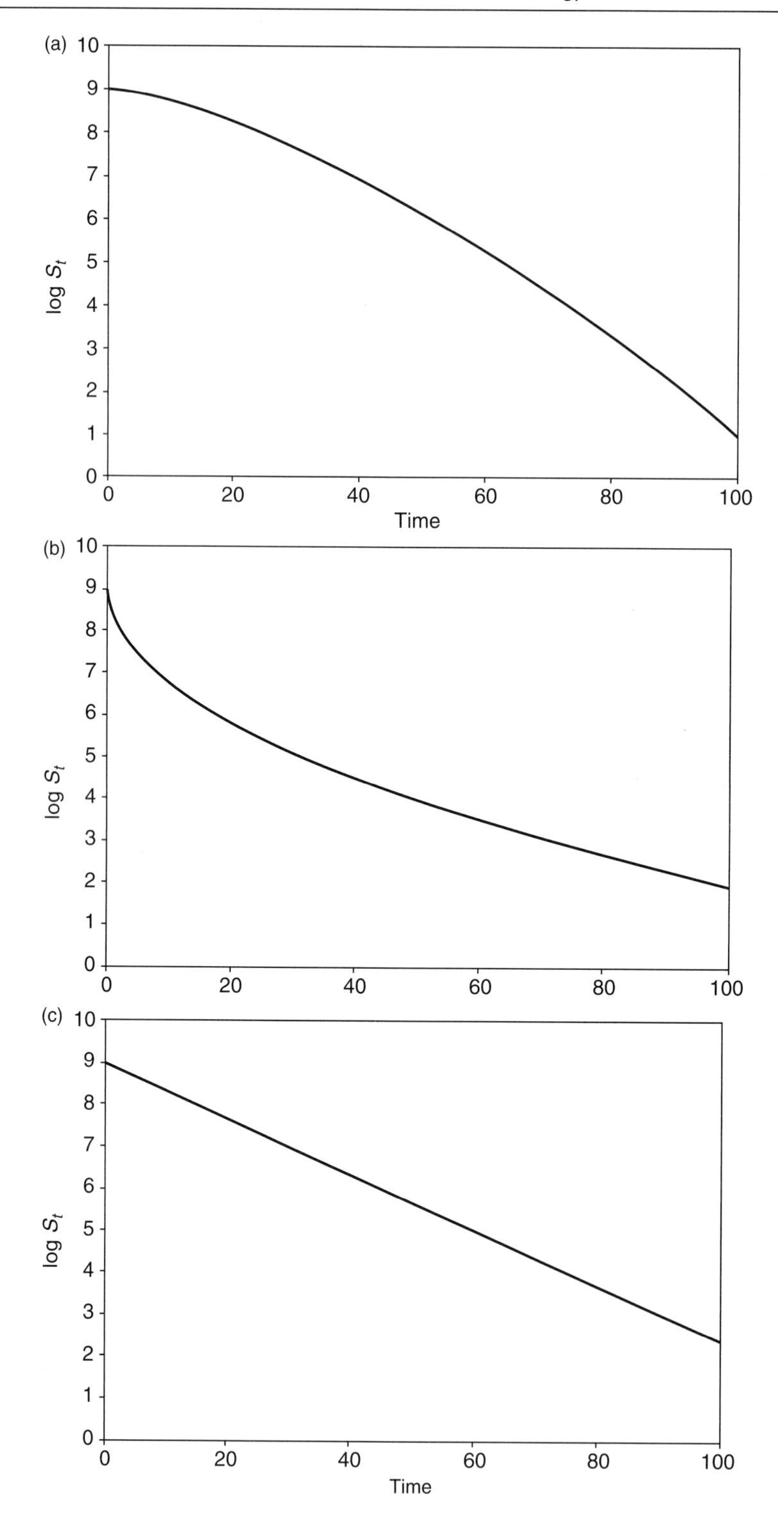


**Figure 18.6** Graphical examples of Weibull models showing downward concavity ($\beta > 1$) (a), upward concavity ($\beta < 1$) (b) and straight line ($\beta = 1$) (c).

(vi) convex curves, (vii) convex/concave curves followed by tailing, (viii) biphasic inactivation kinetics and (ix) biphasic inactivation kinetics preceded by a shoulder.

On the basis of the power law, similar to the Weibull model, a nonlinear model can be proposed to represent a shoulder/tailing function, such as:

$$\log \frac{N}{N0} = -\frac{t^p}{D} \tag{18.27}$$

where $p$ is the power which takes a concave curve when it is lower than one and a concave curve (shoulder curve) when it is higher than one.

Other modelling approaches use logistic functions, in which their inactivation form is a Fermi equation. For the quantification of sigmoid decay curves, the following expression is used:

$$\log S_t = -\log\{1 - \exp[k(t - t_c)]\} \tag{18.28}$$

where $S_t$ is the survivor ratio at time $t$, $k$ is the reduction or inactivation rate and $t_c$ is the lag time or the period of time corresponding to the shoulder.

Other sigmoid models commonly used for representing for microbial growth can be also applied to describe inactivation patterns, such as the modified Gompertz equation and Baranyi's model, through which both shoulder and tailing can be modelled.

As an example of more complex inactivation model, the model developed by Geeraerd *et al.* (2000) is highlighted, including shoulder/tail inactivation based on the physiological state of cells and the residual population density (tail region).

$$N = \left[(N_0 - N_{res})\exp(-k_{\max}\ t)\frac{\exp(-k_{\max}\ t_L)}{1 + \exp((-k_{\max} t_L) - 1)\exp(-k_{\max} t)} + N_{res}\right] \tag{18.29}$$

where $N$ is the number of microorganisms surviving at time $t$, $N_0$ is the initial microbial load, $k_{max}$ is the maximum specific decay rate, $t_L$ is the time prior to inactivation and $N_{res}$ is the residual population density.

## SECONDARY MODELS

Secondary models describes the effect of environmental factors on a certain kinetic parameter, particularly growth rate and lag time (Figure 18.1). Nonetheless, the latter is less frequent due to lag-specific secondary models being less accurate, as a consequence of this kinetic parameter being strongly influenced by pre-culture conditions (Swinnen *et al.*, 2004). Therefore, the review of models carried out in this section is especially focused on the parameter growth rate, although many of them can be also applied to lag time.

### Square root models

The square root models belong to the family of Bêlehrádek-type models (McMeekin *et al.*, 1993a; Ross, 1993). The typical Bêlehrádek model can be formulated as shown in Equation 18.30, in which the growth rate constant *(k)* increases with the increase of the difference between the temperature ($t$) and the temperature of biological zero ($t_0$), that is the temperature at which there is no growth. The value of $a$ establishes the degree of proportionality between both terms and $d$ determines the magnitude of the increase of the growth rate constant, that is the higher the value of $d$, the higher the value of $k$ for a same difference $t - t_0$. The values of $a$, $t_0$ and $d$ are regression parameters.

$$k = a \cdot (t - t_0)^d \tag{18.30}$$

Ratkowsky *et al.* (1982) proposed a variation of the Bêlehrádek model to explain microbial growth at suboptimal temperatures in which growth rate (i.e. the rate constant in Belehradek model) was square root transformed in order to stabilize variance. The resultant model is the so-called root-square model or Ratkowsky model. In addition, the term 'biological zero' was substituted by minimum temperature ($T_{min}$), which stands for the theoretical temperature at which growth is zero, corresponding with the intercept in the equation (Figure 18.7). The observed minimum temperature supporting growth is often 2 or 3 °C higher than $T_{min}$; hence, to avoid misinterpretation, this term is referred to as a notional or conceptual temperature (Ratkowsky *et al.*, 1983; McMeekin *et al.*, 1993a):

$$\sqrt{\mu_{max}} = b \cdot (T - T_{min}) \tag{18.31}$$

where $\mu_{max}$ is the maximum specific growth rate expressed in log cfu/ml, $b$ is a constant and $T$ is the temperature. The parameter $T_{min}$ is the minimum theoretical temperature at which growth is detected.

Ratkowsky *et al.* (1983) also expanded Equation 18.31 to include the temperature range over the optimal growth temperature, that is the whole biokinetic temperature range (Equation 18.32). In the expanded model, the parameters $c$ and $T_{max}$ are included representing, respectively, the proportionality constant and temperature in the upper part of the range over which growth is not possible (Figure 18.7).

$$\sqrt{\mu} = b \cdot (T - T_{min}) \cdot (1 - \exp(c(T - T_{max}))) \tag{18.32}$$

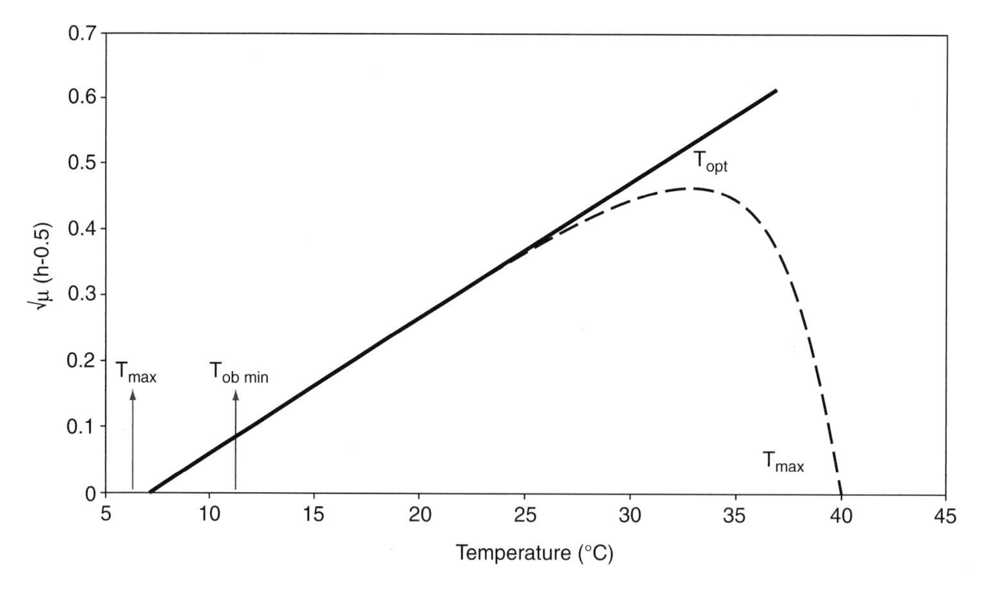


**Figure 18.7** Graphical representaton of the square-root-type model for the whole biokinetic range defined by the notional maximum and minimum temperature, corresponding to $T_{max}$ and $T_{min}$, respectively. The optimal temperatue ($T_{opt}$) is found at the maximum in the graph (optimal growth rate). The observed minimum temperature ($T_{ob\ min}$) is usually 2–3 °C higher than the notinal minimum temperature ($T_{min}$).

This mathematical basis could be extended to include the effect of other environmental factors ($a_w$, pH, $CO_2$, etc.). For example McMeekin *et al.* (1987) proposed a growth model for *Staphylococcus xylosus* in function of temperature and water activity, adding a multiplicative term for the new factor:

$$\sqrt{\mu_{\max}} = b(T - T_{\min})\sqrt{(a_w - a_{w\,\min})} \tag{18.33}$$

In this equation $a_{w\,\min}$ is the theoretical minimum $a_w$ below which growth is not possible

The same example, but for the factor pH, was developed by Adams *et al.* (1991), who found the same multiplicative effect for pH.

$$\sqrt{\mu} = b \cdot (T - T_{\min}) \cdot \sqrt{(pH - pH_{\min})} \tag{18.34}$$

In this equation $pH_{\min}$ is the theoretical minimum *pH* below which growth is not possible.

Also, for the environmental factors $a_w$ and pH, additional terms were suggested accounting for the entire biokinetic range. For example Miles *et al.* (1997) developed a square root model for growth of *Vibrio parahaemolyticus* as a function of temperature and $a_w$ in the whole biokinetic range.

$$\sqrt{\mu} = b \cdot (T - T_{\min}) \cdot (1 - \exp(c(T - T_{\max}))) \cdot \sqrt{(a_w - a_{w\,\min}) \cdot (1 - \exp(d(a_w - a_{w_{\max}})))} \tag{18.35}$$

Here, $a_{w\,\min}$ and $a_{w\,\max}$ correspond to the minimum and maximum water activity below and above which growth is not possible.

By analogy with the model developed by McMeekin *et al.* (1987) (Equation 18.33), Wijtzes *et al.* (2001) expanded the square root model to take into account the whole biokinetic pH range:

$$\sqrt{\mu} = b \cdot (T - T_{\min}) \cdot \sqrt{(a_w - a_{w\,\min})} \cdot \sqrt{(pH - pH_{\min})} \cdot \sqrt{(pH - pH_{\max})} \tag{18.36}$$

Other factors have been included, such as organic acids (e.g. lactic acid), carbon dioxide and phenol (Devlieghere *et al.*, 2000, 2001; Ross *et al.*, 2000; Tienungoon *et al.*, 2000; Gimenez and Dalgaard, 2004). In some cases, the square root model has been applied also to describe lag time (Mataragas *et al.*, 2006; Sant'Ana *et al.*, 2012).

Square root models are widely applied in predictive microbiology and it is considered the most important group of models together the cardinal-type models (Ross and Dalgaard, 2004). Among its strengths, can be highlighted the interpretability of the model parameters ($T_{\min}$, $pH_{\max}$, etc.) which facilitates a more efficient fitting procedure and enables an easier expansion of the model by considering certain parameter constants as if they were cardinal parameters. Another advantage is that square root models can be easily adapted to encompass the whole biokinetic range of environmental factors, as mentioned. From the user side, square root models can be easily applied due to their simplicity, making them more attractive to predictive microbiology practitioners.

## Arrhenius-type models

Arrhenius-type models are considered fundamental in different scientific fields as a means of explaining the effect of temperature on reaction rates of chemical and/or biological processes. A well-known and simplified mathematical form of this typology of equation is:

$$k = A \cdot e^{(E_a/RT)} \tag{18.37}$$

where $k$ is the reaction rate, $A$ is the collision factor, $E_a$ is the activation energy of the reaction, $R$ corresponds to the gas constant (8.314 J/K/ml) and $T$ is the temperature in Kelvin.

The use of this equation as a growth model for bacteria is underpinned by assuming that there is an enzyme in a metabolic pathway that is key to the bacterial specific growth rate, including lower and higher temperatures than the optimum temperature for growth (McMeekin *et al.*, 1993a).

Equation 18.37 can be linearized taking logarithms of both sides yielding the expression:

$$\ln(k) = \ln(A) + \frac{E_a}{RT} \tag{18.38}$$

The main drawback of the simple Arrhenius model is that it describes a straight line when 1/T versus $\ln(k)$ are represented. However, as proved by Ratkowsky *et al.* (1982), both terms are not proportional with real growth data, resulting in a curvilinear relationship when observations for both terms are plotted (i.e. Arrhenius plots).

A modification of the Arrhenius equation was formulated by Eyring (1935) based on the theory of absolute reaction rates:

$$k = KT \cdot e^{(\Delta H^{\ddagger}/RT)} \tag{18.39}$$

here $\Delta H^{\ddagger}$ is the enthalpy difference between the transition state complex and the reactants and $K$ is similar to $A$, but includes steric and entropic effects.

Unlike the model represented by Equation 18.37, this formulation, when the logarithm is taken, does reflect the curvilinear relationship between reaction rate and temperature often observed in microorganisms.

In both Arrhenius-type models for bacterial growth, $K$, $A$, and $E_a$ are regression parameters.

These basic equations have been modified by numerous authors including mechanistic and/or empirical modifications in order to better describe the effect of temperature on microbial growth in the entire biokinetic temperature range. In spite of the improvements introduced in Arrhenius-type models, this type of model still has not found a wider application in predictive microbiology. Although the purpose of this chapter is not to carry out a detailed review of this type, for which it is recommended that the excellent reviews by McMeekin *et al.* (1993b) and Ross and Dalgaard (2004) are consulted, some examples of modified Arrhenius-type equations are highlighted here.

The modification by Ross of the Arrhenius model can be representative of a set of other variants focused on explaining bacterial growth as dependent on the kinetic of conforming of a key enzyme in a bottleneck metabolic reaction determining growth. These models make an attempt to model the enzyme activity using thermodynamics of enzymatic shape and conformation. The equation derived by Ross (1999) (Ross and Dalgaard, 2004) is:

$$k = \frac{CT \cdot e^{(\Delta H^{\ddagger}/RT)}}{1 + e^{\left(-n\left(\Delta H^{*} + T\Delta S^{*} + AC_p[(T - T^{*}{}_H) - T\ln(T/T^{*}{}_S)]\right)/RT\right)}} \tag{18.40}$$

here $C$ is a regression parameter, $\Delta H^{\ddagger}$ is the activation enthalpy of the reaction catalysed by the key enzyme, $AC_p$ is the difference in heat capacity per mole amino acid residue between the active and denatured state of the enzyme, $T^{*}{}_H$ is the temperature (K) at which the $AC_p$ contribution to entropy is zero, $\Delta H^{*}$ is the value of enthalpy at $T^{*}{}_H$ per mole amino acid residue, $\Delta S^{*}$ is the value of entropy at $T^{*}{}_S$ per mole amino acid residue and $n$ is the number of amino acid residues in the enzyme (Ross and Dalgaard, 2004).

On the other hand, Davey (1989) proposed a linear Arrhenius model, named the Davey model, to describe growth dependent on temperature: it takes the mathematical form:

$$\ln(k) = C_0 + \frac{C_1}{T} + \frac{C_2}{T^2} + C_3 a_w + C_4 {a_w}^2 \tag{18.41}$$

here $C_0$, $C_1$, $C_2$ and $C_4$ are regression parameters, $T$ and $a_w$ are temperature (K) and water activity and $k$ corresponds to growth rate.

Later, the same authors expanded the model described by Equation 18.41 to include additional environmental factors in a general format of the model (Davey, 1994):

$$\ln(k) = C_0 + \sum_{i=1}^{j} \left(C_{2i-1}V_i + C^2 V_i\right) \tag{18.42}$$

in which $j$ environmental factors, V, are combined in a summation with $C_0$, $C_1$, $C_2$, . . . $C_j$ regression coefficients.

Although less often than with other secondary models, the Davey model or the linear Arrhenius model and its variants have been applied in scientific studies to reflect mould growth (Samapundo *et al.*, 2005; Baert *et al.*, 2007; Silva *et al.*, 2010) and bacterial growth and inactivation (Davey and Daughtry 1995; Cerf *et al.*, 1996; Amos *et al.*, 2001). Other Arrhenius-type models have had more extensive application in predictive microbiology (Gibson *et al.*, 1988; Zwietering *et al.*, 1990; te Giffel and Zwietering, 1999; Fujikawa and Morozumi, 2005). Koutsoumanis and Nychas (2000) and Koutsoumanis *et al.* (2006) applied a modified Arrhenius-type model to describe spoilage microflora bacterial growth in fresh fish and meat products and Koutsoumanis *et al.* (2000) included as an additional environmental factor carbon dioxide for spoilage microflora growth in fish. More recently, Huang *et al.* (2011) carried out a modification of the Arrhenius-type models developed by McMeekin *et al.* (1993a) and Ratkowsky *et al.* (1982) by including a exponent to improve the fit to growth data, being demonstrated in different microorganisms and food matrices:

$$\mu = AT \cdot e^{(\Delta G'/RT)^{\alpha}} \cdot \left[1 - e^{(B(T - T_{\max}))}\right] \tag{18.43}$$

where $\mu$ is growth rate, $AG'$ is equivalent to energy term expressed in J/mol and $\alpha$ is an exponent. $A$, $AG'$, $T_{\max}$, $B$ and $\alpha$ are regression parameters.

## Polynomial models

Polynomial models, also named response surface models (RSM) are empirical models fitted to kinetic data by a linear or nonlinear regression procedure. These models can include multiple environmental factors (variables) the most often being second-order polynomial models in which environmental effects can be described with: first-order terms, second-order terms, and interaction terms as described by Equation 18.44. A hypothetical second-order polynomial model might be:

$$\text{Kinetic parameter} = C_0 + C_1X_1 + C_2X_2 + C_3X_1X_2 + C_4{X_1}^2 + C_5{X_2}^2 + C \tag{18.44}$$

where $C_0$, $C_1$, $C_2$ and $C_4$ are the regression coefficients and $X_1$ and $X_2$ are the environmental factors (e.g. pH, temperature, etc.)

There are numerous scientific works applying polynomial models for growth rate, lag time and maximal population density of pathogenic and spoilage microorganisms in foods and artificial media

(Guerzoni *et al.*, 1994; te Giffel and Zwietering, 1999; Lebert *et al.*, 2000; Pin *et al.*, 2000; Silva *et al.*, 2010). Their wide application is due to both their good fit to experimental data and the simplicity of their mathematical development. However, certain aspects limit the usefulness of this type of model. The main drawbacks are the lack of biological interpretation of regression coefficients and the possible overfitting, which leads to model experimental error. Baranyi *et al.* (1996) indicated that extrapolation should be avoided, since it can lead to erroneous predictions due to the great flexibility shown by polynomial models. According to these authors, the interpolation region is defined by a minimum convex polyhedron (MCP), which can be remarkably smaller than the nominal variable space delimited by the endpoints of the ranges of environmental factors (Ross and Dalgaard, 2004). As a positive aspect, polynomial models are easy to implement in spreadsheet software, even though the narrow interpolation region and the impossibility of carrying out extrapolations reduce the possibility of expanding the model to new ranges or environment factors.

## Cardinal-type models

Cardinal-type models are a set of secondary models widely applied within predictive microbiology. In these models, regression parameters embody a biological meaning, thus facilitating the fitting process (i. e. the starting values) to experimental data or enabling the use of theoretical values from literature or experimental data set to define the model. These models are built on the hypothesis that environmental factors have an independent effect of maximum growth. Among the most significant contribution to this type we highlight the cardinal parameter model introduced by Lobry *et al.* (1991) (Rosso *et al.*, 1993, 1995), which were specific to temperature and were so-called cardinal parameter model with inflection (CPMI) by the authors. On the basis of the original CPMI, other works have developed and applied effective variants including different factors (pH, $a_w$, organic acid, phenol, synergy among factors, competitive microflora, etc.) for different microorganisms (bacteria, moulds and yeasts) and food matrices (Augustin and Carlier, 2000a, 2000b; Gimenez and Dalgaard, 2004; Sanaa *et al.*, 2004; Dantigny *et al.*, 2007; Zuliani *et al.*, 2007; Van Impe *et al.*, 2011; Gougouli and Koutsoumanis, 2012). A general form of the CPM model can be given as (Ross and Dalgaard, 2004):

$$\mu = \mu_{opt} \cdot CM_2(T) \cdot CM_2(a_w) \cdot CM_1(pH) \cdot \prod_{i=1}^{n} \gamma(c_i) \cdot \prod_{j=1}^{n} k_j \tag{18.45}$$

$$\mathrm{CM_n} = \begin{cases} 0, & X \leq X_{\min} \\ \dfrac{(X - X_{\max}) \cdot (X - X_{\min})^n}{(X_{\mathrm{opt}} - X_{\min})^{n-1}\left[(X_{\mathrm{opt}} - X_{\min}) \cdot (X - X_{\mathrm{opt}}) - (X_{\mathrm{opt}} - X_{\max}) \cdot ((n-1) \cdot X_{\mathrm{opt}} + X_{\min} - n \cdot X)\right]}, & X_{\min} < X < X_{\min} \\ 0, & X \geq X_{\max} \end{cases} \tag{18.46}$$

$$\gamma(c_i) = \begin{cases} (1 - c_i/MIC_i)^2, & c_i < MIC_i \\ 0, & c_i \geq MIC_i \end{cases} \tag{18.47}$$

where $X$ represents temperature, water activity or pH. $X_{\min}$ and $X_{\max}$ are, respectively, the values of $X_i$ below and above which growth is not possible, $X_{\mathrm{opt}}$ is the value at which $\mu$ is equal to its optimal value, that is $\mu_{opt}$. $MIC_i$ is the minimal inhibitory concentration of specific compounds above which there is no growth.

The other type of cardinal model, also widely used in predictive microbiology is the gamma (γ) concept, of which a more detail analysis is provided in this section, even although it exhibits important similarities with the above-mentioned CPMI. The gamma concept or model was proposed by Zwietering *et al.* (1992) for which two important principles were invoked:

(i) Environmental factors affect bacterial growth independently. Therefore, their overall effect can be modelled by multiplying the individual effect of each factor.
(ii) The inhibitory effect of environmental factors on growth rate can be measured as the rate of the optimum growth rate (when environmental factors are at the optimal level) and growth rate at a specific environmental condition.

Therefore, the overall inhibitory effect of environmental factors can be made explicit as:

$$\prod_{i=1}^{n} \frac{\mu_i}{\mu_{i\,\mathrm{opt}}} \tag{18.48}$$

in which $i = 1 \ldots . n$ represents the number of environmental factors.

The relative inhibitory effect of each environmental factor can be described by a growth factor or function 'gamma'. The function takes the value of one at optimal conditions and when the inhibitory effect is at its maximum it takes the value zero.

The multiplication of the gamma functions for each environmental factor provides the total accumulated inhibitory effect, so that it can be formulated:

$$\mu = \mu_{\mathrm{opt}} \cdot \gamma(T) \cdot \gamma(pH) \cdot \gamma(a_w) \cdot \gamma(c) \tag{18.49}$$

here the gamma functions $\gamma(T)$, $\gamma(pH)$, $\gamma(a_w)$ and $\gamma(c)$ represent for the inhibitory effect from temperature, pH, $a_w$ and any additional environmental factor, respectively.

The gamma functions in Equations 18.49, 18.50 and 18.51 describe the inhibitory effect of one specific environmental from its optimum value to the maximum and/or minimum value at which inhibition is complete following a specific mathematical relationship, which is elucidated from experimental data.

$$\gamma(T) = \left( \frac{T - T_{\mathrm{min}}}{T_{\mathrm{opt}} - T_{\mathrm{min}}} \right)^2 \tag{18.50}$$

$$\gamma(pH) = \frac{(pH - pH_{\mathrm{min}}) \cdot (pH_{\mathrm{max}} - pH)}{(pH_{\mathrm{opt}} - pH_{\mathrm{min}}) \cdot (pH_{\mathrm{max}} - pH_{\mathrm{opt}})} \tag{18.51}$$

$$\gamma(a_w) = \frac{a_w - a_{w\,\mathrm{min}}}{1 - a_{w\,\mathrm{min}}} \tag{18.52}$$

In relation to its application, Zwietering (2002) has pointed out that the gamma concept produces reasonable results except for those specific cases in which a clear interaction between factors is given, such as, for example between pH and weak acids. In this sense, a more recent study by Biesta-Peters *et al.* (2011) has endorsed the validity of the gamma concept for explaining the effect of pH and $a_w$ so even interactions are observed. Other authors have challenged the gamma concept against possible interaction or synergy among factors; their results seem to support the gamma concept as a suitable secondary model in such situations (Lambert and Bidlas, 2007; Bidlas and Lambert, 2008).

The gamma model is highly applicable in quantitative microbial risk studies, since it can be expanded with new terms and factors. This means that modifications can be made easily, in order to assess new factors, without having to change the whole risk model. One of the characteristics that a quantitative microbial risk studies should possess is that it should be dynamic and flexible, in the sense of including new information and new approaches that enable the process to be improved continually.

## Artificial neuronal networks

Artificial neural networks (ANN) is an artificial intelligence tool that has become increasingly popular in different scientific areas, including those dealing with biological systems. ANN methods have been successfully applied to model growth kinetics and growth/no-growth interfaces of different micro-organisms (e.g. *L. monocytogenes*, *E. coli*) (Schepers *et al.*, 2000; Valero *et al.*, 2007; Oscar, 2009). ANN is inspired in the functioning of neurons in human brain. ANN are empirical models able to explain dependency between explanatory variables and response variables irrespective to both the nonlinearity level between variables and independence and normality assumptions, which are important constrains for other regression methods such as LS. ANN derive arbitrary nonlinear multiparametric discriminant functions directly from experimental data (Almeida, 2002). There are four elements that characterize a neural network: topology, regression method, type of association between inputs and output and how information is presented. ANN models are made up by three kinds of computational neurons: input, hidden and output neurons (Figure 18.8). The input neurons are intended to collect information or signals coming from an external device or sensor while the output neurons are in charge of providing the ANN answer. Finally, hidden neurons are isolated from the exterior and are situated between input and output neurons. Neurons are also known as nodes and they are organized in layers. Connections between nodes are built on a basic function and activation function and the strength of each connection between nodes is represented by a weight which is modified during the learning process. The topology of ANNs can vary

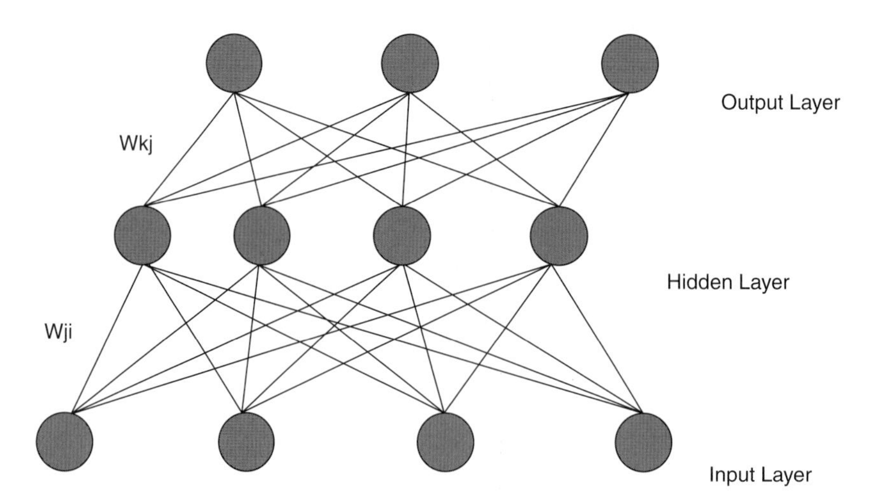


**Figure 18.8** Basic stracture of Artificial Neural Networks based on three layers: Output (k), Hidden (j) and Input (i). Weights assigned to connections between nodes or nerons are represented by W.

depending on the number of layers and nodes used and how they are organized and interconnected in the network. The main advantage of an ANN approach is that it can model complex systems based on multiple equations or mathematical processes. In contrast, ANN models are difficult to handle and interpret as a consequence of their great complexity apart from the fact that practitioners are required to have a thorough knowledge of ANN concepts for optimal application.

### Secondary models for microbial inactivation

The secondary models for microbial inactivation kinetic are mainly based on z-value concept, which consists of assuming a straight line relationship between log D-value and temperature (Equation 18.52). In this case, the reciprocal of the slope is equal to the z-value (Equation 18.53), which is the temperature increase needed to reduce $D$ by a factor 10, so to increase the destruction rate by a factor 10.

$$D = D_{ref} 10^{\left(\frac{T_{ref}-T}{z}\right)} \tag{18.53}$$

$$z = \frac{T_{ref} - T}{\log D - \log D_{ref}} \tag{18.54}$$

where $D_{ref}$ is the D-value at the reference temperature, $T_{ref}$ (the usual $T_{ref}$ is 121.1 °C). D-values could be influenced by the type of organism (strain), treatment temperature, physiological state of cells, pH, fat and $a_w$ content.

Also, the Arrhenius equation has been used to represent the dependency of the inactivation rate on temperature:

$$k = A \cdot e^{(-E_a/RT)} \tag{18.55}$$

where $Ea$ is the activation energy, $R$ the universal gas constant and $T$ is the temperature in degrees Kelvin and A is the collision factor, which is a regression parameter. The Arrhenius model has been used for describing the kinetic behaviour of *E. coli* as a function of temperature, pH and $a_w$ (Cerf *et al.*, 1996).

Another straightforward alternative is to apply multivariate regression to derive polynomial functions as secondary models. This approach allows not only the effect of temperature to be considered but also other environmental factors, such as pH or water activity.

In relation to the Weibull model parameters, secondary models are subjected to the dependency of both parameters $\beta$ and $\alpha$ on temperature. Peleg (2006a) proposed different modelling approaches for considering secondary modelling for such parameters consisting of using a log-logistic function, discrete models or more empirical models employing those functions which best fit to data.

## OTHER PREDICTIVE MODELS AND GROWTH FACTORS

Although most secondary models are specific to growth rate, scientific literature also describes models for lag time and, to a lesser extent, MPD (maximum population density). In the case of lag time, on the basis of the relationship between lag time and growth rate denoted by Baranyi and Roberts (1994), a straightforward approach can be taken to model this parameter (Koseki and Isobe, 2005):

$$lag = h_0/\mu_{max} \tag{18.56}$$

in which $h_0$ is a parameter derived from the primary model proposed by Baranyi and Roberts (1994) defining the 'work to be done' by cells to adapt to the new environment and start to grow (i.e. exponential phase) $h_0 = \ln\left(1 + \frac{1}{q_0}\right) = \mu_{\max}\lambda$. In these cases, $h_0$ is treated as a regression parameter.

Other approaches have been used based on this relationship between both kinetic parameters, enabling the use of secondary models such as square root models and Arrhenius-type models for growth rate (Cardenas *et al.*, 2008; Yoon *et al.*, 2008; Sant'Ana *et al.*, 2012). The downside to this deterministic model and other similar approaches is that they are much less precise and accurate in their predictions than secondary models for growth rate, because lag time is not only influenced by environmental factors but also by pre-incubation conditions (i.e. environmental conditions prior to the experiment) (McMeekin *et al.*, 1993b; Baranyi *et al.*, 1995; Mellefont and Ross, 2003; Swinnen *et al.*, 2004). Hence, alternative approaches have been proposed to better capture lag behaviour, based on stochastic modelling and single-cell based models, even though results are still not as precise as desired. This inability to make reliable predictions, together with the high degree of variability associated with the parameter, makes the use of probability distributions the most appropriate approach (Métris *et al.*, 2005).

Secondary models predicting MPD are also considered in the literature although to a lesser extent than growth rate and lag time secondary models. It should be noted that MPDs obtained in culture medium cannot be extrapolated to foods, since liquid matrices (i.e. plancktonic cells) enable higher population levels to be reached; therefore, this parameter should be derived from food matrices themself or systems simulating solid foods. In scientific literature, polynomial functions have been mostly used to model this parameter (Koseki and Isobe 2005; Zurera-Cosano *et al.*, 2006; Pérez-Rodríguez *et al.*, 2007b; Carrasco *et al.*, 2010).

Other important factors to be considered when modelling growth are the effect of accompanying microbial flora or between-species interaction. Not considering competitive microflora might lead to overestimate growth by between one and two logarithms. The mechanistic approach is a barely explored alternative to account for between-species interaction and, as pointed out by Leroi and De Vuyst (2007), this approach can lead to complex mathematical models with multiple interrelated variables. Therefore, more simplistic approaches have been preferred for quantifying how much the growth of one population is reduced based on the effect of inhibitory substances produced from competitive microorganisms. These inhibitory factors can be included in secondary models as additional terms to account for interaction between bacterial species. The cardinal-type and gamma models have been used for such a purpose (Leroi *et al.*, 2012). Other modelling approach taken by some authors derives from the Jameson effect (Jameson, 1962), which states that only high levels of competitive flora have a significant effect on the growth of pathogens. These models (i.e. primary models) incorporate a deceleration function which reduces the growth rate of one microorganism as the other microorganism population rises (Cornu *et al.*, 2002, 2011; Gimenez and Dalgaard, 2004; Mejlholm and Dalgaard, 2007). The interaction models based on the Lotka–Volterra type model are also valid means of considering interaction between different bacterial populations, which consists of including, in the primary model, as many inhibitory functions as bacterial populations (Vereecken *et al.*, 2000; Vereecken and Van Impe, 2002; Powell, 2004). The reader might obtain a more thorough review of existing models considering between-species interactions by consulting the work by Leroi and De Vuyst (2007).

Another important aspect in terms of kinetic models is the modelling of growth in dynamic conditions, in other words, when the levels of factors change as a function of time (Bernaerts *et al.*, 2004). Swinnen *et al.* (2004) highlighted the need to develop these types of models, since those applied in static conditions can make mistakes in their predictions. However, even though they do exist in the bibliography (Baranyi and Roberts, 1994), they have not yet been applied to quantitative assessment, probably due to their mathematical complexity and the lack of models available for certain microorganisms.

Transfer models are a new class of predictive model aimed at modelling cross-contamination events in food-related environments (Schaffner, 2004; Pérez-Rodríguez *et al.*, 2008). These models essentially describe a physical phenomenon consisting of the transfer of cells from one environment to another, in which one could be a food. Still, the development of transfer models is at an early stage; most of them are empirical models based on the use of transfer rates. Transfer rates (Tr(%)) can be defined as the proportion of cells transferred from one surface to another, which can be mathematically formulated as:

$$Tr(\%) = \frac{cells\ on\ recepient\ surface}{cells\ on\ donnor\ surface} \times 100 \tag{18.57}$$

There is a great variety of transfer models describing cross-contamination during food handling or slicing (from slicer to food and vice versa) (Chen *et al.*, 2001; Montville *et al.*, 2001; Montville and Schaffner, 2003; Vorst *et al.*, 2006a, 2006b; Mylius *et al.*, 2007; Pérez-Rodríguez *et al.*, 2007b). Few transfer models have been built on a more mechanistic basis, such as, for example by considering different transfer compartments and bacterial populations or using a probabilistic approach to account for bacterial transfer (Aziza *et al.*, 2006; Rodríguez *et al.*, 2011; Møller *et al.*, 2012). An important aspect that should be noted when models are put in action is that cross-contamination is an additive phenomenon, in contrast to microbial kinetic in which processes occur logarithmically (i.e. exponential process). Therefore, when transfer models are applied, concentration values should be used in arithmetic scale and not in logarithmic one.

## PROBABILITY MODELS

The probability models, also named growth/no growth models, are intended to define the growth boundaries or/and determine those conditions supporting or not supporting growth. As mentioned by McMeekin *et al.* (2000) growth/no growth models allow quantification of the effects of various hurdles (by applying the hurdle concept) on the probability of growth and the combinations at which the growth is impeded to be defined. The probability models were first proposed to predict the probability of formation of staphylococcal enterotoxin and botulinum toxin (Genigeorgis 1981; Gibson *et al.*, 1987). In the 1990s, this topology experienced a notorious development providing several probability models for different pathogens and exploring new modelling approaches (Presser *et al.*, 1997; Bolton and Frank, 1999; Tienungoon *et al.*, 2000). Probability models have also dealt with spoilage microorganisms such as lactic acid bacteria (Masana and Baranyi, 2000; Zurera-Cosano *et al.*, 2006). Among the different types of probability models, highlighted are:

- Deterministic approach
- Logistic regression
- Artificial neural networks (ANN).

The deterministic approach tends to model specific responses such as, for example the temperature values resulting in no growth as a function of other environmental factors such as water activity or pH. The earliest example of this approach was the model proposed by Pitt (1992); it estimated the temperature/water activity interface for aflatoxin production and *Aspergillus* spp. growth. In this case, the interaction between both environmental factors was modelled to derive the upper and lower temperature limits for growth. Masana and Baranyi (2000) modelled the growth interface for *Brocothrix thermosphacta* by deriving through interpolation the midpoint between growth and no growth observations and fitting a polynomial model. Other works have derived the growth/no growth boundary based on cardinal-type models including interaction terms through which factor combinations yielding no growth are derived and represented in contour plots (Augustin and Carlier, 2000a; Le Marc *et al.*, 2002; Zuliani *et al.*, 2007; Polese *et al.*, 2011).

The use of logistic regression allows the binary outcome used in these models, that is growth and no growth, to be considered as the categorical variable. Using the logistic function and its transformation as the logit function (i.e. log-odd ratio) (Equation 18.56), the logistic regression can describe the probability of a Bernuilli event, in our case the probability of growth (1) and no growth (0):

$$\text{Logit}(P) = \ln(P/1 - P) \tag{18.58}$$

$P$ being the probability of growth defined by a logistic function, as presented in Equation 18.57:

$$P = \frac{e^Y}{1 + e^Y} \quad or \quad \frac{1}{1 + e^{-Y}} \tag{18.59}$$

$$\text{logit}\,(P) = a_0 + \sum_{i=1}^{n} a_i x_i + \sum_{i=j}^{n} a_{ij} x_{ij}^2 + \sum_{i \neq j}^{n} a_{ij} x_i x_j, \quad i,j = 1, \ldots . n \tag{18.60}$$

The first work introducing the use the logistic regression to describe growth/no growth interfaces was that developed by Ratkowsky and Ross (1995). In this case, the square root type function was used. However, subsequent works dealing with logistic regression have preferred to apply polynomial functions, whose general form is represented by Equation 18.58. Both linear and nonlinear regression methods have been applied to derive logistic functions; notwithstanding that the nonlinear approach seems to yield more accurate models, specially when notional or cardinal parameters are included in the model (Ratkowsky, 2002). Among the works applying logistic regression models, the study developed by Valero *et al.* (2009), which reported growth boundaries of *Staphylococcus aureus* considering the effect of temperature, pH and water activity by using a second-order linear logistic model, can be highlighted. Also, McKellar and Lu (2001) and Skandamis *et al.* (2007) published both growth/no growth models for *E. coli* O157:H7 under different environmental conditions (temperature, pH, acetic acid, etc.) and including pre-culture conditions (acid adaptation).

ANN models have been demonstrated to be a suitable approach to model growth probability as its high flexibility and accurate. Different ANN methods have been applied, such as probabilistic neural network (PNN) (Hajmeer and Basheer, 2002) and product unit neural network (PUNN) (Valero *et al.*, 2007). More recently, classification algorithms have been introduced to model growth probability (Fernández-Navarro *et al.*, 2010). In this case, a radial basis function neural network (RBFNN) is applied on growth/no growth data in order to create a multiclassification model that was able to predict the probability of belonging at three different classes: growth, growth transition and no growth.

To assess the performance of probability models different statistics have been used, such as the Hosmer–Lemeshow statistic based on comparing the observed and the expected number of events by using the $\chi 2$ Pearson coefficient. Lower values of the Hosmer–Lemeshowstatistic indicate a better fit. The proportion concordant/discordant/tied and its representation in a receiver operating characteristic curve (ROC) allow the discrimination capacity of the model to be established. The closer the value of ROC is to one, the greater is the discrimination capacity. The Pearson residuals is other index which measures the difference between observed and predicted events, taking into account the number of observations.

## MODEL VALIDATION

Validation can be defined as the process whereby the capacity of a model to predict the behaviour of the real system is assessed. In predictive microbiology, most models are generated in artificial culture media, since it enables multiple environmental conditions to be assayed in a cost-effective manner, so reducing time and human resources. Hence, predictive models should be applied provided a validation process confirms the suitability of the model to predict microbial behaviour in foods. The validation process is often applied to predictions from secondary models and, especially, to growth rate as it is importance to food safety. Validation is divided into two steps, internal and external validation. The former refers to those mathematical processes aimed at ascertaining if the proposed model is able to describe observations from the same system, that is observations coming from the experiment used to derive the model. This type of validation is widely evident in predictive microbiology, since it is first necessary to confirm the correct choice of the model function (García-Gimeno *et al.*, 2002; Zurera-Cosano G. *et al.*, 2004). The latter process, the external validation, is determinant in relation to the applicability of the model, since in this step the capacity of the model to generalize is assessed. In other words, in this process, the model is challenged against independent data sets, which can be obtained from literature or from different experiments, in many cases performed on real food matrices. Thus, for example a predictive model generated for a specific microorganism in a modified culture medium such as triptone soy broth (TSB) could be validated for a certain food or food category such as milk, cooked meat and so on (Whiting and Buchanan, 1994; te Giffel and Zwietering, 1999; Ross *et al.*, 2000). Predictive microbiology applies different statistical indexes for internal and external validation. The goodness-of-fit indexes based on the proximity between observations and predictions are also used for validation purposes. They are mainly the coefficient of determination ($R^2$) and the root mean square error (RMSE). The former informs on the proportion of the total variability explained by the model, thus the closer the $R^2$ is to one, the better the model represents observations. The RMSE is a standardized measure of model residuals (i.e. least squares) that can be used to evaluate how well the model describes observations. A low RMSE value means better adequacy of the model to describe data.

The confidence intervals or prediction limits (e.g. 95%) are also relevant when models are subjected to a validation process. By considering them, models can be evaluated regarding whether or not observations fall within the prediction error displayed by the model. Models can be satisfactorily validated if most observations are within model prediction intervals.

The validation indexes bias factor ($B_f$) and accuracy factor ($A_f$) were proposed by Ross (1996) to assess if predictive models can correctly describe independent observations obtained from real food matrices.

$$B_f = 10^{\left[\sum \log(gpred/gobs)/n\right]} \quad (18.61)$$

$$A_f = 10^{\left[\sum |\log(gpred/gobs)|/n\right]} \quad (18.62)$$

where *gobs* and *gpred* represent observed and predicted growth rate and *n* stands for the number of validation data.

The bias factor is simply an average ratio of discrete model predictions to observations. This index determines whether the model is fail-safe, fail-dangerous, or perfect. A value of one means that observations are equally distributed above and below predictions, values $<1$ means that predicted values are lower than observations (i.e. fail-dangerous model), whereas values $>1$ indicate that predicted values are higher than observations (i.e. fail-safe model). The acceptable bias factor value for a predictive model can be 0.75–1.25. The accuracy factor is the average of absolute values of the ratio of predictions to observations, thus informing how close predictions are to observations. A value of one indicates perfect

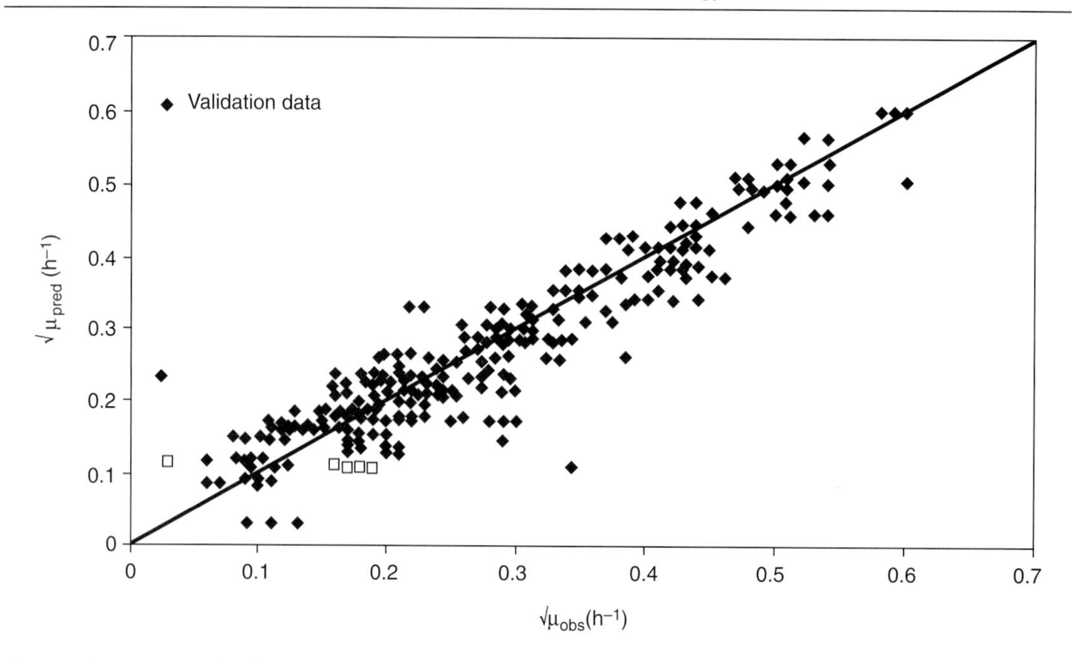


**Figure 18.9** An example of validation plot consisting of representing pairs of predictied and observed values of squared root growth rate (h$^{-1}$) around the equivalence line (i.e. line indicating perfect concordance between prediction and observations). If most of the validation data are above the equivalence line, the model is considered fail-safe while if they are below, the model is fail-dangerous.

concordance whereas a value of two would mean that, on average, predictions are two factors of difference with respect to observations. These validation indexes have been widely used in validation studies (te Giffel and Zwietering, 1999; Nerbrink *et al.*, 1999; Ross *et al.*, 2000; Devlieghere *et al.*, 2001). The evaluation of external validation indexes is usually combined with a visual analysis by plotting observations versus predictions in a scatter plot with a line representing perfect concordance between predictions and observations (i.e. equivalence line), as shown in Figure 18.9. If data points are mostly situated above the equivalence line, the model is considered fail-safe, since it overpredicts growth rate which is a more conservative approach and preferred from a food safety standpoint. In contrast, if data points are situated below the equivalence line, the model is considered fail-dangerous, since the model is underpredicting growth rate.

## TERTIARY MODELS AND DATA BASES

Tertiary models are the integration of secondary and primary models in computer programs providing a user-friendly interface for making predictions. Tertiary models are a means of translating the complex predictive models into easy-to-use tools to be applied by nonexpert end-users, including stakeholders, risk managers and the predictive microbiology beginner.

The most widespread tertiary models are the Pathogen Modeling Program (PMP) and ComBase Predictor. The former, the PMP program, was developed by the US Department of Agriculture (USDA) in the 1990s and has been continuously upgraded over the years, incorporating new models and graphical features and reaching its seventh version, which can be freely downloaded (http://portal.arserrc.gov/PMP.aspx). The PMP is mainly aimed at pathogenic bacteria, including inactivation, survival, growth and probability models (Figure 18.10). One of the main characteristic is the use of the Gompertz

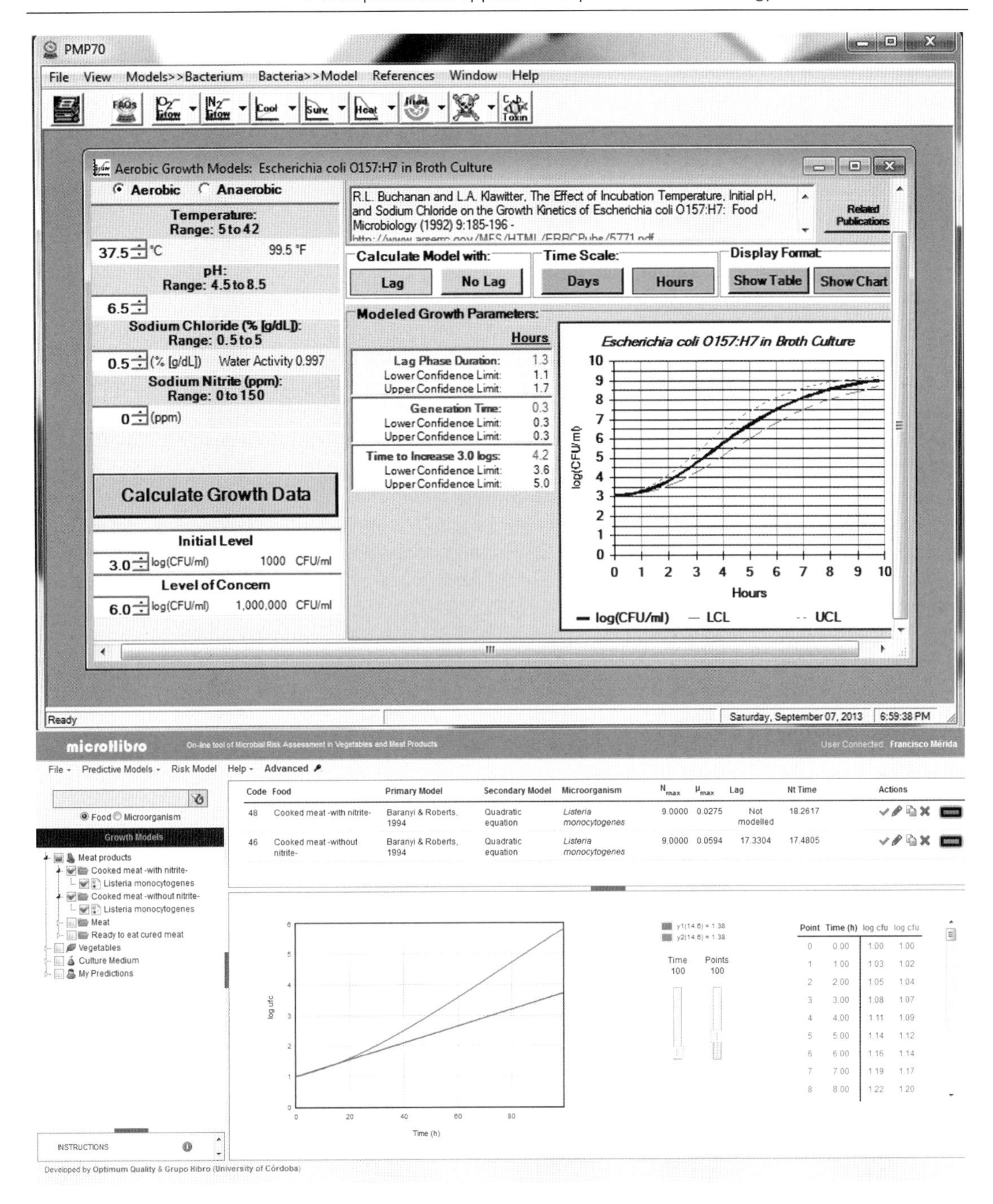


**Figure 18.10** Screen prints of the tertiary models Pathogen Modeling Program (PMP) (on the top). Courtesy Pathogen Modeling Program (PMP), USDA Agricultural Research Service, http://portal.arserrc.gov/PMP.aspx and Microhibro (www.microhibro.com) (on the bottom). Courtesy of Microhibro www.microhibro.com. (See colour version of this figure in colour plate section.)

equation as the underlying primary model for predictions. In this application, users introduce the point-estimate values for the considered environmental factors and the application returns the attendant kinetic parameters (lag, growth rate and maximum density population) together with a graphical representation of the growth, including its confidence limits. Also, PMP includes an interesting function that allows estimating the time needed to reach a level of concern established by the user. Recently, the PMP has been translated into an online version of the application, incorporating new models encompassing transfer models or cross-contamination models.

The ComBase Predictor is an online application developed by the Institute of Food Research (Norwich, UK) that allows prediction of growth and inactivation of different spoilage and pathogenic bacteria to be made. The underlying models are based on kinetic data generated in culture broths; this reduces its applicability and necessitates a previous validation process, especially when predictions are applied to solid food matrices. An important feature of this online software is that it allows predictions to be made in dynamic temperature conditions by introducing time–temperature profiles. The application can be freely accessed by registering at the ComBase website (www.combase.cc).

The Seafood Spoilage and Safety Predictor (SSSP) software is an application developed by the DTU National Institute of Aquatic Research, which is specially focused on growth models for seafood products considering spoilage bacteria (e.g. *Photobacterium phosphoreum* and *Shewanella putrefraciens*) and more recently pathogenic microorganisms (i.e. *L. monocytogenes*). The application incorporates the relative rate of spoilage models for seafood products. With these models, users can predict shelf life at different statistic or dynamic temperature conditions based on a prior knowledge of product deterioration at a specific temperature. The underlying secondary models are validated for the different considered food matrices; however, it also includes default models to be defined by users based on cardinal parameters (i.e. cardinal-type models). The SSSP software can be freely downloaded at the website of DTU National Institute of Aquatic Research (http://sssp.dtuaqua.dk). The latest version corresponds to SSSP v. 3.1, which is translated to several language including English, Spanish and Chinese.

Over the last few years, numerous tertiary applications and predictive microbiology software tools have been developed; some of them, such as Symprevius and Microbial Risk Viewer (MRV), are presented in Table 18.1. All are the consequence of the need to transfer predictive models to predictive microbiology practitioners in an applicable way. Predictive models are not only mathematical functions interpreting reality but rather tools to be applied. The application of predictive models relies on the ability of making them available once they are published and validated. However, models are quite inaccessible and long time periods elapse until they reach end-users. The integration of computational elements into modelling is crucial to provide an applicability dimension to predictive models. In this sense, the online application of predictive microbiology named Microhibro developed by the University of Córdoba makes an attempt to confer this character to predictive models by means of expert systems. The application Microhibro is freely available at http://www.microhibro.com for which a previous registration is required including different user levels (Figure 18.10). Microhibro allows predictions to be made for inactivation and growth of different microorganisms in different food matrices and culture media. Models can be easily implemented and stored in the database for subsequent use, public or private, in quantitative microbial risk assessment studies or stand-alone prediction systems.

The need for gathering and organizing the enormous amount of information on microbial response in foods has led to researchers to promote the development of a unified data base on predictive microbiology. The result of this initiative has been ComBase (Combined, or Common i.e. joint, Data Base of microbial responses to food environments) which was launched at the 4th International Conference on Predictive Modeling in Foods, Quimper, France, June 2003. The database was first developed by the Institute of Food Research (IFR), Norwich, UK, and over the last few years new research centres have joined the ComBase consortium, increasing data input and incorporating new functionalities. ComBase is an online database that allows kinetic data to be retrieved for different

**Table 18.1** List of available tertiary models and predictive microbiology software.

| Application | Availability/location | Source or developer | Main features |
|---|---|---|---|
| Pathogen Modeling Program (PMP) | http://portal.arserrc.gov/PMP.aspx | US Department of Agriculture (USDA, USA) | Growth, inactivation and survival models for pathogenic bacteria |
| ComBase Predictor | http://www.combase.cc | Institute of Food Research (IFR, England) | Growth and inactivation models under static and dynamic conditions |
| Simprevius | http://www.symprevius.net | INRA (French National Institute of Agricultural Research) | Commercial software for predictive microbiology and risk assessment |
| Seafood Spoilage and Safety Predictor (SSSP) | http://sssp.dtuaqua.dk | DTU National Institute of Aquatic Research (Denmark) | Shelf-life predictive models for seafood products |
| Microbial Risk Viewer (MRV) | http://mrv.nfri.affrc.go.jp/Default.aspx#/About | NFRI (National Food Research Institute, Japan) | Microbial growth/no growth and kinetic models derived from ComBase data base. |
| Listeria Control Model TM | http://lcm.purac.com/index.php | PURAC™ | Predictive models for Listeria monocytogenes as a function of PURAC preservative formulations |
| UGPM (Unified Growth Prediction Model) | http://www.aua.gr/psomas | Psomas *et al.*, 2011, 2012 | Predictive models under dynamic temperature conditions |
| Microhibro | http://www.microhibro.com | University of Córdoba | E-predictive models and quantitative risk assessment expert system |
| ICRA (Interactive online Catalogue on Risk Assessment) | http://icra.foodrisk.org/ | National Institute for Public Health and Environment (RIVM, The Netherlands), DTU (Denmark), and the Joint Institute for Food Safety and Applied Nutrition (JIFSAN, USA) | Web tool offering a dynamic model catalogue for existing microbial risk |

microorganisms in different food matrices by using an advanced search engine including different search parameters (environmental conditions, food type, microorganism, etc.). The database is fed by contributions from researchers all over the world. More detailed description is available in Baranyi and Tamplin (2004) and on the website http:\\www.combase.cc.

# QUANTITATIVE MICROBIAL RISK ASSESSMENT

## Risk analysis framework

In the last few decades, a novel framework has been proposed as a basis to derive food standards at national and international levels, so-called risk analysis, which is made of three components: risk assessment; risk management and risk communication. The interaction between the three components is the basis to guide food policy and risk management concerning chemical, physical and biological hazards (Figure 18.11) (FAO/WHO, 2006). The definitions of these components are given by the FAO/WHO (1995). Risk assessment is 'the qualitative and/or quantitative evaluation of the nature of the adverse effects associated with biological, chemical and physical agents, which may be present in food' (FAO/WHO, 1995). Risk assessment is carried out in four steps: hazard identification; hazard characterization, exposure assessment and risk characterization. The foundations and concepts on how to conduct a microbiological risk assessment (MRA) were first presented in the document entitled 'Principles and Guidelines for the Conduct of Microbiological Risk Assessment (Alinorm

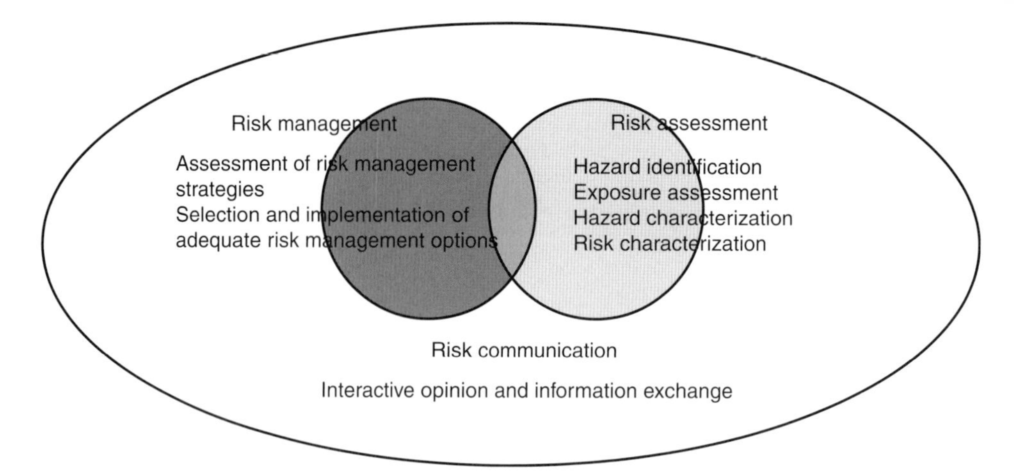


**Figure 18.11** Graphical representation of interaction between the elements of Risk Analysis (FAO/WHO, 2006). (Food and Agriculture Organisation of the United Nations, (2006), Food Safety Risk Analysis: A Guide for National Food Safety Authorities, FAO Food and Nutrition Papers-87, http://www.who.int/foodsafety/publications/micro/riskanalysis06/en/)

99/13A)', which was adopted in June 1999 (FAO/WHO, 1999). In this document, specific definitions for the four steps are provided:

(i) Hazard identification. 'The identification of biological agents capable of causing adverse health effects and which may be present in a particular food or group of foods'.
(ii) Hazard characterization. 'The qualitative and/or quantitative evaluation of the nature of the adverse health associated with the hazard'.
(iii) Exposure assessment. 'The qualitative and/or quantitative evaluation of the likely intake of a biological agent via food, as well as exposure from other sources if relevant'.
(iv) Risk characterization. 'The process of determining the qualitative and/or quantitative estimation, including attendant uncertainties, of the probability of occurrence and severity of known or potential adverse health effects in a given population based on hazard identification, hazard characterization and exposure assessment'.

Risk assessment was adopted by member states of the World Trade Organization (WTO) as the framework providing scientific basis for ensuring fair commerce, avoiding commercial barriers, and guaranteeing consumer health through the General Agreement on Tariffs and Trade (GATT) and Sanitary and the Phytosanitary (SPS).

## Types and classification of microbial risk assessment

From then on, risk assessment has been applied as a science-based tool to evaluate microbiological risk in foods within different contexts: industrial, governmental and scientific. Microbial risk assessment (MRA) can be carried out in different manners according to the nature of the variables that are considered. Thus, we can distinguish, in first instance, qualitative and quantitative MRA studies. The former refers to studies in which variables are qualitatively defined, which normally lead to use of categorical variables using 'qualitative levels' to describe them (e.g. low, medium and high levels). However, the quantitative studies are preferred since a more accurate estimate can be achieved, in which variables are described by numeric values, thereby yielding quantitative estimate of risk. Results from this approach can be more accurately interpreted by risk managers and, therefore, measures can be better derived. Quantitative microbial risk assessment (QMRA) studies can be conducted using either point-estimate values or probability distributions. In the first case, each variable is described by a single value which is usually the mean of the variable or another representative statistic such as, 5th and 95th percentile, mode or median (Lammerding and Fazil, 2000).

The variables considered in QMRA are related to prevalence and concentration of the microorganism and attendant factors such as temperature, pH, time, serving size and so on, which are, in fact, part of the exposure assessment step through which the exposure level to the food-borne pathogen at the time of consumption is estimated. All these variables include an important source of variability and uncertainty, since in many cases risk is derived from different possible scenarios for the different considered variables or factors. For example storage temperatures during food refrigeration usually vary over a wide range, including temperatures supporting and not supporting growth. This fact will greatly influence risk. Therefore, in order to take into consideration this inherent variability, probability distributions are used in QMRA. By including probability distributions, a more complete and accurate estimate of risk can be obtained, since all possible scenario combinations are contemplated. This information helps risk managers make better decisions concerning risk mitigation strategies, control measures and food regulation. Probabilistic or stochastic QMRA studies necessitate the application of different mathematical techniques to operate with probability distributions. The most used technique is Monte Carlo analysis, which is a set of numerical methods that can be operated with any type of distribution. Its application has been facilitated thanks to commercial software that incorporates Monte Carlo analysis methods in a user-friendly and easy-to-use interface, such as @Risk (Palisade ©, Newfield, NY) or Analytica (Lumina Decision Systems, Inc., Los Gatos, CA). Also, Bayesian methods have been satisfactorily applied in probabilistic QMRA (Hald *et al.*, 2004; Delignette-Muller *et al.*, 2006); however, the use of this technique requires advanced knowledge and more complex software such as, for example WinBUGS (Imperial © College School, MRC). Other more general statistical software, such as MATLAB (The MathWorks Inc., Natick, MA) or R statistical package (http://www.R-project .org/), can be used for such purposes.

The distinction between variability and uncertainty is an important aspect which should be addressed when a probabilistic QMRA studies are conducted (Pérez-Rodríguez *et al.*, 2007a). Firstly, define both concepts. Variability can be defined as the natural variation of the target system or population (i.e. special, temporal or inter-individual differences) while uncertainty is the lack of knowledge on the system or population, the consequence of analytical or technological limitations which are usually inherent to a human being. Therefore, uncertainty can be reduced by increasing observations and measurements on the system or improving the analytical method. In turn, variability cannot be eliminated, since it is a property of the system or population. This differentiation between both concepts has important implications to decision makers, since having a clear understanding on which fraction of variation corresponds to natural variability and which part is a consequence from lack of information (i.e. experimental error, data gaps, modelling assumptions, etc.) allows to better value the impact of any decision in addition to knowing which aspects should be further studied to reduce uncertainty (Cullen and Frey, 1999). From that, if suitable information is available, probabilistic QMRA studies separating variability and uncertainty are preferred. In such cases, QMRA studies are so-called second-order or two-dimensional to differentiate them from those QMRA studies in which both components are considered jointly, which are referred as to as a first-order model or one-dimensional. In second-order QMRA studies, both components are characterized by two different probability spaces, which means that distinct probability distributions are used to define the variability components and uncertainty components within the same variable, which should be simulated separately. In spite of the need to carry out second-order QMRA, it is not always possible or indispensable, as in many cases, data or information are not suitable for that purpose or because the goal of the QMRA study can be accomplished by using a first-order approach, which usually implies lower human and technical resources and, therefore, less time to obtain results.

## Predictive models in QMRA

In QMRA, there are two steps, hazard characterization and exposure assessment, which are carried out based on the use of predictive models. On the one hand, hazard characterization relies on the application of dose-response models, intended to estimate the probability of getting ill or probability of infection as a function of the ingested hazard dose (Lammerding and Fazil, 2000). These models embody an important component of uncertainty, since models and epidemiological and experimental data are very limited and scarce. An extensive review of existing dose-response models applied to food-borne pathogens has been presented by Haas *et al.* (1999). Regarding their application, Pérez-Rodríguez *et al.* (2007b) and Pérez-Rodríguez and Zwietering (2011) studied different quantitative methods to integrate dose-response models in QMRA, establishing methodological and conceptual differences for individual risk and public health (i.e. population risk).

Quantitative exposure assessment is another step which entails an important modelling load. This step has as its aim to describe the transmission routes of target microorganisms along the food chain from farm to fork or along a specific segment of the food chain (e.g. from retail to consumption). By doing so, the exposure level of consumers to a specific microbiological hazard can be estimated. However, in many cases, information is scarce or data are not available since there are food steps or processes in which surveys and experimentation are not feasible, such as, for example when an inactivation process is applied, reducing microbial load to levels below the quantification and detection limit, or when microbiological analysis should be conducted at the moment of consumption but are not affordable from a practical point of view. In these cases, predictive models are essential to predict how the different process and transmission conditions affect microorganism transmission and, thus, the exposure level (Lammerding and Paoli, 1997).

In the case of exposure assessment, predictive models are usually applied by taking a systems approach in which significant technological and microbial processes affecting prevalence and concentration of the target microorganism are modelled. Estimates of microbial contamination are later introduced in the dose-response model to obtain an estimate of the risk associated with the microorganism. For example if the aim is to evaluate transmission of pathogen in cooked ham from retail to consumption, risk assessors should first consider those steps in which microorganism levels can be affected. In this case, it might be hypothesized that refrigeration and cross-contamination occurring at both stages (i.e. retail and home) can be key elements to be modelled. Then, adequate predictive models should be chosen. In this case, growth models (combination of a primary model and a secondary model) and transfer models should be applied considering as inputs the initial levels of microorganisms, that is to say, prevalence and concentration when cooked ham is delivered to retail.

Although at first glance QMRA could seem a noncomplex work, factors and processes can be multiple in addition to the lack of information, which in many cases hampers modelling work and reduces accuracy. From that, QMRA is considered a difficult field in which risk assessors should take a multidisciplinary approach and possess specific skills to deal with lack of information, general microbiology concepts, predictive modelling, mathematical techniques and the wide perspective on the food sector. Nonetheless, the main problem affecting QMRA studies is the lack of a standardized method, since it should not be forgotten that QMRA is proposed as a science-based tool to support decisions regarding food risk at international and national levels. Therefore, it is crucial that methodology is harmonized in order to obtain comparable results and, therefore, risk policies. Many attempts have been made to derive a methodological basis for QMRA procedures; however, still many aspects remain undefined. The first study dealing with QMRA is the work by Cassin *et al.* (1998) in which the so-called process risk model (PRM) approach was proposed to conduct probabilistic QMRA studies. This is mainly based on describing the risk process throughout manufacturing, handling and food consumption. Then, doses (from an exposure assessment step) are input into the dose-response model in order to estimate the health risk associated with a food. van Gerwen *et al.* (2000) developed a stepwise approach to conduct QMRA studies in which QMRA is performed in a hierarchical form providing better quantitative detail to those variables most impacting risk. Another modelling approach, the so-called modular process risk model (MPRM), was proposed by Nauta (2001) and has been further developed in subsequent works by the same author (Nauta 2005; Nauta *et al.*, 2005; Havelaar *et al.*, 2008). Basically, this approach proposes that any step throughout the food chain can be described mathematically through six basic processes: growth, inactivation, mixing, partitioning, removal and cross-contamination. Through the application of six types of models, an accurate estimation can be made of the levels of exposure to a pathogen from a specific food process. The MPRM approach has been basis of an expert opinion by the European Commission (2003) aimed at laying the foundations of EA methodology in a framework of QMRA. More recently, a compendium of mathematical tools for QMRA has been carried out commissioned by ILSI (International Life Sciences Institute) (Bassett *et al.*, 2012). This work is focused on different aspects concerning QMRA foundation, predictive models, software tools and interpretation of QMRA results. However, in spite of these methodological advances, QMRA is still an incipient field which should be further developed and harmonized in the future.

Predictive models such as inactivation, growth and transfer models are the basis for performing QMRA studies. In some cases, specific models can be applied, such as behavioural models describing the handler or consumer behaviour in food environments. Generally, predictive model QMRA should be easy to update, so that QMRA can be easily modified by adding new variables or information, as they becomes available, or when new risk manager question are arisen. Therefore, cardinal-type model or square root models can be adequate candidates for QMRA, since they allow incorporation of new terms or even, in some cases like in the cardinal-type models (including the gamma concepts), models can be defined by using theoretical values. The capacity of extrapolation is an important limitation of models when applied to QMRA studies; polynomial secondary models should be avoided since, in addition to being less updatable, the capacity of extrapolation is quite limited in comparison to others models. With regards to inactivation primary models, although most of scientific literature is concerned with log-linear inactivation models, it is recommended that shoulder and tail effects are considered in QMRA by using Weibull models and two-phase and three-phase inactivation models, since these models can adapt to the different microbial inactivation patterns. Growth/nongrowth models are complementary to kinetic models. For example a growth/nongrowth model would be used to determine whether the conditions of a certain storage phase (temperature, storage time, pH, etc.) permit growth or not. If the result is growth, the kinetic model would then be applied in order to estimate the concentration following storage time. In the case of nongrowth, the kinetic model would not be applied and, therefore, the concentration value following storage would be the same as the initial value. The use of lag secondary models is less frequent due to the great uncertainty associated with this kinetic parameter as mentioned in previous sections; hence, QMRA studies tend not to include lag time, thereby providing more conservative risk estimates. Nevertheless, if adequate data are available a stochastic approach could be taken by defining probability distributions reflecting the possible lag times for the target microorganism and conditions. Cross-contamination/transfer and survival models for microorganisms on (inert and food) surfaces are also needed in QMRA. Cross-contamination is an important factor in the increase of pathogen prevalence along the different steps from farm to fork and, therefore, in final risk. Cross-contamination models should be accompanied by survival models to estimate living cells on surfaces available for cross-contamination. The combination of both types of models in QMRA leads to more accurate estimate of bacterial transfer to foods (Pérez-Rodríguez *et al.*, 2008).

## FUTURE TRENDS IN PREDICTIVE MICROBIOLOGY

It is obvious that predictive microbiology has made significant advances over the last few years, providing useful and reasonably accurate models, However, predictive models still lack a mechanistic basis. The mechanistic approach is not itself a goal but rather a means of undertanding microbial phenomena and proposing more accurate and universal models. Limitations inherent to the current modelling approach, based on a macroscopic observation of the system, could be overcome if a closer look at cell level is taken, gaining insight into microbial population complexities (population heterogenity, response to stress conditions, quorum sensing, competition, etc.) and generating models with better application to real foods (i.e. application outside of the interpolation range). It is, therefore, time that predictive microbiology moves towards a microscopic level, looking at a cell level and underlying biochemical process governing microbial response.

One of the most challenging areas faced by predictive microbiology is obtaining reliable models for lag time, which, as already mentioned, is subjected to important variability sources. The study of lag phase at single cell level is increasingly being addressed by reserachers by applying innovative analysis methods enabling observations of individual cells to be obtained (e.g. image analysis coupled with microscopy) (Guillier *et al.*, 2006; Niven *et al.*, 2006). These studies are intended to model invidual cells during cell division cycles, estimating the heterogenity of bacterial population and yielding more reliable pictures on how population lag time is produced under different environmental conditions. This information, together with a better knowledge of biochemical mechanisms in cells, is expected to yield more mechanistic models (McKellar, 2008; Huang, 2011; Olofsson and Ma, 2011).

In addition, predictive microbiology can take advantage of the enormous amount of omic information in addition to applying concepts from systems biology as basis to develop top-down models (or mechanistically-based models) (Westerhoff and Hofmeyr, 2005; Hertog *et al.*, 2011). Indeed, some attempts, though scarce, have been made in the last few year to associate certain metabolic or physiological substances with observed kinetic parameters (Poschet *et al.*, 2005; Van Impe *et al.*, 2005). Genome-scale models based on a complete representation of the matabolic network in a cell system are proposed as a mechanistic approach to attain more reliable and universal models (Stelling, 2004; Orth *et al.*, 2011). Different techniques applied in systems biology have also been highlighted as crucial for obtaining Genome-scale models, such as metabolic flux analysis or network analysis, and preliminary attempts in this novel field have been published by researchers in predictive microbiology (Brul *et al.*, 2008; Métris *et al.*, 2011; Van Impe *et al.*, 2013). Therefore, systems biology should be considered as a valid and promising *track* to be followed by predictive microbriology in the next years, focusing on a microscopic and biochemical level as a means to improve predictive models.

## ACKNOWLEDGEMENTS

This work was supported by MICINN AGL2008-03298/ALI, the Excellence Projects AGR-01879 and P08-CTS-3260 (Junta de Andalucía), the EU project from the Seventh Framework Program KBBE 222738 Food, Agriculture and Fisheries, and Biotechnology and the Research Group AGR-170 HIBRO of the 'Plan Andaluz de Investigación, Desarrollo e Innovación' (PAIDI).

## REFERENCES

Adams, M.R., Hartley, A.D. and Cox, L.J. (1989) Factors affecting the efficacy of washing procedures used in the production of prepared salads. *Food Microbiology* **6**, 69–77.

Almeida J.S. (2002) Predictive non-linear modeling of complex data by artificial neural networks. *Current Opinion in Biotechnology* **13**, 72–76.

Amos S.A., Davey K.R. and Thomas C.J. (2001) A comparison of predictive models for the combined effect of UV dose and solids concentration on disinfection kinetics of *Escherichia coli* for potable water production. *Process Safety and Environmental Protection* **79**, 174–182.

Augustin J. and Carlier V. (2000a) Modelling the growth rate of *Listeria monocytogenes* with a multiplicative type model including interactions between environmental factors. *International Journal of Food Microbiology* **56**, 53–70.

Augustin J.C. and Carlier V. (2000b) Mathematical modelling of the growth rate and lag time for *Listeria monocytogenes. International Journal of Food Microbiology* **56**, 29–51.

Augustin J.C., Rosso L. and Carlier V. (1999) Estimation of temperature dependent growth rate and lag time of *Listeria monocytogenes* by optical density measurements. *Journal of Microbiological Methods* **38**, 137–146.

Aziza F., Mettler E., Daudin J.-J. and Sanaa M. (2006) Stochastic, compartmental, and dynamic modeling of cross-contamination during mechanical smearing of cheeses. *Risk Analysis* **26**, 731–745.

Baert K., Valero A., De Meulenaer B. *et al.* (2007) Modeling the effect of temperature on the growth rate and lag phase of *Penicillium expansum* in apples. *International Journal of Food Microbiology* **118**, 139–150.

Baranyi J. and Roberts T.A. (1994) Review paper:A dynamic approach to predicting bacterial growth in food. *International Journal of Food Microbiology* **23**, 277–294.

Baranyi J. and Roberts T.A. (1995) Mathematics of predictive food microbiology. *International Journal of Food Microbiology* **26**, 199–218.

Baranyi J. and Roberts T.A. (2000) Principles and application of predictive modeling of the effects of preservative factors on microorganisms. In: *The Microbiological Safety and Quality of Food* (eds B.M. Lund, A.C. Baird-Parker and G.W. Gould), pp. 342–358. Aspen Publishers, Inc., Gaithersburg, MD.

Baranyi J. and Tamplin M.L. (2004) ComBase: a common database on microbial responses to food environments. *Journal of Food Protection* **67**, 1967–1971.

Baranyi J., Roberts T.A. and McClure P. (1993) A non-autonomous differential equation to model bacterial growth. *Food Microbiology* **10**, 43–59.

Baranyi J., Robinson T.P., Kaloti a and Mackey B.M. (1995) Predicting growth of *Brochothrix thermosphacta* at changing temperature. *International Journal of Food Microbiology* **27**, 61–75.

Baranyi J., Ross T., McMeekin T.A. and Roberts T.A. (1996) Effects of parameterization on the performance of empirical models used in 'predictive microbiology'. *Food Microbiology* **13**, 83–91.

Bassett J., Nauta M., Lindqvist R. and Zwietering M.H. (2012) *Tools for Microbiological Risk Assessment*. ILSI Europe Report Series, ILSI Europe, Brussels, Belgium.

Bernaerts K., Dens E., Vereecken K. *et al.* (2004) Concepts and tools for predictive modeling of microbial dynamics. *Journal of Food Protection* **67**, 2041–2052.

Bidlas E. and Lambert R.J.W. (2008) Quantification of hurdles: predicting the combination of effects – Interaction vs. non-interaction. *International Journal of Food Microbiology* **128**, 78–88.

Biesta-Peters E.G., Reij M.W., Zwietering M.H. and Gorris L.G.M. (2011) Comparing nonsynergy gamma models and interaction models to predict growth of emetic Bacillus cereus for combinations of pH and water activity values. *Applied and Environmental Microbiology* **77**, 5707–5715.

Bigelow, W.D. (1921) The logarithmic nature of thermal death time curves. *Journal of Infectious Diseases* **27**, 528–536.

Bigelow W.D. and Esty J.R. (1920) The thermal death point in relation to typical thermophylic organisms. *Journal of Infectious Diseases* **27**, 602–617.

Bolton L.F. and Frank J.F. (1999) Defining the growth/no-growth interface for Listeria monocytogenes in Mexican-style cheese based on salt, pH, and moisture content. *Journal of Food Protection* **62**, 601–609.

Brul S., Mensonides F.I.C., Hellingwerf K.J. and Teixeira de Mattos M.J. (2008) Microbial systems biology: new frontiers open to predictive microbiology. *International Journal of Food Microbiology* **128**, 16–21.

Buchanan R.L., Whiting R.C. and Damert W.C. (1997) When is simple good enough: a comparison of the Gompertz, Baranyi, and three-phase linear models for fitting bacterial growth curves. *Food Microbiology* **14**, 313–326.

Cardenas F., Giannuzzi L. and Zaritzky N. (2008) Mathematical modelling of microbial growth in ground beef from Argentina. Effect of lactic acid addition, temperature and packaging film. *Meat Science* **79**, 509–520.

Carrasco E., Pérez-Rodríguez F., Valero A. *et al.* (2010) Risk assessment and management of *Listeria monocytogenes* in ready-to-eat lettuce salads. *Comprehensive Reviews in Food Science and Food Safety* **9**, 498–512.

Cassin M.H., Lammerding A.M., Todd E.C. *et al.* (1998) Quantitative risk assessment for *Escherichia coli* O157:H7 in ground beef hamburgers. *International Journal of Food Microbiology* **41**, 21–44.

Cerf O., Davey K.R. and Sadoudi a. K. (1996) Thermal inactivation of bacteria – a new predictive model for the combined effect of three environmental factors: temperature, pH and water activity. *Food Research International* **29**, 219–226.

Chen Y., Jackson K.M., Chea F.P. and Schaffner D.W. (2001) quantification and variability analysis of bacterial cross-contamination rates in common food service tasks. *The Hand* **64**, 72–80.

Cornu M., Kalmokoff M. and Flandrois J.-P. (2002) Modelling the competitive growth of *Listeria monocytogenes* and *Listeria innocua* in enrichment broths. *International Journal of Food Microbiology* **73**, 261–274.

Cornu M., Billoir E., Bergis H. *et al.* (2011) Modeling microbial competition in food: application to the behavior of *Listeria monocytogenes* and lactic acid flora in pork meat products. *Food Microbiology* **28**, 639–647.

Cullen A.C. and Frey H.C. (1999) *Probabilistic Techniques in Exposure Assessment.A Handbook for Dealing with Variability and Uncertainty in Models and Inputs*. Plenum, New York.

Dalgaard P., Ross T., Kamperman L. *et al.* (1994) Estimation of bacterial growth rates from turbidimetric and viable count data. *International Journal of Food Microbiology* **23**, 391–404.

Dantigny P., Marín S., Beyer M. and Magan N. (2007) Mould germination: data treatment and modelling. *International Journal of Food Microbiology* **114**, 17–24.

Davey K.R. (1989) A predictive model for combined temperature and water activity on microbial growth during the growth phase. *Journal of Applied Microbiology* **67**, 483–488.

Davey K.R. (1994) Modelling the combined effect of temperature and pH on the rate coefficient for bacterial growth. *International Journal of Food Microbiology* **23**, 295–303.

Davey K.R. and Daughtry B.J. (1995) Validation of a model for predicting the combined effect of three environmental factors on both exponential and lag phases of bacterial growth: temperature, salt concentration and pH. *Food Research International* **28**, 233–237.

Delignette-Muller M.L., Cornu M., Pouillot R. and Denis J.-B. (2006) Use of Bayesian modelling in risk assessment: application to growth of *Listeria monocytogenes* and food flora in cold-smoked salmon. *International Journal of Food Microbiology* **106**, 195–208.

Devlieghere F., Geeraerd A.H., Versyck K.J. *et al.* (2000) Shelf life of modified atmosphere packed cooked meat products: addition of Na-lactate as a fourth shelf life determinative factor in a model and product validation. *International Journal of Food Microbiology* **58**, 93–106.

Devlieghere F., Geeraerd A.H., Versyck K.J. *et al.* (2001) Growth of *Listeria monocytogenes* in modified atmosphere packed cooked meat products: a predictive model. *Food Microbiology* **18**, 53–66.

Dupont C. and Augustin J.-C. (2009) Influence of stress on single-cell lag time and growth probability for Listeria monocytogenes in half Fraser broth. *Applied and Environmental Microbiology* **75**, 3069–3076.

Esty J.R. and Meyer K.F. (1922) The heat resistance of spores of *B. botulinus* and related anaerobes. *Journal of Infection Diseases* **31**, 650–663.

European Commission (2003) Risk assessment of food borne bacterial pathogens: Quantitative methodology relevant for human exposure assessment (final report). http://ec.europa.eu/food/fs/sc/ssc/out308_en.pdf (last accessed 28 June 2013).

Eyring H. (1935) The activated complex in chemical reactions. *Journal of Chemical Physics* **3**, 107–115.

FAO/WHO (Food Agriculture Organization/World Health Organization) (1999) *RISK ASSESSMENT OF MICRO-BIOLOGICAL HAZARDS IN FOODS*. Report of the Joint FAO/WHO Expert Consultation Geneva, Switzerland.

FAO/WHO (Food Agriculture Organization/World Health Organization) (1995) Application of risk analysis to food standards. Report of the joint FAO/WHO expert consultation. FAO/WHO, Geneva, Switzerland.

FAO/WHO (Food Agriculture Organization/World Health Organization) (2006). Food Safety Risk Analysis – A Guide for National Food Safety Authorities, FAO Food and Nutrition Papers-87, FAO, Rome, Italy. http://www.who.int/foodsafety/publications/micro/riskanalysis06/en/ (last accessed 28 June 2013).

Fernández-Navarro F., Valero A., Hervás-Martínez C. *et al.* (2010) Development of a multi-classification neural network model to determine the microbial growth/no growth interface. *International Journal of Food Microbiology* **141**, 203–212.

Fujikawa H. and Morozumi S. (2005) Modeling surface growth of *Escherichia coli* on agar plates. *Applied and Environmental Microbiology* **71**, 7920–7926.

García-Gimeno R.M., Hervás-Martínez C. and De Silóniz M.I. (2002) Improving artificial neural networks with a pruning methodology and genetic algorithms for their application in microbial growth prediction in food. *International Journal of Food Microbiology* **72**, 19–30.

Geeraerd A.H., Herremans C.H. and Van Impe J.F. (2000) Structural model requirements to describe microbial inactivation during a mild heat treatment. *International Journal of Food Microbiology* **59**, 185–209.

Geeraerd A.H., Valdramidis V.P. and Van Impe J.F. (2005) GInaFiT Geeraerd and Van Impe Inactivation model Fitting Tool Requirements of GInaFiT. *International Journal of Food Microbiology* **102**, 95–105.

Genigeorgis C.A. (1981) Factors affecting the probability of growth of pathogenic microorganisms in foods. *American Veterinary Medicine Association* **179**, 1410–1417.

van Gerwen S.J., te Giffel M.C., van't Riet K. *et al.* (2000) Stepwise quantitative risk assessment as a tool for characterization of microbiological food safety. *Journal of Applied Microbiology* **88**, 938–951.

Gibson A.M., Bratchell N. and Roberts T.A. (1988) Predicting microbial growth: growth responses of salmonellae in a laboratory medium as affected by pH, sodium chloride and storage temperature. *International Journal of Food Microbiology* **6**, 155–178.

Gibson A.M., Bratchell N. and Roberts T.A. (1987) The effect of sodium chloride and temperature on the rate and extent of growth of *Clostridium botulinum* type A in pasteurized pork slurry. *The Journal of Applied Bacteriology* **62**, 479–490.

te Giffel M.C. and Zwietering M.H. (1999) Validation of predictive models describing the growth of *Listeria monocytogenes*. *International Journal of Food Microbiology* **46**, 135–149.

Gimenez B. and Dalgaard P. (2004) Modelling and predicting the simultaneous growth of Listeria monocytogenes and spoilage micro-organisms in cold-smoked salmon. *Journal of Applied Microbiology* **96**, 96–109.

Gougouli M. and Koutsoumanis K.P. (2012) Modeling germination of fungal spores at constant and fluctuating temperature conditions. *International Journal of Food Microbiology* **152**, 153–161.

Guerzoni M.E., Lanciotti R., Torriani S. and Dellaglio F. (1994) Growth modelling of *Listeria monocytogenes* and *Yersinia enterocolitica* in food model systems and dairy products. *International Journal of Food Microbiology* **24**, 83–92.

Guillier L., Pardon P. and Augustin J.-C. (2006) Automated image analysis of bacterial colony growth as a tool to study individual lag time distributions of immobilized cells. *Journal of Microbiological Methods* **65**, 324–334.

Gurtler J.B., Marks H.M., Jones D.R. *et al.* (2011) Modeling the thermal inactivation kinetics of heat-resistant *Salmonella* Enteritidis and Oranienburg in 10 percent salted liquid egg yolk. *Journal of Food Protection* **74**, 882–892.

Haas C.N., Rose J.B. and Gerba C.P. (1999) *Quantitative Microbial Risk Assessment*. John Wiley & Sons, Inc., New Jersey.

Hajmeer M. and Basheer I. (2002) A probabilistic neural network approach for modeling and classification of bacterial growth/no-growth data. *Journal of Microbiological Methods* **51**, 217–226.

Hald T., Vose D., Wegener H.C. and Koupeev T. (2004) A Bayesian approach to quantify the contribution of animal-food sources to human salmonellosis. *Risk Analysis* **24**, 255–269.

Havelaar A.H., Evers E.G. and Nauta M.J. (2008) Challenges of quantitative microbial risk assessment at EU level. *Trends in Food Science and Technology* **19**, S26–S33.

Hertog M.L.A.T.M., Rudell D.R., Pedreschi R. *et al.* (2011) Where systems biology meets postharvest. *Postharvest Biology and Technology* **62**, 223–237.

Hills B.P. and Wright K.M. (1994) A new model for bacterial growth in heterogeneous systems. *Journal of Theoretical Biology* **168**, 31–41.

Huang L. (2011) A new mechanistic growth model for simultaneous determination of lag phase duration and exponential growth rate and a new Bělehdrádek-type model for evaluating the effect of temperature on growth rate. *Food Microbiology* **28**, 770–776.

Huang L., Hwang A. and Phillips J. (2011) Effect of temperature on microbial growth rate-mathematical analysis: the arrhenius and eyring-polanyi connections. *Journal of Food Science* **76**, E553–E560.

Jameson J. (1962) A discussion of the dynamics of Salmonella enrichment. *Journal of Hygiene* **60**, 193–207.

Juneja V.K. and Marks H.M. (2003) Mathematical description of non-linear survival curves of *Listeria monocytogenes* as determined in a beef gravy model system at 57.5 to 65 °C. *Innovative Food Science and Emerging Technologies* **4**, 307–317.

Koseki S. and Isobe S. (2005) Prediction of pathogen growth on iceberg lettuce under real temperature history during distribution from farm to table. *International Journal of Food Microbiology* **104**, 239–248.

Koutsoumanis K. and Nychas G.J. (2000) Application of a systematic experimental procedure to develop a microbial model for rapid fish shelf life predictions. *International Journal of Food Microbiology* **60**, 171–184.

Koutsoumanis K.P., Taoukis P.S., Drosinos E.H. and Nychas G.E. (2000) Applicability of an Arrhenius model for the combined effect of temperature and $CO_2$ packaging on the spoilage microflora of fish. *Applied Environmental Microbiology* **66**(8), 3528.

Koutsoumanis K., Stamatiou A., Skandamis P. and Nychas G.E. (2006) Development of a microbial model for the combined effect of temperature and pH on spoilage of ground meat, and validation of the model under dynamic temperature conditions. *Society* **72**, 124–134.

Lambert R.J.W. and Bidlas E. (2007) A study of the Gamma hypothesis: predictive modelling of the growth and inhibition of *Enterobacter sakazakii*. *International Journal of Food Microbiology* **115**, 204–213.

Lammerding A.M. and Fazil A. (2000) Hazard identification and exposure assessment for microbial food safety risk assessment. *International Journal of Food Microbiology* **58**, 147–157.

Lammerding A.M. and Paoli G.M. (1997) Quantitative risk assessment: an emerging tool for emerging foodborne pathogens. *Emerging Infectious Diseases* **3**, 483–487.

Lebert I., Robles-Olvera V. and Lebert A. (2000) Application of polynomial models to predict growth of mixed cultures of *Pseudomonas* spp. and Listeria in meat. *International Journal of Food Microbiology* **61**, 27–39.

Le Marc Y., Huchet V., Bourgeois C.M. *et al.* (2002) Modelling the growth kinetics of Listeria as a function of temperature, pH and organic acid concentration. *International Journal of Food Microbiology* **73**, 219–237.

Leroi F., Fall P.A., Pilet M.F. *et al.* (2012) Influence of temperature, pH and NaCl concentration on the maximal growth rate of *Brochothrix thermosphacta* and a bioprotective bacteria *Lactococcus piscium* CNCM I-4031. *Food Microbiology* **31**, 222–228.

Leroy F. and De Vuyst L. (2007) Modelling microbial interactions in foods. In: *Modelling Microorganisms in Food* (eds S. Brul, S.J.C.van Gerwen and M.H. Zwietering), pp. 214–224. CRC Press, Boca Raton, FL.

Lindqvist R. (2006) Estimation of *Staphylococcus aureus* growth parameters from turbidity data: characterization of strain variation and comparison of methods. *Applied and Environmental Microbiology* **72**, 4862–4870.

Lobry J.R., Rosso L. and Flandrois J.-P. (1991) Parameter confidence limit in non-linear model. *Binary* **3**, 86–93.

Masana M.O. and Baranyi J. (2000) Growth/no growth interface of *Brochothrix thermosphacta* as a function of pH and water activity. *Food Microbiology* **17**, 485–493.

Mataragas M., Drosinos E.H., Vaidanis A. and Metaxopoulos I. (2006) Development of a predictive model for spoilage of cooked cured meat products and its validation under constant and dynamic temperature storage conditions. *Journal of Food Science* **71**, M157–M167.

McKellar R.C. (1997) A heterogeneous population model for the analysis of bacterial growth kinetics. *International Journal of Food Microbiology* **36**, 179–186.

McKellar R.C. (2008) Correlation between the change in the kinetics of the ribosomal RNA rrnB P2 promoter and the transition from lag to exponential phase with Pseudomonas fluorescens. *International Journal of Food Microbiology* **121**, 11–17.

McKellar R.C. and Lu X. (2001) A probability of growth model for *Escherichia coli* O157:H7 as a function of temperature, pH, acetic acid, and salt. *Journal of Food Protection* **64**, 1922–1928.

McMeekin T., Olley J. and Ross T. (1993a) *Predictive Microbiology: Theory and Application*. John Wiley & Sons Ltd, Chichester.

McMeekin T.A., Olley J.N., Ross T. and Ratkowsky D.A. (1993b) *Predictive Microbiology: Theory and Application*. Research Studies Press, Baldock, UK.

McMeekin T.A., Chandler R.E., Doe P.E. *et al.* (1987) Model for combined effect of temperature and salt concentration/water activity on the growth rate of *Staphylococcus xylosus*. *Journal of Applied Microbiology* **62**, 543–550.

McMeekin T.A., Presser K.A., Ratkowsky D. *et al.* (2000) Quantifying the hurdle concept by modelling the bacterial growth/no growth interface. *International Journal of Food Microbiology* **55**, 93–98.

Mejlholm O. and Dalgaard P. (2007) Modeling and predicting the growth of lactic acid bacteria in lightly preserved seafood and their inhibiting effect on *Listeria monocytogenes*. *Journal of Food Protection* **70**, 2485–2497.

Mellefont L.A. and Ross T. (2003) The effect of abrupt shifts in temperature on the lag phase duration of *Escherichia coli* and *Klebsiella oxytoca*. *International Journal of Food Microbiology* **83**, 295–305.

Membré J.-M., Leporq B., Vialette M. *et al.*, (2005) Temperature effect on bacterial growth rate: quantitative microbiology approach including cardinal values and variability estimates to perform growth simulations on/in food. *International Journal of Food Microbiology* **100**, 179–186.

Métris A., Le Marc Y., Elfwing A. *et al.* (2005) Modelling the variability of lag times and the first generation times of single cells of E. coli. *International Journal of Food Microbiology* **100**, 13–19.

Métris A., George S. and Baranyi J. (2011) Modelling osmotic stress by flux balance analysis at the genomic scale. *International Journal of Food Microbiology*, 6–11.

Miles D.W., Ross T., Olley J. and McMeekin T.A. (1997) Development and evaluation of a predictive model for the effect of temperature and water activity on the growth rate of *Vibrio parahaemolyticus*. *International Journal of Food Microbiology* **38**, 133–142.

Møller C.O.A., Nauta M.J., Christensen B.B. *et al.* (2012) Modelling transfer of *Salmonella* Typhimurium DT104 during simulation of grinding of pork. *Journal of Applied Microbiology* **112**, 90–98.

Montville R. and Schaffner D.W. (2003) Inoculum size influences bacterial cross contamination between surfaces. *Society* **69**, 7188–7193.

Montville R., Chen Y. and Schaffner D.W. (2001) Glove barriers to bacterial cross-contamination between hands to food. *Journal of Food Protection* **64**, 845–849.

Mylius S.D., Nauta M.J. and Havelaar A.H. (2007) Cross-contamination during food preparation: a mechanistic model applied to chicken-borne Campylobacter. *Risk Analysis* **27**, 803–813.

Nauta M.J. (2001) *Modular process risk model structure for quantitative microbiological risk assessment and its application in an exposure assessment of Bacillus cereus*. REPFED, Bilthoven, The Netherlands.

Nauta M.J. (2005) Microbiological risk assessment models for partitioning and mixing during food handling. *International Journal of Food Microbiology* **100**, 311–322.

Nauta M.J., van der Fels-Klerx I. and Havelaar A. (2005) A poultry-processing model for quantitative microbiological risk assessment. *Risk Analysis* **25**, 85–98.

Nerbrink E., Borch E., Blom H. and Nesbakken T. (1999) A model based on absorbance data on the growth rate of *Listeria monocytogenes* and including the effects of pH, NaCl, Na-lactate and Na-acetate. *International Journal of Food Microbiology* **47**, 99–109.

Niven G.W., Fuks T., Morton J.S. *et al.* (2006) A novel method for measuring lag times in division of individual bacterial cells using image analysis. *Journal of Microbiological Methods* **65**, 311–317.

Olofsson P. and Ma X. (2011) Modeling and estimating bacterial lag phase. *Mathematical Biosciences* **234**, 127–131.

Orth J.D., Conrad T.M., Na J. *et al.*, (2011) A comprehensive genome-scale reconstruction of *Escherichia coli* metabolism. *Molecular Systems Biology* **7**, 535.

Oscar T.P. (2009) General regression neural network and Monte Carlo simulation model for survival and growth of salmonella on raw chicken skin as a function of serotype, temperature, and time for use in risk assessment. *Journal of Food Protection* **72**, 2078–2087.

Peleg M. (2006a) Isothermal and nonisothermal bacterial growth in a closed habitat. In: *Advanced Quantitative Microbiology for Foods and Biosystems* (ed. F.M. Clydesdale), pp. 205–241. CRC Press, Boca Raton, FL.

Peleg M. (2006b) Isothermal microbial heat inactivation. In: *Advanced Quantitative Microbiology for Foods and Biosystems* (ed. F.M. Clydesdale), pp. 1–48. CRC Press, Boca Raton, FL.

Pérez-Rodríguez F. and Zwietering M.H. (2011) Application of the central limit theorem in microbial risk assessment: high number of servings reduces the coefficient of variation of food-borne burden-of-illness. *International Journal of Food Microbiology* **153**, 413–419.

Pérez-Rodríguez F., van Asselt E.D., Garcia-Gimeno R.M. *et al.* (2007a) Extracting additional risk managers information from a risk assessment of *Listeria monocytogenes* in deli meats. *Journal of Food Protection* **70**, 1137–1152.

Pérez-Rodríguez F., Valero A., Todd E. *et al.* (2007b) Modeling transfer of *Escherichia coli* O157:H7 and *Staphylococcus aureus* during slicing of a cooked meat product. *Meat Science* **76**, 692–699.

Pérez-Rodríguez F., Valero A., Carrasco E. *et al.* (2008) Understanding and modelling bacterial transfer to foods: A review. *Trends in Food Science and Technology* **19**, 131–144.

Pin C., Baranyi J. and de Fernando G.G. (2000) Predictive model for the growth of *Yersinia enterocolitica* under modified atmospheres. *Journal of Applied Microbiology* **88**, 521–530.

Pitt R.E. (1992) A descriptive model of mold growth and afratoxin formation as affected by environmental conditions. *Journal of Food Protection* **56**, 139–146.

Polese P., Del Torre M., Spaziani M. and Stecchini M.L. (2011) A simplified approach for modelling the bacterial growth/no growth boundary. *Food Microbiology* **28**, 384–391.

Poschet F., Vereecken K.M., Geeraerd a H. *et al.* (2005) Analysis of a novel class of predictive microbial growth models and application to coculture growth. *International Journal of Food Microbiology* **100**, 107–124.

Powell M. (2004) Considering the complexity of microbial community dynamics in food safety risk assessment. *International Journal of Food Microbiology* **90**, 171–179.

Presser K.A., Ratkowsky D.A. and Ross T. (1997) Modelling the growth rate of Escherichia coli as a function of pH and lactic acid concentration. *Applied and Environmental Microbiology* **63**, 2355–2360.

Psomas A.N., Nychas G.-J., Haroutounian S.A. and Skandamis P. (2012) LabBase: Development and validation of an innovative food microbial growth responses database. *Computers and Electronics in Agriculture* **85**, 99–108.

Psomas A.N., Nychas G.-J., Haroutounian S.A. and Skandamis P.N. (2011) Development and validation of a tertiary simulation model for predicting the growth of the food microorganisms under dynamic and static temperature conditions. *Computers and Electronics in Agriculture* **76**, 119–129.

Ratkowsky D.A. (2002) Some examples of, and some problems with, the use of nonlinear logistic regression in predictive food microbiology. *International Journal of Food Microbiology* **73**, 119–125.

Ratkowsky D.A. and Ross T. (1995) Modelling the bacterial growth/no growth interface. *Letters in Applied Microbiology* **20**, 29–33.

Ratkowsky D.A., Lowry R.K., McMeekin T.A. *et al.* (1983) Model for bacterial culture growth rate throughout the entire biokinetic temperature range. *Journal of bacteriology* **154**, 1222–6.

Ratkowsky D.A., Olley J., McMeekin T.A. and Ball A. (1982) Relationship between temperature and growth rate of bacterial cultures. *Journal of Bacteriology.* **149**, 1–5.

Richards F.J. (1959) A Flexible Growth Function for Empirical Use. *Journal of experimental botany* **10**, 290–300.

Roberts T.A. and Jarvis B. (1983) Predictive modelling of food safety with particular reference to *Clostridium botulinum* in model cured meat systems. In: *Food Microbiology: Advances and Prospects* (eds T.A. Roberts and F.A. Skinner), pp. 85–95. Academic Press, New York.

Roberts T.A., Gibson A.M. and Robinson A. (1981) Prediction of toxin production by *Clostridium botulinum* in pasteurised pork slurry. *Journal of Food Technology* **16**, 337–355.

Rodríguez F.P., Campos D., Ryser E.T. *et al.* (2011) A mathematical risk model for *Escherichia coli* O157:H7 cross-contamination of lettuce during processing. *Food Microbiology* **28**, 694–701.

Ross T. (1993) Belehrádek-type models. *Journal of Industrial Microbiology* **12**, 180–189.

Ross T. (1996) Indices for performance evaluation of predictive models in food microbiology. *Journal of Applied Bacteriology* **81**, 501–508.

Ross T. (1999) Assessment of a theoretical model for the effects of temperature on bacterial growth rate. In: *Refrigeration Science and Technology Proceedings*, Quimper, France, pp. 64–71. International Institute of Refrigeration, Paris, France.

Ross T. and Dalgaard P. (2004) Secondary models. In: *Modelling Microbial Responses in Food* (eds R.C. McKellar and X. Lu), pp. 63–150. CRC Press, Boca Raton, FL.

Ross T., Dalgaard P. and Tienungoon S. (2000) Predictive modelling of the growth and survival of Listeria in fishery products. *International Journal of Food Microbiology* **62**, 231–245.

Rosso L., Lobry J.R. and Flandrois J.P. (1993) An unexpected correlation between cardinal temperatures of microbial growth highlighted by a new model. *Journal of Theoretical Biology* **162**, 447–463.

Rosso L., Lobry J.R., Bajard S. and Flandrois J.P. (1995) Convenient model to describe the combined effects of temperature and pH on microbial growth. *Applied and Environmental Microbiology* **61**, 610–616.

Samapundo S., Devlieghere F., De Meulenaer B. *et al.* (2005) Predictive modelling of the individual and combined effect of water activity and temperature on the radial growth of *Fusarium verticilliodes* and F. proliferatum on corn. *International Journal of Food Microbiology* **105**, 35–52.

Sanaa M., Coroller L. and Cerf O. (2004) Risk assessment of listeriosis linked to the consumption of two soft cheeses made from raw milk: Camembert of Normandy and Brie of Meaux. *Risk Analysis* **24**, 389–399.

Sant'Ana A.S., Franco B.D.G.M. and Schaffner D.W. (2012) Modeling the growth rate and lag time of different strains of *Salmonella enterica* and *Listeria monocytogenes* in ready-to-eat lettuce. *Food Microbiology* **30**, 267–273.

Schaffner D.W. (2004) Models – What comes after the next generation? In: *Modelling microbial response in food* (eds R.C. McKellar and X.L. Xu), pp. 303–312. CRC Press, Boca Raton, FL.

Schepers A., Thibault J. and Lacroix C. (2000) Comparison of simple neural networks and nonlinear regression models for descriptive modeling of *Lactobacillus helveticus* growth in pH-controlled batch cultures. *Enzyme and Microbial Technology* **26**, 431–445.

Silva a R., Sant'Ana a S. and Massaguer P.R. (2010) Modelling the lag time and growth rate of Aspergillus section Nigri IOC 4573 in mango nectar as a function of temperature and pH. *Journal of Applied Microbiology* **109**, 1105–1116.

Skandamis P.N., Stopforth J.D., Kendall P.A. *et al.* (2007) Modeling the effect of inoculum size and acid adaptation on growth/no growth interface of *Escherichia coli* O157:H7. *International Journal of Food Microbiology* **120**, 237–249.

Stelling J. (2004) Mathematical models in microbial systems biology. *Current Opinion in Microbiology* **7**, 513–518.

Stringer S.C., George S.M. and Peck M.W. (2000) Thermal inactivation of *Escherichia coli* O157:H7. *Symposium series (Society for Applied Microbiology)*, 79S-89S.

Swinnen I.A.M., Bernaerts K., Dens E.J.J. *et al.* (2004) Predictive modelling of the microbial lag phase: a review. *International Journal of Food Microbiology* **94**, 137–159.

Tienungoon S., Ratkowsky D.A., Mcmeekin T.A. and Ross, T. (2000) Growth limits of *Listeria monocytogenes* as a function of temperature, pH, NaCl, and lactic acid. *Applied Environmental Microbiology* **66**(11), 4979.

Valero A., Pérez-Rodríguez F., Carrasco E. *et al.* (2006) Modeling the growth rate of *Listeria monocytogenes* Using absorbance measurements and calibration curves. *Journal of Food Science* **71**, M257–M264.

Valero A., Hervás C., García-Gimeno R.M. and Zurera G. (2007) Product unit neural network models for predicting the growth limits of *Listeria monocytogenes*. *Food Microbiology* **24**, 452–464.

Valero A., Pérez-Rodríguez F., Carrasco E. *et al.* (2009) Modelling the growth boundaries of *Staphylococcus aureus*: Effect of temperature, pH and water activity. *International Journal of Food Microbiology* **133**, 186–194.

Van Impe J.F., Poschet F., Geeraerd A.H. and Vereecken K.M. (2005) Towards a novel class of predictive microbial growth models. *International Journal of Food Microbiology* **100**, 97–105.

Van Impe J.F., Vercammen D. and Van Derlinden E. (2011) Developing next generation predictive models: a systems biology approach. *Procedia Food Science* **1**, 965–971.

Van Impe J.F., Vercammen D. and Van Derlinden E. (2013) Toward a next generation of predictive models: A systems biology primer. *Food Control* **29**, 336–342.

Vereecken K.M. and Van Impe J.F. (2002) Analysis and practical implementation of a model for combined growth and metabolite production of lactic acid bacteria. *International Journal of Food Microbiology* **73**, 239–250.

Vereecken K., Dens E.J. and Van Impe J. (2000) Predictive modeling of mixed microbial populations in food products: evaluation of two-species models. *Journal of Theoretical Biology* **205**, 53–72.

Verhulst P.F. (1838) Notice sur la loi que la population poursuit dans son accroissement. *Correspondance Mathématique et Physique* **10**, 113–121.

Vorst K.L., Todd E.C.D. and Ryser E.T. (2006a) Transfer of *Listeria monocytogenes* during slicing of turkey breast, bologna, and salami with simulated kitchen knives. *Journal of Food Protection* **69**, 2939–2946.

Vorst K.L., Todd E.C.D. and Rysert E.T. (2006b) Transfer of *Listeria monocytogenes* during mechanical slicing of turkey breast, bologna, and salami. *Journal of Food Protection* **69**, 619–626.

Watier D., Dubourguier H.C., Leguerinel I. and Hornez J.P. (1996) Response surface models to describe the effects of temperature, pH, and ethanol concentration on growth kinetics and fermentation end products of a Pectinatus sp. *Applied and Environmental Microbiology* **62**, 1233–1237.

Westerhoff H.V. and Hofmeyr J.S. (2005) What is systems biology? From genes to function and back. *Systems Biology* **13**, 4–9.

Whiting R.C. and Buchanan R.L. (1994) Microbial modeling. *Food Technology* **48**, 113–120.

Wijtzes T., Rombouts F.M., Kant-Muermans M.L. *et al.* (2001) Development and validation of a combined temperature, water activity, pH model for bacterial growth rate of *Lactobacillus curvatus*. *International Journal of Food Microbiology* **63**, 57–64.

Yoon K.S., Min K.J., Jung Y.J. *et al.* (2008) A model of the effect of temperature on the growth of pathogenic and nonpathogenic *Vibrio parahaemolyticus* isolated from oysters in Korea. *Food Microbiology* **25**, 635–641.

Zuliani V., Lebert I., Augustin J.-C. *et al.* (2007) Modelling the behaviour of *Listeria monocytogenes* in ground pork as a function of pH, water activity, nature and concentration of organic acid salts. *Journal of Applied Microbiology* **103**, 536–550.

Zurera-Cosano G., Castillejo-Rodriguez A.M., Garcia-Gimeno R.M. and Rincon-Leon F. (2004) Performance of response surface and davey model for prediction of *Staphylococcus aureus* growth parameters under different experimental conditions. *Journal of Food Protection* **67**, 1138–1145.

Zurera-Cosano G., García-Gimeno R.M., Rodríguez-Pérez R. and Hervás-Martínez C. (2006) Performance of response surface model for prediction of *Leuconostoc mesenteroides* growth parameters under different experimental conditions. *Food Control* **17**, 429–438.

Zwietering M.H. (2002) Quantification of microbial quality and safety in minimally processed foods. *International Dairy Journal* **12**, 263–271.

Zwietering M.H., Jongenburger I., Rombouts F.M. and van 't Riet K. (1990) Modeling of the bacterial growth curve. *Applied and Environmental Microbiology* **56**, 1875–1881.

Zwietering M.H., Wijtzes T., de Wit J.C. and Van't Riet K. (1992) A decision support system for prediction of the microbial spoilage in foods. *Journal of Food Protection* **55**, 973–979.

Zwietering M.H., de Wit J.C. and Notermans S. (1996) Application of predictive microbiology to estimate the number of *Bacillus cereus* in pasteurised milk at the point of consumption. *International Journal of Food Microbiology* **30**, 55–70.

# 19 Statistical approaches for the design of sampling plans for microbiological monitoring of foods

**Ursula Andrea Gonzales-Barron[1], Vasco Augusto Pilão Cadavez[1] and Francis Butler[2]**

[1] *CIMO Mountain Research Centre, School of Agriculture (ESA) of the Polytechnic Institute of Bragança (IPB), Bragança, Portugal*

[2] *School of Biosystems Engineering, Agriculture and Food Science Centre, University College Dublin, Belfield, Dublin 4, Ireland*

## ABSTRACT

Sampling and testing for microorganisms in foods is a risk management strategy used to evaluate whether a food safety system achieves the appropriate level of control. Sampling plans can be derived to meet the consumer's and/or producer's quality requirements, and have been traditionally designed using classical acceptance sampling theory. This chapter describes in detail and illustrates the methodologies to derive the different type of sampling plans used for microbiological criteria in foods; namely, the two-class (based on prevalence, on concentrations, and with an enrichment step) and three-class attributes sampling plans and the variables sampling plans. The validity of the assumption that the log microbial concentration is normally distributed among food units with a variance that is approximately stable batch to batch is questioned; and, within this context, new modelling trends based on more realistic assumptions are examined that consider the clustering of bacteria, the intrinsic variability among food batches and the use of past monitoring microbial data.

## INTRODUCTION

In recent years, considerable advances have been made in establishing procedures for enhancing the management of microbiological food safety (Stringer, 2005). In many areas of the food chain, microbiological safety is the major risk concern, which has led to a much greater focus on public health and methods for establishing clear health targets. Given the difficulty of using public health goals such as an appropriate level of protection (ALOP) to establish control measures, the risk-based concepts of food safety objectives (FSOs) and performance objectives (POs) were introduced to provide meaningful guidance to food safety management in practice. It is evident that specific POs need to be selected in the food chain that can be linked directly to improvements in public health, such that public health goals begin to drive the performance requirements of the food safety management chain. These performance requirements are regulated through the control measures operated by the good hygiene practices (GHP), good manufacture practices (GMP) and hazard analysis and critical control points (HACCP) systems at diverse points of the food chain (Figure 19.1) (Gorris, 2005). The performance criterion (PC) describes the overall effect of the control measures on the hazard level at a

*Mathematical and Statistical Methods in Food Science and Technology*, First Edition.
Edited by Daniel Granato and Gastón Ares.

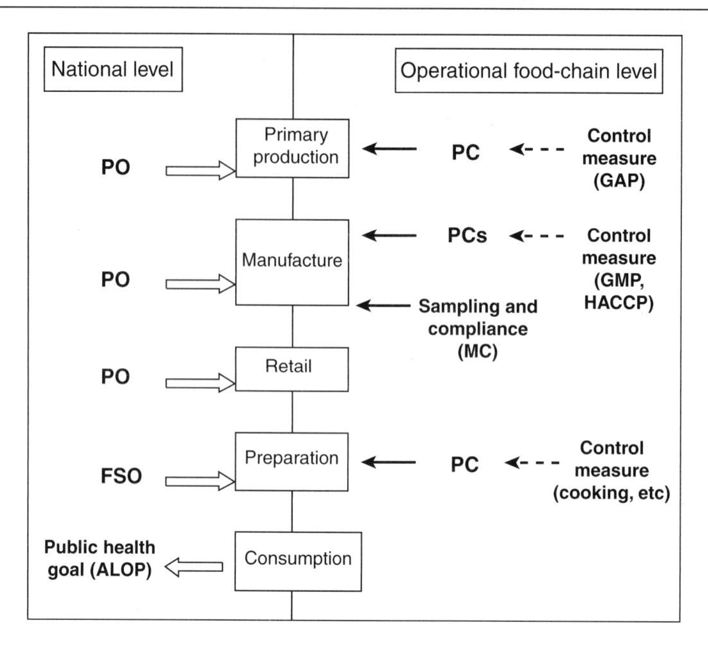


**Figure 19.1** Schematic diagram of the positions of the risk-based microbiological metrics at national level and operational level along an imaginary food chain. The operational level measures are embedded in the food safety management systems operated in the chain, such as GAP, GHP, GMP, HACCP.

step, and is in general decided on by food safety managers at key points in the design of the production of a food in a supply chain. Said otherwise, PCs are the specific operational, supply chain measures at (a) specific step(s) that result in meeting the objective for that step, the PO (Figure 19.1). Currently, the link between public health goal and microbiological targets – in some cases ascertained by microbial risk assessment models – does not exist for all food commodities. Nevertheless, guidance for most food commodities is provided by 'compliance levels' or 'acceptance criteria' in the form of standards, guidelines or specifications. This chapter describes the current considerations for the design of acceptance sampling plans used in the food industry to meet the level of protection of predefined food safety objectives and performance objectives. The statistical basis and assumptions of the different sampling plans, namely, the classical two-class attributes sampling plan, the newly proposed two-class with enrichment, the three-class attributes sampling plan and the variables sampling plan, are revised. The construction of the operating characteristic curve, an important tool for the assessment of the performance of a sampling plan, is illustrated with examples for each of the types of sampling plans. Finally, the chapter addresses the weakness of current assumptions and methodologies, and discusses in depth new modelling trends for the design of more efficient sampling plans, such as the use of past monitoring microbial data, the use of heterogeneous Poisson distributions to better characterize spatial clustering of bacteria, and the allowance for the batch-to-batch variability in microbial concentrations.

## NATURE OF MICROBIOLOGICAL CRITERIA

A microbiological criterion can be used to define the microbiological quality of raw materials, food ingredients and end products at any stage in the food chain, or can be used to evaluate or compare the stringency of alternative food control systems and product and process requirements. The establishment of a microbiological criterion is useful for verifying that food control systems are implemented correctly (ICMSF, 2002). With basis on the knowledge a microbiological criteria are derived, they can be broadly categorized as GHP-based, hazard-based or risk-based:

- *GHP-based microbiological criteria* are generally developed from empirical scientific knowledge and experience and relate to food hygiene. They are, for example, used for verification that hygiene conditions have been applied.
- *Hazard-based microbiological criteria* are developed from scientific knowledge of a likely level of control of a microbiological hazard at a step or series of steps in a food chain and can be validated as to their efficacy in hazard control. There is an expectation of consumer protection but the actual degree of protection will be unknown. They are, for example, used for the verification of the performance of HACCP systems and for lot-by-lot acceptance.
- *Risk-based microbiological criteria* are developed from risk assessment or specific knowledge of the likely levels of consumer protection that will result. They have a quantitative base and should be able to be validated against a level of consumer protection. They can, for example, be used for verification that a PO has been met or by the use of an appropriate risk assessment model to show that the ALOP has been achieved.

A microbiological criterion should be established only when there is a need and when it can be shown to be effective and practical for the stated purpose. Nevertheless, any microbiological criterion should include the ten following components:

(i) the microorganism, toxins or metabolite and the reason for selection;
(ii) the food, process or environment to where the criterion applies;
(iii) a sampling plan defining the number of samples to be taken and the size of the analytical unit;
(iv) the analytical method to be used to detect and/or quantify the microorganism(s) or their toxins/metabolites;
(v) the specific point(s) in the food chain where the microbiological criterion should be applied;
(vi) the microbiological limit(s) considered appropriate to the food at the specified point(s) of the food chain;
(vii) the sampling plan defining the number and size of samples to be taken;
(viii) the number of samples that should conform to the microbiological limit(s);
(ix) any indication of the statistical performance of the sampling plan; and
(x) the actions to be taken when the criterion is not met.

Within the risk analysis context, microbiological criteria based on within-batch testing are meant to provide a statistically-sound means for determining whether the FSO/PO targets are being achieved. As in any sampling plan, two types of errors can result from decisions based on the results from the test samples. Thus, the derivation of a sampling plan needs to consider either of the following errors in order to reduce risks to a minimum yet without the need for excessive sampling:

(i) The probability that an ‘acceptable batch’ will be rejected by the sampling scheme (Type I or $\alpha$ error). This error is also called producer’s risk as it may bring about economic losses.
(ii) The probability that a ‘bad batch’ is accepted (Type II or $\beta$ error). This error is also called consumer’s risk as ultimately it may involve adverse health effects.

## TYPES OF SAMPLING PLANS USED IN MICROBIOLOGICAL CRITERIA IN FOODS

The microbiological criteria for foodstuffs, laid down in Regulation EC No 1441 (EC, 2007), cover two types of sampling plans, which in classical acceptance sampling theory are classified as by attributes and by variables (Duncan, 1986). Table 19.1 compiles the parameters defining each of these types of sampling plans within the context of microbiological criteria.

A simple way to decide whether to accept or reject a food batch may be based on some microbiological test performed on several sample units. For pathogens, this will usually be a test for the presence (positive) or absence (negative) of the microorganism. If instead concentrations of microorganisms are measured, the result of the individual sample can be assigned to a particular attribute class by determining whether it is above (positive) or below (negative) some preset concentration. The decision making process of a *two-class attributes sampling plan* is essentially defined by two numbers. The first, denoted as $n$, determines the number of sample units that have to be drawn independently and randomly from the batch. The second number, denoted as $c$, is the maximum allowable number of defective sample units. In microbiological testing, the term *defective* implies that the individual sample contains more than the specified number of microorganisms $m$, when tested by an appropriate test, or in the case of a presence/absence test, that the target microorganism is detected. In the former case of grouped quantitative data, there is one microbiological limit, denoted by $m$, which separates good quality from defective quality. In this case, the maximum allowable number of sampling units exceeding this limit is given by $c$, which is usually set to zero for pathogens (Jarvis, 2008). Two-class attributes sampling plans are used for more stringent microbiological criteria testing pathogens or toxins in foods, and the microbiological limit $m$ is commonly expressed as 'absence in a certain sample weight' (for

**Table 19.1** Types of acceptance sampling plans currently used to express microbiological criteria in foods.

| Sampling plan | Parameters | Definition | Uses |
|---|---|---|---|
| By attributes | | | |
| a) Two-class | $c$ | Maximum number of sample units giving microbial concentrations over $m$ for the batch to be 'accepted' | Stringent sampling plans to assess presence of pathogens in foods. Unsatisfactory results should lead to batch rejection. |
| | $n$ | Number of samples to test or sample size | |
| | $m$ | Microbiological limit, commonly expressed as absence in certain sample weight | |
| b) Three-class | $c$ | Maximum number of sampling units giving microbial concentrations between $m$ and M and still allow the batch to be 'accepted' | Less stringent sampling plans to assess hygiene indicators. Unsatisfactory results should lead to improvements in production hygiene. |
| | $n$ | Sample size | |
| | $m$ | Lower microbiological limit of an individual sample | |
| | M | Upper microbiological limit of an individual sample. If any sample gives results exceeding M, the batch is rejected | |
| By variables | $n$ | Sample size | Sampling plans more informative than by attributes but of limited use. Unsatisfactory results should lead to improvements in production hygiene. |
| | $m$ | Lower microbiological limit. If the sample's average is below $m$, the batch is considered of *satisfactory* hygiene. | |
| | M | Lower microbiological limit. If the sample's average falls between $m$ and M, the batch is considered of *acceptable* hygiene. Beyond M, the batch is unsatisfactory. | |

instance, the safety criterion for milk powder is based on absence of *Salmonella* in 25 g ($m$) in all of the five individual samples taken ($c = 0$, $n = 5$)).

In situations where decisions are not based on results of presence-absence tests but on quantitative analytical results, three-class attributes sampling plans can be applied as an alternative to two-class plans working with data grouped according to a single microbiological limit, $m$. In a three-class sampling plan, the quality of food batches can be divided in three groups. The marginally defective sample is defined as one that contains a number of microorganisms lower than a specified upper limit $M$ but a greater number than a lower (acceptable) specified limit $m$. However, sample results above the upper concentration $M$ are unacceptable (or defective) and usually a batch is rejected if the test from any sample unit exceeds $M$ (Dahms, 2004). Jarvis (2008) argues that a marginally defective grouping may be considered in order to make some allowance for variation in the distribution of microorganisms in the food and for the imprecision associated with colony count procedures.

In a sampling plan by variables, microbiological measurements are made on a series of samples taken from a batch. A batch is considered of acceptable quality when the mean value of the microbial counts from the individual samples does not exceed a microbiological limit $m$. As such plans make full use of microbial counts, rather than ascribing them to categories or classes, variables plans can be more useful under some conditions than attributes plans. Although the formulation of the decision rule ($n$, $m$) in a variables sampling plan may be more complex than for attributes sampling plans, this procedure has the advantages of using a smaller sample size and that the definition of a conforming batch and the desired confidence in decision making become more transparent (Dahms, 2004).

## CLASSICAL ACCEPTANCE SAMPLING THEORY

The performance of a given sampling plan as well as its design can be assessed by the construction of 'operating characteristic' (OC) curves. OC curves are used as an important tool to measure the performance of a sampling plan, allowing the assessment of the relative efficiencies of various potential sampling plans. The derivation of OC curves for both attributes and variables sampling plans have been traditionally based on classical acceptance sampling theory (Duncan 1986), and is a graphical representation that relates the probability of accepting a batch, based on the number of food units tested, to the mean quality of the batch.

In attributes sampling plans (i.e. each unit in the sample is judged to be either conforming or nonconforming), the OC curve plots the cumulative probability of acceptance (*Pa*) conditional to the percentage of defective sample units in a batch (the batch fraction defective). However, in food microbiology, OC curves for attributes sampling plans have been produced translating the proportion defective of a batch to the mean quality characteristic, this is converting prevalence into an estimate of the mean concentration of microorganisms in the batch, by assuming that the underlying statistical distribution of a microorganism in a batch is lognormal (Hildebrandt *et al.*, 1995; Dahms and Hildebrandt, 1998; Legan *et al.*, 2001; Dahms, 2004; Van Schothorst *et al.*, 2009). In a variables sampling plan (i.e. the quality characteristic of each unit in a sample is measured), the OC curve similarly displays the discriminatory power of a sampling plan with *Pa* plotted against the mean microbial concentration of the batch; nevertheless, it is constructed in a different way. While the primary advantage of variables sampling plans is that the same OC curve can be obtained with a smaller sampling size than would be required by an attributes sampling plans, its use is very limited, almost certainly due to lack of research in this field.

An OC curve, which assesses the performance of indistinctively a sampling plan by attributes or variables, is shown in Figure 19.2. Interest focuses on two points of the OC curve. The acceptable quality level (AQL) defines the quality (mean microbial concentration) of good batches that the purchaser/consumer is prepared to accept most of the time (set at the probability $P_1$). Hence, the producer's risk (type I

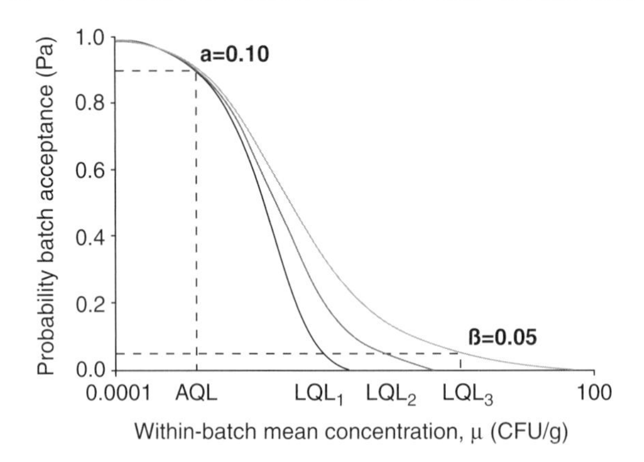


**Figure 19.2** Operating characteristic curves showing the points of interest defining an acceptance sampling plan: the acceptable quality level (AQL) at a producer's risk $\alpha$ and the limiting quality level (LQL) at a consumer's risk $\beta$. The steeper the curve the higher the discriminatory power of the sampling plan.

error or $\alpha$) of rejecting good quality batches is $1-P_1$. The producers are interested in determining the batch quality that would need to be achieved, so that there is a high probability that batches would be accepted, and adjust their production processes accordingly (Van Schothorst *et al.*, 2009). The limiting quality level (LQL) point denotes the quality of poor batches that the purchaser/consumer wishes to reject as often as possible (at the probability $P_2$). Hence, the consumer's error (type II error or $\beta$) of accepting bad quality batches is $1-P_2$. It is assumed that the producer is operating at a fallout level that is considerably better than the LQL. Specifications of either the AQL or LQL points and their probabilities of acceptance make it possible to design sampling plans; and setting either the AQL or the LQL implies the other.

OC curves can be used to evaluate the influence that parameters of the microbiological criteria ($n$, $m$, $c$), and the mean and standard deviation of the underlying batch distribution have on the efficiency of the microbiological testing programme. This information quantifies the confidence that we can have that an unacceptable or defective batch will be rejected. If one were able to test every unit of food within the batch, the OC curve would change from 100% probability of acceptance to a 100% probability of rejection exactly at the quality level that distinguishes an acceptable from a defective batch. At the other extreme, taking a single sample, particularly if negative, has virtually no ability to discriminate between conforming and nonconforming batches. Increasing the number of samples ($n$) examined is one of the primary means for increasing the ability of a sampling plan to discriminate acceptable from defective batches. A higher sample size will produce a steeper OC curve, indicating a higher discriminatory power of the sampling plan (Figure 19.2).

## Design of two-class attribute sampling plans (n, c), (n, c, m) and (n, c, absence in w)

### *Based on microbial prevalence (n, c)*

In its simplest form, an OC curve can be constructed based on the true within-batch prevalence ($d$) of the target microorganism as the quality measure. In that case, the probability of accepting a batch ($Pa$) after sampling with a regime ($n$, $c$) reduces to a cumulative binomial problem.

$$Pa = P(x \leq c) = \sum_{C=0}^{c} \binom{n}{c} d^c (1-d)^{n-c} \tag{19.1}$$

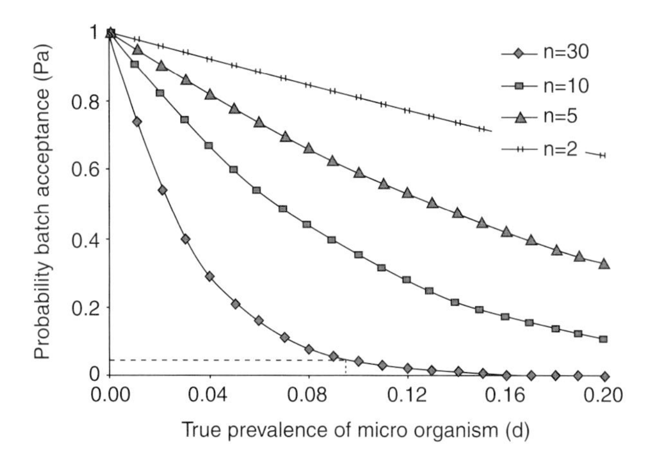


**Figure 19.3** Operating characteristic curves for two-class attributes sampling plans based on prevalence.

In stringent sampling plans, normally there is no allowance for positive samples (i.e. the value of $c$ is set to zero); hence $Pa$ becomes:

$$Pa = P(x = 0) = (1 - d)^n \tag{19.2}$$

The OC curve is then simply the plot of the probability of accepting a batch ($Pa$) of a given microbial prevalence under the sampling plan ($c = 0, n$). Observing the OC curves of Figure 19.3, we can see that, if negative results are obtained on 30 samples, and the true prevalence is 1%, the likelihood of accepting the batch is 74% (conversely, the probability of rejection is 26%), but if only five samples were tested, the probability of accepting the batch is 95%. Similarly, if the batch contained 10% true defective units, then the likelihood of product acceptance after testing ten sample units with negative results will be only 35% but if two samples are tested, the probability of acceptance is 81%.

In setting a criterion for the number of samples to be tested, an assumption is made about the incidence of true defectives (this is, the proportion of units produced in a batch which exceeds a given microbiological limit) that are acceptable. For example, a sampling plan $n = 30$, $c = 0$ has an AQL of $d = 0.095$ at a $\beta$ risk of 5% (Figure 19.3). This means that such a sampling plan would give a confidence of *at least* 95% of rejecting batches having 9.5% or more defective units. Thus, for any given true prevalence in a batch, and assuming that the distribution of contaminated units among the batch is random, the sample size that would be necessary for attain a given confidence can be determined:

$$n = \frac{\log(Pa)}{\log(1 - d)} \tag{19.3}$$

Using the same binomial expression (Equation 19.2), it is possible to estimate confidence limits for the true prevalence of contamination at a given confidence level $(1 - \alpha)$, when all sample units test negative (Jarvis, 2007). The confidence limits are bounded by 0 and $d_u$, where $d_u$ is chosen so that if the true prevalence were $d_u$, then the probability of the observation of no positive results $P(x = 0)$ in $n$ samples is equal to $\alpha$. Thus, rearranging Equation 19.2:

$$d_u = 1 - \sqrt[n]{\alpha} \tag{19.4}$$

For example, if no positive samples were found from a sample size $n = 10$, we can estimate that the true prevalence of contamination lies within 0 and 0.26 ($d_u$) with a 95% confidence ($\alpha = 0.05$).

From the estimate of the upper confidence limit of the true prevalence, it is also possible to approximate an upper confidence limit for the mean microbial concentration ($\mu$) in the batch tested. Assuming that the microbial cells are randomly distributed among food units, and that the test procedure is sufficiently sensitive to detect the microorganism if present, the probability $P(x=0)$ of a negative results on any sample is given by the Poisson probability of finding zero organisms:

$$P(x = 0) = \exp(-w\mu) \tag{19.5}$$

where $w$ is the sample weight. Hence, following the previous example, if 10 samples of $w=25$ g with negative results ($d_u = P(x=0) = 0.26$) are tested, the upper limit of the mean microbial concentration estimated using Equation 19.5 would be 0.012 cell/g; meaning that we have a 95% confidence that the mean microbial concentration lies within 0–12 cells/kg.

## Example 19.1

As discussed earlier, the sampling plans established in microbiological criteria should be designed to satisfy or test compliance with a pre-established PO. The following example is an illustration of the derivation of a sampling plan to test *Salmonella* spp. in frozen chicken, where a PO has been formulated as proportion of defectives in a batch. Suppose that the PO is formulated as: 'not more than 15% chicken carcasses in a batch may test positive for *Salmonella*', and that the confidence that a defective batch is rejected is set at 95% probability. Replacing in Equation 19.3 the true microbial prevalence $d$ by 0.15 and the probability of accepting the batch $Pa$ by 0.05, the sample size $n$ equals 18.5. Thus, a batch of frozen chicken will be compliant to the PO if none of the 19 samples randomly withdrawn tests positive for *Salmonella*. Considering that the analytical sample unit ($w$) is 5 g of neck skin, it can further be said that 95% of the times that a sampled batch is labelled as compliant, the within-batch mean *Salmonella* concentration ($\mu$) in frozen chicken is not higher than 0.38 CFU/g (Equation 19.5).

### *Based on microbial concentration (n, c, m)*

When the result of a microbiological analysis is given in a quantitative manner (CFU/g or MPN/g), the batch quality may be then described by a statistical distribution of the concentration of microorganisms among the food units produced in a batch (instead of the true prevalence measure). For this, an assumption must firstly be made regarding the statistical distribution of the microorganisms in a batch (i.e. this is also referred to as within-batch distribution of variability). Traditionally, the lognormal distribution (normal distribution of the logarithm of the microbial concentrations) has been assumed although, as is discussed later, this assumption is more valid for high contamination than for low contamination (Gonzales-Barron and Butler, 2011a). In the absence of available data, a lognormal distribution is often assumed and a default value for the standard deviation applied. Van Schothorst *et al.* (2009) used standard deviation values of 0.2 log CFU/g to describe a rather homogeneous spread of microbes in food within a batch (e.g. for liquid food with a high degree of mixing); 0.4 log CFU/g for food of intermediate homogeneity (e.g. ground beef); and 0.8 log CFU/g for an inhomogeneous food (e.g. solid food). Higher standard deviation values may be used for larger inhomogeneity although in such cases it is rather advisable to consider another family of distributions, such as the heterogeneous Poisson, which characterize far better high microbial clustering and high proportion of zero counts (Gonzales-Barron *et al.*, 2010).

Apart from having some information on the within-batch standard deviation, it is necessary to know the detection limit of the analytical procedure, as its value is normally assigned to the microbiological limit $m$, which is a measure that separates a conforming sample unit (microorganism nondetected by the

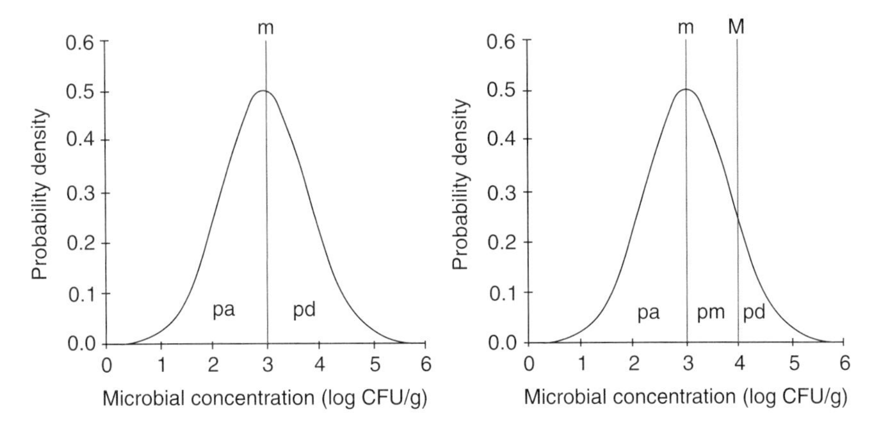


**Figure 19.4** Estimation of the proportion of acceptable food units ($p_a$) and defective food units ($p_d$) for two-class sampling plan (left); and the proportion of marginally acceptable units ($p_m$) for a three-class sampling plan (right), given a lognormal distribution of the microbial concentration in a batch with mean $\mu = 3.0$ log CFU/g and standard deviation $\sigma = 0.8$ log CFU/g.

microbiological protocol) from a defective sample unit (microorganism detected) (Figure 19.4, left). A plate counting technique able to quantify *Listeria monocytogenes* above 10 CFU/g has a $m = 1$ log CFU/g, while a 1/0.1/0.01 g three-tube MPN procedure with a lower limit of detection of 0.3 CFU/g has a $m = -0.52$ log CFU/g. If the analytical procedure was sensitive to extremely low levels of contamination, batches that should be accepted would have samples testing positive and be erroneously rejected (Whiting *et al.*, 2006)

Assuming that bacteria are lognormally distributed in the product and that the standard deviation is known, an OC curve can be established by using the statistical distribution of microbial concentration to calculate the proportion of acceptable ($p_a$) and defective ($p_d$) food units in a batch (Hildebrandt *et al.*, 1995; Legan *et al.*, 2001; Dahms, 2004). The *probability of extracting one defective sample* ($p_d$) from the batch can be used to determine the probability of accepting the batch after testing (*Pa*) in the usual way by a cumulative binomial distribution. For a given within-batch microbial mean concentration $\mu$, the area under the probability density function above the microbiological limit $m$ is used the define the value of proportion defective ($p_d$) for a two-class sampling plan (Figure 19.4, left). The value of $p_d$ is calculated as $1 - p_a$, where $p_a$ is the cumulative probability of the normal distribution ($\mu$, $\sigma$) evaluated at $m$.

The proportion of acceptable units $p_a$ can be calculated in Microsoft® Excel using the 'Normdist' function, $p_a$ = Normdist ($m$, $\mu$, $\sigma$, 1). The next step is then to calculate the probability of accepting the batch (*Pa*) after the sampling plan ($n$, $c$, $m$) assuming a binomial process where $p_d$ becomes the probability of success (i.e. finding a positive sample). The value of *Pa* can be computed as, *Pa* = Binomdist ($p_d$, $n$, $c$, 1). This procedure, repeated for a range of mean microbial values $\mu$, will yield different *Pa* values, which coupled will produce the OC curve of the sampling plan ($n$, $c$, $m$). Notice that in this way OC curves in terms of mean concentrations are developed by fixing the standard deviation $\sigma$, and then increasing the mean of the normal distribution through a range of values. This implies that $\sigma$ is assumed invariable for a little or highly contaminated batch.

From an OC curve, two important measures can be determined: the mean microbial concentration at which the batch will be accepted with 95% probability (AQL) and the mean concentration at which the batch will be rejected with 95% probability (LQL). Reading off the AQL and LQL values from an OC curve may be imprecise on steeply sloping curves. As a quicker and more precise alternative, the 'goal seek' feature of Microsoft® Excel may be used to find the mean microbial concentrations that correspond to $Pa = 0.95$ and $Pa = 0.05$, respectively. Figure 19.5 illustrates OC curves derived for

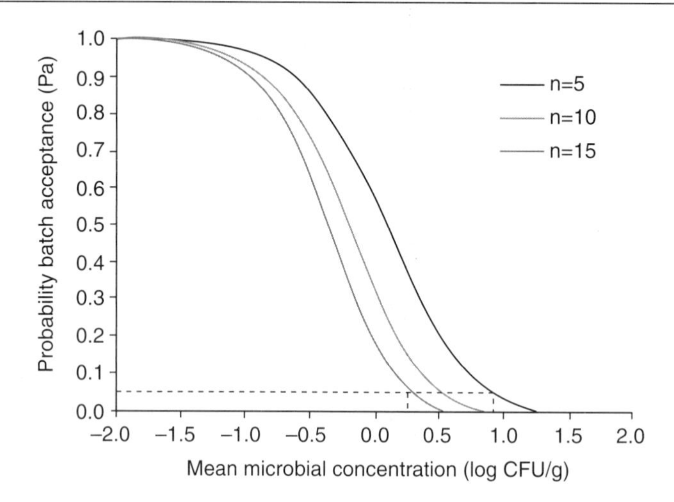


**Figure 19.5** Operating characteristic curves for two-class attributes sampling plans of microbiological limit m = 1 log CFU/g and different sampling size n, assuming a within-batch lognormal distribution of microbial counts with standard deviation $\sigma = 0.8$ log CFU/g.

three sampling plans with microbiological limit 10 CFU/g ($m = 1$ log CFU/g) and different sample size $n$. They were constructed assuming that the expected standard deviation of the lognormal distribution representing the microbial counts among food units in a batch is $\sigma = 0.8$ log CFU/g. Compliance with a sampling plan ($n = 5$, $c = 0$, $m = 10$ CFU/g) ensures that batches with mean contamination equal or higher than 0.9 log CFU/g (LQL = 8 CFU/g) have at least 95% probability of being rejected. Batches with lower levels of contamination would have lower chance of being rejected and greater chance of being accepted (Figure 19.5). For comparison, a sample size $n = 15$ (producing a LQL = 0.24 log CFU/g) implies that now batches of lower contamination level (LQL = 1.7 CFU/g) will have at least 95% chance of being rejected (Figure 19.5). A greater sample size translates into a more discriminatory and stringent sampling plan, which is characterized by a steeper OC curve. Finally, notice that an increased sample size has a greater effect on the LQL value than on the AQL value.

The performance of an attributes sampling plan is also dependent on the validity of the underlying assumption for the frequency distribution, especially with regard to its standard deviation $\sigma$. Errors in $\sigma$ can greatly affect the target AQL value at which the industry intends to operate (Figure 19.6). For instance, if the heterogeneity of the microorganisms among food units was underestimated by assuming $\sigma = 0.6$ instead of an 'actual but unknown' 1.0, the sampling plan ($n = 5$, $c = 0$, $m = 1.5$ log CFU/g) would wrongly lead the producer to target its mean microbial concentration in a batch (AQL) to a maximum of 0.1 log CFU/g (instead of −0.8 log CFU/g). At 0.1 log CFU/g, the producer would expect that 95% of the batches will be positively accepted, when in fact that quality level will only provide 65% confidence (Figure 19.6). A higher standard deviation means that the mean microbial concentration that must be achieved so as to not exceed the PO/FSO must be decreased; and that a greater sample size should be tested to assess compliance to a given safety target. Whiting *et al.* (2006) indicate that taking a larger sample, homogenizing or compositing samples may reduce the variation between samples within the batch, particularly in nonhomogeneous foods. This may reduce the standard deviation of the samples and will allow a shift in the distribution of the LQL batch to a higher batch mean and $m$ value that may be easier to detect. Nonetheless, with prior experience substantiating the assumption of the standard deviation (i.e. availability of historical monitoring microbial data), attributes sampling plans can still be used to assess mean microbiological concentrations in batches of food.

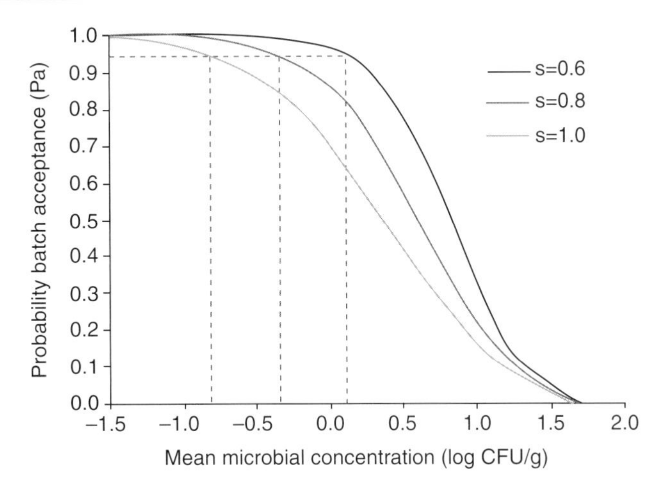


**Figure 19.6** Effect of the within-batch standard deviation ($\sigma$) on the performance of the sampling plan n = 5, c = 0, m = 1.5 log CFU/g.

## Example 19.2

When designing a sampling plan from knowledge of the FSO/PO in terms of maximum microbial concentration, a second requirement is the proportion (95%, 99%, 99.9%, etc.) of the distribution of possible concentrations that must satisfy the test limit so that FSO/PO is met. Suppose that the wish is to derive a sampling plan to ensure compliance of *Listeria monocytogenes* levels in cold-smoked salmon to a hypothetical maximum concentration at consumption of FSO = 2.3 log CFU. Assuming that during the shelf-life of two weeks from the point of manufacture to the point of consumption when stored at 5°C or below, there is a maximum increase of 0.6 log CFU/g; the required PO would then be 1.7 (2.3–0.6). Furthermore, to ensure that the PO would be met by 99% of the food units in such a 'limiting quality' batch, the mean concentration of the LQL batch should be 2.33 × $\sigma$ below the calculated PO (Figure 19.7). Thus, assuming that the within-batch standard deviation is $\sigma = 0.8$, the

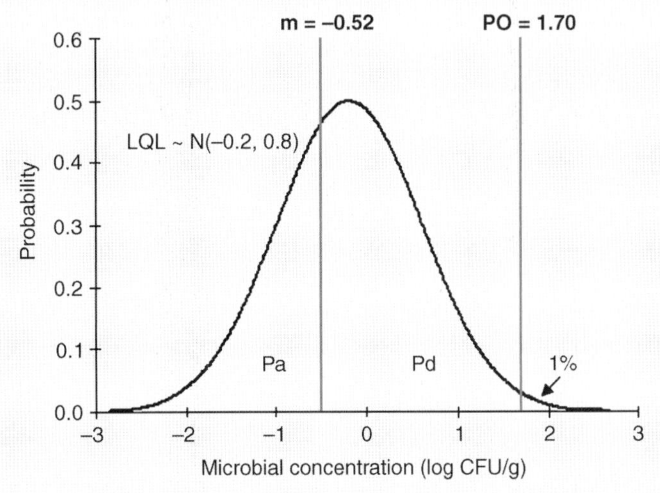


**Figure 19.7** A LQL distribution used to design a sampling plan of *Listeria monocytogenes* in cold-smoked salmon ensuring that the hypothetical PO of 1.70 log CFU/g would be met by 99% of the food units in the batch.

mean microbial concentration of the LQL batch is 1.7 − 2.33 × 0.8 = −0.2 log CFU/g. Batches with mean contamination levels higher than LQL = Normal (−0.2, −0.8) should be rejected at least 95% of the times ($Pa = 0.05$). If the three-tube MPN procedure has a limit of quantification of 0.3 *Listeria monocytogenes*/g, the $m$ value can be set at −0.52 log MPN/g. The next step is then to calculate the proportion $p_a$ of acceptable food units in the LQL batch, which is estimated as the cumulative probability evaluated at $m$. In Microsoft® Excel notation, this is $p_a$ = Normdist ($m$, $\mu$, $\sigma$,1) = Normdist (−0.52, −0.2, −0.8, 1) = 0.345. The value of $p_d$ will then be 0.655. Replacing in Equation 19.3, where now $d = p_d$, n = log $Pa$/(log(1 − $p_d$)) = 2.8 ~ 3. Hence, a sampling plan (n = 3, c = 0, m = −0.52 log CFU/g) would ensure that whenever the microbial concentration in a batch exceeds the PO, the batch would be rejected with not less than 95% confidence.

### *Based on microbial concentration with an enrichment step (n, c, w)*

Van Schothorst *et al.* (2009) modified the approach proposed by Legan *et al.* (2001) for a two-class attribute testing where a specified number of analytical units are cultured via enrichment and then assessed for presence/absence rather than enumeration. If microbial cells were evenly distributed in the samples and were present at the level of one cell per sample unit, not every sample would be expected to be positive for growth: some samples selected at random would contain one or more cells, and produce a positive result, while others not. Thus, when testing absence of a microorganism in an analytic unit (i.e. $m$ = absence in 25 g), it is necessary to consider the consequences of sampling *coincidences* (i.e. the detection of a cell in a set of samples even when such detection is highly improbable based on the mean concentration of the organism in the batch) on the interpretation of the results of analytical methods. The probability of detecting cells ($P_{detect}$) by randomly sampling from a well-mixed system can be described by a Poisson distribution. Re-arranging Equation 19.5:

$$P_{\text{detect}} = P(x > 0) = 1 - \exp(-wC) \tag{19.6}$$

In other words, if the concentration of the cells in the sample ($C$) is perfectly homogeneous, the sampling and enumeration method will sometimes overestimate or underestimate the concentration. However, the aim is to determine by sampling whether the mean concentration in the batch is such that less than 1% of the batch exceeds the PO. As long as the distribution of microbial concentrations is known, it is possible to calculate the *overall probability of detecting a cell* from any sample drawn from a batch. This is because the overall probability of obtaining a positive sample ($p_d$) is the product of the probability of that concentration occurring in the batch, and the probability of detecting a cell in the sample based on the weight of the analytical unit ($w$) and the concentration of cells in the sample ($C$). If the distribution of microbial concentrations can be assumed to be lognormal, the probability of sampling any particular concentration is given by the lognormal distribution, and is combined with the Poisson sampling process to calculate the probability that a cell is present in the sample. Mathematically, this is expressed by the (mixed) Poisson-lognormal distribution:

$$p_d = \int_{-\infty}^{\infty} P_{normal}(\log C, \mu, \sigma) \cdot P_{\text{detect}} d \log C = \int_{-\infty}^{\infty} P_{normal}(\log C, \mu, \sigma) \cdot (1 - \exp(-wC)) d \log C \tag{19.7}$$

Once the probability of extracting one defective sample ($p_d$) from the batch is calculated, the probability of accepting the batch ($Pa$) is calculated from the binomial process ($Pa$ = Binomdist ($p_d$, $n$, $c$, 1)

| | A | B | C | D | E | F | G |
|---|---|---|---|---|---|---|---|
| 1 | **Construction of an OC curve for a sampling plan of the type (n, c, w)** | | | | | | |
| 2 | | | | | | | |
| 3 | INPUT VALUES | | | Weight = | 25 | g | |
| 4 | Sigma = | 0.4 | log CFU/g | n = | 10 | | |
| 5 | | | | c = | 0 | | |
| 6 | | | | | | | |
| 7 | Mean log | Normal | Poisson | flag | $p_s$ | $p_d$ | Pa |
| 8 | -2 | **=RiskNormal(A8,$B$4)** | **=RiskPoisson((10^B8)*$E$3)** | **=IF(C8=0,1,0)** | **=RiskMean(D8)** | **=1-E8** | **=BINOMDIST($E$5,$E$4,F8,1)** |
| 9 | -1.9 | -2.184243244 | 0 | 1 | 0.725 | 0.274 | 0.040 |
| 10 | -1.8 | -2.167406964 | 0 | 1 | 0.669 | 0.330 | 0.018 |
| 11 | -1.7 | -1.644919527 | 0 | 1 | 0.603 | 0.396 | 0.006 |
| 12 | -1.6 | -1.931329466 | 0 | 1 | 0.594 | 0.405 | 0.005 |
| 13 | -1.5 | -1.841656364 | 1 | 0 | 0.525 | 0.474 | 0.001 |
| 14 | -1.4 | -1.612998431 | 0 | 1 | 0.404 | 0.595 | 0.000 |
| 15 | ... | | | | | | |

**Figure 19.8** An overview of a spreadsheet with @Risk® add-in used to construct by simulation operating characteristic curves of attributes sampling plans by enrichment (n, c, w) based on the Poisson-lognormal distribution.

in Microsoft® Excel notation). Unfortunately, the Poisson-lognormal distribution does not have a closed form from which $p_d = 1 - P(x = 0)$ can be readily estimated (as in the Poisson distribution, Equation 19.6); yet an easy way to solve for Equation 19.7 is through Monte Carlo simulation. Figure 19.8 shows a spreadsheet for the construction of OC curves for attributes sampling plans with an enrichment step, which uses @Risk software to estimate by simulation $p_d$ values for a range of within-batch mean microbial concentrations. The formulae for the Poisson-lognormal distribution are indicated in Figure 19.8. Alternatively, OC curves for the sampling plans with enrichment (Van Schothorst *et al.*, 2009) can be assessed by using a downloadable spreadsheet file developed by ICMSF (2009), based on the assumptions of the lognormal distributions of microbes in foods and the fixed within-batch standard deviation. This application can also be used to assess the performance of the simplest two-class and three-class attributes sampling plans proposed by Legan *et al.* (2001).

## Example 19.3

Suppose that the requirement is to determine the appropriate sampling plan for *Salmonella* in ice cream that is compliant with a hypothetical PO of −1.4 log CFU/g (0.04 CFU/g), and that there is the constraint that the maximum number of units microbiologically analysed should not exceed ten. Fixing $n = 10$, the problem reduces to find the optimal sample weight $w$. Assuming that the concentration of *Salmonella* among ice cream units produced in a batch follows a lognormal distribution with standard deviation $\sigma = 0.4$, OC curves for different sample weights ($w$) were derived by simulation using the spreadsheet shown in Figure 19.8. To ensure that a batch of ice cream in which more than 1% of the units have a concentration higher than −1.4 log CFU/g (PO) would be rejected with 95% confidence, it is first necessary to calculate the LQL. The mean concentration of the LQL batch is LQL $= -1.4 - 2.33 \times 0.4 = -2.33$ log CFU/g, and should be rejected with at least 95% confidence ($Pa = 0.05$). Analysis of the OC curves constructed by simulation (Figure 19.9) indicates that a sample weight of 50 g would ensure this level of safety (i.e. notice that for this OC curve, the AQL for $Pa = 0.05$ is ~2.3). While a sample weight of 25 g will be insufficient to assure the PO level of safety, the sample weight of 100 g will only lead to a stricter (lower) PO. Furthermore, the defined sampling plan ($n = 10$, $c = 0$, absence in $w = 50$ g) demands that the producer should target its production at a safer level so that good batches be accepted at least 95% of the times. From Figure 19.9, the extrapolated AQL is −4.15 log CFU/g,

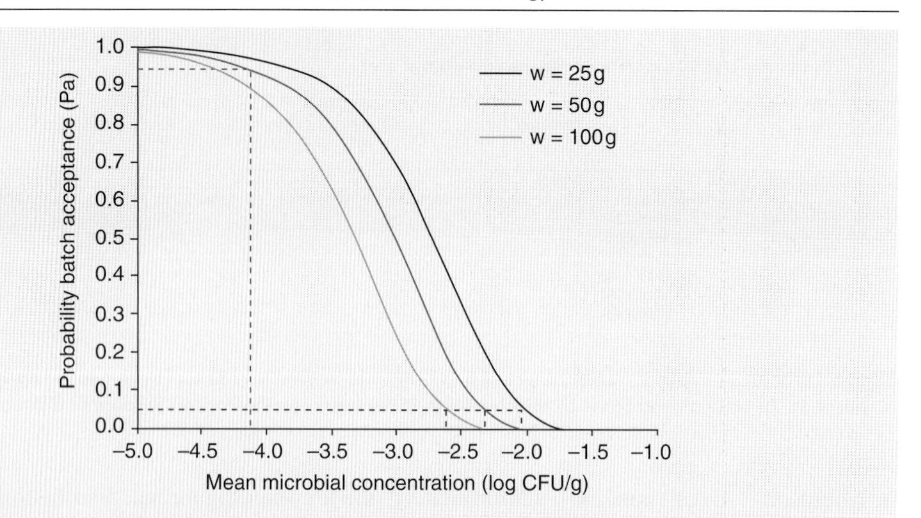


**Figure 19.9** Effect of the sample weight (w) on the performance of a sampling plan testing absence of *Salmonella* in n = 10 ice cream samples, assuming a within-batch standard deviation $\sigma = 0.4$.

meaning that the producer should aim to produce ice cream batches of mean *Salmonella* concentration levels below −4.15 log CFU/g (<7 CFU/100 kg).

## Design of three-class attribute sampling plans (n, c, m, M)

In two-class sampling plans, the quantitative data (CFU/g) is converted into attributive data by two steps: firstly, a contamination limit $m$ is fixed, and then the transformation itself is carried out by assigning all test results to groups according to their being above or below the previously determined limit. On the one hand, transforming a piece of quantitative information into an attribute criterion avoids the difficulty of having to determine the variability; but, on the other hand, the level of information is lowered considerably by this manipulation: one can derive neither the actual amount of variation within the batch nor how far away the single results are from the limit (Hildebrandt *et al.*, 1995). In order to retain more of the information contained in the quantitative data, Bray *et al.* (1973) developed the three-class attribute sampling plan. In this sampling plan, one random sample of $n$ units is tested for two different limits at the same time. The lower microbiological limit $m$ is equivalent to the upper limit of a good manufacturing practice (GMP), and the other is the upper microbiological limit $M$, which marks the borderline beyond which the quality is no longer acceptable. Thus, this plan discerns three classes of contamination: the acceptable range from 0 to $m$, the marginally acceptable class from $m$ to $M$; and the defective quality class above $M$. An acceptance number $c$ may be assigned to each of the two limits $m$ and $M$, although in microbiological criterion the value assigned to $M$ is always zero. A batch will be rejected if at least one individual sample exceeds the limit $M$ and/or if more than $c$ individual samples range above the limit $m$.

Statistical analysis demonstrates that for batches with mean contamination levels near to $m$, the frequency of batch rejection due to a single sample result exceeding $M$ increases if the distance between $m$ and $M$ decreases and/or if the sample variation increases (Dahms and Hildebrandt, 1998). Actually, when a limit $M$ is introduced into microbiological sampling plans, the batch heterogeneity is taken into account. The distance between $m$ and $M$ contains information on the maximum level of variability that is accepted when the three-class sampling plan is applied. Therefore, the design of such sampling plans could be based on knowledge about production technologies leading to values for $m$ and $M$ that take into account the maximum level of heterogeneity under conditions of good manufacturing practice. Especially with regard to nonpathogenic microorganisms, such as total bacterial count or hygiene indicators, which represent no health risk for the consumer, it is hardly recommendable to reject a batch still meeting GMP requirements solely because of a single sample result exceeding $M$. Dahms and Hildebrandt (1998) proposed a methodology to estimate the optimal distance between $m$ and $M$ based on the variability between samples.

As in a two-class plan, the performance of a three-class sampling plan can be evaluated assuming a normal distribution for the log-concentrations of microorganisms in food and a constant standard deviation. For a given within-batch normal distribution ($\mu, \sigma$) of microbial concentrations, the area below the density function below $m$ defines the proportion of acceptable units ($p_a$). The area between $m$ and $M$ defines the value for the proportion of the marginally-acceptable units ($p_m$), and the area above $M$ defines the value for the proportion of defective food units ($p_d$) (Figure 19.4, right). In Microsoft® Excel notation, the proportion acceptable and marginally-acceptable for a given batch can be calculated as $p_a = \text{Normdist}(m, \mu, \sigma, 1)$, and $p_m = \text{Normdist}(M, \mu, \sigma, 1) - p_a$. Finally, the proportion defective can be estimated as $p_d = 1 - p_a - p_m$. Since the acceptable number of units above $M$ is normally set to zero (i.e. $c$ marginally acceptable units are allowed between $m$ and $M$, but not above $M$), the multinomial distribution expression used to calculate the probability of accepting the batch ($Pa$) can be simplified to:

$$Pa = \sum_{i=0}^{C} \binom{n}{i} p_a^{\,n-i} p_m^{\,i} \quad (19.8)$$

For example, the probability of accepting a batch ($Pa$) of characteristics $p_a = 0.7$, $p_m = 0.2$, $p_d = 0.1$, after the sampling plan $n = 10$, $c = 2$ can be estimated using Equation 19.8 as:

$$Pa = P(c = 0) + P(c = 1) + P(c = 2)$$

$$Pa = \frac{10!}{10!0!}(0.7)^{10}(0.2)^0 + \frac{10!}{9!1!}(0.7)^9(0.2)^1 + \frac{10!}{8!2!}(0.7)^8(0.2)^2 = 0.213.$$

## Example 19.4

A producer wishes to establish a three-way sampling plan to verify the hygiene level in the production of minced beef using counts of *Escherichia coli* as hygiene indicator. From past experience the producer has knowledge that the limits of 1.5 log CFU/g ($m$) and 2.5 log CFU/g ($M$) take into account the maximum level of heterogeneity within a batch under conditions of GMP, although the average *E. coli* concentration from the samples normally do not exceed ~5 CFU/g (AQL~0.70 log CFU/g). Assuming that the producer does not wish to exceed a maximum concentration of 3 log CFU/g for the end of the process, and that the average within-batch standard deviation characterizing the variability in microbial counts among sample units is $\sigma = 0.6$, an optimal three-way sampling plan can be determined. As in previous examples, the LQL value is calculated assuming that in a *just-acceptable batch* no more than 1% of the units produced can exceed the maximum concentration of 3 log CFU/g. Thus, LQL = 3 – 2.33 × 0.6 = 1.6 log CFU/g. Thus, it is required that batches of minced beef with mean *E. coli* concentrations higher than 1.6 log CFU/g should be spotted with 95% confidence by sampling. The problem reduces to determine a three-way sampling plan producing an OC curve that meets the consumer's requirement (LQL = 1.6, $Pa = 0.05$) and ideally also the producer's requirement (AQL = 0.70, $Pa = 0.95$). For an upper microbiological limit $M = 2.5$ log CFU/g and a sampling size $n = 8$, OC curves were produced at varying marginally-acceptable units $c$ (Figure 19.10).

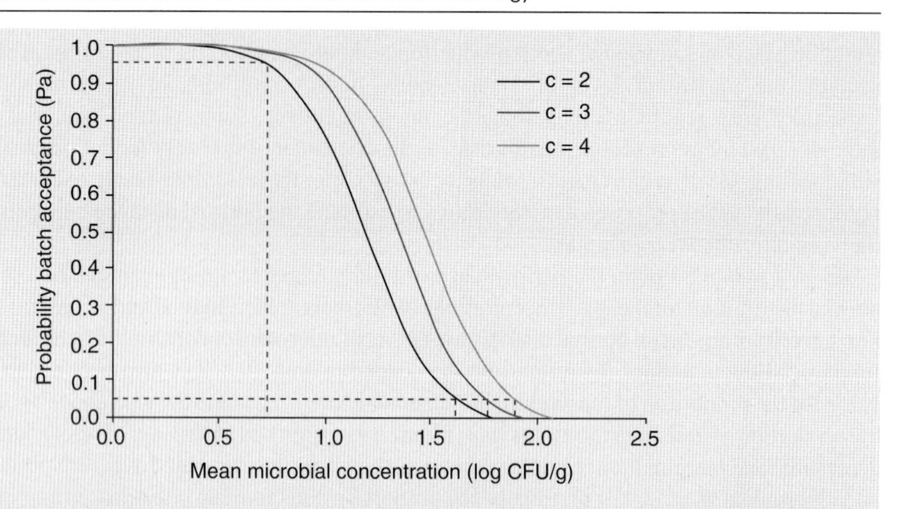


**Figure 19.10** Operating characteristic curves for three-class attributes sampling plans of microbiological limits m = 1.5 log CFU/g, M = 2.5 log CFU/g, n = 8 and different marginally acceptable units c, assuming a within-batch lognormal distribution of microbial counts with standard deviation $\sigma = 0.6$ log CFU/g.

The smaller the *c* value, the more stringent the sampling plan becomes, causing a shift of the OC curve to the left. It was found from this analysis that a *c* value of 2 meets both the consumer's and producer's requirements. Hence, a sampling plan (n = 8, c = 2, m = 1.5 log CFU/g, M = 2.5 log CFU/g) ensures that the producer tests its hygiene level against GMP standards. When none of the tested units are above the *m* level, the producer will have 95% confidence that batches with mean *E. coli* concentrations not greater than 0.70 log CFU/g are being manufactured. However, in batches that may come out of control, there is a tolerance that up to two units out of the eight samples may have microbial concentrations between 1.5 and 2.5 log CFU/cm$^2$, which minimizes the risk of a false alarm for a hygiene indicator that does not represent health risk. A noncompliant batch, however, should give rise to improvements in production hygiene and in the selection and/or origin of raw materials.

## Design of variables sampling plans (n, m)

When sufficient historical information is available to identify the distribution, variables sampling plans may be more cost effective because all the information in the test results is used to support decision making. Variables plans are generally more flexible than three-class sampling plans, have more discriminating power, and they respond generally in a more consistent manner to the true microbial quality of the batch. A special advantage of the variables sampling plans is that a clear distinction can be made between decisions for safety/quality and GMP (Kilsby *et al.*, 1979). This distinction is not made in three-class sampling schemes. Nevertheless, the variables sampling plans are useful only when the distribution of the microbial concentrations within a batch is well known. Ideally, the distribution can be lognormal, although a variables sampling plan can be derived with basis on other distributions (Gonzales-Barron *et al.*, 2012; 2013).

To design a variables sampling plan, the first decision to make is to define what makes a batch unacceptable. A batch of food is defined as unacceptable if more than a proportion $p_0$ of the units produced has microbial concentrations exceeding a critical concentration *C* (Figure 19.11). It is at this stage that either safety (risk-based) or GMP specifications are set. Based on these values and information on the expected standard deviation $\sigma$ within batches, the mean concentration of the limiting quality batch

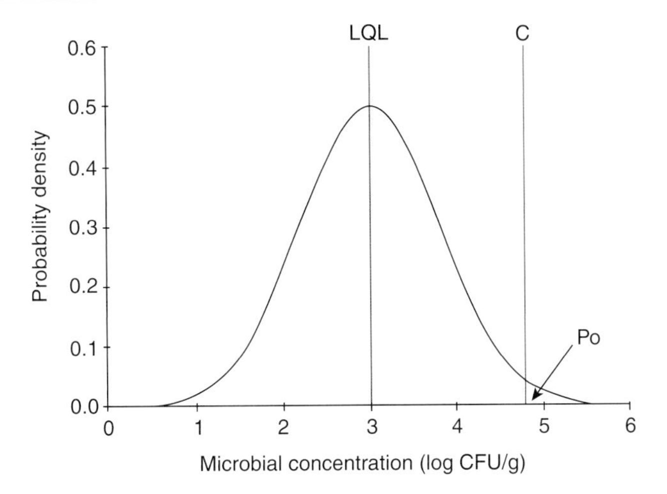


**Figure 19.11** A normal distribution of log microbial concentrations representing a limiting quality batch of LQL mean, defined by the maximum proportion $p_0$ of the batch that can be accepted with concentrations above the critical limit C for a variables sampling plan.

(LQL) can be found as LQL $= C - z(p_0) \times \sigma$. The second decision is to choose the maximum risk to accept a defective batch ($\beta$). This corresponds to $1 - \beta$ being the desired confidence to reject a defective batch (i.e. a batch that exceeds the tolerance criterion defined by $C$ and $p_0$). Thus, a batch of mean concentration higher than the LQL must be rejected with '$1 - \beta$' confidence. On the basis of the sample size $n$, it is necessary to decide whether the percentage of potential values beyond $C$ in a batch is likely to exceed the value $p_0$. The problem consists then of finding the decision rule $m$ (maximum mean log of the samples or microbiological limit) and $n$ (sample size) using the prescriptions above. When acceptable batches are tested, the mean of the samples is expected to be lower than the limit mean concentration m derived from $C$ and $p_0$. Because the within-batch distribution of the log microbial concentrations is assumed to be normal, the distribution of the samples' mean will also be a normal distribution with $\sigma' = \sigma/\sqrt{n}$. As it is on this samples' mean distribution that the limit $m$ is measured, the values of $m$ and $n$ should be chosen so that:

$$\Pr\left(Normal\left(LQL, \frac{\sigma}{\sqrt{n}}\right) > m\right) = 1 - \beta \tag{19.9}$$

As different ($m$, $n$) pairs can meet the consumer's requirement above, the equality can only be solved by keeping one variable fixed (for example, assuming a suitable $n$) and solving for the other ($m$). Thus, for a fixed sample size $n$, $m$ can be solved as $m =$ LQL $- z(\beta) \times \sigma/\sqrt{n}$. The whole procedure can be illustrated by the following example: Suppose that a batch should be rejected with 95% confidence ($\beta = 0.05$) if more than $p_0 = 10\%$ of the food units exceed a critical concentration $C = 2.0$ log CFU/g. For the probabilities of $p_0$ and $\beta$, the $z$ values are $z(p_0) = 1.28$ and $z(\beta) = 1.64$. Assuming that the standard deviation of the microbial concentrations among units is $\sigma = 0.8$, the mean of the LQL batch can be calculated as, LQL $= 2 - 1.28 \times 0.8 = 0.976$. Thus, fixing the sample size to $n = 5$, the microbiological limit $m = 0.976 - 1.64 \times 0.8/\sqrt{5} = 0.389$ log CFU/g, and for a sample size $n = 8$, the microbiological limit $m = 0.976 - 1.64 \times 0.8/\sqrt{8} = 0.510$ log CFU/g. If the mean log of the individual results from the individual samples exceeds that $m$ limit, the batch should be then rejected. Other microbiological limits can be estimated for different sample sizes.

In cases, where there is also a known requirement on the side of the producer (i.e. the AQL mean concentration of a batch that should be accepted with '$1 - \alpha$' confidence), solving for the decision criterion ($m$, $n$) will only lead to one solution, since now two preconditions need to be satisfied (i.e. two equations and two variables):

$$\begin{aligned} \Pr\left(Normal\left(AQL, \frac{\sigma}{\sqrt{n}}\right) < m\right) &= 1 - \alpha \\ \Pr\left(Normal\left(LQL, \frac{\sigma}{\sqrt{n}}\right) > m\right) &= 1 - \beta \end{aligned} \tag{19.10}$$

## Example 19.5

The construction of OC curves for assessing the performance of variables sampling plans is simpler than for attributes sampling plans. Assuming that the log microbial concentrations among food units in a batch distribute as a normal distribution ($\mu$, $\sigma$), given a sampling plan ($n$, $m$), the probability of batch acceptance $Pa$ is determined as the cumulative probability of the sample's mean distribution below the microbiological limit $m$. In Microsoft® Excel notation, this probability is calculated as $Pa =$ Normdist ($m$, $\mu$, $\sigma/\sqrt{n}$, 1). The OC curve is a plot of $Pa$ values calculated for a range of within-batch mean concentrations $\mu$. Figure 19.12 presents OC curves constructed in this way for a fixed microbiological limit $m = 0.4$ log CFU/g and three sample sizes $n = 3$, 5 and 8, assuming a constant within-batch standard deviation $\sigma = 0.8$ log CFU/g. If the safety requirement was to reject with 95% confidence ($Pa = \beta = 0.05$) batches having more than 10% ($p_0$) of the food units surpassing the contamination critical limit of 2 log CFU/g ($C$), the mean of the LQL batch can be estimated as LQL$= 2 - 1.28 \times 0.8 = 0.976$. Then, the problem is to find a decision criterion ($m$, $n$) that meets the condition (Equation 19.9)

$$\Pr\left(Normal\left(0.976, \frac{0.8}{\sqrt{n}}\right) < m\right) = 0.05$$

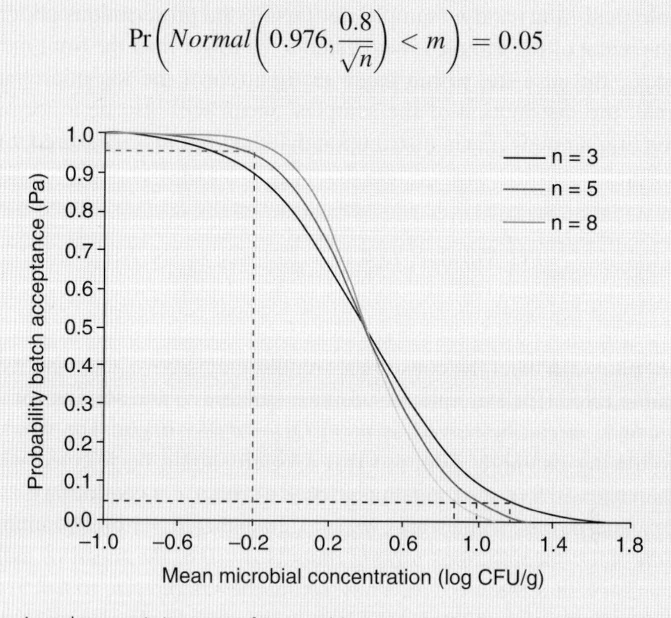


**Figure 19.12** Operating characteristic curves for variables sampling plans of microbiological limit m = 0.4 log CFU/g, and different sample size n, assuming a within-batch lognormal distribution of microbial counts with standard deviation $\sigma = 0.8$ log CFU/g.

or, in other words, to find an OC curve that contains or approximately passes the LQL point. This can be done by first keeping $n$ or $m$ fixed. Figure 19.12 shows OC curves constructed for a fixed microbiological limit $m = 0.4$ log CFU/g and different sample sizes. It can be noticed that the higher the sample size, the higher the discriminatory power of the sampling plan, and the lower the misclassification errors (i.e. hence the steeper the curve). The OC curve for $n = 3$ indicates that such a sample size will not be sufficient to achieve that degree of safety as LQL at $Pa = 0.05$ is 1.16 log CFU/g. Similarly, a sample size $n = 8$ will lead to a conservative sampling plan with a LQL = 0.86, which in the long run will reject batches that still meet the safety criterion. The OC curve for $n = 5$ presents a LQL = 0.97, which is very close to the safety requirement at a confidence of rejection of 95%. At this level of consumer's protection, the producer will seek to produce food batches of mean microbial concentrations lower than 0.63 CFU/g (AQL = − 0.2 log CFU/g in Figure 19.12).

## FUTURE TRENDS

The statistical methods proposed for the design of sampling plans to test microorganisms in foods have been entirely based on quality control statistics (Duncan, 1986); the reasons as to why two simplifying assumptions have been traditionally adopted are: that the log microbial concentrations among food units produced in a batch can be represented by a normal distribution and that its measure of spread (i.e. the standard deviation for a normal distribution) is constant among batches of production. Furthermore, from these assumptions, it is implicitly understood that the normal distribution is adequate for any contamination level (i.e. low counts or high counts) and that the measure of spread (i.e. within-batch standard deviation) is independent of the contamination level. Nevertheless, recent work has demonstrated that in many cases the normality assumption does not hold, especially when modelling low microbial concentrations (Gonzales-Barron *et al.*, 2010; Gonzales-Barron and Butler, 2011a), and that the within-batch measure of spread is, in fact, associated with the within-batch mean concentration (Gonzales-Barron and Butler 2011b; Gonzales-Barron *et al.*, 2012; 2013). This dependence, a phenomenon that has been observed in many microbial data sets, is very likely to be an effect of the physical agglomeration of bacterial cells (Mussida *et al.*, 2013). This is of significance in the design of sampling plans, since the derivation of OC curves should ideally take into account the shifts in the measure of spread for the acceptance probability calculation as movement is along the x-axis values of within-batch mean concentration. The importance of the choice of the most appropriate distribution to represent bacterial clustering resides on its critical effect on the acceptance probability for typical microbiological criteria and on the design of sampling plans. Gonzales-Barron and Butler (2011b) have shown that the Poisson-gamma assumption allowing for clustering produces stricter sampling plans than both the lognormal and the Poisson distribution for the same level of safety (LQL or PO).

Unlike the manufacturing process of nonfood items that, when under control, have been proven to operate with only chance (or nonassignable) causes of variation that are an intrinsic part of the process, and therefore lead to a stable variance (Montgomery, 2009), in the production of food items the batch-to-batch variability in the microbiological quality of the food product is expected to be more unstable due to phenomena inherent to the biological systems (i.e. microbial growth and death, usual microbial variability in raw food materials due to environmental factors, difference among serotypes and serovars, cross-contamination, etc.). To account for the effect of the batch-to-batch variability, Gonzales-Barron *et al.* (2012; 2013) developed a novel approach to model the association between the within-batch mean and the within-batch spread measure using a random-effects regression model based on the Poisson-gamma distribution characterizing low and high microbial counts from many batches of production. A model of such association is a joint model of within-batch and between-batch variability in microbial counts. Thus, to derive an OC curve, the uncertainty around the within-batch spread measure for a given within-batch mean is propagated by simulation to the probability of batch acceptance. In this way, OC curves are obtained with confidence intervals representing the uncertainty about the spread measure arising from the between-batch variability. The OC curves obtained as such lead to more conservative sampling plans.

In classical acceptance sampling theory, the sampling plans are designed basically on a pre-established level of safety (that could be the AQL/LQL at a risk $\alpha/\beta$, or alternatively a tolerance criterion ($C$, $p_0$), both in accordance to a PO/FSO) and with little information about the process itself. Using the methods from classical acceptance sampling, the lognormality of microbial concentrations will be presumed and the only piece of information from the process itself will be, at most, some estimate of the 'constant' standard deviation, if not an assumption. Nevertheless, when there is availability of past monitoring data, sampling plans should ideally be designed making use of this information, in particular when they are derived on the producer's risk and there is evidence that hygiene and safety measures in the production process have long been under control (Gonzales-Barron *et al.*, 2013). In such cases, the objective of the sampling plan would be to verify that the safety of the production process is under control; and mathematically it should be able to discern the outlying 'out-of-control' batches. As proposed by Gonzales-Barron *et al.* (2012), when microbial data are available, the within-batch and between-batch variability in contamination can be modelled jointly and this spectrum of possible batches (i.e. universe of contaminated batches) can be used to estimate the misclassification risks ($\alpha$ and $\beta$) with confidence intervals for the all possible microbiological limits and sample sizes. Selecting the sampling plan will be then a matter of finding the optimal trade-off between risks and taking into account risk management considerations. In this way, taking into account the results of sample units that were created by the process itself will result in sampling plans that are more informative and dynamic (Gonzales-Barron *et al.*, 2013). Sampling plans obtained using past information are also amenable to being updated as new sampling data are obtained, which is unarguably its better advantage.

## CONCLUDING REMARKS

While no practical amount of sampling and testing for microorganisms in foods can on its own assure food safety, microbiological testing is still a common tool used to evaluate whether a food safety risk management system provides the appropriate level of control (FSO, PO). The sampling plans used to test microorganisms can be assigned to two groups depending on the type of microbiological test results. For presence/absence tests, the sampling plans can be designed based on the simple two-class type ($n$, $c$, $m$) or the two-class by enrichment ($n$, $c$, $w$), while for quantitative analytical results the sampling plans that can be designed are of three-class by attributes ($n$, $c$, $m$, $M$) or by variables ($n$, $m/M$). Despite considerable research that has been devoted to attributes sampling plans, little attention has been paid to variables sampling plans despite their advantage of being more informative and requiring less sampling for the same level of protection. The sampling plans – derived either to meet the producer's or consumer's safety requirements – have traditionally been designed using classical acceptance sampling theory, which assumes that the measure of the quality characteristic (i.e. log concentration of target microorganism) distributes in the product (i.e. food) follow a normal distribution with a variance that is approximately stable batch to batch. Recent concerns in relation to bacterial physical agglomeration have prompted questioning as to whether these oversimplifying assumptions may lead to an inefficient design of sampling plans. To address this problem, lately there have been some efforts in the investigation of the adequacy of other statistical distributions to better represent bacterial clustering and a situation originated thereof, the high proportion of zero counts in samples. It has been found that the Poisson-gamma or negative binomial distribution is more suitable than both the lognormal and the

Poisson-lognormal distributions when dealing with clustered microbial data consisting of many zero counts. Thus, the design of statistically-sound sampling plans should be based on historical microbial data from many production batches and new methodologies, if possible Bayesian, should be developed that are based on more realistic assumptions taking into account bacterial clustering, variable measures of spread conditional to the mean contamination level, and the intrinsic variability among batches of production.

## ACKNOWLEDGMENTS

Dr Ursula Gonzales-Barron wishes to acknowledge the financial support provided by the Portuguese Foundation for Science and Technology (FCT) through the award of a five-year Investigator Fellowship (IF) grant for Career Development (IF/00570).

Professor Francis Butler wishes to acknowledge the Food Institutional Research Measure (FIRM) administered by the Irish Department of Agriculture, Fisheries and Food.

## REFERENCES

Bray, D.F., Lyon, D.A. and Burr, I.W. (1973) Three-class attribute plans in acceptance sampling. *Technometrics* **15**, 575–585.

Dahms, S. (2004) Microbiological sampling plans – statistical aspects. *Mitteilungen aus Lebensmitteluntersuchung und Hygiene* **95**, 32–44.

Dahms, S. and Hildebrandt, G. (1998) Some remarks on the design of three-class sampling plans. *Journal of Food Protection* **61**(6), 757–761.

Duncan, A.J. (1986) *Quality Control and Industrial Statistics*, 5th edn. Richard D Irwin, Inc., Homewood, IL.

EC (2007) Commission Regulation (EC) No. 1441/2007 of 5 December 2007 amending Regulation (EC) No 2073/2005 on microbiological criteria for foodstuffs. *Official Journal of the European Communities* **L22**, 12–29, Available: http://eur-lex.europa.eu/LexUriServ/LexUriServ.do?uri=OJ:L:2007:322:0012:0029:EN:PDF (last accessed 28 June 2013).

Gonzales-Barron, U. and Butler, F. (2011a) A comparison between the discrete Poisson-Gamma and Poisson-Lognormal distributions to characterise microbial counts in foods. *Food Control* **22**, 1279–1286.

Gonzales-Barron, U. and Butler, F. (2011b) Characterisation of within-batch and between-batch variability in microbial counts in foods using Poisson-gamma and Poisson-lognormal regression models. *Food Control* **22**, 1268–1278.

Gonzales-Barron, U., Kerr, M., Sheridan, J. and Butler, F. (2010) Count data distributions and their zero-modified equivalents as a framework for modelling microbial data with a relatively high occurrence of zero counts. *International Journal of Food Microbiology* **136**, 268–277.

Gonzales-Barron, U., Lenahan, M., Sheridan, J. and Butler, F. (2012) Use of a Poisson-gamma model to assess the performance of the EC process hygiene criterion for *Enterobacteriaceae* on Irish sheep carcasses. *Food Control* **25**, 172–183.

Gonzales-Barron, U., Zwietering, M. and Butler, F. (2013) A novel derivation of a within-batch testing regime based on a Poisson-gamma model characterising low microbial counts in foods. *International Journal of Food Microbiology* **161**, 84–96.

Gorris, L.G.M. (2005) Food safety objective: An integral part of food chain management. *Food Control* **16**, 801–809.

Hildebrandt, G., Bohmer, L. and Dahms, S. (1995) Three-class attributes plans in microbiological quality control: a contribution to the discussion. *Journal of Food Protection* **58**(7), 784–790.

ICMSF (International Commission on Microbiological Specifications for Foods) (2009) Microbiological sampling plans: A tool to explore ICMSF recommendations. Version 2.05. [Online] Available: http://www.icmsf.iit.edu/main/sampling_plans.html (last accessed 29 June 2013)

ICMSF (International Commission on Microbiological Specifications for Foods) (2002) *Microorganisms in Foods 7: Microbiological Testing in Food Safety Management*. Kluwer Academic/Plenum Publishers, New York.

Jarvis, B. (2007) On the compositing of samples for qualitative microbiological testing. *Letters in Applied Microbiology* **45**, 592–598.

Jarvis, B. (2008) *Statistical Aspects of the Microbiological Examination of Foods*, 2nd edn. Elsevier, London.

Kilsby, D.C., Aspinall, L.J. and Baird-Parker, A.C. (1979) A system for setting numerical microbiological specifications for foods. *Journal of Applied Bacteriology* **46**, 591–599.

Legan, J.D., Vandeven, M.H., Dahms, S. and Cole, M.B. (2001) Determining the concentration of microorganisms controlled by attributes sampling plans. *Food Control* **12**, 137–147.

Montgomery, D.C. (2009) *Statistical Quality Control: A Modern Introduction*, 6th edn. John Wiley & Sons, Inc., Hoboken, New Jersey.

Mussida, A., Gonzales-Barron, U. and Butler, F. (2013) Effectiveness of sampling plans by attributes based on mixture distributions characterising microbial clustering in foods. *Food Control* **34**, 50–60.

Stringer, M. (2005) Food safety objectives – role in microbiological food safety management. *Food Control* **16**, 775–794.

Van Schothorst, M., Zwietering, M.H., Ross, T. *et al.* (2009) Relating microbiological criteria to food safety objectives and performance objectives. *Food Control* **20**, 967–979.

Whiting, R.C., Rainosek, A., Buchanan, R.L. *et al.* (2006) Determining the microbiological criteria for lot rejection from the performance objective or food safety objective. *International Journal of Food Microbiology* **110**, 263–267.

# 20 Infrared spectroscopy detection coupled to chemometrics to characterize foodborne pathogens at a subspecies level

**Clara C. Sousa[1] and João A. Lopes[2]**

[1] *REQUIMTE, Laboratório de Microbiologia, Departamento de Ciências Biológicas, Faculdade de Farmácia, Universidade do Porto, Porto, Portugal*

[2] *REQUIMTE, Laboratório de Análises Químicas e Físico-Químicas, Departamento de Ciências Químicas, Faculdade de Farmácia, Universidade do Porto, Porto, Portugal*

## ABSTRACT

According to the World Health Organization, foodborne and waterborne diseases kill about 2.2 million people annually and are an important cause of mortality and morbidity. Amongst the most dangerous foodborne diseases and foodborne pathogens are: bovine spongiform encephalopathy (BSE), *Campylobacter*, *Shigella*, *Listeria monocytogenes*, *Escherichia coli* infections and Salmonellosis. To trengthen surveillance systems and effective characterization methods of these pathogens are a public health priority. Infrared spectroscopy (near infrared or mid infrared spectroscopy) is currently one of the most effective, quick and low-cost techniques for detection of these foodborne pathogens in a nondestructive way. Due to the huge amount of data generated by spectroscopic techniques, appropriate mathematical tools are needed to extract the relevant information for the analysis. The development of computers hardware and increasingly sophisticated chemometric tools place spectroscopy on top in bacterial identification and characterization of foodborne pathogens.

## INTRODUCTION

According to the World Health Organization, foodborne and waterborne diseases kill about 2.2 million people annually and are an important cause of mortality and morbidity. The main causes of these diseases are foodborne pathogens such as *Campylobacter jejuni*, *Clostridium botulinum*, *Clostridium perfringens*, *Cryptosporidium parvum*, Norovirus, *Salmonella*, *Staphylococcus aureus*, Giardia lamblia, *Escherichia coli* O157:H7, *Bacillus cereus*, *Shigella*, *Yersinia enterocolitica* and *Listeria monocytogenes*. To reduce the incidence and economic consequences of foodborne diseases includes strength surveillance strategies, assuring the safety of food from production to final consumption and high throughput methodologies to identify foodborne pathogens.

Spectroscopic methodologies such as those based on infrared spectroscopy are some of the main techniques used in food safety control for bacterial characterization due to their effectiveness, low cost, speed and minimal or no sample preparation, and they avoid destruction. These techniques have been used in the food industry for different purposes, ranging from qualitative and quantitative analysis of a category or a specific bioactive compound (Wu *et al.*, 2009), determination of antioxidant activity

*Mathematical and Statistical Methods in Food Science and Technology*, First Edition.
Edited by Daniel Granato and Gastón Ares.

(Versari *et al.*, 2010) to food quality assurance, detection of adulterants and food safety determinations, including the characterization of foodborne pathogens (Alvarez-Ordóñez *et al.*, 2011). The first applications of these methodologies were performed to access food chemical composition with special emphasis in molecular architecture of food proteins. More recently, these methods have been applied to quantitative analysis of food ingredients (Ishiguro *et al.*, 2006; Hashimoto and Kameoka, 2008), to characterize stress response in foodborne pathogenic bacteria by food processing technologies and as a method to characterize foodborne pathogens even at a sub-species level (Preisner *et al.*, 2010, 2012).

One of the biggest issues of spectroscopic techniques concerns data analysis, particularly the large volume of data generated by these methods. The application of chemometric approaches as experimental design or multivariate data analysis allows the optimization of the analytical process and data analysis, extracting the meaningful information and, therefore, improving the quality of the results. Previous studies on microbial identification applied both unsupervised and supervised methods to represent information from hyper-spectral data, namely principal component analysis, factor analysis, linear discriminant analysis, hierarchical cluster analysis, partial least squares regression and artificial neural networks, among others (Mobley *et al.*, 1996; Goodacre *et al.*, 1998; Udelhoven *et al.*, 2000). Challenges associated with these chemometric tools are to assess the accuracy (bias) and also the precision. Both should be minimized in order to develop a robust model. Errors are normally estimated with RMSECV (root mean square error of cross validation: internal validation error) or RMSEP (root mean square error of prediction: external validation error). Even when the experimental sampling and spectroscopic procedure are well established there are sources of variability that must be kept to a minimum. Calibration transfer has been extensively analysed and is known to be a problem (Bakeev and Kurtyka, 2005). Removal of physical effects caused by light scattering, baseline drift and random noise has an impact on the performance of the microbial characterization process. These are typically compensated for by adequate spectra pre-processing (e.g. spectra normalization, filtering, applying derivatives, standard normal variate, multiplicative scattering correcting) (Martens *et al.*, 2003). A criterion for wavelength (or wave number) selection is normally to prevent noninformative parts of the spectra from being included in the chemometric model (regarding the specific objective). These factors are very important in developing robust methodologies for bacteria characterization, yet requiring a great deal of attention.

## INFRARED SPECTROSCOPY

### Fundamentals

Infrared radiation (IR) was firstly discovered by William Herschel, an English astronomer, during one of his experiments in 1800. Spreading sunlight with a small glass prism into its components, Hershel found different temperatures associated to each colour and, moreover, the highest temperature was beyond the red portion of the visible spectrum. This new kind of radiation was named infrared radiation (meaning 'below red') because it has a lower frequency than red light.

Within the electromagnetic spectrum, infrared radiation lies from the visible region to the microwave region (from 780 to $1 \times 10^6$ nm) and is usually divided into three regions: NIR – near infrared radiation (12 500–4000 $cm^{-1}$), MIR – mid infrared radiation (4000–400 $cm^{-1}$) and FIR – far infrared radiation (400–10 $cm^{-1}$). Invisible to the human eye, this radiation can be felt in the form of heat and can lead to some changes in systems subjected to it.

At the atomic level, infrared energy can be absorbed, causing alterations in rotational-vibrational movements through a change in the dipole moment. Vibrational movements of covalently bonded atoms can be classified as stretching, which involves changes in the bond length (symmetric or asymmetric), and bending, corresponding to changes in bond angles, designated scissoring, rocking, wagging and twisting. These vibrations are characteristic of a chemical bond and their frequency is used to study organic compounds through the recording of a spectrum containing all the frequencies of absorption. It should be stressed that spectral IR absorbance bands can only reflect information of molecular functional groups and cannot identify a specific chemical compound.

According to the region of infrared radiation used to obtain the spectrum, infrared spectroscopy get different designations. Near infrared spectroscopy (NIRS) or Fourier transform near infrared spectroscopy (FT-NIRS) if Fourier transform is applied, covers the range between the visible and mid infrared regions. Fourier transform is an algorithm developed by Fourier in which any mathematical function can be expressed as the sum of sinusoidal waves with different frequencies. In addition to the information on chemical structures that an infrared spectra can provide, quantitative information can be obtained through the application of Beer's law, which relates concentration to absorbance.

The spectral absorption bands in NIRS are the result of overtones or combinations of the fundamental ones; their intensity is dependent of the composition and on the change in the dipole moment of the chemical bonds (Bokobza, 1998). These combinations and overtones leads to complex spectra with broad bands that are difficult to interpret, making it impossible to assign specific features to specific chemical components. Some of these issues are surpassed with appropriate mathematical and statistical treatment, as will be shown later.

Despite these limitations, NIRS still is a spectroscopic technique used frequently, mainly due to its speed, low cost, nondestructive characteristic, little or no sample preparation and, nowadays, with the development of fibre optics, due to its portability.

The earliest known applications of NIRS date back to the 1950s (White *et al.*, 1957). However, it is only relatively recently that NIRS has started to be used for qualitative and quantitative analysis in agriculture (Norris 1996; Roggo *et al.*, 2004) and the food (Tøgersen *et al.*, 2003; Ng *et al.*, 2011), chemical (Larrechi and Callao, 2003), oil (Blanco *et al.*, 2001) and pharmaceutical (Kim *et al.*, 2011) industries as well as in microbiology to detect and identify bacteria (Rodriguez-Saona *et al.*, 2001; Tito *et al.*, 2012).

When dealing with mid infrared radiation the designation mid infrared spectroscopy, MIRS, is used. The most common type of MIRS is Fourier transform mid infrared spectroscopy, FT-MIRS, and is usually designated by FTIR spectroscopy (FTIRS).

The first available MIR spectrometers consisted of an infrared beam dispersed by a grating or a prism onto a slit that blocks all but a narrow range of frequencies from reaching the detector; the resolution was dependent on the width of the slit. The result was a low intensity spectrum making the method highly insensitive (Figure 20.1a). The introduction of an interferometer allowed a simultaneous monitoring of a broad band of vibrational frequencies. The most common is the 'Michelson interferometer', a device that

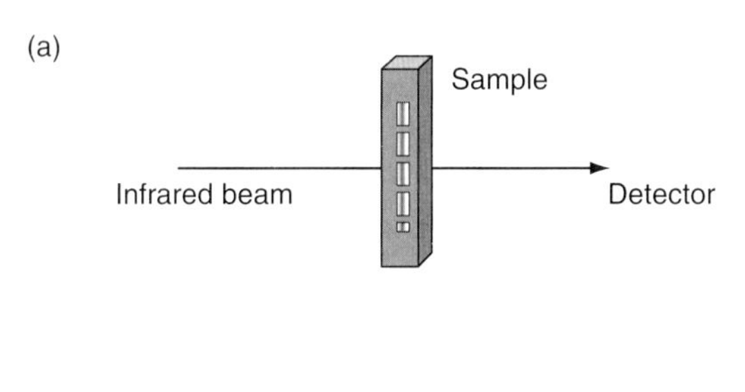


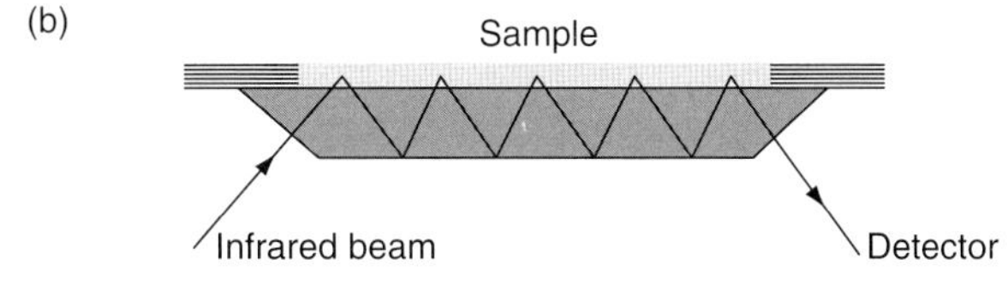


**Figure 20.1** Fourier-transform infrared sampling principle: (a) transmission and (b) attenuated total reflectance (ATR).

divides the infrared beam into two paths and then recombines both after the introduction of a path difference. The apparatus consists in two perpendicular mirrors, one fixed and the other that can move perpendicularly to its plane. Between these two mirrors there is a radiation splitter where the infrared beam is partially reflected to the fixed mirror and partially transmitted to the movable one. The two beams return to the splitter and are sent to the detector and combined to obtain the spectrum.

The available infrared techniques appropriate for food science research until 1993 were transmittance and reflectance spectroscopy, and mainly for near infrared wavelengths. For mid infrared, samples had to be incorporated in a potassium bromide matrix; this procedure was not always suitable for food samples, giving rise to irreproducible results. The introduction of attenuated total reflectance (ATR) overcame issues related with sample preparation and samples could be analysed after just removing water. With an ATR accessory (Figure 20.1b) an internally reflected infrared beam is directed onto an optically dense crystal with high refractive index at a certain angle. The resulting evanescent wave enters the crystal and contacts with the sample, a few microns (0.5–5 $\mu$m), being attenuated.

## Bacteria characterization by IR spectroscopy

Infrared spectroscopy has been applied with great success for bacterial classification (Naumann, 2000), identification and detection allowing discrimination at different taxonomic levels (Gómez *et al.*, 2003; Essendoubia *et al.*, 2005; Savic *et al.*, 2008). The use of infrared spectroscopy in this context had already been reported by Rouf (Rouf and Stokes, 1962) and Norris (Norris, 1959) in the late 1950s. However, due to the limited performance specifications of IR spectrometers up to the 1960s (time, sensitivity, reproducibility) reports on IR applications in microbiology were almost nonexistent. The development and cost reduction of modern spectrometers (improvement of spectrometer specifications such as the development of interferometry IR spectroscopy) associated with low-cost computers and effective multivariate statistical analysis boosted the number of applications of vibrational spectroscopy in the microbiology field. This technique nowadays presents very attractive advantages for microbiological classification and identification: fast (requiring virtually no sample processing), nondestructive, general, multipurpose (e.g. detection, enumeration, classification, identification) and allows discrimination at different taxonomic levels (genus, species and subspecies) (Naumann *et al.*, 1991). It was only after Naumann's studies that this technique started to be used for bacterial identification. Naumann and coworkers have shown that biomolecular absorbance of infrared (IR) radiation is a bacteria 'fingerprint', directly related to cellular composition (Naumann *et al.*, 1991).

Typically, an infrared spectrum of a bacterial strain shows broad and overlapped bands rather than sharp peaks, especially if intact cells are used (all cell components contribute to the final spectrum). The broad overlaid absorbance bands reflect the total composition of the cells (proteins, membranes, cell wall, nucleic acids, carbohydrates); it is very difficult to attempt a direct assignment of specific chemical groups (Maquelin *et al.*, 2002). However, the spectrum carries a great amount of potentially useful information despite of this limitation. Generally, the study of a particular strain is based on one or more spectral regions (Figure 20.2) and not the specific peaks characteristic of some functional group (Table 20.1). Analyses considering very restricted spectral regions are useful if the targets are very specific biomolecules (Alvarez-Ordóñez *et al.*, 2011). Despite this complexity, some spectral ranges are considered to contain more specific information about particular cell components (Naumann *et al.*, 1991) (Figure 20.2) and can be used for bacterial cell identification. The most important mid infrared characteristic peaks in biological matrices are summarized in Table 20.1 (Naumann, 2001; Maquelin *et al.*, 2002; Burgula *et al.*, 2007).

The spectral region between 3000 and 2800 cm$^{-1}$ (region $W_1$, in Figure 20.2) corresponds to the fatty acid region dominated by C–H stretching vibrations of $-CH_3$ and $>CH_2$ functional groups; it is, therefore, dominated by the spectral characteristics of fatty acid chains as phospholipids and some amino

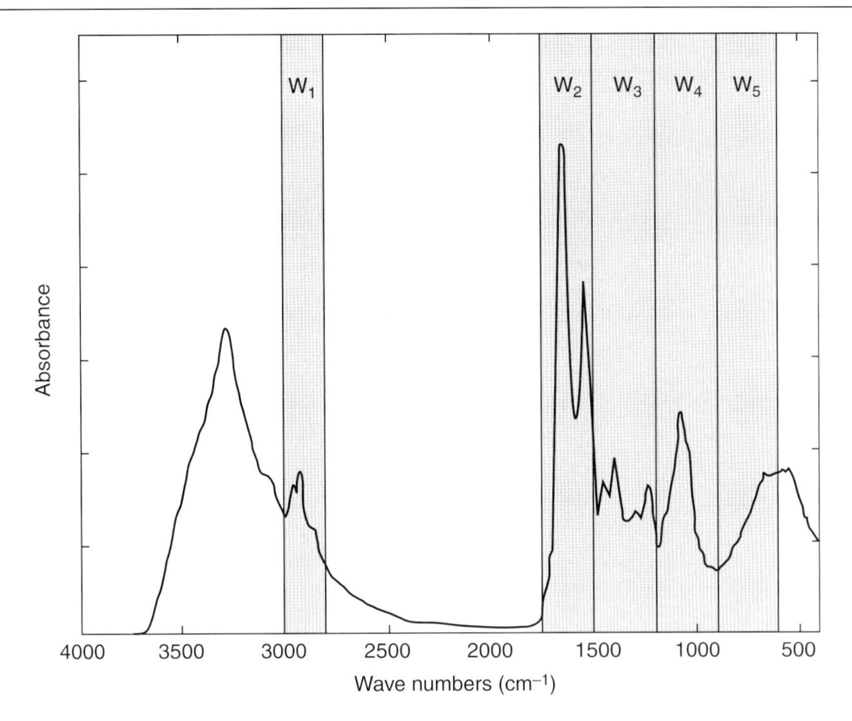


**Figure 20.2** FT-MIR spectrum of an *Escherichia coli* strain measured by ATR with indication of major spectral regions (region $W_1$: fatty acids; region $W_2$: proteins (amides I and II); region $W_3$: mixed region (phospholypids/DNA/RNA); region $W_4$: polysaccharides and region $W_5$: 'fingerprint' region).

acid side chains. Deformation modes of these functional groups are also found in the region between 1470 and 1350 cm$^{-1}$. The sharp peaks between 1800 and 1500 cm$^{-1}$ (region $W_2$, in Figure 20.2) correspond mainly to amide I and amide II vibrations of peptides and proteins and are the most intense bands in the spectra of nearly all bacterial samples. The amide I and amide II bands indicate the predominance of α- or β-structures present in proteins. Information in this region can also be obtained from bands near 1740 cm$^{-1}$, resulting from >C=O stretching vibrations of the ester functional groups in lipids. The spectral range from 1500 to 1200 cm$^{-1}$ ($W_3$) is usually referred as the mixed region, as it contains information about fatty acid bending vibrations, proteins and phosphate-carrying compounds, predominantly from >$CH_2$ and –$CH_3$ bending modes. Around 1400 cm$^{-1}$, a characteristic feature can also be observed and is usually attributed to symmetric stretching vibrations of –$COO^-$ functional groups of amino acid side chains or free fatty acids. In the $W_4$ region (1200–900 cm$^{-1}$), vibrations are mainly due to carbohydrates, leading to a complex sequence of peaks due to C—O—C and C—O—P stretching vibrations of the various oligo and polysaccharides. The range from 900 to 700 cm$^{-1}$ ($W_5$) is the true fingerprint region and holds very specific, weak spectral patterns from aromatic ring vibrations of AAA (tyrosine, tryptophan and phenylalanine) and nucleotides.

Besides these spectral ranges generally associated to carbohydrates, proteins and lipids there are other regions, or specific wave numbers, that could be associated to vibrations of some particular functional groups, amino acids or nucleic acids. Glycogen contributes to the bands at 1024, 1078 and 1151 cm$^{-1}$. Nucleic acids mainly account for bands in the range 990–950 cm$^{-1}$ and 1080–1010 cm$^{-1}$, usually attributed to P=O stretching of nucleic acids. The band near 1220 cm$^{-1}$ is due to the phosphodiester functional groups of DNA/RNA polysaccharide backbone structures while bands in the range of 1400–1380 cm$^{-1}$ are attributed to the symmetric stretching of $COO^-$ of amino acid side chains. Through this

**Table 20.1** Major infrared bands of biomolecules.

| Wave number ($cm^{-1}$) | Spectral region | Assignment |
|---|---|---|
| ~3500 | | O–H stretching of hydroxyl groups |
| ~3200 | | N–H stretching in resonance with overtone |
| ~2955 | $W_1$ | $CH_3$ asymmetric stretching |
| ~2930 | $W_1$ | $CH_2$ asymmetric stretching |
| ~2918 | $W_1$ | $CH_2$ asymmetric stretching |
| ~2898 | $W_1$ | C–H stretching in methylene groups |
| ~2870 | $W_1$ | $CH_3$ symmetric stretching |
| ~2850 | $W_1$ | $CH_2$ symmetric stretching |
| ~1740 | $W_2$ | C=O stretching of esters |
| ~1715 | $W_2$ | RNA C=O stretching |
| ~1680–1715 | $W_2$ | C=O stretching |
| ~1695 | $W_2$ | Amide I bands |
| ~1685 | $W_2$ | Antiparallel pleated sheets |
| ~1675 | $W_2$ | Proteins β-turns vibrations |
| ~1655 | $W_2$ | DNA C=O stretching; N–H bending; RNA C=O stretching Amide I of α-helical structures |
| ~1637 | $W_2$ | Amide I of β-pleated sheet structures |
| ~1550–1520 | $W_2$ | Amide II |
| ~1515 | $W_2$ | Tyrosine band |
| ~1485 | $W_3$ | $(CH_3)_3N^+$ asymmetric bending |
| ~1400 | $W_3$ | C=O symmetric stretching of $COO^-$ |
| ~1310–1240 | $W_3$ | Amide III |
| ~1250–1220 | $W_3$ | P=O asymmetric stretching of $>PO_2^-$ |
| ~1200–900 | $W_4$ | C–O, C–C stretching, C–O–H, C–O–C deformation |
| ~1090–1085 | $W_4$ | P=O symmetric stretching of $>PO_2^-$ |
| ~720 | $W_5$ | C–H rocking of $>CH_2$ |
| ~900–600 | $W_5$ | Fingerprint region |

information it is possible to perform some spectral band assignment. The difficulty of interpreting complex IR spectra and the application to the analysis of closely related strains with very similar spectra, is that it requires empirical multivariate mathematical methods (chemometrics) that are able to extract spectral features with very low a priori knowledge (Naumann, 2001).

## CHEMOMETRICS

There are several multivariate data analysis techniques (chemometric techniques) that can be applied to analyse vibrational spectra in the microbiological field (Naumann *et al.*, 1991) Vibrational spectra contain a large number of correlated variables (the wave numbers). To reduce the number of correlated variables it is necessary to describe the data variability in a few uncorrelated variables containing only the relevant information. The method of excellence to reduce variables is principal component analysis (PCA). The most common multivariate methods used in quantitative analysis regression are multiple linear regression (MLR), principal component regression (PCR) and partial least squares regression (PLSR). When the spectra have nonlinear relations with the target property, nonlinear calibration methods, such as locally weighted regression (LWR) or artificial neural networks (ANN), can be used (Goodacre *et al.*, 1998).

Multivariate classification methods, also known as pattern recognition methods, are subdivided in supervised and nonsupervised learning algorithms (Miller, 2000). Nonsupervised methods do not require a priori knowledge about structure in the data. Analysis can be performed by simple visualization techniques such as PCA or using hierarchical methods. For supervised methods some structure in the data must be known. Usually, samples are measured with some alternative technique. These methods can be used, for example to classify samples or to estimate some biochemical parameter. Methods such as linear discriminant analysis (LDA), quadratic discriminant analysis (QDA), soft independent modelling of class analogies (SIMCA), Kohonen networks or PLS discriminant analysis (PLSDA) are some of the most commonly used supervised classification methods.

## Spectral processing methods

Interfering spectral parameters, such as light scattering, path length variations and random noise, resulting from variable physical sample properties or instrumental effects, call for mathematical corrections. These are so-called data pretreatments and are applied prior to multivariate modelling in order to reduce, eliminate or standardize their impact on the spectra (Bakeev, 2005; Reich, 2005). To remove base line drifts, methods such as multiplicative scatter correction (MSC) and standard normal variate (SNV) are often used (Miller, 2000). Derivatives can be used to improve resolution and remove base line offsets but, because derivation amplifies the noise, derivatives are frequently applied with smoothing algorithms, such as Taylor and Savitzky-Golay (SG) (Naes *et al.*, 2002; Bakeev, 2005). Chemometrics methods often rely on covariance or correlation matrices (e.g. PCA and PLSR), which requires that spectral data should be centred or scaled.

## Unsupervised methods

PCA is one of the most widely used methods in chemometrics for spectral analysis since it is a method for data compression and data reduction The usefulness of this method resides in the fact that multivariate data can be well described in a more workable set of variables that contain almost all the information or variability in the original data (Naes *et al.*, 2002). Parallel factor analysis (PARAFAC) or Tucker3 are methods for modelling three-way or higher order data. These methods are based on a decomposition that can be compared to the bilinear method PCA (Bro, 1997; Bro and Kiers, 2003). Hierarchical cluster analysis (HCA) helps to identify similarities between spectra using the distances. Ward's algorithm and the average linkage correlation are the most commonly used cluster analysis algorithms for microbial identification (Gaul and Meinhardt-Wollweber, 2011; Yongnian *et al.*, 2012). Dendrograms are used to show HCA results, unveiling the similarities between spectra of bacteria (Helm *et al.*, 1991; Naumann *et al.*, 1991).

## Supervised methods

The objective of PLS regression is to establish a model for the analysis of unknown samples (determination of a physical or chemical property from the spectrum) (Geladi and Kowalski, 1986; Miller, 2000). The PLS algorithm can be extended to handle several dependent variables simultaneously (Naes *et al.*, 2002). The multiblock PLS (MBPLS) is an extension of the PLS method and consists of the separation of the predictors into subsets or blocks, according to a meaningful criterion or process knowledge (Westerhuis *et al.*, 1998; 103. Westerhuis and Smilde, 2001; Brás *et al.*, 2005; Felício *et al.*, 2005). Partial least squares discriminant analysis (PLSDA) is a method based on PLS for dealing with discrimination problems. The PLSDA model can be used to develop chemometric calibration models for bacteria discrimination. The multivariate PLSDA model is based on the well-known PLS model (Geladi and Kowalski, 1986). The specificity of the PLSDA model is that it is used to classify samples (Alsberg *et al.*, 1998).

## Models validation and optimization

Commonly the mean square error (MSE) and the root mean square error (RMSE) of the predictions are used as an estimator of supervised model accuracy (e.g. PLSR or PLSDA). These indicators measure the difference between the actual values $y_i$ and the predicted values $\hat{y}_i$. The RMSE can be defined as, where $N$ is the number of samples:

$$RMSE = \sqrt{\frac{\sum_{i=1}^{N} (\hat{y}_i - y_i)^2}{N}} \qquad (20.1)$$

The $y$ can be a concentration of some molecule but can only codify the assignment to some group, as in discriminant analysis (e.g. LDA or PLSDA). In supervised models it is very important to properly validate model results. The root mean square error obtained for a calibration set (RMSEC) cannot be used as a model fit indicator. The difference between RMSEC and the true model prediction error can be very large (Naes *et al.*, 2002).

Validation is a crucial process in developing multivariate methods. Not taking into account this procedure may lead to model overfitting. Overfit happens when a model over-adjusts the calibration data, being unable to provide good estimations in the presence of new data (Hawkins, 2004). Therefore, model errors must be estimated from unseen data sets (validation data). Available data are often partitioned in two sets: one for calibration and one for validation. The prediction error (RMSEP) is the simplest test that can be carried out to validate a model. The cross-validation is often used when the available data set is small and cannot be partitioned (Miller, 2000). The root mean square error of cross-validation (RMSECV) is calculated as RMSE but the estimated predictions are obtained by sequentially leaving out small data portions from the calibration set. The cross-validation may be performed using different methods (e.g. leave-one-out, contiguous blocks, venetian blinds and random subsets) (Naes *et al.*, 2002). A suitable procedure to optimize discriminant regression models is based on the division of the original data set into a training set and a validation set (Preisner *et al.*, 2010). Samples are divided randomly but the following two rules must apply: (i) the proportion of classes should be maintained in both calibration and validation sets (to ensure a correct balance between calibration and validation data sets) and (ii) replicates (if existing) of each sample must all be assigned to the calibration or validation set (to prevent overfitting). The model yielding the lowest validation error should be kept. However, the results could depend on a particular training-validation division. Therefore, the entire procedure should be repeated a given number of times (e.g. 1000 times).

### *Outliers*

One of the reasons why a sample can be an outlier is if it belongs to a population other than the normal samples. Another reason can be a malfunction of the instrument, giving erroneous or misleading signals affecting one or all $x$-variables. The third case is when there are errors in y caused by reference method failure or transcription error. Thus, this sample will not fit into the regression model obtained from the rest of the data. Outliers can be detected in component models (e.g. PCA/PLS) by analysing the Q and $T^2$ statistics (Naes *et al.*, 2002). The $Q$ statistic is simply the sum-of-squares of each row (sample) of $E$. This statistic indicates how well each sample fits the model. It is a measure of the difference, or residual, between a sample and its projection into the A components retained in the model. The $T^2$ is the sum of normalized squared scores and is a measure of the variation in each sample within the model (e.g. PCA, PLS). The $T^2$ values are a distance measure of the scores or loadings with respect to their respective origins

### *Wavelength selection*

Variable selection, or more specifically wavelength selection, is typically mandatory in multivariate models based on infrared spectra in the context of microorganism analysis (Balabin and Smirnov, 2011). Genetic algorithms (GA) are classes of evolutionary optimization methods that can be used as a search strategy in large multivariate problems for which there are many possible solutions (Michalewicz, 1997; Mitchell, 1999). These problems include the optimization of multivariate models or selection problems. A typical selection problem is the search for the most appropriate wavelengths to be used for a particular purpose (Jouan-Rimbaud *et al.*, 1995; Arakawa *et al.*, 2011). Note that the number of possible combinations of wavelengths is too high for an exhaustive search. These algorithms are intended to allow a fast convergence to a near-optimal solution without completing all possible combinations. Another example of the use of GAs is the selection of appropriate samples for multivariate models calibration (Michalewicz, 1997).

### *Robustness analysis*

Despite the existence of many successful applications of bacteria characterization with FTIR in the literature, reports of successful field implementation of FTIRS on a routine basis hardly exist (Naumann, 2000; Oust *et al.*, 2004). This is mainly due to the lack of robustness of the developed models. Robust estimates of the accuracy and precision of regression models are crucial for any method to be adopted as routine (Preisner *et al.*, 2007). Jackknifing (as proposed by Quenouille, 1956) and bootstrapping (as proposed by Efron, 1979) are the two resampling techniques with greatest use amongst the chemometrics community (Quenouille, 1956; Efron, 1979; Sokal and Rohlf, 1981). The attractiveness of jackknifing and bootstrapping is that they provide a significant and previously unattainable type of information: estimates of accuracy and dispersion statistics of unknown or poorly known distribution (Diaconis and Efron, 1983).

Bootstrapping is a nonparametric resampling technique that can be used to estimate statistical parameters of some population. Mean and standard deviation statistics are some of the parameters that can be estimated with this technique. Bootstrapping is based on a repetitive estimation of some statistic by changing the population individuals used to estimate that statistic. Within the framework of multivariate regression methods, such as PCA or PLS, one application of bootstrapping is the assessment of statistical significance of model coefficients (regression coefficients). The bootstrapping algorithm may be applied as a variable selection technique, as previously proposed by Lazraq *et al.* (2003). Bootstrapping can equally be applied to other nonregression models. Within the PCA framework this methodology may assess the uncertainty in principal components (or scores). The bootstrap method coupled to the leave-one-out concept yields different estimates for each sample score. Standard deviation for each sample score can then be used to assess uncertainty. For example in a two-dimensional score map, samples are actually represented by a cloud of points (where the cloud area represents uncertainty). For classification purposes this representation is more likely to give an idea of clusters than observing single estimates for scores. A method to estimate the uncertainty in PCA scores is given in Preisner *et al.* (2008).

## INFRARED RADIATION AND FOOD TECHNOLOGY: A BRIEF REVIEW

The use of infrared radiation in food technology is becoming a routine method, being one of the most used resources in food safety determinations. Samarakoon and coworkers (Samarakoon *et al.*, 2012) showed that infrared radiation can be used with antibacterial potential for drying citrus press-cakes. This methodology, when compared with the freeze drying method, showed stronger antibacterial activity against pathogenic bacteria. The inactivation of microorganisms is another issue in food processing and preservation and could be performed by using infrared radiation (Lu *et al.*, 2011). Heating by IR radiation is advantageous in respect of the heating time, uniform heating, reduced quality losses, it uses simple and compact equipment and represents significant energy saving (Rastogi, 2012). When used to obtain spectral data, infrared radiation was applied to analyse dairy products (De Marchi *et al.*, 2007), meat (Ellis *et al.*, 2002), fish (Guillen *et al.*, 2004), edible oils (Marigheto *et al.*, 1998), cereals (Aït Kaddour *et al.*, 2008), sugar and honey (Bertelli *et al.*, 2007), fruit and vegetables (Holland *et al.*, 1998) and coffee (Kemsley *et al.*, 1995), amongst others, and to identify and characterize foodborne pathogens at different taxonomic levels (Amiel *et al.*, 2001; Al-Qadiri *et al.*, 2006; Lamprell *et al.*, 2006; Preisner *et al.*, 2010).

## Use of infrared spectroscopy in food technology

Real-time spectroscopic techniques such as infrared spectroscopy (NIRS or MIRS) have gained in popularity and become one of the most used techniques for a wide range of analysis in several fields. Despite the lower sensitivity of NIRS, this technique has been preferable in the dairy industry; most packaging stuffs are transparent to NIR radiation, allowing a totally nondestructive analysis. Being a fast and nondestructive methodology, with the support of rapid and effectiveness multivariate calibration tools, infrared spectroscopy has high potential in process monitoring, allowing the optimization of production, saving money and time and improving food quality. NIRS and MIRS have shown also high potential in the determination of geographical origin and adulteration. Recently, Sun *et al.* (2012) showed that NIRS, combined with chemometric tools, can be used as a rapid and effective method to discriminate the geographical origin of lamb meat. Another popular application of infrared spectroscopy, NIRS or MIRS, is in the determination of vitamin C in food. Yang (Yang and Irudayaraj, 2002) showed that several techniques based on infrared radiation (NIR, Fourier transform near-infrared (FT-NIR), Fourier transform infrared-attenuated total reflectance (FTIR-ATR), diffuse reflectance (DRIFTS), Fourier transform infrared-photoacoustic (FTIR-PAS)) can be used to quantify vitamin C in powdered mixtures and solutions. Other applications of infrared spectroscopy, summarized in Table 20.2, include:

**Table 20.2** Some applications of infrared spectroscopy in food technology.

| Aim | Spectroscopic technique | Chemometric method | References |
|---|---|---|---|
| Emulsion stabilizing properties of sunflower proteins | FTIR | — | Burnett *et al.*, 2002 |
| Denaturation of meat proteins | FTIR μ-spectroscopy and imaging | PCA, MSC, PLS | Kirschner *et al.*, 2004 |
| Imaging of microstructure of cereal tissues | FTIR μ-spectroscopy | — | Yu *et al.*, 2004 |
| Conformational study of globulin from rice | FTIR | — | Ellepola *et al.*, 2005 |
| Protein and phytate content of soybean | FTIR | — | Ishiguro *et al.*, 2006 |
| Effect of brine salting on Atlantic salmon | FTIR | PCA, PLSR | Bocker *et al.*, 2008 |
| Nitrogen concentration in tomato leaves | VIS-NIR | PLS, RMSEC, RMSECV | Ulissi *et al.*, 2011 |
| Food adulteration and contamination | NIR, FTIR-ATR | PLS | Ellis *et al.*, 2012 |
| Quality protein maize | NIR | PLS | Rosales *et al.*, 2011 |
| Acid content in vinegar | NIR | PLS, ANN | Chen *et al.*, 2012 |
| Determination of vitamin C | NIR, FT-NIR, FTIR-ATR | PLS | Yang and Irudayaraj, 2002 |
| Classification of geographical origins of lamb meat | NIR | PCA, D-PLS, LDA, PLSR | Sun *et al.*, 2012 |
| Determination of glucosinolates in broccoli | NIR | PLSDA | Hernández-Hierro *et al.*, 2012 |
| Antioxidant activity in bamboo leaves | NIR | PLS, ANN | Wu *et al.*, 2012 |
| Quality control of corn steep liquor | NIR | PLSDA | Xiao *et al.*, 2012 |
| Sugar level in snack products | NIR, MIR | PLSR | Wang *et al.*, 2012 |
| Cheese properties | NIR | PLS | Oca *et al.*, 2012 |
| Antioxidant content and activity in food | FTIR, NIR | PLS | Lu and Rasco, 2012 |

determination of glucosinolates in broccoli (Hernández-Hierro *et al.*, 2012), determining the antioxidant activity indices (Lu and Rasco, 2012; Wu *et al.*, 2012), prediction of weight of fat percentage, dry matter, protein and fat/dry matter contents in cheese (Oca *et al.*, 2012), quality control of fermentation raw materials for food additives and fine chemicals, especially in corn steep liquor (Xiao *et al.*, 2012), and the determination of sugar levels in snack products (Wang and Rodriguez-Saona, 2012).

## Use of IR spectroscopy to characterize foodborne pathogens

Bacterial contamination still represents a big issue in food safety. The emergence of new subtypes of foodborne pathogens and the importance of tracking outbreaks create the need for quick and effective methods for subtyping these pathogens. Infrared spectroscopy has been used in several works with proven success in subtyping foodborne pathogens, such as *Escherichia coli* (Wang *et al.*, 2010; Davis *et al.*, 2012a), *Salmonella* spp. and *Listeria monocytogenes* (Davis and Mauer, 2011), among others.

Davis and coworkers (Davis *et al.*, 2012a) proved that FTIRS, coupled with multivariate statistical analysis, is a suitable, quick ($\leq$16 h) and economical procedure to subtyping *E. coli* O157:H7 with compared accuracy to multiple-locus variable number tandem repeat analysis (MLVA) typing. This methodology also proved to be able to differentiate between live and dead *E. coli* O157:H7 cells (Davis *et al.*, 2012b) and cells subjected to various inactivation treatments (heat, salt, UV, antibiotics and alcohol). With the application of canonical variate analysis (CVA) it was possible to differentiate between the spectra of live and dead cells. This method was also used to differentiate between the spectra of the differentially treated cells. Clear separation was obtained between clusters of spectra of bacteria exposed to the different inactivation treatments. Another study of Davis *et al.* (Davis *et al.*, 2010) reports the effectiveness of FTIRS for detection, differentiation and quantification of *E. coli* O157:H7 isolated from ground beef. Spectral data were validated trough standard plate counts. PLS and CVA were applied with success, proving the usefulness of this technique for rapid detection and differentiation of pathogens in complex foods.

Despite the efforts to avoid bacterial contamination by even more effective washing systems, pathogen internalization still remains a challenge in food safety. Wang *et al.* (2010) used FTIRS with attenuated total reflectance for quantification of *E. coli* K-12 internalized in baby spinach. The strains were inoculated into the spinach leaves and, after confirmation by scanning electron microscopy (SEM), FTIRS analyses were performed. The obtained concentrations of *E. coli* K-12 by FTIRS agreed well with the concentrations determined by plate counting, demonstrating that FTIRS can identify and quantify *E. coli* even when internalized in vegetables.

*Salmonella* spp. is another important foodborne pathogen due not only to the large number of foodborne illnesses caused by this pathogen but also due to its severity. FTIR-ATR spectroscopy proved to be capable of distinguishing between live and dead cells of *S. typhimurium* and *S. enteritidis* (Sundaram *et al.*, 2012). Spectral features of *Salmonella* spectra were analysed using PCA and 100% of accuracy was obtained for serotype differentiation.

Other studies proved that FTIRS has the potential as a routine technique to detect *S. typhimurium* in a short time with lower cost, with sensitivity and accuracy (Koluman *et al.*, 2012), and to differentiate *S. enterica* serovar *enteritidis* phage types (Preisner *et al.*, 2010) and *S. enterica* serotypes (Preisner *et al.*, 2012). *L. monocytogenes* is usually subtyped at the serotype and halotype level by multilocus genotyping (MLGT) and pulsed field gel electrophoresis (PFGE). However, these are laborious and expensive methods. Davis and Mauer (2011) proved that FTIRS, with appropriated chemometric tools, is once more able to subtype this pathogen at the serotype and halotype level with 96.6% and 91.7% of correct identifications, respectively.

# INFRARED APPLICATIONS IN FOODBORNE PATHOGEN ANALYSIS

The following sections describe a series of applications of infrared spectroscopy (NIRS and MIRS) considering, essentially, the objectives of food microbiological contamination detection and foodborne pathogens identification and discrimination at different taxonomic levels. These examples are divided according to the type of radiation involved: NIRS (12 500–4000 $cm^{-1}$) and MIRS (4000–400 $cm^{-1}$). Due to the low absorptivity of NIR radiation this spectroscopy has more potential to be used directly in food matrices, despite the possibility of using MIRS coupled with an ATR probe for this. The NIRS is more adequate to deal with samples with high percentages of water (e.g. milk). On the other hand, MIRS is more adequate to measure pure cultures giving strong signals and providing a better spectroscopic fingerprint.

## Bacterial samples processing

Samples to be used in infrared spectroscopy techniques require minimal processing. Usually they are inoculated overnight in a regular culture medium with no special requirement at 37°C for a specific period of time, which can be different according to the microorganism. The incubation time should ensure that the microorganism has reached the stationary phase. All replicates should be acquired in the same strictly conditions.

For FTIR-ATR spectroscopy, colonies are directly transferred from the agar plates to the ATR crystal, followed by drying in the optical surface in a thin film. For transmittance measurements, a loop of cells is collected from the agar plates and suspended in sterilized water. Aliquots of the bacterial suspensions are transferred in triplicate to a ZnSe sample wheel. Prior to analysis, samples are dried at 45°C for 45 minutes in an oven until a transparent film is obtained. Temperature and drying time should be optimized according to the microorganism; mentioned values constitute only an example. FTIR spectra are acquired usually from 4000 to 400 $cm^{-1}$ with a resolution and number of scans according the choice of the operator. Between each isolate measurement, a background should be acquired. For NIRS, cells can be suspended in sterile saline solution and the optical density adjusted by measuring the absorbance at a wavelength of 600 nm ($OD_{600}$). According to McFarland standards, at this wavelength optical density ($OD_{600}$) has a correlation with cellular concentration. Dilutions could be made in order to achieve the needed concentrations.

## Applications with NIRS

The nature of NIR radiation makes it possible to analyse food matrices directly. Due to the strong water absorptions in the NIR range some parts of the spectrum are masked if aqueous solutions are analysed. Nevertheless, it is possible to use NIRS to evaluate microbial contamination in samples containing a high percentage of water. An example is given here to demonstrate the application of NIRS in assessing microbial contamination using saline solutions as the medium.

Strains of three different genera were analysed by NIRS: gram-negative *Escherichia coli* and *Salmonella enterica* and gram-positive *Staphylococcus aureus*. Cells were grown in TSA medium and suspended in sterile saline solution. The $OD_{600}$ of each suspension was adjusted to 0.132 (equivalent to $1.5 \times 10^8$ CFU $ml^{-1}$), according to the McFarland standards. Tenfold dilutions were made to obtain cell suspensions at different concentrations, with the highest and lowest concentrations approximately $10^8$ CFU $ml^{-1}$ and 10 CFU $ml^{-1}$, respectively.

The procedure for NIR spectral acquisition used two sampling accessories: a temperature controlled transmittance accessory for 1 ml clear borosylicate glass vials (8 mm path length) and a transflectance dip-probe (4 mm path length). In vials, the aliquots were deposited in clear glass vials and placed in the adapter for measuring. For the probe measurements, the probe was immersed in 25 ml centrifuge

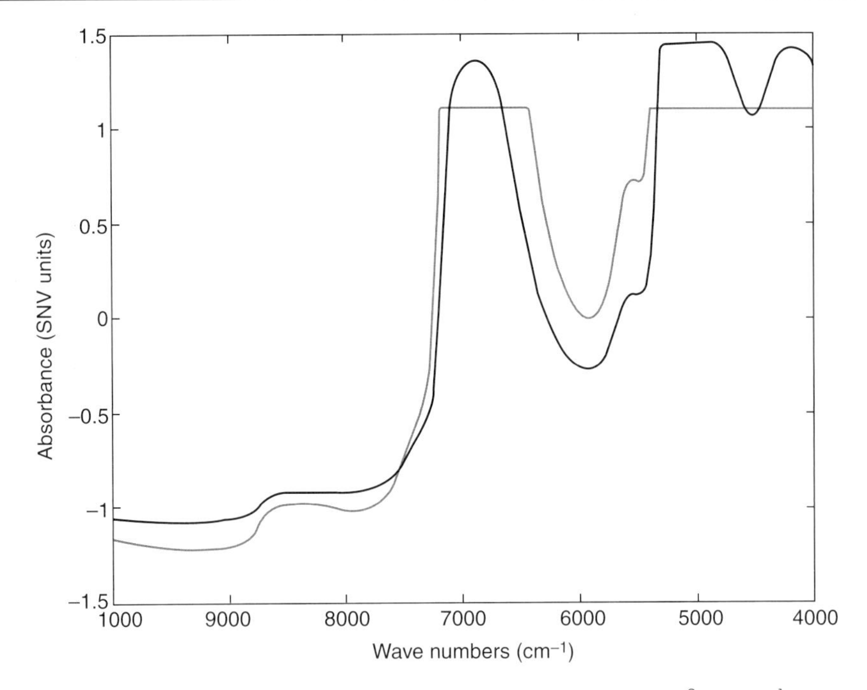


**Figure 20.3** NIR spectra obtained from suspended *E. coli* at a concentration of $1 \times 10^8$ CFU ml$^{-1}$ dissolved in a saline solution acquired with 8 mm path length temperature-controlled borosylicate vials (grey) and 4 mm path length transflectance dip-probe (black).

tubes containing the diluted samples. Spectra were recorded in the 10 000–4000 cm$^{-1}$ range using an ABB FT-NIR FTLA 200 spectrometer operating at 8 cm$^{-1}$ resolution. Figure 20.3 compares the average spectra obtained for *E. coli* at a concentration of $10^8$ CFU ml$^{-1}$ acquired with both sampling accessories. It is evident that when using the vials accessory (8 mm path length) there is saturation in the regions corresponding to the water bands strongly absorbing NIR radiation (around 6870 cm$^{-1}$ for the first overtone and 5200 cm$^{-1}$ for the combination region of O–H in water). The overflow problem was partially solved when the path length was reduced to half, as can be observed when the transflectance dip-probe (4 mm path length) is used. Saturated bands should not be used in subsequent spectral analysis. The temperature controlled accessory is a major reason to prefer the vials accessory, since temperature control was not ensured as efficiently for the probe. Temperature variations are known to affect considerably the NIR spectra, namely when aqueous solutions are measured.

### *Discrimination at the genus level*

The possibility of discrimination of the three different genera using NIRS in this matrix was assessed using PCA. Experiments were replicated using six independent plates for each bacterium to allow statistical assessment of results. The NIRS ability to discriminate between different bacteria species was assessed using spectral data acquired from samples in vials and using the dip-probe at concentration cells of $10^8$ CFU ml$^{-1}$ for all bacteria. Spectral data from 6500 to 5500 cm$^{-1}$ were pre-processed with SNV, Savitzky-Golay (width filter of 15 points, second-degree polynomial, first derivative) and mean centring. Models generated for the two situations show that total discrimination is possible in both cases (Figures 20.4a and b).

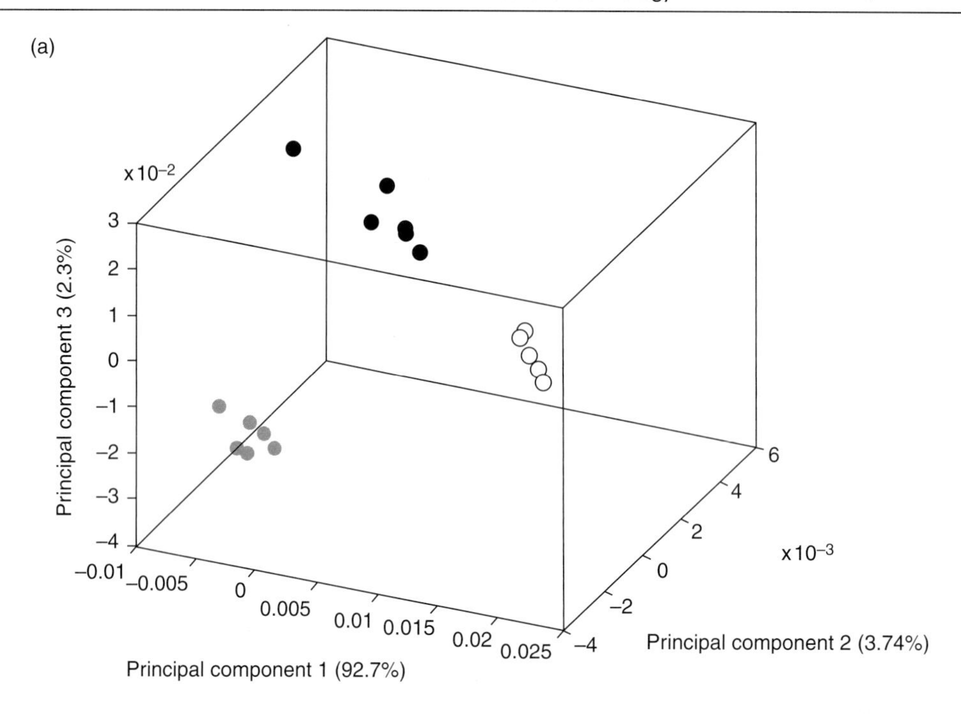


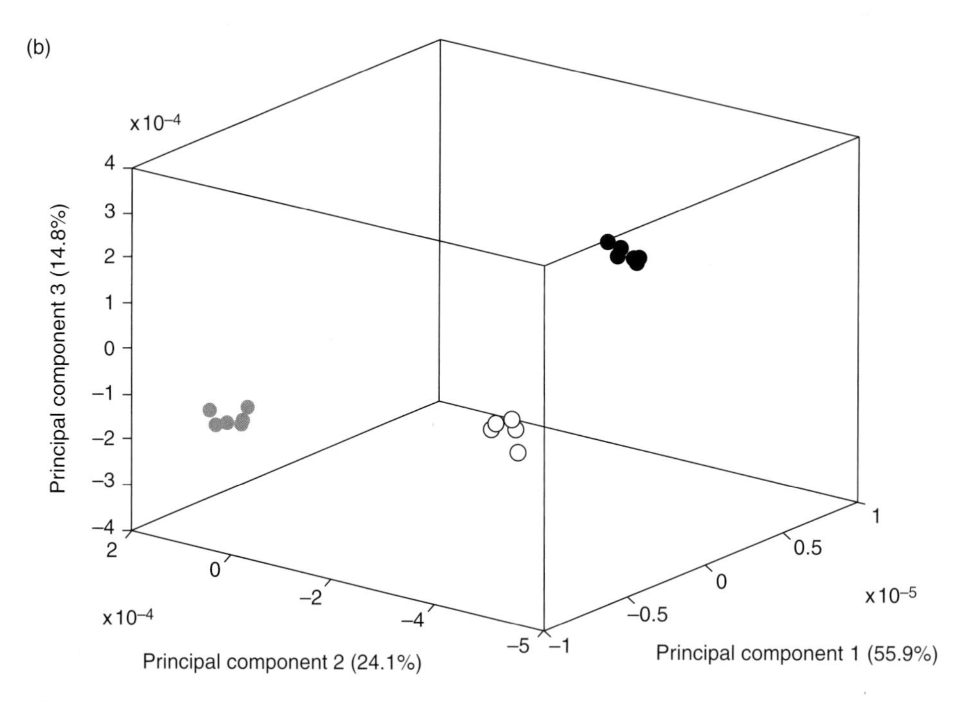


**Figure 20.4** Three dimensional score plot for a three component PCA model made with FT-NIR spectra collected from samles in (a) vials and (b) using the dip-probe (● *Escherichia coli*; ● *Salmonella enterica*; ○ *Staphylococcus aureus*).

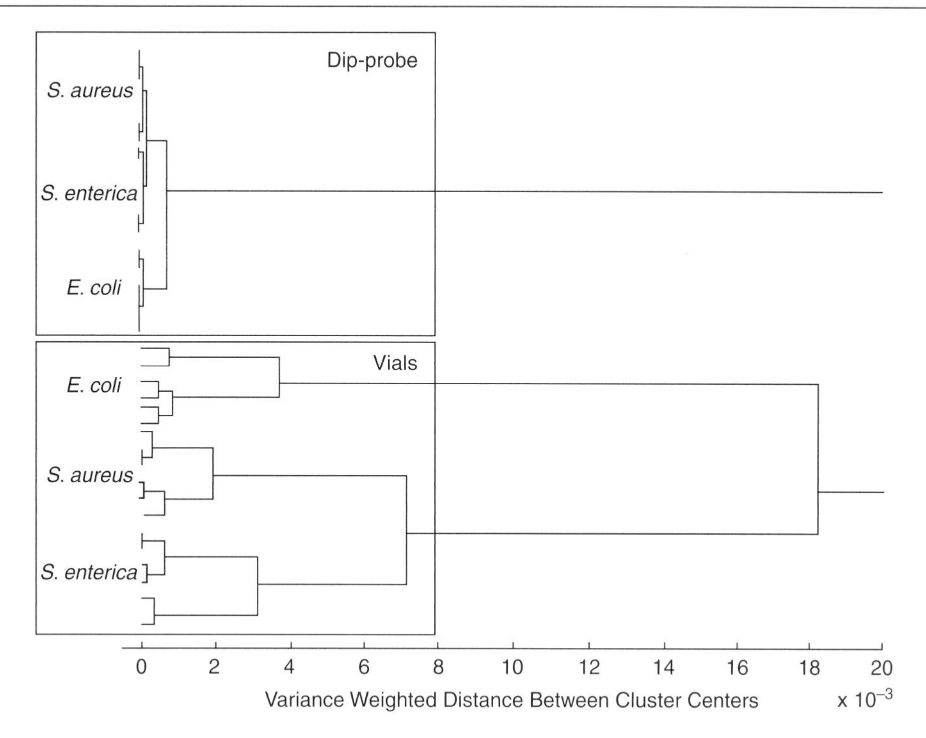


**Figure 20.5** Dendrogram obtained with the first principal component (97% of total variance) of the PCA model generated for the FT-NIR spectra of three genera obtained with the two sampling accessories (Ward's algorithm and Euclidean distance).

To investigate the relative separation between the three genera when using different sampling accessories, the two data sets were merged and a single PCA was performed considering the same wavelength region and the same spectral processing. A HCA was then performed using the first principal component of the PCA model (Ward's algorithm and Euclidean distance). The resulting dendrogram shows that spectra obtained with the dip-probe are substantially more consistent than spectra obtained with the vials accessory (Figure 20.5). Looking at the PCA model loadings it is possible to verify that the first loading (the major component allowing the discrimination) has a broad peak centred at 5685 $cm^{-1}$. This peak corresponds to the maximum slope of the original spectra centred at 5685 $cm^{-1}$. No further features were observed promoting the discrimination. This result shows that although the NIR spectra of saline solutions contaminated with bacteria was enough to discriminate between these three genera, it can be expected that going further in terms of taxonomy would present a very difficult task for NIRS in this type of samples. In this sense, the MIR spectrum is much more informative, thus allowing better separations at the subspecies level.

## Contamination detection

The NIRS data were used additionally to assess the possibility of this technique to detect contamination at increasingly smaller concentrations levels of bacterial cells. Eight different dilutions were prepared (from $10^8$ CFU $ml^{-1}$ to 10 CFU $ml^{-1}$) considering three replicates from each bacteria in three different days. Six spectra were acquired for each sample yielding a total of 576 spectra. Additionally, spectra of a saline solution (bacteria free) were acquired, also considering six replicates. This analysis was performed

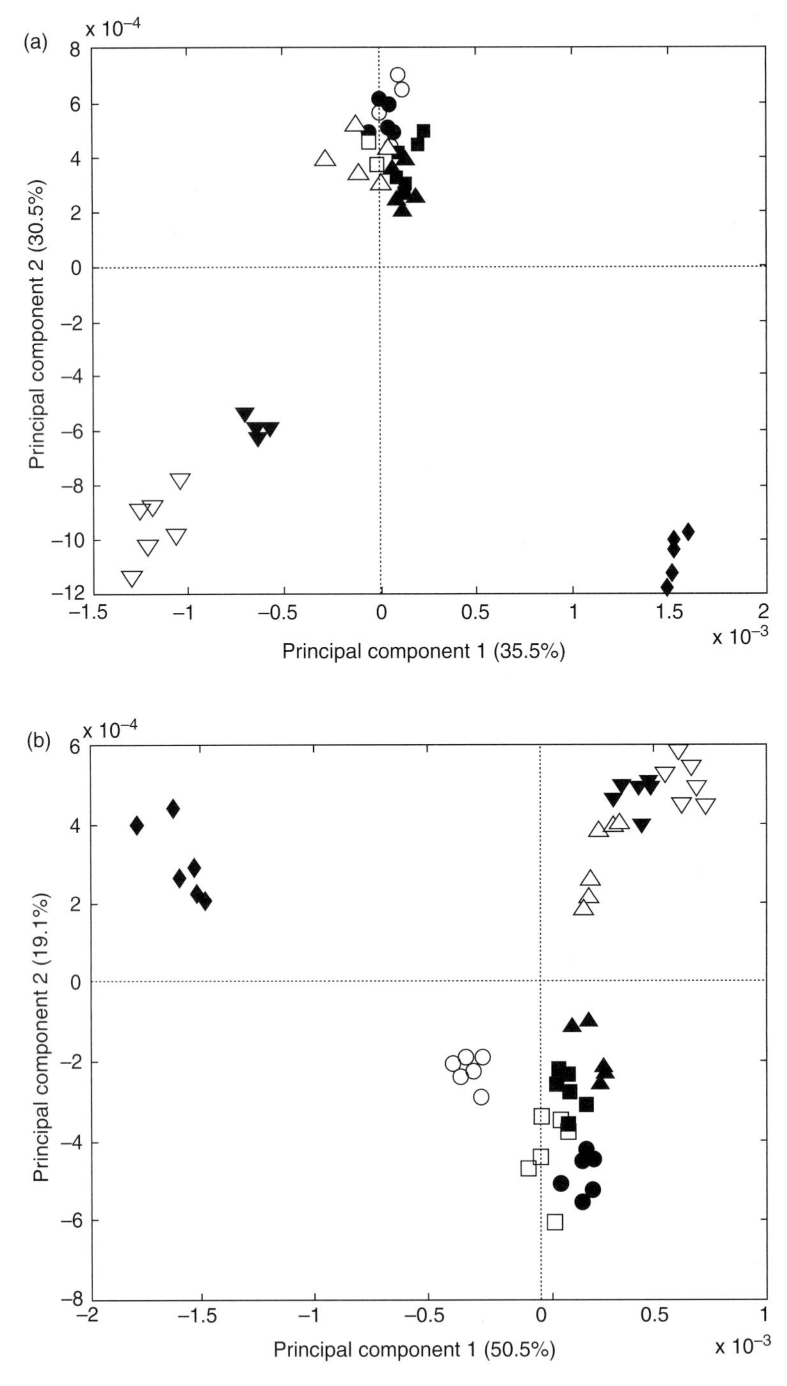


**Figure 20.6** Scores of PCA models obtained for a saline solution and saline solutions contaminated with bacteria concentrations (a - *Escherichia coli*. b - *Salmonella emerica*, c - *Staphylococcus aureus*) from $10^8$ CFU ml$^{-1}$ to 10 CFU ml$^{-1}$ in 10-fold dilution steps (○ $10^8$ CFU ml$^{-1}$, ● $10^7$; CFU ml$^{-1}$, □ $10^6$ CFU ml$^{-1}$, ■ $10^5$ CFU ml$^{-1}$, ▲ $10^4$ CFU ml$^{-1}$, △ $10^3$ CFU ml$^{-1}$, ▼ $10^2$ CFU ml$^{-1}$, ▽ 10 CFU ml$^{-1}$, ◆ saline solution).

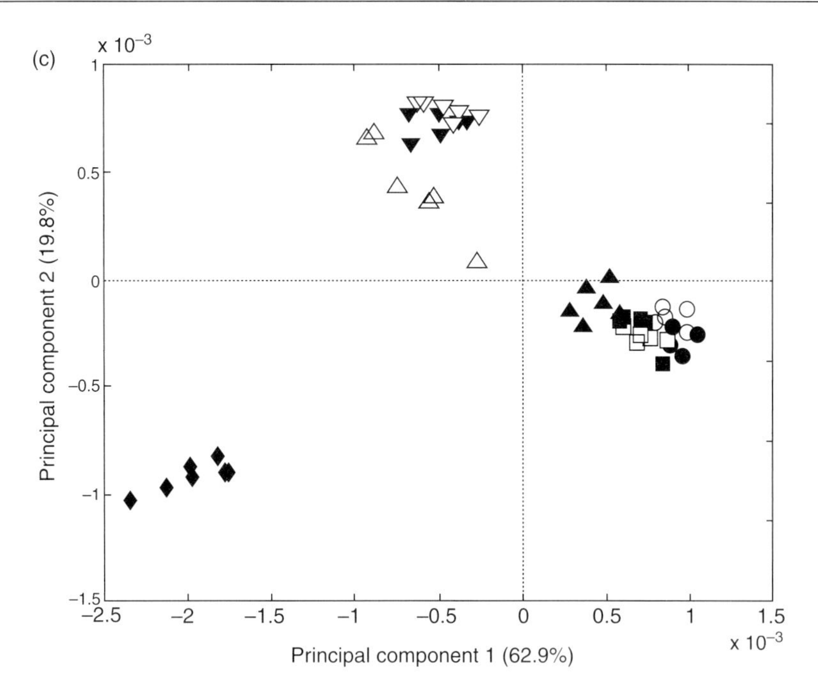


**Figure 20.6** *(Continued)*

using the dip-probe. A first analysis was performed with PCA using the above mentioned spectral region and spectral processing. Distinct models were calibrated for each genus. In each model the spectra obtained for the weight dilutions and the saline solution were considered. The goal was to evaluate at which dilution level the method loses the ability to discriminate a contaminated from the noncontaminated saline solution.

The scores obtained for each genus are presented in Figure 20.6. The evaluation of the three different models yields very similar conclusions. It can be observed that the models have problems discriminating cell concentrations above $10^5$ CFU $ml^{-1}$. In all situations, clusters formed by concentrations $10^5$–$10^8$ CFU $ml^{-1}$ can be observed. However, close inspection of these clusters shows that there is still a pattern in these clusters. It is observable that the different concentrations within these clusters can actually be separated along some principal component axis (dependent on the particular model). It is also evident that after some level of dilution the spectra appear increasingly distinct. In the three cases, the 10–100 CFU $ml^{-1}$ levels are distinct from the rest (in the *S. enterica* case the concentration level $10^3$ CFU $ml^{-1}$ was also very distinct from the higher concentrations). The saline solution was in all cases perfectly separated from the bacteria-contaminated saline solutions, indicating the very good ability of NIRS to make the identification of contamination.

These results opens the door to very interesting applications, namely those in the pharmaceutical industry, especially the manufacturing of sterile products for which very tight control of bacterial contamination is required. In more complex situations or matrices (such as food matrices) the presence of other chemical substances and suspended solids may naturally influence the ability to detect contamination. In the present situation, NIRS was able to detect the presence of bacteria via the light scattering phenomena generated by the presence of a suspended material in the samples. In food matrices the presence of bacteria may additionally influence the chemical composition of the matrices, thus allowing predictability the NIR spectra to capture contamination indirectly.

## Applications with MIRS

Bacteria discrimination at different levels (genus, species and subspecies) is illustrated here along with the challenges faced as the taxonomic level decreases and the spectral similarity increases. Foodborne pathogenic bacteria may have very distinct MIR spectral patterns. Although the same biological building blocks are present, phenotypic characteristics yield very different patterns over the spectra. Additionally, the characteristics of the growth medium, metabolic state and other factors contribute substantially for the spectral features. Mid infrared spectroscopy is designated by FTIRS in the following sections due to the use of Fourier-transform equipment.

### *Discrimination at the genus level*

To illustrate the MIR spectral differences present in different foodborne pathogens, strains of the genera *Bacillus*, *Escherichia*, *Staphylococcus*, *Salmonella* and *Listeria* were measured in reflectance by FTIR-ATR. Colonies from agar plates were harvested and directly transferred onto the ATR crystal, followed by drying at the optical surface in a thin film. FTIR spectra were collected in the 4000–400 $cm^{-1}$ range using a PerkinElmer Spectrum BX FTIR System spectrophotometer with a PIKE Technologies Gladi ATR accessory. The spectra resulted of the average of 32 consecutive scans at 4 $cm^{-1}$ resolution. Between each sample measurement a background was acquired with the crystal cleaned with ethanol and dried. Figure 20.7 shows the average spectra of these five different foodborne bacteria. Major differences in spectra are visible, leading to the expectation that discrimination at this level is absolutely straightforward not even requiring any chemometric method. Peak ratios are very distinct for the different genera. Some features are very characteristic, for example the peak at 1738 $cm^{-1}$, only observable for the genus *Bacillus* resulting from the production of poly-β-hydroxybutyrates (PHB). A very simple PCA model restricted to the 1750–900 $cm^{-1}$ spectral region allows the observation of clear differentiation between the five foodborne pathogens (Figure 20.8). Discrimination at the species level has equally been demonstrated with the MIRS technique. Spectral differences among species are generally less pronounced than between genera. Moreover, while for genera discrimination any of the

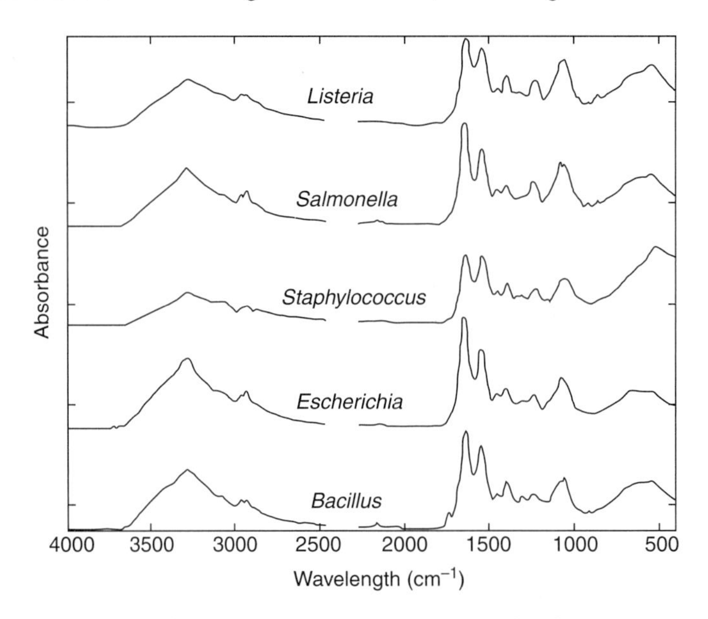


**Figure 20.7** Raw FTIR-ATR spectra of bacteria of the genera *Bacillus, Escherichia, Staphylococcus, Salmonella* and *Listeria*.

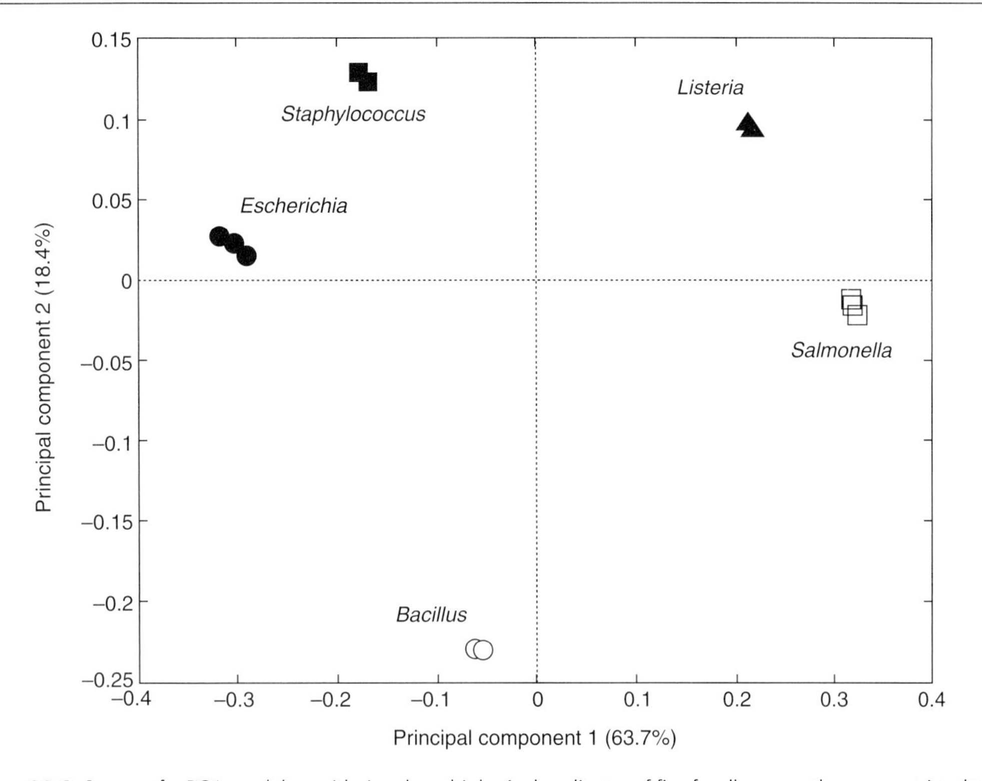


**Figure 20.8** Scores of a PCA model considering three biological replicates of five foodborne pathogens restricted to the 1750–900 $cm^{-1}$ spectral region (spectra processed with Savitzky-Golay filtering (first derivative, 9-points window size and second-order polynomial and first derivative), SNV and mean centring.

highlighted spectral regions is enough, to discriminate between species localized spectral regions are typically required. Therefore, a wavelength selection is required.

### *Discrimination at the species level*

A total of 105 isolates of multiresistant *Acinetobacter baumanni*, vancomicin-teicoplanin resistant *Enterococcus faecium* (VRE) and meticillin resistant *Staphylococcus aureus/epidermidis* (MRSA) was selected to illustrate the differences observed between genera and species. The samples set consisted of 49 isolates of *A. baumannii*, 28 isolates of *E. faecium*, and 28 isolates of *S. aureus/epidermidis* typed by pulse field gel electrophoresis (Preisner *et al.*, 2008). Samples were run in triplicate and analysed in transmittance mode. Spectra were collected over the wave number range of 4000 to 600 $cm^{-1}$ with a resolution of 4 $cm^{-1}$. Models based on PCA were built considering the different spectral regions as defined before in this chapter. Spectra were processed using Savitzky–Golay filtering (first derivative, nine-point window size and second-order polynomial and first derivative), SNV and mean centring. Before developing each model, the spectral region was restricted for the specific wavelengths.

Score plots obtained for each model (first versus second principal component) are shown in Figure 20.9a to 29.e, with each figure corresponding to a specific wavelength region. The analysis of each model shows that it is possible to discriminate between genera using the first two components independently of the selected region (the exception here is for the region between 1500 and 1200 $cm^{-1}$, where the first two components do not discriminate two genera well). It is possible to verify that the differences between the two *Staphylococcus* species are generally less evident. Additionally, the

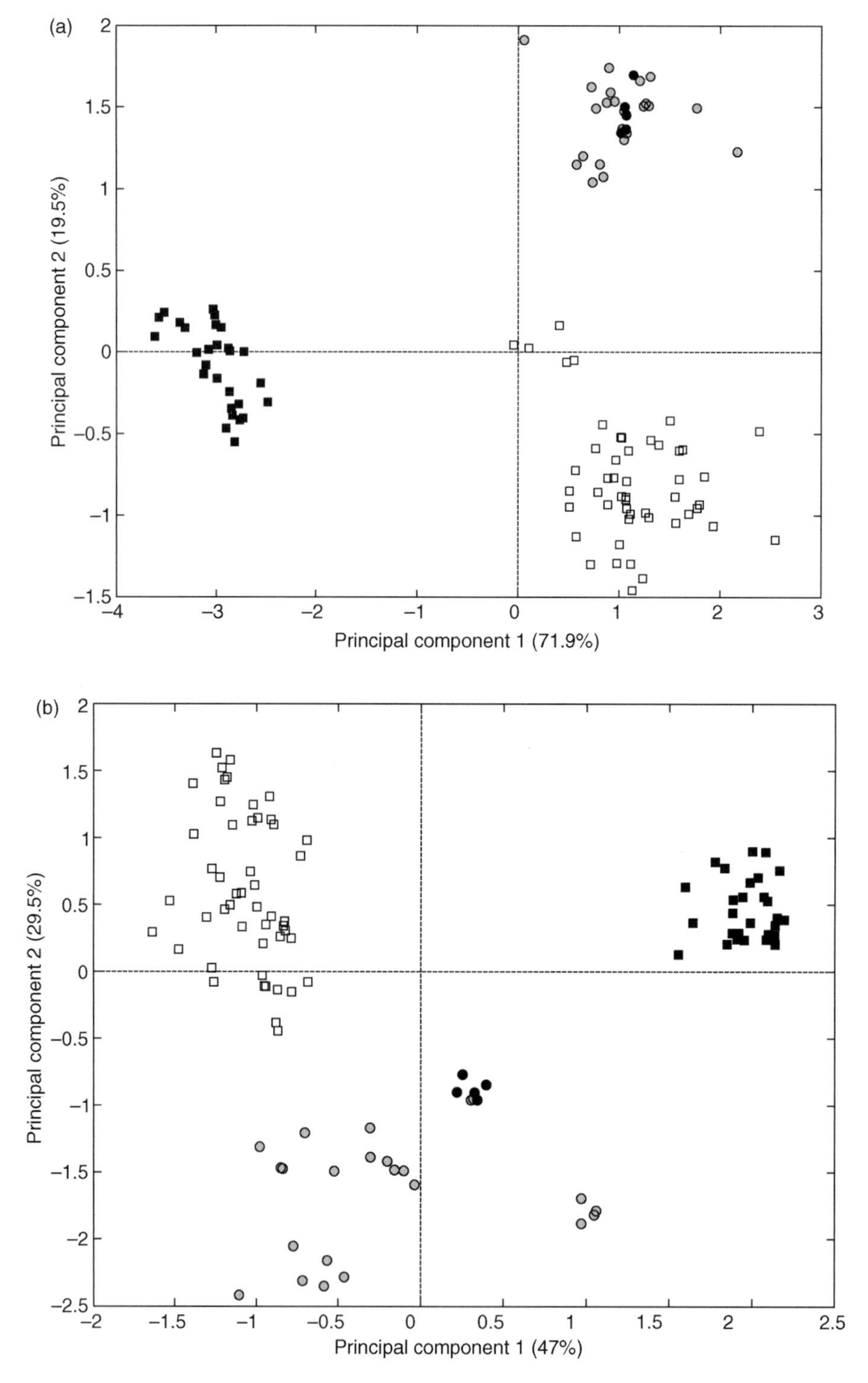


**Figure 20.9** Scores of principal component analysis models obtained using different wavelength sections of transmission FT-IR spectra corresponding to *Staphylococcus aureus* (○), *Staphylococcus epidermidis* (●), *Acinetobacter baumannii* (□) and *Enterococcus faecium* (■). The wavelength sections were the (a) lipids (3000–2820 $cm^{-1}$), (b) proteins (1750–1500 $cm^{-1}$), (c) phospholipids (1500–1200 $cm^{-1}$), (d) carbohydrates (1200–900 $cm^{-1}$) and (e) fingerprint (900–600 $cm^{-1}$) regions.

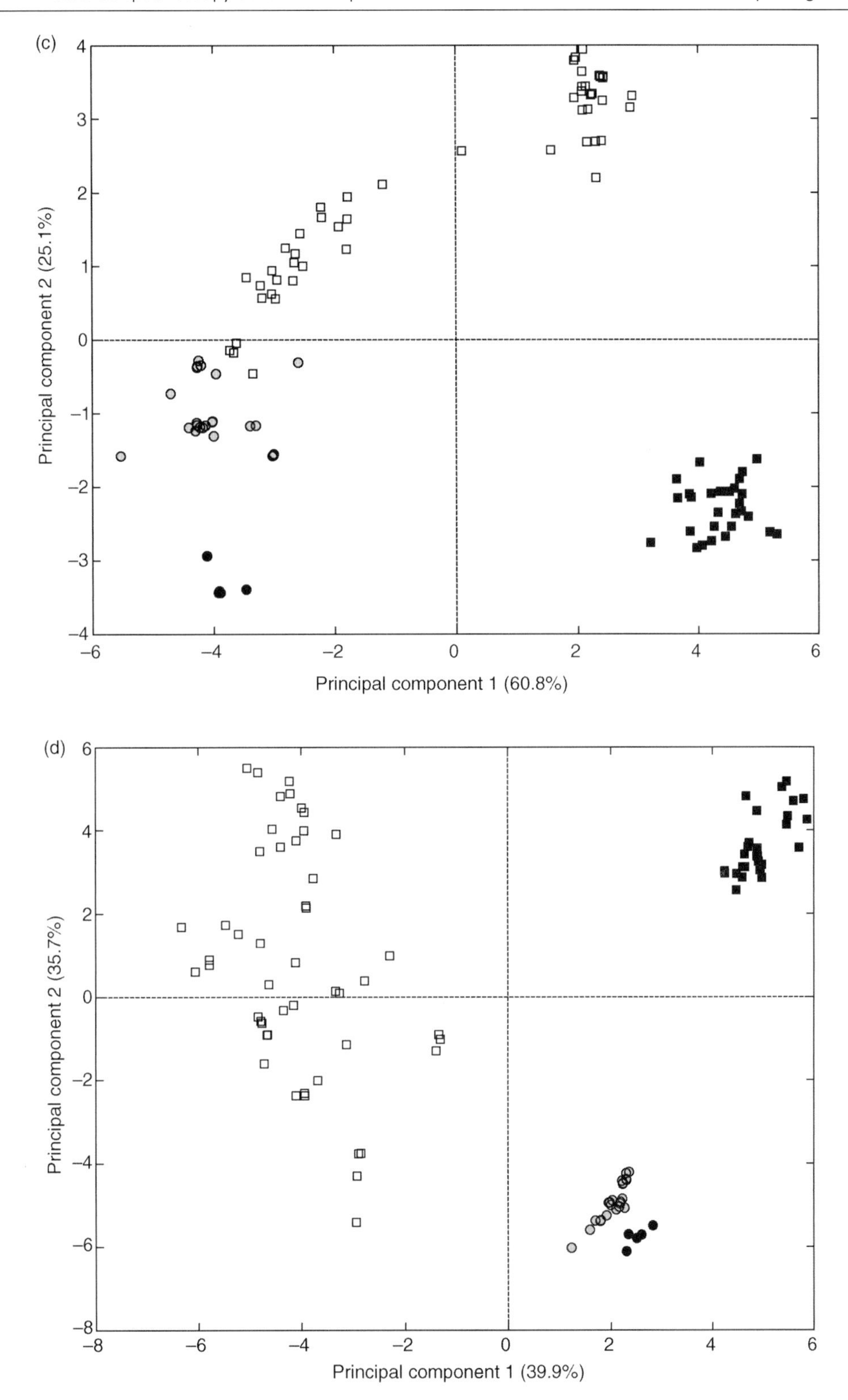


**Figure 20.9** *(Continued)*

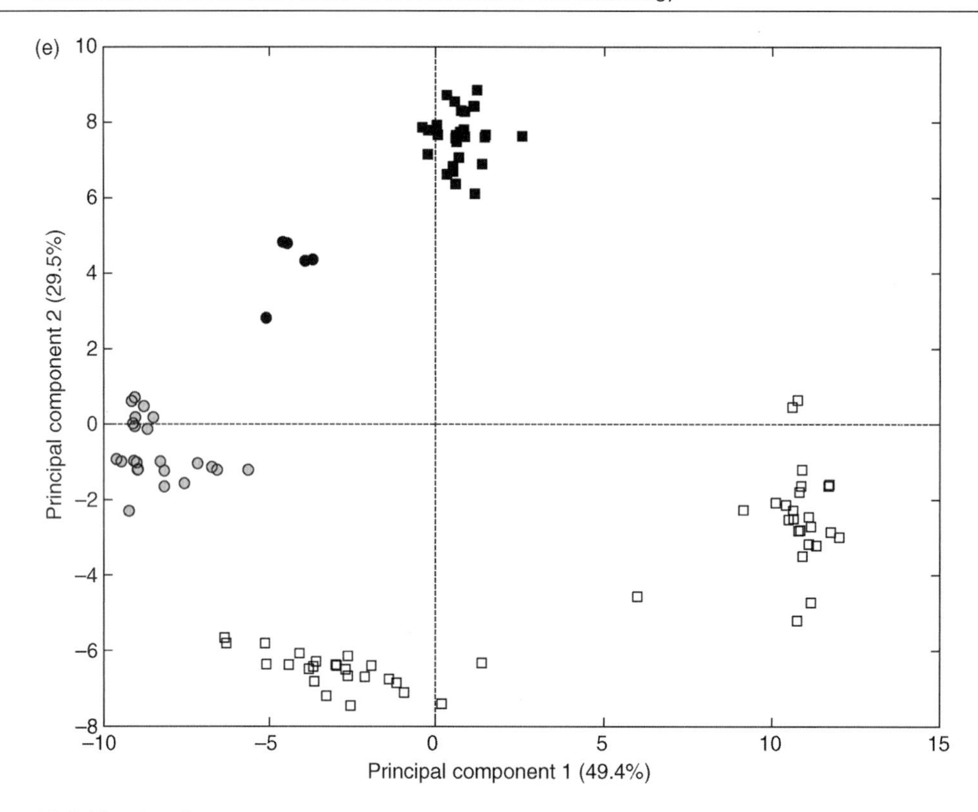


**Figure 20.9** *(Continued)*

differentiation between the species is only possible between 1500 and 600 cm$^{-1}$ corresponding to three regions. Note that the region between 1200 and 900 cm$^{-1}$ has often been considered for this discrimination and, in this case, it seems the most appropriate, giving a clear separation and also a good level of clustering. It should be stated here that the selection of the spectral window to discriminate between species should be made taking into account previous knowledge about biochemical and phenotypic differences between the species.

### *Discrimination at the subspecies level*

Discrimination at the subspecies level presents a challenge because typically IR spectra are very much similar. In this section two examples are used to demonstrate the techniques used to circumvent the high similarity between spectra. The first example consists of the discrimination of serotypes of *Salmonella enterica* (Preisner *et al.*, 2012). The first approach considers the discrimination of serogroups and then within each serogroup the different serotypes are discriminated. The second example demonstrates the discrimination of phage types of *S. enterica* servovar Enteritidis (Preisner *et al.*, 2010).

The possibility to discriminate at this level is not possible using simple methods such as PCA or HCA. Features must be extracted using supervised approaches. Among these approaches, PLS regression is a method that conjugates simplicity (it is a linear method after calibration) and performance.

FTIRS together with chemometric techniques were used to discriminate closely related *S. enterica* serotypes. Intact cells of 26 different serotypes belonging to serogroups B, C1, C2–C3 and D1 were

**Table 20.3** Performance of the FTIRS based PLSDA models for discrimination between the four *S. enterica* serogroups considering different spectral regions indicating the optimal number of latent variables for each model.

| Region | Region ($cm^{-1}$) | Latent variables | Correct predictions (%) | Standard deviation (%) |
|---|---|---|---|---|
| All | 4000–2600 1900–600 | 10 | 99.9 | 0.3 |
| Lipids | 3000–2800 | 8 | 91.0 | 4.6 |
| Proteins | 1700–1500 | 11 | 73.7 | 6.0 |
| Phospholipids | 1500–1200 | 8 | 96.6 | 3.5 |
| Polysaccharides | 1200–900 | 6 | 100.0 | 0.0 |
| Fingerprint | 900–600 | 7 | 98.59 | 1.7 |

Adapted from Preisner *et al.* (2012).

examined by FTIRS in transmittance mode and classified based on the agglutination pattern reactions using the Kauffmann–White classification scheme. Strains were prepared according to Preisner *et al.* (2012). Samples were run in triplicate and analysed by FTIRS in transmittance mode (4000–600 $cm^{-1}$ with a resolution of 4 $cm^{-1}$, collecting and averaging 64 scans and converting to absorption spectra). A PLSDA regression method was adopted to build the discrimination model. The model validation was described previously. The application of different spectral regions to the discrimination of the four serogroups yielded the results expressed in Table 20.3. Using FTIR spectra in the 1200–900 $cm^{-1}$ range (outer membrane polysaccharides) allowed the correct classification of all isolates according to the four analysed serogroups. A discrimination analysis applied within each serogroup demonstrated that it was also possible to differentiate between the different serotypes. Note that some parts of the spectra, especially the amides I and amides II regions (proteins), do not allow a good discrimination.

The analysis of the PLSDA model loadings revealed that the major discrimination region was located between 1000 and 960 $cm^{-1}$. Discrimination within each serogroup was found to be possible also. However, the application of PLSDA regression to small data sets (when classes have a limited number of elements) does not ensure statistical validation, therefore the simple PCA modelling was used here. Serogroups C1 and C2–C3 were selected. Each serogroup in this database contained three species with a sufficiently high number of isolates to ensure modelling robustness. Figure 20.10 shows the scores of PCA models developed for serogroups C1 and C2–C3. In both cases it is possible to verify that isolates cluster according to the species, even using an unsupervised technique, showing that the major differences observed between FTIR spectra in the 1200–900 $cm^{-1}$ region are related to the species.

The discrimination of five closely related *S. enterica* serotype Enteritidis phage types, phage type PT1, PT1b, PT4b, PT6 and PT6a is also described here (Preisner *et al.*, 2010). The FTIR spectra of these strains was acquired considering the entire bacterial spectra and the outer membrane protein extract as described in Preisner *et al.* (2010). Phage types were interpreted on the basis of the lysis patterns, using the method of Ward (Ward *et al.*, 1987). Intact cells and outer membrane protein (OMP) extracts from bacterial cell membranes were subjected to FTIRS analysis in transmittance mode. Spectra obtained from intact cells and OMP extracts are substantially different. Figure 20.11 compares the spectra obtained from intact cells and OMP extracts of two *S. enterica* and *E. coli* strains. Bands in the lipids region (3000–2800 $cm^{-1}$) are clearly amplified for OMP extracts. Bands centred at 1308, 1395 and 1457 $cm^{-1}$, characteristic of vibrations of phospholipids, are also more intense for OMP extracts as expected. Partial least squares discriminant analysis was used to develop calibration models based on pre-processed FTIR spectra (Savitzky-Golay filter with nine points, second-order polynomial and second-order derivative and mean centring). Models considering the entire spectra and the OMP were considered; the results are displayed in Table 20.4 as confusion matrices. The analysis based on OMP extracts provided greater separation between the phage type PT1-PT1b, PT4b and PT6-PT6a groups than

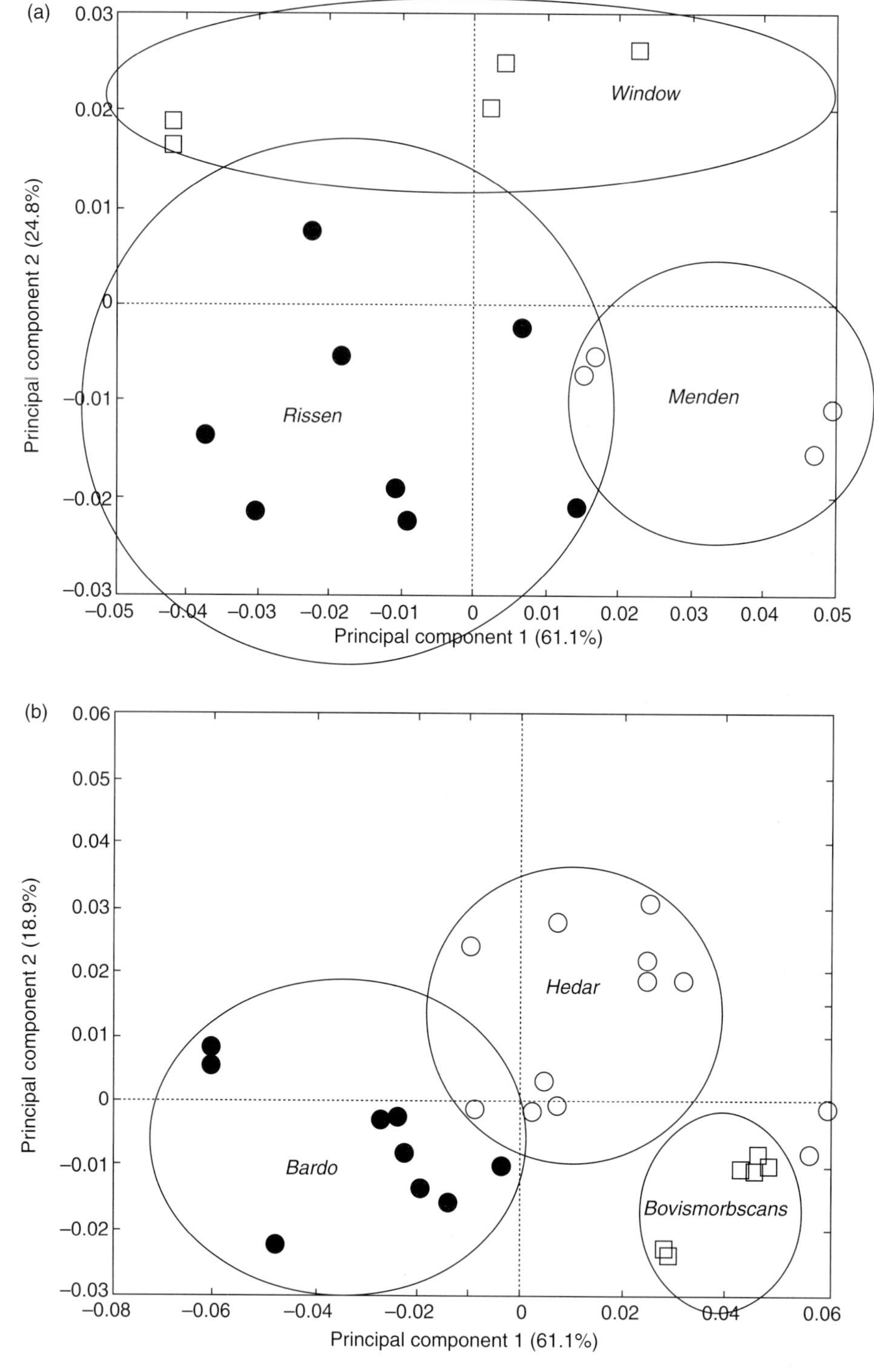


**Figure 20.10** Scores of PCA models for serogroups (a) C1 and (b) C2–C3 restricted to the 1200–900 $cm^{-1}$ spectral region. Spectra were processed with Savitzky–Golay filtering (9-points window size, second-order polynomial and first derivative), SNV and mean-centring.

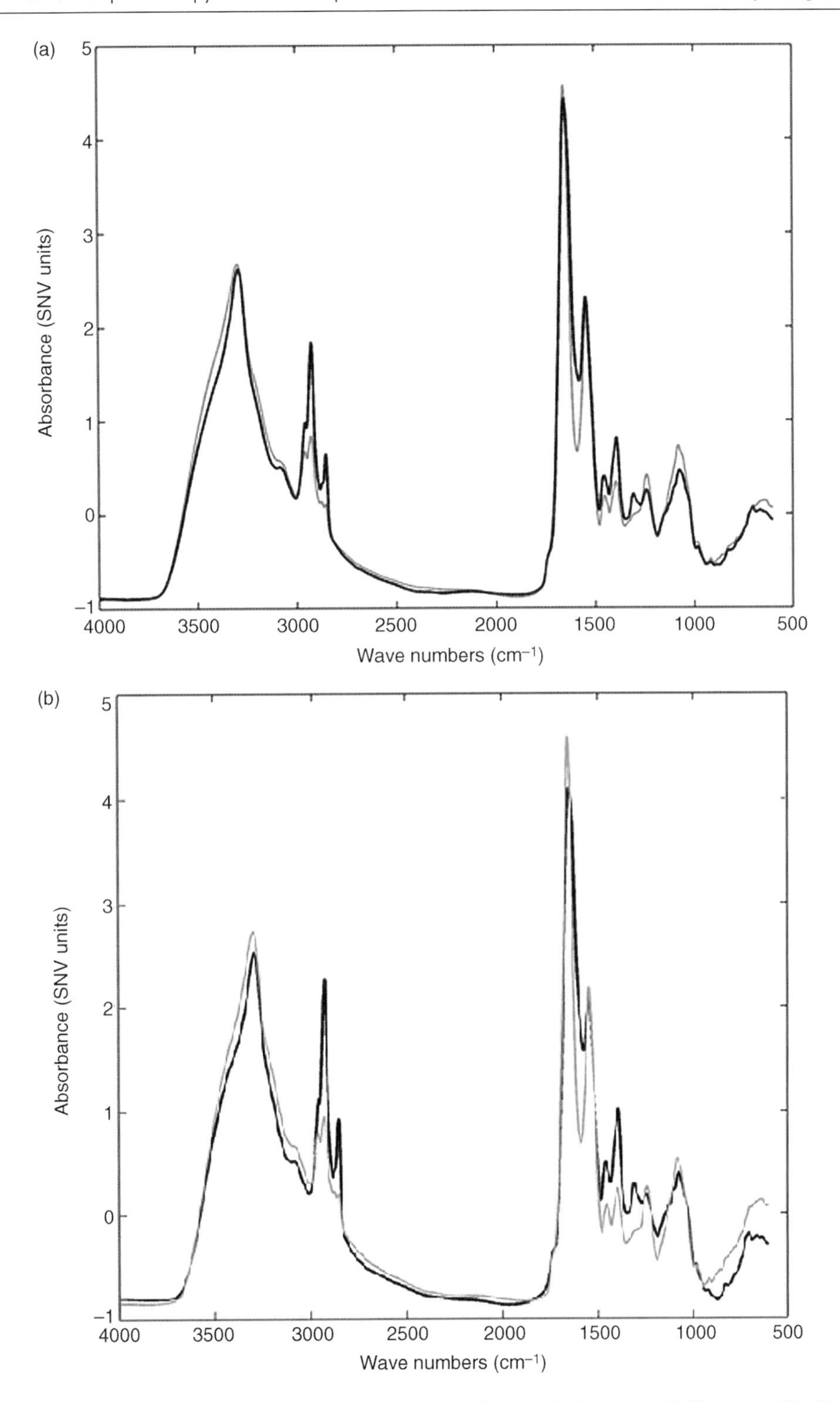


**Figure 20.11** Comparison between FTIR spectra obtained from intact cells (grey) and OMP extracts (black) for (a) a *Salmonella enterica* strain and (b) an *Escherichia coli* strain.

**Table 20.4** Confusion matrices for *S. enterica* intact cell- and OMP extract-based PLSDA discrimination models for the analysed phage types (values are in %).

| FTIR Method | Reference method | | | | | | | | | | | |
|---|---|---|---|---|---|---|---|---|---|---|---|---|
| | Intact cells | | | | | | OMP extract | | | | | |
| | PT1 | PT1b | PT4b | PT6 | PT6a | Total | PT1 | PT1b | PT4b | PT6 | PT6a | Total |
| PT1 | 10.0 | 1.1 | 1.8 | 0.9 | 0.4 | 14.2 | 19.6 | 0.4 | 0.0 | 0.0 | 0.0 | 20.0 |
| PT1b | 1.3 | 18.2 | 5.3 | 0.9 | 6.4 | 32.2 | 1.8 | 21.1 | 0.0 | 0.0 | 0.0 | 22.9 |
| PT4b | 6.7 | 3.3 | 9.1 | 5.1 | 1.6 | 25.8 | 0.0 | 0.0 | 21.4 | 0.0 | 0.0 | 21.4 |
| PT6 | 2.0 | 1.1 | 2.7 | 6.2 | 2.9 | 14.9 | 0.0 | 0.0 | 0.0 | 19.3 | 5.0 | 24.3 |
| PT6a | 0.0 | 2.9 | 1.1 | 0.2 | 8.7 | 12.9 | 0.0 | 0.0 | 0.0 | 2.1 | 9.3 | 11.4 |
| Total | 20.0 | 26.7 | 20.0 | 13.3 | 20.0 | 100.0 | 21.4 | 21.4 | 21.4 | 21.4 | 14.3 | 100.0 |



the intact cell analysis. When these three phage type groups were considered, the method based on OMP extract FTIR spectra was 100% accurate. Moreover, when complementary local models that considered only the PT1-PT1b and PT6-PT6a groups were developed the level of discrimination increased. PT1 and PT1b isolates were differentiated successfully with the local model using the entire OMP extract spectrum (98.3% correct predictions), whereas the accuracy of discrimination between PT6 and PT6a isolates was 86.0%.

## Uncertainty estimation of infrared-based chemometrics models

An example of the estimation of uncertainty is given here for PCA scores and a PLS regression vector. The selected data corresponds to a collection of *Staphylococcus aureus* (23 isolates) and *Staphylococcus epidermidis* (5 isolates). These isolates were measured by FTIRS in transmittance mode as previously described in this section. In this situation the number of isolates per species is unbalanced, which makes model construction a harder task (especially to validate results). If the goal is to devise a method for discrimination either based on PCA or PLS the statistical validity of model results must be ensured. In the first situation the uncertainty in scores estimation is attempted, while in the second case the standard deviation for the regression vector is estimated so that confidence intervals can be devised (Preisner *et al.*, 2008). Both models consider the same spectral range between 1200 and 900 $cm^{-1}$, as this was the best spectral region for the discrimination and the same pre-processing. Because the discrimination is between two classes (the species) the PLSDA model needs in practice only one regression vector (the other is the symmetric). Results were obtained by bootstrapping the models 1000 times.

In Figure 20.12a the resampled models generating the black dots allow the definition of statistical confidence bounds around each particular sample for PCA. It is clear that the discrimination is achieved at a very high level of probability (low p-value). The same result would not be obtained, for example when using the entire spectra. Strong overlap would be verified between samples of both species (not shown). Regarding the PLSDA model regression vector it is clear that for some wave numbers the regression vector value cannot be assumed to be different from zero (Figure 20.12b). Then, for this level of significance (~95% corresponding to ±2 standard deviations) these wavelengths should be discarded from the model, since their statistical significance is null. This strategy allows a wavelength selection and increased model robustness against unexpected spectral variations not related with the modelled problem. These resampling strategies were very helpful as they allowed estimation of uncertainty in this type of models with no specific a priori knowledge about data

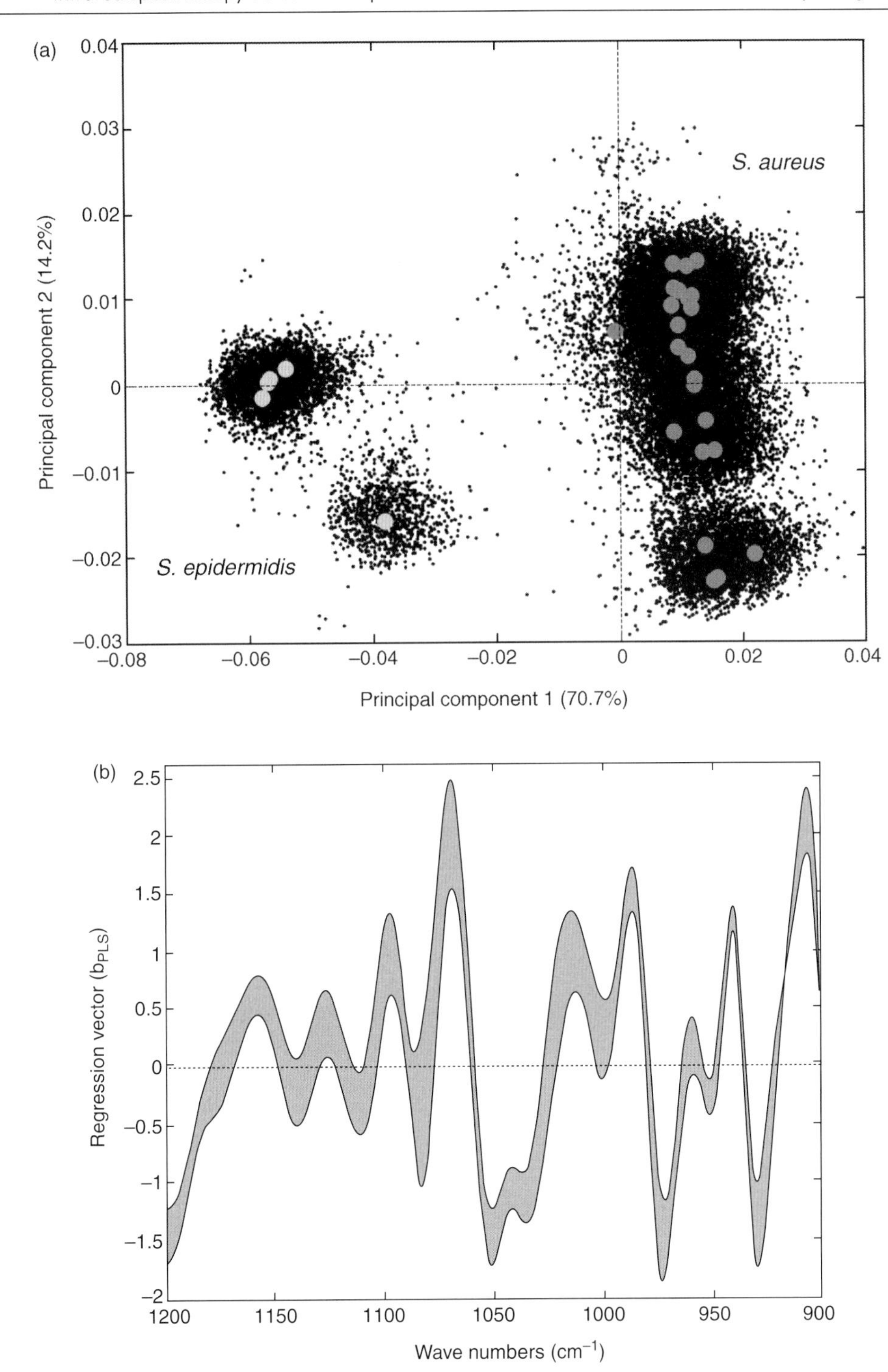


**Figure 20.12** Uncertainty estimation for PCA scores (a) and PLS regression vector boundaries (b) for models discriminating between *Staphylococcus aureus* and *epidermidis* species. The shadow region corresponds to the estimated PLS regression vector $\pm 2$ standard deviations.

distributions. The bootstrap method allows a cloud of estimations to be obtained for each sample in PCA and, therefore, enhances the clusters definition for different type of pathogens. This was observed for the analysed data set where clusters were positively identified. Bootstrapping demonstrated to be a valid alternative to estimate bias and variance for nonsupervised (PCA) and supervised models (PLSDA).

## PERSPECTIVES

Despite of the substantial achievements in this field of research, a series of developments are expected in the near future. These developments will be driven not only by developments in hardware (e.g. spectrometers, optics, probes) and processing algorithms but also by the need for field implementation of the current achievements.

- *Underpinning science.* A better understanding of spectral features obtained for different foodborne bacteria must be achieved. Standard procedures for developing foodborne analysis methods based on infrared spectroscopy are required. Migrating from the case-to-case situation to a more systematic approach of developing these methods is still something to achieve.
- *Subspecies discrimination levels.* Developments of new applications in terms of taxonomic discrimination are also expected using different sampling approaches. The possibility to identify microbial resistance with these straightforward methods requires a substantial effort not only in terms of the type of instrumentation and spectral processing but also in terms of bacteria samples processing (e.g. adequate growth procedures, effects of growth media).
- *Field/routine method applications.* It is expected that the consolidation of the current achievements will allow the transference of these methods for an increasing number of real-life applications. Applications range from microbiological diagnosis (health agencies) to field analysis (rapid, *in situ*) of foods and real-time diagnosis of bacterial contamination. This process requires proper generation of spectral libraries, including a systematic way of analysing samples for these libraries. Miniaturization of spectrometers plays a special role in this process.
- *Methodologies validation.* Validation of these methods for practical use is still a challenge in this area. The assessment of figures-of-merit for these methods is absolutely essential to promote validation upon the respective food protection governmental agencies. Methods validation comprises the previous point idea (large spectral databases with a samples universe wide enough) and also application of multivariate methods for assessing figures-of-merit.
- *Variability understanding.* Understanding the infrared spectral matrix effects of bacteria isolated in geographically distinct sites or from distinct sources (e.g. clinical, environmental) is still under investigated. Often these show substantial spectral differences. The same applies for similar food products belonging to different geographic origins. These differences should ideally be mitigated for in an appropriate universal methodology for practical use.
- *Mixed cultures.* Reported applications of bacteria discrimination rely on pure cultures. The extension of these methods to the analysis of mixed cultures is currently a challenge. The use of sampling methods such as micro-FTIR spectroscopy presents a possible solution but there are problems related with limitations of the radiation (wavelength). Resourcing to synchrotron radiation or confocal Raman spectroscopy (Schuster *et al.*, 2000), single cell analysis is possible, though (Holman *et al.*, 2010; Batard *et al.*, 2011). The effects of microsampling on spectra require a suitable approach, since the spectral reproducibility reduces considerably.
- *New methodologies and sampling devices.* Instrumental achievements and the production of cost-effective infrared instruments with better optical characteristics is also expected to happen in the near

future. Increase in optical resolution and micro-infrared spatial resolution (synchrotron radiation) may allow applications to be developed.

- *Competing spectroscopic methods.* The comparison of the achievements of infrared radiation (NIRS and MIRS) with concurrent techniques, also rapid and potentially applicable *in situ*, is required. Techniques like Raman (Rösch *et al.*, 2005) and Maldi-TOF/mass spectroscopy (Lay, 2001; Veloo *et al.*, 2011) are expected to be increasingly used in this field. Use of synchrotron radiation sources and microprobes is also increasingly reported (Dumas *et al.*, 2007).

## ACKNOWLEDGMENTS

Clara C. Sousa thanks the Fundação para a Ciência e a Tecnologia in Portugal for the award of her post-doctoral scholarship (BPD/70548/2010). This work was also supported by Fundação para a Ciência e a Tecnologia in Portugal through grant number PEst-C/EQB/LA0006/2011.

João A. Lopes thanks FSE (Fundo Social Europeu) and MCTES (Ministério da Ciência, Tecnologia e Ensino Superior) for the financial support through the POPH-QREN program.

Both authors would like to acknowledge the contribution over the past years of many colleagues that worked in collaboration in this field. José C. Menezes (Technical University of Lisbon, Portugal), Ornella E. Preisner (Technical University of Lisbon, Portugal), Jorge Machado (INSA, Portugal), Raquel Giomar (INSA, Portugal), Raquel T. Silva (REQUIMTE, Portugal), Luís Meirinhos-Soares (INFARMED I.P., Portugal), Carla P. Coutinho (REQUIMTE, Portugal) and Mafalda C. Sarraguça (REQUIMTE, Portugal) are greatly acknowledged for their contributions to the results presented in this chapter.

## REFERENCES

Aït Kaddour, A., Mondet, M. and Cuq, B. (2008) Application of two-dimensional cross-correlation spectroscopy to analyse infrared (MIR and NIR) spectra recorded during bread dough mixing. *J Cereal Sci* **48**, 678–85.

Al-Qadiri, H.M., Lin, M.S., Cavinato, A.G. and Rasco, B.A. (2006) Fourier transform infrared spectroscopy, detection and identification of *Escherichia coli* O157:H7 and *Alicyclobacillus* strains in apple juice. *Int J Food Microbiol* **111**, 73–80.

Alsberg, B.K., Kell, D.B., and Goodacre, R. (1998). Variable selection in discriminant partial least-squares analysis. *Anal Chem* **70**(19), 4126–33.

Alvarez-Ordóñez, A., Mouwen, D. J. M., López, M. and Prieto, M. (2011). Fourier Transform infrared spectroscopy as a tool to characterize molecular composition and stress response in foodborne pathogenic bacteria. *J Microbiol Meth* **84**, 369–78.

Amiel, C., Mariey, L., Denis, C. *et al.* (2001) FTIR spectroscopy and taxonomic purpose: Contribution to the classification of lactic acid bacteria. *Lait* **81**, 249–55.

Arakawa, M., Yamashita, Y. and Funatsu, K. (2011) Genetic algorithm-based wavelength selection method for spectral calibration. *J Chemom* **25**(1), 10–19.

Bakeev, K. A. (ed.) (2005) *Process Analytical Technology – Spectroscopy Tools and Implementation Strategies for the Chemical and Pharmaceutical Industries*. Blackwell, Oxford.

Bakeev, K.A. and Kurtyka, B. (2005) Sources of measurement variability and their effect on the transfer of near infrared spectral libraries. *J. Near Infrared Spectrosc* **13**(6), 339–48.

Balabin, R.M. and Smirnov, S.V. (2011) Variable selection in near-infrared spectroscopy: Benchmarking of feature selection methods on biodiesel data. *Anal Chim Acta* **692**, 63–72.

Batard, E., Jamme, F., Montassier, E. *et al.* (2011) Synchrotron radiation infrared microspectroscopy to assess the activity of vancomycin against endocarditis vegetation bacteria. *J Microbiol Methods* **85**(3), 235–238.

Bertelli, D., Plessi, M., Sabatini, A.G. *et al.* (2007) Classification of Italian honeys by mid-infrared diffuse reflectance spectroscopy (DRIFTS). *Food Chem* **101**, 1582–87.

Blanco, M., Maspoch, S., Villarroya, I. *et al.* (2001) Geographical origin classification of petroleum crudes from near-infrared spectra of bitumens. *Appl Spectrosc* **55**, 834–39.

Bocker, U., Kohler, A., Aursand, I.G. and Ofstad, R. (2008) Effects of brine salting with regard to raw material variation of Atlantic salmon (Salmo salar) muscle investigated by Fourier transform infrared. *J Agric Food Chem* **56**, 5129–37.

Bokobza, L. (1998) Near infrared spectroscopy. *J Near Infrared Spectrosc* **6**, 3–17.

Bras, L.P., Bernardino, S.A., Lopes J.A. and Menezes, J.C. (2005) Multiblock PLS as an approach to compare and combine NIR and MIR spectra in calibrations of soybean flour. *Chemom Intel Lab Syst* **75**(1), 91–9.

Bro, R. (1997) PARAFAC. Tutorial and applications. *Chemom Intel Lab Syst* **38**(2), 149–71.

Bro, R. and Kiers, H.A.L. (2003) A new efficient method for determining the number of components in PARAFAC models. *J Chemom* **17**(5), 274–86.

Burgula, Y., Khali, D., Kim, S. *et al.* (2007) Review of mid-infrared Fourier transform-infrared spectroscopy applications for bacterial detection. *J Rapid Methods Autom Microbiol* **15**(2), 146–75.

Burnett, G.R., Rigby, N.M., Mills, E.N.C. *et al.* (2002) Characterization of the emulsification properties of 2S albumins from sunflower seed. *J Colloid Interface Sci* **247**, 177–85.

Chen, Q., Ding, J., Cai, J., Zhao, J. (2012) Rapid measurement of total acid content (TAC) in vinegar using near infrared spectroscopy based on efficient variables selection algorithm and nonlinear regression tools. *Food Chem* **135**(2), 590–95.

Davis, R. and Mauer, L.J. (2011) Subtyping of *Listeria monocytogenes* at the haplotype level by Fourier transform infrared (FT-IR) spectroscopy and multivariate statistical analysis. *Int J Food Microbiol* **150**(2–3) 140–49.

Davis, R., Irudayaraj, J., Reuhs, B.L. and Mauer, L.J. (2010) Detection of *E. coli* O157:H7 from ground beef using Fourier transform infrared (FT-IR) spectroscopy and chemometrics. *Food Sci* **75**(6), M340–46.

Davis, R., Paoli, G. and Mauer, L.J. (2012a) Evaluation of Fourier transform infrared (FT-IR) spectroscopy and chemometrics as a rapid approach for sub-typing *Escherichia coli* O157:H7 isolates. *Food Microbiol* **31**(2), 181–90.

Davis, R., Deering, A., Burgula, Y. *et al.* (2012b) Differentiation of live, dead and treated cells of *Escherichia coli* O157:H7 using FT-IR spectroscopy. *J Appl Microbiol* **112**(4), 743–51.

De Marchi, M., Dal Zotto, R., Cassandro, M. and Bittante, G. (2007) Milk coagulation ability of five dairy cattle breeds. *J Dairy Sci* **90**, 3986–92.

Diaconis, P. and Efron, B. (1983) Computed-interactive methods in statistics. *Sci Am* **248**, 116–30.

Dumas, P., Sockalingum, G.D. and Sulé-Suso, J. (2007) Adding synchrotron radiation to infrared microspectroscopy: what's new in biomedical applications? *Trends Biotechnol* **25**(1), 40–44.

Efron, B. (1979) Bootstrap methods: another look at Jackcknife. *Ann Stat* **7**, 1–26.

Ellepola, S.W., Siu, M.C. and Ma, C.Y. (2005) Conformational study ofglobulin from rice (Oryza sativa) seeds by Fourier-transform infrared spectroscopy. *Int J Biol Macromol* **37**, 12–20.

Ellis, D.A., Brewster, V.L., Dunn W.B. *et al.* (2012) Fingerprinting food: current technologies for the detection of food adulteration and contamination. *Chem Soc Rev* **41**, 5706–27.

Ellis, D.I., Broadhurst, D., Kell, D.B. *et al.* (2002) Rapid and quantitative detection of the microbial spoilage of meat using FT-IR spectroscopy and machine learning. *Appl Environ Microbiol* **68**, 2822–28.

Essendoubia, M., Toubasb, D., Bouzaggoua, M. *et al.* (2005) Rapid identification of *Candida* species by FT-IR microspectroscopy. *Biochim Biophys Acta* **1724**, 239–47.

Felicio, C.C., Bras, L.P., Lopes, J.A. *et al.* (2005) Comparison of PLS algorithms in gasoline and monitoring with MIR and NIR. *Chemom Intel Lab Syst* **78**(1–2), 74–80.

Gaul, T.W. and Meinhardt-Wollweber, M. (2011) Hierarchical cluster analysis (HCA) of microorganisms: an assessment of algorithms for resonance Raman spectra. *Appl Spectrosc* **65**, 165–73.

Geladi, P. and Kowalski, B. R. (1986). Partial least-squares regression: A tutorial. *Anal Chim Acta* **185**(1), 1–17.

Goodacre, R., Adaoin, E., Timmins, M. *et al.* (1998) Rapid identification of urinary tract infection bacteria using hyperspectral whole-organism fingerprinting and artificial neural networks. *Microbiol* **144**, 1157–70.

Gómez M.A.M., Pérez, M.A.B., Gil, F.J.M. *et al.* (2003) Identification of species of *Brucella* using Fourier transform infrared spectroscopy. *J. Microbiol Meth* **55**, 121–31.

Guillen, M.D., Ruiz, A. and Cabo, N. (2004) Study of the oxidative degradation of farmed salmon lipids means of Fourier transform infrared spectroscopy. Influence of salting. *J. Sci Food Agric* **84**, 1528–34.

Hashimoto, A. and Kameoka, T. (2008) Application of infrared spectroscopy to biochemical, food, and agricultural processes. *Appl Spectrosc Rev* **43**, 416–51.

Hawkins, D.M. (2004) The problem of overfitting, *J Chem Inf Comput Sci* **44**, 1–12.

Helm, D., Labischinski, H., Schallehn, G. and Naumann, D. (1991) Classification and identification of bacteria by Fourier-Transform Infrared Spectroscopy. *J Gen Microbiol* **137**, 69–79.

Hernández-Hierro, J.M., Valverde, J., Villacreces, S. *et al.* (2012) Feasibility Study on the use of visible-near-infrared spectroscopy for the screening of individual and total glucosinolate contents in broccoli. *J Agric Food Chem* **60**(30), 7352–58.

Holland, J.K., Kemsley, E.K. and Wilson, R.H. (1998) Use of Fourier Transform infrared spectroscopy and chemometrics for the detection of adulteration of strawberry purees. *J Sci Food Agric* **76**, 263–69.

Holman, H.N., Bechtel, H.A., Hao, Z. and Martin, M.C. (2010) Synchrotron IR spectromicroscopy: chemistry of living cells. *Anal Chem* **82**(21), 8757–65.
Ishiguro, T., Ono, T., Wada, T. *et al.* (2006) Changes in soybean phytate content as a result of field growing conditions and influence on tofu texture. *Biosci Biotechnol Biochem* **70**, 874–80.
Jouan-Rimbaud, D., Massart, D., Leardi, R. and De Noord, O.E. (1995) Genetic algorithms as a tool for wavelength selection in multivariate calibration. *Anal Chem* **67**(23), 4295–301.
Kemsley, E.K., Ruault, S. and Wilson, R.H. (1995) Discrimination between coffea-arabica and coffea-canephora variant robusta beans using infrared-spectroscopy. *Food Chem* **54**, 321–26.
Kim, S., Kano, M., Nakagawa, H. and Hasebe, S. (2011) Estimation of an active pharmaceutical ingredient content using locally weighted partial least squares and statistical wavelenght selection. *Int J Pharm* **421**(2), 269–74.
Kirschner, C., Ofstad, R., Skarpeid, H.J. *et al.* (2004) Monitoring of denaturation processes in aged beef loin by Fourier transform infrared microspectroscopy. *J Agric Food Chem* **52**, 3920–29.
Koluman A, Celik G. and Unlu T. (2012) *Salmonella* identification from foods in eight hours: A prototype study with *Salmonella Typhimurium*. *Iran J Microbiol* **4**(1), 15–24.
Lamprell, H., Mazerolles, G., Kodjo, A. *et al.* (2006) Discrimiation of *Staphylococcus aureus* strains from different species of *Staphylococcus* using Fourier transform infrared (FT-IR) spectroscopy. *Int. J Food Microbiol* **108**, 125–29.
Larrechi, M.S. and Callao, M.P. (2003) Strategy for introducing NIR spectroscopy and multivariate calibration techniques in industry. *Trends Anal Chem* **22**, 634–40.
Lay J.O.Jr., (2001) MALDI-TOF mass spectrometry of bacteria. *Mass Spectrom Rev* **20**(4), 172–194.
Lazraq, A., Cléroux, R. and Gauchi, J-P. (2003) Selecting both latent and explanatory variables in the PLS1 regression model. *Chemom Intel Lab Syst* **66**, 117–26.
Lu, X., Liu, Q., Wu, D. *et al.* (2011) Using of infrared spectroscopy to study the survival and injury of *Escherichia coli* O157:H7, *Campylobacter jejuni* and *Pseudomonas aeruginosa* under cold stress in low nutrient media. *Food Microbiol* **28**(3), 537–46.
Lu, X. and Rasco, B.A. (2012) Determination of antioxidant content and antioxidant activity in foods using infrared spectroscopy and chemometrics: a review. *Crit Rev Food Sci Nutr* **52**(10), 853–75.
Maquelin, K., Kirschner, C., Choo-Smith, L.P. *et al.* (2002) Identification of medically relevant microorganisms by vibrational spectroscopy. *J Microbiol Meth* **51**, 255–71.
Marigheto, N.A., Kemsley, E.K., Defernez, M. and Wilson, R.H. (1998) A comparison of mid-infrared and Raman spectroscopies for the authentication of edible oils. *J Am Oil Chem Soc* **75**, 987–92.
Martens, H., Nielsen, J.P. and Engelsen, S.B. (2003) Application to near-infrared tansmission analysis of powder mixtures. *Anal Chem* **75**(3), 394–404.
Michalewicz, Z. (ed.) (1997) *Genetic Algorithms + Data Structures = Evolution Programs*. Springer Verlag, Germany.
Miller, C. E. (2000) Chemometrics for on-line spectroscopy applications – theory and practice. *J Chemom* **14**(5–6), 513–28.
Mitchell, M. (ed.) (1999) *An Introduction to Genetic Algorithms*. The MIT Press, Cambridge, MA.
Mobley, P., Kowalski, B., Workman, J. and Bro, R. (1996) Review of chemometrics applied to spectroscopy. *Appl Spectrosc Rev* **31**, 347–68.
Naes, T., Tomas, I., Fearn, T. and Tony, D. (2002) *A User-Friendly Guide to Multivariate Calibration and Classification*. NIR Publications, Chichester.
Naumann D. (2000) Infrared spectroscopy in microbiology. In: *Encyclopedia of Analytical Chemistry*. (ed. R.A. Meyers), John Wiley & Sons Ltd, Chichester. pp. 102–131.
Naumann, D. (2001) FT-infrared and FT-Raman spectroscopy in biomedical research. *Appl Spectrosc Rev* **36**(2–3), 239–98.
Naumann, D., Helm, D. and Labischinski, H. (1991) Microbiological characterization by FT-IR spectroscopy. *Nature* **351**, 81–2.
Ng, C.L., Wehling, R.L. and Cuppett, S.L. (2011) Near-infrared spectroscopy determination of degradation in vegetable oils used to fry various foods. *J Agric Food Chem* **59**(23), 12286–9D.
Norris, K.P. (1959) Infra-red spectroscopy and its application to microbiology. *J Hyg (Lond)* **57**, 326–45.
Norris, K.H. (1996) History of NIR. *J. Near Infrared Spectrosc* **4**, 31–7.
Oca, M.L., Ortiz, M.C., Sarabia, L.A. *et al.* (2012) Prediction of Zamorano cheese quality by near-infrared spectroscopy assessing false non-compliance and false compliance at minimum permitted limits stated by designation of origin regulations. *Talanta* **99**, 558–65.
Oust, A., Møretrø, T., Kirschner, C. *et al.* (2004) FT-IR spectroscopy for identification of closely related lactobacilli. *J Microbiol Methods* **59**, 149–62.
Preisner, O., Lopes, J.A., Guiomar, R. *et al.* (2007) Fourier transform infrared (FT-IR) spectroscopy in bacteriology: towards a reference method for bacteria discrimination. *Anal Bioanal Chem* **387**, 1739–48.
Preisner, O.E., Lopes, J.A. and Menezes J.C. (2008) Uncertainty assessment in FT-IR spectroscopy based bacteria classification models, *Chemom Intel Lab Syst* **94**, 33–42.

Preisner, O., Guiomar, R., Machado, J. *et al.* (2010) Application of Fourier transform infrared spectroscopy and chemometrics for differentiation of *Salmonella enterica* serovar *enteritidis* phage types. *Appl Environ Microbiol* **76**(11), 3538–44.

Preisner, O., Menezes, J.C., Guiomar, R. *et al.* (2012) Discrimination of *Salmonella enterica* serotypes by Fourier transform spectroscopy. *Food Res Int* **45**, 1058–64.

Quenouille, M. (1956) Note on bias in estimation. *Biometrika* **43**, 353–60.

Rastogi, N.K. (2012) Recent trends and developments in infrared heating in food processing. *Crit Rev Food Sci Nutr* **52**(9), 737–60.

Reich, G. (2005) Near-infrared spectroscopy and imaging: Basic principles and pharmaceutical applications. *Adv Drug Deliv Rev* **57**, 1109–43.

Rodriguez-Saona, L.E., Khambaty, F.M., Fry, F.S. and Calvey, E.M. (2001) Rapid detection and identification of bacterial strains by Fourier transform near infrared spectroscopy. *J Agric Food Chem* **49**(2), 574–79.

Roggo, Y., Duponchel, L. and Huvenne, J.P. (2004) Quality evaluation of sugar beet (Beta vulgaris) by Near-Infrared spectroscopy. *J Agric Food Chem* **52**(5), 1055–61.

Rosales, A., Galicia, L., Oviedo, E.X. *et al.* (2011) Near-infrared reflectance spectroscopy (NIRS) for protein, tryptophan, and lysine evaluation in quality protein maize (QPM) breeding programs. *J Agric Food Chem* **59**, 10781–86.

Rösch, P., Harz, M., Schmitt, M. *et al.* (2005) Chemotaxonomic identification of single bacteria by micro-Raman spectroscopy: application to clean-room-relevant biological contaminations. *Appl Environ Microbiol* **71**(3), 1626–37.

Rouf, M.A. and Stokes, J.L. (1962) Isolation and identification of the sudanophilic granules of *Sphaerotilus natans*. *J Bacteriol* **83**(2), 343–47.

Samarakoon, K., Senevirathne, M., Lee, W.W. *et al.* (2012) Antibacterial effect of citrus press-cakes dried by high speed and far-infrared radiation drying methods. *Nutr Res Pract* **6**(3), 187–94.

Savic, D., Jokovic, N. and Topisirovic, L. (2008) Multivariate statistical methods for discrimination of lactobacilli based on their FTIR spectra. *Sci Technol* **88**, 273–90.

Schuster, K.C., Urlaub, E. and Gapes J.R. (2000) Single-cell analysis of bacteria by Raman microscopy: spectral information on the chemical composition of cells and on the heterogeneity in a culture. *J Microbiol Methods* **42**(1), 29–38.

Sokal, R.R. and Rohlf, F.J. (1981) *Biometry: The Principles and Practice of Statistics in Biological Research*, 2nd edn. W.H. Freeman Co., New York.

Sun, S., Guo, B., Wei, Y. and Fan, M. (2012) Classification of geographical origins and prediction of δ(13)C and δ(15)N values of lamb meat by near infrared reflectance spectroscopy. *Food Chem* **135**(2), 508–14.

Sundaram, J., Park, B., Hinton, A.Jr. *et al.* (2012) Classification and structural analysis of live and dead Salmonella cells using Fourier transform infrared spectroscopy and principal component analysis. *J Agric Food Chem* **60**(4), 991–1004.

Tito, N.B., Rodemann, T. and Powell, S.M. (2012) Use of near infrared spectroscopy to predict microbial numbers on Atlantoic salmon. *Food Microbiol* **32**(2), 431–36.

Tøgersen, G., Arnesen, J.F., Nilsen, B.N. and Hildrum, K.I. (2003) On-line prediction of chemical composition of semi-frozen ground beef by non-invasive NIR spectroscopy. *Meat Sci* **63**(4), 515–23.

Udelhoven, T., Naumann, D. and Schmitt, J. (2000) Development of a Hierarchical Classification System with Artificial Neural Networks and FT-IR Spectra for the Identification of Bacteria. *App Spectrosc* **54**(10), 1471–79.

Ulissi, V., Antonucci, F., Benincasa, P. *et al.* (2011) Nitrogen concentration estimation in tomato leaves by VIS-NIR non-destructive spectroscopy. *Sensors* **11**(6), 6411–24.

Veloo, A.C.M., Welling, G.W. and Degener, J.E. (2011) The identification of anaerobic bacteria using MALDI-TOF MS. *Anaerobe* **17**(4), 211–2.

Versari, A., Paola Parpinello, G., Scazzina, F. and Del Rio, D. (2010). Prediction of total antioxidant capacity of red wine by Fourier transform infrared spectroscopy. *Food Control* **21**, 786–89.

Wang, J., Kim, K.H., Kim, S. *et al.* (2010) Simple quantitative analysis of *Escherichia coli* K-12 internalized in baby spinach using Fourier transform infrared spectroscopy. *Int J Food Microbiol* **144**(1), 147–51.

Wang, T. and Rodriguez-Saona, L.E. (2012) Rapid determination of sugar level in snack products using infrared spectroscopy. *J Food Sci* **77**(8), C874–9.

Ward, L.R., De Sa, J. and Rowe, B. (1987) A phage-typing scheme for Salmonella enteritidis. *Epidemiol Infect* **99**, 291–294.

Westerhuis J.A. and Smilde, A.K. (2001) Deation in multiblock PLS. *J Chemom* **15**(5), 485–93.

Westerhuis, J.A., Kourti, T. and MacGregor, J.F. (1998) Analysis of multiblock and hierarchical PCA and PLS models. *J Chemom* **12**(5), 301–21.

White, C.S., Watkins, L.C. and Fletcher, E.E. (1957) Emission spectroscopy in analysis of respiratory gases. IV. Calibration characteristics of oxygen emission in the near infrared. *J Aviat Med* **28**(4), 406–16.

Wu, D., Chen, X., Shi, P. *et al.* (2009). Determination of α-linolenic acid and linoleic acid in edible oils using near-infrared spectroscopy improved by wavelet transform and uninformative variable elimination. *Anal Chim Acta* **634**, 166–71.

Wu, D., Chen, J., Lu, B. *et al.* (2012) Application of near infrared spectroscopy for the rapid determination of antioxidant activity of bamboo leaf extract. *Food Chem* **135**(4), 2147–56.

Xiao, X., Hou, Y., Du, J. *et al.* (2012) Determination of main categories of components in corn steep liquor by near-infrared spectroscopy and partial least-squares regression. *J Agric Food Chem* **60**(32), 7830–35.

Yang, H. and Irudayaraj, J. (2002) Rapid determination of vitamin C by NIR, MIR and FT-Raman techniques. *J Pharm Pharmacol* **54**(9), 1247–55.

Yongnian N., Gu Y. and Kokot S. (2012) Interpreting analytical chemistry data: recent advances in curve resolution with the aid of chemometrics. *Anal Lett* **45**, 933–48.

Yu, P., McKinnon, J.J., Christensen, C.R. and Christensen, D.A. (2004) Using synchroton transmission FTIR micro-spectroscopy as a rapid, direct and non-destructive analytical technique to reveal molecular microstructural-chemical features within tissue in grain barley. *J Agric Food Chem* **52**, 1484–94.

# Section 4

# 21 Multivariate statistical quality control

**Jeffrey E. Jarrett**
*University of Rhode Island, Kingston, RI, USA*

## ABSTRACT

Better methods for quality monitoring are suggested with the application of multivariate methods in food and science and technology. In particular, the construction and use of industrial quality control methods are examined in both the prevention of diseases and the quality performance of food and beverage providers, especially in the area of controlling and improving quality output. The implementation is suggested of modern multivariate applications of quality control techniques common in industrial applications in food science and technology sector to better the health of the public and maintain a great reputation for a firm's food and beverage output.

## INTRODUCTION

The notion that statistical quality control (SQC) has no part in the provision of food science and technology is dubious and probably absurd. Although the term ***quality of care*** may have numerous definitions, we are almost certain in the SQC industry as to the meaning of the terms quality control and improvement. Today, much of the food and beverage industry is relying on quality control techniques that have changed since the early development of SQC in manufacturing operations, especially in the communications industries. Applications of SQC in recent years in activities in the food and beverage industry have been described by Bharati and McGregor (1998), Bakshi (1998), Figen *et al.* (2005), Lachenmeier (2007) and Philips *et al.* (2006). SQC has also been applied by health care institutions to identify improper treatment and disease prevention strategies (A complete treatment of this subject is given by Al-Assaf and Schmele, 1993.)

Decision makers from other fields of application recommend standard control charts for use in monitoring and improving a hospital's performance. Some examples include the monitoring of infection rates, waiting times for diagnostic procedures and the rates of patient falls. These applications developed originally for industrial statistical process control include Benneyan (1998a, 1998b), Lee and McGreevey (2002), and Benneyan *et al.* (2003). Books authored by Cary (2003), Hart and Hart (2002), Manciu *et al.*, (2006), Knapp and Miller (1983) and Morton (2005) illustrate how to apply standard statistical process control (SPC) charts in the monitoring of health care services and products. Our purpose in this study is to explain the need to improve these applications in the food science and technology arena to produce both quality control and improvement in this very important and vital area. At this time, we distinguish the differences between monitoring chronic and infectious problems that may occur in food and beverage preparation. Methods for monitoring this latter type of problem require the use of advanced time series models to account for seasonal effects. (details of these methods can be found elsewhere VanBrackle and Williamson, 1999). Finally, general multivariate methods that have other goals, such as

*Mathematical and Statistical Methods in Food Science and Technology*, First Edition.
Edited by Daniel Granato and Gastón Ares.

the determination of factors that distinguish subsets of a population, are not considered. Methods such as cluster analysis to find similar groups in a large sample of data are not considered here.

## AN EXAMPLE IN FOOD SCIENCE

The primary objective of any process control or monitoring system is to minimize the frequency of false signals or actions, and to rapidly detect external sources of process variation. In food science this means studies concerning all technical aspects of foods, beginning with harvesting or slaughtering, and ending with cooking and consumption. Often this is an ideology commonly referred to as 'from field to fork'. The conventional Shewhart subgroup control model achieves the objective of quality control and improvement by providing successive process observations that are serially independent (not autocorrelated). Unfortunately, many service industry processes including food and beverage preparation follow a first-order, positive autoregression. Some suggest that the existence of autoregression in the process violates the basic assumptions of the Shewhart model and alternative models for dealing with autoregression. This discussion is left to Montgomery and Mastrangelo (1991), and the fixed limit chart of Alwan (1992), West *et al.* (2002), and to West and Jarrett (2004). All have made contributions to assisting the construction of quality control and improvement in this area by refining the original Shewhart process.

Standard quality control charts may have applications in food and beverage preparation. For example one may apply the cumulative sum (CUSUM) chart for any specified underlying probability distribution, which characterizes the model generating data on the application of a particular food and beverage service. This very practical and useful technique based on standard statistical practice does provide a method by which decision makers can evaluate the quality of the service they provide. However, Steiner and Haass (2000) described the use of a risk-adjusted application based on meaningful performance criteria, which has optimality properties with respect to its ability to detect process shifts (Spiegelhalter, 2004). The standard CUSUM chart does not easily recognize these process shifts. In addition, Rogers *et al.* (2004) stated that other methods not described here might provide equally beneficial performance. This debate on the use of one such standard method of quality control and improvement (CUSUM Charts) is not the focus of this chapter. I do recognize, however, the benefits of implementing simple industrial quality methods in the food and beverage sector and its supply chain.

The purpose here is not to argue the robustness of CUSUM charts but to encourage the use of industrial tools such as multivariate control charts and to investigate further the service applications of SQC to monitor and possibly improve the quality of food and beverage services. The desire for improved services and monitoring is a well-established principle in the general field of decision making. Practitioners and researchers in industrial statistics have additional important contributions to the theory and further development of procedures to monitor and improve food science and technology applications, with the purpose of improving specific and general care of the population of consumers of these products.

The most active research areas in statistical process control include multivariate methods, effect of estimation error, short-run methods, autocorrelated data, variable sampling methods, economic design methods, change point estimation and engineering process control, as well as the use of nonparametric methods. In this chapter, the concern is with multivariate methods. We begin by discussing the benefits of multivariate analysis for process control as the multivariate analogue of a previously designed univariate solution.

In the next sections, how improve health monitoring in the health care supply chain may be improved by the use of multivariate quality control charts as opposed to standard techniques currently in use is explored. This is particularly important because food science and technology applications involve two or more related variables in the diagnoses of problems resulting from inattention to quality care in the preparation of food and beverage products.

## CURRENT APPLICATION OF STANDARD SHEWHART CONTROL CHARTS AND ITS FAILURE

One should not display data for two or more related measurements, that is simple Shewhart control charts, for a variety of reasons. A *multivariate* quality control chart shows how several variables jointly influence a process or outcome. For example multivariate control charts can be used to investigate how one therapeutic action coupled with another action may affect the quality of a prepared product to be consumed later. Similar examples exist in other fields, such as producing Silicon rods, industrial gases and semiconductors, and in the designing of software to manufacture semiconductors. If the data include correlated variables, the use of separate control charts is misleading because the variables both jointly and individually affect the process. If separate single variable quality control charts are used in a multivariate situation, the error of rejecting a true null hypothesis (Type I error) and the probability of a point correctly plotted in control are not equal to their expected values. That is, the plot is incorrect, which may lead to a false signal. The distortion of these values increases with the number of measurement variables. In complex production and operations processes the numbers of variables are many.

Multivariate quality control charts (Hotelling, 1947; Jackson, 1956, 1959, 1985) have several advantages over creating multiple univariate charts for the same business situation:

(i) The actual control region of the related variables in the bivariate case is elliptical.
(ii) You can maintain a specific probability of a Type I error (the $\alpha$ risk).
(iii) The determination of whether the process is out of or in control is a single control limit.

On the other hand, decision makers have more difficulty interpreting multivariate charts than interpreting the classic Shewhart control chart. For example the scale on multivariate charts is unrelated to the scale of any of the variables, and an out-of-control signal does not reveal which variable (or combination of variables) causes the signal.

Often, whether to use a univariate or multivariate chart is determined by constructing and interpreting a correlation matrix of the pertinent variables. If the correlation coefficients are greater than 0.1, the variables can be assumed to correlate and it is appropriate to construct a multivariate quality control chart.

The development of information technology enables the collection of large-size databases with high dimensions and short sampling time intervals at low cost. Computational complexity is now relatively simple for online computer-aided processes. In turn, monitoring results by automatic procedures produces a new focus for quality management. The new focus is on fitting the new environment. MQC requires methods to monitor multivariate and serially correlated processes existing in new commercial practice.

Illustrations of processes, which are both multivariate and serially correlated, are numerous in the production of food and beverage products and their associated services, as well as in industrial practice. In the service industries, the correlation among processes are serial, due to the inertia of human behaviours, and cross-sectional (i.e. across one time period) because of the interactions among various human actions and activities. This is particularly true for the prevention of infectious diseases arising from subquality preparation services. These subquality services give rise to disease and the use of substances that correlate with each other and other factors in both the cure and prevention of disease. For example the number of interventions in food preparation in a facility in a hospital may be serially dependent and related to (i) the waiting time for service provided by a nearby alternative facility and (ii) the cost and convenience of transport from one facility to another facility at a greater distance from the original facility. Furthermore, the latter factors are also autocorrelated and cross-sectionally correlated to each other. The management and span of control problems in food and beverage product preparation facilities relate unit preparation services to internal economic factors, such as inventory, extent of insurance, labour and materials costs, and environmental factors, such as outputs, input prices, specific demands and the relevant economy of food science and technology services. These problems are multivariate and serially correlated because one factor at one point in time is associated with other factors at other points in time (past, present and future).

Multivariate quality control (MQC) emphasizes the properties of control for decision making while it ignores the complex issues of process parameter estimation. Estimation is less important for Shewhart control charts for serially independent processes because the effects of different estimators of process parameters are nearly indifferent to the criterion of *average run length* (ARL). For processes having serial correlation, estimation becomes the key to correct construction of control charts. Adopting workable estimators is then an important issue.

In the past, researchers studied SQC for serially correlated processes and SQC for multivariate processes separately. Research on quality control charts for correlated processes focused on univariate processes. Box and Hill (1974) and Berthouex *et al.* (1978) noticed and discussed the correlated observations in production processes. Alwan and Roberts (1988) proposed a general approach to monitor residuals of univariate autocorrelated time series, where the systematic patterns are filtered out and the special changes are more exposed. Other studies include Jarrett and Pan (2007), Montgomery and Friedman (1989), Harris and Ross (1991), Montgomery and Mastrangelo (1991), Maragah and Woodall (1992), Wardell *et al.* (1994), Lu and Reynolds (1999), West *et al.* (2002) and West and Jarrett (2004). Others (English and Sastri, 1990; Pan and Jarrett, 2004) suggested *state space methodology* for the control of autocorrelated process.

In Alwan and Roberts's (1988) approach, they separated time series data into two parts that are monitored in two charts. One is the common-cause chart and the other is the special-cause chart. The common cause chart essentially accounts for the process's systematic variation that is represented by an autoregressive-integrated-moving-average (ARIMA) model, while the special cause chart is for detecting assignable causes that can be assigned in the residual of the ARIMA model. That is, one designs the special cause chart as a Shewhart-type chart to monitor the residuals filtered and whitened from the autocorrelated process (with certain or estimated parameters).

A quick way to see the advantages of MQC is to superimpose univariate control charts on top of each other and create a graph of all the points of each control chart in an area of space. This is shown in Figure 21.1.

Figure 21.1 shows a scatter of multivariate data composed of two variables. The chart indicates individual control limits for each variables' respective univariate chart in the control rectangle. This particular pattern shows that the process is in-control for each individual variable, as the data points fall within the control rectangle (Mastrangelo *et al.*, 1996; Tracy *et al.*, 1992). However, when the variables correlate (as they often do when from the same process), superimposing univariate charts is not a useful method of monitoring processes, as it does not capitalize upon their correlation and the probability of both charts simultaneously plotting in control is not $1 - \alpha$. If a process is in-control, the probability of $p$ means that the plot is in control is $(1 - \alpha)\,p$. Thus, the joint probability of a type I error is much larger: $(1 - \alpha)\,p$ (Alt, 1982, 1984; Jackson, 1985).

Multivariate quality control (or process monitoring), originally developed by Hotelling (1947), applied his procedures to data on allied bombsites during the Second World War. Subsequently, others (Hicks, 1955; Jackson, 1956, 1959, 1985; Crosier, 1988; Hawkins, 1991, 1993; Lowry *et al.*, 1992;

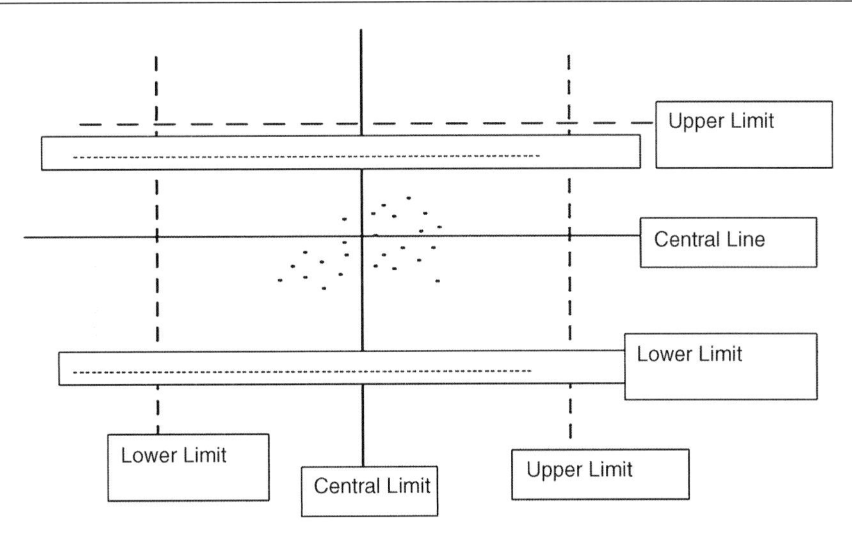


**Figure 21.1** Hotelling T chart analysis. (Conclusion: No points out of control. This will not be the case when compared with Univariate charts.)

Lowry and Montgomery, 1995; Pignatiello and Runger, 1991; Tracy *et al.*, 1992; Alt, 1985, among others) produced applications in a wide variety of management settings where two or more variables interrelate. In these settings, the size of the databanks is millions of individual records on hundreds of variables. Monitoring quality control and improvement methods of the univariate type are either ineffective or misleading in these environments. Hence, in recent years the implementation of multivariate process monitoring procedures has developed rapidly.

In another example, a food processer interested in monitoring the procession to quality product output (the store shelf for instance) randomly samples five units for 20 days for three correlated variables (i.e. *length* to take positive results, therapeutic material *supplier 1* and therapeutic *supplier 2*). By randomly sampling five items each day, the effects are measured and analysed. Because the length of the process and raw materials supplies are correlated, one creates a Hotelling $\mathbf{T}^2$ control chart to monitor the measurements for the three variables (length, supplier 1 and supplier 2).

Thus, the $\mathbf{T}^2$ control chart is created to monitor the mean measurements for the three variables simultaneously, which uses univariate procedures.

In Figure 21.2, the multivariate control contains only one control limit although the median line prints in this version. The $\mathbf{T}^2$ control chart shows two out-of-control points: 8 and 14. Point 8 is out of control due to length and Suppler 1. Point 14 is out of control due to Supplier 2. At this time, the quality manager needs to investigate what special causes may have affected length and supplier 1 in one sample and supplier 2 in another sample. If the quality manager printed three univariate control charts, no points would have been out of control. The obvious advantage of multivariate monitoring can be comprehended from the above results.

In Figure 21.3 it can be noted that no points are out of control. This is the multivariate form of the S chart. It can be used it to simultaneously monitor the process variability of two or more related process characteristics. For example the length can be monitored at the same time as the data from suppliers 1 and 2. By so doing, it can be determined if these variables jointly remain constant over the course of food production or quality of food preparation service.

If we choose to analyse the same data to control the mean and variation in the length, variable results follow.

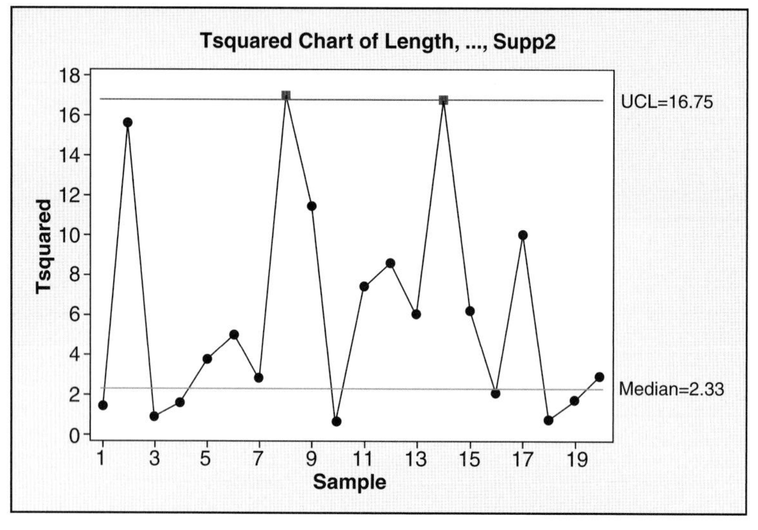


Test Results for Tsquared Chart of Length,..., Supp2

| | Point | Variable | P-Value |
|---|---|---|---|
| Greater Than UCL | 8 | Length | 0.0022 |
| | 14 | Supp1 | 0.0299 |
| | | Supp2 | 0.0002 |

**Figure 21.2** $T^2$ (Tsquared) of three variables.

The mean and standard deviation charts drawn by customary methods appear in Figure 21.4. The mean chart shows an out-of-control signal at point 8. However, the standard deviation chart shows no point out-of-control. Although not shown here the ange chart also does not show out-of-control signals.

How do we interpret the various results of the four control charts drawn by both the univariate and multivariate methods? The univariate results indicate one out-of-control point in the mean only at

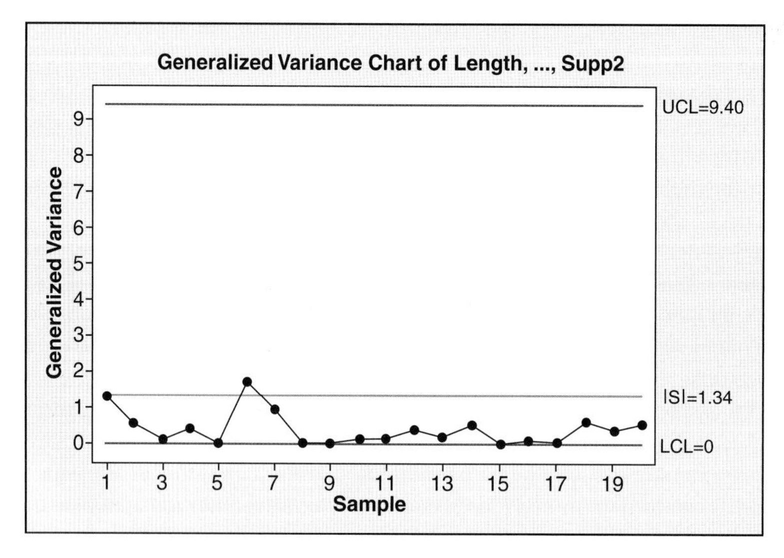


**Figure 21.3** Generalized variance chart. (Note: No points are above the UCL. Since LCL = zero, they cannot be smaller than the LCL.)

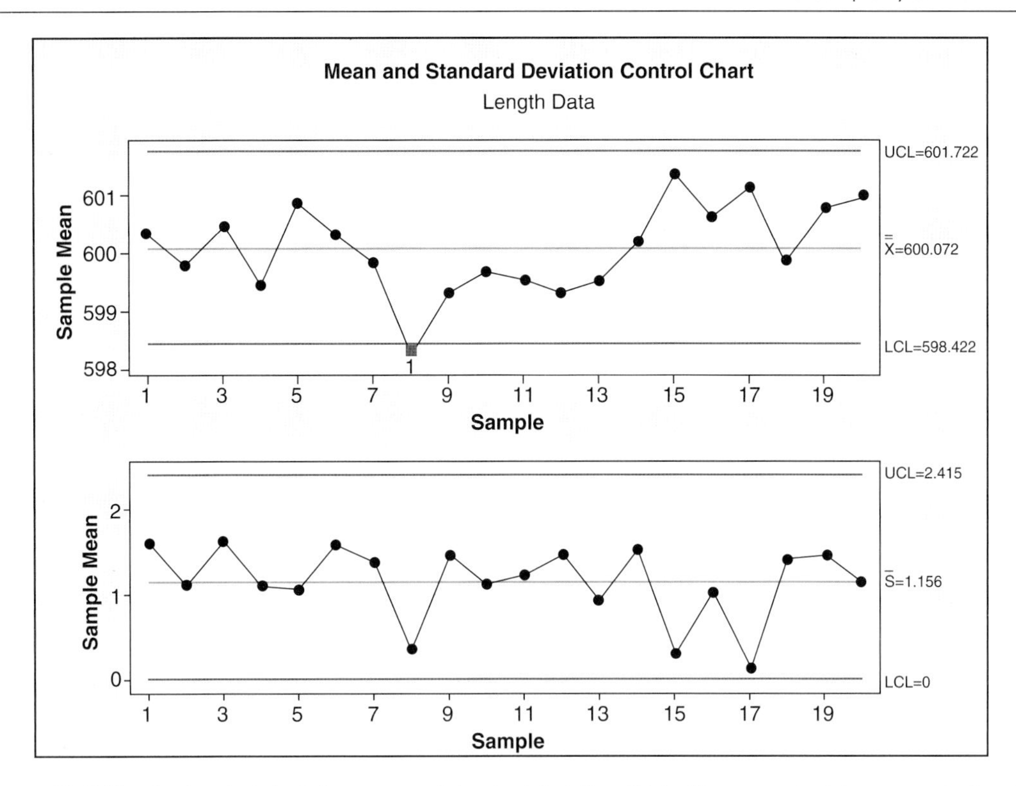


**Figure 21.4** Shewhart control charts for mean and standard deviation. (Note: The mean chart shows an out-of-control signal at point 8. However, the standard deviation chart shows no point out-of-control. Although not shown here the range chart also does not show out-of-control signals.)

point 8. The multivariate methods indicate that at point 8 the length variable is out of control and so is the variable for supplier 1. Hence, we associate the out-of-control signal in the mean length at point 8 to the out-of-control signal for supplier 1 at the same point. In addition, there is an out-of-control signal emanating from supplier 2. We locate the *p-values* for the out-of-control signals at the bottom of Figure 21.2 and note that these are very small probabilities (0.0002–0.0299). Besides the 'six sigma control' tests described here, there are many other tests for *special causes* associated with these control charts. At this point, it can be concluded that multivariate quality control does provide the additional evidence necessary for management to control and improve the quality of health production processes.

This chapter discusses the control chart that uses vector autoregressive (VAR) models and combines Alwan and Roberts's residual chart and the traditional Multivariate Hotelling $T^2$ chart to monitor multivariate serially correlated processes. We view the scheme as a generalization of Alwan and Roberts's special cause approach to multivariate cases. Detailed here are the guideline and procedures of the construction of VAR residual charts. Molnau *et al.* (2001) produced a method for calculating *average run length* (ARL) for multivariate exponentially weighted moving average charts. Mastrangelo and Forrest (2002) simulated a VAR process for statistical process control (SPC) purposes. The general study on VAR residual charts has not been reported., although most of the methods used here on VAR model estimation and its approximation to vector autoregressive-moving-average (ARMA) models exist in the literature of econometrics and mathematical statistics. A general consideration and summarization for control chart construction using this model does not appear in SPC research, so this paper should still be useful for multivariate quality control practice. More recent algorithms for producing control charts developed by Jarrett and Pan (2007) may also apply to food and beverage service quality control, leading to exciting new areas for research as well as improved quality of output.

## SUMMARY AND CONCLUSION

The conventional multivariate control charts described by example above assume that the in-control samples are independently and identically distributed (i.i.d.) over time while the variables correlate. When each variable is normal, these variables have a multivariate normal distribution. For the individual measurements (sample size $r = 1$), the distribution of Phase I follows Beta distribution (Tracy *et al.*, 1992; Weirda, 1994; Sullivan and Woodall, 1996). As long as the assumption fits the physical applications of quality control and improvement, the multivariate method will provide managers with more informative and better results when monitoring output of processes.

In the future, new developments in multivariate application in quality control will be observed. Newer methods for monitoring variation in output processes include control charts that use vector autoregressive (VAR) models and combine Alwan and Roberts's residual chart and the traditional multivariate Hotelling $T^2$ chart to monitor multivariate serially correlated processes. We view this scheme as a generalization of Alwan and Roberts's special cause approach to multivariate cases. The guideline and procedures for the construction of VAR residual charts have been detailed in this chapter. Molnau *et al.* (2001) produced a method for calculating *average run length* (ARL) for multivariate exponentially weighted moving average charts. Mastrangelo and Forrest (2002) simulated a VAR process for SQC purposes. However, the general study on VAR residual charts has not been reported before. Specifically, our interest is health care provision, although most of the methods used in this paper on VAR model estimation and its approximation to vector autoregressive-moving-average (ARMA) models exist in econometrics and mathematical statistics literature. We need only apply these methods in food and beverage product output. A general consideration and summarization for control chart construction using this model does not appear in SQC research. Most importantly, these applications are useful for multivariate quality control practice in the food science and technology industry.

## REFERENCES


Al-Assaf, A.F. and Schmele, J. (1993) *The Textbook of Total Quality in Healthcare*. CRC Press, Danvers, MA.

Alt, F.B. (1985) Multivariate quality control. In: *Encyclopedia of Statistical Sciences*. Vol. 6 (eds N. L. Johnson and S. Kotz). John Wiley & Sons, Inc., New York.

Alwan, L.C. (1992). Effects of Autocorrelation on Control Charts. *Communications in Statistics – Theory and Methods* **21**, 1025–1049.

Alwan, B.M. and Roberts, H.V. (1988) Time-series modeling for detecting level shifts of autocorrelated processes. *Journal of Business and Economics Statistics* **6**, 87–96.

Bakshi, B.R. (1998) Multiscale PCA with application to multivariate statistical process monitoring. *AIChE Journal* **44**, 1596–1610.

Benneyan, J.C. (1998a) Statistical quality control methods in infection control and hospital epidemiology, Part 1: Introduction and basic theory. *Infection Control and Hospital Epidemiology* **19**, 194–214.

Benneyan, J.C. (1998b) Statistical quality control methods in infection control and hospital epidemiology, Part 2: Chart use, statistical properties, and research issues. *Infection Control and Hospital Epidemiology* **19**, 265–277.

Benneyan, J.C., Lloyd, R.C. and Plsek, P.E. (2003) Statistical process control as a tool for research and healthcare improvement. *Quality and Safety in Health Care* **12**, 458–464.

Berthouex, P. M., Hunter, E. and Pallesen, L (1978) Monitoring sewage treatment plants: some quality control aspects. *Journal of Quality Technology* **10**, 139–149.

Bharati, M.H. and MacGregor, J.F. (1998) Multivariate image analysis for real-time process monitoring and control. *Industrial & Engineering Chemistry Research* **37**, 4715–4724.

Box, G.E.P. and Hill, W.J. (1974) Correcting inhomogeneity of variance with power transformation weighting, *Technometrics* **15**, 359–389.

Carey, R.G. (2003) *Improving Healthcare with Control Charts: Basic and Advanced SPC Methods and Case Studies*. ASQ Quality Press, Milwaukee, WI.

Crosier, R.B. (1988) Multivariate generalizations of cumulative sum quality-control schemes, *Technometrics* **30**(3), 291–303. doi: 10.2307/1270083.

English, J.R. and Sastri, T. (1990) Enhanced quality control in continuous flow processes. *Computers and Industrial Engineering* **19**, 258–262.

Figen (Kosebalaban), T., Cinar A. and Schlesser, J.E. (2005) HACCP with multivariate process monitoring and fault diagnosis techniques: application to a food pasteurization process. *Food Control* **16**, 411–422.

Harris, T.J. and Ross, W. H. (1991) Statistical process control procedures for correlated observations. *Canadian Journal of Chemical Engineering* **69**, 48–57.

Hart, M.K. and Hart, R.F. (2002) *Statistical Process Control for Health Care*. Duxbury Press, Pacific Grove, CA.

Hawkins, D.M. (1991) Multivariate quality control based on regression adjusted for variables. *Technometrics* **33**, 61–75.

Hawkins, D.M. (1993) Regression adjustment for variables in multivariate quality control. *Journal of Quality Technology* **25**(3), 37–43.

Hicks, S.D. (1955) Natural History Survey of Coppermine Northwest Territories, 1951. *The Canadian Field Naturalist* **69**, 162–166.

Hotelling, H. (1947) Multivariate quality control. In: *Techniques of Statistical Analysis* (eds Eisenhart, Hastay and Wallis). McGraw-Hill.

Jackson, J.E. (1956) Quality control methods for two related variables. *Industrial Quality Control* **12**(7), 4–8.

Jackson, J.E. (1959) Quality control methods for several related variables. *Technometrics* **1**, 359–377.

Jackson, J.E. (1985) Multivariate quality control. *Communications in Satistics – Theory and Methods* **14**, 2657–2688.

Jarrett, J.E. and Pan, X. (2007) Monitoring variability and multivariate autocorrelated process. *Journal of Applied Statistics* **34**, 459–469.

Knapp, R.G. and Miller, M.C.III (1983) Monitoring two or more indices of health care. *Evaluation and Health Professions* **6**, 465–482.

Lachenmeier, D.W. (2007) Rapid quality control of spirit drinks and beer using multivariate data analysis of Fourier transform infrared spectra. *Food Chemistry* **10**, 825–832.

Lee, K. and McGreevy, C. (2002) Using control charts to assess performance measurement data. *Journal on Quality Improvement* **28**, 90–101.

Lowry, C.A.W., Woodall, C.W. and Rigdon, S.E. (1992) A multivariate exponentially weighted moving average control chart. *Technometrics* **34**, 46–53.

Lowry, C. A. and Montgomery, D.C. (1995) A review of multivariate charts. *IIE Transactions* **27**, 800–810.

Lu, C.W., and M.R. Reynolds (1999) Control charts for monitoring the mean and variance of autocorrelated processes. *Journal of Quality Technology* **31**, 259–274.

Manciu, A, Popa, M., Mitrea, D. and Capatina, D. (2006) Parameters monitoring solutions for the quality of water used in healthcare units. In: *Automation, Quality and Testing, Robotics. 2006 IEEE International Conference on Robotics and Automation, 2*. IEEE pp. 457–462.

Maragah, H.O. and Woodall, W.H. (1992) The effect of autocorrelation on the retrospective X-chart. *Journal of Statistical Computation and Simulation* **40**(1), 29–42.

Mastrangelo, C.M. and Forrest, D. R. (2002) Multivariate autocorrelated processes: data and shift generation. *Journal of Quality Technology* **34**, 216–220.

Molnau, W.E., Montgomery, D. C., and Runger, G.C. (2001) Statistically constrained economic design of the multivariate exponentially weighted moving average control chart. *Quality and Reliability Engineering International* **17**(1), 39–49.

Montgomery, D.C. and Friedman, J.J. (1989) Statistical process control in a computer-integrated manufacturing environment. In: *Statistical Process Control in Automated Manufacturing* (eds J.B. Kates and N.F. Hunele). Marcel Dekker, Inc., New York.

Montgomery, D.C. and Mastrangelo, C.M. (1991) Some statistical process control methods for autocorrelated data. *Journal of Quality Technology* **23**(3), 179–193.

Morton, A. (ed.) (2005) Methods for Hospital Epidemiology and Healthcare Quality Improvement. Available at http://www.eicat.com/.

Pan, X. and J. Jarrett (2004) Applying state space into SPC: monitoring multivariate time series. *Journal of Applied Statistics* **31**(4), 397–418.

Phillips, K.M., Patterson, K. Y, Rasor, A.S. *et al.* (2006) Quality-control materials in the USDA National Food and Nutrient Analysis Program (NFNAP). *Analytical and Bioanalytical Chemistry* **384**, 1341–1355.

Pignatiello J.J., Jr., and Runger, G.C. (1991) Adaptive sampling for process control. *Journal of Quality Technology* **22**, 173–186.

Rogers S.J., Williams C.S. and Roman G.C. (2004) Myelopathy in Sjogren's syndrome: role of nonsteroidal immunosuppressants. *Drugs* **64**(2), 123–132.

Spiegelhalter, D.J. (2004) Incorporating Bayesian Ideas into Health-Care Evaluation. *Statistical Science* **19**(1), 156–174.

Steiner, H. and Haass, C. (2000) Intramembrane proteolysis by presenilins. *National Review Molecular Cell Biology* **1**(3), 217–224.

Sullivan, J.H. and Woodall, W.H. (1996) A review of multivariate charts. *Journal of Quality Technology* **28**, 261–264.

Tracy, N.D., Young, J.C. and Mason, R.L. (1992) Multivariate quality control charts for individual observations. *Journal of Quality Technology* **24**, 88–95.

VanBrackle, L. and Williamson G.D. (1999) A study of the average run length characteristics of the National Notifiable Disease Surveillance System. *Statistics in Medicine* **18**, 3309–3319.

Wardell, D.G., Moskowitz, H. and R.D. Plante (1994) Run-length distribution of special-cause control charts for correlated processes. *Technometrics* **36**(1), 3–17.

West, D. and Jarrett, J. (2004) The impact of first order positive autoregression on process control. *International Journal of Business and Economics* **3**, 29–37.

West, D., Delana, S. and Jarrett, J. (2002) Transfer function modeling of processes with dynamic inputs. *Journal of Quality Technology* **34**, 315–321.

Wierda, S.J. (1994) Multivariate statistical process control: recent results and directions for future researches. *Stastica Neerlandica* **48**, 147–168.

# 22 Application of neural-based algorithms as statistical tools for quality control of manufacturing processes

**Massimo Pacella**[1] **and Quirico Semeraro**[2]

[1] *Dipartimento di Ingegneria dell'Innovazione, Università del Salento, Lecce, Italy*
[2] *Dipartimento di Meccanica, Politecnico di Milano, Milan, Italy*

## ABSTRACT

The use of neural networks began to be applied because the traditional control charts used for monitoring manufacturing process, in some cases, did not provide the possibility of correctly and quickly signalling the existing causes of variation. In today's manufacturing environment, neural networks present increasing usefulness for implementing the automation of statistical process control. This chapter targets issues on the use of neural networks for quality control of manufacturing processes, concerning the way of operation of each network model, the network's architecture and the results provided. Applications of neural networks for pattern recognition and for detection of mean and/or variance shifts in process are discussed. Comparisons between the performances of the neural approach and those of traditional control charts are also presented. Results prove that the neural network model is a useful alternative to the existing control schemes.

## INTRODUCTION

A process can be viewed as a set of causes and conditions that repeatedly come together to transform inputs into outputs. The inputs might include people, materials or information. The outputs include products, services, behaviour or people. Figure 22.1 shows a general model of a process. In any process, no matter how well designed it is, variation is encountered. Two sources of variation can be considered: *common* and *special* causes (Montgomery, 2008).

The first source of variation (common causes) is the result of numerous unremarkable changes that may occur in the process. This kind of variation is, to some extent, inevitable without a revision of the production procedure. When only common causes are in effect, a process is considered to be in a natural state (i.e. 'in control'). If the variation due to common causes is small compared to the requirements of customers, and if the process is 'on target', this poses no problems. The process is capable of meeting the demands.

However, the process may be affected by external sources of variation, which are upsetting to the natural functioning of the process (special causes). New methods and different machines introduced into the process, or changes in the measurement instruments and in the turnover of the labour force, are common examples of special causes of variation. Special causes of variation are not part of the process and occur only accidentally; therefore, the process is also said to be in an unnatural state (i.e. 'out of

*Mathematical and Statistical Methods in Food Science and Technology*, First Edition.
Edited by Daniel Granato and Gastón Ares.

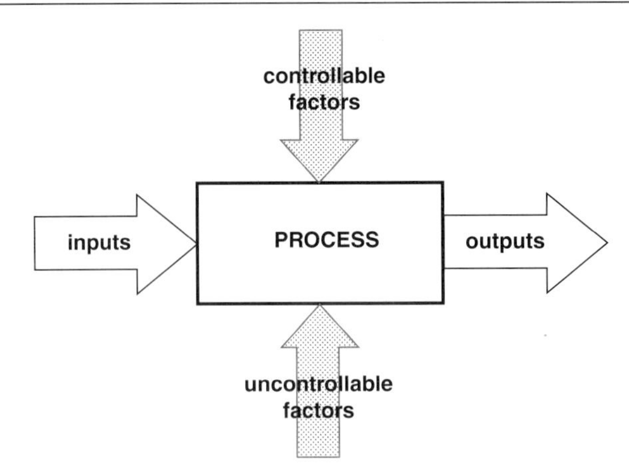


**Figure 22.1** General model of a process.

control'). When a special cause of variation is present, it may have a large effect on the outcomes of the process. If removal is possible, a special cause can usually be eliminated without revising the process, thereby improving the quality of the outcomes of the process.

Typically, the responsibility for reducing the effect of special causes of variation lies on a different management level than the reduction of common causes of variation. In many cases, an operator can be instructed to recognize and remove special causes of variation, whereas reducing the effect of common causes of variation is the responsibility of the owner of the process. It is important for an operator to know whether special causes are present, so that action to remove this cause of variation can be undertaken. However, it is even more important to know when to leave the process alone when only common causes of variation are affecting process outputs. What happens in practice is that operators try to counteract the effect of common causes of variation as if it was a special cause of variation (MacGregor, 1990). Trying to counteract the effect of the variation due to common causes is in many cases like intervening in a stable system, thereby increasing instead of reducing the variation. It is therefore of critical importance to be able to distinguish situations where only common causes of variation affect the outcomes of a process, from situations where special causes are also present.

A process is said to be 'statistically in control' if only common causes are affecting the outputs and their effect is to some extent predictable. The 'predictability' means that it is possible to determine limits that bound statistically the process outputs. The statistical predictability is the basis for the control chart, a tool that can be used to distinguish between situations where only common causes of variation affect the outcomes of a process, and situations where special causes are present also. In its original form, the control chart is a simple time plot of a sequence of sample statistics. Statistics of interest (like the mean, range and standard deviation) are computed for each sample in order to summarize the information contained within each subgroup. The points in the plot are compared to control limits, which indicate the bandwidth of the variation due to common causes.

A control chart is exemplified in Figure 22.2. The centre line is the overall average of the sample statistic, while the upper and lower control limits are drawn at a distance of plus and minus k-times the standard deviations from the centre line. Figure 22.2 shows a sequence of 25 sample statistics, which wander randomly above and below the centre line. The sample statistics plotted on the graph are compared to limits that represent the bandwidth of the variation due to common causes. In Figure 22.2, the sample statistics are contained within the control limits, with the exception of samples with index 22 and 23.

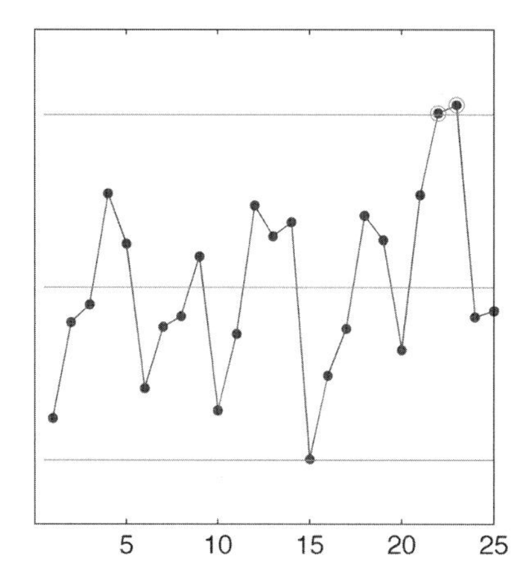


**Figure 22.2** An example of control chart that shows a sequence of 25 sample statistics.

Generally, a point outside the control limits is considered an 'alarm', as it indicates that more variation is present than can be attributed to the effect of common causes of variation. However, due to the random nature of the observations, there is a small probability that an alarm is encountered while the process is statistically in control. Such a signal is referred to as a 'false alarm'. A control chart must be sensitive enough to detect the effect of special causes of variation but must not generate too many false alarms. In practice, a balance between these two must be struck by determining the width of the control limits.

Dr Walter A. Shewhart, an engineer at the Bell Telephone Laboratories Inc., where he was faced with the issue of obtaining good quality in the mass production of interchangeable equipment for the rapidly expanding telephone system in the 1920s, developed the control chart technique. Actually, statistical process control (SPC), that is the collection of statistical techniques and graphical displays (as control charts) to enable a process to be monitored and special causes of variability to be eliminated, may be considered to have begun with the pioneering work of Shewhart (1931).

Later, different kinds of control chart procedures were developed with the aim of improving the performance of control charts. As a first example: the Shewhart control chart is inappropriate to signal timely small or moderate changes of a process. Page (1954) proposed the cumulative sum (CUSUM) procedure to overcome this weakness of Shewhart control charts. Furthermore, Roberts (1959) focused on overcoming the weakness of the Shewhart procedure and proposed the exponentially weighted moving average (EWMA) procedure. CUSUM and EWMA enhanced the performance capability of the control charts. Lucas and Saccucci (1990) evaluated CUSUM and EWMA procedures and concluded that the performance properties of the procedures are the same. As a second example: the decision to signal an alarm in the Shewhart control chart is based on the current sample statistic, ignoring information from previous ones. To increase its detection performance, several supplementary stopping rules (also known as run rules) have been proposed (Western Electric, 1956). For instance, an alarm is obtained when eight successive points fall below or above the centre line, eight increasing or decreasing points in a row, and fourteen points in a row alternate in direction, increasing then decreasing. Often, a combination of different run rules is used to improve the detection performance of the Shewhart control chart. However, in practice, run rules are used with care in actual applications because they lead to an increase in the false alarm rate (Champ and Woodall, 1987).

Before computers, the implementation and maintenance of control charts were manually performed by human operators, and hence they were subjected to human error. Nowadays, with the rapid movement towards computer-integrated manufacturing the automation of SPC implementation is considered unavoidable. With the availability of advanced data collection systems, new methods have been developed to accommodate rapid data input rates and, in general, to take advantage of the high capabilities of modern soft computing methods. Computer procedures have been developed to implement, or at least help, human operators to carry out the various quality control tasks automatically.

Recently, artificial intelligence techniques have been widely investigated as promising tools for the automation of SPC implementation. What makes these algorithms popular is their ability to learn from experience and to handle uncertain and complex information in a competitive and quality demanding environment.

Among the useful artificial intelligence techniques, neural networks (NNs) have the ability to learn a specific knowledge, to adapt it to new situations and to provide reliable classifications and approximations of data (Haykin, 1998). NNs have received a great deal of attention in a wide variety of applications where statistical methods are usually employed. NNs are used for classification and regression problems because of their ability to elaborate large amounts of data in real time as well as for their capacity for handling noisy data measurements, requiring no assumption about the statistical distribution of the monitored data.

The combination of control chart and NNs began to be applied, in many papers, in the late 1980s, because the traditional control charts used for monitoring manufacturing processes, in some cases, did not provide the possibility of correctly and quickly signalling the existing causes of variation. In today's manufacturing environment, NNs present increasing usefulness for the automation of SPC implementation. As such, their use has grown rapidly in the recent years.

This chapter targets issues on the use of NNs in SPC, concerning the way of operation of each network model, the network's architecture and the results provided. Specifically, the next section gives a brief overview of machine learning theory. This is followed by a review of the literature on the general topic of NNs for the automation of SPC implementation. In particular, applications of NNs for pattern recognition and for detection of mean and/or variance shifts in process are discussed. Subsequently, a specific NN for process monitoring is presented by providing a basic description of the approach. Finally, conclusions are given along with directions of future research.

## AN INTRODUCTION TO MACHINE LEARNING

Consider a measurement system that is capable of acquiring in reasonable time large amounts of data from a manufacturing process. Data are supposed to contain some useful information one wants to learn about the process. The construction of a model for the observed data, often referred to as the learning procedure, encodes the patterns, that is any type of rule or dependency structure present in the data.

In actual applications, data are often noisy and the dependency expressed by patterns is not deterministic. In addition, observed data give a partial view of the process. The aim of learning is to identify inherent patterns, that is patterns that are exhibited by subsequent data collected from the process and not only by the observed data. A learning procedure that is able to identify inherent patterns generalizes well. The generalization property makes it possible to infer with some confidence the missing information of a partially observed state of the process. In contrast, if the learning procedure returns patterns that are present in the observed data but absent from other data collected from this process, it is said to be overfitting. When overfitting occurs, inference of missing information might be completely erroneous.

Designing automatic learning procedures for tackling practical pattern recognition tasks is often viewed as the subject of the machine learning research field. Usually, in this research field it is not sufficient to have good estimators from learning theory if their computational cost is excessive, for example because of the size of the data set and/or of the high flexibility of the models. The fitted model should be useful to be exploited for classification and/or prediction tasks. Usefulness of the learning procedure should be measured by a proper evaluation, which can be assessed by a quantitative criteria (e.g. in a classification task, by the proportion of correct classifications).

There are essentially two types of learning procedure, referred to as supervised and unsupervised tasks, which are briefly discussed in the following subsections. For a more complete overview of learning theory, there are many standard references, such as the books of Vapnik (2000) and Bishop (2006). The first part of the book of Bishop (1995) presents the issue of pattern recognition.

## Supervised and unsupervised learning

In supervised learning, pairs of vectors, input values and target values are used. The goal of supervised learning is to derive from the data set the dependency of targets on inputs, such that the model is capable of returning target predictions for new inputs. In practice, an initial model produces a value for the output vector, which differs from the value of the target vector. This difference is called error and based on this, as well as on a specific learning procedure, the adjustment of the model parameters occurs, until this error is decreased enough to fall under a predetermined limit.

When one has some knowledge of the structures that should be present in the data, a supervised learning procedure can be adopted. However, in practical applications there is no such prior knowledge, either because it is too complex to characterize the structures in the data or because the observed data are significantly stochastic. This is typically the case of manufacturing processes, where observed data are noisy, dependencies are highly stochastic and there is no simple physical rule to represent them.

When one has little knowledge about the patterns present in the data, looking for clusters in the observed data is a good starting point. The second type of learning task is the unsupervised one, also known as clustering. The goal of clustering is to identify inherent separations in the data. One could view clustering as a classification problem without label information. Data belonging to a cluster should be close to one another, while data from different clusters should be far apart. While in the case of supervised learning, adjusting a model requires the definition and the minimization of a loss function, for unsupervised learning the loss function is a measure of the similarity of samples within each cluster. Different natural clustering would be found just by changing the similarity measure. Setting the number of clusters is also an important issue, as it might depend on the similarity measure.

## Neural networks

NNs are parallel computing algorithms typically thought of as black boxes. Such computer algorithms are used to learn a specific knowledge, to adapt it to new situations and to provide reliable classifications and approximations of data (Haykin, 1998). A NN implements learning procedures, and hence it can be exploited to identify the fundamental patterns in data sets.

The adjective neural is because these algorithms simulate in a very simplified form the ability of brain neurons to process information. The principle is to combine in a network simple processing functions, which are called neurons or nodes, linked by weighted connections. The function of the synapse, the structure responsible for storing information in the brain, is modelled by a modifiable weight, which is associated to each connection between two neurons. Within each neuron, all the weighted input signals are summed up and a signal is then produced as output. In particular, the output signal is computed as the response of an activation function on the summation of the weighted input signals of that neuron. Two basic activation functions are the sigmoid and hyperbolic function. The sigmoid function gives outputs in $[0, 1]$ and the hyperbolic function gives outputs in $[-1, 1]$.

Neurons set in parallel form a layer of the network, while the output signal of a neuron is fed to the neurons in the subsequent layer as input signal. A NN is composed of successive layers, namely the input layer, one or more hidden layers and the output layer. In general, all neurons of a layer have the same

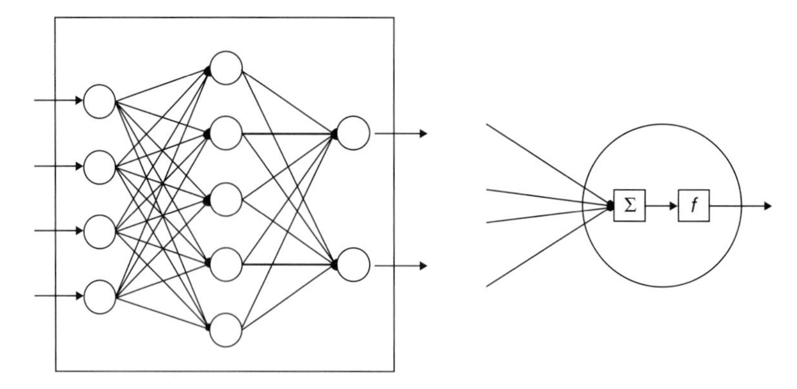


**Figure 22.3** Schematic structure of a three-layer neural network (left) and model of a neuron with four input connections and an activation function f (right).

activation function and are fully connected to the neurons of adjacent layers. Depending on the way that the nodes of various layers are connected, two categories of NNs are usually considered: feedforward and recurrent NNs. In a feedforward NN, the nodes of a layer are not connected with the nodes of a previous layer and the nodes of each layer are not connected among them. The flow of information is unidirectional. In a recurrent NN, the nodes of a layer are connected with the nodes of a previous layer and the nodes of each layer are connected among them. The feedforward NNs are commonly used because of their simple form, the lesser time they need for training and their long-term memory.

Figure 22.3 graphically exemplifies a three-layer feedforward NN. The input layer contains four nodes used as input data for four data points. The output layer consists of two nodes. The single hidden layer of the network contains 5 nodes. Also depicted in Figure 22.3 is the model of a neuron, with four input connections, where each connection is associated to a specific weight.

In the following subsections, two NN models are summarized, namely: the multilayer perceptron (MLP) and adaptive resonance theory (ART) networks. The MLP network represents the common model used for supervised learning tasks, while the ART network is related to unsupervised tasks.

## Supervised learning: the MLP NN

MLP is a flexible feedforward NN model configured by setting the number of layers, the number of neurons in each layer and the types of activation functions. The training stage of MLP corresponds to the optimization of the different layers of weights in order to accomplish a supervised learning task. The basic idea of this method is to determine the 'proportion' of the error that corresponds in the weights of each node. Firstly, using the input and target values, the error for the nodes of the output layer is calculated and then this error is used to calculate the errors in the last hidden layer. This process is repeated retrospectively to the first layer and based on the so-called 'back propagation' of the error, a calculation of contribution of each weight of nodes in the total error takes place. Afterwards, the errors that were calculated for the nodes of each layer are used in order to adjust the weights of each node and the process is repeated until the total error becomes smaller than a predetermined value. Because of this training algorithm, the MLP NN is also referred to as a back-propagation neural network.

The success of MLP is due to the approximation property of this model. Indeed, theoretical results (Hornik *et al.*, 1989) showed that a MLP with one hidden layer of nonlinear processing functions (followed by a linear one) is capable of approximating 'any measurable function to any desired degree of accuracy'. According to the results of Hornik *et al.* (1989) '. . . any lack of success in applications must arise from inadequate learning, insufficient numbers of hidden units or the lack of a deterministic relationship between input and target'.

The main issue with MLP is how to develop a network of appropriate size for capturing the underlying patterns in the data. Although MLP theory suggests that more hidden nodes typically lead to improved accuracy in approximating a functional relationship, they also cause the problem of overfitting (the network fits the observed data very well but generalizes very poorly). Therefore, determination of how many hidden layers and hidden neurons to use is often chosen through experimentation or by trial-and-error.

A common approach to tackle the overfitting problem is to divide the set of observed data into three subsets, namely, training, validation and testing data. The training and validation parts are used for model building with the last one used for model evaluation. In particular, the training set is used for computing and updating iteratively the network weights. During training, the error on the validation set is monitored, while this set of data is not used for updating the network weights. The validation error normally decreases during the initial phase of training, as does the training set error. However, when the network begins to overfit the training data, the error on the validation set begins to rise, while the training set error continues to decrease. When the validation error increases for a specified number of iterations, the training is stopped (early stopping). The network with the best performance on the validation set is then evaluated on the testing data set.

Different weight elimination and node pruning methods have also been proposed in the literature for building the optimal architecture of MLP (Reed, 1993; Roy *et al.*, 1993; Murata *et al.*, 1994; Wang *et al.*, 1994; Cottrell *et al.*, 1995; Schittenkopf *et al.*, 1997), even if none of these methods can guarantee the best solution for every possible situation. The basic idea with these methods is to find a parsimonious model that fits the data well. Generally, a parsimonious model not only gives adequate representation of the data but also has the more important generalization capability.

In general, the overfitting problem is more likely to occur in a NN model if it presents a large parameter set to be estimated. On the other hand, if the number of parameters in a network is much smaller than the total number of samples in the training set, then there is little or no chance of overfitting. Furthermore, overfitting does not apply to some learning paradigms, such as the unsupervised ones (e.g. the ART network) because they are not trained using an iterative process.

## Unsupervised learning: the ART network

ART is a computer algorithm able to cluster input vectors that resemble each other according to the stored prototypes. ART can adaptively create a new cluster corresponding to an input if this specific pattern is not similar to any existing prototype.

In a physical system, when a small vibration of a proper frequency produces a large amplitude vibration, it is defined as resonance. Actually, ART gets its name due to the fact that information, that is the output of neurons, reverberates back and forth between two layers, namely F1 (the comparison layer) and F2 (the recognition layer), which are fully connected by weights. On the one hand, the comparison layer (F1) acts as a feature detector that receives external input, on the other hand the recognition layer (F2) acts as a category classifier that receives internal patterns (the general ART model is depicted in Figure 22.4).

The application of a single input vector leads to a set of activity that the network develops in the so-called resonant state, and which produces different top-down templates (prototypes) from layer F2 to F1. Each template is associated with one of the cluster nodes in the layer F2.

The orienting subsystem of the ART model is responsible for generating a reset signal to the recognition layer when the bottom-up input pattern and the top-down template mismatch according to a vigilance criterion. This signal, if sent, will cause either a different cluster to be selected or, if no more

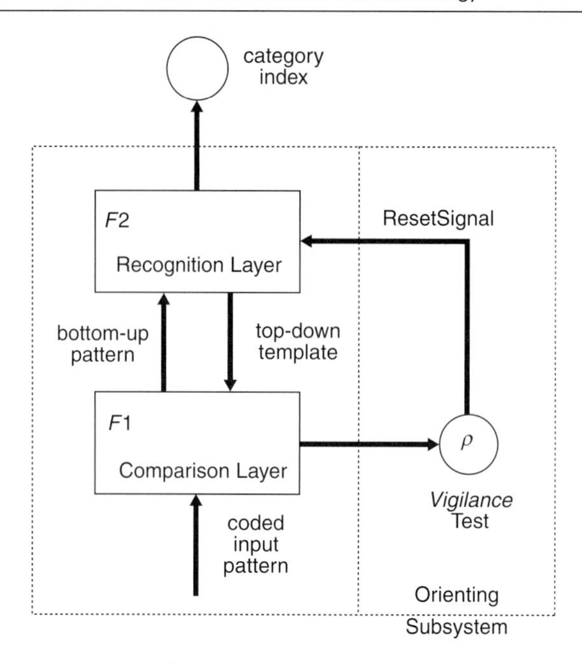


**Figure 22.4** The schematic representation of the ART NN.

cluster is available, the end of the resonance state. During training, the formerly coded template associated with the cluster node that represents the best match to the current bottom-up input, will be modified to include the input features. If there are no clusters that match to the current bottom-up input, a new one is initialized with the incoming pattern. This is called the vigilance test and is incorporated in the orienting subsystem of the ART network. The ART model allows controlling the vigilance test, that is the degree of similarity of patterns placed in the same cluster. The similarity depends on the vigilance parameter $\rho$, where $0 < \rho < 1$. If $\rho$ is small, the result is inclined to a coarse categorization. On the other hand, if $\rho$ is chosen to be close to one, many finely divided clusters are formed.

The ART model operates in a plastic mode (i.e. a continuous and cumulative training mode) as long as new patterns are presented to it. In fact, this type of model was firstly introduced for solving the stability/plasticity problem, that is to provide a method by which a NN can incrementally learn new patterns without forgetting old knowledge. During training, the ART categorizes input patterns of data into clusters with similar features, and when it is confronted by a new input it produces a response that indicates which cluster the pattern belongs to (if any). Detecting whether an input vector resembles the natural categories formed during training is the function of the matching algorithm.

Among different ART models, the Fuzzy ART is considered henceforth. It inherits the design features of other ART models and incorporates computations from fuzzy set theory by which it can cluster analog patterns. A detailed description of Fuzzy ART can be found in the papers of Huang *et al.* (1995), Georgiopoulos *et al.* (1996, 1999) and Anagnostopoulos and Georgiopouolos (2002).

## NEURAL NETWORKS FOR QUALITY MONITORING

In this section, an overview of the literature in NNs for quality monitoring is given. We focus on applications of NNs for monitoring univariate quality characteristics only. In the case of univariate process control, the approaches proposed in the literature can be classified into two categories: control chart pattern recognition and unnatural process behaviour detection. These approaches are briefly reviewed in the following two subsections.

## Control chart pattern recognition

Control chart pattern recognition provides a mechanism for identifying different types of predefined 'patterns', that is fluctuation of observations plotted on a univariate control chart. The recognized patterns then serve as the primary information for identifying the causes of unnatural process behaviour. In nearly all the approaches proposed in the literature, a natural pattern fluctuates randomly and its observations follow the underlying distribution of the statistics in question (usually normal distribution). An unnatural pattern, on the other hand, may fluctuate too widely or narrowly, or follow a certain path (such as trend or cyclic).

Many articles have been published about the combination of NNs and SPC in pattern recognition. Hwarng and Hubele (1993a, 1993b) carried out extensive studies on pattern recognition by training a MLP in order to detect some basic unnatural pattern on a univariate control chart. These patterns are upward trends, downward trends, cycle patterns, systematic patterns, stratification patterns, mixtures patterns, sudden upward shifts and sudden downward shifts. Anagun (1998) proposed a three-layered MLP for the recognition of the same unnatural patterns as Hwarng and Hubele (1993a, 1993b), comparing the effectiveness of raw data against feature-based representation.

Guh and Tannock (1999) investigated the feasibility of MLP to identify concurrent patterns (where more than one pattern exists together, which may be associated with different assignable causes). The authors used a four-layer MLP model with an input layer of 16 neurons (used to input from 16 consecutive sample data points), an output layer of four neurons, and two hidden layer each of 13 neurons. They found that once the number of hidden neurons exceeded 13, performance was not enhanced while the total training time was increased. They also observed that there is no established theoretical method to determine the optimal configuration of a MLP model, thus most of the design parameters must be determined empirically.

Generally, the size of a MLP network, that is the number of the model parameters, increases as the number of the input vector elements increases. In the literature, data-window sizes from 16 to 64 observations were used. Even with 16 data-window observations, the need to train hundreds of weights to classify unnatural patterns is a normal requirement.

Guh and Hsieh (1999) presented a control system composed of several inter-connected MLP networks both to recognize the unnatural patterns and to estimate their parameters. Perry *et al.* (2001) proposed an NN model that consisted of two four-layered MLP NNs. The first NN was used for the identification of four specific out-of-control situations. If one of these situations existed, a second NN was used for the recognition of unnatural patterns in the data (upward trends, downward trends, cycle patterns, systematic patterns and mixtures patterns). Guh (2002) tried to explore the effect of non-normality on the performance of NN-based models for control chart pattern recognition. The NN-based models for pattern recognition perform well in a non-normal distribution environment in terms of recognition accuracy and speed.

Guh (2005) proposed a hybrid-learning based model integrating MLPs and decision trees as an effective identification system of patterns in the series of process quality measurements. This approach showed high performance in detecting and recognizing some unnatural patterns on the control charts. Guh (2010) presented a NN-based model for the simultaneous recognition of both mean and variance control chart patterns. The author studied only the shift, trend and cycle pattern. The simulation results proved that the proposed NN model could effectively recognize not only the single mean and variance control chart patterns but also the mixed patterns in which the mean and variance occur concurrently.

In most of the approaches proposed in the literature, coding schemes were applied on the quality measurements after standardization. In the coding process, the measured variable range was divided into

*N* zones (where the width of each zone was pre-specified), each returning an integer code. The objective of the coding process was to reduce the effect of the noise in the input data before the data were presented to the neural network while retaining the main features in the data. The choice of the coding zone width was critical for MLP classification performance. Too small gradations might not be able to detect the important features in the data due to the effect of random noises. On the other hand, if gradations were too large, the true process variations might be lost (Cheng, 1997; Guh and Tannock, 1999).

## Unnatural process behaviour detection

Generally, control charts may have insufficient ability to determine quickly the occurrence of a fault in the process. Therefore, many researchers have proposed the application of NNs in detection of mean and/ or variance shifts in process.

Pugh (1989, 1991) reported the first application of a MLP model for mean shift detection in a manufacturing process. The author also compared the performance of the implemented NN to that of the Shewhart control charts. Smith (1994) implemented a NN model to signal shifts in means or variance in an X-bar and R control charts. This NN model was trained to replace combined X-bar and R charts and it comprised six NNs (three NNs for large shifts and three NNs for small shifts) that differed by the number of input and hidden nodes.

Chang and Ho (1999) used a NN to discover shifts in variances in two steps. The first part is a NN that decides whether the pattern is in or out of control. The second part provides a coded value for the shift magnitude. Ho and Chang (1999) proposed a combined NN approach to monitor simultaneously mean and variance shifts. A comparative study proved that their NN model performance was superior to the performance of the traditional control charts such as X-bar and R, CUSUM and EWMA in the majority of situations for individual observations, as well as in the case of subgroup size of five observations. They also proved that their proposed NN model was superior than that of Smith (1994). Similarly, the performance of the MLP model in detecting changes in process mean was found to be superior to that of a combined Shewhart-CUSUM control scheme in (Cheng 1995). Cheng and Cheng (2001) proposed a MLP model to monitor exponential mean shifts.

Some studies also reported the use of NNs for mean process shifts in the presence of autocorrelation in the data. This problem has more influence on the foods, chemicals, papers and woods industries. Indeed, autocorrelation is present in the data generated by most continuous and batch process operations, as the value of a parameter depends upon its previous value. Traditional control charts have large weaknesses in recognizing shifts in process, in the presence of autocorrelation in the data, as these charts result in a large number of false out-of-control signals.

Chiu *et al.* (2001) proposed NN models to identify various shifts of standard deviation in process parameter values from time series models with varying values of the autoregressive parameter. Each NN consisted of different number of hidden nodes. Hwarng (2004) proposed a MLP to recognize various shifts of standard deviations in correlated data with various values of autoregressive parameter. The author noted that the NN outperformed competitive control charts in all cases, except when a high level of autocorrelation with large shifts occurred. Hwarng (2005) also proposed a neural-based identification system for both mean shift and correlation parameter change in a time series model. Pacella and Semeraro (2007) proposed a modified MLP model, which employs feedback connections between layers, the Elman's NN (Elman, 1990), for discovering mean shifts in the case of serially correlated data.

From the results cited above, it can be concluded that NNs are at least comparable to the traditional control charts and they can be successfully used for the detection and determination of mean and/or variance shifts in process. It can also be concluded that MLP algorithms are the most widely used NNs for the determination of mean and variance shifts in process.

## ART NN FOR UNNATURAL PROCESS BEHAVIOUR DETECTION

Differently from the above-mentioned approaches, Al-Ghanim (1997) proposed an unsupervised neural-based system capable of signalling any unnatural change (not just a shift in the mean/variance) in the behaviour of a manufacturing process. Specifically, the binary implementation of the ART network was trained on a set of natural data in order to cluster them into groups with similar features. After training, the ART network provides an indication that a change in process outputs has occurred when the series of process data does not fit to any of the learned categories. Al-Ghanim's pioneering methodology, however, did not have a degree of sensitivity comparable to that of other NNs (such as the MLP). The drawback can be mainly ascribed to the binary coding of the ART algorithm, as it is a less flexible way of using process data than a method based on graded continuous number encoding.

Subsequent researches extended Al-Ghanim's methodology and presented outperforming ART-based approaches for unnatural behaviour detection (Pacella *et al.*, 2004a, 2004b; Pacella and Semeraro, 2005). In particular, a simplified Fuzzy ART algorithm, which does not require binary coding of data, was presented. In Pacella *et al.* (2004a), the NN was trained using a series of process natural output data. In Pacella *et al.* (2004b), it was shown that the training set could even be limited to a single vector whose components are equal to the process nominal value. In the post-training phase, Fuzzy ART compares input vectors to learned clusters and produces a signal if the current input does not fit to any of the learned categories.

These Fuzzy ART-based approaches have been showed to achieve similar performances in signalling a sustained change of process mean with those of the CUSUM control chart. At the same time, they are also capable of detecting a wide set of potential unnatural changes that cannot be addressed by a sole CUSUM chart. Since it can model different control strategies simultaneously, the Fuzzy ART network can be exploited as the sole tool for signalling a generic modification in the state of the process, so it provides a powerful diagnostic tool for detecting assignable causes in actual industrial applications.

An additional advantage of the Fuzzy ART network is that its responses to input stimulus can be easily explained, in contrast to other NNs, where typically it is more difficult to realize why an input produces a specific output. By using this feature, Pacella and Semeraro (2005) provided a criterion for deciding on the values of ART parameters that should be used in order to achieve predefined monitoring performance for the process control case at hand. This represents an important advantage of the Fuzzy ART network, as in many cases the selection method, for values of the network parameters, is based on a trial-and-error approach where different values are used and evaluated. A trial-and-error approach for NN implementation has two main drawbacks: firstly, it is time consuming, especially when NNs have slow convergence rates, and, secondly, because it leaves the user with developing empirically, for the process control case at hand, the relationship between NN performances and its parameters.

The remainder of this section provides the reader with an illustrative example on how to use a Fuzzy ART network for process monitoring. A comparison study between the Fuzzy ART performances and those of common SPC tools is also presented.

### Fuzzy ART implementation

The neural-based control system accepts as input the process outputs and produces a binary response, which in turn signals whether the process is in-control or out-of-control.

The Fuzzy ART is schematically depicted in Figure 22.2. It consists of two major subsystems: the attentional and the orienting subsystem. Three fields of nodes denoted as F0, F1 and F2 compose the attentional subsystem. The F0 layer contains $M$ neurons. The number of nodes in the F1 field is equal to $2M$, while the number of neurons in the F2 layer is equal to (or greater than) the number of clusters formed in the training phase. On the other hand, the orienting subsystem consists of a single node called the reset node. The output of the reset node, which depends on the vigilance parameter $\rho$, affects the nodes in the F2 layer.

Two pre-processing stages of process output data take place before they are presented to the Fuzzy ART network. The first one transforms the temporal series of process outputs into $M$-dimensional vectors. The second one takes an $M$-dimensional input vector and transforms it into a $M$-dimensional vector whose components fall into the interval [0,1].

In practice, the approach to apply the Fuzzy ART network for quality monitoring applications can be summarized in four steps.

(i) *Selection of the neurons number M.* This number influences the performance of the Fuzzy ART network. A strategy is to select $M$ according to the minimum magnitude in the process behaviour that is important to signal (lower magnitude requires higher window sizes). In the most diffused literature, the parameter $M$ is referred to as the window size of the monitoring approach. The most recent $M$ observations are collected from the process to form the input vector for the Fuzzy ART network. The need to arrange the series of quality measurements as $M$-dimensional vectors implies that the system provides no indication on the process state during the first $M - 1$ time intervals. Once the first $M$ data are collected, a new $M$-dimensional input vector for the Fuzzy ART network can be implemented whenever a new quality measurement becomes available (also known as a single-step moving window).

(ii) *Data normalization into the range [0, 1].* This step is needed as Fuzzy ART only accepts input ranging between zero and one. To this aim, a simple linear scaling transform is implemented from the set of possible values of process observations into the range $[0, 1]$. The set of possible process outputs can be considered ranging from $\mu - l$ to $\mu + l$, where $\mu$ represent the nominal value and $l$ is a proper saturation limit for the deviations from $\mu$. This type of coding introduces no distortion to the variable distribution as there is a one-to-one relationship between the original and 'normalized' values in $[0, 1]$.

(iii) *Training of the NN.* The Fuzzy ART is trained on the process nominal value $\mu$, that is the training list is composed by the single vector with $M$ components equal to $\mu$. Such a training method represents the least supervised approach practically possible by using a single training vector.

(iv) *Tuning of the NN (choice of the parameter* $\rho$*).* In the tuning phase, no more weight adaptations or cluster creations are allowed, and vectors from a tuning list are presented to the NN in order to check the performance of different settings of the vigilance parameter $\rho$. Tuning vectors are examples of natural (in-control) data patterns, which are obtained using either real or simulated data. A possible strategy is to select $\rho$ in order to maintain the false alarm rate about equal to a predefined value. It results that higher vigilance imposes a stricter matching criterion to the natural template learned in the training phase (this results in higher false alarm rates); on the contrary, lower vigilance tolerates greater mismatches (this results in lower false alarm rates).

It is worth noting that a monotonically increasing relation between vigilance and false alarm rate exists. Thus, when natural process data are either available or they can be simulated, the vigilance parameter can be tuned (offline) through a very simple approach, as higher vigilance parameters cause higher false alarm rates and *vice versa*. The objective of tuning is to assess the NN false alarm rate for different trial values of the vigilance parameter. This approach is quite common in quality control, as an example, it resembles the designing phase of a statistic-based control chart (e.g. the choice of parameter h for a CUSUM control chart). Given the monotonically increasing relation between vigilance and false alarm rate, a simple computer program, such as a binary search method, which iteratively divides the space of admissible values for the vigilance (i.e. the interval of real numbers between 0 and 1) in subintervals of decreasing length, can be also used to find the vigilance value.

**Table 22.1** Fuzzy ART Tuning phase. Type I error rates.

| M | l | ρ | False alarm rate | False alarm rate 95% Confidence Interval |
|---|---|---|---|---|
| 35 | 4 | 0.8631 | 0.288% | [0.257%, 0.319%] |
| 45 | 4 | 0.8676 | 0.261% | [0.224%, 0.298%] |
| 50 | 4 | 0.8693 | 0.265% | [0.236%, 0.294%] |
| 55 | 4 | 0.8707 | 0.278% | [0.247%, 0.309%] |
| 65 | 4 | 0.8731 | 0.261% | [0.232%, 0.290%] |
| 35 | 6 | 0.9087 | 0.285% | [0.254%, 0.316%] |
| 45 | 6 | 0.9117 | 0.259% | [0.223%, 0.295%] |
| 50 | 6 | 0.9128 | 0.263% | [0.234%, 0.292%] |
| 55 | 6 | 0.9138 | 0.279% | [0.247%, 0.311%] |
| 65 | 6 | 0.9154 | 0.262% | [0.233%, 0.291%] |
| 35 | 8 | 0.9315 | 0.283% | [0.252%, 0.314%] |
| 45 | 8 | 0.9338 | 0.262% | [0.225%, 0.299%] |
| 50 | 8 | 0.9346 | 0.263% | [0.234%, 0.292%] |
| 55 | 8 | 0.9354 | 0.287% | [0.255%, 0.319%] |
| 65 | 8 | 0.9366 | 0.269% | [0.239%, 0.299%] |

Despite this, in Pacella and Semeraro (2005) an analytical criterion was also presented to select the Fuzzy ART parameters ($M, l, \rho$) that should be used to obtain a predefined false alarm rate. In particular, Pacella and Semeraro (2005) derived a probabilistic model of the Fuzzy ART to estimate the effect of the three parameters on the performance of quality monitoring procedure. Statistical methods were used to derive analytically bound limits for monitoring performances and, hence, for deciding on the values of network parameters that should be used to achieve a specific performance in terms of false alarm rate.

By using the results (Pacella and Semeraro 2005), Table 22.1 reports combinations of parameters $M, l, \rho$ for the Fuzzy ART network to be used in order to obtain a false alarm rate about equal to that of the standard Shewhart three-sigma control chart (0.273%). The results in Table 22.1 are based on the assumption that the outcomes of the in-control process are normally distributed, with zero mean and constant standard deviation equal to one. Similar results can be obtained for different values of the false alarm rate.

## Comparison study

Process monitoring denotes the use of a control system that can cyclically check the desired stable state of the process. Its properties can be described in terms of probabilities; in fact, process monitoring parallels statistical hypothesis testing ($H_0$: the process is in a natural state; $H_1$: the process is in an unnatural state). As with every statistical test, errors of Type I and Type II can occur with probabilities $\alpha = P\{H_1|H_0\}$ and $\beta = P\{H_0|H_1\}$, respectively.

Henceforth, it is assumed that the outcomes of the in-control process (natural state) are normally distributed, with zero mean and constant standard deviation equal to one. The Type II error rates presented by the Fuzzy ART network are estimated for process mean changes of 1.0, 1.5 and 2.0 units of standard deviations and then they are compared to those of three SPC benchmarks. The Fuzzy ART network was implemented as follows.

- It has $M = 50$ neurons in the F0 layer, 100 neurons in the field F1 and a single node in the F2 layer.
- The linear scaling transform is considered to normalize the input data into the range [0, 1]. $l = 6$ is the saturation limit used for the deviations of the quality measurement from the process nominal value.

- The Fuzzy ART NN was trained on the process nominal value $\mu = 0$.
- To compare the NN to any traditional charting technique it is required that performances must be identical when the process is in a natural state (Type I error rates). This serves to provide an unbiased comparison when the process drifts to unnatural states. Hence, the vigilance parameter of the Fyzzy ART NN was in turn adjusted in order to give a comparable performance in terms of the Type I error rate ($\hat{\alpha}_{nn}$) to that of a predefined SPC benchmark ($\hat{\alpha}_{cc}$).

Comparisons of the Fuzzy ART performances to those of a control chart benchmark are based on Type II error rates, which have been experimentally estimated by introducing two controlled disturbance signals: systematic variation and shift. For each disturbance signal and each magnitude setting, Type II error point and interval estimators were assessed on 50 batches of 2000 independent simulation runs. The following control charts have been selected.

- Bilateral cumulative summation (CUSUM) control chart with parameters $k = 0.5$ and $h = 4.7749$. Estimated Type I error rate $\hat{\alpha}_{cc} = 0.269\%$.
- Shewhart control chart with Western Electric run rules (Western Electric Statistical Quality Control Handbook, 1956). Estimated Type I error rate $\hat{\alpha}_{cc} = 1.115\%$.
- Shewhart control chart with Western Electric run rules and four additional sensitizing rules (Nelson, 1984). Estimated Type I error rate $\hat{\alpha}_{cc} = 1.617\%$.

A one-step moving window of size $M$ was exploited for Type II error estimation. Numerical results and comparisons are discussed in the following sections for each of the three SPC benchmarks considered.

### CUSUM control chart

Table 22.2 compares Type I and Type II errors of the CUSUM schema $k = 0.5$ and $h = 4.7749$ (Type I error $\hat{\alpha}_{cc} = 0.269\%$) to those of the Fuzzy ART network with vigilance parameter $\rho = 0.9128$ (Type I error $\hat{\alpha}_{nn} = 0.263\%$). The values of the CUSUM parameters $k$ and $h$ have been set for signalling a shift of one standard deviation in the mean with a false alarm rate about equal to that of the standard Shewhart three-sigma control chart (0.273%).

In order to confirm the statistical significance of the difference between NN and control chart performances, the t-based confidence intervals (coverage 95%) have also been presented in the same

**Table 22.2** Performance comparison between Fuzzy ART and CUSUM control chart (simulation results, 50 sets of 2000 data).

| | **CUSUM $k = 0.5$ $h = 4.7749$** | **Fuzzy ART $M = 50$ $\rho = 0.9128$** | **Comparison Neural Network vs Control Chart** | | |
|---|---|---|---|---|---|
| *Natural* | $\hat{\alpha}_{cc}$ | $\hat{\alpha}_{nn}$ | $(\hat{\alpha}_{nn} - \hat{\alpha}_{cc})_-$ | $\hat{\alpha}_{nn} - \hat{\alpha}_{cc}$ | $(\hat{\alpha}_{nn} - \hat{\alpha}_{cc})_+$ |
| | 0.269% | 0.263% | −0.050% | −0.006% | 0.038% |
| *Sys. Var.* | $\hat{\beta}_{cc}$ | $\hat{\beta}_{nn}$ | $(\hat{\beta}_{nn} - \hat{\beta}_{cc})_-$ | $\hat{\beta}_{nn} - \hat{\beta}_{cc}$ | $(\hat{\beta}_{nn} - \hat{\beta}_{cc})_+$ |
| *1.0* | 99.487% | 67.645% | −32.741% | −31.842% | −30.943% |
| *1.5* | 99.158% | 0.000% | −99.225% | −99.158% | −99.091% |
| *2.0* | 98.424% | 0.000% | −98.508% | −98.424% | −98.340% |
| *Shift* | $\hat{\beta}_{cc}$ | $\hat{\beta}_{nn}$ | $(\hat{\beta}_{nn} - \hat{\beta}_{cc})_-$ | $\hat{\beta}_{nn} - \hat{\beta}_{cc}$ | $(\hat{\beta}_{nn} - \hat{\beta}_{cc})_+$ |
| *1.0* | 0.019% | 67.989% | 67.045% | 67.970% | 68.895% |
| *1.5* | 0.000% | 0.012% | 0.000% | 0.012% | 0.030% |
| *2.0* | 0.000% | 0.000% | 0.000% | 0.000% | 0.000% |

table. The columns marked as $\hat{\beta}_{nn} - \hat{\beta}_{cc}$ gives the difference between the Type II error point estimators. The lower limit of the t-based confidence intervals is reported in the column labelled as $(\hat{\beta}_{nn} - \hat{\beta}_{cc})_-$, while the upper limit in the column labeled as $(\hat{\beta}_{nn} - \hat{\beta}_{cc})_+$.

The NN performance is better (i.e. smaller Type II errors) than that of the CUSUM chart for signalling systematic variations of the process mean. On the other hand, the NN has a worse performance if compared to the CUSUM chart (i.e. higher Type II errors) for shifts of 1.0 unit of standard deviation. For higher shifts, the performances are similar.

The results of Table 22.2 prove that the CUSUM schema cannot be adopted as the sole tool for signalling a generic modification in the state of the process (e.g. it performs poorly in signalling alarms for a systematic variation of the mean, while it performs better for a constant shift of the mean). On the other hand, the Fuzzy ART NN recognizes different kinds of change with the same capability. Indeed, the NN performances in tackling systematic variations and shifts of the mean are similar for each level of magnitude.

### *Shewhart control chart with Western Electric run rules*

Table 22.3 compares Type II errors of the Shewhart chart with the three Western Electric run rules (two of three consecutive points outside the ±2-sigma limits; four of five consecutive points beyond the ±1-sigma limits; a run of eight consecutive points on one side of the centre line), to those of the NN.

While the simultaneous tests proposed in the Western Electric Statistical Quality Control Handbook (1956) improve the performance of the Shewhart control chart in recognizing changes of the process mean, they do so at the cost of increases in false alarm rates. Therefore, a higher value of the vigilance parameter ($\rho = 0.9168$) has been adopted in order to obtain a NN false alarm rate that is comparable to that presented by the benchmark.

The results of Table 22.3 prove that the NN achieves better performances (lower Type II error rates) than those of the SPC benchmark in recognizing any disturbance signals.

### *Shewhart control chart with Western Electric run rules and sensitizing rules*

Table 22.4 compares Type II errors of the Shewhart control chart with seven run rules, to those given by the Fuzzy ART NN with vigilance parameter $\rho = 0.9179$. The run rules implemented in the SPC benchmark are the standard three tests described in Western Electric (1956) and four additional sensitizing rules proposed by Nelson (1984): six points in a row steadily increasing or decreasing; fifteen points in a row within the ±1-sigma limits; fourteen points in a row alternating up and down; eight points in a

**Table 22.3** Performance comparison between Fuzzy ART and Shewhart control chart with standard Western Electric (1956) run rules (simulation results, 50 sets of 2000 data).

| | **Shewhart WE RRs** | **Fuzzy ART M = 50 ρ = 0.9168** | **Comparison Neural Network vs Control Chart** | | |
|---|---|---|---|---|---|
| *Natural* | $\hat{\alpha}_{cc}$ | $\hat{\alpha}_{nn}$ | $(\hat{\alpha}_{nn} - \hat{\alpha}_{cc})_-$ | $\hat{\alpha}_{nn} - \hat{\alpha}_{cc}$ | $(\hat{\alpha}_{nn} - \hat{\alpha}_{cc})_+$ |
| | 1.115% | 1.133% | −0.075% | 0.018% | 0.111% |
| *Sys. Var.* | $\hat{\beta}_{cc}$ | $\hat{\beta}_{nn}$ | $(\hat{\beta}_{nn} - \hat{\beta}_{cc})_-$ | $\hat{\beta}_{nn} - \hat{\beta}_{cc}$ | $(\hat{\beta}_{nn} - \hat{\beta}_{cc})_+$ |
| *1.0* | 95.635% | 42.181% | −54.654% | −53.454% | −52.254% |
| *1.5* | 87.731% | 0.000% | −87.901% | −87.731% | −87.561% |
| *2.0* | 72.220% | 0.000% | −72.536% | −72.220% | −71.904% |
| *Shift* | $\hat{\beta}_{cc}$ | $\hat{\beta}_{nn}$ | $(\hat{\beta}_{nn} - \hat{\beta}_{cc})_-$ | $\hat{\beta}_{nn} - \hat{\beta}_{cc}$ | $(\hat{\beta}_{nn} - \hat{\beta}_{cc})_+$ |
| *1.0* | 80.820% | 43.423% | −38.506% | −37.397% | −36.288% |
| *1.5* | 42.902% | 0.008% | −43.419% | −42.894% | −42.369% |
| *2.0* | 8.335% | 0.000% | −8.609% | −8.335% | −8.061% |

**Table 22.4** Performance comparison between Fuzzy ART and Shewhart control chart with standard Western Electric (1956) and Nelson (1984) sensitizing run rules (simulation results, 50 sets of 2000 data).

| | Shewhart WE + SR RRs | Fuzzy ART $M = 50$ $\rho = 0.9179$ | Comparison Neural Network vs Control Chart | | |
|---|---|---|---|---|---|
| *Natural* | $\hat{\alpha}_{cc}$ | $\hat{\alpha}_{nn}$ | $(\hat{\alpha}_{nn} - \hat{\alpha}_{cc})_-$ | $\hat{\alpha}_{nn} - \hat{\alpha}_{cc}$ | $(\hat{\alpha}_{nn} - \hat{\alpha}_{cc})_+$ |
| | 1.671% | 1.653% | −0.135% | −0.018% | 0.099% |
| *Sys. Var.* | $\hat{\beta}_{cc}$ | $\hat{\beta}_{nn}$ | $(\hat{\beta}_{nn} - \hat{\beta}_{cc})_-$ | $\hat{\beta}_{nn} - \hat{\beta}_{cc}$ | $(\hat{\beta}_{nn} - \hat{\beta}_{cc})_+$ |
| 1.0 | 87.888% | 34.829% | −54.326% | −53.059% | −51.792% |
| 1.5 | 65.761% | 0.000% | −66.193% | −65.761% | −65.329% |
| 2.0 | 32.188% | 0.000% | −32.652% | −32.188% | −31.724% |
| *Shift* | $\hat{\beta}_{cc}$ | $\hat{\beta}_{nn}$ | $(\hat{\beta}_{nn} - \hat{\beta}_{cc})_-$ | $\hat{\beta}_{nn} - \hat{\beta}_{cc}$ | $(\hat{\beta}_{nn} - \hat{\beta}_{cc})_+$ |
| 1.0 | 80.255% | 35.669% | −45.727% | −44.586% | −43.445% |
| 1.5 | 42.546% | 0.005% | −43.017% | −42.541% | −42.065% |
| 2.0 | 8.265% | 0.000% | −8.521% | −8.265% | −8.009% |

row on both sides beyond the ±1-sigma limits. The use of four additional run rules increases the false alarm rate of the SPC benchmark, thus a higher value of the vigilance parameter has been adopted in this case.

The NN has better performances (lower Type II error rates) than those of the SPC chart when recognizing the disturbance signals, for each magnitude level considered in the test.

## Discussion

From the experimental results and comparisons, it is fair to conclude that the ART-based control system is superior to (or in par with) several SPC charts in terms of Type II error rates. In particular, test comparisons show that the proposed method is a good control procedure for tackling different kinds of alteration in the process mean. For example, the Fuzzy ART network possesses superior detection capability against fluctuations of the process mean (systematic variations) than the CUSUM test, while it presents a comparable (or slightly worse) ability in signalling constant shifts. At the same time, the network outperforms Shewhart control charts with a set of run rules and sensitizing rules.

Simulation results prove that the proposed approach can model different control strategies simultaneously: for example, those of a CUSUM and of a Shewhart control chart with run rules, which were designed to recognize different kinds of change in the process structure (steady shifts of moderate magnitude and sudden fluctuation of high magnitude in the process mean, respectively). Indeed, the Fuzzy ART network can be potentially adopted to signal any types of unnatural pattern, so it provides a powerful diagnostic tool for detecting assignable causes in real processes.

Thus, the main advantage of the neural-based approach over traditional ones is that it can be exploited as the sole tool for signalling a generic modification in the state of the process. Indeed, the NN can be useful when starting processing of new products, or with a new installed process, for which no prior knowledge of the unnatural changes are available in advance in order to design a proper control strategy.

# CONCLUSIONS

In this chapter, issues referred to the use of NNs in SPC have been discussed, concerning the way of operation of each network model, the network's architecture and the results provided.

After a brief overview of machine learning and neural network theory, papers referring to pattern recognition using NNs have been discussed. Traditional Shewhart control charts do not provide pattern information of the data because only the most recent point is used to decide the status of the process. The results of using NNs for this purpose showed that these networks are, in many cases, the most reliable tools for the recognition of a pattern (trend, cycle, stratification etc.) in the data.

Furthermore, papers that proposed NNs for the detection and determination of mean and variance shifts in a process have been discussed. The results showed that in a number of cases, the proposed NNs outperform the Shewhart control charts. For small mean and variance shifts, they have at least equal performance with other control charts such as CUSUM and EWMA charts.

The last section focused on an ART-based network for process monitoring by providing a basic description of such a network. The analysed neural approach is mainly intended for identifying unnatural process behaviour by detecting changes in the state of the process. This method is quite simple to implement and the training set can be limited to a single ideal pattern. Details of the proposed neural-based control schema have been discussed while the performances of the control system reported in the chapter are shown to be superior to (or in par with) several traditional control charts in terms of Type II error rates.

Finally, a few studies have been presented in the literature concerning the use of NNs in multivariate control charts. In fact, traditional multivariate control charts under some conditions do not provide sufficient ability for monitoring more than one quality characteristics simultaneously. For this reason, NNs should be investigated. The application of NNs in multivariate SPC may improve the ability to automate the monitoring of industrial processes, detect and recognize more quickly and accurately the status of the process. Therefore, more studies should be performed in the near future in order to optimize the use of NNs in multivariate SPC.

## REFERENCES


Al-Ghanim, A. (1997) An unsupervised learning neural algorithm for identifying process behavior on control charts and a comparison with supervised learning approaches. *Computers and Industrial Engineering* **32**, 627–639.

Anagnostopoulos, G.C. and Georgiopouolos, M. (2002) Category regions as a new geometrical concepts in Fuzzy-ART and Fuzzy-ARTMAP. *Neural Networks* **15**, 1205–1221.

Anagun, A.S. (1998) A neural network applied to pattern recognition in statistical process control. *Computers and Industrial Engineering* **35**, 185–188.

Bishop, C.M. (1995) *Neural Networks for Pattern Recognition*. Oxford University Press, New York.

Bishop, C.M. (2006) *Pattern Recognition and Machine Learning*. Springer, New York.

Champ, C.W. and Woodall, W.H. (1987) Exact results for Shewhart control charts with supplementary run rules. *Technometrics* **29**, 393–399.

Chang, S.I. and Ho, E.S. (1999) A two-stage neural network approach for process variance change detection and classification. *International Journal of Production Research* **37**, 1581–1599.

Cheng, C.S. (1995) A multi-layer neural network model for detecting changes in the process mean. *Computers and Industrial Engineering* **28**, 51–61.

Cheng, C.S. (1997) A neural network approach for the analysis of control chart patterns. *International Journal of Production Research* **35**, 667–697.

Cheng, C.S. and Cheng, S.S. (2001) A neural network-based procedure for the monitoring of exponential mean. *Computers and Industrial Engineering* **40**, 309–321.

Chiu, C.C., Chen, M.K. and Lee, K.M. (2001) Shifts recognition in correlated process data using a neural network. *International Journal of System Science* **32**, 137–143.

Cottrell, M., Girard, B., Girard, Y. *et al.* (1995) Neural modeling for time series: a statistical stepwise method for weight elimination. *IEEE Transactions on Neural Networks* **6**, 1355–1364.

Elman, J.L. (1990) Finding structure in time. *Cognitive Science* **14**, 179–211.

Georgiopoulos, M., Fernlund, H., Bebis, G. and Heileman, G.L. (1996) Order of search in Fuzzy ART and Fuzzy ARTMAP: effect of the choice parameter. *Neural Networks* **9**, 1541–1559.

Georgiopoulos, M., Dagher, I., Heileman, G.L. and Bebis, G. (1999) Properties of learning of a Fuzzy ART variant. *Neural Networks* **12**, 837–850.

Guh, R.S. (2002) Robustness of the neural network based control chart pattern recognition system to non-normality. *International Journal of Quality and Reliability Management* **19**, 97–112.

Guh, R.S. (2005) A hybrid learning-based model for on-line detection and analysis of control chart patterns. *Computers and Industrial Engineering* **49**, 35–62.
Guh, R.S. (2010) Simultaneous process mean and variance monitoring using artificial neural networks. *Computers and Industrial Engineering* **58**, 739–753.
Guh, R.S. and Hsieh, Y.C. (1999) A neural network based model for abnormal pattern recognition of control charts. *Computers and Industrial Engineering* **36**, 97–108.
Guh, R.S. and Tannock, J.D.T. (1999) Recognition of control chart concurrent patterns using a neural network approach. *International Journal of Production Research* **37**, 1743–1765.
Haykin, S. (1998) *Neural Networks. A Comprehensive Foundation*, 2nd edn. Macmillan, New York.
Ho, E.S. and Chang, S.I. (1999) An integrated neural network approach for simultaneous monitoring of process mean and variance shifts – a comparative study. *International Journal of Production Research* **37**, 1881–1901.
Hornik, K., Stinchcombe, M. and White, H. (1989) Multilayer feedforward networks are universal approximators. *Neural Networks* **2**, 359–366.
Huang, J., Georgiopoulos, M. and Heileman, G.L. (1995) Fuzzy ART proprieties. *Neural Networks* **8**, 203–213.
Hwarng, H.B. (2004) Detecting process mean shift in the presence of autocorrelation: A neural-network based monitoring scheme. *International Journal of Production Research* **42**, 573–595.
Hwarng, H.B. (2005) Simultaneous identification of mean shift and correlation change in AR(1) model. *International Journal of Production Research* **43**, 1761–1783.
Hwarng, H.B. and Hubele, N.F. (1993a) Back-propagation pattern recognizers for X-bar control charts: methodology and performance. *Computers and Industrial Engineering* **24**, 219–235.
Hwarng, H.B. and Hubele, N.F. (1993b) X-bar control chart pattern identification through efficient off-line neural network training. *IIE Transasctions* **25**, 27–40.
Lucas, J.M. and Saccucci, M.S. (1990) Exponentially weighted moving average control schemes: properties and enhancements. *Technometrics* **32**, 1–12.
MacGregor, J.F. (1990) A different view of the funnel experiment. *Journal of Quality Technology* **22**, 255–259.
Montgomery, D.C. (2008) *Introduction to Statistical Quality Control*, 6th edn. John Wiley & Sons, Inc., New York.
Murata, N., Yoshizawa, S. and Amari, S. (1994) Network information criterion-determining the number of hidden units for an artificial neural network model. *IEEE Transactions Neural Networks* **5**, 865–872.
Nelson, L.S. (1984) The Shewhart control chart-tests for special causes. *Journal of Quality Technology* **16**, 237–239.
Pacella, M., Semeraro, Q. and Anglani, A. (2004a) Manufacturing quality control by means of a Fuzzy ART network trained on natural process data. *Engineering Applications of Artificial Intelligence* **17**, 83–96.
Pacella, M., Semeraro, Q. and Anglani, A. (2004b) Adaptive resonance theory-based neural algorithms for manufacturing process quality control. *International Journal of Production Research* **40**, 4581–4607.
Pacella, M. and Semeraro, Q. (2005) Understanding ART-based neural algorithms as statistical tools for manufacturing process quality control. *Engineering Applications of Artificial Intelligence* **18**, 645–662.
Pacella, M. and Semeraro, Q. (2007) Using recurrent neural networks to detect changes in autocorrelated processes for quality monitoring. *Computers and Industrial Engineering* **52**, 502–520.
Page, E.S. (1954) Continuous inspection schemes. *Biometrika* **41**, 100–115.
Perry, M.B., Spoerre, J.K. and Velasco, T. (2001) Control chart pattern recognition using back propagation artificial neural networks. *International Journal of Production Research* **39**, 3399–3418.
Pugh, G.A. (1989) Synthetic neural networks for process control. *Computers and Industrial Engineering* **17**, 24–26.
Pugh, G.A. (1991) A comparison of neural networks to SPC charts. *Computers and Industrial Engineering* **21**, 1–4.
Reed, R. (1993) Pruning algorithms – a survey. *IEEE Transactions on Neural Networks* **4**, 740–747.
Roberts, S.W. (1959) Control chart tests based on geometric moving averages. *Technometrics* **1**, 239–250.
Roy, A., Kim, L.S. and Mukhopadhyay, S. (1993) A polynomial time algorithm for the construction and training of a class of multilayer perceptrons. *Neural Networks* **6**, 535–545.
Schittenkopf, C., Deco, G. and Brauer, W. (1997) Two strategies to avoid overfitting in feedforward networks. *Neural Networks* **10**, 505–516.
Smith, A.E. (1994) X-bar and R control chart interpretation using neural computing. *International Journal of Production Research* **32**, 309–320.
Shewhart, W.A. (1931) Statistical method from an engineering viewpoint. *Journal of the American Statistical Association* **26**, 262–269.
Vapnik, V.N. (2000) *The nature of statistical learning theory*, 2nd edn. Springer, New York.
Wang, Z., Di Massimo, C., Tham, M.T. and Morris, A.J. (1994) A procedure for determining the topology of multilayer feedforward neural networks. *Neural Networks* **7**, 291–300.
Western Electric ( (1956) *Statistical Quality Control Handbook*. AT&T, Princeton, NJ.

# 23 An integral approach to validation of analytical fingerprinting methods in combination with chemometric modelling for food quality assurance

**Grishja van der Veer**[1], **Saskia M. van Ruth**[2] **and Jos A. Hageman**[3]

[1] *RIKILT Institute of Food Safety, Wageningen UR, Wageningen, The Netherlands*
[2] *Product Design and Quality Group, Wageningen UR and RIKILT Institute of Food Safety, Wageningen UR, Wageningen, The Netherlands*
[3] *Plant Sciences Group, Wageningen UR, Wageningen, The Netherlands*

## ABSTRACT

Analytical fingerprinting in combination with chemometric modelling provides a powerful approach in the framework of food quality assurance. This recently emerging approach allows verifying claims regarding a whole range of food quality characteristics that were previously difficult or impossible to determine. These include claims regarding the nutritional composition, product typicality and method of production as well as the geographical origin of food. As such, the fingerprinting approach can for example be used to verify claims of fish being from a wild origin, or of eggs being organic.

Since the fingerprinting approach differs from the classical analytical approach for food quality assurance, the existing protocols for validation of classical analytical methods are not directly applicable. To further standardize and harmonize the fingerprinting approach on an international level, an extension to existing validation protocols is required. This chapter provides a first step towards a further standardization by providing a tentative strategy for integral validation of the fingerprinting approach.

## INTRODUCTION

Quality is defined in DIN ISO 9000 (2005-12) as 'the totality of features and characteristics of product or service that bear on its ability to satisfy stated or implied needs'. In food science the term *quality* not only refers to characteristics such as appearance, taste and flavour but also includes claims about nutritional composition, product typicality, production method, as well as geographical origin. Such claims involve, for example the food being described as halal or kosher, or being produced by an organic farming management system, or being from a specific designated area of origin. These claims are difficult – or impossible – to determine by classical analytical methods or by sensory panels. Analytical fingerprinting in combination with chemometric modelling is a newly emerging approach to verify such claims

*Mathematical and Statistical Methods in Food Science and Technology*, First Edition.
Edited by Daniel Granato and Gastón Ares.

regarding food quality independently of a paper trail and physical control (Charlton *et al.*, 2002; Møller *et al.*, 2005; Bertelli *et al.*, 2010; van Ruth *et al.*, 2010; Tres *et al.*, 2012).

Analytical fingerprinting can be loosely described as an approach in which a characteristic pattern, which consists of a set of chemical or physical attributes – or markers – is used to determine certain quality characteristics or properties of a sample. This characteristic pattern – or fingerprint – can, for example consist of a mass spectrum or a chromatogram, or specific sections hereof (Figure 23.1). In principle, therefore, any analytical technique that is able to produce results for several chemical, biochemical or physical markers in a single run could be used for fingerprinting purposes. Analytical techniques that are often used in food quality assurance include different types of mass spectrometry (e.g. LC-MS, PTR-MS), chromatography (e.g. GC, HPLC) and spectroscopy (e.g. NIRS, NMR) as well as other techniques (e.g. microarray techniques, gel electrophoresis).

Because of the large amount of information that is often contained in these fingerprints, interpretation and modelling of the data requires advanced chemometric techniques that are able to deal with the data's

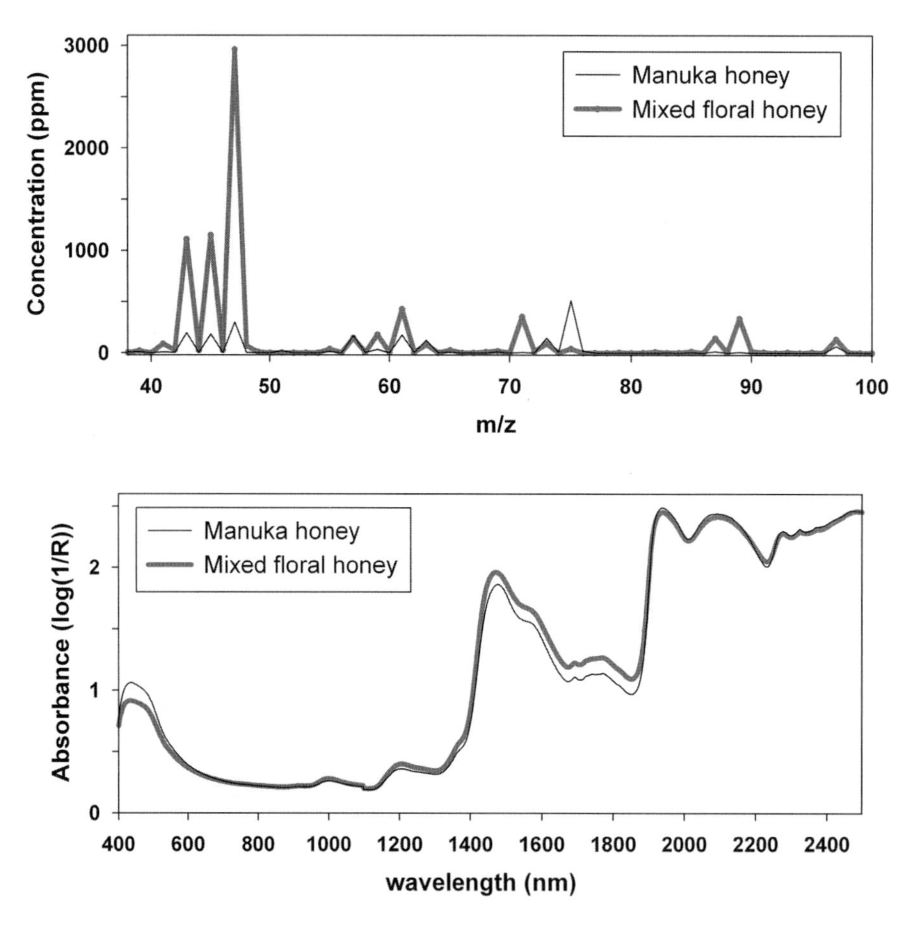


**Figure 23.1** (a) Proton mass transfer mass spectrometry (PTR-MS) fingerprint of two honey samples showing the difference between Manuka honey from New Zealand and mixed floral honey from Australia in terms of volatile compounds. The example shows that most volatile compounds occur in both types of honey, yet their respective proportions are different. (b) Near infrared spectroscopy (NIRS) fingerprint of the same two honey samples showing the differences in terms of physicochemical characteristics as observed by NIR spectroscopy (including part of visible spectrum).

high dimensional nature. In this integrated fingerprinting–classification approach, a multivariate classification model is 'trained' to recognize those samples that meet the required quality characteristics and to distinguish them from those who do not based on the fingerprints of a set of reference samples of both groups. The trained classification model then allows the compliance of future samples to be tested at a certain level of confidence.

The general concept behind the fingerprinting–classification modelling approach is quite different from the classical approach. In general, classical analytical chemistry is concerned with determination of the concentration (quantitative) or the presence (qualitative) of an analyte and, as such, the approach is targeted in a sense that it is known what specific analyte or set of analytes are being measured, either directly or indirectly. In contrast, knowledge about the exact identity or nature of the analytes or markers is not required in the fingerprinting approach, as long as their collective pattern – or fingerprint – is characteristic for the quality characteristic that is to be determined by the method. This so-called untargeted approach is especially useful in cases where the information is simply not, or only partly, available, which can be the case, for example in more advanced types of spectroscopy and mass spectrometry.

Secondly, the fingerprinting method is based on a reference data set that contains samples from all relevant groups/classes that need to be discerned. This requires sampling of a representative – and often large – set of reference samples. This reference set then plays a comparable role, as certified reference materials and/or in-house standards do in the classical approach, in the sense that it is used to 'calibrate' or 'train' the chemometric model. Since this approach requires an extensive database of sample measurements, it is sometimes also referred to as the 'database approach'.

Yet another difference between the classical analytical approach and the fingerprinting approach is that the latter is usually specifically developed for a single combination of food commodity and questions regarding its quality (e.g. Are these eggs organic? Is this orange juice fresh or from concentrate?). This stems from the fact that each type of food has its own typical fingerprint pattern and each quality characteristic is reflected by a different set of markers depending on the commodity. As such, the method needs to be calibrated specifically for each commodity/quality characteristic.

As can be seen the most importance differences between both approaches lay in the fact that the fingerprinting method (i) works for untargeted or semi-target fingerprinting data as well, whereas the classical approach applies typically to target data only, (ii) needs to be calibrated (trained) by an extensive set of reference samples, and (iii) needs to be calibrated for a specific combination of food commodity/quality question.

Another typical aspect of the fingerprinting approach involves the dimensionality of the data, which is often high (>1000 variables) to very high (>100 000 variables). In this case, the number of samples is considerably smaller than the amount of variables. This forms a problem for some classification techniques, such as linear discriminant analysis (LDA), which would therefore require feature selection or reduction prior to modelling. Most chemometric techniques are, however, able to deal with such high dimensional data.

Since this approach to fingerprinting differs from the classical analytical approach for food quality assurance in a number of ways, existing protocols for validation are not directly applicable in the fingerprinting context. Therefore, an extension to existing validation protocols is required to further standardize and harmonize the fingerprinting approach on an international level. As a first review of this emerging topic, this chapter provides a set of tentative guidelines to validate methods based on the fingerprinting approach.

A significant part of the applications of chemometrics in analytical chemistry falls in the general framework of classification, and the discussion is therefore limited to the application of supervised classification models. Moreover, the focus is on in-house validation, since most fingerprinting methods will first be developed and validated as a single laboratory method.

# THE FINGERPRINTING – CLASSIFICATION MODELLING APPROACH

## General concept

The basic starting point for developing and validating a new analytical method is to define its goal and scope in terms of what the method should be able to test for. In the case of a method that is based on a combination of fingerprinting and classification modelling, this first step moreover includes defining the populations/classes that need to be discerned/identified by the method. In the case of food this requires a definition of the populations/classes in terms of production method, brand, animal breed or crop variety and so on. Since a method developed, for example for European beef might not be applicable for beef from South America, the geographical extent for which the method is valid should be decided upon as well.

In practice, the development of a method based on analytical fingerprinting in combination with classification modelling generally involves the following steps:

(i) *Sampling:* Collection of reference samples that need to representatively cover the variation present in the populations/classes that need to be distinguished from each other.
(ii) *Fingerprinting analysis:* Determination of the fingerprints of the reference samples using the relevant fingerprinting technique.
(iii) *Data pretreatment:* Pretreatment of the fingerprinting data using data pre-processing (e.g. mean centring), transformation (e.g. smoothing) or aggregation.
(iv) *Feature selection/reduction (optional):* Reduction of the dimensionality of the data before further modelling.
(v) *Building a classification model:* Selection of the optimal classifier and model parameter setting.
(vi) *Model validation:* Determination of model performance using internal and external model validation procedures.

For the purpose of this chapter the focus is mainly on the model validation step; further information about model building aspects and well as data pretreatment and feature selection is available elsewhere (Massart *et al.*, 1998; Vandeghinste *et al.*, 1998; Duda *et al.*, 2001; Bishop, 2006; Webb and Copsey, 2011). In the remainder of this section only some key issues in chemometrics that are relevant for understanding the later text are reviewed.

## Classification methods

Classification in chemometrics and pattern recognition refer to classifying a set of objects, or samples, into a number of pre-existing and known categories or classes. This is referred to as supervised classification, as opposed to unsupervised classification in which no information about the categories or classes to which the objects belong is involved.

There are many different supervised classification algorithms (classifiers) available, including, for example linear and quadratic discriminant analysis (LDA and QDA), k-nearest neighbour (kNN), support vector machines (SVM), artificial neural networks (ANN), classification and regression trees (CART), partial least squares discriminant analysis (PLS-DA), soft independent modelling of class analogy (SIMCA) and so on.

A general distinction can be made between soft and hard classification methods, also referred to as class-modelling and pure classification. Soft techniques, such as SIMCA and UNEQ, focus on modelling the analogies among the elements of a class rather than on discriminating among the different classes, and each class is modelled separately. As these techniques build frontiers between each class and the rest of the universe, three situations can arise in case of binary classification: an object can be assigned to a single class, to two classes or to no class at all (outlier).

Hard or pure classification, which applies to most other classification techniques, focuses on discriminating between different groups; these techniques divide the hyperspace in as many regions as the number of groups, so that an object is always assigned to only one class. Hard classification in general provides better discrimination between the classes and is, therefore, often preferred in the framework of food quality assurance. This however requires careful screening for outliers prior to data modelling and prediction.

Apart from the distinction between hard versus soft modelling techniques, other relevant distinctions include linear versus nonlinear methods and parametric versus nonparametric methods. In terms of choosing a classifier it should be noted that most classification tasks can be performed using linear methods, such as LDA and PLS-DA, whereas nonlinear methods, such as SVM and ANN, are rarely needed (Beruetta *et al.*, 2007). The choice of a parametric (e.g. LDA) or nonparametric (e.g. kNN) method furthermore depends on whether a specific underlying probability distribution is assumed or not.

Another aspect that could be taken into account in the choice of classifier is the sensitivity of the method to overfitting. Overfitting takes place if the model learns the idiosyncrasy of the data and the noise is modelled as well. As a result, the model loses its generalization ability (i.e. ability to correctly predict the class membership of new samples/objects). LDA, CART, ANN, SVM and UNEQ, especially, appear sensitive to overfitting, which can also be an issue in PLS-DA when too many latent components are selected (Berrueta *et al.*, 2007).

Although the above mentioned arguments can be used to choose the most optimal classifier according to the modeller's demand (soft/hard method) and the characteristics of the data (linear/nonlinear, fixed distribution/free distribution), in practice this choice is based on an empirical selection procedure in which the performance of a number of classifiers is compared and evaluated (Duda *et al.*, 2001). How to evaluate the performance of such classifiers is further detailed in the next section.

## Model performance

There are various measures that quantify the performance of a classification model. One obvious measure is the percentage, or fraction, of good predictions, which is also referred to as the correct classification rate, or accuracy of the model. This is a very general measure of model performance that does not include any information about how well the individual classes can be model. For this purpose, a zero and alternative hypothesis (or hypotheses in case of a multiclass problem) need to be defined. In the context of for food quality assurance the zero and alternative hypothesis could, for example look like:

$H_0$: The food sample belongs to the population that meets the quality criterion.
$H_1$: The food sample does not belong to the population that meets the quality criterion.

Herein the quality criterion could be any physical or chemical quality aspect related to a food safety or authenticity question. Furthermore, the samples that belong can be regarded as the 'positives' and the samples that do not the 'negatives'.

It is common practice to summarize model performance in a so-called 'confusion matrix' (Table 23.1), which shows how many samples are being classified as positive and negative, as well as how many of these are correctly classified as true positives (TP) or true negatives (TN), and how many are falsely classified as false positives (FP) or false negatives (FN). The chance of erroneously classifying a truly negative sample as a positive is referred to as the $\alpha$ error (chance of a Type I error), whereas classifying a truly positive sample as a negative is referred to as the $\beta$ error (chance of a Type II error).

**Table 23.1** Example of a 2 × 2 contingency table (also known as a 'confusion matrix').

| | | True class | |
|---|---|---|---|
| | | Positive | Negative |
| **Predicted class** | Positive | True positive ($p = 1 - \alpha$) | False positive ($p = \alpha$) |
| | Negative | False negative ($p = \beta$) | True negative ($p = 1 - \beta$) |

From the confusion matrix in Table 23.1, a number of measures for the performance of the model can be established, such as the sensitivity (also referred to as the true positive rate), which is defined as TP/(TP + FN), and the specificity (also referred to as the false negative rate), which is defined as TN/(TN + FP). An overview of related measures for model performance can be found in Fawcett (2004).

## Model validation

In chemometrics, validation refers to establishing the predictive ability of a model by using one or more validation data sets. In general, two types of validation can be distinguished:

(i) External validation using an independent validation set.
(ii) Internal validation using random subsets of the data for validation.

External validation is based on collecting a new and independent set of samples – the external validation set – which is used to challenge the classification model. This approach provides the most direct way of model validation, as the validation set is sampled under truly different conditions reflecting, for example different sample locations and different production periods and so on. External validation is, therefore, the preferred method to assess the generalization ability of the model, and hence its ability to correctly predict the identity of new samples.

Using the confusion matrix as predicted from the external validation set, the correct classification rate of the external validation procedure ($CCR_{EV}$) can be determined and can be used as a measure of the model performance. When additional information is required regarding the performance for predicting each class separately, one could moreover determine the model sensitivity and specificity from the confusion matrix of the validation predictions.

External validation often involves a considerable investment in terms of time and resources, and before these validation samples are collected an internal validation should be applied to check whether the model has sufficient prediction power for the purpose of the method. During internal validation, a model is built by using a subset of the data (training set), after which the prediction power of the model is determined by predicting the class memberships of the samples that were left out (test set). This procedure is then repeated N times, where N depends on the resampling approach used to select the training and the test set.

Resampling can be done using various methods including different forms of cross-validation. The simplest form of cross-validation is based on a leave-one-out (LOO) resampling scheme. In this approach, one sample is left out from model calibration and only used for prediction, which is then repeated for each sample. However, especially for small data sets LOO-CV tends to overfit the model (Baumann, 2003); related resampling schemes, including leave-k-out cross-validation or k-fold cross-validation, are generally preferred over LOO-CV.

In bootstrapping, which provides an alternative resampling strategy, a test set is created by randomly selecting $n' < n$ samples/objects from the training set with replacement (Efron, 1979). This procedure is then independently repeated N times to yield N bootstrap sets. The classification results from each individual set can then be established using for example the correct classification rate ($CCR_i$). The individual $CCR_i$ can then be averaged to derive a general measure of the model performance. Since in bootstrapping N is only confined by available computer resources, one has more control over the estimate and variance of the statistic, which improves with increasing N. This is a distinct advantage of bootstrapping, especially for small sample sizes.

Internal validation is not only used to estimate the model performance *an sich* but is often also used to optimize a classification algorithm in terms of the parameter settings, such as the number of neighbours in kNN or the number of latent components in PLS-DA. In this case, the free parameter is varied and each time the model performance is evaluated by resampling. The optimal value for the parameter is then determined by making a fair trade-off between the error of prediction and the chance of overfitting the model by increasing the value for the parameter.

Having optimized a classification model in terms of model parameters, a similar procedure can be used to compare the performance of different classification algorithms and to select the most optimal classifier. As explained by the no-free-lunch theorem, there is no single classifier that works best on all given problems and even selection of the optimal classifier for a specific problem is not trivial. Another approach to this issue is to use a combination of classifiers instead of just one (ensemble methods or mixture-of-expert models). Further discussion about these topics is available elsewhere (Duda *et al.*, 2001; Webb and Copsey, 2011).

## CLASSICAL APPROACH TO METHOD VALIDATION

In analytical chemistry, the goal of validation is to confirm and provide objective evidence that particular requirements for a specific intended use are fulfilled (e.g. ISO 17025 and EC Directive 2002/657/EC). In this section how method validation is performed in a classical analytical sense is reviewed shortly and how this would apply to a fingerprinting context is discussed. Since most fingerprinting techniques will initially be developed as a single laboratory method, the discussion is limited to in-house method validation.

For the purpose of this chapter the EC Directive 2002/657/EC guidelines are followed as an open access example of an in-house validation protocol. Single laboratory or in-house validation can also be performed according to alternative methods (Gowik, 1998; Jülicher *et al.*, 1998). For inter-laboratory studies, further reference is made to methods established by ISO or the IUPAC.

In-house method validation in a classical analytical context usually proceeds according to the general scheme shown in Figure 23.2. The first step herein involves defining the goal of the method. In the case of the classical approach, this could be, for example a method for determining the concentration of aflatoxins in a food matrix. In the fingerprinting approach for food quality assurance, the goal could be, for example to discern different brands of virgin olive oil, or a test for the adulteration of milk powder. Since the method is developed specifically for a certain type of commodity, the commodity of interest should be defined in detail, as should the different (sub)populations – or classes – that need to be discerned (e.g. different brands of whiskey, different tomato varieties, organic beef versus conventional beef, etc.). Since the reference samples are collected from a certain area or region, this description should involve the geographical extent of the method (target area) as well.

The second step then involves selecting the appropriate method for validation. This step applies to analytical methods and analytes for which international validation protocols are available. Since this is not (yet) the case for the fingerprinting methods as described here, this step can be ignored for the moment.

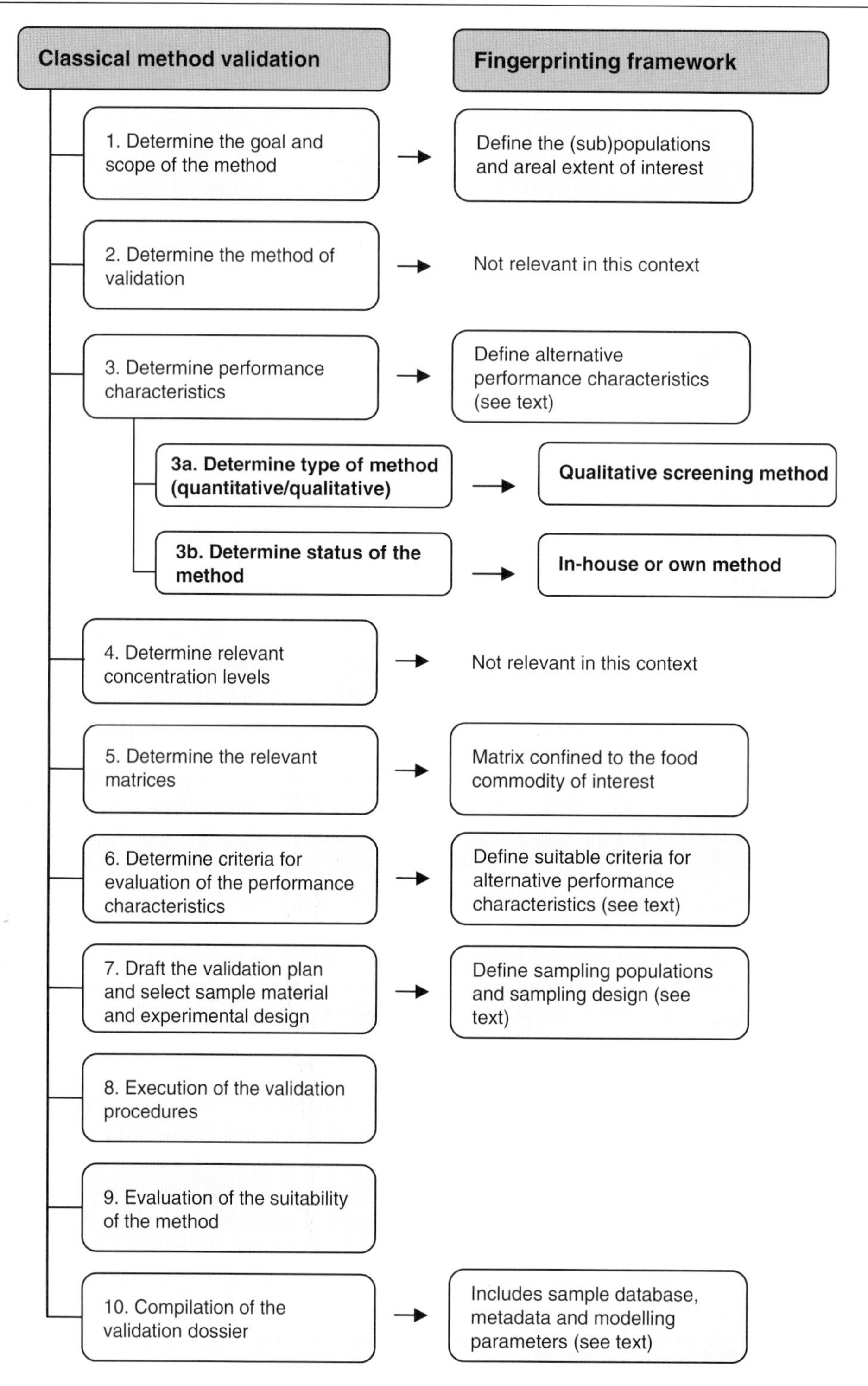


**Figure 23.2** General approach for method validation in a classical analytical context and adaptations required in the context for validation of fingerprinting methods.

**Table 23.2** Overview of the performance characteristics that are mentioned in Council Directive 2002/657/EC.

| Performance characteristic | Definition |
|---|---|
| Repeatability | Precision under repeatability conditions. Repeatability conditions means conditions where independent test results are obtained with the same method on identical test items in the same laboratory by the same operator using the same equipment. |
| Reproducibility | Precision under reproducibility conditions. Reproducibility conditions means conditions where test results are obtained with the same method on identical test items in different laboratories with different operators using different equipment. |
| Within-laboratory reproducibility | Precision obtained in the same laboratory under stipulated (predetermined) conditions (concerning e.g. method, test materials, operators, environment) over justified long time intervals. |
| Decision limit ($CC_{\alpha}$) | The limit at and above which it can be concluded with an error probability of $\alpha$ that a sample is noncompliant. |
| Detection capability/limit ($CC_{\beta}$) | The smallest content of the substance that may be detected, identified and/or quantified in a sample with an error probability of $\beta$. In the case of substances for which no permitted limit has been established, the detection capability is the lowest concentration at which a method is able to detect truly contaminated samples with a statistical certainty of $1-\beta$. In the case of substances with an established permitted limit, this means that the detection capability is the concentration at which the method is able to detect permitted limit concentrations with a statistical certainty of $1-\beta$. |
| Trueness/recovery | Trueness means the closeness of agreement between the average value obtained from a large series of test results and an accepted reference value. Trueness is usually expressed as bias. Recovery means the percentage of the true concentration of a substance recovered during the analytical procedure. It is determined during validation, if no certified reference material is available. |
| Precision | Precision means the closeness of agreement between independent test results obtained under stipulated (predetermined) conditions. The measure of precision usually is expressed in terms of imprecision and computed as standard deviation of the test result. |
| Specificity | The ability of a method to distinguish between the analyte being measured and other substances. This characteristic is predominantly a function of the measuring technique described, but can vary according to class of compound or matrix. |
| Applicability/ruggedness (minor changes) | The susceptibility of an analytical method to changes in experimental conditions which can be expressed as a list of the sample materials, analytes, storage conditions, environmental and/or sample preparation conditions under which the method can be applied as presented or with specified minor modifications. For all experimental conditions which could in practice be subject to fluctuation (e.g. stability of reagents, composition of the sample, pH, temperature) any variations which could affect the analytical result should be indicated. |
| Stability | The susceptibility of an analytical method as a result of sample storage and analyses. |

*Source*: Commission Decision 2002/657/EC, 2002.
Note that the International Union of Pure and Applied Chemistry (IUPAC) prefers using the term selectivity instead of specificity.

In the third step, the relevant performance characteristics need to be selected. An overview of the performance characteristics mentioned in EC Directive 2002/657/EC is provided in Table 23.2. A number of these characteristics are required for validation irrespective of the type of method, which include: repeatability, reproducibility and within-laboratory reproducibility. Depending on the type of method, a number of additional performance characteristics are required. Herein a distinction is made between qualitative methods on the one hand and quantitative methods on the other hand. In this respect, the fingerprinting – classification modelling approach can be described as a qualitative method providing yes/no answers with a certain level of confidence. A further distinction is made between screening methods, i.e. rapid methods to detect suspicious samples at a given level of confidence, and confirmatory methods, which allow confirming the presence or absence of a substance beyond any reasonable doubt.

Since the fingerprinting method is based on a reference data set and hence the quality of the performance strongly depends on the completeness of the reference data set in terms of spatial and temporal coverage. It is questionable therefore whether the method can be applied as a confirmatory method, moreover because the model rarely reaches >99% correct classification. As such, the fingerprinting technique should first of all be regarded as a qualitative screening method. The relevant performance criteria for such methods include the specificity, detection limit, applicability/ruggedness as well as stability. Together with the criteria for acceptance of the performance they will be further discussed in Validation of the Fingerprinting – Classification Approach section.

Under step 3b (Figure 23.2), the status of the method is mentioned. This refers to whether the method can (i) claim conformity with a reference method, or (ii) claim to be identical to a reference method or (iii) whether it concerns an own/in-house method. Although fingerprinting methods are being used as a rapid alternative for various methods, think for example of near-infrared spectroscopy (NIRS) to determine the water content in various food matrices, the focus of this chapter is on development and validation of fingerprinting methods for which no reference or alternative methods exist yet. As such, the status of a new fingerprinting method would be 'own' or 'in-house'.

Step 4 of the validation scheme involves determination of the relevant concentration levels (Figure 23.2). As discussed in the section 'General concept', however, the concentration of an analyte is not directly relevant in a fingerprinting context, especially when the data is of an untargeted or semi-targeted nature. Since the fingerprinting approach as discussed here focuses on discerning between different (sub)populations by classification modelling, the working range of the method – or the extent to which it is valid – is largely determined by the (sub)populations included in the reference database.

In step 5, the relevant matrices for which the method is valid are to be determined. Since the fingerprinting methods as discussed here are specifically developed for a single type of food commodity, the analytical matrix in this approach is largely confined by the food commodity of interest. This in contrast to the classical analytical approach, where the matrix is usually defined more broadly.

After having defined the matrix, the criteria for acceptance of the performance of the method should be looked up or otherwise determined. In the Council Directive both general criteria (i.e. method independent) as well as method-specific criteria and requirements are provided for a number of performance characteristics. Yet for other characteristics, especially for stability, it is more difficult to provide generic criteria, as they depend strongly on the analyte and matrix involved. In these cases it depends on the requirements of the end user of the method to determine whether the performance of the method is sufficient.

Having thought over and decided upon the previous steps, a validation plan should be drafted after which the validation procedures can be started (step 7). The validation plan usually describes, in addition to the topics that were already mentioned under steps 1–6, the choice of sample material as well as the experimental design. In the classical approach to method validation the sample material includes certified reference materials, in-house standards and (fortified) blanks as well as (spiked) field samples. During validation these samples are analysed according to a specified experimental design, which is used to efficiently determine the various performance characteristics by testing for a number of influencing factors at the same time, including different storage conditions, spiking levels, measurement days or operators. In most cases some sort of factorial design is used (see, for example the Youden approach described in EC Directive 2002/657/EC). In this design, the number of repetitions for each factor is also defined.

With respect to both the sample material as well as the experimental design, the differences between the fingerprinting and the classical approach are again obvious. As mentioned previously, the fingerprinting approach relies on an extensive set of reference samples of all (sub)populations or classes that are relevant to the scope and goal of the method. Therefore, instead of preparing the relevant sample material in the laboratory, samples have to be collected during various sampling campaigns. As such, setting up an appropriate sampling design is an important aspect of the experimental design required for the fingerprinting approach. This topic is further worked out in the sections 'Sample design' and 'Validation sampling'.

The last two steps in the general approach to in-house validation include evaluation of the suitability of the method and compilation of validation dossier. The suitability of the method is simply assessed by comparing the values found for the different performance characteristics during the validation procedure with the predefined criteria. The outcome of this comparison, as well as the previously discussed steps, should be laid down in a validation dossier. In the framework of a fingerprinting method the validation dossier should include a sample database containing information about the samples and sample locations as well as the fingerprinting data. Furthermore, to ensure model conformance and interoperability, all details concerning the final classification model should be recorded in a standardized way, for example using predictive model mark-up language (PMML). This will be further discussed in section 'Validation dossier'.

## VALIDATION OF THE FINGERPRINTING – CLASSIFICATION APPROACH

In the previous section the classical analytical approach to method validation was reviewed and its applicability in the framework of chemical/physical fingerprinting and classification modelling explored. From this discussion it is clear that validation of such fingerprinting methods requires a different validation approach that should allow for integral validation of the analytical as well as chemometric aspects of the method. In this section such an alternative validation approach is discussed and tentative guidelines provided for dealing with different aspects such as the experimental design, performance characteristics and criteria for acceptance.

For the remainder of this section it is assumed that our goal is to validate a fingerprinting method that is being newly developed, for example a method to discern organic coffee from conventional coffee by using direct-injection MS as a fingerprinting method. As such, the scope and goal of the method have been roughly defined and the feasibility has been tentatively assessed during a pilot study. Moreover, it is assumed that standard operating procedures (SOP) for the fingerprinting analysis are available and regular quality control is being applied as a part of proper laboratory practice.

### Sampling design

Probably the most important aspect of developing and validating a new fingerprinting method concerns the sampling design and strategy. This not only defines the (geographical) scope of the method but in the end also the quality of the performance statistics as well as the generalizability of the method. Since it is of utmost importance to develop and validate the method on the basis of samples whose characteristics are genuine, samples should be collected directly from the producers (i.e. not from indirect source such as shops or retailers). Setting up a sampling strategy usually proceeds according to the following steps:

#### *Define the target subpopulations as well as the target area*

The subpopulations refer to the different classes that need to be modelled by the classifier. In case of the example of organic versus conventional coffee it should, for example be decided which coffee varieties and which brands are to be included in both subpopulations. Moreover, the geographical extent of the method should be decided upon; should the method be valid for only coffee produced in central Africa, or should it apply to all coffee producing regions on Earth? This would than define the target area as well.

### *Define the sampling design*

Define the sampling design by deciding how the samples are allocated within the target areas, for example according to a simple random, stratified, systematic, cluster or accidental design (Cochran, 1977; Särndal *et al.*, 1992). Since, however, the target population consists of two or more subpopulations, it seems natural to apply stratification at this level. This step involves first of all a delineation of the occurrence of both subpopulations within the target area, either in geographical space or otherwise.

In the case of coffee, for example this would include delineation of all organic coffee production regions as well as all conventional coffee production regions within the target area. This information might not be readily available in a geo-spatial format, or only at a limited resolution, such as, for example the number of organic and conventional coffee plantations within an area or country; hence, it will not always be possible to provide an exact representation of the strata in a geographical sense. Nevertheless, the stratified approach improves the accuracy and efficiency of estimation and it might be optimal to further substratify the two subpopulations, for example according to different geographical regions or provinces, or according to areas with a different geology, soil conditions or climate characteristics.

Since food is often produced in geographically confined areas (farms, plantations, domains, etc.), a clustered sampling approach within each of the strata will be most optimal in terms of cost of sampling and the potential bias introduced by the clustering. In this set-up various producers or production areas (clusters or subunits) are selected at random within each of the strata, from which a predetermined number of samples are to be collected. Obviously, a fair trade-off should be found between the number of clusters and the number of samples within each cluster, which requires knowledge about the number of producers/production areas as well as their respective size/production numbers. An example of such a stratified clustered sampling design is shown in Figure 23.3.

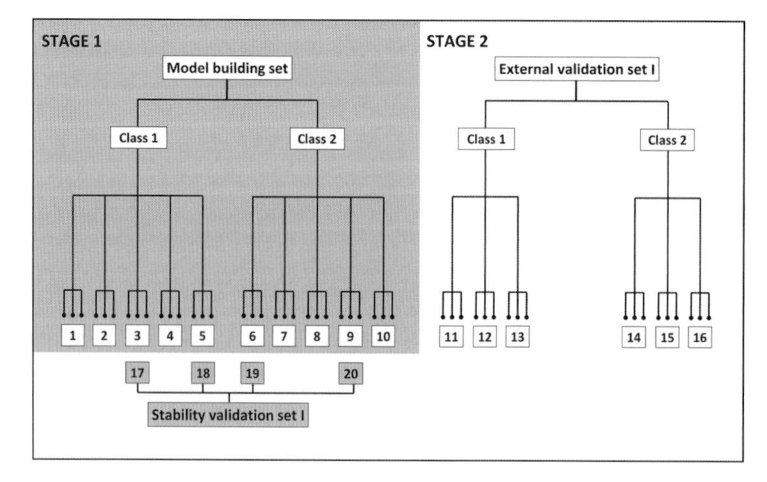


**Figure 23.3** Example of a stratified clustered sampling design showing the design for stage 1 (model building) and stage 2 (validation stage) for validation of a binary classification model. For simplicity, the number of samples is kept much lower than would be required for actual model building and validation.

*Define the sample size*

The sampling design should be further defined by choosing an appropriate sample size, that is the total number of sample locations. Obviously, the total number of samples will greatly determine the final quality of the classification model and one of the major questions is: 'How many samples are required to build a realistic model that has sufficient generalization ability?' Unfortunately, the answer to this question is not trivial and depends on many factors, including the choice of classifier, the variation within the populations of interest and the extent up to which differences between the groups are reflected by the chemical fingerprinting method ('effect size') as well as the stability of these differences over time. Since the approach involves chemometric classification on high dimensional data, moreover, the choice of classifier as well as of the number of variables included in the model are of influence.

When there is no information about the variation in the data available beforehand, the proper amount of samples required for building a model can only be guessed. In the case that preliminary data are available from both populations, for example from a pilot study, power analysis could be used to determine the minimum amount of samples required for model development. Power analysis is based on the effect size, which is usually calculated as the difference between the means of the two population normalized by the standard deviation of one population (assuming equal variance) (Cohen, 1988). Next, the effect size is used to calculate the amount of samples needed from both populations using a $t$-test with a predefined power and $\alpha$ error. An extension of this approach for a multivariate case, which is based on the Mahalanobis distance or related distance metric, has been described in Morse (1999), Olejnika and Alginab (2000) and Guo *et al.* (2010). Alternative suggestions have been provided by, for example Dobbin and Simon (2007).

Although power analysis or alternative sample size estimations can give some guidelines on the amount of samples needed for model building, one should be cautious when the outcome suggests that less than about 100 samples are required for model building. Such sparse models might lead to initial overfitting of the data and hamper proper feature selection. It should be noted in this context that the EC directive already foresees 80 measurements in the experimental design for establishing a number of performance characteristics.

Lastly, it should be noted that multivariate classification methods can be very sensitive to large unbalances in the data (Berrueta *et al.* 2007) and it is, therefore, advisable to have roughly an equal amount of samples in each of the two subpopulations. In terms of sample allocation this means that the number of sample locations in each of the major strata should preferably be equal. This requirement strictly relates to the modelling aspect of the method, since in cases where the occurrence of both populations is very different, this approach would lead to an unbalanced sampling scheme and inherent introduction of further sampling bias. One way around this issue is to use classifiers that are more robust to unbalanced classes. This topic deserves further attention, however.

*Define the sampling procedures*

The sampling procedures refer to all proceedings regarding the collection, storage and transport of the sample material to the laboratory. On this level it should be decided how much sample material is collected from an individual lot or batch, and whether the material should be pooled. Also, temporary storage of sample material should be well thought over, as most food products have a limited shelf life. Again, these issues depend strongly on the food commodity of interest and general guidelines are difficult to provide.

## Validation sampling

The previously mentioned recommendations regarding the sampling strategy and design apply to both the initial sampling stage, which is needed to build and calibrate the model, as well as any of the further validation stages. As will be seen in the next section, the proposed validation scheme involves repeated (re-)sampling of both locations that were previously not included in the data set (external validation sampling) as well as of sample locations that were already including for model building in an earlier stage ('stability validation' sampling). In most cases it will be most efficient to collect both types of validation samples during the same sampling campaign.

In practice, therefore, two sampling stages are required for the first cycle: (i) collection of samples for model building/calibration and (ii) collection of samples for external validation and for validation of stability (Figure 23.3). If the performance characteristics (see next section) meet the pre-specified criteria, then both the external validation set as well as the 'stability validation' set can be added to the initial classification model. Next, the values for the various performance characteristics should be updated using the aggregated data set. In order to keep to model operational and up-to-date, the second sampling stage should be repeated at regular intervals (stage 3, stage 4, . . . stage X) whereby noticing that sample locations previously used for external validation should be added to the set of locations for the 'stability validation' set for the following cycle.

When, however, some or all of the performance characteristics do not meet pre-specified criteria, one could either decide to improve the fingerprinting method (i.e. analytical aspects) or the chemometric aspects (e.g. data pre-processing, feature selection) of the method. If the method then still fails to meet the criteria one should either decide to start the whole procedure from the start by including all data into a newer model that would then again have to be validated according to previously mentioned procedures, or to start an alternative approach to tackle the problem.

As for model building stage, the question arises how many samples are required for each validation cycle? Having collected the set of reference samples, a multivariate approach to a power test (Morse, 1999; Olejnika and Alginab, 2000; Guo *et al.*, 2010) or other approaches could be applied to determine the required sample numbers for external model validation as well as for validation of the stability. From a more practical point of view, it can be argued that both validation sets should consist of at least a 50–60 samples of each class. In this way the performance characteristics that are based on the external and stability validation sets can be determined with sufficient resolution/precision. In this context also see the discussion in Brereton (2006).

## Alternative performance characteristics

Another important aspect of the validation procedure for fingerprinting methods concerns the translation of the classical performance characteristics in Table 23.2 into a fingerprinting–classification modelling context. As will be clear from the previous sections, this will not so much involve a strict mathematical translation but more a conceptual translation and formalisation of existing validation procedures.

Here, alternative performance characteristics that can be assessed by the classification model using the validation sets are discussed. In this case, the analytical error of the fingerprinting method is included in the overall error but not explicitly assessed. Issues relating to the quality of the measurements itself, including, for example calibration of the instrument, blank corrections, replications and inclusion of internal standards, are part of good laboratory practice and should be dealt with in the standard operating procedures (SOP).

As mentioned before, the fingerprinting approach as described here should be regarded as a qualitative method in a sense that it provides yes/no answers. This implies that both the trueness/recovery as well as the precision are not required for validation according to the EC Directive, and that only the repeatability, (within-laboratory) reproducibility, detection and decision limit, specificity, applicability/ruggedness and stability are relevant in this context.

### *Repeatability*

Repeatability refers to the precision under similar analytical conditions (Table 23.2) and is usually determined by repeated analysis of a set of fortified matrices/samples at different concentration levels. The precision is then expressed as a coefficient of variation, which can be compared to a set of predefined criteria at different concentration levels (EC Directive 2002/657/EC). Since the repeatability is determined under similar operating conditions it represents a minimum value for the analytical error of the method. The performance of the classifier during estimation (i.e. using all samples in the model building set to estimate the correct classification rate) can also be regarded as an optimistic estimator of the true error of the model; we suggest to use the correct classification rate determined by estimation ($CCR_E$) as an analogue to the coefficient of variation of the repeatability in the classical approach. Herein, it is assumed that the $CCR_E$ is determined using the optimal model parameter settings as determined during the model building stage.

### *Reproducibility and within laboratory reproducibility*

Reproducibility refers to the precision as determined by measurements in different laboratories and by different instruments, which requires participation in collaborative studies according to, for example ISO 5725-2. Since the focus is on in-house validation with an own method, this topic is beyond the scope of the discussion. The within-laboratory reproducibility is, however, of relevance. The definition of the within-laboratory reproducibility is similar to that of the repeatability, only in that the conditions including operator, temperature, pH and so on are now varied throughout the experiment. As such, the coefficient of variation of the within-laboratory reproducibility can be regarded as a more relevant measure of the analytical error of the method, as it reflects the conditions during application of the method for routine analysis more realistically. Similar to the coefficient of variation of the repeatability, it is therefore suggested to use the correct classification rate as determined by external validation ($CCR_{EV}$) as an analogue to the coefficient of variation of the within-laboratory reproducibility as measure of error.

### *Decision and detection limit*

The definition of both these performance characteristics is provided in Table 23.2. The decision limit ($CC_\alpha$) is the concentration at and below which the sample can be declared compliant (analyte absent or present at a concentration lower than a certain maximum concentration or limit) with a confidence level of $(1 - \alpha)$. Similarly, the detection capability, or detection limit ($CC_\beta$), can be seen as the concentration above which a substance is present above a certain limit with a confidence level of $1 - \beta$. Both measures are to safeguard that future statements about compliance or noncompliance of a measurement are associated with known uncertainties.

In practice it is determined either according to ISO 11843 or by analysing at least 20 blank materials, or fortified blanks in case of a permitted limit of the analyte, per matrix. According to Council Directive 2002/657/EC three times the signal-to-noise ratio can be used as decision limit, or in the case of an analyte with a permitted limit the concentration at the permitted limit plus 1.64 times the corresponding standard deviation ($\alpha = 5\%$). $CC_\beta$ can be determined in a similar way (ISO 11843; Council Directive 2002/657/EC).

In terms of the previously formulated hypothesis about a food quality characteristics (see section 'Model performance'), the measurements that fall above the $CC_\alpha$ can be regarded as 'positives', with an $\alpha$ chance of them being false positives, and those that fall below the $CC_\beta$ would be referred to as the 'negatives', with a $\beta$ chance of them being false negatives. For measurements that fall above the decision limit but have an estimated concentration between $CC_\alpha$ and $CC_\beta$ no statements can be made and, from a statistical point of view, the result remains unclassified (Antignac *et al.*, 2003).

Since the $CC_\alpha$ and $CC_\beta$ refer to concentrations of a single analyte, these concepts cannot be directly translated to a multivariate setting in which, moreover, concentrations are not really relevant. Irrespective of this, information regarding the uncertainty in terms of compliancy statements can be readily retrieved from the model performance characteristics mentioned in an earlier section. Recalling that specificity $= 1 - \alpha$ and sensitivity $= 1 - \beta$ it seems intuitive to 'translate' the $CC_\alpha$ to the specificity and the $CC_\beta$ to the sensitivity of the method as determined by the external validation procedure.

To prevent further confusion with the term specificity as used in a classical validation context, they will be denoted $\text{specificity}_{EV}$ and $\text{sensitivity}_{EV}$ respectively. The $\text{sensitivity}_{EV}$ as well as $\text{specificity}_{EV}$ can be determined from the confusion matrix of the external validation prediction.

### *Specificity*

In the context of the analytical methods described in Council Directive 2002/657/EC, the specificity refers to the ability of the method to discriminate between the analyte and any other compounds (Table 23.2). In practice, the specificity is determined by repeated analysis of blanks and blanks fortified with potentially interfering substances. According to the Directive 'Therefore, potentially interfering substances shall be chosen and relevant blank samples shall be analysed to detect the presence of possible interferences and to estimate the effect of the interferences . . . '

Because the fingerprinting approach does not focus on specific analytes/compounds but on the differences between two or more groups that are based on a set of markers or fingerprint, any marker compound in the fingerprint could potentially interfere with the model results. The most obvious form of such interference can be caused by cross-contamination of samples from the different classes, for example during sample preparation, yet chances of cross-contamination by similar type commodities should also not be ruled out. Since chances of this happening during the whole development and validation procedure of a fingerprinting-based method are considerable, it is for now assumed that this effect is implicitly accounted for by the model and no additional alternatives are required.

### *Applicability/ruggedness (minor/major changes)*

In a classical analytical context, the applicability/ruggedness of a method is related to its susceptibility to variations in the experimental conditions (Table 23.2). To evaluate ruggedness (minor changes) according to Council Directive 2002/657/EC a pre-determined number of spiked samples as well as a blank sample are measured under varying experimental conditions, such as pH, time of preparation, device setting and so on. In case of ruggedness (major changes) the analytical method should be tested under different experimental conditions, which include, for example different species, different matrices or different sampling conditions. Using a factorial design, the importance of these changes can be evaluated using, for example the Youden approach as proposed by Council Directive 2002/657/EC.

In the fingerprinting approach, testing for different experimental conditions is extensively done during external model validation by collecting a different set of samples from different locations, farms and/or producers. Since also the analytical conditions will be different during the measurement of the validation set compared to those during analysis of the calibration set, this effect is implicitly accounted for by these alternative validation procedures. Moreover, the effect of a different sample matrix is not so relevant, since the fingerprinting approach is usually developed for a specific commodity. It can, therefore, be concluded that as such there is no direct need for a further alternative to the ruggedness test in a multivariate fingerprinting setting.

*Stability*

Stability refers to susceptibility of an analytical method as a result of sample storage (Table 23.2) and is usually tested by repeated analysis of a set of incurred samples (if available), or matrices fortified with the analyte. For testing the stability of an analyte in a matrix, the EC Directive suggests a scheme in which a set of sample materials is analysed at $T = 0$ (fresh) and after one, two, four and 20 weeks while stored at least at $-20°C$ or lower if required.

Especially when dealing with fresh food commodities that have a limited shelf-life, determination of the stability is obviously of importance. In case of the fingerprinting approach, however, it is even more relevant to know the stability of the model performance. Since the model is based on an extensive set of reference samples, re-sampling and re-analysis of a subset of these samples would provide a means to determine the temporal stability of the model. This would require, besides extensive sampling to collect an external validation set, additional re-sampling of locations that have already been sampled for the calibration set. This was already mentioned in the section 'Validation sampling'. In an analogous way to the $CCR_{EV}$ determined from the external validation set, the error of stability can then be expressed as the $CCR_{SV}$ using the 'stability validation' set.

*Permutation test*

In addition to the 'translated' performance characteristics discussed above, an essential addition in the context of validation of classification models involves the permutation test. This test evaluates whether the specific classification of the samples in the different classes is significantly better than any other random classification (Good, 2000; Golland *et al.* 2005; Ojaja and Garriga, 2010). Although mentioned as last in this section, the permutation test should preferably be conducted before proceeding with the external validation procedures since the outcome could indicate that the model first needs to be improved.

Ojaja and Garriga (2010) describe two simple procedures, of which the first one (permuted labels) is probably the most relevant in this context. In practice, the test is performed by randomization of the labels, or class membership, of the samples and evaluation of the performance of the model build on these randomized classification scheme by calculation of a p-value according to:

$$p_{\text{perm}} = \frac{1 + \sum_{i}^{N} \Phi(CCR_{\text{perm,i}} > CCR_E)}{N} \tag{23.1}$$

Where $\Phi$ is the indicator function, $CCR_{\text{perm,i}}$ is the CCR of the i-th permutated set and $CCR_E$ is the CCR determined during estimation (see under repeatability) and N is the total number of repetitions. As such the p-value of the permutation test gives the fraction of occurrences in which the classifier performs better for the randomized set than for the original data set. Note that the CCR has been used as a measure to compare the performance of the classifier for each repetition and hence a right-tailed test is used. However, any other measures that evaluate the classifier performance, such as, for example the leave-one-out cross-validation error, are equally valid for this approach.

## Alternative criteria for acceptance

In the previous section a set of alternative performance characteristics were defined for validation of a method based on fingerprinting and classification modelling. In this section the appropriate criteria for acceptance of these characteristics are defined. A number of these characteristics can be determined from the model building set or calibration set ($CCR_E$, p-value of permutation test), whereas others require external validation sampling ($CCR_{EV}$, $CCR_{EV}$, specificity$_{EV}$ and sensitivity$_{EV}$). For practical purposes it is wise to first evaluate those characteristics that do not require external validation sampling, since in the event that these characteristics do not meet the pre-set criteria further external validation is redundant at this stage.

### *Correct classification rate of cross-validation ($CCR_{CV}$)*

As discussed in section 'Model validation', the correct classification rate of cross-validation ($CCR_{CV}$) can be determined by iteratively building a model using a subset of the data (training set), after which prediction power of the model is determined by predicting the class memberships of the samples that were left out (test set). When the number of samples in the test set and the number of iterations are sufficiently large, say 20 or more samples in the test set and 100 or more iterations, a distribution of the correct classification rate of cross-validation can be established.

For the purpose of method validation it is proposed to apply this distribution of errors to define the criteria for the previously defined performance characteristics. Since in bootstrapping the number of iterations N is determined by the modeller, it is proposed to use bootstrapping as the resampling method for cross-validation in this context (Figure 23.4).

Having determined the empirical distribution of individual $CCR_i$ established during each iteration, a number of measures can be established from it such as the median, 25th and 5th percentile. These values, or any other moments of the distribution, can then be taken as an lower limit for the correct classification rate of estimation ($CCR_E$), correct classification rate of the prediction using the stability validation set ($CCR_{SV}$) and the correct classification rate of the prediction using the external validation set ($CCR_{EV}$) respectively (Figure 23.4). As mentioned, other sets of percentiles of the distribution can be used as well, but it is recommended to follow the same order since it is expected $CCR_E < CCR_{SV} < CCR_{EV}$. At last it should be noticed that the actual values of the $CCR_E$, $CCR_{SV}$ and $CCR_{EV}$ should of course match the purpose of the method.

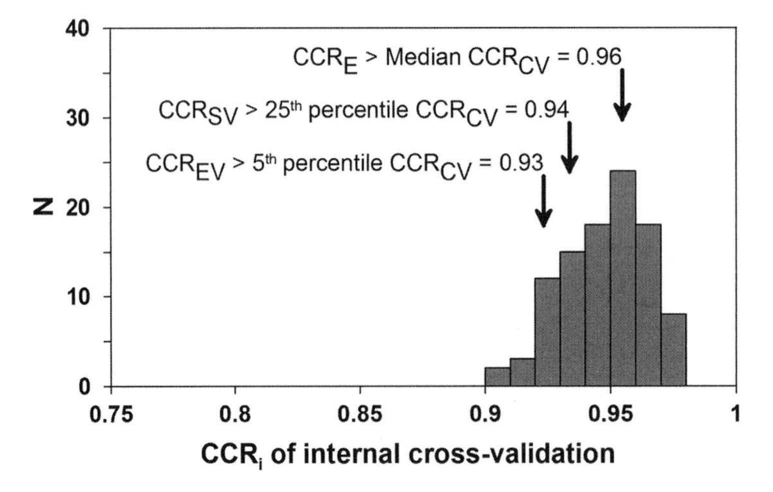


**Figure 23.4** Distribution of the correct classification rate after a 100 bootstrap realizations during internal validation of a PLS-DA model using 25% of the data as a test set. The multivariate artificial data set used for this purpose consists of two rather well separated clusters with 200 samples in each of the clusters and 20 randomly generated variables. The criteria which are to be met by the various performance characteristics, which are established from the distribution of $CCR_i$, are furthermore indicated ($CCR_E$, $CCR_{SV}$, $CCR_{EV}$).

### *Sensitivity$_M$ and specificity$_M$*

The sensitivity$_{EV}$ and specificity$_{EV}$ are calculated from the prediction of the external validation set (section 'Alternative performance characteristics') and provide information about how well the model can predict the individual classes. Obviously, it is up to the developer of the method to decide if the sensitivity$_{EV}$ and specificity$_{EV}$ are sufficiently high for the purpose of the method. However, the Council Directive 2002/657/EC mentions that 'only those analytical techniques, for which it can be demonstrated in a documented traceable manner that they are validated and have a false compliant rate of $<5\%$ ($\beta$-error) at the level of interest shall be used for screening purposes'. In the fingerprinting–classification approach this implies that the sensitivity$_{EV}$ should be larger than 95%, which seems a fair requirement and is adopted here as well. For practical purposes the same threshold could be used for the specificity$_{EV}$, although this again depends on the requirements of the end-user of the method.

### *P-value of permutation test*

As discussed in the previous section, the p-value of the permutation test based on permutating the class labels ($p_{perm}$) measures how likely the observed accuracy would be obtained by chance. Therefore, if the p-value is small enough, for example $\alpha = 0.05$, it can be said that the value of the error in the original data is indeed significantly small and as a result the classifier can be regarded significant under the given null hypothesis (i.e. the null hypothesis can be rejected). As mentioned before, this test should preferably be performed during the earliest stages of model validation since it provides important insight of the fundamental existence of a structure.

## Validation dossier

The validation dossier basically entails a detailed description of all the previously mentioned steps, including sample collection, fingerprinting analysis, data pre-processing, modelling and validation. To ensure model conformance and interoperability, all details concerning the data pretreatment, outlier detection procedures and, possibly, feature selection steps should be recorded in a standardized way. A common standard for complex statistical models is the predictive model mark-up language (PMML) developed by the Data Mining Group (http://www.dmg.org/). The mark-up language provides a way for applications to define models related to predictive analytics and data mining and to share those models between PMML-compliant applications.

Since the fingerprinting–classification approach is based on a large set of reference samples, the validation dossier should include a sample database containing the relevant details about the samples as well as the sample measurements (fingerprints). In addition, a clear description of the data contained in the database should be provided in the metadata. As for the other aspects of the validation procedure, the validation dossier and sample database require a regular update after each stage of (re-)sampling.

# SUMMARY AND CONCLUDING REMARKS

This chapter describes an integral approach to validation of analytical fingerprinting methods focussing specifically on food quality assurance. In the field of food quality assurance, fingerprinting methods are increasingly used either as a rapid alternative to more expensive 'classical' methods, as well as to verify food quality claims that were previously difficult to determine, including nutritional composition, production method and product typicality as well as the geographical origin.

The integral approach to validation of methods that are based on a combination of analytical fingerprinting techniques and chemometric classification modelling, as described here, is based on a

**Table 23.3** Overview of classical performance characteristics for which alternative performance characteristics have been formulated, as well as their respective criteria for acceptance as proposed in this chapter.

| Performance characteristic (classical) | Alternative characteristic | Criteria for acceptance |
|---|---|---|
| Repeatability | $CCR_E$ as determined by prediction of the model building set | $CCR_E > \text{median}(CCR_{CV})$ |
| Within-laboratory reproducibility | $CCR_{EV}$ as determined by external model validation | $CCR_{EV} > \text{5th percentile}(CCR_{CV})$ |
| $\text{Sensitivity}_A$ | None; Not relevant in this context since matrix is confined | — |
| Decision limit ($CC_\alpha$) | $\text{Specificity}_{EV}$ as determined by external model validation | $\text{Specificity}_{EV} > 0.95$ |
| Detection limit/capability ($CC_\beta$) | $\text{Sensitivity}_{EV}$ as determined by external model validation | $\text{Sensitivity}_{EV} > 0.95$ |
| Ruggedness | None; Sufficient variation included in the model building and validation sets | — |
| Stability | $CCR_{SV}$ as determined by re-sampling locations already included in the previous model building set | $CCR_{SV} > \text{25th percentile}(CCR_{CV})$ |
| Additional characteristic | p-value as determined by a permutation test | $p_{perm} < 0.05$ |

conceptual translation of the procedures laid down in existing protocols for 'classical' method validation in existing validation procedures in a multivariate context (Table 23.3). The approach and guidelines provided here are generic in a sense that they would also apply to commodities other than food, ranging from industrial wood and biomass to illegal drugs and medicines. Moreover, the guidelines are exemplary as well as tentative in a sense that many valid – or more advanced – alternatives exist to the proposed sampling strategy, performance characteristics and criteria for acceptance.

At the basis of the validation procedure for the fingerprinting–classification approach lies a repeated (re-)sampling scheme in which food samples are collected from the relevant (sub)populations for continuous external validation and validation of model stability over time. This is required since the fingerprinting approach is intrinsically heuristic in a sense that it depends on a reference data set of fingerprints of field samples, which may or may not change their characteristics over time (e.g. as a result of different climatic conditions or slight changes in the production method). Moreover, by including previous validation sets in the sample set and re-running the classification model on the aggregated data set, the generalizability of the method increases as more and more of the variation within the populations is included in the model.

Whether to include validation samples to construct an updated model depends on whether the criteria for acceptance are met by the relevant performance characteristics as suggested in Table 23.3. If during any of the (re-)sampling stages these criteria are not met, and this is not caused by the presence of outliers, a set of newer validation samples should be collected to see how well they fit the criteria for acceptance. In case the newest validation set also does not meet the criteria it should be decided either to restart the validation procedure from the beginning using the first noncompliant validation set as a starting point for model building and the latest set for further validation, or to adjust the fingerprinting method if possible.

So far it has been assumed that the fingerprinting data sets are of an untargeted nature and the exact (bio)chemical or physical identity of the markers is unknown, for example because multiple compounds can have the same mass or can absorb energy at the same wavelength. In other cases, this information is simply not (yet) available because the compound is unknown. The benefit of this approach is that it saves time and money because data does not need to be interpreted in a compositional way. However, not knowing the identity of the markers can provide a serious limitation for anticipating changes of the fingerprint patterns in the sample population as well as possible disturbances due to contamination or otherwise. As such, it might be desirable to apply the fingerprinting approach in an untargeted mode as an exploratory method to first identify which (bio)chemical or physical markers are responsible for a certain food quality characteristic. Provided that these patterns are sufficiently robust over time, the nature of the discriminating markers should than be revealed using an alternative analytical method.

The need for repeated (re-)sampling and model validation is required until it can be safely assumed that the markers that are responsible for making a distinction between two or more classes are sufficiently robust and their identity is resolved. So far, however, in most studies that describe a specific fingerprinting method for food quality assurance, only one stage of external validation sampling is applied. In other cases only internal validation is applied and the external validation stage is omitted. Although it might be argued that in some case it can be safely assumed that the model is sufficiently robust over time and the identity of the markers is known an understood, most of these studies are merely proofs of concept.

To further validate these fingerprinting methods and to harmonize the methods between laboratories additional validation sampling and model evaluation stages are required. Although several issues have to be further worked out, for example the required sample sizes for validation as well as the choice of classifier in relation to the occurrence of samples in the relevant populations, it is hoped that this chapter will provide a first onset to further intergration of the fingerprinting–classification approach in an validation and accreditation framework.

## REFERENCES


Antignac, J.P., Le Bizec, B., Monteau, F. and Andre, F. (2003). Validation of analytical methods based on mass spectrometric detection according to the '2002/657/EC' European decision: guideline and application. *Analytica Chimica Acta* **483**, 325–334.

Baumann, K. (2003). Cross-validation as the objective function for variable-selection techniques. *Trends in Analytical Chemistry* **22**, 395–406.

Berrueta, L.A., Alonso-Salces, R.M. and Héberger, K., 2007. Supervised pattern recognition in food analysis. *Journal of Chromatography A* **1158**, 196–214.

Bertelli, D., Lolli, M., Papotti, G. *et al.*2010. Detection of honey adulteration by sugar syrups using one-dimensional and two-dimensional high-resolution nuclear magnetic resonance. *Journal of Agricultural and Food Chemistry* **58**(15), 8495–8501.

Bishop, C.M. (2006). *Pattern Recognition and Machine Learning*. Springer Science + Business Media (ISBN 978-0387310732).

Brereton, R.G. (2006). Consequences of sample size, variable selection, and model validation and optimisation, for predicting classification ability from analytical data. *Trends in Analytical Chemistry* **25**(11), 1103–1111.

Charlton, A.J., Farrington, W.H.H. and Brereton, P.B. (2002). Application of $^1$H NMR and multivariate statistics for screening complex mixtures: quality control and authenticity of instant coffee. *Journal of Agricultural and Food Chemistry* **50**(11), 3098–3103.

Cochran, W. G. (1977). *Sampling Techniques*, 3rd edn. John Wiley & Sons, Inc., New York (ISBN 978-0-471-16240-7).

Cohen, J. (1988). *Statistical Power Analysis for the Behavioral Sciences*, 2nd edn. Lawrence Erlbaum Associates, New Jersey.

Commission Decision 2002/657/EC, Commission Decision of 12 August 2002 implementing Council Directive 96/23/EC concerning the performance of analytical methods and the interpretation of results. http://eur-lex.europa.eu/LexUriServ/LexUriServ.do?uri=OJ:L:2002:221:0008:0036:EN:PDF (last accessed 1 July 2013).

DIN EN ISO 9000 (2005-12). Quality management systems – Fundamentals and vocabulary (ISO 9000:2005); Trilingual version EN ISO 9000:2005.

Dobbin, K.K. and Simon, R.M. (2007). Sample size planning for developing classifiers using high-dimensional DNA microarray data. *Biostatistics* **8**(1), 101–117.

Duda, R.O., Hart, P.E. and Stork, D.G. (2001) *Pattern Classification*, 2nd edn. John Wiley & Sons, Inc., New York (ISBN 0-471-05669-3).

Efron, B. (1979). Bootstrap methods: another look at the jackknife. *The Annals of Statistics* **7**(1), 1–26.

Fawcett, G. (2004). ROC Graphs: Notes and Practical Considerations for Researchers. *Pattern Recognition Letters*, **27**(8), 882–891.

Golland, P., Liang, F., Mukherjee, S. and Panchenko, D. (2005). Permutation tests for classification. *Lecture notes in Computer Science* **3559**, 501–515.

Good, P.I. (2000) *Permutation Tests: A Practical Guide to Resampling Methods for Testing Hypotheses*, 2nd edn. Springer (ISBN 038798898X).

Gowik, P., Jülicher, B. and Uhlig, S. (1998) Multi-residue method for non-steroidal anti-inflammatory drugs in plasma using high performance liquid chromatography-photodiode-array detection. Method description and comprehensive in-house validation. *Journal of Chromatography* **716**, 221–232.

Guo, Y., Graber, A., McBurney, R.N. and Balasubramanian, R. (2010) Sample size and statistical power considerations in high-dimensionality data settings: a comparative study of classification algorithms. *BMC Bioinformatics* **11**, 447–466.

ISO 17025 , 1999. General requirement for the competence of calibration and testing laboratories.

ISO 5725-2 Part 2 : 1994. Basic method for the determination of repeatability and reproducibility of a standard measurement method.

ISO 11843 : 1997. Capability of detection – Part 1: Terms and definitions, Part 2: Methodology in the linear calibration case Part 2: Methodology in the linear calibration case.

Jülicher, B., Gowik, P. and Uhlig, S. (1998) Assessment of detection methods in trace analysis by means of a statistically based in-house validation concept. *Analyst* **120**, 173–179.

Massart, D.L., Vandeginste, B.G.M., Buydens, L.M.C. and De Jong, S. (1998) *Handbook of Chemometrics and Qualimetrics: Part A*. Elsevier, Amsterdam, The Netherlands (ISBN 0-444-82853-1).

Møller, J.K.S., Catharino, R.R. and Eberlin, M.N. (2005) Electrospray ionization mass spectrometry fingerprinting of whisky: immediate proof of origin and authenticity. *Analyst* **130**, 890–897.

Morse, D.T. (1999) Minsize2: a computer program for determining effect size and minimum sample size for statistical significance for univariate, multivariate, and nonparametric tests. *Educational and Psychological Measurement* **59**, 518–531.

Ojala, M. and Garriga, G.C. (2010). Permutation tests for studying classifier performance. *Journal of Machine Learning Research* **11**, 1833–1863.

Olejnika, S. and Alginab, J. (2000) Measures of effect size for comparative studies: applications, interpretations, and limitations. *Contemporary Educational Psychology* **25**(3), 241–286.

Särndal, C.-E., Swensson, B. and Wretman, J. (1992). *Model Assisted Survey Sampling*. Springer-Verlag (ISBN 0-387-40620-4).

Tres, A., van der Veer, G., Perez-Marin, M.D. *et al.* (2012) Authentication of organic feed by near infrared spectroscopy combined with chemometrics. *Journal of Agricultural and Food Chemistry* **60**(33), 8129–8133.

Vandeginste, B. G. M., Massart, D. L., Buydens, L. M. C. *et al.* (1998) Supervised pattern recognition. In: *Handbook of Chemometrics and Qualimetrics: Part B*. Elsevier, Amsterdam, The Netherlands (ISBN 0-444-82853-2).

Van Ruth, S.M., Villegas, B., Rozijn, M. *et al.* (2010) Prediction of the identity of fats and oils by their fatty acid, triacylglycerol and volatile compositions using PLSDA. *Food Chemistry* **118**, 948–955.

Webb, A.R. and Copsey, K.D. (2011) *Statistical Pattern Recognition*, 3rd edn. John Wiley & Sons, Inc., Hoboken, NJ (ISBN 978-0-470-68227-2).

# 24 Translating randomly fluctuating QC records into the probabilities of future mishaps

**Micha Peleg[1], Mark D. Normand[1] and Maria G. Corradini[2]**

[1] *Department of Food Science, Chenoweth Laboratories, University of Massachusetts, Amherst, MA, USA*

[2] *Department of Food Science, Rutgers, The State University of New Jersey, New Brunswick, NJ, USA*

## ABSTRACT

A QC chart typically depicts a random time series fluctuating around a characteristic value. This is because an individual entry is determined by the interplay of many factors, most unknown or undocumented, which usually balance each other but not exactly. There is a probability, however, that the effects of several factors will coincide to raise or lower the entry to a level outside the permitted range. If the entries are independent having no trend or periodicity, the probability of such an event can be estimated from their distribution. This can be demonstrated with industrial records, by comparing the estimated frequencies of these events with those observed in fresh data. The estimation procedure has been automated and posted as freeware on the Internet. Free software is also available for demonstrating that if the factors' effects were multiplicative, the entries' distribution would be approximately lognormal.

## INTRODUCTION

Quality control (QC) or quality assurance (QA), together with HACCP, is an integral part of industrial food manufacturing and a standard practice in almost every industry. Much of it is based on compiling the results of physical and chemical measurements and/or microbial counts during processing or in the final product. The record is usually in the form of tabulated or graphical data collected periodically or from successive batches or lots. Law sometimes requires keeping such data, to serve as a record of the plants' compliance with safety standards and other regulations. Almost invariably, the compiled records have the appearance of a randomly fluctuating time series with or without a trend. A trend can indicate improvement or deterioration that might require corrective measures. The entries themselves are usually the mean values of several measurements or counts taken from the same or different samples and reported with or without the corresponding standard deviation. QC or QA charts of chemical and physical attributes commonly have two horizontal lines marking the upper and lower boundaries of the permitted fluctuations range, as shown in Figure 24.1. Obviously, when it comes to microbial counts, especially of harmful, potentially harmful or spoilage organisms, only the upper limit is meaningful. In contrast, when viable probiotic cells or spores are concerned it is the other way around, that is a problem would arise only if the count falls *below* the desired level.

*Mathematical and Statistical Methods in Food Science and Technology*, First Edition.
Edited by Daniel Granato and Gastón Ares.

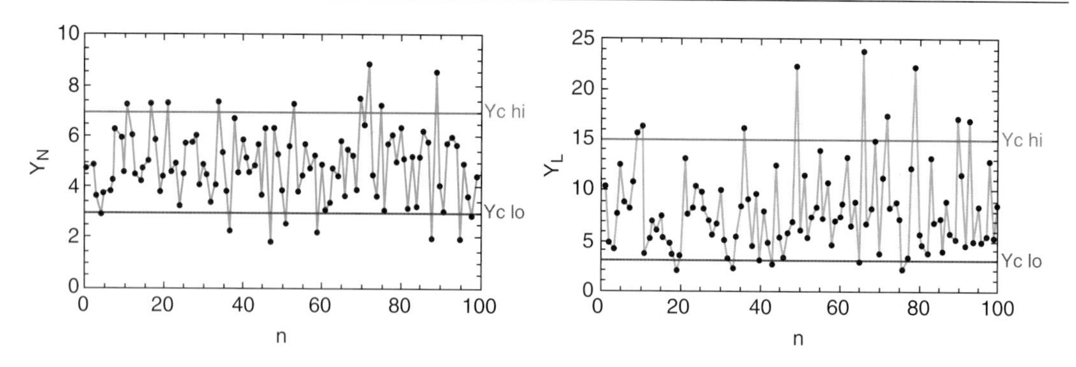


**Figure 24.1** Simulated QC charts with the entries having normal distribution (left) and lognormal distribution (right).

Methods of collecting, charting and interpreting QC data can be found in the general literature on quality control and on the Internet. They include statistical criteria for judging whether the presence of outliers in a QC chart is evidence of a process's instability. These methods are not of concern here; suffice it to say that, by and large, they have been very successful and instrumental in guaranteeing safe food products that meet their established quality, nutritional and durability requirements. Here the focus is only on the not uncommon situations where the QC record indicates a normal or acceptable 'steady state', that is that the system operates or functions routinely and shows no sign of a trend, unusually large oscillation or periodicity. The methods that are presented and discussed were originally developed for predicting the frequencies of microbial outbursts (Horowitz *et al.*, 1999; Nussinovitch and Peleg, 2000; Peleg and Horowitz, 2000; Peleg *et al.*, 2000). But the calculation methods and the concepts on which they are based equally apply to QC records of chemical and physical measurements (Gonzalez-Martinez *et al.*, 2003).

Apart from the entries' units, absolute magnitude and fluctuation range, the main difference, as will be explained, is in the symmetry and other properties of the entries' distribution. The typical scenario that will be addressed is of a system, where the QC record fluctuates within the acceptable limit or limits, and then suddenly, without any warning sign, the recorded microbial count or a component's concentration by far exceeds or falls below the acceptable or allowed level (Figure 24.1). In most cases, such an odd occurrence can be traced to an accident, equipment failure or human error for example, and is dealt with by *ad hoc* measures. Sometimes, however, excessively high counts cannot be traced to any specific cause and their appearance in the record remains puzzling. The same can be observed in physical or chemical properties. Although the effect on these is rarely as dramatic as in microbial counts, it can still result in an unacceptable product that would have to be withdrawn, reprocessed or destroyed.

Ideally, to eliminate mishaps of either kind a model that could predict the exact time of their occurrence will have to be developed. But as will be shown, the development of such a model any time soon seems very unlikely, except of course where there is obvious trend. But what can be estimated on the basis of a stationary record is the *probability* that a mishap of a given magnitude will happen in the future. This probability, in turn, can be directly translated into the future frequency of a mishap occurrence, even if it has not yet happened. In this chapter, the underlying concept and calculation method are described with emphasis on the mathematical principles and procedures rather than on any specific biological, physical or chemical causes that are responsible for the fluctuations, their magnitude and frequency.

## WHERE DO THE FLUCTUATIONS COME FROM?

A recorded microbial count, for example or a food's moisture content, say, is determined by many factors, some totally unknown or undocumented. In this chapter, only their manifestation will be dealt with, with a focus on the mathematical/statistical principles and procedures, rather than on the biological, physical or chemical factors themselves. It can be safely assumed that some of the factors tend to increase the microbial count or other physical or chemical property, while others tend to decrease it. This is true for both fresh and processed foods. Usually, the random effects of the two groups of factors cancel each other but not exactly, hence the fluctuations around the mean.

This can be demonstrated in the following way. Suppose there are $n$ factors $f_1, f_2, f_3, \ldots, f_n$, all known or could be guessed, that affect a microbial count $N$, given that it has started with a value $M$. If so then:

$$N = M \cdot f_1 \cdot f_2 \cdot f_3 \cdot f_4 \cdot f_5 \cdot \ldots \cdot f_k \tag{24.1}$$

Notice that if a factor $f_i$ belongs to the first group (i.e. causes an increase) it will have a value bigger than one ($f_i > 1$) and if it belongs to the second group (i.e. causes a decrease) it will have a value smaller than one ($f_i < 1$).

Of course, all the factors $f_i$'s, let alone their value, are rarely if ever known exactly. But an expert would be able not only to identify the main factors but also to come up with reasonable estimates of their effects' magnitudes. Thus, an expert using Equation 24.1 as a formula could estimate the number $N$, with reasonable accuracy. The reason why this estimation method frequently works is that if all the estimates are reasonable, it is unlikely that all or most of them will be much too high or too low. More likely, the errors in one direction will cancel out the errors in the other direction, albeit not exactly, but sufficiently to render an acceptable estimate of $N$. This is an example of the application of what has been dubbed the 'Fermi Solution' (von Baeyer, 1993). As already stated in previous works (Peleg *et al.*, 2007), 'reasonable estimate' is not a scientific term. But it is not the same as a wild guess, especially when given by an expert, a principle that underlies several risk assessment methods.

The Expanded Fermi Solution (Peleg *et al.*, 2007) is based on the notion that the initial number and factors' effects should be entered as the lower and upper limits of their *probable ranges*, that is as $M_{min}$ and $M_{max}$, $f_{1min}$ and $f_{1max}$, $f_{2min}$ and $f_{2max}$, and so on. When the factors and their boundaries are identified, a series of several hundreds of Monte Carlo runs is performed with different combinations of $M$ and the factors values, each chosen randomly within its respective range. In each such run $j$, $N_j$ is calculated with Equation 24.1 as a model and its value logged. (The corresponding $M$'s and the factors' random values are all assumed to have uniform distribution within their respective ranges, which reflects 'maximum ignorance'.) The Expanded Fermi Solution was originally developed for risk assessment and can serve as a model to generate series of $N$ values reminiscent of real industrial microbial count records (Peleg *et al.*, 2012), as shown in Figure 24.2. Not unexpectedly, the distribution of the $N$'s so generated is approximately lognormal (Peleg *et al.*, 2007). The approximate lognormality is a manifestation of the *central limit theorem* (CLT), which states that under certain conditions the means of equal size samples of random numbers have approximately normal distribution. In our case, however, the simulated $N$'s are calculated with Equation 24.1 and therefore for each run $j$

$$\log N_j = \log M_j + \log f_{1j} + \log f_{2j} + \log f_{3j} + \ldots + \log f_{nj} \tag{24.2}$$

where $j$ is the simulation's (iteration's) index ($j = 1, 2, 3, \ldots$). Each $\log N_j$ is the sum of $n$ random numbers, because both $M_j$ and the corresponding $f_{ij}$'s are all chosen randomly, which makes their logarithms random numbers too. Thus, according to the CLT the distribution of the $\log N_j$'s generated by the Monte Carlo simulations ought to be approximately normal. And, if the $\log N_j$'s have approximately normal distribution, that of the $N_j$'s themselves should be approximately lognormal as shown in the figure. An interested reader can generate his or her own simulated records with the Wolfram Demonstration 'Simulating Microbial Count Records with An Expanded Fermi Solution Model' (http://demonstrations.wolfram.com/SimulatingMicrobialCountRecordsWithAnExpandedFermiSolutionMo/). The Wolfram CDF

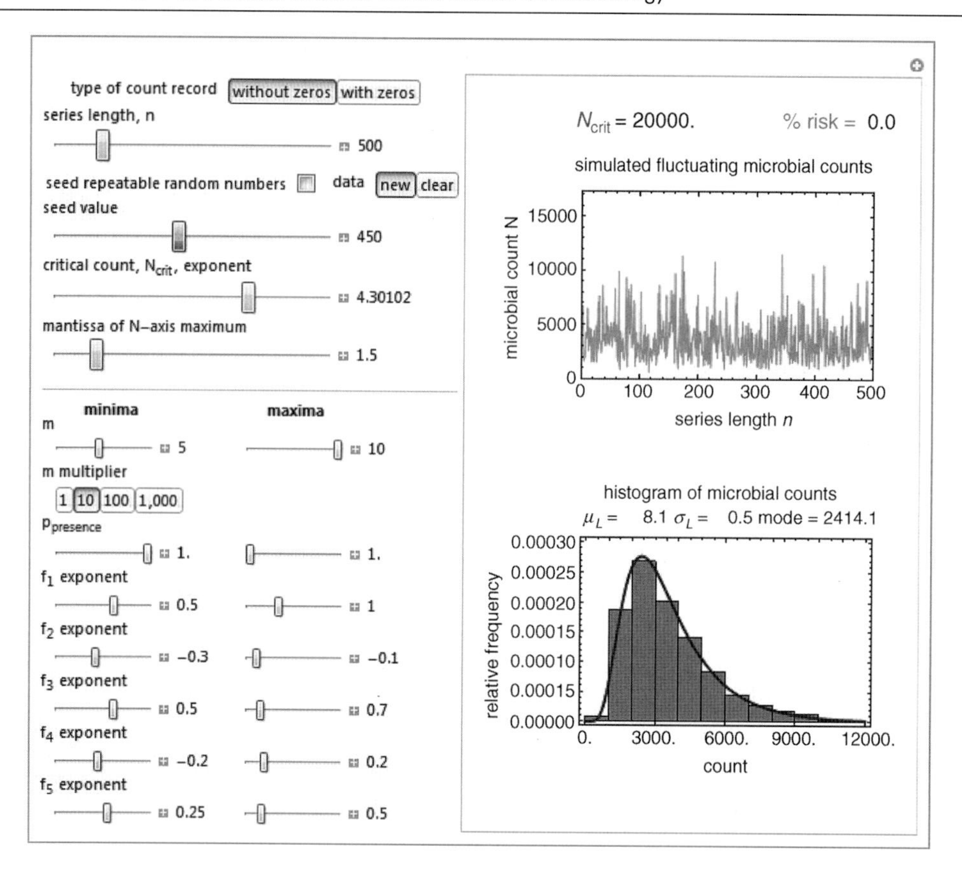


**Figure 24.2** The screen display of the Wolfram Demonstration that simulates random QC charts without zero entries using the Expanded Fermi Solution as a model. Notice that all the parameters can be entered either numerically or by moving a slider.

Player that runs the Demonstration (and over 9000 other Demonstrations to date) is a freely downloadable program courtesy of Wolfram Research.

The emergence of lognormally distributed entries in simulated and real microbial count records is consistent with that microbial counts can reach very large numbers but never become negative. The lognormal distribution is a flexible model that can describe both highly skewed and almost symmetric distributions. But although it is consistent with both simulated and actual microbial count records (Peleg *et al.*, 2000; Nussinovitch *et al.*, 2000) and lends support to the suggested explanation of the fluctuations origin, the lognormal model neither proves that the factors' effects must be multiplicative nor is a reason to exclude alternative explanations or different distribution functions (Corradini *et al.*, 2001b). Also, the lognormal model is totally inappropriate for microbial records containing a substantial number of zero entries, a point that is returned to later, or records having negative entries, such as those charting deviations from an expected value.

An alternative to the above explanation is that each factor only adds or subtracts a certain amount from the measured and reported quantity. Since subtraction is the same as addition of a negative value the combined effect can be called 'additive'. In this case the quantity or number $N$ will be:

$$N = M + f_1 + f_2 + f_3 + f_4 + f_5 + \ldots + f_k \tag{24.3}$$

where any factor $f_i$ can be either positive or negative number, that is adding or subtracting from the value of $M$, respectively.

Here, too, the Expanded Fermi Solution can be used, assigning $M$ and each $f_i$ within the boundaries of its probable range. In this case, however, the CLT applies to the $f_i$'s themselves and not their logarithms. In contrast with the lognormal model, the normal distribution model is suitable for records having negative entries. However, the normal distribution model's applicability is limited to entries that are symmetric or approximately symmetric. This might not be the case with microbial counts in raw foods where asymmetry is idiosyncratic, stemming from the fact that microbial populations can reach a very large size but can never become negative.

In reality, the effects of some factors can be additive while that of others multiplicative or interrelated in several different ways (Peleg *et al.*, 2012). For many industrial records, especially if short or imperfect, it would be expected that the lognormal or normal distribution model would be adequate. If not, other distribution functions could be tried and compared (Corradini *et al.*, 2001a, 2001b, and below).

## Extracting the probabilities of mishaps of different magnitudes

### *The basic assumptions*

Under normal conditions, as already stated, it is expected that the effects of factors that tend to raise the entry value will be cancelled by the effects of the factors that tend to lower it, and vice versa, albeit not exactly. But if some or all the effects vary randomly, there is a probability that several or many of the factors that affect the entry in one direction will coincide, in which case the result will be an entry of unusually high or unusually low magnitude, perhaps exceeding the permitted level, that is resulting in a mishap. This should apply to both microbial count records and physical or chemical attributes composition, except that the magnitude of the deviation would be expected to be much higher in the former than in the latter. The probability of such a coincidence can be estimated from the entries' *distribution's mean and standard deviation* provided that (Peleg and Horowitz, 2000):

(i) The entries are independent and the record is stationary having no trend or periodicity. (Stationary here means that the record will have the same mathematical properties if started at a different entry.)
(ii) The probability of a deviation from the characteristic mean value diminishes with the deviation's magnitude.
(iii) The conditions of the monitored system remain practically unchanged (i.e. equipment and raw materials are not modified or replaced, formulation and/or procedure are the same, etc.)

### *Tests for independence*

Randomness is an intuitively clear concept but its establishment in a given set of data is not a simple matter (because of the possibility that there is a hidden order not yet discovered, for example). For our practical purpose, it would be sufficient to treat a time series as being composed of independent entries, that is 'random', if it passes a statistical test for randomness. The simplest test for periodicities, albeit qualitative, is the 2D Packard-Takens plot. For a given series $a_1, a_2, \ldots . a_i, \ldots . a_{n-1}, a_n$ (see examples in Figure 24.3), it is the plot of $a_{i+lag}$ vs. $a_i$ where the 'lag' can be 1 or 2 or 5, for example. (The 3D version is produced in the same manner except that the three axes are $a_i$, $a_{i+lagj}$, $a_{i+lagk}$.) Where the series is monotonic and smooth, so too will be its Takens-Packard plot. If the series has periodicities, the plot will be an elaborate graphical object, frequently very aesthetically appealing (Figure 24.3). [The interested reader can generate periodic time series and their Packard-Takens 2D and 3D plots with the

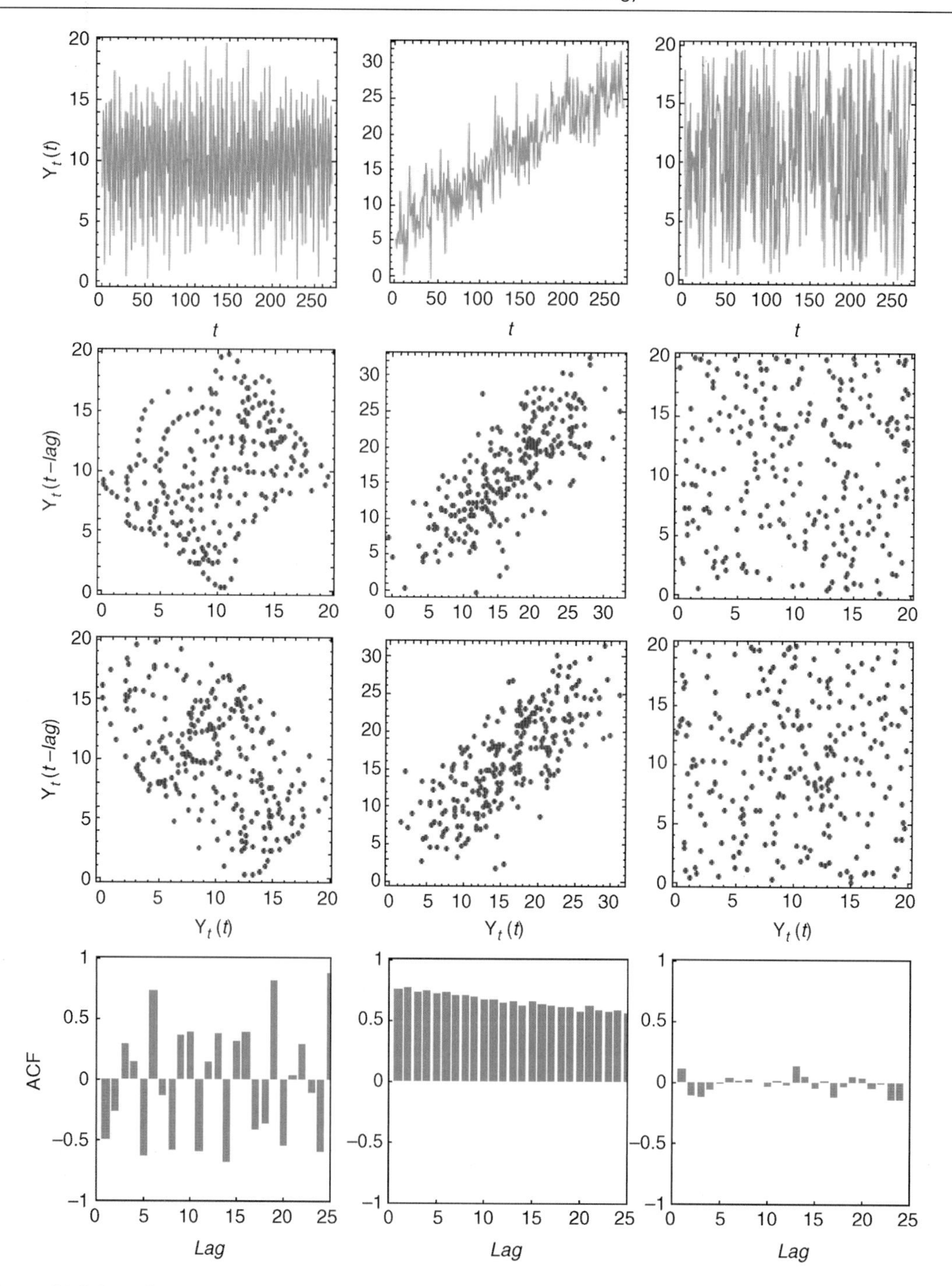


**Figure 24.3** Periodic, monotonic and random time series (left, middle and right, respectively), their Packard-Takens plots for lag 1 and 3, and corresponding autocorrelations functions (ACFs).

freely downloadable Wolfram Demonstration (http://demonstrations.wolfram.com/2DAnd3DPackard-TakensAutocorrelationPlotsOfSinusoidalFunction/).] And, as also shown in Figure 24.3 if the series is random so will be the appearance of the Packard-Takens plot.

A most convenient quantitative measure of periodicity is the *autocorrelation function* (ACF), the calculation of which is a standard option in commercial statistical software. It can also be easily programmed for use in general purpose mathematical software. To generate the ACF of a series $a_1$, $a_2$, . . . . $a_i$, . . . . $a_{n-1}$, $a_n$, plots of $a_{i+1}$ versus $a_i$ (lag 1), $a_{i+2}$ versus $a_i$ (lag 2), $a_{i+3}$ versus $a_i$ (lag 3) and so on are created and their linear regression coefficient, $r$, calculated. The ACF is the $r$ versus lag relationship, which is usually presented as a plot as shown in Figure 24.3. A large (significant) positive $r$ means positive correlation while a large (significant) negative $r$ negative correlation for the particular lag. As demonstrated in the figure, a highly correlated time series shows a pattern and many high significant $r$-values, while a series of independent entries (randomly generated) a featureless plot and small insignificant $r$ values for all lags. [A different possible pattern that can emerge in a QC record will show the prominence of a particular lag and its multiples, for example 7, 14, 21, etc.; an example is shown in Figure 24.4. In the shown hypothetical case, had the entries been microbial counts plotted against the production date, the ACF would raise the suspicion of a potentially weekly sanitary problem, probably associated with weekends. Identifying such periodicity is a good reason in itself to create and examine the ACF of QC data.] The absolute magnitude of an $r$ is a measure of how strongly entries separated by a particular distance are correlated and whether this correlation is significant by statistical

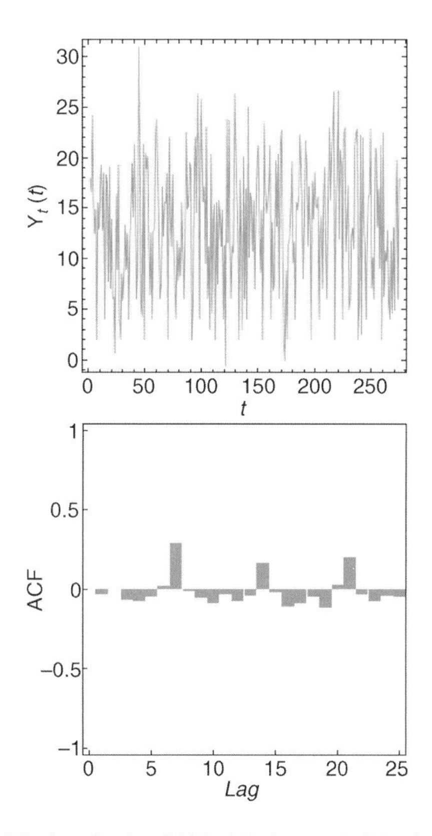


**Figure 24.4** Simulated 'random' QC chart having 'hidden' 7-days periodicity (top), and its autocorrelation function (ACF).

criteria. However, it has been shown in the analysis of industrial microbial count records that the existence of weakly correlated entries, such as those of the product made in successive days (as compared with samples produced weeks apart), even if statistically significant, has only a minor effect on the predicted probabilities of an 'outburst' (Peleg *et al.*, 2000; Corradini *et al.*, 2003; Peleg, 2006).

### Parametric versus nonparametric distributions

In principle, the distribution of entries in a QC or QA record can be quantified by a *parametric* or *nonparametric distribution*. A nonparametric distribution presents the actual frequencies of the entries' magnitudes and can be presented in a tabulated form or graphically as a histogram. It involves no assumptions, except perhaps that an observed frequency of an entry having or exceeding a given magnitude can be converted into the probability that it will be observed in the future. The nonparametric approach would be most useful for cases where the record, even if sufficiently long, does not suggest any practical mathematical model. Obviously, the nonparametric distribution is the easiest to determine and it can be used for records of any kind, including having zero and/or negative entries. But it cannot be used to estimate the probability of encountering any value exceeding the highest or falling bellow lowest value already observed.

A parametric distribution is one of several distribution functions such as the already mentioned normal or lognormal distribution function. The use of a parametric model requires the identification of the proper distribution function and determination of its parameters from the record at hand. The great advantage of a parametric model, apart from being concise, is that at least in principle it can be used to estimate the probabilities of encountering entries having a magnitude not yet observed.

### Parametric distribution functions

The most likely candidates, and convenient, distribution functions for describing a symmetric fluctuation pattern in a QC record are the normal distribution function

$$f(X) = \frac{1}{\sigma\sqrt{2\pi}} \exp\left[-\frac{(X-\mu)^2}{2\sigma^2}\right] \tag{24.4}$$

where $X$ is the entry's value and $\mu$ and $\sigma$ are the mean and standard deviation, and for an asymmetric pattern, the lognormal distribution function

$$f(X) = \frac{1}{\sigma_L\sqrt{2\pi}} \exp\left[-\frac{(\log X-\mu_L)^2}{2\sigma_L^2}\right] \tag{24.5}$$

where $\mu_L$ and $\sigma_L$ are the logarithmic mean and standard deviation.

Nevertheless, a particular pattern can sometimes be also better described by other parametric distribution functions, such as the Laplace (symmetric), logLaplace, Weibull, extreme value or beta (Corradini *et al.*, 2001b).

For a chosen distribution function, the probability, $P$, of *exceeding* a value, $X_c$, can be calculated from its cumulative version, that is

$$P(X \geq X_c) = \int_{X_c}^{\infty} f(X)dX \tag{24.6}$$

or

$$P(X \geq X_c) = \int_{\log X_c}^{\infty} f(\log X) d \log X \tag{24.7}$$

Similarly, the probability of encountering a value *smaller* than a given $X_c$

$$P(\mathrm{X} \leq X_c) = \int_{0}^{X_c} f(X) dX \tag{24.8}$$

or

$$P(\mathrm{X} \leq X_c) = \int_{0}^{\log X_c} f(\log X) d \log X \tag{24.9}$$

Notice that once the distribution's parameters have been determined the calculation of $P(X \geq X_c)$ or $P(\mathrm{X} \leq X_c)$ can be done using standard functions of statistical and mathematical software.

The estimated number of values *exceeding* the chosen $X_c$ in a future record of length $m$ is therefore:

$$\text{Estimated Number} = m * P(X \geq X_c) \tag{24.10}$$

and of *smaller* than $X_c$

$$\text{Estimated Number} = m * P(X \leq X_c) \tag{24.11}$$

The abscissa of a QC or QA chart is frequently a lot number, entered in a sequential order, or time. In the first case, an estimated number calculated using Equations 24.10 or 24.11 refers to encountering entries $X \geq X_c$, or $\mathrm{X} \leq X_c$, in $m$ successive lots regardless of time. In the second case, if the values are entered at constant time intervals, $m$ can represent time, for example a week, a month, a year and so on. Either way, depending on the safety or economic implications of $X \geq X_c$ or $\mathrm{X} \leq X_c$, the estimated number would constitute crucial information on the decision on whether to take remedial action and of what kind (Peleg, 2006). The probabilities or frequencies calculated using the above model could be also used to assess the efficacy of corrective measures or the impact of changes, such as equipment replacement or formulation modification. Expressing the effect not in terms of the mean's rise or fall but as the increase or decrease of the probability or expected frequency of future mishaps seems to be a more efficient way to translate QC data into risk (Peleg, 2006, 2009). As has been shown by Nussinovitch *et al.* (2000), ranking milk sources by the mean microbial count is not the same as ranking them on the basis of the probability that the count will surpass the allowed level.

Estimated numbers calculated with Equations 24.10 and 24.11 can also be used to test and validate the model with existing QC data. When the available record is long enough, it can be divided into two (or more) sections. Once done, and the entries' distribution in one of the halves has been determined, it can be used to estimate the expected frequencies of high and low entries of different magnitudes. These can be compared with the actual frequencies in the other half of the record. Satisfactory agreement between the estimated and observed frequencies will serve as evidence that the model works. Failure to estimate the frequencies correctly will be a good reason to reject the model and search for an alternative, be it another parametric distribution function or if none found resorting to the nonparametric option.

### Choosing a distribution function

The shape of the entries' or their logarithms' histogram usually suggests the *type of distribution function* to consider: symmetric or skewed, concave (bell shaped) such as normal or lognormal, or convex (Laplace), continuous or truncated, unimodal or bimodal and so on. Once a candidate distribution function has been considered, its compatibility with the data at hand can be tested. This can be done by producing a Q-Q plot and examining its linearity. A familiar example is testing the linearity of a plot on a probability paper to establish a distribution's normality. The principle applies to other distribution functions. [In Mathematica® (Wolfram Research, Champaign, IL) and probably other advanced mathematical and statistical software, the Q-Q plot can be generated using standard functions of the program.] When examining actual QC records of foods the plot is rarely perfectly linear. Also, the test itself might to some extent be subjective or inconclusive. Still, the method can be used effectively to screen contemplated distribution functions and to discard those that are clearly inferior or inappropriate (Corradini *et al.*, 2001a, 2001b, 2002; Hadas *et al.*, 2004).

### The distribution parameters' determination

Once a distribution function has been chosen, it parameters can be estimated in three ways: by nonlinear regression, the method of moments (MM) and maximum likelihood estimation (MLE). Since this distribution function is to be used to estimate the probability of rare events, its important or relevant part is not the centre but the tail or tails. Most standard regression procedures are based on minimizing the mean square error. Therefore, compensation for deviations in the centre could have a dramatic effect on the tail(s) where the values are very small in comparison. For this reason, estimating the distribution coefficients by regression is not a useful option. In the method of moments (MM) it is assumed that the mean, $\mu$, and standard deviation, $\sigma$, of the underlying distribution are the same as those of the sample, in our case the examined record, that is that $\mu = \bar{x}$ and $\sigma = s$, respectively. In the maximum likelihood estimation (MLE), the sought parameters are those that would have produced the sample (the observed record) with the highest probability. Therefore, the MLE method is generally considered more rigorous or reliable. In the case of the normal and lognormal distribution, the MM and MLE are the same, that is the parameters calculated by the simpler method of moments are also those that would have produced the record with the highest probability. The situation is different in other distribution functions, such as the Laplace or logLaplace, for example. Unlike the parameters estimation with the MM, which is straightforward, their calculation by the MLE method requires trial and error starting with guessed values, which can frequently be time consuming with no guarantee of a result. However, using the MM calculated parameters as the initial guesses can shorten the process considerably and will almost always guarantee a result (Corradini *et al.*, 2001b).

Comparison of frequencies of high microbial counts in foods and wash water records using parameters calculated by the MM and MLE method has shown that MLE had not always rendered consistently better predictions and, even when they were closer to the actual values, the difference was not large. In light of the imperfections of most food records it seems that the application of the MLE method might not be necessary (Peleg, 2006). As already stated, the issue does not arise at all, of course, when the normal or lognormal distribution is used as the fluctuations model.

### Truncated distributions

Records of pathogens, bacteria of faecal origin and other special microorganisms in food and water can have a substantial number of zero entries. The same is probably true for the concentration of chemical contaminants or counts of foreign objects, rodents' hair and insect parts (e.g. in grains or flour) and so on.

The reason can be total absence or failure to detect them, both subject to the same stochastic considerations that determine the entries' magnitudes. A program to simulate such records, therefore, has to include the presence probability as one of the control parameters. The screen display of the previously described Wolfram Demonstration set for producing zero entries is shown in Figure 24.5.

In all cases, where $f(X<0)=0$ and $f(X=0)>0$, the common parametric distribution functions, such as the normal, lognormal, Weibull or Laplace in their standard form, cannot serve as a model. However, parametric distributions having the range $-\infty < X < \infty$, such as the normal or Laplace distribution function, can still be used albeit in a truncated form, which for our purpose can be written as:

$$f_{\text{truncated}}(X) = \frac{f(X)}{\int_0^{\infty} f(X)dX} \tag{24.12}$$

where $f(X)$ is the distribution function in its probability density function (PDF) form.

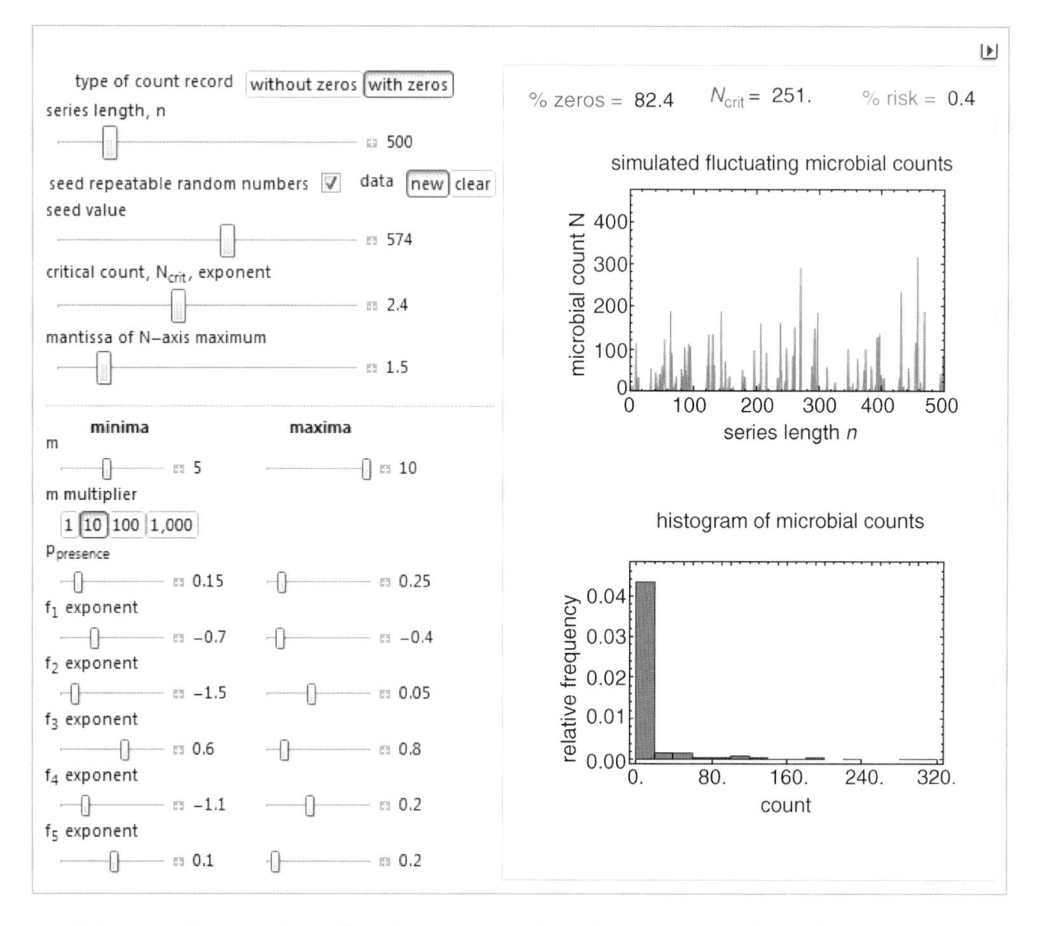


**Figure 24.5** Screen display of the Wolfram Demonstration simulating random QC charts with a substantial number of zero entries using the Expanded Fermi Solution as a model. Notice that all the parameters can be entered either numerically or by moving a slider.

The probability of encountering a value exceeding a given value $X_c$ would therefore be:

$$P(X \geq X_c) = \frac{\int_{X_c}^{\infty} f(X)dX}{\int_{0}^{\infty} f(X)dX} \tag{24.13}$$

Notice that $X$ in the two equations can represent either the entries themselves or be any of their suitable transform, such as the entry raised to a given power (Hadas *et al.*, 2004; Peleg, 2006). Testing the data's Q-Q plot will reveal whether a transform is needed and help in choosing its kind. Calculation of $P(X \geq X_c)$ with a truncated parametric distribution function $f_{\text{truncated}}(X)$ can now be done with existing commands of the latest version of Mathematica®, and perhaps other advanced programs. It can also be done with a program especially written using general-purpose mathematical software. An example of simulated QC record having a substantial number of zeroes and corresponding histogram is given in Figure 24.6.

### *Complications*

All the above refers to records that pass the test for independence but this is not always the case. We are not concerned here with cases where the ACF shows an occasional small but statistically significant $r$-value or several such values, typically for lags 1 and 2. It seems that the method is sufficiently robust so

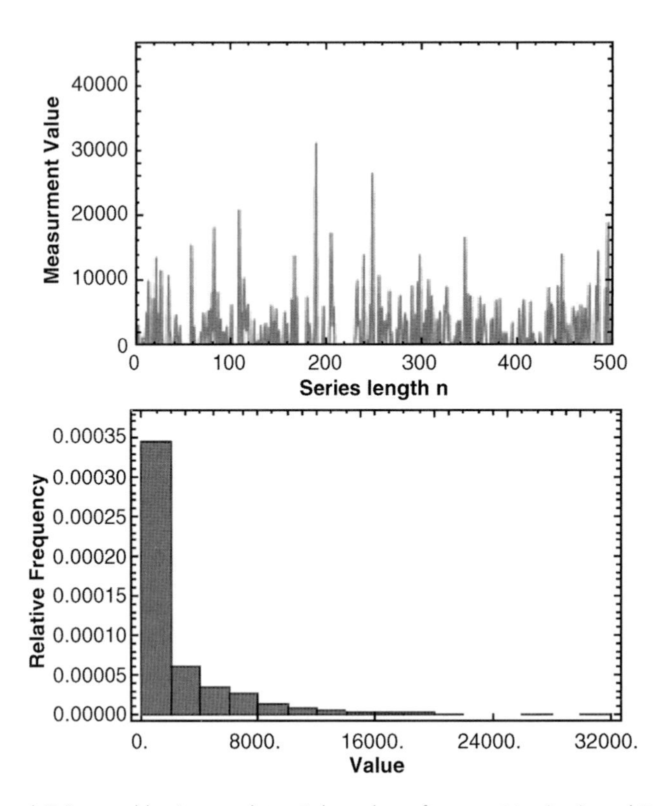


**Figure 24.6** A simulated QC record having a substantial number of zero entries (top), and its histogram (bottom).

that such an occurrence has little or no practical impact on the predicted frequencies. What is of concern is systematic violation of the independence assumption that can render the method totally inoperable. One example is the result of inconsistent rounding, which can be encountered in microbial count records. It happens when different people round the count to the nearest 10 or 100, say. Thus, counts that would be otherwise spread over a range may appear as having the same value, so distorting their true distribution. This can be overcome by *de-rounding* the record, that is by replacing the rounded entries by random numbers drawn within the pertinent range (Corradini *et al.*, 2001a; Peleg, 2006). The record will also be distorted if some entries are rounded while others are not. Something similar happens when small counts are reported as $<10$, for example. Thus, if assigned either zero or ten, the result will be an artificial abundance of zeros or tens, again distorting the actual distribution. This artifact can be corrected by replacing the $<10$ entries by a random number between 0 and 10 (Corradini *et al.*, 2001a; Peleg, 2006).

Certain records may show a clear trend or the existence of inherent periodicity, which will be clearly reflected in the autocorrelation function, as demonstrated in Figure 24.7. In such a case, the concept can still be applied and the described procedures used by switching the search from finding the probability of encountering entries exceeding a given absolute value to finding the probability of exceeding a given *ratio* (relative to the previous value). All that is needed is to convert the original record into a new time series whose members are in the form $Y_{n+1}/Y_n$. As shown in the figure, this transformation may allow the record to pass the test of independence as judged by the autocorrelation function (Corradini *et al.*, 2002).

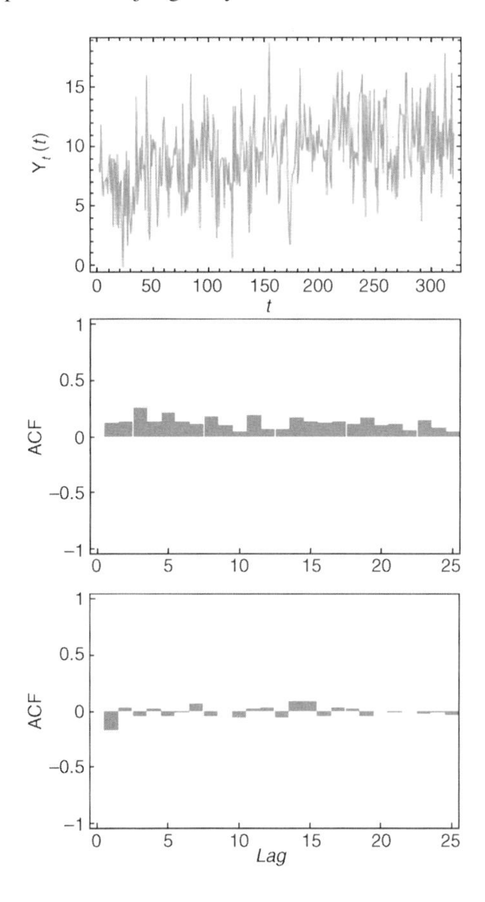


**Figure 24.7** A simulated QC record having a clear trend (top) and its autocorrelation functions (ACFs) before and after the entries being converted into the ratios $Y_{n+1}/Y_n$ (middle and bottom, respectively).

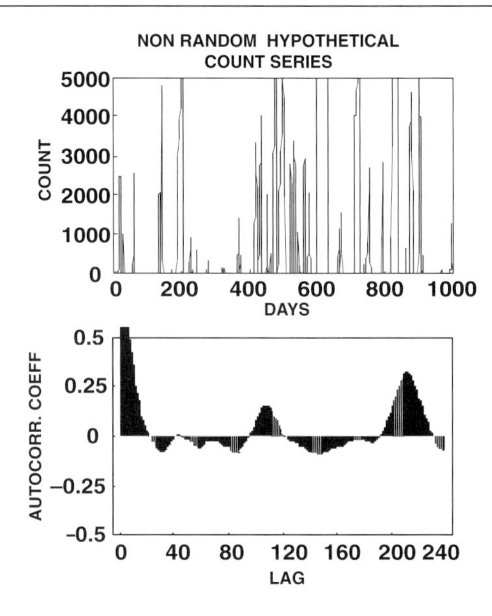


**Figure 24.8** Simulated QC record having peaks of random height and duration (top) and its corresponding autocorrelation function, ACF (bottom). Engel *et al.* (2001). Reproduced with permission from John Wiley & Sons.

An altogether different kind of complication arises when the record contains evidence of lingering episodes, such as peaked microbial outbursts spread over several entries or time, as shown in Figure 24.8. In principle, the same can happen in other kinds of contamination if the source is not completely eliminated prior to the next sample testing. The existence of peaks having duration is easy to detect in the raw QC chart. It will also be clearly reflected in the shape of the ACF (Figure 24.8). This is because during the rise and fall of the recorded entity, the successive entries are highly correlated. The peaks can have randomly varying magnitude (height) and duration (width), a manifestation of the stochastic character of the fluctuations. Here, not only many random factors influence the rise onset but also how fast and how long it will last. The same is true for the eventual decline timing and its rate. Unlike the case of rounded record, though, the remedy is not the data manipulation but invoking a totally different model. Engel *et al.* (2001a, 2001b) describe such a model. It consists of an *algorithm* rather than an algebraic equation and accounts for two modes of behaviour: during and between outbursts, ruled by different sets of probabilities. The model is yet to be fully validated with actual data but it can already offer qualitative predictions that can be verified. Being in the form of an algorithm that contains if statements and calls for random numbers generation, its parameters cannot be determined by any of the already mentioned conventional methods (regression, MM and MLE). But as has been shown with computer simulations they can be estimated by a specially devised mathematical method (Engel *et al.*, 2001a, 2001b).

## ANALYSIS OF RANDOMLY FLUCTUATING QC RECORDS

To generate a time series of random numbers with a normal distribution having a mean $\mu$ and standard deviation $\sigma$ one can use the simple model:

$$X_n = \mu + \sigma R_n \tag{24.14}$$

where $R_n$ is a random number between zero and one ($0 < R_n < 1$) picked from a uniform distribution.

To generate a time series of random numbers with a lognormal distribution having a logarithmic mean $\mu_L$ and a logarithmic standard deviation $\sigma_L$, the model is:

$$X_n = Exp(\mu_L + \sigma_L R_n) \tag{24.15}$$

Or if the logarithm base is 10, as is quite common in presenting microbial counts, the model will become:

$$X_n = 10^{\mu_{L10} + \sigma_{L10} R_n} \tag{24.16}$$

The Wolfram Demonstration 'Failure Probabilities from Quality Control Charts (http://demonstrations.wolfram.com/FailureProbabilitiesFromQualityControlCharts/) uses the above algorithms to simulate QC records with entries having a normal or lognormal distribution (two of its screen displays are shown in Figures 24.9 and Figure 24.10). The user can choose the model with a bar setter and the series length and distribution parameters with sliders on the screen or by entering them numerically. The user can also enter, numerically or with sliders, the permitted upper limits. Once done, or any of the sliders moved, the Demonstration automatically generates and draws a new random record for the modified settings, counts and displays number of entries outside the permitted range. The Demonstration also calculates and displays the probabilities of encountering such values analytically from the corresponding distribution function, and on the basis of their frequency as observed in the simulated record. [The program also offers the option to fix the random numbers generator's seed in order to create a reproducible record.]

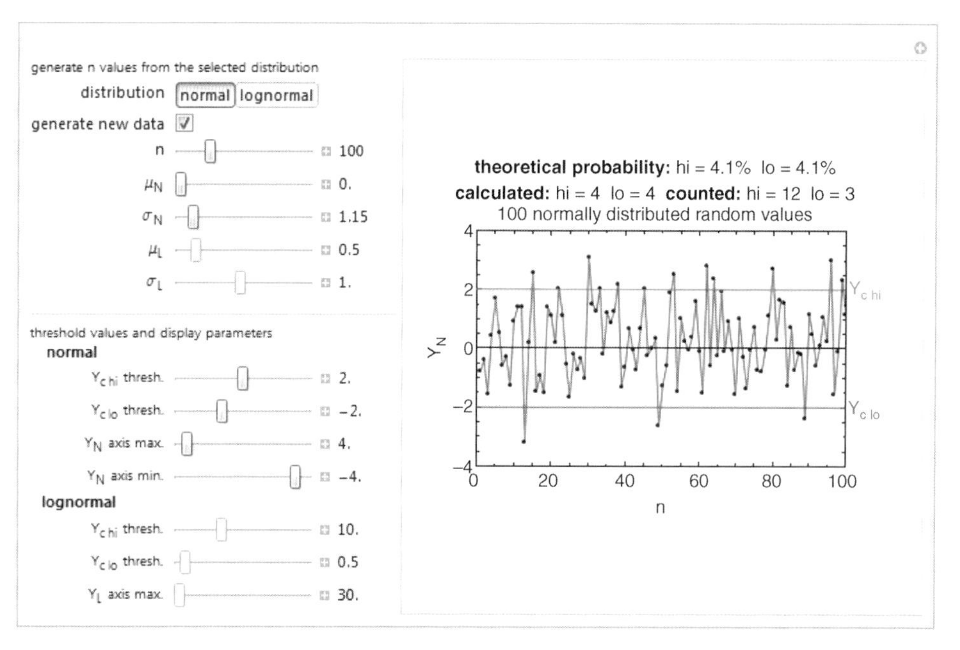


**Figure 24.9** Screen display of the Wolfram Demonstration that simulates random QC charts using the normal distribution as a model with chosen values of the mean and standard deviation (which can be those of an actual record). The Demonstration also estimates the probabilities of encountering entries outside a range specified by the user through its upper and lower bounds, $N_{c\ hi}$ and $N_{c\ lo}$, respectively. Notice that all the parameters can be entered either numerically or by moving a slider.

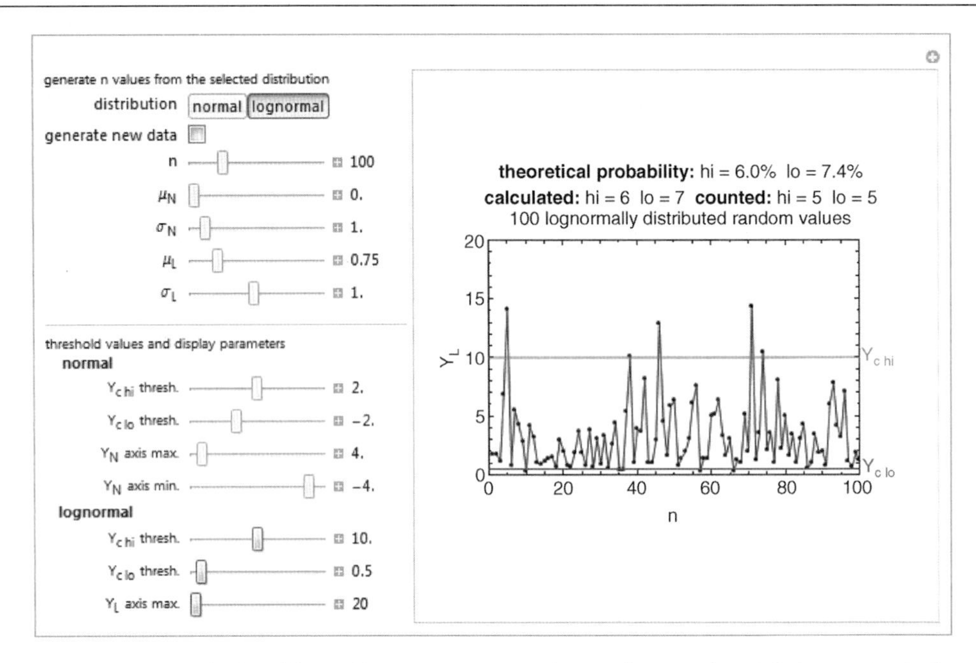


**Figure 24.10** The screen display of the Wolfram Demonstration that simulates random QC charts using the lognormal distribution as a model with chosen values of the mean and standard deviation (which can be those of an actual record). The Demonstration also estimates the probabilities of encountering entries outside a range specified by the user through its upper and lower bounds, $N_{c\,hi}$ and $N_{c\,lo}$, respectively. Notice that all the parameters can be entered either numerically or by moving a slider.

The original purpose of the Demonstration was to present the concept visually and provide a learning aid to quality control personnel in the form of an interactive program. Its great advantage is that it allows the user, or trainee, to examine numerous hypothetical scenarios in a very short time by moving the sliders and assess their potential quality and safety implications instantaneously. But the Demonstration can also be used as a rapid tool to estimate the probabilities of mishaps from actual QC records, if their entries' mean and standard deviation, or logarithmic mean and standard deviation have been calculated separately, and it can be assumed that the record follows Equations 24.14 or 24.15 as a model. But since the Demonstration does not allow pasting the actual record for examination, it gives no indication on whether the method's prerequisites have actually been met.

All this can be achieved by using the MS Excel spreadsheet "Estimating the Probability of High Microbial Counts Using Excel" (http://people.umass.edu/aew2000/MicCountProb/microbecounts.html), which too is a freely downloadable program. It too was originally developed for analysing microbial count records and calculating the probabilities of encountering an outburst, hence the name. But it can also be used for other types of QC records as long as they satisfy the normal or lognormal model's assumptions. The program allows the user to paste (and edit) his or her actual record and calculates the probabilities that a count will exceed up to five different levels chosen by the user. Part of the spreadsheet's screen display is shown in Figure 24.11. As can be seen in Figure 24.11, the pasted record is plotted and its autocorrelation function calculated and displayed too. This provides visual indication of whether the individual entries are really independent, or approximately independent, for the method to work. Also displayed graphically are the original entries' histogram and that of their logarithms, and the Q-Q plots of the normal and lognormal distributions. Together with the calculated skewness coefficient they help the user to decide whether to use the normal or lognormal distribution

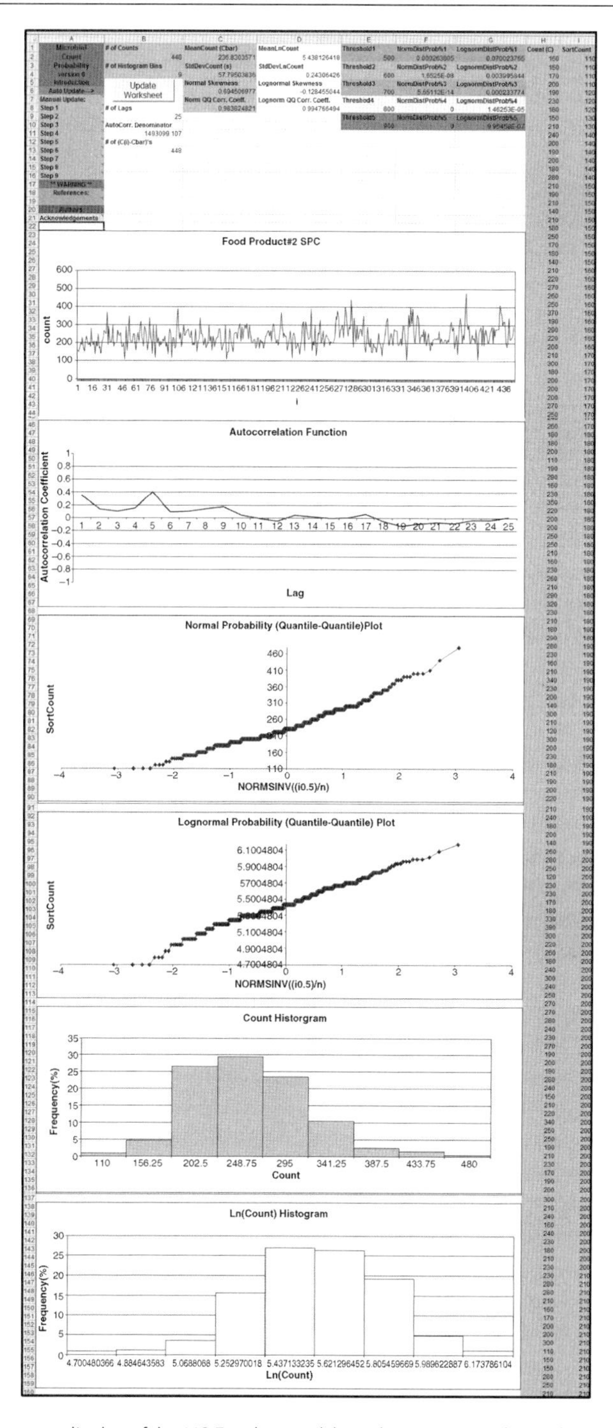


**Figure 24.11** Part of the screen display of the MS Excel spreadsheet that estimates the probabilities of exceeding five levels of chosen critical values form an actual record pasted by the user. It plots the record and its autocorrelation function (ACF) and uses the normal and lognormal distribution models simultaneously. To help decide between them, the program plots and analyses the two Q-Q plots and displays the counts and their logarithms' histograms. Normality will be manifested by a symmetric histogram of the entries and lognormality by a symmetric histogram of their logarithms.

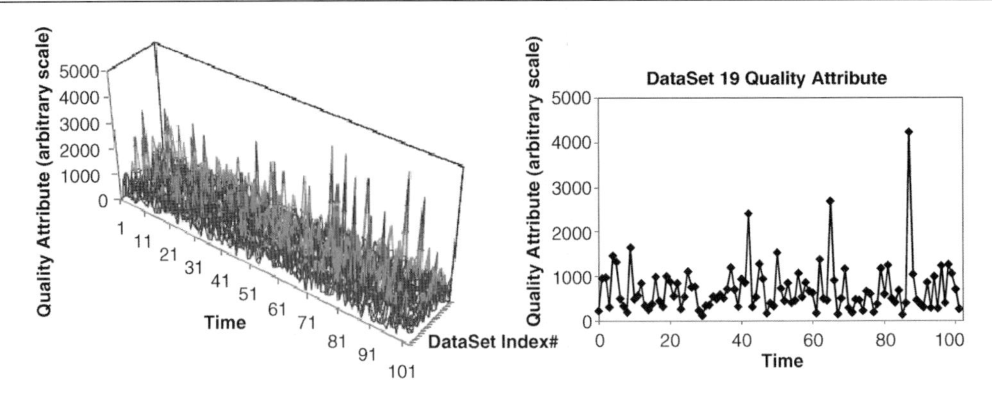


**Figure 24.12** Simulated microbial outbursts, or other industrial mishaps, happening in space and time produced by the MS Excel spreadsheet 'Modeling Microbial Population Explosions Using Excel' (PC version only). Corradini *et al.* (2009). Reproduced with permission from Springer Science + Business Media.

model for the probabilities calculations. The spreadsheet is set to automatically calculate only probabilities of exceeding the five chosen values. A modified version that calculates the probabilities of entries *falling* below five chosen values or any number of critical values for this matter is also available in the internet (See http://people.umass.edu/aew2000/ProbOfExtinction/ProbOfExtinction.html).

## Mapping the temporal and geographical distribution of mishaps

In certain cases, notably pathogens outbursts, there is interest not only in their occurrence pattern in time but also in space (Corradini *et al.*, 2009). Free software for simulating scenarios where each site has its own random fluctuating pattern can be found at http://www-unix.oit.umass.edu/~aew2000/MicPopExpl/MicPopExplModel.html. An example of the program's output is given in Figure 24.12. It is based on the model of Engel *et al.* (2001a, 2001b), which as has already mentioned is in an algorithmic rather than an algebraic form. This allows it to include in the model formulation not only a 'common' fluctuations generator, such as Equation 24.15, but also components, such as a detection level, explosion threshold, the probabilities of the counts rising and falling when the population is in an explosive regime. The model, which is fully stochastic, calls for three random number generators and uses 'If statements' to initiate and terminate a regime.

The space can be conveniently represented by a single axis on which locations are assigned numerical values, such as those used in the zip code by the US Mail Service. The software also calculates rudimentary statistical properties that can be used to compare the frequencies and magnitudes of outbursts generated with different model parameter combinations. At this time, the model and program can only be used to generate hypothetical scenarios and examine the resulting pattern. It is not impossible that in the future, when large databases become available, patterns of the kind shown could also be used to quantify actual microbial outbursts and other industrial mishaps and translated into risk.

# CONCLUDING REMARKS

The focus of this chapter has been methods developed or elaborated at the University of Massachusetts at Amherst for extracting *additional* useful information from quality control data of foods and water to that obtained by traditional methods. The statistical principles of QC data analysis and interpretation are well known and established (Bain and Engelbardt, 1992; Rice, 1995) and so are the procedures of their applications in the food industry (FAO, 1988, 1991; Hubbard, 1996). Therefore, no attempt has been made to reintroduce them. Needless to say, they are most useful, especially in identifying an existing problem and revealing a perilous trend. That can indicate that a process is under control and the process safe. Although sometimes a problem, sampling and testing related issues have not been discussed. The underlying assumption has been that the numbers in the record are correct and representative. Consequently, any predictions made with the described method only refer to results obtained by the original procedure.

What has been shown is that, at least in principle, it is possible to translate a randomly fluctuating record, however determined, into the probabilities or frequencies of future desirable or undesirable events. This makes the described methods, and the software developed for their application, most powerful tools in industrial risk assessment. They can be used both in routine analysis of QC data and to assess the potential quality or safety implications of changes in raw material or procedures when they occur. They can also be used to quantify the consequences, improvements or corrective measures in terms of lowering the risk of future mishaps. When needed, the described methods can be used to rank suppliers not in terms of mean performance but in terms of the expected frequencies of future problems. As already mentioned, the order need not be the same. In other words, a source ranked superior based on its mean performance can actually be inferior as far as the frequency of product failures is concerned!

Gathering QC data, especially when chemical assays or microbial counts are involved, has a considerable cost associated with materials, labour, equipment, storage space and administration. The little extra effort that implementation of the methods described in this chapter require are miniscule in comparison. Therefore, it seems to us as wasteful that QC data obtained at a high cost are commonly discarded without further analysis. And, if as already claimed (Peleg, 2009), even one recall could be avoided, this extra effort would be more than justified. The peace of mind that would come with the knowledge that a process or product is at a low risk of an unexpected failure, would be another reason to implement the described methods.

The reliability of the calculated risk estimates primarily depends on the validity of their underlying assumptions. Thus, it is imperative that the record be always examined to reveal significant violations. Examined records so far have indicated that minor violations of the condition of independence, or an imperfect fit of the chosen parametric distribution function, need not have a dramatic effect on the estimates quality. Yet, as already suggested (Corradini *et al.*, 2001b), once in doubt, one could try several models and choose the one rendering the most conservative estimates.

## ACKNOWLEDGEMENTS

Contribution of the Massachusetts Agricultural Experiment Station at Amherst.

A significant part of the work described in this chapter was also supported by the USDA-NRICGP 9802797 grant.

## REFERENCES

Bain, L.J. and Engelbardt, M. (1992) *Introduction to Probability and Mathematical Statistics*, 2nd edn. PWS-KENT, Duxbury, MA.

Corradini, M.G., Horowitz, J., Normand, M.D. and Peleg, M. (2001a) Analysis of the fluctuating pattern of *E. coli* counts in the rinse water of an industrial poultry plant. *Food Research International* **34**, 565–572.

Corradini, M.G., Normand, M.D., Nussinovitch, A. *et al.* (2001b) Estimating the frequency of high microbial counts in commercial food products using various distribution functions. *Journal of Food Protection* **64**, 674–681.

Corradini, M.G., Engel, R., Normand, M.D. and Peleg, M. (2002) Estimating the frequency of high microbial counts from records having a true or suspected trend or periodicity. *Journal of Food Science* **67**, 1278–1285.

Corradini, M.G., Normand, M.D. and Peleg, M. (2009) On modeling the temporal and geographical distribution of pathogens' outbursts and other industrial mishaps. *Food Engineering Reviews* **1**, 3–15.

Engel, R., Normand, M.D., Horowitz, J. and Peleg, M. (2001a) A model of microbial contamination of a water reservoir. *Bulletin of Mathematical Biology* **63**, 1025–1040.

Engel, R., Normand, M.D., Horowitz, J. and Peleg, M. (2001b) A qualitative probabilistic model of microbial outbursts in foods. *Journal of the Science of Food and Agriculture* **81**, 1250–1262.

FAO (1988) Introduction to food sampling. Manuals of food quality control (no.9). Food and Agriculture Organization of the United Nations, Rome, Italy.

FAO (1991) Quality assurance in the food control microbiological laboratory. Manuals of food quality control (no.12). Food and Agriculture Organization of the United Nations, Rome, Italy.

Gonzalez-Martinez, C., Corradini, M.G. and Peleg, M. (2003) Probabilistic models of food microbial safety and nutritional quality. *Journal of Food Engineering* **56**, 135–142.

Hadas, O., Corradini, M.G. and Peleg, M. (2004) Statistical analysis of the fluctuating counts of fecal bacteria in the water of lake Kinneret. *Water Research* **38**, 79–88.

Horowitz, J., Normand, M.D. and Peleg, M. (1999) On modeling the irregular fluctuations in microbial counts. *Critical Reviews in Food Science* **39**, 503–517.

Hubbard, N.R. (1996) *Statistical Quality Control for Food Industry*, 2nd edn. Chapman & Hall, New York.

Nussinovitch, A. and Peleg, M. (2000) Analysis of the fluctuating patterns of microbial counts in frozen industrial food products. *Food Research International* **33**, 53–62.

Nussinovitch, A., Curasso, Y. and Peleg, M. (2000) Analysis of the fluctuating microbial counts in commercial raw milk – A case study. *Journal of Food Protection* **63**, 1240–1247.

Peleg, M. (2001) Interpretation of the irregularly fluctuating microbial counts in commercial dairy products. *International Dairy Journal* **12**, 255–262.

Peleg, M. (2006) *Advanced Quantitative Microbiology for Food and Biosystems: Models for Predicting Growth and Inactivation*. CRC Press, Boca Raton, FL.

Peleg, M. (2009) Applying statistical methods to reduce product recalls. *Food Technology* **63**, 100.

Peleg, M. and Horowitz, J. (2000) On estimating the probability of aperiodic outbursts of microbial populations from their fluctuating counts. *Bulletin of Mathematical Biology* **62**, 17–35.

Peleg, M., Nussinovitch, A. and Horowitz, J. (2000) Interpretation and extraction of useful information from irregular fluctuating industrial microbial counts. *Journal of Food Science* **65**, 740–747.

Peleg, M., Normand, M.D., Horowitz, J. and Corradini, M.G. (2007) An Expanded Fermi Solution for risk assessment. *International Journal of Food Microbiology* **113**, 92–101.

Peleg, M., Normand, M.D. and Corradini, M.G. (2012) A study of the randomly fluctuating microbial counts in foods and water using the expanded Fermi solution as a model. *Journal of Food Science* **71**, R63–R71.

Rice, J.A. (1995) *Mathematical Statistics and Data Analysis*, 2nd edn, pp. 321–328. Brookside Publishing Company, Duxbury, MA.

von Baeyer, H.C. (1993) *The Fermi Solution: Essays on Science*. Random House, New York, NY.

# 25 Application of statistical approaches for analysing the reliability and maintainability of food production lines: a case study of mozzarella cheese

**Panagiotis H. Tsarouhas**

*Department of Standardization & Transportation of Products – Logistics, Alexander Technological Educational Institute of Thessaloniki, Katerini, Greece*

## ABSTRACT

The reliability and maintainability (R&M) analysis of automated mechanical equipment of a mozzarella cheese production line for a period of twenty-four months was carried out at machine and entire line level. The most important characteristics of failure data and the most important failures were identified; also carried out was the determination of the theoretical distributions parameters that best fit the failures data. Furthermore, the reliability, maintainability, failure rate and repair rate models of the production line for all machines and the entire production line were calculated. The analysis could prove to be a useful tool both to assess the current conditions and to predict the R&M for upgrading the operations management policies of the production line. This methodology can also be used in the cheese industry sector by the machinery manufacturers and the manufacturers of cheese products to improve the design and operation management of their production lines.

## INTRODUCTION

The global economy is forcing food companies to modernize their operations through increased mechanization and automation. Thus, as food equipment is becoming more complex and sophisticated, its cost is increasing rapidly. To meet production targets, food companies are increasingly demanding better equipment reliability. High costs motivate seeking engineering solutions to reliability problems for reducing financial expenditure, enhancing reliability, satisfying customers with on-time deliveries through increased equipment availability and by reducing costs and problems arising from products that fail easily (Barringer, 2000). Reliability and maintainability (R&M) are two fundamental measures that can be used to estimate the effectiveness of the production system.

There is a vast literature on R&M, including many articles and books. Ansell and Phillips (1989) analysed practical problems in the statistical analysis of reliability data. The failures of a system are not scheduled and can be considered as a random variable. Since failure cannot be prevented entirely, it is important to minimize both the probability of its occurrence and the impact of failures when they do occur (Blischke and Murthy, 2003). de Castro and Cavalca (2003) presented an availability optimization problem of an engineering system assembled in a series configuration that has the redundancy of units

*Mathematical and Statistical Methods in Food Science and Technology*, First Edition.
Edited by Daniel Granato and Gastón Ares.

and maintenance teams as optimization parameters. Hajeeh and Chaudhuri (2000) have worked on the reliability and availability assessment of reverse osmosis, assessing the performance of reverse osmosis plants in the Arabian Gulf region by analysing failure behaviour and downtime patterns. Smith and Smith (2002) have presented a method of calculating system availability and reliability probability distributions using permutations of inseparable system failure and restore data sets.

According to Mobley (2002), the maintenance costs are a major part of the total operating costs of all manufacturing or production plants; depending on the specific industry, maintenance costs can represent between 15 and 60% of the cost of the goods produced. Lewis and Steinberg (2001) reported that, for example maintenance-related costs account for approximately 30–50% of direct mining costs. Hauge (2002) reported that the two requirements to be met for the preventive maintenance of a component to be appropriate are: firstly, preventive maintenance when the component deteriorates with time and, secondly, that the cost of preventive maintenance must be less than the cost of corrective maintenance. Preventive maintenance is widely considered an effective strategy for reducing the number of system failures, thus lowering the overall maintenance cost (Okogbaa and Peng, 1996). Biswas *et al.* (2003) have given a methodology to calculate the availability of a periodically inspected system, maintained under an imperfect repair policy. Francois and Noyes (2003) displayed a methodology for the evaluation of maintenance strategies by taking into consideration the effect of certain variables on the dynamic of maintenance, its structure and its context of evolution.

For food machinery, manufacturers' reliability engineering is one of the most important fields that define the quality of the products. When an unexpected failure occurs, the failed equipment stops and forces most of the line upstream of the failure to operate without processing material. The negative impact of failures on the actual production rate of automated production lines adds pressure on food products manufacturers to assess and improve the reliability of their production lines (Liberopoulos and Tsarouhas, 2005).

In literature there are a limited number of case studies on reliability and maintainability (R&M) for food production lines. Liberopoulos and Tsarouhas (2002) presented a case study of the speeding up a croissant production line by inserting an in-process buffer in the middle of the line to absorb some of the downtime, based on the simple assumption that the failure and repair times of the workstations of the lines are ruled by exponential distributions. The parameters of these distributions were computed based on actual data collected over ten months. Percy *et al.* (1997) reported that empirical evidence indicates that sets of failure times typically include ten or fewer observations, which emphasizes the need to develop methods dealing adequately with small data sets (of course the larger the data set, the more precise the statistical analysis). Tsarouhas *et al.* (2009a, 2009b, 2009c) developed the reliability and maintainability analysis of strudel, feta cheese and juice bottling production lines at machine, workstation and entire line level; descriptive statistics of the failure and repair data were carried out and the best fitness index parameters were determined, and the reliability and hazard rate modes for all workstations and production lines (strudel, feta cheese and juice bottling) were calculated. Moreover, in other studies Tsarouhas *et al.* (2010a, 2010b) estimated the reliability and maintainability analysis of bread and beer packaging production lines based on field failure data.

In this chapter the application of statistical approaches of failure data for analysing the R&M of a mozzarella production line (MPL) are presented. The analysis includes the computation of the most important characteristics of the failure data, the identification of the most important failures and the computation of the parameters of the theoretical distributions that best fit the failure data. The reliability and hazard rate models at machine and production line level that can be a useful tool for food and engineering to assess the current conditions, and to predict reliability for upgrading the operation management (i.e. maintenance policy, spare parts, reducing the delivery time to the customers etc.) of the line, were calculated. The aim of this case study was to provide a valid reliability and maintainability modelling for food product machinery manufacturers, which target optimizing the design and operation of their production lines at peak reliability, thereby increasing their productivity and availability.

## COLLECTION AND ANALYSIS OF FAILURE DATA

The collection of the failure data is a necessary step for the reliability estimation and improvement in the process of mozzarella cheese production. The field of failure data recorded for the production line must be for the longest possible period. Moreover, the quality of field failure data must be unbiased and recorded from maintenance staff, that is mechanics, electricians and so on. The records can include the failure mode or modes that occurred during the production process per shift, the action taken to repair the failure and the report of the exact time between the failures and the time to repair a failure. The failure data are often recorded in time units of minutes or hours.

The flow diagram that describes the methodology for the reliability modelling of a repairable system used as a framework to analyse the failure data of the production lines is displayed in Figure 25.1. Similar

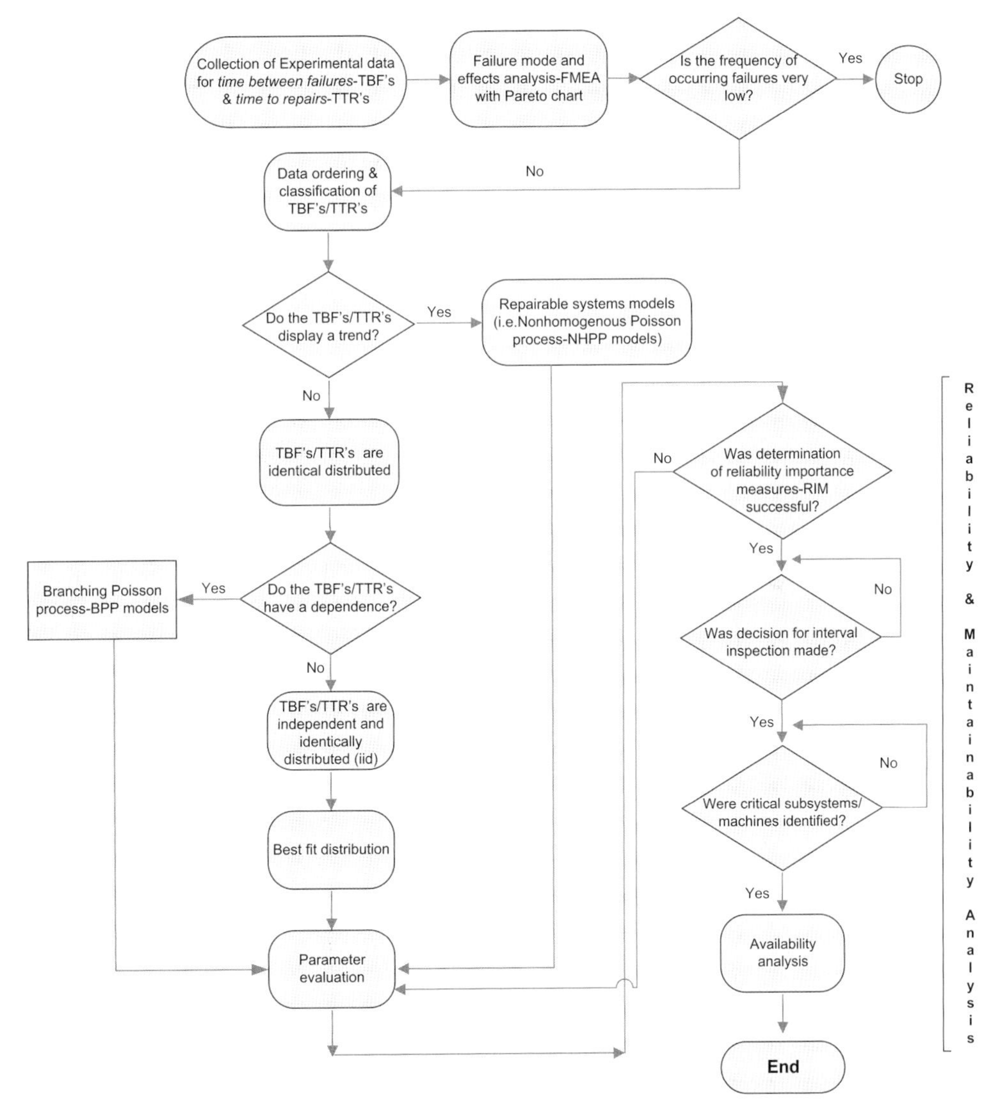


**Figure 25.1** Reliability/Maintainability analysis process of a repairable system. Adapted from Ascher and Feingold (1984); Barabady and Kumar (2008).

flow diagrams are shown in the literature (Coetzee, 1997; Louit *et al.*, 2009). Initially, the failure data with significant consequence are identified. For this reason, it is appropriate to use a Pareto chart. Then, after the collection, sorting and classification of the data comes the validation of the assumption of the independent and identically distributed (iid) nature of the time-between-failure (TBF) and time-to-repair (TTR) data of each machine. If the assumption that the data are iid is not valid, then a nonstationary model such as the nonhomogeneous Poisson process (NHPP) must be fitted (Ascher and Feingold, 1984; Kumar and Klefsjo, 1992). A functional form that has been most commonly applied to repairable systems is the NHPP model based on the power law process (PLP) (Rigdon and Basu, 2000). Then, the fitting of TBF and TTR data for machines and faults with a theoretical probability distribution is applied. The next step is the estimation of the reliability and maintainability parameters at machine and line level with a best-fit distribution. Finally, the identification of critical machines and faults and formulation of a better maintenance policy to improve reliability is presented.

## THE MOZZARELLA PROCESSING LINE

The mozzarella production line (MPL) consists of several machines in series; each machine may have several failure modes. The line has a common transfer mechanism and a common control system. The transfer of material from one machine to another is performed automatically by mechanical means (i.e. conveyor belts; Figure 25.2).

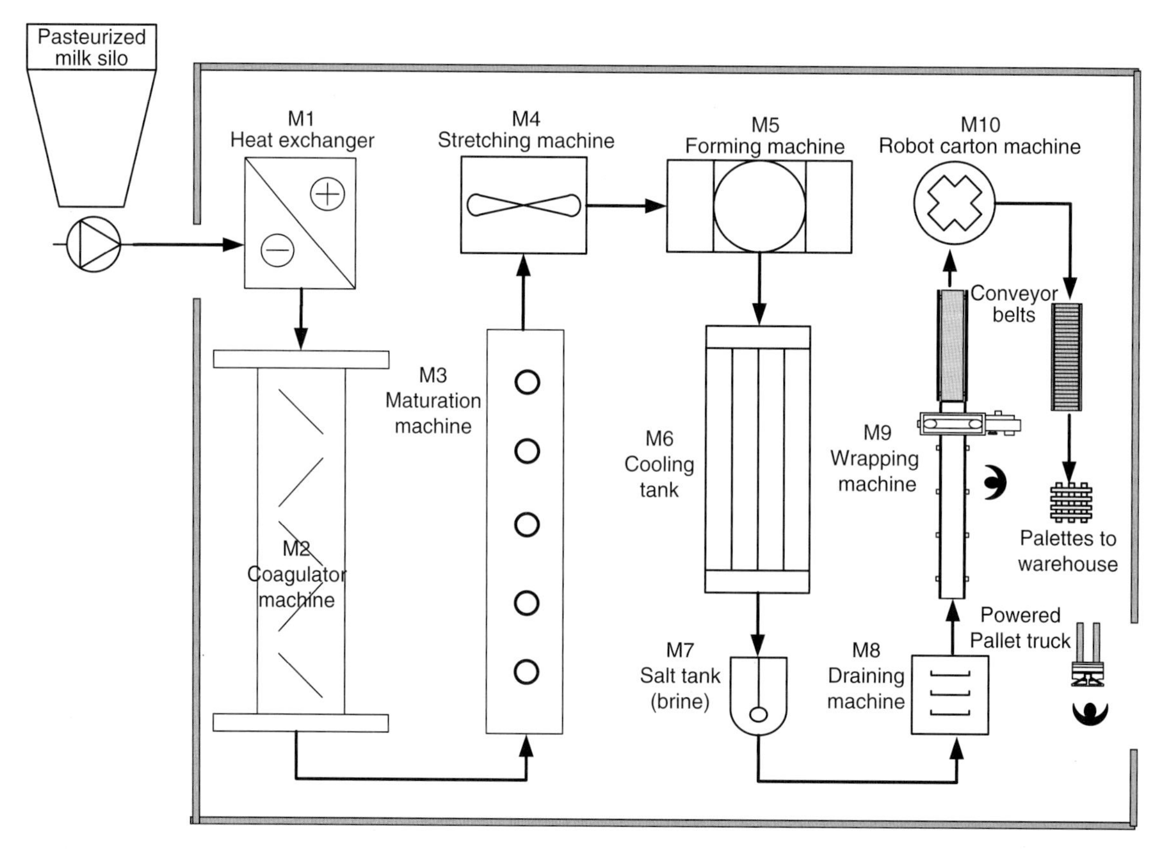


**Figure 25.2** Schematic presentation of the mozzarella production line.

The process flow of the mozzarella production line is as follows: The pasteurized-standardized milk for mozzarella production remains in silos for pasteurized milk (100 tonnes). The milk, before its transfer to the coagulator, is heated to 72°C for 15 seconds by means of heat exchanger (M1). As soon as the milk enters the next stage, coagulation by citric acid solution takes place (coagulator machine, M2). Coagulating enzymes or rennet is the crude preparation or extract of the abonasum (stomach of a young calf) consisting of two main enzymes, rennin and pepsin. Rennin is an extremely powerful clotting enzyme responsible for rapid clotting without much proteolysis, whereas pepsin induces proteolysis resulting in bitterness in cheese (De, 1980; Tsarouhas *et al.*, 2009a). In this stage, the milk is heated to 37°C for 60 min (Morea *et al.*, 1998). Then, the curd is pumped into the maturation machine (M3) where the mozzarella mixture is extruded through metal dies and, subsequently, cooled in large vats containing refrigerated water. The ripening of the curd lasts from 90–120 min until the pH reaches $5.3 \pm 0.1$. The process of stretching stands for a significant heat treatment of the fresh curd by adding hot water at 65°C for 15 min (stretching machine, M4). The ripened cheese mixture is then processed in multiple-screw extruders for longitudinal orientation of fibres and draining the water from the mozzarella particles (Ferrari *et al.*, 2003).

The next stage is the forming machine (M5), which pressurizes the mozzarella into metal moulds. The length of time required at M5 is 15 min, where the formed pats of cheese are already hot (50°C). Then, the mozzarella pats need to be cooled further in a cooling tank (M6) for 60 min to maintain their shape. In this stage, the mozzarella reaches a temperature of 12°C.

By the end of the cooling stage, the cheese is transported to the salt tank (M7) where it remains in brine (salt solution) for 30 min. This step is important for proper product conservation and to avoid the growth of any microorganisms or germs during storage. Then, the mozzarella pats are removed from the salt solution to be dried by the draining machine (M8), and transported to the wrapping machine (M9). The mozzarella is packaged either in plastic (under vacuum) or in salt solution. The final products leaving the mozzarella processing line are loaded onto a conveyor belt. From there, they are picked up by a robot carton machine (M10) and packed in cartons. The filled cartons are placed on a different conveyor belt that leads them to a worker who stacks them onto palettes and transfers them to the finished goods warehouse (refrigerators).

The above stages describe the entire mozzarella production line that is subject to failures. However, besides the failures of the equipment that may be characterized as 'internal failures', there are also failures due to the exterior environment that affect the entire line. These failures are described within the frame of this study as 'external failures' and can be defined as an eleventh dummy machine (M11) (Liberopoulos and Tsarouhas, 2005). This machine has four external failure modes: there can be shortages of the electric power, water, gas and air supply. Failures at M11 are not that frequent but are still important because they considerably affect the entire line. The most significant of these failures is the electric power generator shortage that temporarily supplies the system with electricity in case of an electric power outage.

The total cycle time of the mozzarella production line amounts to 4 h and 30 min.

## OPERATIONS MANAGEMENT OF THE MOZZARELLA LINE

The mozzarella production line (MPL) operates in three 8-h shifts during each workday, including weekends, 7 days per week with no pauses. The maintenance policy that is used is corrective maintenance. When a failure occurs, the maintenance staff that is responsible for the proper operation of the line performs the necessary corrective maintenance operations to repair the failure. Corrective maintenance comprises actions taken to restore a failed component or machine to the operational state. The actions involve repair or replacement of all failed components necessary for successful operation of the production line. Corrective maintenance actions are unscheduled actions intended to restore the line from a failed state into a working state. The maintenance staff (mechanics and electricians) is also responsible for keeping hand-written records of the failures per shift.

The field data of 1889 failures recorded by the maintenance staff of the mozzarella production line for a period of 761 continuous working days (about two years) were analysed. The records included the failures that had occurred per shift, the action taken to repair the failure, the down time and the exact time of failure. Therefore, available are the exact time that the machine failed, the exact time-between-failure (TBF) and the corresponding time-to-repair (TTR) of this failure. The TBF of equipment is defined as the time that elapses from the moment the equipment goes up and starts operation after a failure until the moment it goes down again and stops operation due to a new failure. The TTR of failed equipment was defined as the time that elapses from the moment the equipment stops functioning until the moment it starts operating again. Both TBF and TTR were recorded in minutes.

Table 25.1 gives the total number of failures, total TBF and total TTR at workstation level for the MPL. Over this working period, the line operated a total of 1 095 840 min, out of which for 968 980 min the line operated without failures and for the remaining 126 860 min the line was under repair. Therefore, 88.42% $\left(or\,\frac{968980}{1095840}\times 100\right)$ of the total operating time the line was functioning properly, whereas the remaining 11.58% $\left(or\,\frac{126840}{1095840}\times 100\right)$ of the total operating time the line was under repair.

The failure frequency of each failure mode was evaluated by means of a Pareto chart (Figure 25.3). This chart resulted from an analysis of the high rank and occurrence of failures, indicating the number of failure occurrences per failure mode of the total failure occurrence. The most frequent failures are observed at the coagulator machine (M2), amounting to 34.3% of all the failures. The second most frequent failures are at the wrapping machine (M9), accounting for 23.1% of all the failures; while the stretching machine (M4) failures are ranked in third position with 10.2% of all the failures. Therefore, these machines (M2, M9, and M4) are accountable for two-thirds (67.6%) of the total failure occurrence on the mozzarella production line.

## TREND TEST AND SERIAL CORRELATION ANALYSIS FOR FAILURE DATA

After collection, sorting and classification of the data, the validation of the assumption for the independent and identically distributed (iid) nature of the TBF and TTR data of each device must be identified. Thus, the null hypothesis $H_0$: *No-trend in data* (homogeneous Poisson process) and the alternative hypothesis $H_1$: *Trend in data* (non-homogeneous Poisson process) are considered. Moreover, the test statistic $X^2$ is chi-square distributed with $2(n-1)$ degrees of freedom, *df* (Department of Defense, 1981). The $X^2$ statistic is calculated from the experimental failure data whereas the $x^2_{a,df}$ can be determined from the chi-square distribution given the degrees of freedom. If the statistic $X^2 > x^2_{a,df}$ then the null hypothesis is plausible, otherwise the null hypothesis is rejected and the alternative hypothesis $H_1$ is accepted. The validation of the trend for the TBF and TTR at machine and line level are displayed in Table 25.2 and at $a = 5\%$ level of significance the $H_0$ is not rejected, except for the M3, M4, M7, M9 and the entire line for TBFs.

In addition, the correlation of the failure data should be identified. A serial correlation diagram represents the sketching of $\hat{\rho}_k$ against lag $k$, where $\hat{\rho}_k$ are the correlation coefficients and lag $k$ are the lag-time periods separating the ordered data. Correlation coefficients, $\hat{\rho}_k$, range in value from $-1$ (a perfect negative relationship) up to $+1$ (a perfect positive relationship). A value of zero indicates the absence of a linear relationship that is no-correlation. A high level of correlation is implied by a correlation coefficient that is greater than 0.5 in absolute terms, that greater than 0.5 or less than $-0.5$. A mid-level of correlation is implied if the absolute value of the coefficient is greater than 0.2 but lower than 0.5. A low level of correlation is implied if the absolute value of the coefficient is less than 0.

Figures 25.4 and Figure 25.5 show the serial correlation diagrams of the TBF and TTR at machine and line level for the mozzarella line where the correlation coefficients are calculated for lags that range from 1 to 10 ($k = 1, 2, 3, \ldots, 10$). The outcome is that there is a lack of correlations for the TBFs and TTRs.

**Table 25.1** Total number of failures, total TBF and total TTR at machine level for the mozzarella production line.

| | M1 | M2 | M3 | M4 | M5 | M6 | M7 | M8 | M9 | M10 | M11 | Line |
|---|---|---|---|---|---|---|---|---|---|---|---|---|
| Number of failures | 43 | 647 | 163 | 193 | 83 | 95 | 39 | 47 | 437 | 27 | 115 | 1889 |
| Total TBF | 1 093 070 | 1 051 480 | 1 084 205 | 1 082 790 | 1 089 585 | 1 089 375 | 1 093 230 | 1 092 585 | 1 065 750 | 1 093 870 | 1 091 325 | 968 980 |
| Total TTR | 2770 | 44 380 | 11 635 | 13 050 | 6255 | 6465 | 2610 | 3255 | 30 090 | 1970 | 4515 | 126 860 |

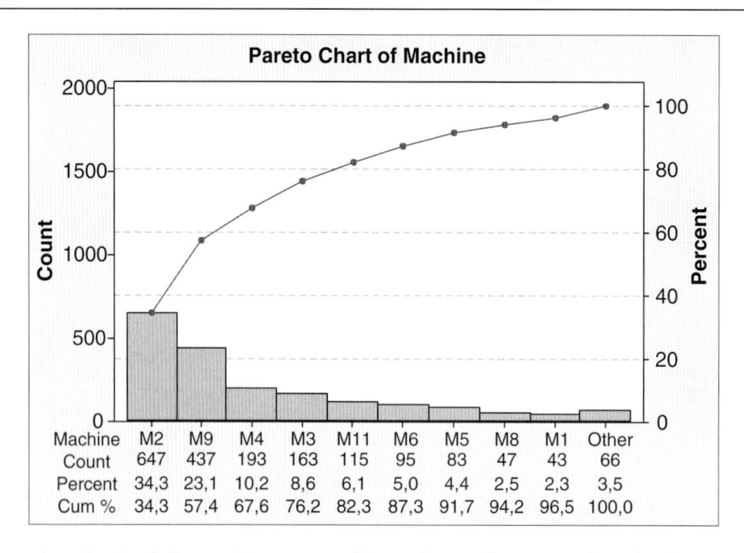


**Figure 25.3** Pareto chart for the failures of the mozzarella production line at machine level.

Therefore, from the trend test and the serial correlation test it is obvious that the failure data for all the machines and the entire line, except for the maturation machine (M3), the stretching machine (M4), the salt tank (M7), and the wrapping machine (M9) and at line level for TBF of the MPL, are free from the presence of trends and serial correlations.

**Table 25.2** Calculation of the test statistic $X^2$ for TBF and TTR at machine and line level.

| Unit | Variable | *df* | $X^2$ statistic | $x^2_{a,df}$ | Decision for $H_0$ |
|---|---|---|---|---|---|
| M1 | TBF | 86 | 90.46 | 65.6233 | Not rejected |
| | TTR | 84 | 80.47 | 63.8763 | Not rejected |
| M2 | TBF | 1294 | 1262.07 | 1211.47 | Not rejected |
| | TTR | 1292 | 1250.2 | 1209.54 | Not rejected |
| M3 | TBF | 326 | 274.51 | 285.167 | rejected |
| | TTR | 324 | 321.63 | 283.296 | Not rejected |
| M4 | TBF | 386 | 309.66 | 341.463 | rejected |
| | TTR | 384 | 362.25 | 339.581 | Not rejected |
| M5 | TBF | 166 | 174.71 | 137.209 | Not rejected |
| | TTR | 164 | 159.23 | 135.39 | Not rejected |
| M6 | TBF | 190 | 200.75 | 159.113 | Not rejected |
| | TTR | 188 | 172.45 | 157.282 | Not rejected |
| M7 | TBF | 78 | 57.02 | 58.6539 | rejected |
| | TTR | 76 | 71.77 | 56.9198 | Not rejected |
| M8 | TBF | 94 | 106.64 | 72.6398 | Not rejected |
| | TTR | 92 | 78.56 | 70.8816 | Not rejected |
| M9 | TBF | 874 | 795.07 | 806.386 | rejected |
| | TTR | 872 | 839.06 | 804.465 | Not rejected |
| M10 | TBF | 54 | 73.81 | 38.1162 | Not rejected |
| | TTR | 52 | 51.06 | 36.4371 | Not rejected |
| M11 | TBF | 230 | 255.1 | 195.895 | Not rejected |
| | TTR | 228 | 244.15 | 194.049 | Not rejected |
| Line | TBF | 3778 | 3600.26 | 3636.17 | rejected |
| | TTR | 3776 | 3707.16 | 3634.2 | Not rejected |

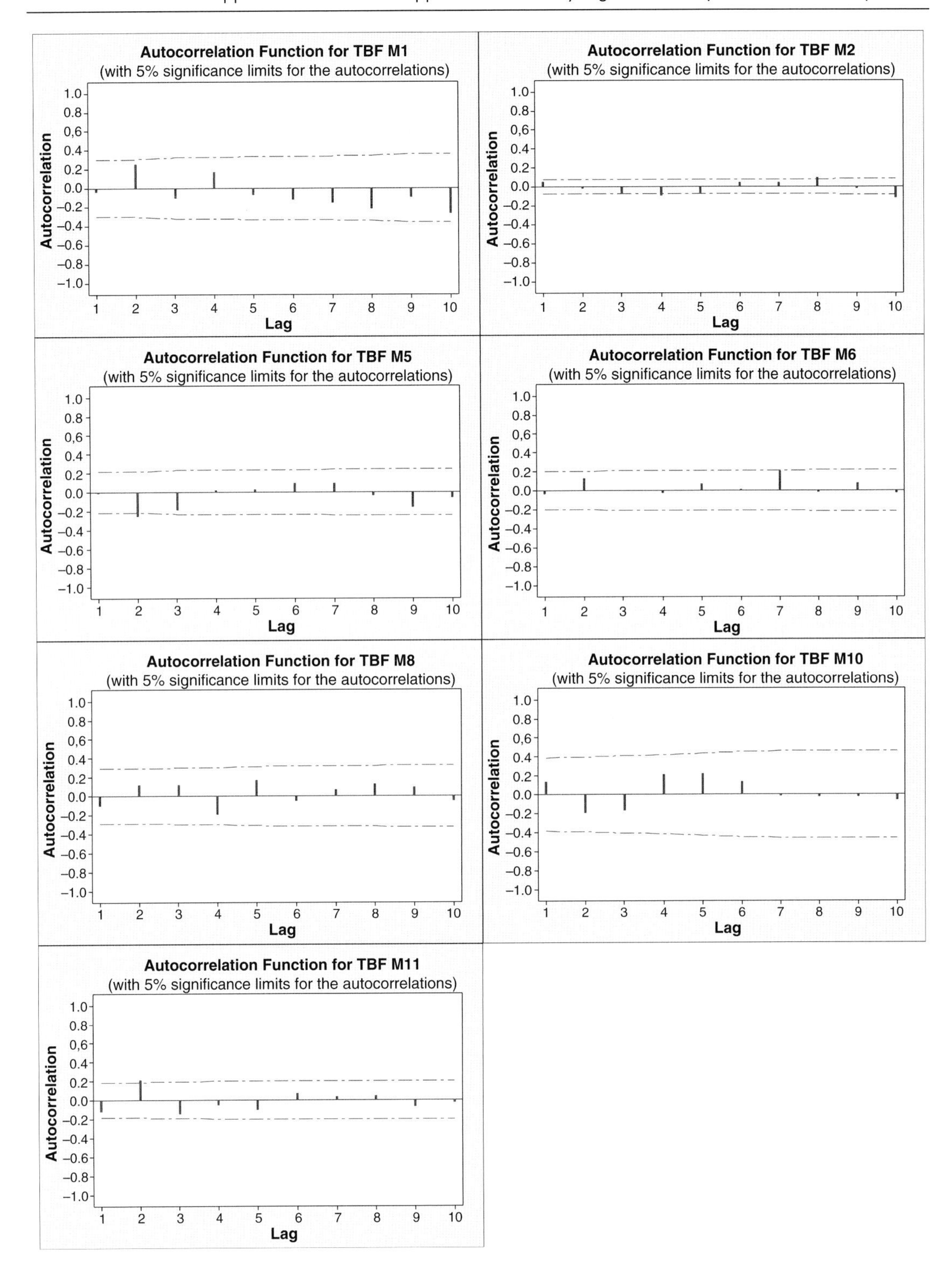


**Figure 25.4** Correlation diagrams of time-between-failure (TBF) at machine and line level for the mozzarella production line.

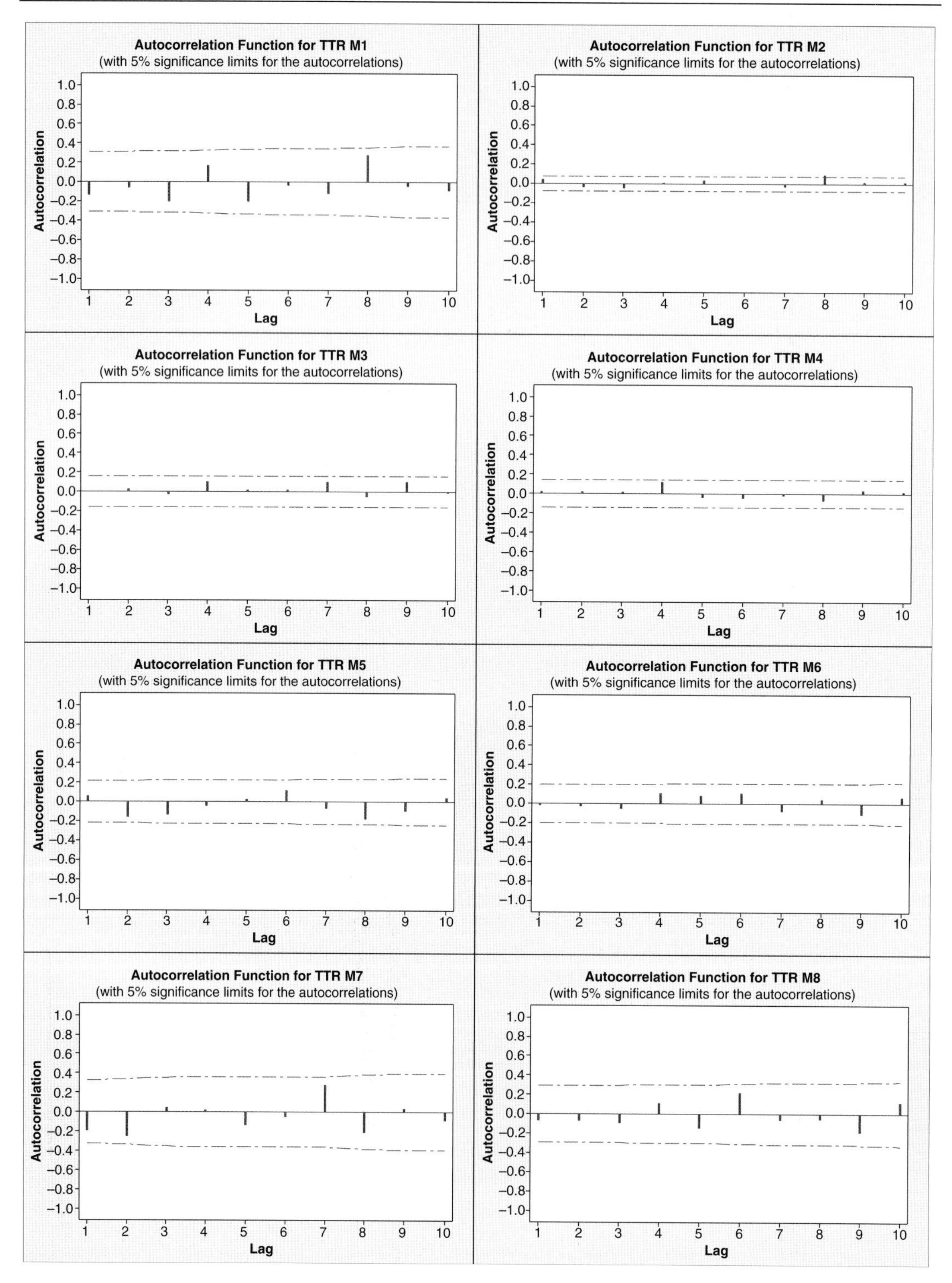


**Figure 25.5** Correlation diagrams of time-to-repair (TTR) at machine and line level for the mozzarella production line.

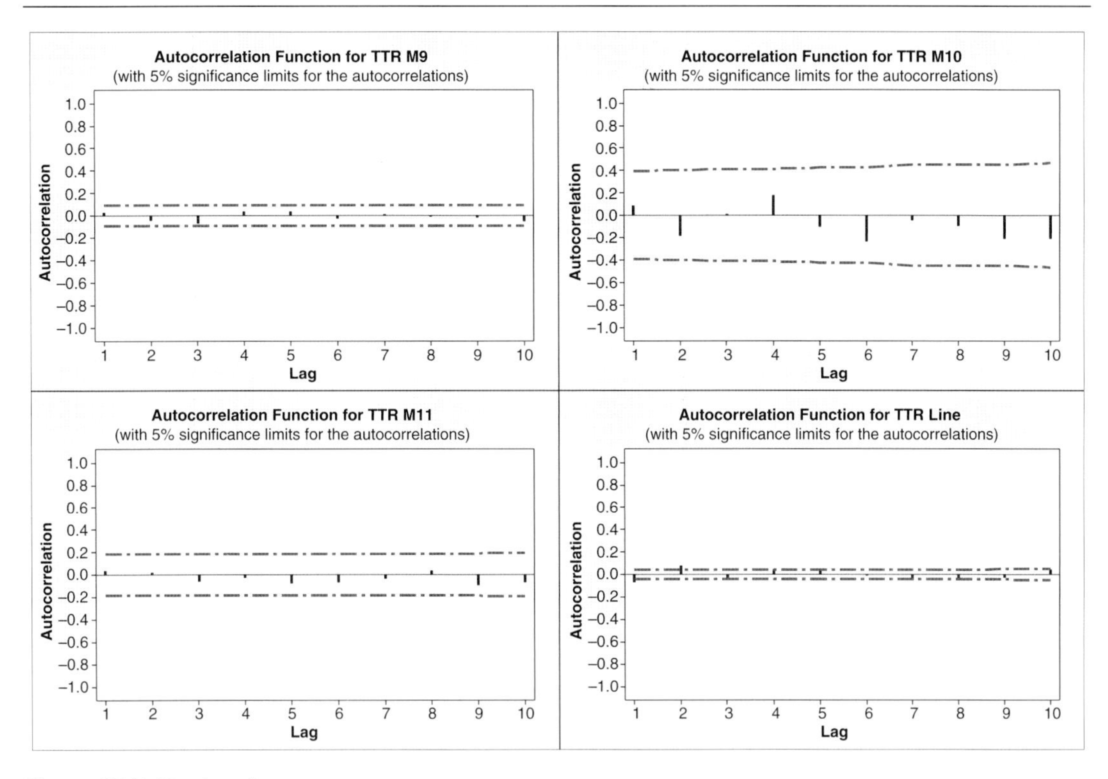


**Figure 25.5** *(Continued)*

# RELIABILITY ANALYSIS

Reliability is the probability that a machine or system will perform a required function, under stated conditions, for a stated period of time, $t$. Thus, reliability is the probability of nonfailure in a given period of time. If $T$ is the continuous random variable that represents the time-between-failure (TBF) of the system with $T \geq 0$, then the reliability can be expressed as (Wang and Pham, 2006):

$$R(t) = P(T \geq t) \tag{25.1}$$

If $F(t)$ is the unreliability of the system, then:

$$R(t) + F(t) = 1 \tag{25.2}$$

Thus, the unreliability is the probability that a failure occurs before time $t$:

$$F(t) = 1 - R(t) = P(T < t) \tag{25.3}$$

$F(t)$ is also called the cumulative distribution function of the failure distribution.

In reliability theory, the hazard or failure rate function is indicated as:

$$\lambda(t) = f(t)/R(t) \tag{25.4}$$

where $f(t)$ is the probability density function of the failure distribution and is defined by:

$$f(t) = \frac{dF(t)}{dt} = -\frac{dR(t)}{dt} \tag{25.5}$$

The expected or mean value of $f(t)$ is the mean time between failure ($MTBF$); it is given by:

$$MTBF = \int_0^{\infty} t \cdot f(t)dt = \int_0^{\infty} R(t)dt \tag{25.6}$$

The MPL consists of eleven machines in series. In this case, all the included machines of the system should be functioning to maintain the required operation of the system. A failure of any one machine of the system will cause failure of the whole system. Thus, the reliability of the entire production line can be calculated by:

$$R_{sys}(t) = R_1(t)^* R_2(t)^* \ldots ^* R_{11}(t) = \prod_{i=1}^{11} R_i(t) \tag{25.7}$$

To identify the distributions of the trend-free failure data between several theoretical distributions (i.e. Weibull, lognormal, exponential, loglogistic, normal and logistic distribution), the maximum likelihood estimation method was used per candidate distribution and its parameters assessed by applying a goodness-of-fit test – Anderson–Darling. The Anderson–Darling statistics of several theoretical distributions for TBF based on failure data of the machine level are summarized in Table 25.3. A smaller statistic value indicates that the distribution fits the data better, that is in heat exchanger (M1) column for TBF the lowest value is 1.404, which belongs to the Weibull distribution. Furthermore, the cooling tank (M6), and the draining machine (M8) are loglogistically distributed. The Weibull distributions are the best fit for the forming machine (M5) and the robot carton machine (M10). The coagulator machine (M2) follows the exponential distribution, whereas the dummy machine (M11) is lognormal distributed.

In Figure 25.6 the hazard functions and the evaluation of parameters for TBF at machine and line level are displayed. The hazard rate functions at wrapping machine (M9) and the robot carton machine (M10) were shown to increase, as they did for the entire production line. Therefore, the probability of failure is relatively high for an extended period of time. This means that the current maintenance policy is not adequate for those machines. The rest of the machines displayed decreasing failure rate, except the coagulator machine M2 that presents a constant failure rate.

In Table 25.4 the reliability for all the machines and the entire production line for different time intervals is calculated. The following observations can be made: (i) the reliability of the line, in one hour (60 min) of operation is 85.66%, in 8-h or 480 min (a shift) of operation it is 37.01% and in 24-h or 1440 min (a working day) of operation it is 6%; (ii) the highest reliabilities are reported at the robot carton machine (M10), the salt tank (M7) and at the heat exchanger (M1); and (iii) the lowest reliabilities are recorded at the coagulator machine (M2), the wrapping machine (M9) and at the stretching machine (M4).

The reliability diagram for all machines and the entire line is plotted in Figure 25.7. Its usefulness resides in the estimation of the optimal maintenance interval with the expected reliability level; that is, to

**Table 25.3** The Anderson–Darling statistics for time-between-failure (TBF) at machine level for the mozzarella line.

| | | | T | B | F | | |
|---|---|---|---|---|---|---|---|
| **Distribution** | **M1** | **M2** | **M5** | **M6** | **M8** | **M10** | **M11** |
| Weibull | 1.404* | 6.489 | 1.762* | 0.999 | 0.846 | 0.686* | 1.121 |
| Lognormal | 1.958 | 22.941 | 3.949 | 0.917 | 3.086 | 0.921 | 0.584* |
| Exponential | 1.64 | 6.488* | 2.376 | 1.018 | 0.891 | 1.284 | 1.16 |
| Loglogistic | 1.674 | 17.346 | 2.706 | 0.394* | 0.826* | 0.699 | 0.659 |
| Normal | 1.85 | 18.997 | 2.512 | 8.4 | 2.978 | 1.271 | 8.783 |
| Logistic | 1.814 | 12.89 | 2.455 | 5.479 | 2.126 | 1.309 | 6.328 |

* indicates the best-fit of the failure data between the candidate distributions.

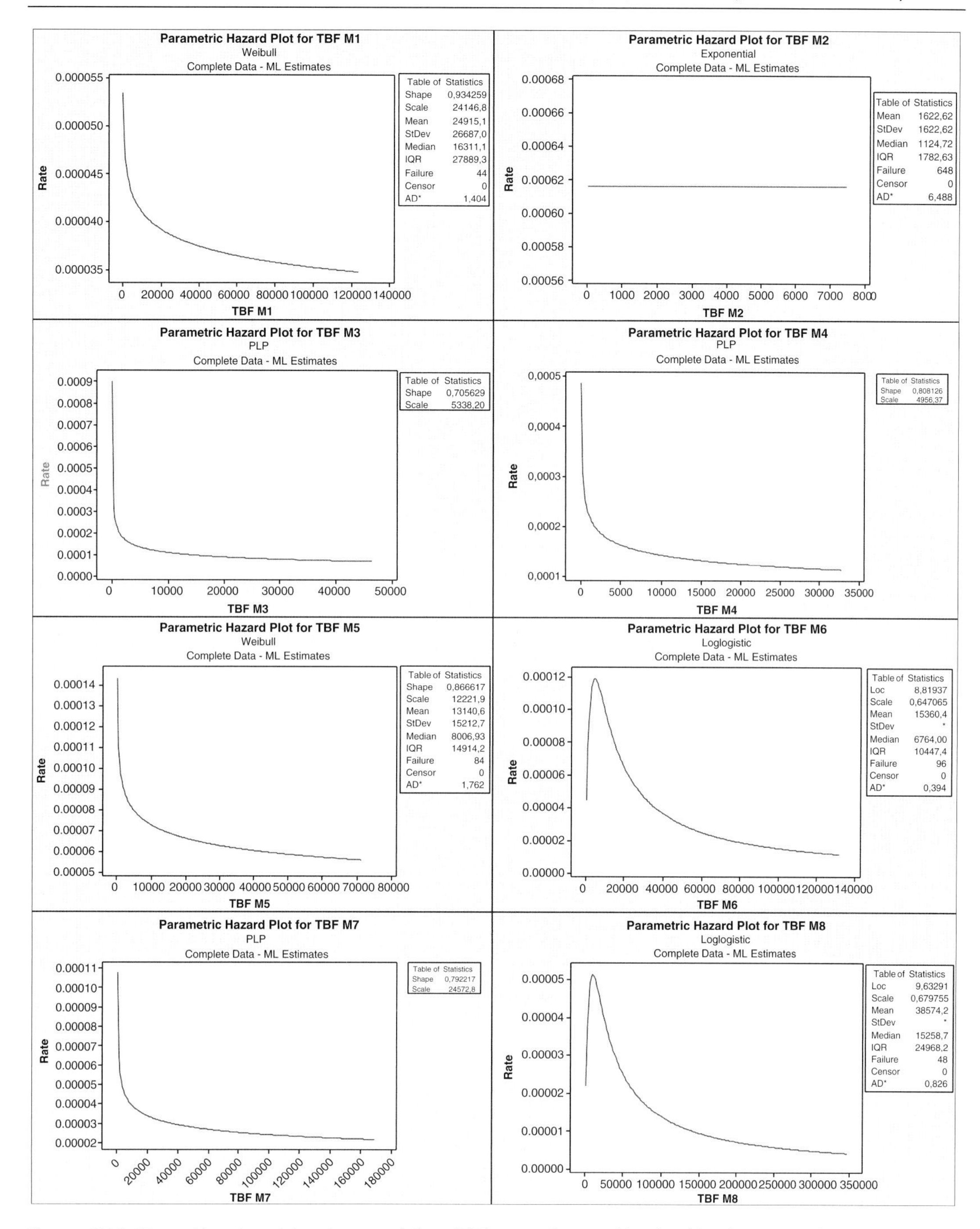


**Figure 25.6** Hazard function of time-between-failure (TBF) at machine and line level for the mozzarella production line.

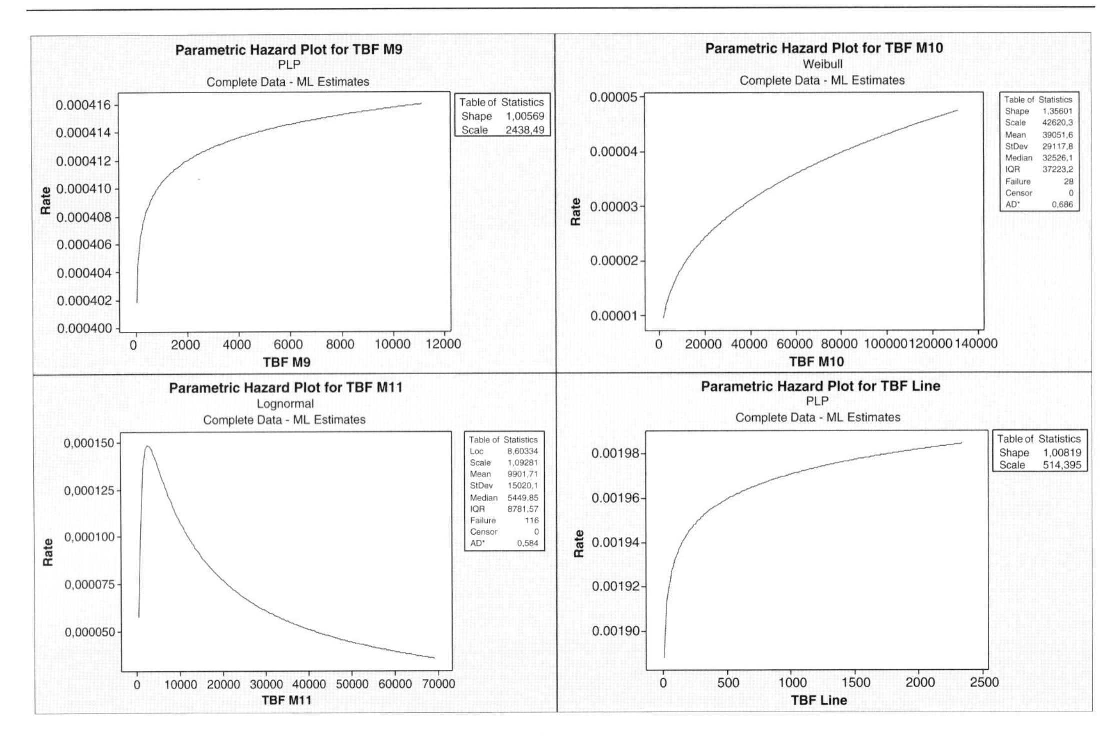


**Figure 25.6** *(Continued)*

achieve 90% reliability for the robot carton machine (M10), the maintenance must be carried out within 8640 min or 144 hrs. Moreover, for $Rel_{Line}$ (600) = 0.30, which means that the system will not fail for 600 min or 10 h of operation with only a 0.30 probability.

## MAINTAINABILITY ANALYSIS

Maintainability is the probability that a failed machine or system will be restored to operational effectiveness within a given period of time, *t*, when the repair action is performed in accordance with the prescribed procedures. Maintainability is the probability of completing the repair at a given time. If $T_r$ is the continuous random variable representing the time-to-repair (TTR) of the system, having a probability density function of $r(t)$, then according to Ben-Daya *et al.* (2009) the maintainability is:

$$M(t) = P(T_r \leq t) \tag{25.8}$$

The repair rate function is given by:

$$\lambda_r(t) = r(t)/(1 - M(t)) \tag{25.9}$$

The mean time-to-repair (*MTTR*) is the average time it takes to restore a system to operational status after it has failed to function, and can be calculated by:

$$MTTR = \int_0^{\infty} (1 - M(t))dt \tag{25.10}$$

**Table 25.4** Reliability for all machines and the entire mozzarella production line for different time intervals (in minutes).

| Time | Rel M1 | Rel M2 | Rel M3 | Rel M4 | Rel M5 | Rel M6 | Rel M7 | Rel M8 | Rel M9 | Rel M10 | Rel M11 | Rel Line |
|---|---|---|---|---|---|---|---|---|---|---|---|---|
| 1 | 0.99992 | 0.999384 | 0.997659 | 0.998968 | 0.999713 | 1 | 0.999668 | 1 | 0.999608 | 1 | 1 | 0.99492926 |
| 5 | 0.999638 | 0.996923 | 0.992731 | 0.996217 | 0.998843 | 0.99999 | 0.998811 | 0.99999 | 0.998022 | 1 | 1 | 0.98130019 |
| 10 | 0.999309 | 0.993856 | 0.988173 | 0.993385 | 0.997891 | 0.99996 | 0.997941 | 0.99998 | 0.996033 | 0.99999 | 1 | 0.9669525 |
| 30 | 0.998073 | 0.981681 | 0.9745 | 0.984003 | 0.994543 | 0.99977 | 0.995092 | 0.9999 | 0.988073 | 0.99995 | 1 | 0.91837858 |
| 60 | 0.996321 | 0.963698 | 0.958748 | 0.972158 | 0.990073 | 0.99933 | 0.991515 | 0.99971 | 0.976196 | 0.99986 | 0.99998 | 0.85664043 |
| 120 | 0.992982 | 0.928714 | 0.933604 | 0.951762 | 0.981973 | 0.99804 | 0.985352 | 0.9992 | 0.952776 | 0.99965 | 0.99976 | 0.75290519 |
| 240 | 0.986631 | 0.86251 | 0.894005 | 0.917074 | 0.967374 | 0.99429 | 0.97477 | 0.99778 | 0.907436 | 0.99911 | 0.99786 | 0.59048189 |
| 360 | 0.980535 | 0.801025 | 0.861434 | 0.886804 | 0.953958 | 0.98937 | 0.96538 | 0.99598 | 0.864128 | 0.99846 | 0.99355 | 0.46676062 |
| 480 | 0.974609 | 0.743923 | 0.832995 | 0.859355 | 0.941311 | 0.98351 | 0.956713 | 0.99387 | 0.822809 | 0.99772 | 0.9869 | 0.37014973 |
| 600 | 0.968816 | 0.690892 | 0.807439 | 0.833997 | 0.929243 | 0.97688 | 0.948561 | 0.99151 | 0.78341 | 0.99692 | 0.97826 | 0.29401012 |
| 720 | 0.963134 | 0.641641 | 0.784073 | 0.810307 | 0.917643 | 0.96959 | 0.940809 | 0.98893 | 0.745856 | 0.99606 | 0.968 | 0.23373818 |
| 960 | 0.952042 | 0.553422 | 0.7423 | 0.766903 | 0.895582 | 0.95335 | 0.92623 | 0.98319 | 0.675972 | 0.99418 | 0.94396 | 0.14794278 |
| 1440 | 0.930736 | 0.411704 | 0.672529 | 0.691912 | 0.854949 | 0.91612 | 0.899729 | 0.9699 | 0.555014 | 0.98994 | 0.88837 | 0.05948554 |
| 2880 | 0.871826 | 0.1695 | 0.523626 | 0.524734 | 0.751452 | 0.7891 | 0.832787 | 0.92077 | 0.306611 | 0.97444 | 0.72027 | 0.00397298 |
| 4320 | 0.818456 | 0.069784 | 0.422621 | 0.408651 | 0.666272 | 0.66661 | 0.777019 | 0.86488 | 0.169084 | 0.95612 | 0.58418 | 0.00027805 |
| 5760 | 0.769427 | 0.02873 | 0.348148 | 0.32332 | 0.593909 | 0.56176 | 0.728427 | 0.8074 | 0.093136 | 0.93587 | 0.4798 | 2.0419E-05 |
| 7200 | 0.724071 | 0.011828 | 0.290818 | 0.258661 | 0.531428 | 0.47588 | 0.685135 | 0.75118 | 0.051258 | 0.91421 | 0.39942 | 1.5694E-06 |
| 8640 | 0.681934 | 0.00487 | 0.245459 | 0.208691 | 0.476924 | 0.40654 | 0.646037 | 0.69777 | 0.028191 | 0.8915 | 0.33663 | 1.2579E-07 |
| 10080 | 0.64267 | 0.002005 | 0.208873 | 0.169522 | 0.429034 | 0.35057 | 0.610396 | 0.64792 | 0.015495 | 0.86801 | 0.28681 | 1.0469E-08 |
| 11520 | 0.606004 | 0.000825 | 0.17893 | 0.138484 | 0.386728 | 0.30515 | 0.577682 | 0.60192 | 0.008513 | 0.84396 | 0.24669 | 9.0098E-10 |
| 12960 | 0.571706 | 0.00034 | 0.154141 | 0.113673 | 0.349194 | 0.26797 | 0.547497 | 0.55977 | 0.004675 | 0.81952 | 0.21397 | 8.0069E-11 |
| 14400 | 0.539578 | 0.00014 | 0.133427 | 0.093698 | 0.315777 | 0.23726 | 0.519527 | 0.52129 | 0.002566 | 0.79485 | 0.18697 | 7.3074E-12 |
| 15840 | 0.509449 | 0.000058 | 0.115982 | 0.077518 | 0.28594 | 0.21164 | 0.493518 | 0.48625 | 0.001408 | 0.77007 | 0.16445 | 6.879E-13 |
| 17280 | 0.481167 | 0.000024 | 0.101192 | 0.064342 | 0.259233 | 0.19008 | 0.469263 | 0.45438 | 0.000773 | 0.74528 | 0.14549 | 6.6213E-14 |
| 18720 | 0.454598 | 0.00001 | 0.08858 | 0.053564 | 0.235275 | 0.17176 | 0.446586 | 0.42537 | 0.000424 | 0.72058 | 0.12941 | 6.5467E-15 |
| 20160 | 0.429618 | 0.000004 | 0.077773 | 0.044712 | 0.213743 | 0.15607 | 0.42534 | 0.39897 | 0.000232 | 0.69604 | 0.11565 | 6.3176E-16 |
| 21600 | 0.406119 | 0.000002 | 0.068469 | 0.037415 | 0.194359 | 0.14254 | 0.405396 | 0.37489 | 0.000127 | 0.67174 | 0.10381 | 7.7588E-17 |
| 23040 | 0.384 | 0.000001 | 0.060427 | 0.031379 | 0.176882 | 0.13078 | 0.386642 | 0.35292 | 0.00007 | 0.64773 | 0.09355 | 9.7487E-18 |
| 24480 | 0.363169 | 0 | 0.053452 | 0.026373 | 0.161102 | 0.12049 | 0.368982 | 0.33283 | 0.000038 | 0.62408 | 0.08462 | 0 |
| 25920 | 0.343542 | 0 | 0.047382 | 0.022208 | 0.146837 | 0.11144 | 0.352328 | 0.31443 | 0.000021 | 0.60081 | 0.07679 | 0 |
| 27360 | 0.325041 | 0 | 0.042083 | 0.018736 | 0.133927 | 0.10343 | 0.336605 | 0.29754 | 0.000011 | 0.57797 | 0.06991 | 0 |
| 28800 | 0.307596 | 0 | 0.037445 | 0.015833 | 0.122231 | 0.0963 | 0.321743 | 0.28201 | 0.000006 | 0.55559 | 0.06383 | 0 |

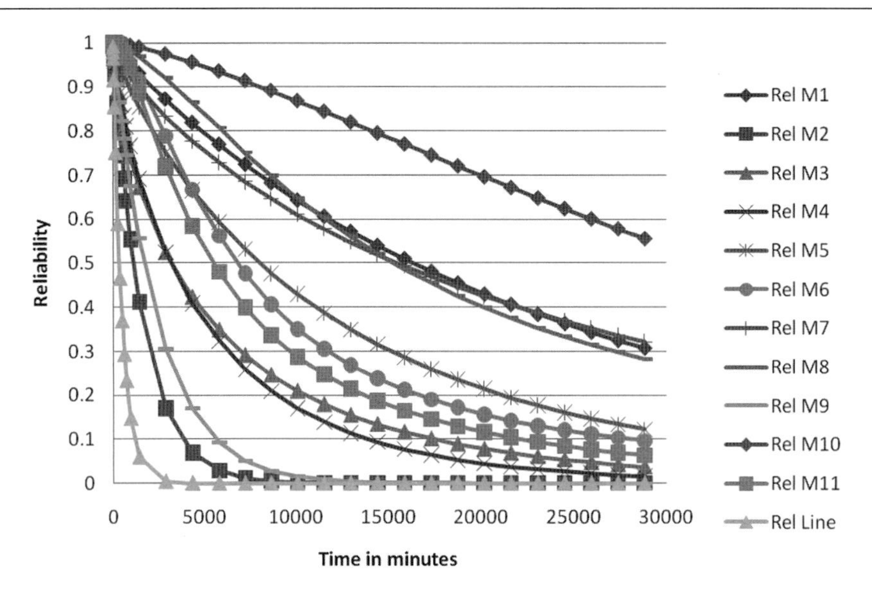


**Figure 25.7** Reliability diagram for all the machines and the entire mozzarella production line.

Maintainability analysis is used to identify any weaknesses in maintenance operation on the production line. As mentioned in Table 25.2, the trend test of TTR, with $a = 5\%$ level of significance, for all the machines and the entire MPL was presented. It is observed that the $H_0$: *No-trend in data* (homogeneous Poisson process) is not rejected. Thus, further analysis as the test of serial correlations was needed (Figure 25.5). As no trend and serial correlation were identified for all the machines and the line, then the iid assumption for the TTR was verified.

In order to identify the distributions of the repair data between several theoretical distributions, using the maximum likelihood estimation method was used and its parameters assessed by applying a goodness-of-fit test – Anderson–Darling. The Anderson–Darling statistics of several theoretical distributions for TTR based on repair data of the machine and line level for MPL are summarized in Table 25.5. A smaller statistic value indicates that the distribution fits the data better, that is in the heat exchanger (M1) column for TTR the lowest value is 0.964, which belongs to the loglogistic distribution.

In Table 25.6 the maintainability is calculated for all machines and the entire MPL for different time intervals. From Table 25.6 the following conclusions can be derived: (i) $Main_{Line}(100) = 0.90$, which means that there is a 90% chance that any failure in the MPL will be repaired within 100 min; (ii) the maintainability must initially be improved on the forming machine (M5) and the robot carton machine (M10), as well as for the entire production line; (iii) there is a 100% chance that any failure in the MPL will be repaired within $t > 210$ min.

**Table 25.5** The Anderson–Darling statistics for TTR at machine and line level for the mozzarella production line.

| | | | | | T | T | R | | | | | |
|---|---|---|---|---|---|---|---|---|---|---|---|---|
| **Distribution** | **M1** | **M2** | **M3** | **M4** | **M5** | **M6** | **M7** | **M8** | **M9** | **M10** | **M11** | **Line** |
| Weibull | 1.297 | 7.389 | 3.059 | 2.861 | 1.509 | 1.541 | 1.434 | 2.114 | 7.773 | 1.174 | 12.735 | 34.847 |
| Lognormal | 1.234 | 17.877 | 3.812 | 6.298 | 3.057 | 2.784 | 1.528 | 1.377 | 11.151 | 1.094 | 10.164 | 74.426 |
| Exponential | 9.800 | 120.104 | 35.923 | 34.562 | 17.311 | 17.084 | 10.349 | 10.057 | 90.016 | 6.464 | 13.576 | 307.813 |
| Loglogistic | 0.964* | 11.348 | 3.096 | 4.480 | 1.981 | 2.113 | 1.340 | 0.983* | 7.469 | 1.049* | 9.061* | 47.157 |
| Normal | 1.241 | 6.969 | 2.964 | 2.732* | 1.329* | 1.537 | 1.266* | 2.044 | 7.147 | 1.162 | 22.024 | 25.295 |
| Logistic | 1.141 | 6.367* | 2.765* | 2.838 | 2.391 | 1.428* | 2.279 | 1.444 | 6.018* | 1.147 | 17.444 | 18.623* |

* indicates the best-fit of the repair data between the candidate distributions.

**Table 25.6** Maintainability for all machines and the entire mozzarella production line for different time intervals (in minutes).

| Time | Main M1 | Main M2 | Main M3 | Main M4 | Main M5 | Main M6 | Main M7 | Main M8 | Main M9 | Main M10 | Main M11 | Main Line |
|---|---|---|---|---|---|---|---|---|---|---|---|---|
| 1 | 0,0000 | 0,0087 | 0,0042 | 0,0044 | 0,0015 | 0,0107 | 0,0001 | 0,0000 | 0,0059 | 0,0000 | 0,0011 | 0,0134 |
| 5 | 0,0000 | 0,0115 | 0,0057 | 0,0069 | 0,0024 | 0,0140 | 0,0002 | 0,0000 | 0,0079 | 0,0000 | 0,0419 | 0,0173 |
| 10 | 0,0000 | 0,0162 | 0,0084 | 0,0118 | 0,0045 | 0,0195 | 0,0006 | 0,0001 | 0,0116 | 0,0000 | 0,1741 | 0,0238 |
| 15 | 0,0005 | 0,0229 | 0,0124 | 0,0193 | 0,0079 | 0,0272 | 0,0016 | 0,0006 | 0,0169 | 0,0002 | 0,3459 | 0,0328 |
| 20 | 0,0022 | 0,0323 | 0,0182 | 0,0306 | 0,0134 | 0,0379 | 0,0038 | 0,0026 | 0,0246 | 0,0011 | 0,5039 | 0,0449 |
| 25 | 0,0072 | 0,0454 | 0,0268 | 0,0469 | 0,0219 | 0,0525 | 0,0086 | 0,0080 | 0,0357 | 0,0037 | 0,6276 | 0,0612 |
| 30 | 0,0191 | 0,0634 | 0,0392 | 0,0696 | 0,0348 | 0,0722 | 0,0180 | 0,0198 | 0,0515 | 0,0099 | 0,7182 | 0,0830 |
| 35 | 0,0430 | 0,0878 | 0,0570 | 0,0999 | 0,0532 | 0,0987 | 0,0349 | 0,0420 | 0,0737 | 0,0225 | 0,7834 | 0,1116 |
| 40 | 0,0848 | 0,1204 | 0,0822 | 0,1388 | 0,0785 | 0,1334 | 0,0630 | 0,0790 | 0,1045 | 0,0452 | 0,8304 | 0,1484 |
| 45 | 0,1492 | 0,1631 | 0,1171 | 0,1870 | 0,1122 | 0,1779 | 0,1065 | 0,1344 | 0,1460 | 0,0823 | 0,8648 | 0,1948 |
| 50 | 0,2369 | 0,2170 | 0,1642 | 0,2443 | 0,1551 | 0,2333 | 0,1681 | 0,2087 | 0,2005 | 0,1368 | 0,8904 | 0,2513 |
| 60 | 0,4546 | 0,3594 | 0,3014 | 0,3823 | 0,2694 | 0,3757 | 0,3470 | 0,3978 | 0,3502 | 0,2984 | 0,9247 | 0,3926 |
| 70 | 0,6577 | 0,5318 | 0,4863 | 0,5373 | 0,4151 | 0,5433 | 0,5694 | 0,5894 | 0,5368 | 0,4948 | 0,9457 | 0,5545 |
| 80 | 0,7985 | 0,6969 | 0,6751 | 0,6868 | 0,5736 | 0,7017 | 0,7713 | 0,7376 | 0,7135 | 0,6686 | 0,9593 | 0,7056 |
| 90 | 0,8824 | 0,8232 | 0,8202 | 0,8105 | 0,7210 | 0,8230 | 0,9051 | 0,8357 | 0,8426 | 0,7924 | 0,9686 | 0,8219 |
| 100 | 0,9299 | 0,9041 | 0,9092 | 0,8985 | 0,8379 | 0,9019 | 0,9699 | 0,8963 | 0,9201 | 0,8710 | 0,9751 | 0,8988 |
| 120 | 0,9727 | 0,9748 | 0,9797 | 0,9803 | 0,9630 | 0,9729 | 0,9987 | 0,9559 | 0,9815 | 0,9477 | 0,9834 | 0,9705 |
| 150 | 0,9917 | 0,9969 | 0,9980 | 0,9994 | 0,9986 | 0,9964 | 1,0000 | 0,9852 | 0,9981 | 0,9838 | 0,9899 | 0,9958 |
| 180 | 0,9969 | 0,9996 | 0,9998 | 1,0000 | 1,0000 | 0,9995 | 1,0000 | 0,9940 | 0,9998 | 0,9939 | 0,9933 | 0,9994 |
| 210 | 0,9987 | 1,0000 | 1,0000 | 1,0000 | 1,0000 | 0,9999 | 1,0000 | 0,9973 | 1,0000 | 0,9973 | 0,9953 | 0,9999 |
| 240 | 0,9993 | 1,0000 | 1,0000 | 1,0000 | 1,0000 | 1,0000 | 1,0000 | 0,9986 | 1,0000 | 0,9987 | 0,9965 | 1,0000 |
| 270 | 0,9997 | 1,0000 | 1,0000 | 1,0000 | 1,0000 | 1,0000 | 1,0000 | 0,9992 | 1,0000 | 0,9993 | 0,9973 | 1,0000 |
| 300 | 0,9998 | 1,0000 | 1,0000 | 1,0000 | 1,0000 | 1,0000 | 1,0000 | 0,9995 | 1,0000 | 0,9996 | 0,9979 | 1,0000 |
| 360 | 0,9999 | 1,0000 | 1,0000 | 1,0000 | 1,0000 | 1,0000 | 1,0000 | 0,9998 | 1,0000 | 0,9999 | 0,9986 | 1,0000 |

The maintainability of the MPL graphically at machine level and entire line level is shown in Figure 25.8. It is estimated that for the optimal repair level with the expected maintainability, that is to achieve 95% maintainability for the dummy machine (M11), the repair must be carried out within 70 min.

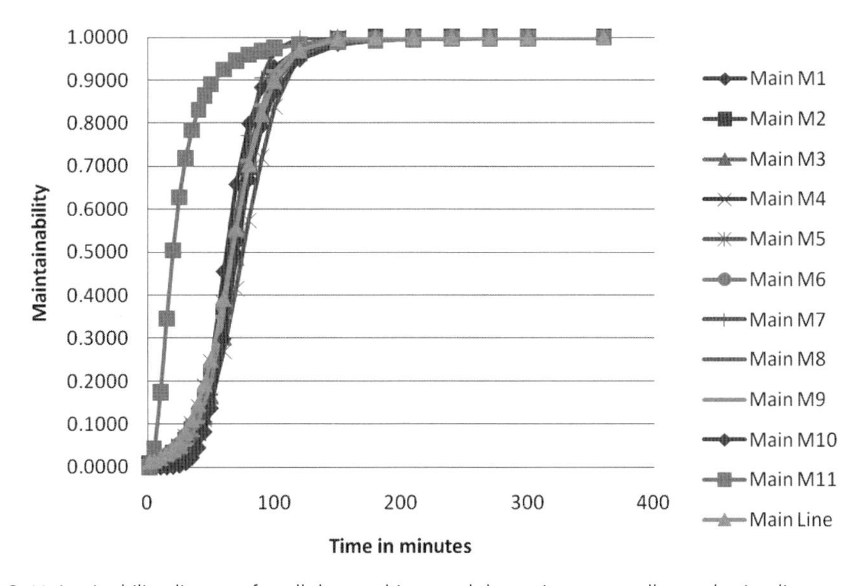


**Figure 25.8** Maintainability diagram for all the machines and the entire mozzarella production line.

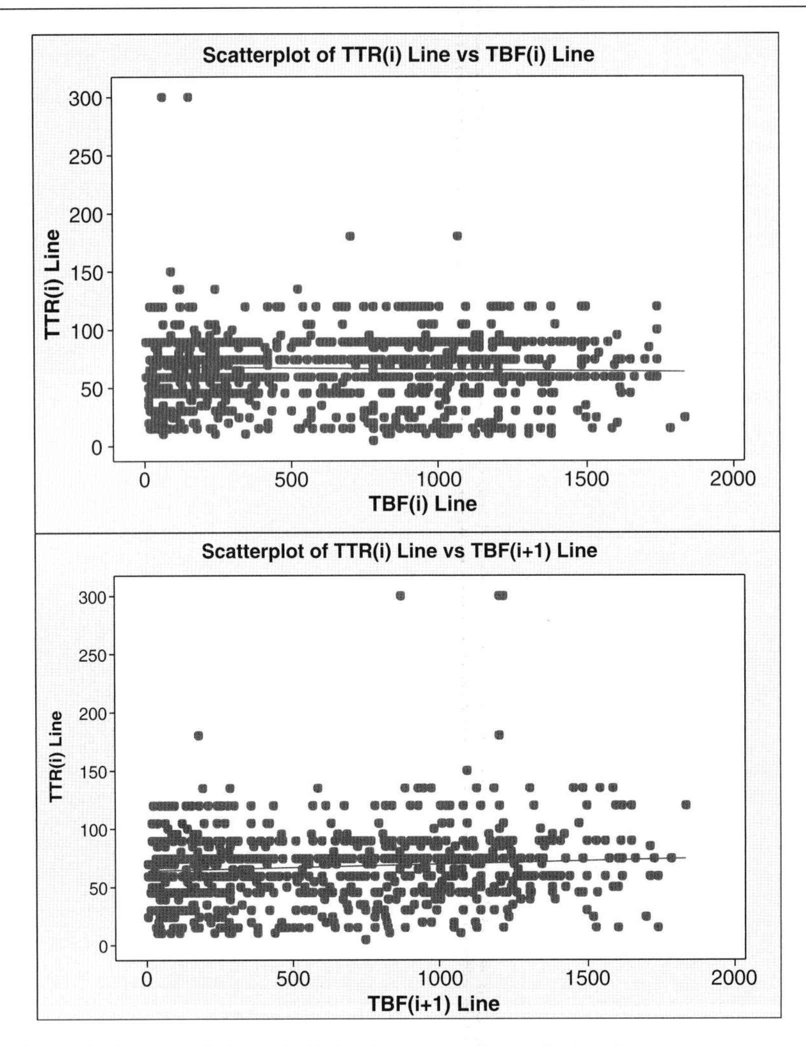


**Figure 25.9** Relationship between TBFs and TTRs for the mozzarella production line.

## RELATIONSHIP BETWEEN TBFs AND TTRs

To explore the relationship between TBFs and TTRs, the scatter plots of TBF against repair times are shown. The first scatter plot shows the relationship between a $TTR_{(i)}$ and the previous $TBF_{(i)}$, whereas the second scatter plot shows the relationship to the following $TBF_{(i+1)}$ (Figure 25.9). From Figure 25.9 it is evident that there is a slightly positive correlation between a $TTR_{(i)}$ and the following $TBF_{(i+1)}$; therefore, the longer the time it takes to repair a failure, the longer the time until the next failure.

## CONCLUSIONS

The main research findings can be summarized as follows:

(i) The operating time of the mozzarella production line was 88.42% and for the remaining 11.58% of the total operating time the line is under repair because of equipment failures.
(ii) The coagulator machine (M2), the wrapping machine (M9) and the stretching machine (M4) are accountable fortwo-thirds (67.6%) of the total failure occurrence on the mozzarella production line.
(iii) The hazard rate functions at the wrapping machine (M9) and the robot carton machine (M10) were shown to increase, as were those of the entire production line, meaning that the current maintenance policy is not adequate for those machines.
(iv) The lowest reliabilities are recorded at the coagulator machine (M2), the wrapping machine (M9) and at the stretching machine (M4).
(v) The maintainability must initially be improved on the forming machine (M5) and the robot carton machine (M10), as well as in the entire production line.
(vi) There is a slightly positive correlation between a $TTR_{(i)}$ and the following $TBF_{(i+1)}$; therefore, the longer the time it takes to repair a failure, the longer the time until the next failure.

Consequently, the statistical analysis of failure/repair data may estimate the cons of the production line, thereby intervening to improve the operation management and increase the performance and the production rate of the line. Moreover, the reliability analysis is very useful for deciding maintenance intervals, and for planning and organizing maintenance.

## REFERENCES


Ansell, J.I. and Phillips, M.J. (1989) Practical problems in the statistical analysis of reliability data. *Applied Statistics* **38**(2), 205–247.

Ascher, H. and Feingold, H. (1984) *Repairable System Reliability*. Dekker, New York.

Barabady, J., & Kumar, U. (2008). Reliability analysis of mining equipment: A case study of crushing plan at Jajarm Bauxite mine in Iran. *Reliability Engineering and System Safety* **93**, 647–653.

Barringer, P.E. (2000) *Reliability Engineering Principles*. Barringer & Associates, Humble, TX, USA.

Ben-Daya, M., Duffua, O.S., Raouf, A. *et al.* (2009) *Handbook of Maintenance Management and Engineering*. Springer-Verlag, NY, USA.

Biswas, A., Sarkar, J. and Sarkar, S. (2003) Availability of a periodically inspected system, maintained under an imperfect-repair policy. *IEEE Transactions on Reliability* **52**(3), 311–318.

Blischke, W.R. and Murthy, D.N.P. (2003) *Case Studies in Reliability and Maintenance*. John Wiley & Sons, Inc., Hoboken, NJ, USA.

Coetzee, J.L. (1997) The role of NHPP models in the practical analysis of maintenance failure data. *Reliability Engineering and System Safety* **56**, 161–168.

de Castro, H.F. and Cavalca, K.L. (2003) Availability optimization with genetic algorithm. *International Journal of Quality & Reliability Management* **20**(7), 847–863.

De, S. (1980). *Outlines of Dairy Technology*. Oxford University Press, New Delhi, India.

Department of Defense (1981) *Relibility Growth Management*. Military Handbook MIL-HDBK-189, Department of Defense, Washington, DC.

Ferrari, E., Gamberi, M., Manzini, R. *et al.* (2003) Redesign of the Mozzarella cheese production process through development of a micro-forming and stretching extruder system. *Journal of Food Engineering* **59**, 13–23.

Francois, P. and Noyes, D. (2003) Evaluation of a maintenance strategy by the analysis of the rate of repair. *Quality and Reliability Engineering International* **19**(2), 129–148.

Hajeeh, M. and Chaudhuri, D. (2000) Reliability and availability assessment of reverse osmosis. *Desalination* **130**, 185–92.

Hauge, B.S. (2002) Optimizing intervals for inspection and failure-finding tasks. In: *Proceedings of the Annual Reliability and Maintainability Symposium* (Seattle, WA), pp. 14–19. IEEE, New York, NY.

Kumar, U. and Klefsjo, B. (1992) Reliability analysis of hydraulic system of LHD machine using the power low process model reliability. *Engineering and System Safety* **35**(3), 217–224.

Lewis, M.W. and Steinberg, L. (2001) Maintenance of mobile mine equipment in the information age. *Journal of Quality in Maintenance Engineering* **7**(4), 264–274.

Liberopoulos, G. and Tsarouhas, P. (2002) Systems analysis speeds up Chipita's food processing line. *Interfaces* **32**(3), 62–76.

Liberopoulos, G. and Tsarouhas, P. (2005) Reliability analysis of an automated pizza processing line. *Journal of Food Engineering* **69**, 79–96.

Louit, D.M., Pascual, R. and Jardine, A.K.S. (2009) A practical procedure for the selection of time-to-failure models based on the assessment of trends in maintenance data. *Reliability Engineering and System Safety* **94**, 1618–1628.

Mobley, R.K. (2002) *An Introduction to Predictive Maintenance*, 2nd edn. Butterworth Heinemann, Boston, MA, USA.

Morea, M., Baruzzi, F., Cappa, F. and Cocconcelli, P.S. (1998) Molecular characterization of the *Lactobacillus* community in traditional processing of Mozzarella cheese. *International Journal of Food Microbiology* **43**, 53–60.

Okogbaa, O.G. and Peng, X. (1996) Methodology for preventive maintenance analysis under transient response, In: *Proceedings of the Annual Reliability and Maintainability Symposium* (Durham, NC), pp. 335–340. IEEE, New York, NY.

Percy, D.F., Kobbacy, K.A.H. and Fawzi, B.B. (1997) Setting preventive maintenance schedules when data are sparse. *International Journal of Production Economics* **51**, 223–234.

Rigdon, S.E. and Basu, A.P. (2000) *Statistical Methods for the Reliability of Repairable Systems*. John Wiley & Sons, Inc., New York, USA.

Smith, J.B. and Smith, W.B. (2002) Probabilistic assessment of availability from system performance data. In: *Proceedings of the Annual Reliability and Maintainability Symposium*, pp. 569–576. IEEE, New York, NY.

Tsarouhas, H.P. and Arvanitoyannis, S.I. (2010a) Reliability and maintainability analysis of bread production line. *Critical Review in Food Science and Nutrition* **50**(4), 327–343.

Tsarouhas, H.P. and Arvanitoyannis, S.I. (2010b) Assessment of operation management for beer packaging line based on field failure data: a case study. *Journal of Food Engineering* **98**(1), 51–59.

Tsarouhas, P., Varzakas, T. and Arvanitoyannis, I. (2009a) Reliability and maintainability analysis of strudel production line with experimental data; a case study. *Journal of Food Engineering* **91**, 250–259.

Tsarouhas, H.P., Arvanitoyannis, I. and Varzakas, T. (2009b) Reliability and maintainability analysis of cheese (Feta) production line in a Greek medium-size company: A case study. *Journal of Food Engineering* **94**, 233–240.

Tsarouhas, H.P., Arvanitoyannis, S.I. and Ampatzis, D.Z. (2009c) A case study of investigating reliability and maintainability in a Greek juice bottling medium size enterprise (MSE). *Journal of Food Engineering* **95**, 479–488.

Wang, H. and Pham, H. (2006) *Reliability and Optimal Maintenance*. Springer-Verlag, NJ, USA.

# Index



*Mathematical and Statistical Methods in Food Science and Technology*, First Edition.
Edited by Daniel Granato and Gastón Ares.